# AGRICULTURAL POWER AND MACHINERY

# AGRICULTURAL POWER AND MACHINERY


**CLINTON O. JACOBS**
Department of Agricultural Education
University of Arizona

**WILLIAM R. HARRELL**
Department of Agricultural Education
Sam Houston State University

**Glen C. Shinn**
Consulting Editor
Department of Agricultural and
Extension Education
Mississippi State University



**GREGG DIVISION**
**McGRAW-HILL BOOK COMPANY**

New York Atlanta Dallas St. Louis San Francisco
Auckland Bogotá Guatemala Hamburg Lisbon
London Madrid Mexico Montreal New Delhi Panama Paris
San Juan São Paulo Singapore Sydney Tokyo Toronto

Sponsoring Editor: **Roberta Moore**
Editing Supervisor: **Patricia Nolan**
Design and Art Supervisor: **Patricia F. Lowy**
Production Supervisor: **Priscilla Taguer**
Cover Designer: **Patricia F. Lowy**

**Library of Congress Cataloging in Publication Data**
Jacobs, Clinton O., (date)
Agricultural power and machinery.

Includes index.
1. Agricultural machinery. 2. Power transmission.
I. Harrell, William R., (date). II. Title.
S675.J26 1982 631.3 82-7322
ISBN 0-07-032210-4 AACR2

**Agricultural Power and Machinery**



5 6 7 8 9 0 SEMSEM 8 9

ISBN 0-07-032210-4

# CONTENTS

# PREFACE

*Agricultural Power and Machinery* is designed for students preparing for employment in occupations in the agricultural power and machinery industry. The text provides instruction in basic principles of power and machinery operation, maintenance, service, repair, and management.

The text is divided into units and chapters that can be studied in the order presented or in the order that is best suited to a particular course or curriculum. Each unit is introduced with an occupational matrix that lists the competencies that students will acquire through mastery of the material in that unit. The matrix indicates the relative importance of each competency to the major categories of occupations in agricultural mechanics, so that students may assess how the material relates to their individual career goals.

Each chapter is introduced with an overview of the concepts to be covered and a list of behavioral goals so that students will understand the learning expectations. The chapters in Unit I, "An Orientation to Agricultural Power and Machinery," provide an overview of the industry and the major categories of occupations in agricultural mechanics. Unit II, "The Small Internal-Combustion Engine," contains chapters on basic principles of the small internal-combustion engine and service and repair of the small engine. Unit III, "Power and Power Transmission Components," begins with an overview of the basic principles of power and then covers specific components of power transmission, such as clutches, transmissions, and final drives; power transmission by belts, chains, and power takeoff drives; and hydraulic power transmission. Unit IV, "The Agricultural Tractor," Unit V, "Soil Preparation and Crop Production Equipment," and Unit VI, "Harvesting and Handling Agricultural Products," all cover maintenance, service, repair, and adjustment of specific kinds of agricultural equipment. Unit VII, "Agricultural Power and Machinery Management," contains chapters on determining machinery costs and selecting and managing machinery both on and off the farm.

At the end of each chapter is a list of questions called "Thinking It Through." These questions may be used for individual study, as a basis for class discussion, or as a means for the instructor to evaluate student progress.

The *Agricultural Power and Machinery* text is accompanied by an activity guide and a teacher's manual and key. The activity guide is designed to be used by the student as an integral part of the instructional program. It contains exercises that provide students an opportunity to test their knowledge of the information learned from the text. It also has many jobs that require performance of service, repair, maintenance, and adjustment of machinery in the shop or in the field. Each job is referenced to the corresponding material in the text.

The teacher's manual and key contains information on how to set up a program of instruction in agricultural mechanics, suggested answers to the end-of-chapter questions in the text, and answers to the activity guide exercises.

The *Agricultural Power and Machinery* program presents a comprehensive package of materials for a one- or two-year program in agricultural mechanics. It is based on nationally identified competencies that will enable students to acquire the knowledge and skills necessary for a broad range of jobs in on-farm or off-farm occupations related to agricultural mechanics.

The authors would like to express their sincere appreciation to the many agricultural educators, both professors of higher education and high school instructors, who reviewed the project during its development. Appreciation also goes to the following people for their contributions to the project: Ronald A. Brown, Professor, Mississippi State University; Weston McCoy, Photographer, Sam Houston State University; and Karl Collins, Service Representative, Sperry-New Holland, Arlington, Texas. We are also grateful to the faculties of Sam Houston State University and The University of Arizona for their help and support, and special acknowledgment is given to Myra Jacobs and to Ramona, Ginger, and Glena Harrell for their help in typing the manuscript, and for their general support throughout the development of the project.

Clinton O. Jacobs
William R. Harrell

# UNIT I AN ORIENTATION TO AGRICULTURAL POWER AND MACHINERY

| **Competencies** | Agricultural Machinery Service Manager | Agricultural Machinery Parts Person/ Manager | Agricultural Tractor and Machinery Mechanic | Agricultural Machinery Setup and Delivery Mechanic | Agricultural Tractor and Machinery Mechanic's Helper | Small-Engine Mechanic | Farmer, Grower, and Rancher |
|---|---|---|---|---|---|---|---|
| Describe the role of mechanization in modern agriculture | Essential | Essential | Essential | Essential | Essential | Essential | Essential |
| Identify employment opportunities for persons trained in power and machinery | Desirable | Desirable | Desirable | Desirable | Desirable | Desirable | Desirable |
| List skills needed for employment in power and machinery | Essential | Essential | Essential | Essential | Essential | Essential | Desirable |
| Identify personal qualities as they relate to a career in power and machinery | Essential | Essential | Essential | Essential | Essential | Essential | Desirable |
| Determine training for job entry and advancement in power and machinery | Essential | Essential | Essential | Essential | Essential | Essential | Desirable |

**Essential**

**Desirable**

**Not necessary**

In modern agriculture, machines are used to reduce the hours of labor required to produce agricultural products. Machines have been developed that enable agricultural producers to increase the number of acres that they manage or operate.

Agricultural producers use tractors to operate the equipment that is used to till, plant, fertilize, and harvest crops. Modern tractors range in size from 10 to 15 horsepower (hp) [7.5 to 11 kilowatts (kW)] to over 400 hp [over 300 kW]. Grain combines with headers up to 30 feet [9.1 meters] in width are used to harvest grain and seed crops. Hay and other forage crops can be harvested and stored with modern equipment without any manual handling. Cotton, peanuts, fruits, vegetables, and nut crops are harvested almost entirely by specialized harvesting equipment.

The development of the machines used in modern agriculture has brought new demands on farmers, ranchers, and other agricultural workers. They must be familiar with energy and mechanical forces and have mechanical skills. It is important that they be able to select machines for specific jobs and know how to safely operate, maintain, and repair them.

There is a great demand for persons to manufacture, sell, and service the machines used in agricultural production. This is an important part of the power and machinery industry. It includes the equipment and tractor companies, the farm equipment and supply dealers, the independent repair and maintenance shops, the research and development centers, and all other services that are necessary to support production agriculture's needs for machines and equipment.

This unit will examine the importance of agricultural power and machinery to the agricultural industry and describe the many jobs available for people trained in power and machinery. It will also point out many of the basic skills needed to attain a job in this field.

Upon completion of this unit, you should know about many of the career opportunities available in agricultural power and machinery. In addition, you should know what personal qualities and characteristics you will need to succeed in these careers.

# CHAPTER 1 AGRICULTURAL POWER AND MACHINERY

## CHAPTER GOALS

In this chapter your goals are to:

- Explain the importance of power and machinery in modern agriculture
- Identify areas of employment in agricultural power and machinery
- Identify entry requirements for employment in the various jobs available in agricultural power and machinery
- Identify characteristics and working conditions for the areas of employment in agricultural power and machinery

A few years ago the Future Farmers of America (FFA) used the slogan "Agriculture Is More than Farming" as a national theme. As a student of vocational agriculture, you know that this statement is true and that millions of workers are employed in agribusiness occupations.

It is true that the number of people employed in agricultural production has declined as improved technology and machines have increased the productive capability of farmers and ranchers. It is also true that the number of skilled employees needed to manufacture, sell, service, and operate agricultural machinery and equipment has increased.

In this chapter, the role of agricultural power and machinery in modern agriculture is discussed. Career opportunities, job entry requirements, job characteristics, and working conditions are also discussed.

## THE AGRICULTURAL POWER AND MACHINERY INDUSTRY

The production of food and fiber to meet the needs of the population is a basic requirement of any civilization. For many centuries the majority of the population in most countries, including the United States, was engaged in producing food and fiber. Draft animals such as horses, oxen, and water buffalo were used to produce the power necessary to operate simple machines.

The transition from animal power to mechanical power began early in the twentieth century. The first tractors were steam-powered. It is generally considered that the steam-traction tractor was first produced successfully in the United States in 1876. Since that time gasoline, kerosene, tractor fuel, liquefied petroleum (LP) gas, and diesel engines have been developed to power agricultural tractors and equipment. The agricultural industry uses a wide variety of machinery and equipment. The machines used in modern agriculture have been developed through research and field testing. Agricultural equipment manufacturers, state agricultural experiment stations, private research centers, farmers, ranchers, and other agricultural producers have worked together to meet the needs of modern agriculture.

The development of power farming has had a tremendous effect on the American way of life. In 1860 it required 57.7 hours (h) of work to produce 1 acre [0.4 hectare (ha)] of wheat. By 1914, 15.2 h of labor were required to produce 1 acre of wheat. Today less than 1.5 h of labor are required to produce 1 acre of wheat with the modern equipment and technology used in agricultural production in the United States. Similar advances have been made in the production of all other crop and livestock enterprises.

At the beginning of the twentieth century one farmer produced enough food and fiber for around seven people. The modern American farmer produces enough food and fiber for more than 70 people. This has brought about a reduction in the number of workers needed to produce food and fiber, while the total production of agricultural products in the United States has increased by about 80 percent.

Modern agriculture is the largest industry in the United States. Approximately 4 million farmers, ranchers, and agricultural workers are actively engaged in producing food and fiber. Another 18 to 20 million American workers are engaged in processing, transporting, storing, and marketing agricultural products or providing goods and services to farmers and ranchers. These are agribusiness occupations and combined with the agricultural producers amount to approximately 28 percent of the total labor force in the United States.

Farmers, ranchers, and other agricultural producers provide a large market for equipment. Equipment dealers sell and service new and used agricultural equipment such as tractors, harvesting and handling equipment, and tillage and planting equipment. They provide employment for thousands of individuals interested in sales, service, repair, parts, and customer relations.

## EMPLOYMENT OPPORTUNITIES

Power and machinery are an integral part of modern agriculture. Never before have the employment opportunities been greater for young men and women who have an interest in learning about the service, repair, and operation of agricultural equipment. A brief description of some of the major employment areas may make it easier for you to select a career and establish educational goals.

## Equipment Manufacturers

Agricultural equipment manufacturers provide excellent employment opportunities for young men and women. They provide job opportunities in manufacturing, research and development, customer relations, sales, education, and distribution (Fig. 1-1).

Many of the jobs are centralized near the companies' manufacturing centers. A background in agriculture and training in agricultural power and machinery are recommended but not essential for employment. Some jobs require a degree in agricultural engineering, mechanized agriculture, or business administration. Others may require a degree in agricultural power and machinery from a two-year vocational-technical school or a high school diploma.

**Branch Centers** Major companies have branch or regional service, distribution, and sales centers located throughout the United States. These centers provide services to the dealers located in their region. They employ salespeople and service people to work with local dealers to help them sell and service their products.

A branch serviceperson or salesperson is usually an individual who has had work experience in the

(*a*) (*c*) (*b*) (*d*)

**Fig. 1-1.** Equipment manufacturers provide many employment opportunities: (*a*) Machine operator. (*b*) Training mechanics for dealers. (*c*) Customer service and distribution. (*d*) Design. (*Deere & Co.*)

factory center or at a dealership or who has a college degree in mechanized agriculture or agricultural engineering.

## Equipment Dealers

There are over 36,000 agricultural equipment dealers in the United States (Fig. 1-2). Equipment dealers sell and service agricultural equipment. Dealers provide employment opportunities for people interested in sales, customer relations, management, parts, and service.

Dealerships provide good employment opportunities for people who enjoy working with machinery and have the natural ability to work well with their hands. An agricultural background is good but is not essential.

A person does not have to be an expert mechanic to be hired by a dealer. There is actually a shortage of qualified personnel to fill the jobs that are available. However, people interested in working for a dealer can improve their employment opportunities by taking high school vocational agriculture courses in power and machinery or cooperative part-time training and by taking post-high school courses at a community college or university. A high school diploma is required by most dealers.

Most people enter the dealership as a mechanic's helper, setup person, or helper in the parts department. After employment, training is obtained on the job and through training programs sponsored by the dealer and the manufacturers.

In large dealerships, the mechanics usually specialize in specific areas. There is a great demand for mechanics specializing in diesel fuel systems, hydraulics, electrical systems, air conditioning, and areas such as combines or hay balers.

The opportunity for job advancement is very good in most dealerships. It usually requires about 3 years to become an experienced mechanic and to earn mechanics' wages. However, this is one of the fastest-growing phases of agricultural industry, and there is unlimited advancement potential. An experienced mechanic can become a head mechanic or a service manager. Other jobs are available in areas such as sales, management, and customer service.

It is not impossible to become a dealer. Some dealers offer proven employees a chance to buy into the dealership. They may do this as an incentive to keep employees, or they may be ready to retire and want to pass the dealership on to someone they know has the ability to maintain it.

## Production Agriculture

Farmers and ranchers employ people to operate, maintain, and repair the tractors and other machines they use to produce agricultural products. Many larger producers maintain well-equipped shops so that they can perform routine repairs and preventive maintenance on their equipment.

# SKILLS AND COMPETENCIES NEEDED FOR JOB ENTRY

It is evident that there are many jobs available for young people interested in a career in agricultural

**Fig. 1-2.** Agricultural equipment dealerships provide excellent employment opportunities for people trained in power and machinery. (*Ford Tractor Operations*)

power and machinery. Following are some descriptions of specific jobs and skills that they require.

## Mechanic's Helper

Many job titles fall under the category of mechanic's helper. Three commonly used job titles in this category are setup mechanic, welder, and painter. A job as a mechanic's helper is a good way to enter the field with limited education and work experience.

**Setup Mechanic** The major duties of a setup mechanic are to unpack and assemble new machines. The job requires that a person be able to use hand tools, hoisting equipment, and precision measuring tools. It will also be necessary to read service manuals and setup instructions (Fig. 1-3).

Other duties may include the cleanup and disassembly of equipment that is to be repaired in the dealer's shop. Many times a setup mechanic will assist in repairing equipment and tractors and in cleaning parts.

The setup mechanic may have some janitorial duties in the shop and may be called on at times to assist a service person making repairs on equipment at a customer's farm or ranch.

**Fig. 1-3.** A setup mechanic must be able to use service manuals and special tools to assemble and adjust new equipment. (*Ford Tractor Operations*)

**Welder** Some dealer shops do fabrication work on equipment. Workers may install bumpers and brush guards and perform other custom work. They may also operate a portable welder and make repairs on equipment at the customer's farm or ranch. It is not uncommon for a dealer to employ a young person who has developed welding skills in vocational agriculture.

**Painter** Another common job performed in many dealer shops is the painting of equipment for customers and for resale. The skills developed by a young person who has painted equipment and projects in vocational agriculture are sufficient for obtaining a job as an agricultural equipment painter.

As with the setup mechanic, the welder or painter may do other jobs around the shop. All jobs offer a young person who is willing to work and learn new skills an opportunity to advance.

## Mechanics

Most mechanics entered the profession as a mechanic's helper. Advancement is much quicker if a person has some training before becoming a helper. This training may be obtained in a high school class specializing in agricultural power and machinery, at a community college or vocational-technical school, or in short courses offered at many universities.

Skilled mechanics are specialists in the machinery handled by the dealership or service center. It is common for mechanics to specialize in hydraulics, diesel fuel systems, power trains, or agricultural equipment such as combines, hay balers, or cotton pickers (Fig. 1-4).

Mechanics may do and supervise repair jobs such as overhauling an engine or rebuilding a combine. They must be able to use the special tools and equip-

**Fig. 1-4.** A mechanic working on a tractor in a dealership. (*Deere & Co.*)

ment required to make necessary repairs and adjustments. It is essential that they be able to read service manuals and determine the condition of components and parts.

Mechanics and service people must be able to meet the public. When a customer brings in a tractor or other equipment for service, the employees must be able to discuss the problem with the customer in order to determine the service that is wanted and needed.

It is essential that mechanics continue their education. Many young mechanics attend night classes to gain additional training and to specialize in an area such as diesel fuel injection. Dealers and manufacturers continually offer short courses to train mechanics when new products are marketed or other changes occur.

## Parts People

Service shops and agricultural producers depend on the parts department to supply the parts needed to keep agricultural equipment operating. Most dealerships have a parts manager and one or more parts helpers.

**Parts Helpers** Parts helpers must be familiar with the equipment serviced by a dealer (Fig. 1-5). When a mechanic, farmer, or rancher needs a specific part, the parts helper must look it up in a partsbook or on microfiche to determine the part number. Then the parts helper will find the stock number and check to see if the part is in stock. If the part is not in stock, it is usually ordered from the manufacturer or a wholesaler.

Parts helpers receive parts and place them in stock. They usually assist in taking inventory and ordering parts. When equipment is still under manufacturer's warranty, the parts helper may make warranty adjustments.

Since invoices, charges, credits, parts orders, and other paper work are major duties of a parts helper, a good background in math, business, bookkeeping, and English is very helpful. Also, a parts helper must have a pleasant personality and like to work with people.

Work as a mechanic's helper or a farm worker is good training for acquiring a job as a parts helper. Also, vocational agriculture training in machinery or specialized training at a community college will help an individual get the necessary training.

**Parts Managers** In addition to the duties performed by a parts helper, the parts manager is responsible for managing the parts department. It is the responsibility of the managers to maintain the parts department and keep an inventory of parts that are needed by the mechanics and the customers.

Many parts managers entered the dealership as a parts helper. It is not uncommon for parts managers to become the manager or part owner of the dealership. Also, some are employed as area sales or service people by equipment companies.

**Fig. 1-5.** The parts person must be familiar with equipment and must be able to use catalogs and microfiche as well as maintain good customer relations. *(Jane Hamilton-Merritt)*

## Service Managers

Most service shops have a service manager. Service managers are responsible for the operation of the service shop. A service manager must be able to meet customers as they bring equipment in for service. Service managers are usually responsible for troubleshooting, scheduling jobs, keeping shop records, and supervising the work of several mechanics and mechanic's helpers.

As a rule, service managers work their way up by starting as a mechanic or mechanic's helper. Therefore, the job entry requirements are the same as those for a mechanic. Service managers must continue their education by attending short courses provided by the equipment manufacturers.

## Salespeople

Many equipment dealers hire salespeople to sell new and used agricultural equipment. The salesperson may spend much time visiting customers in order to sell equipment (Fig. 1-6).

It is essential that a salesperson know the equipment that is being sold. Many salespeople have worked as mechanics and parts people. A salesper-

**Fig. 1-6.** The salesperson must have good skills in human relations and understand the needs of the agricultural industry.

son must also be able to meet people and have good communications skills.

A salesperson needs a thorough knowledge of agriculture and needs a background in machinery management. Sales jobs are usually acquired after a person has several years of experience in a dealership.

### Manufacturer's Representatives

Manufacturers of agricultural equipment provide employment opportunities for those individuals who have gained experience by working for a dealer or for those who have a college degree in agricultural engineering or mechanized agriculture.

A manufacturer's representative may work in product sales, customer relations, or service. Manufacturers provide continuous training for their employees, and the employees become experts in their area of responsibility.

As a rule, manufacturer's representatives work out of a branch office and have a specific region that they work in. These jobs usually require extensive travel.

## SUMMARY

Power and machinery are an integral part of the agricultural industry. Machines have enabled American farmers and ranchers to produce an abundance of food and fiber efficiently. The use of power has also created many new employment opportunities for young men and women trained in the service, repair, and operation of agricultural equipment.

People selecting a career in power and machinery have a good choice of employment opportunities. A person just out of high school may choose to enter the industry as a mechanic's helper or a parts helper and receive on-the-job training. However, many choose to obtain additional training before seeking employment. People with more training usually advance much faster and obtain better jobs.

There is a special need for people trained in hydraulics, diesel fuel systems, air conditioning, and other specialty areas. As a general rule, mechanics become specialists in an area after they have a few years of experience.

## THINKING IT THROUGH

1. What is the largest industry in the United States?
2. What percent of the jobs in the United States are directly related to agriculture?
3. How can people planning to go to work for an equipment dealer improve their employment opportunities?
4. What employment opportunities are available for a person with a high school diploma and vocational agriculture training?
5. Using your own community, find out the types of jobs that are available locally in power and machinery.
6. Pick an area in power and machinery you are interested in. Visit people employed in that area and ask them about their job. You might ask the following questions:
   (a) What are the future employment opportunities?
   (b) What skills are necessary to become employed?
   (c) What is the nature of the job?
   (d) What are the physical requirements for the job?
   (e) What are the possibilities for career advancement?
   (f) What is the average starting salary?
   (g) How many hours per week are worked?

# CHAPTER 2 CAREER PLANNING IN AGRICULTURAL POWER AND MACHINERY

## CHAPTER GOALS

In this chapter your goals are to:

- Assess your personal interests as they relate to a career in agricultural power and machinery
- Assess your personal qualities as they relate to a career in agricultural power and machinery
- Determine the educational requirements for job entry into the career of your choice

It is not always easy to make a career choice. However, it is important for a person to set career goals as soon as possible, so that education and training can be tailored to provide the understanding and skills required for employment and success.

Career planning requires people to become aware of themselves. It is important that the career chosen fits the person. A career choice should be based on personal characteristics such as interests, aptitude, ability, self-concept, and desire for education and advanced training.

A person must also learn as much as possible about the careers available. Most careers are a series of related jobs—job clusters—with success in one job leading to another. Each advancement usually results in more responsibility and a better salary.

## SELF-EVALUATION

Selecting an occupation is one of the most important considerations in the life of a young adult. The life expectancy of an 18-year-old male is 70.19 years. An 18-year-old female has a life expectancy of 72.95 years. This indicates that the average person can expect to live about 50 years after graduating from high school.

Since the major portion of a person's life is spent as a working adult, it is important to become familiar with a broad field of occupations before making a choice.

### Determining Interests

Individual interests should play a major role in the selection of a career. It is important that you examine yourself to determine your likes and dislikes.

School guidance counselors use interest inventories to help identify areas of interest. You can also ask yourself questions that will help you make career choices that fit your individual interests.

**Outdoor Interests** As a general rule agricultural workers spend considerable time outdoors. Even employment in an equipment dealership usually entails some outside work (Fig. 2-1). Other jobs such as the operator of agricultural equipment or a field service person require much outside work.

**Mechanical Interests** A person with mechanical interests enjoys working with tools and equipment (Fig. 2-2). Agricultural equipment service people, tractor mechanics, agricultural engineers, and equipment operators are all directly involved with machines and tools.

**Persuasive Interests** People with persuasive interests like to meet and deal with people. Sales, parts, and service people, as well as dealers and factory service representatives, all deal directly with people.

**Social-Service Interests** Social-service interests indicate a desire to help people. Vocational agricultural teachers and agricultural extension agents usually have an interest in helping others. There are opportunities for individuals to teach agricultural power and machinery in high schools, community colleges, and universities. Equipment specialist positions with state extension services also provide employment opportunities for people with high social-service interests.

**Fig. 2-1.** Mechanic servicing an air conditioner on a customer's tractor. *(Ford Tractor Operations)*

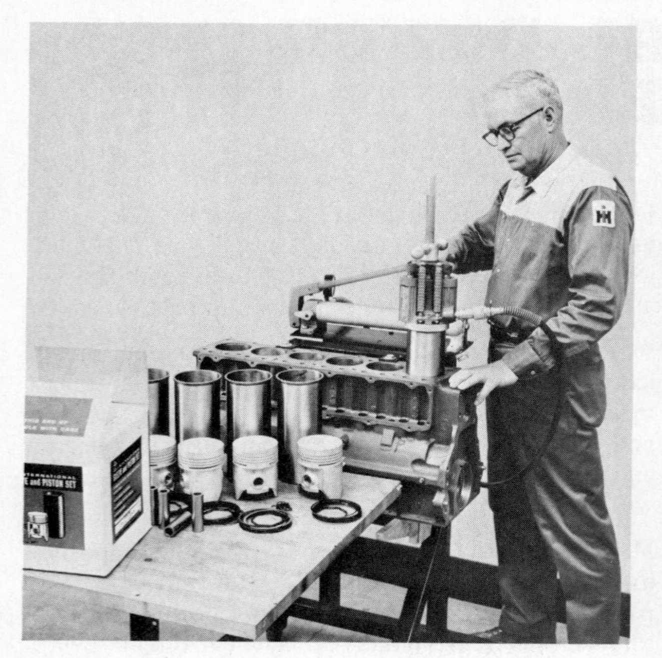

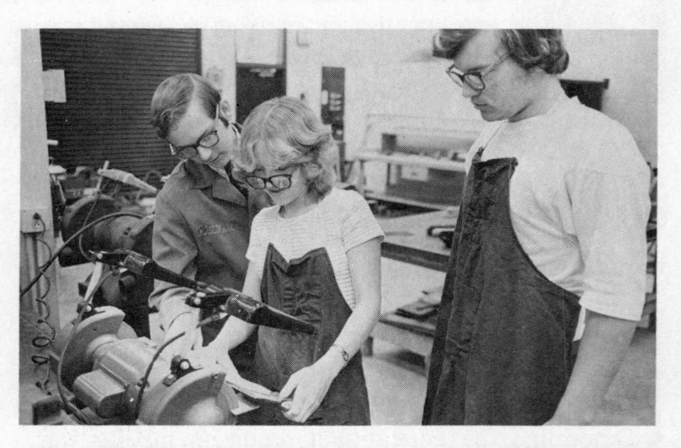

**Fig. 2-2.** There are good career opportunities for people interested in working with tools and equipment. (*International Harvester Co. and Future Farmers of America*).

## Determining Personal Qualities

You should realize that you as an individual are unique and distinct. You are the only person in the world like yourself. Not only do you have special interests, but you also have your own abilities and personal characteristics.

**Ability** Ability refers to a person's capacity to perform or do a job skillfully. A person may develop abilities or skills through practice and hard work or may possess a natural talent for the skills or occupation.

School grades are indicators of ability to perform in certain subject matter areas. Experiences in other activities such as the Future Farmers of America (FFA) contests, shop projects, supervised occupational experience (SOE) programs, and summer jobs may be used to identify other abilities.

A person planning a career in power and machinery should have the ability to work with tools and machinery. Training received in vocational agriculture can help develop these skills.

**Physical Makeup** Some occupations require more strength and stamina than others. People training for careers as mechanics or machinery service persons should have the strength and stamina to handle heavy objects. Long working hours may also be required when the harvest or planting season is in full swing. Other jobs such as that of a parts person will require less physical strength and stamina.

**Educational Aspirations** The choice of a career should include plans to secure the education needed to obtain employment and achieve success. It is not a realistic goal to want to become a design engineer for a major equipment company without planning to earn a degree in agricultural engineering. However, many occupations such as parts person or mechanic can be attained with vocational agriculture training and a high school diploma.

**Self-Concept** Self-concept is the image that individuals have of themselves. It is important that people choose occupations in which they feel comfortable.

An occupation should allow individuals to engage in activities that they find interesting and challenging. Think about your high school classes and other activities and decide which ones provide satisfaction. This can help you identify situations in which you feel comfortable.

## PREPARING FOR JOB ENTRY

Occupations also have varied characteristics. After examining your personal interests and qualities, you should thoroughly examine the occupations that interest you to determine their specific characteristics.

It is important that your choice be based upon facts and sound thinking. A thorough knowledge of the entry requirements and employment outlook for the occupation you are considering is essential.

### Observe the World of Work

Your chances of planning a career that will be interesting and satisfying will be improved if you observe people at work. There are many ways to observe the world of work. One way is to look around when you have the opportunity to visit a dealership or other agricultural industry. Watch the people at work. Determine the type of work that they are doing. You will see how jobs are performed and what skills are required. You can also note working conditions (Fig. 2-3).

**Supervised Occupational Experience (SOE) Program** The opportunity to participate in the SOE program during high school can be used to explore different careers. Placement in an agricultural equipment dealership as a mechanic's helper, setup person, or parts person can provide valuable career information.

Even though you may be doing simple jobs, you will have an excellent opportunity to observe the professional employees. You can see them as they perform their daily duties and determine if this type of work interests you. You can also determine the type of training that is necessary for job entry and career advancement.

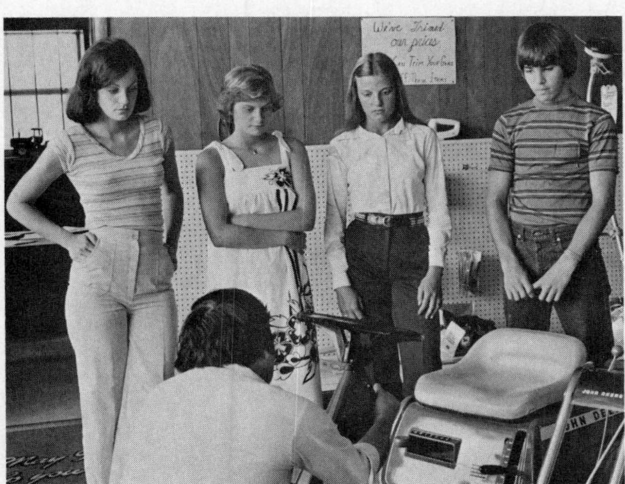

**Fig. 2-3.** Your chances of selecting a career you like can be improved by observing people at work.

### Educational Requirements

A person needs to know as much as possible about an occupation in order to make appropriate educational choices. A high school education is almost essential for job entry and career advancement. Yet each year thousands of young people drop out of high school to go to work. Most of these young people have the ability to finish school, but they quit to take a full-time job. They probably feel that a job with a regular income will solve all their problems.

Unfortunately, those who drop out of high school to go to work are often the very people who are unemployed. Or, many find themselves working at low-paying jobs that offer little or no chance for advancement. They usually find that the lack of a high school education is a real and serious handicap.

Upon completion of high school, you must decide if you are going to enter the world of work or continue your education. If you choose to continue your education, you have several options. You may choose an area vocational-technical school, a trade school, a community college, or a 4-year college or university. Your choice should be based upon your occupational choice.

Your vocational agriculture teacher, school counselor, parents, or friends employed in the occupation you are interested in can often provide meaningful information concerning entry requirements. Also, it is usually possible to visit a prospective employer in order to gain more knowledge about entry requirements.

### Personal Requirements

"Doing to Learn" is a part of the motto of the Future Farmers of America. As a citizen, it is your responsibility to develop the characteristics that will lead to career development. Careful planning is essential. Courses taken during high school should relate to your career choice.

Vocational agriculture offers many learning activities for people planning a career in agricultural power and machinery. Skills are taught in agricultural mechanics classes. Other skills are learned through SOE programs.

Agricultural mechanics contests are conducted on local, area, state, and national levels to allow FFA members to test and perfect their abilities and skills (Fig. 2-4). FFA leadership activities and shows that feature agricultural mechanics projects also allow for personal development. Many people have succeeded in agricultural industry because of their training in vocational agriculture and the FFA.

Desirable personal traits are also very important.

**Fig. 2-4.** The FFA offers many learning activities that help individuals prepare for a career in agricultural power and machinery. *(Future Farmers of America)*

Traits such as loyalty, honesty, dependability, work attendance, courtesy, tact, and personal neatness should be practiced until they become habits. Not only will these habits help you succeed in getting a job, but they will help you advance to better jobs.

## SUMMARY

Selection of an occupation is one of the most important decisions made by young adults. It is very important for people planning careers in agricultural power and machinery to identify their interests and personal characteristics.

Occupations also have varied characteristics, and a thorough knowledge of available occupations is essential. This knowledge can be attained by a purposeful observation of the world of work.

When a decision is reached, a person should plan for job entry. Job entry requirements vary, and the education and training required also vary. Careful planning during high school will help an individual prepare for job entry and advancement.

Good personal traits are important for success in work. Desirable traits should be practiced until they become habits.

## THINKING IT THROUGH

1. Several types of interests are discussed. Select the interests that you have and list occupations that would allow you to fulfill these interests.
2. List the subjects that you have found to be most interesting during high school. Discuss the reasons you like these courses.
3. List the experiences gained during supervised occupational experience and FFA activities that you have found to be most interesting. Explain why you like these activities.
4. Select a career that you are interested in and determine the educational requirements for job entry and advancement.
5. List several desirable personal traits that should be practiced until they become habits.

# UNIT II THE SMALL-INTERNAL COMBUSTION ENGINE

| Competencies | Agricultural Machinery Service Manager | Agricultural Machinery Parts Person/ Manager | Agricultural Tractor and Machinery Mechanic | Agricultural Machinery Setup and Delivery Mechanic | Agricultural Tractor and Machinery Mechanic's Helper | Small-Engine Mechanic | Farmer, Grower, and Rancher |
|---|---|---|---|---|---|---|---|
| Utilize parts and service manuals | Essential | Essential | Essential | Essential | Essential | Essential | Essential |
| Use hand and power tools safely | Essential | Essential | Essential | Essential | Essential | Essential | Essential |
| Troubleshoot engine malfunction | Essential | Desirable | Essential | Essential | Desirable | Desirable | Essential |
| Service and repair ignition system | Desirable | Not necessary | Essential | Essential | Essential | Essential | Desirable |
| Test and maintain cylinder compression | Essential | Not necessary | Essential | Desirable | Essential | Essential | Not necessary |
| Measure and evaluate engine parts | Essential | Essential | Essential | Desirable | Essential | Essential | Not necessary |
| Perform major engine repairs | Essential | Not necessary | Essential | Desirable | Desirable | Essential | Not necessary |
| Perform tuneup operations | Essential | Not necessary | Essential | Essential | Essential | Essential | Desirable |
| Select fuels and lubricants | Essential | Essential | Essential | Essential | Essential | Essential | Desirable |

**Essential**

**Desirable**

**Not necessary**

The modern internal-combustion engine is essential to the production of food and fiber. In fact, the agricultural industry is almost totally dependent upon the power produced by the internal-combustion engine. The farm tractor and auxiliary engines for farm machinery are used to till the soil, plant and cultivate crops, harvest food and fiber, and process products for human or livestock use. Trucks and automobiles depend upon the internal-combustion engine for power to distribute consumable products and make it possible for workers to be transported to jobsites. Likewise, the smaller air-cooled engines power lawn, garden, and other groundskeeping machinery.

Workers in agricultural power and machinery will be applying knowledge and skills in operating, servicing, maintaining, and repairing internal-combustion engines. For example, the machinery service manager must understand the operating components of engines to assist customers in service needs. A machinery parts person will be using knowledge of engines to provide critical repair parts for immediate use. It is extremely important that the agricultural tractor and machinery mechanic and the agricultural machinery setup and delivery mechanic know how to service and maintain engines to achieve proper performance and service life. A knowledge of the principles of engine operation is basic for the salesperson who must know the company product and present it to the prospective customer.

This unit is designed to help you develop a basic understanding of the small internal-combustion engine and to provide a foundation of knowledge to build upon for a career in agricultural power and machinery.

# CHAPTER 3 THE INTERNAL-COMBUSTION ENGINE

## CHAPTER GOALS

In this chapter your goals are to:

- Identify the operating principles of the internal-combustion engine
- Identify the operating characteristics of the two- and the four-stroke-cycle engine
- Identify the basic parts of the internal-combustion engine
- Compare the operation of the spark-ignition gasoline engine and the compression-ignition diesel engine
- Define the three basic terms which are used to describe engine specification characteristics
- Describe the operation of the valve systems used in two-cycle and four-cycle engines
- Describe the operating principles of carburetion and fuel injection systems
- Identify the operating principles of the flywheel magneto-, the battery-, and the solid-state-ignition system

Have you ever stopped to think how many different types of equipment which you use every day are power-operated? Do you know whether they are driven by a "motor" or an "engine"? In using a power tool or an elevator, you may have started an electric motor by working a switch. A *motor* must be supplied with energy from an external source in order to operate. Electric motors change power from an electric generator into motion. A hydraulic motor converts the energy developed by the flow of oil from a pump into rotating force.

The term *engine* identifies a machine which is able to produce power, independent of an external source of energy. An *internal-combustion* engine is capable of converting heat, developed by the burning of a fuel within a combustion chamber, into rotating force. When supplied with the proper fuel, it powers trucks, tractors, and many other agricultural machines, both large and small.

Small *air-cooled* internal combustion engines are an important power source in agriculture. Usually these engines are of single- or two-cylinder design (Fig. 3-1). They are used as auxiliary power sources for operating spray pumps, grain elevators, rotary tillers, lawn mowers, small tractors, and turf and groundskeeping equipment (Fig. 3-2). They also are used in extremely lightweight machines such as power chain saws and trimmers.

Small engines possess many of the functional operating parts found in larger multicylinder engines. It is therefore important for the person who is preparing for employment as a specialist in power and machinery to realize the necessity for developing competencies in small engines. Skills required to service and repair small engines are generally common to their counterpart, the large engine. Knowledge and skills developed from instruction and work experience with the small engine can be readily transferred to skills needed for work with large engines.

Because of the tremendous increase in demand for small-engine power, the opportunities for employment in small-engine service and repair increase with their use. For example, one of the major small-engine manufacturers has a production capacity of 32,000 engines per day! Large numbers of engines are used in rural areas for production agricultural operations.

(*a*)

(*b*)

(*c*)

**Fig. 3-1.** (*a*) and (*b*) Engines classified as horizontal crankshaft construction. (*c*) Engine classified as vertical crankshaft construction. (*Briggs & Stratton Corp.*)

(a) (b) (c) (d)

**Fig. 3-2.** Small engines power many types of groundskeeping equipment and small tractors. (*a*) Hand mower. (*The Toro Company*) (*b*) Lawn tractor. (*Ariens Company*) (*c*) Chain saw. (*d*) Trimmer/cutter. (*The Toro Company*)

Much agriculture in urban locations centers on the need for skilled people to provide maintenance and repair services for lawn, garden, and groundskeeping machinery. Once you have become experienced in the operation of a small-engine service center, you may find the opportunity and have the desire to own your own business or advance to other occupations that require knowledge of and experience with small engines and their application.

## PRINCIPLES OF ENGINE OPERATION

Two basic types of internal-combustion engines are in present use. The first is the *piston* engine. It is often called a ''reciprocating'' type because the piston moves up and down in a cylinder in a reciprocating action (Fig. 3-3). The second type of engine is the *rotary*. There are two types of rotary engines: the turbine engine, which is used in jet aircraft, and the Wankel, which has limited use in automobiles. The piston engine is the type used in present-day agricultural equipment.

A reciprocating engine transforms the up-and-down action of the piston sliding in its cylinder into a rotating motion. This is accomplished by the use of a crankshaft. A connecting rod joins the piston to the crankshaft (see Fig. 3-4).

### Basic Engine Parts

An internal-combustion piston engine consists of five basic parts: the cylinder block, piston, connecting rod, crankshaft, and flywheel.

**Cylinder Block** The cylinder block (shown in Fig. 3-4) includes the cylinder and a housing, or crankcase, for the crankshaft. The block is usually constructed of high-quality cast iron or an aluminum alloy. In an air-cooled engine, circular fins are cast into the outside surface of the cylinder in order to expose more surface to the air for cooling. The inner bore (hole) of the cylinder is machined to a precise diameter. A cover called a *cylinder head* is attached to the open end of the cylinder to form a combustion chamber. The joining surfaces of cylinder and head are sealed, using a cylinder-head gasket. The crankcase portion of the block serves as a bearing support for the crankshaft and as a storage chamber for lubricating oil in some engines.

**Piston** The piston is the sliding part of the combustion chamber. It performs the important function of transmitting heat energy into straight-line motion. The piston is nearly always constructed of a lightweight alloy. It is accurately fitted to the cylinder to permit freedom of movement. Piston rings are fitted into the ring grooves to provide a positive sealing surface between the cylinder wall and piston. The rings nearest the top of the piston are called *compression rings*. During the combustion process these rings prevent the escape of high-pressure gases past the piston and assist in reducing the sliding friction to a mini-

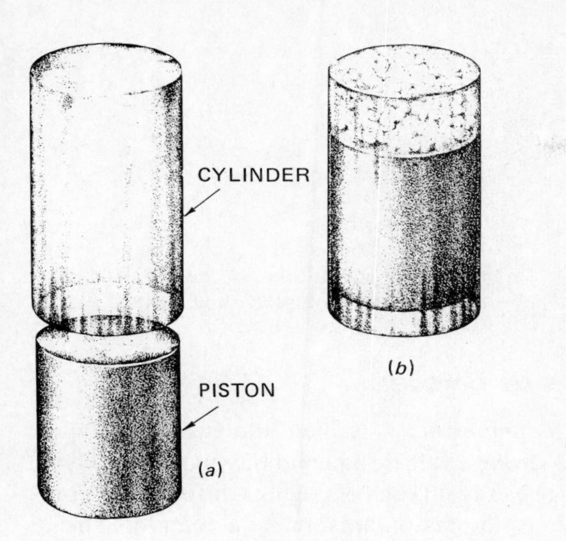


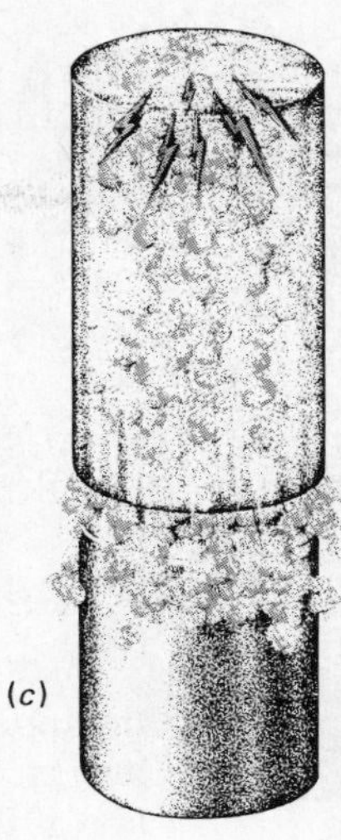


**Fig. 3-3.** In a piston engine the piston slides in the cylinder ("reciprocates"). (*a*) The cylindrical piston fits snugly into the cylinder. (*b*) When the piston is pushed up into the cylinder, air is trapped and compressed. (*c*) The increase of pressure as the gasoline vapor and air mixture is ignited pushes the piston out of the cylinder. (*Crouse/Anglin, Small Engine Mechanics, 2nd ed. Copyright © 1980 by McGraw-Hill, Inc. Reproduced with permission.*)

mum. The lower rings are *oil rings* and are placed in wider grooves below the compression rings. The oil rings are designed to control the amount of oil film on the cylinder wall. A piston pin is fitted into two bearing surfaces in the skirt of the piston. It is retained in the piston by lock rings at each end of the pin.

**Connecting Rod** A connecting rod (Fig. 3-4) is used to transfer the sliding force of the piston to the crankshaft. It is constructed from a cast-aluminum alloy or forged steel. The small end of the connecting rod is connected to the piston by the piston pin. The large end of the connecting rod provides the bearing surface to rotate the crankshaft.

**Crankshaft** The purpose of the crankshaft (Fig. 3-4) is to convert the reciprocating action of the piston into rotary motion and force. Crankshafts are manufactured using highly refined nodular cast iron or drop-forged steel. The offset section of the shaft is the *crank throw*. The crank forms a lever against which the force of the piston is exerted. The action develops a rotating, twisting action which is transferred to the exposed end of the crankshaft, called the *power takeoff*.

**Flywheel** A flywheel is attached to one end of the crankshaft. The function of the flywheel is to even out the power flow to the crankshaft by resisting any change in speed of rotation. In small single-cylinder engines, the flywheel is proportionately much heavier than the type used on multicylinder engines. This is because fewer power strokes are used to turn the crankshaft through a revolution. A heavy flywheel can "store" more energy from the power stroke to rotate the crankshaft through non-power-producing strokes of the piston. The flywheel of most small engines is equipped with fan blades for moving air over the cylinder and cylinder head for cooling (Fig. 3-4). Many small air-cooled engines utilize the flywheel to house the magnets for the magneto-ingnition system. Some manufacturers design the flywheel to also serve as a generator for the electrical system.

## Engine Operation

All spark-ignition engines work in the following ways:

- A mixture of fuel and air is drawn into the combustion chamber of the cylinder.
- The mixture is then compressed into a much smaller volume.
- The compressed gases are ignited at the correct position of the piston in the cylinder. The burning

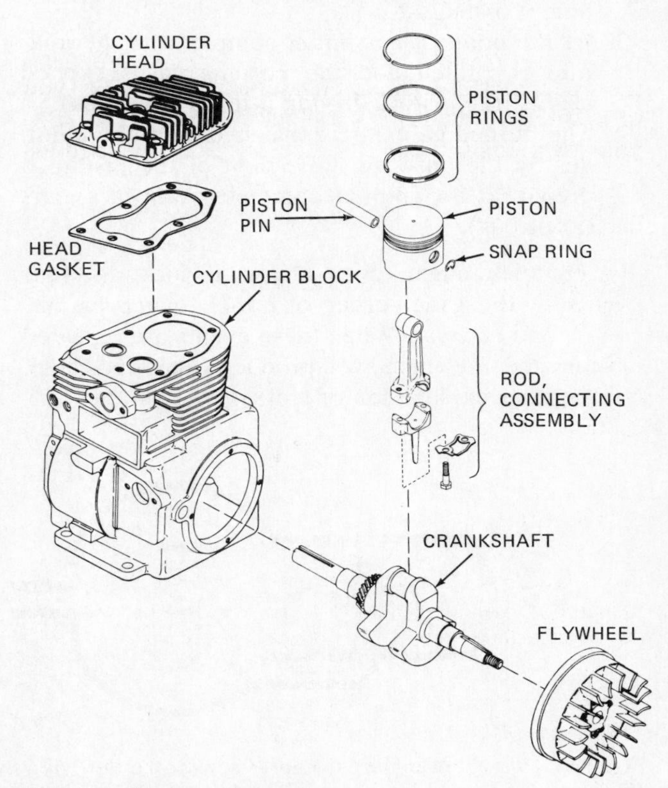


**Fig. 3-4.** Basic engine parts.

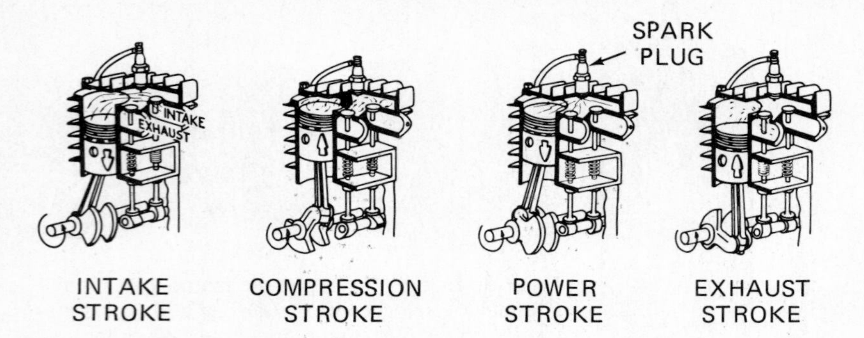


**Fig. 3-5.** Four-stroke-cycle engine operation. Two revolutions of the crankshaft are required to complete the four strokes. (*Briggs & Stratton Corp.*)

gases then expand, pushing the piston down with great force.

- Near the end of the downward movement of the piston a valve opens to allow the burned gases to escape from the cylinder, and readies it for repetition of the above events.

The first successful engine to operate in the manner described in the above sequence of events was developed in Germany by Dr. N. A. Otto in 1876. The engine required four strokes of the piston to complete the events in the proper sequence. Consequently, the engine was classified as operating on the *four-stroke-cycle* principle as follows:

1. A combustible air-fuel mixture is drawn through the carburetor and into the combustion chamber of the cylinder by the inward movement of the piston in the cylinder (INTAKE).
2. The mixture is compressed by the outward movement of the piston to a much smaller volume (COMPRESSION).
3. At the point of maximum compression the mixture is ignited and the burning gases expand against the piston, driving it inward (POWER).
4. The burned gases are exhausted from the cylinder by the outward movement of the piston to prepare it for another complete cycle of events (EXHAUST).

A complete engine cycle always includes the four events in the exact order of *intake, compression, power,* and *exhaust.* When these events are repeated in sequence, the engine will produce work (see Chap. 6 for further explanation of work).

## Types of Cycles

Piston engines are classified into two basic types: the four-stroke-cycle design and the two-stroke-cycle design. These two types of engines differ in the number of times the piston must make a complete stroke in the cylinder to complete the four-event cycle. A stroke is one complete movement of the piston in the cylinder.

**Four-Stroke-Cycle Engine** In the four-stroke-cycle engine, the four basic events are completed with four strokes of the piston in the cylinder. For this reason it is commonly referred to as a four-cycle engine (see Fig. 3-5). The four strokes require two revolutions of the crankshaft. On the *intake* stroke the piston is moved in toward the crank, creating a vacuum in the cylinder. The vacuum causes a combustible mixture of fuel and air from the carburetor to be forced into the cylinder through the open intake valve by atmospheric pressure. Near the completion of the intake stroke of the piston, the intake valve closes, sealing the cylinder. As the piston moves out toward the cylinder head, the mixture is compressed into a much smaller volume. This is called the *compression* stroke. As soon as the mixture is ignited, combustion (burning) begins and the expanding gases drive the piston inward toward the crankshaft for the *power* stroke. Near the end of this stroke the exhaust valve opens, allowing burned gases to start their escape. As the piston moves toward the head, the remaining gases are forced out the open exhaust valve. This stroke, the *exhaust,* clears the cylinder for the start of the next cycle.

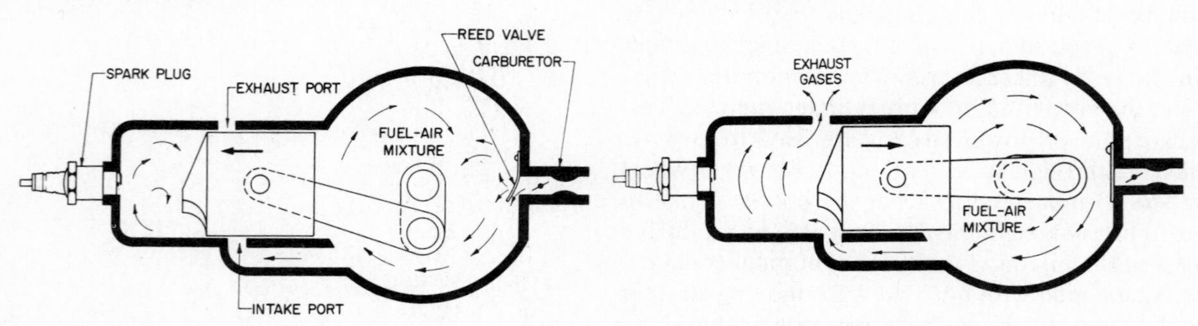


**Fig. 3-6.** Two-cycle engine. Compression stroke (left view) with crankcase vacuum opening the reed valve. Power, exhaust, and intake occur on downstroke of piston. (*Tecumseh Products Co.*)

**Two-Stroke-Cycle Engine** A two-stroke-cycle engine completes all four of the events with two strokes of the piston in the cylinder. Thus, all events are accomplished in one revolution of the crankshaft. Generally this engine is simply referred to as a two-cycle engine. The mechanism used to accomplish two-cycle operation is illustrated in Fig. 3-6. As the piston moves toward the head of the cylinder, it closes off *ports,* or valves machined in the side of the cylinder, and compresses a combustible mixture in the combustion chamber. At the same time a vacuum in the crankcase causes a new mixture to enter the dry-sump-type crankcase. When ignited, the burning mixture forces the piston toward the crankcase. Near the end of the stroke the sliding piston opens an exhaust port and allows the burned gases to excape. At nearly the same moment the piston moves past and opens an intake port on the opposite side of the cylinder. The inward movement of the piston compresses the mixture in the crankcase, which rushes through the intake port opened by the piston. The fuel mixture is now in the combustion chamber. *Compression* and *power* are developed on two separate strokes, while the *exhaust* and *intake* events take place at the time the piston is near the inward position in the cylinder. Since no oil is stored in the crankcase, the engine can be operated in any position. This fact plus the simple construction and light weight makes the two-cycle engine particularly adaptable to handheld portable tools, such as the chainsaw, and special groundskeeping equipment, including trimmers, mowers, and blowers.

## Comparison of Gasoline and Diesel Engines

Engines are also classified by the method used to ignite the air-fuel mixture in the combustion chamber. The two methods are spark-ignition and compression-ignition. Engines which burn gasoline or natural or liquefied petroleum (LP) gas use an electric spark to ignite the fuel in the combustion chamber. The compression-ignition engine is commonly referred to as a *diesel,* in honor of Dr. Rudolph Diesel who in 1894 developed the first successful engine of this type.

Most diesel engines used in modern agricultural machinery are the four-stroke-cycle type. These engines do not differ greatly in design and operation from the spark-ignition type except that they are constructed to withstand the higher compression and combustion pressures which are developed. The major difference is that a four-stroke diesel engine does not require a carburetor to mix the air and fuel. An injection pump and nozzle are used to place the proper amount of fuel into the combustion chamber of the cylinder.

The intake, compression, power, and exhaust strokes of the spark-ignition and diesel engine are compared in Fig. 3-7, diagrams *a* to *d*. The intake stroke occurs for both engines when the piston moves downward in the cylinder. The intake valve opens at the start of this stroke so that atmospheric

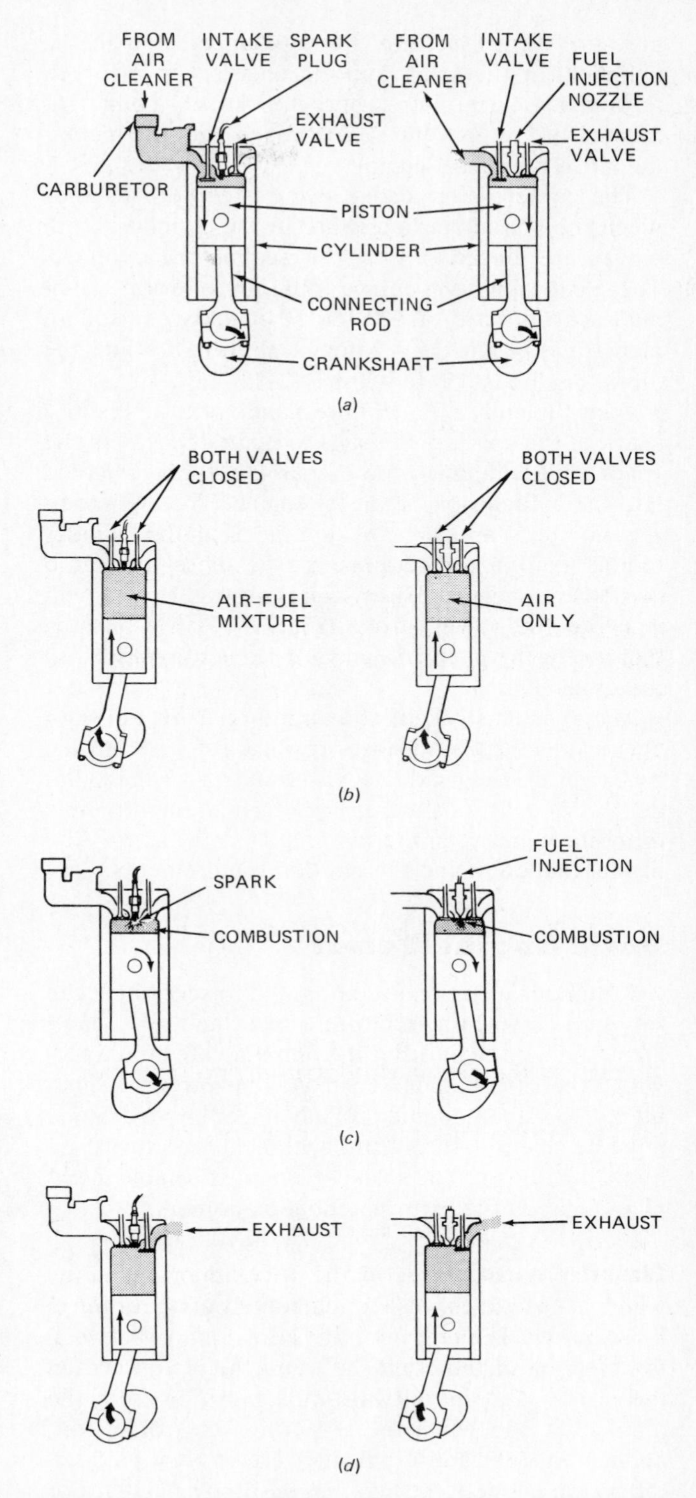


**Fig. 3-7.** Four-stroke-cycle spark-ignition gasoline engine (left) and compression-ignition diesel engine (right) compared. The four cycles include (*a*) intake, (*b*) compression, (*c*) power, and (*d*) exhaust.

pressure forces air into the vacuum created in the cylinder. In the spark-ignition engine, air is forced through the carburetor, where it picks up a combustible mixture of fuel. No fuel is mixed with the incoming air in the diesel engine.

The compression stroke occurs for both engines when the piston starts upward in the cylinder. Both valves are closed (Fig. 3-7*b*), sealing the cylinder. The spark-ignition engine has an average compression pressure of 100 to 150 pounds per square inch (psi) [689 to 1034 kilopascals (kPa)] while the diesel engine is 350 to 450 psi [2413–3103 kPa].

Near the end of the compression stroke an electric spark at the spark plug starts to ignite the fuel in the spark-ignition engine. In comparison, fuel is injected into the cylinder of the diesel engine. The high-compression pressures have heated the air in the cylinder to the ignition temperature of the diesel fuel, and combustion begins. The piston is now starting downward on the power stroke (Fig. 3-7*c*). The burning fuel forces the piston downward, generating force on the crankshaft.

The exhaust stroke is shown in Fig. 3-7*d*. For both engines, as the piston starts upward, it forces burned gases out the open exhaust valve and clears the cylinder for the start of another cycle. Thus, the principle differences between the two engines are the method of fuel induction and the ignition requirements.

## Basic Engine Terms

A number of terms are commonly used which the service person in agricultural power and machinery must understand in order to communicate effectively with other mechanics. To make the proper adjustments, the small-engine parts manager and mechanic must be able to interpret the technical literature used by maufacturers. The salesperson must be able to explain characteristics to potential customers.

**Dead Center** One of the very important terms which is used to express fundamentals of operation is *dead center*. The term identifies the relation between the position of the crankshaft and the piston. When the piston is at its outward-most position from the crankshaft and when the centerline axes of piston, connecting rod, and crankshaft are exactly parallel, the piston is said to be at *top dead center* (TDC). For example, TDC is located at the top of the circle, or 0°, in Fig. 3-8. When the crankshaft is rotated exactly 180° so that the piston, connecting rod, and crankshaft are aligned, the piston is at *bottom dead center* (BDC).

To identify the location of the piston in the cylinder in relation to the crankshaft, the centerline positions of dead centers are further divided into quarters (Fig. 3-9). Thus, as the crankshaft rotates clockwise, it is identified in its relation to the dead centers. If the crankshaft is rotating in the first quadrant of 0 to 90°, it is in the position referred to as *after top dead center* (ATDC). When rotating in the 90 to 180° quadrant, the position is referred to as *before bottom dead center* (BBDC). The remaining positions are identified as follows:

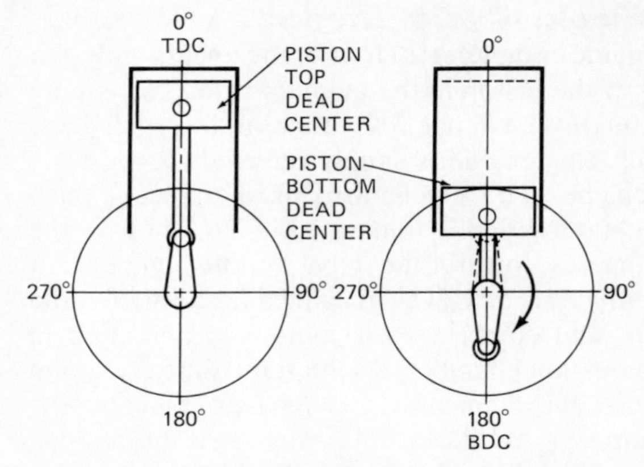

**Fig. 3-8.** Piston positions. Top dead center (TDC) and bottom dead center (BDC).

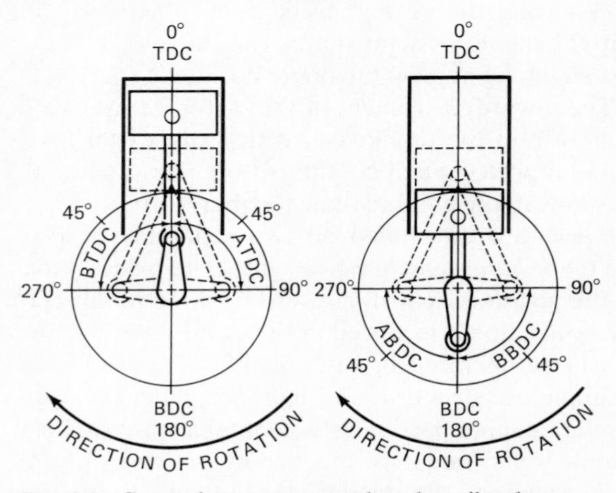

**Fig. 3-9.** Several terms are used to describe the position of the crankshaft in relation to TDC and BDC.

- The crankshaft positions between 180 and 270° are known as *after bottom dead center* (ABDC).
- The crankshaft rotating in the 270 to 360° quadrant is located in the *before top dead center* (BTDC) position.

These terms are used extensively to identify the position of the crankshaft in relation to the piston movement in the cylinder.

**Piston Displacement** The worker in agricultural power and machinery will need to be able to understand manufacturers' literature and identify the characteristics of engines for comparison purposes. *Piston displacement,* one of the important terms, identifies the volume of air which a piston moves (displaces) in one complete stroke. It is a term that is used to describe the size of an engine, and is expressed in cubic inches ($in^3$), cubic centimeters ($cm^3$),

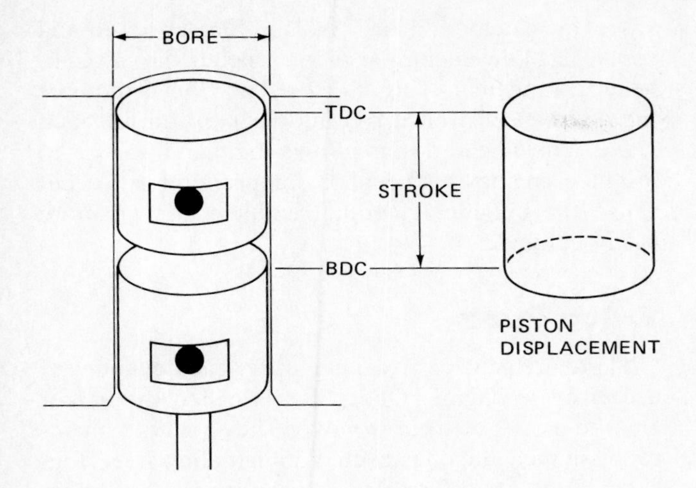


**Fig. 3-10.** The diameter of the cylinder is expressed as the bore, while the stroke is determined by length of piston travel. (*Crouse/Anglin, Motorcycle Mechanics. Copyright © 1982 by McGraw-Hill, Inc. Reproduced with permission.*)

or liters (L). Piston displacement has been determined by the manufacturer by using the following dimensions of the engine:

**Bore** The term *bore* identifies the measurement of the cylinder diameter in inches (in) or millimeters (mm) (Fig. 3-10). By knowing the diameter of the cylinder bore, the area of the cylinder in square inches ($in^2$) or square centimeters ($cm^2$) can be determined.

**Stroke** The *stroke* is determined by the number of inches [millimeters] the piston moves in the cylinder from BDC to TDC (Fig. 3-10). Stroke is controlled by the length of the offset (the *throw*) in the crankshaft. The stroke is always twice the length of the throw because the crankshaft turns one-half turn (180°) to complete one stroke of the piston.

It is standard practice to give the dimensions of an engine by listing the bore first. For example, an engine is listed as 2.750 in [69.85 mm] × 2.250 in [57.15 mm]. Respectively, these are the bore and stroke dimensions of the engine.

With this information the manufacturer determines the piston displacement by figuring the volume of a cylinder as follows:

$$V = A \times h$$

where $V$ = volume of the cylinder in cubic inches [cubic centimeters]
$A$ = area of the cylinder in square inches [square centimeters]
$h$ = stroke of the piston in inches [millimeters]

Thus, the piston displacement of the small engine with the measurements listed above is as follows (Fig. 3-10):

$$Pd = \frac{\pi d^2}{4} \times h \times \text{No. of cylinders}$$

where $Pd$ = piston displacement in cubic inches [cubic centimeters]
$d$ = bore of cylinder in inches [millimeters]
$\pi = 3.14$

**Example** Bore = 2.750 in [69.85 mm]
Stroke = 2.250 in [57.15 mm]

$$Pd = \frac{3.14 \times (2.75\ \text{in})^2}{4} \times 2.25\ \text{in} \times 1\ \text{cylinder}$$

$$Pd = 13.36\ \text{in}^3\ [219\ \text{cm}^3]$$

*Note:* For multiple-cylinder engines, multiply the piston displacement for one cylinder by the total number of cylinders. This value will identify total piston displacement.

**Example** If the engine used in the previous example had two cylinders, the total displacement would be

$$13.36\ \text{in}^3 \times 2 = 26.7\ \text{in}^3\ [438\ \text{cm}^3]$$

**Compression Ratio** *Compression ratio* (CR) is defined as the ratio of the total volume (TV) of the cylinder when the piston is at BDC to the cylinder volume when the piston is at TDC. This volume is called the *clearance volume* (ClV). The difference is generally expressed as the ratio of the total cylinder volume to the clearance (combustion chamber) volume. The illustration in Fig. 3-11*a* shows a compression ratio of 6:1.

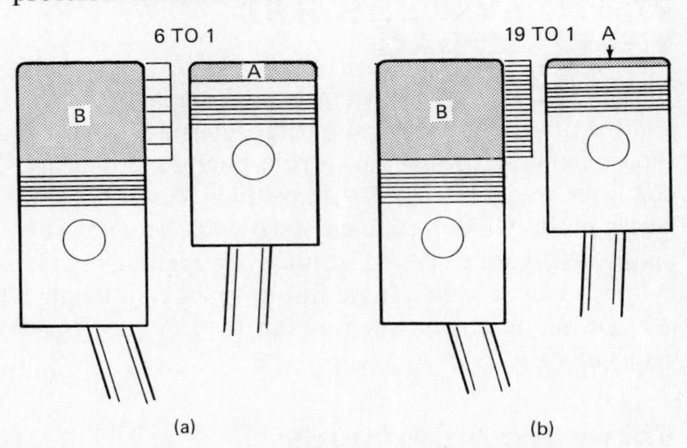


**Fig. 3-11.** (*a*) Compression ratio typical of a gasoline engine. (*b*) Compression ratio typical of a diesel engine. A = clearance volume (ClV); B = total volume (TV). (*Onan Corp.*)

The manufacturer computes the compression ratio for an engine by applying the following formula:

$$CR = \frac{TV}{ClV}$$

The total cylinder volume is obtained by adding the piston displacement volume (PdV) and the clearance volume. Hence, compression ratio is determined by the following formula:

$$CR = \frac{PdV + ClV}{ClV}$$

**Example** The piston displacement of the engine previously calculated was 13.36 in$^3$. If the clearance volume is 2.67 in$^3$, the compression ratio is as follow:

$$CR = \frac{PdV + ClV}{ClV} = \frac{13.36 + 2.67}{2.67} = 6.0:1$$

This ratio means that 6 volumes are compressed into 1 volume on the compression stroke.

Most small air-cooled gasoline engines have a compression ratio of approximately 6:1. The corresponding cylinder compression pressure for this ratio is approximately 100 psi [689 kPa]. In comparison, the compression pressure of the small air-cooled diesel engine with a 19:1 ratio (Fig. 3-11*b*) is approximately 450 psi [3103 kPa]. Some diesel engines have a compression ratio as high as 21:1. This will produce a pressure of approximately 600 psi [4137 kPa] in the cylinder on the compression stroke.

When the manufacturer's literature lists the compression ratio of an engine, the mechanic will understand that it identifies the type of engine and its cylinder compression pressure.

## VALVE SYSTEMS AND VALVE TIMING

The type of valve system used in small internal-combustion engines is primarily determined by the type of engine. Small four-cycle engines utilize the common poppet valve systems like those used in larger multicylinder engines. Two-cycle engines rely upon the piston to cover or uncover openings in the cylinder called *ports*. In addition, two-cycle engines use automatic valves such as the reed or the rotary disk valve.

### Valve Arrangements

Small single-cylinder, four-cycle gasoline engines are usually constructed using the L-head type of valve arrangement (Fig. 3-12*a*). Small diesel and some gasoline engines use the I-head, or valve-in-head, arrangement (Fig. 3-12*b*). The I-head requires the addition of a push rod and rocker arm for each valve. The I-head design allows the manufacturer to produce engines with higher compression ratios because the cylinder clearance volume can be more easily confined.

### Valve Parts

There are generally two valves for each cylinder of a four-cycle engine. One valve is called the *intake* and the other the *exhaust* valve. They are both mushroom-shaped and constructed of high-alloy steel (Fig. 3-13). They are referred to as *poppet* valves because they raise to open and close against a seat during operation. Each valve resembles the other in shape and size. The valve sealing face is accurately ground so that when the valve rests on its seat, it will seal the combustion chamber and prevent escape of the compressed gases. The exhaust valve is constructed of a special alloy steel that will resist the extremely hot exhaust gases whose temperatures may reach 4500 degrees Fahrenheit (°F) [2482 degrees Celsius (°C)].

Valves are held on their seats by spring tension. The opening and closing sequence of the valves is controlled by the *camshaft*. This shaft is equipped with irregular surfaces called *lifting lobes*. The lobes strike a *tappet*, or cam follower, which rides on the lobe and lifts the valve by making contact with the end of the valve stem. This also compresses the *valve spring*, which is used to return the valve to its seat. The camshaft is fitted with a gear which meshes with a smaller gear on the crankshaft. As the crankshaft gear rotates, it drives the camshaft gear. The rate of camshaft rotation is a critical part of valve timing. Each of the valves must complete cycles of opening and closing during two revolutions of the crankshaft. The ratio of rotation, 1:2, is obtained by using a cam gear with twice the number of teeth as the crankshaft gear. When properly meshed, the camshaft rotates one revolution for each two turns of the crankshaft.

### Valve Timing

Each valve must be coordinated to open and close with the position of the piston in the cylinder and the rotating crankshaft. Valve timing is therefore expressed in degrees of crankshaft rotation. The valve timing diagram in Fig. 3-14 is typical for small air-cooled four-cycle engines.

Note that the intake valve starts to open 20° BTDC and closes 50° ABDC. Thus the valve is off its seat for a total duration of 250° of the crankshaft rotation. The exhaust valve opens 50° BBDC and closes 20° ATDC. It also is off its seat for a duration of 250° of

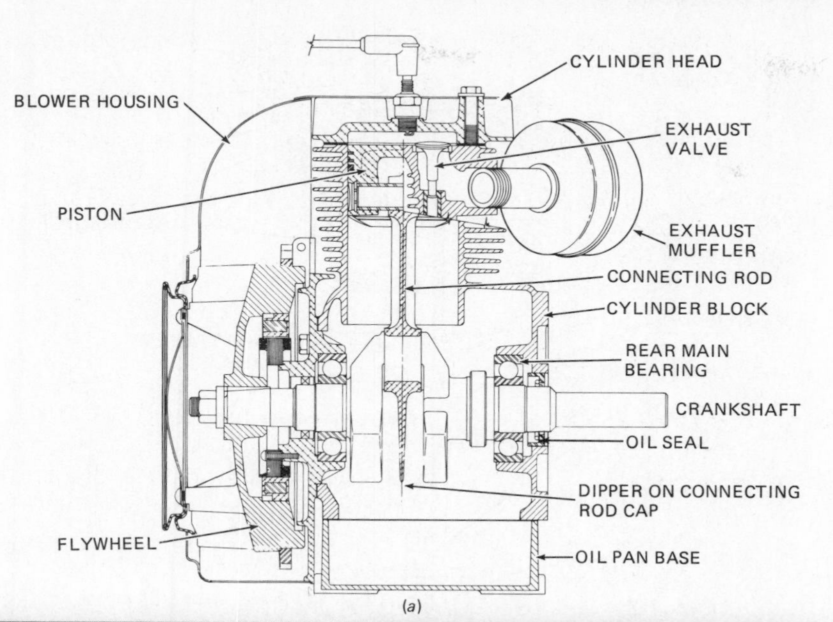


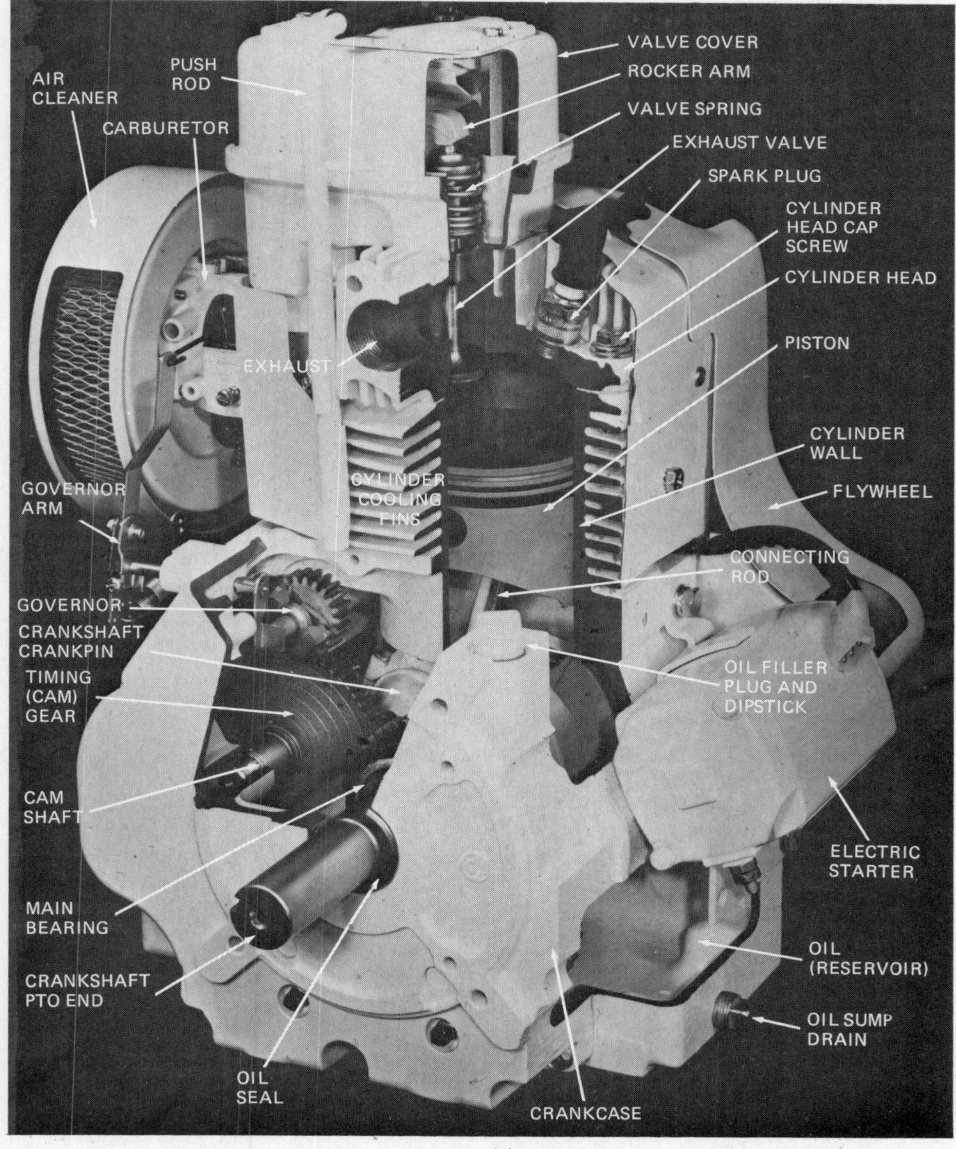


**Fig. 3-12.** (*a*) Cross section of a typical L-head small engine. (*Kohler Co.*) (*b*) Cutaway view of an I-head engine. (*Tecumseh Products Co.*)

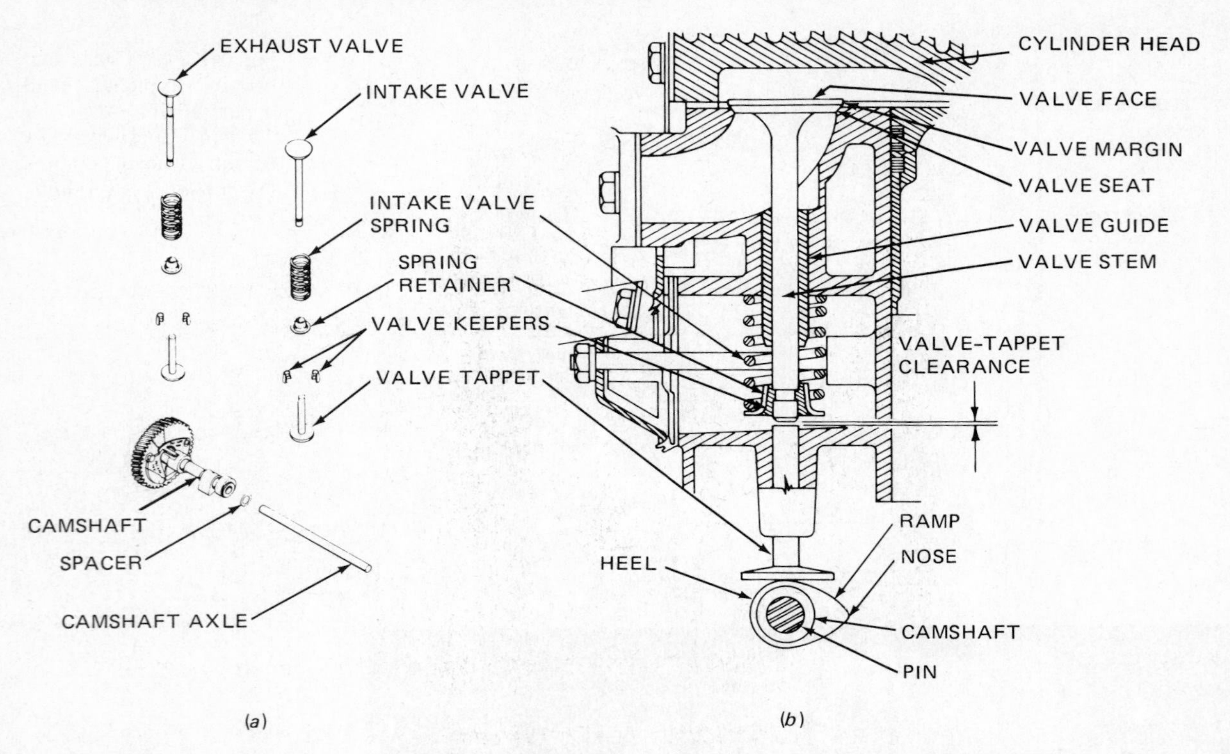


**Fig. 3-13.** Valve parts. (*a*) Camshaft and valve mechanism, often called the valve train. (*b*) Cross section of L-head arrangement in a small air-cooled engine. (*Kohler Co.*)

crankshaft travel. The opening of the intake valve and closing of the exhaust valve occur when the piston is near TDC. During this time both valves are off their seats, and the intake valve's opening "overlaps" the exhaust valve's closing (known as *valve overlap*.) This timing is critical to achieve proper performance from the engine. To permit the mechanic to fit the cam and crankshaft gears properly for correct valve timing, permanent marks are placed on the gears to identify gear teeth which must be meshed (Fig. 3-15).

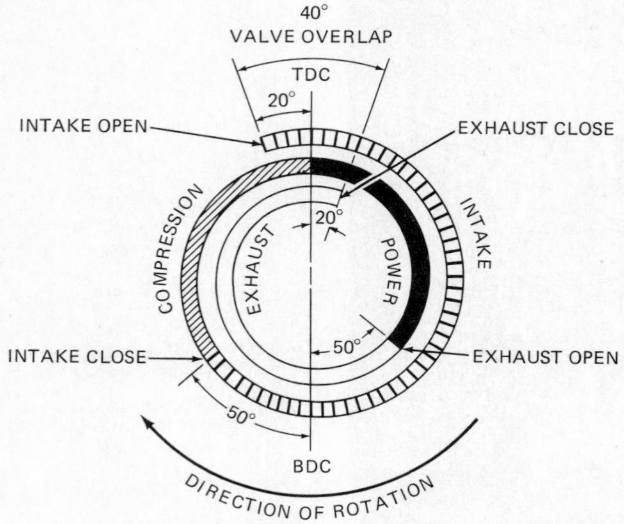

**Fig. 3-14.** Typical valve timing diagram for a small air-cooled four-cycle engine. Both valves are open for a brief period, called valve overlap.

The amount of valve tappet clearance has a pronounced effect upon the valve duration angle (the time, in degrees of crankshaft rotation, that the valve is off its seat). For example, Fig. 3-16 illustrates the effect of normal tappet clearance, too much clearance, and too little clearance on valve timing. A mechanic must realize that setting valve tappet clearance is an important method of setting the valves to operate at the proper time.

## Two-Cycle-Engine Valves

Two-cycle engines are fitted with fixed ports in the cylinder for intake and exhaust (Fig. 3-17). Timing of port opening and closing is dependent upon their location in relation to the piston in the cylinder. A typical valve timing diagram for a two-cycle engine is illustrated in Fig. 3-18.

The major differences in valve operation of two-cycle engines occur in the type of automatic valve used and the method of removing (scavenging) the exhaust gases from the cylinder.

**Reed Valve** A popular automatic valve is the *reed* type. This valve consists of thin strips of spring steel which contact an opening in the crankcase. During intake the vacuum created in the crankcase allows atmospheric pressure to bend the reeds inward, opening the valve (Fig. 3-19). As the piston

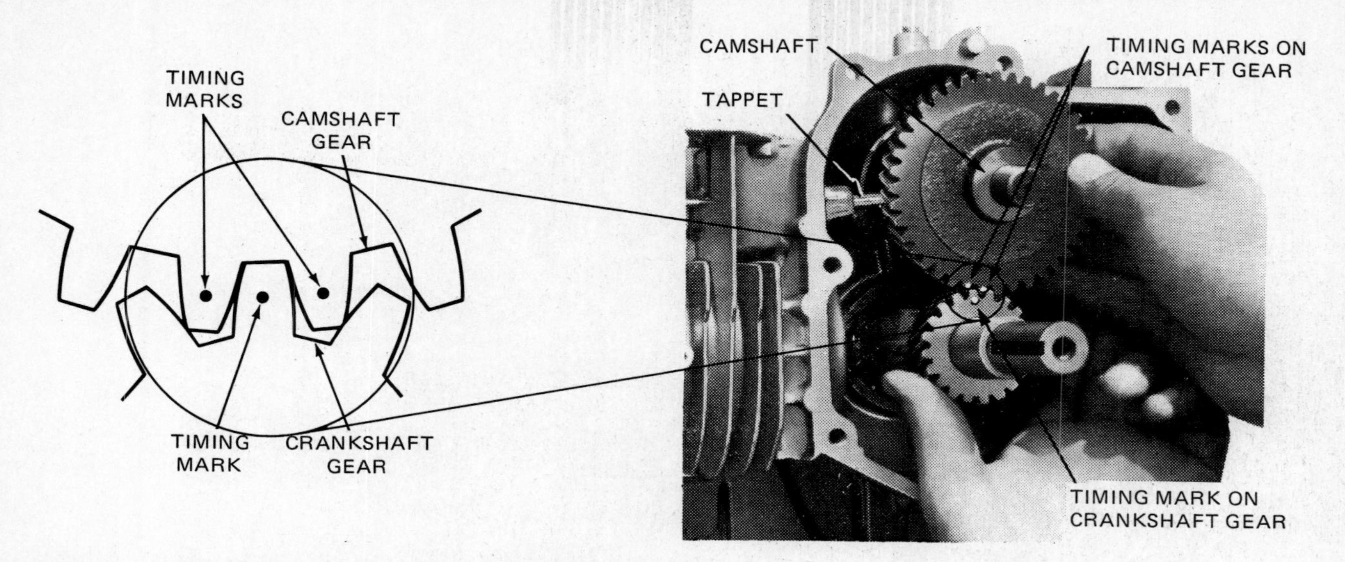


**Fig. 3-15.** A camshaft gear has twice as many teeth as a crankshaft gear. The gear ratio is 2:1 (note the timing marks). (*Teledyne Wisconsin Motor*)

starts toward the crankcase, pressure is developed, causing the reeds to press tightly against the intake opening and seal the crankcase.

**Rotary Valve** The *rotary disk* valve (Fig. 3-20) is a part of the crankshaft. There are two disks: one is stationary on the block; the other rotates with the crankshaft. The disks are machined to a flat surface. They contact each other and form a seal. This is a more positive-acting valve, because valve action occurs when the crankshaft rotation causes openings in each disk to align.

**Third-Port Loop-Scavenged Valve** The *third-port loop-scavenged* valve system operates in a similar manner to other two-cycle engines. The intake port to the cylinder is approximately 90° from the exhaust port. Thus the mixture enters from one side of the cylinder and makes a swirling loop. This action permits a more thorough mixing of fuel mix-

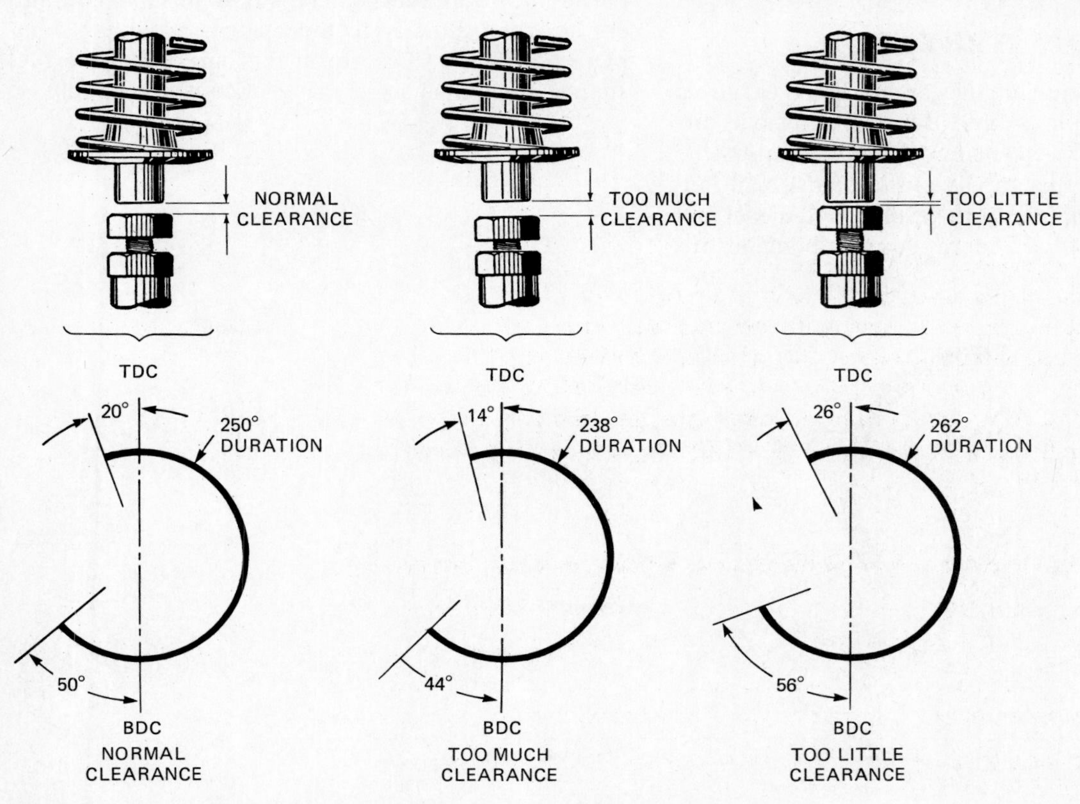


**Fig. 3-16.** Effect of valve tappet clearance on valve angle duration.

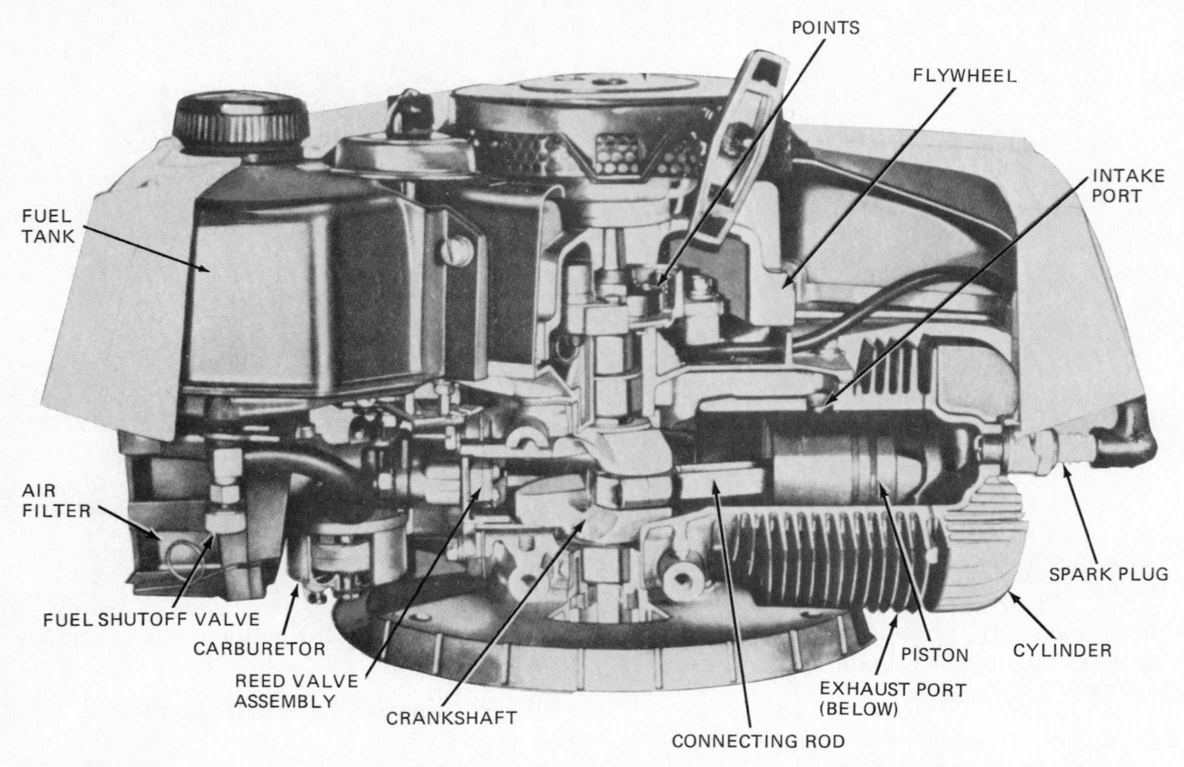


**Fig. 3-17.** Ports (holes in the cylinder wall) form the valves in two-cycle gasoline engines. (*Jacobsen Manufacturing Co.*)

ture in the combustion chamber and less loss of mixture during exhaust (Fig. 3-21).

## FUEL SUPPLY SYSTEM

Internal-combustion engines are designed to burn a petroleum product as a fuel. Because the small air-cooled engine was designed to serve as a portable power source, the fuel supply system must be an integral part of the engine. Both gasoline and diesel fuel are ideal sources of heat energy for small engines for two reasons. (1) A small quantity will produce a tremendous amount of heat energy when properly burned. (2) Both fuels can be stored in small containers attached directly to the engine, providing complete portability. Therefore, the purpose of the fuel supply system is to provide a constant supply of

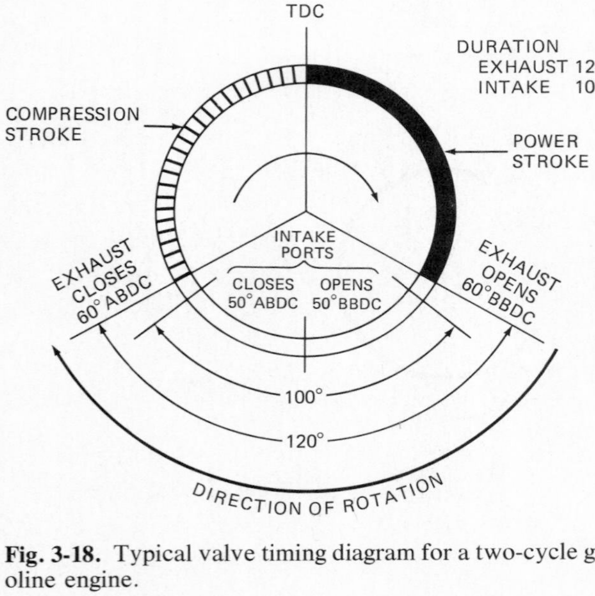

TDC
DURATION
EXHAUST 120°
INTAKE 100°
COMPRESSION STROKE
POWER STROKE
INTAKE PORTS
CLOSES 50° ABDC
OPENS 50° BBDC
EXHAUST CLOSES 60° ABDC
EXHAUST OPENS 60° BBDC
100°
120°
DIRECTION OF ROTATION

**Fig. 3-18.** Typical valve timing diagram for a two-cycle gasoline engine.

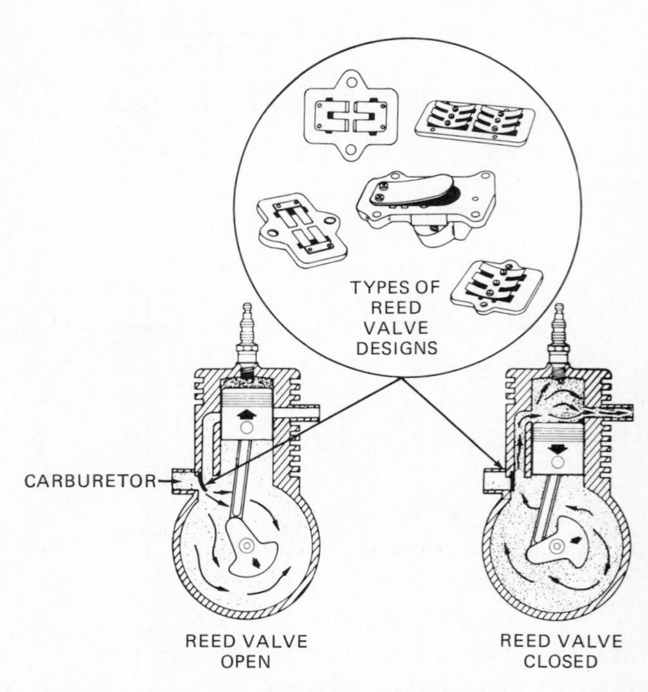


**Fig. 3-19.** Two-cycle engine, cross-scavenged—reed valve design. (*McCulloch Corp. and Tecumseh Products Co.*)

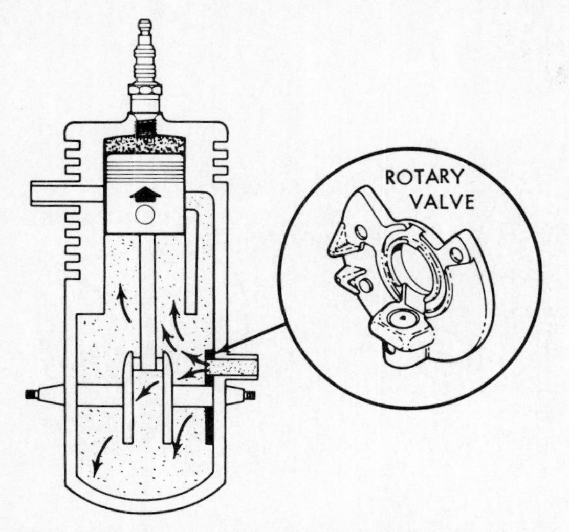


**Fig. 3-20.** Two-cycle engine, cross-scavenged—rotary valve design.

clean fuel to the carburetor of the spark-ignition engine or to the injection pump of the compression-ignition engine where it can be mixed with the proper amount of air to be burned.

## Fuel Systems for Spark-Ignition Engines

The location of the supply tank determines whether the system is *gravity-feed* or *force-feed*. In the gravity-feed system, the fuel tank is higher than the carburetor so that the fuel will flow downward to the carburetor (Fig. 3-22*a*). On engines where the fuel tank is mounted on a level with or below the carburetor, a gravity-feed system will not work. A fuel pump, however, can be used since it will deliver fuel to the carburetor regardless of the relative position of the fuel supply tank to the carburetor. In a force-feed system, a pump is used to lift the fuel against the force of gravity from the supply tank to the carburetor (Fig. 3-22*b*).

In all systems the supply tank must be constructed of a rust-resistant material to prevent contamination of the fuel system by rust. A special coated steel or plastic tank is used. A filler cap containing an atmospheric vent is used to prevent a vacuum from forming in the tank as fuel is removed. Such a vacuum would stop the flow of fuel. The tank dis-

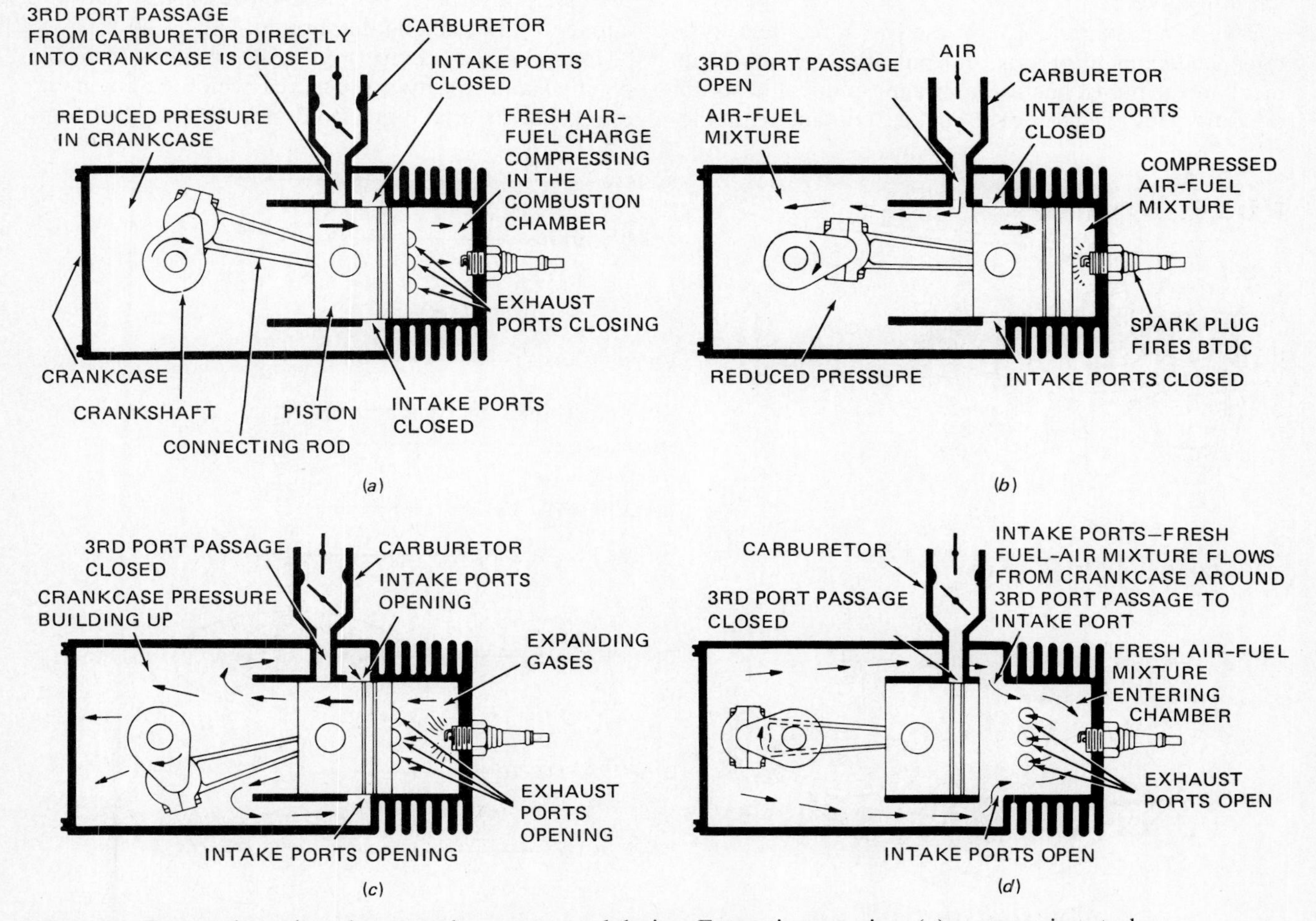


**Fig. 3-21.** Two-cycle engine, three-port loop-scavenged design. Events in operation: (*a*) compression stroke begins as all ports close; (*b*) intake of air-fuel mixture into crankcase through carburetor and start of power stroke; (*c*) power stroke and start of exhaust; (*d*) completion of exhaust and start of intake as air-fuel mixture enters combustion chamber from crankcase. (*Tecumseh Products Co.*)

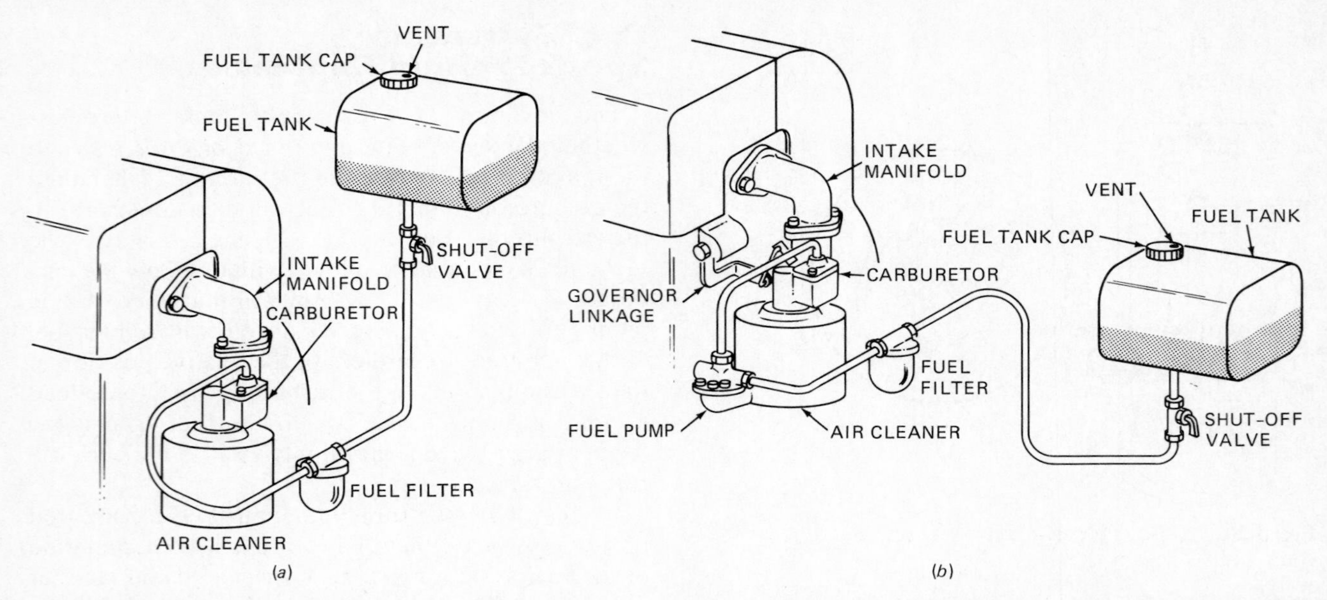


**Fig. 3-22.** Types of fuel systems. (*a*) Gravity system. (*b*) Force-feed system.

charge outlet is equipped with a filter screen and a shutoff valve.

Two types of pumps are used in force-feed systems: mechanical pumps and pulsation pumps. The mechanical pump has a diaphragm connected to an operating lever. The lever makes contact with the cam or crankshaft eccentric operating lobe (Fig. 3-23). The cam causes the lever to actuate the diaphragm, flexing it up and down. A pair of spring-loaded valves serve as intake and discharge ports for the fuel as alternate vacuum and pressure strokes are applied by the flexing diaphragm.

The pulsation pump (Fig. 3-24) operates from the positive and negative pressures which occur in the crankcase or intake manifold of both two- and four-

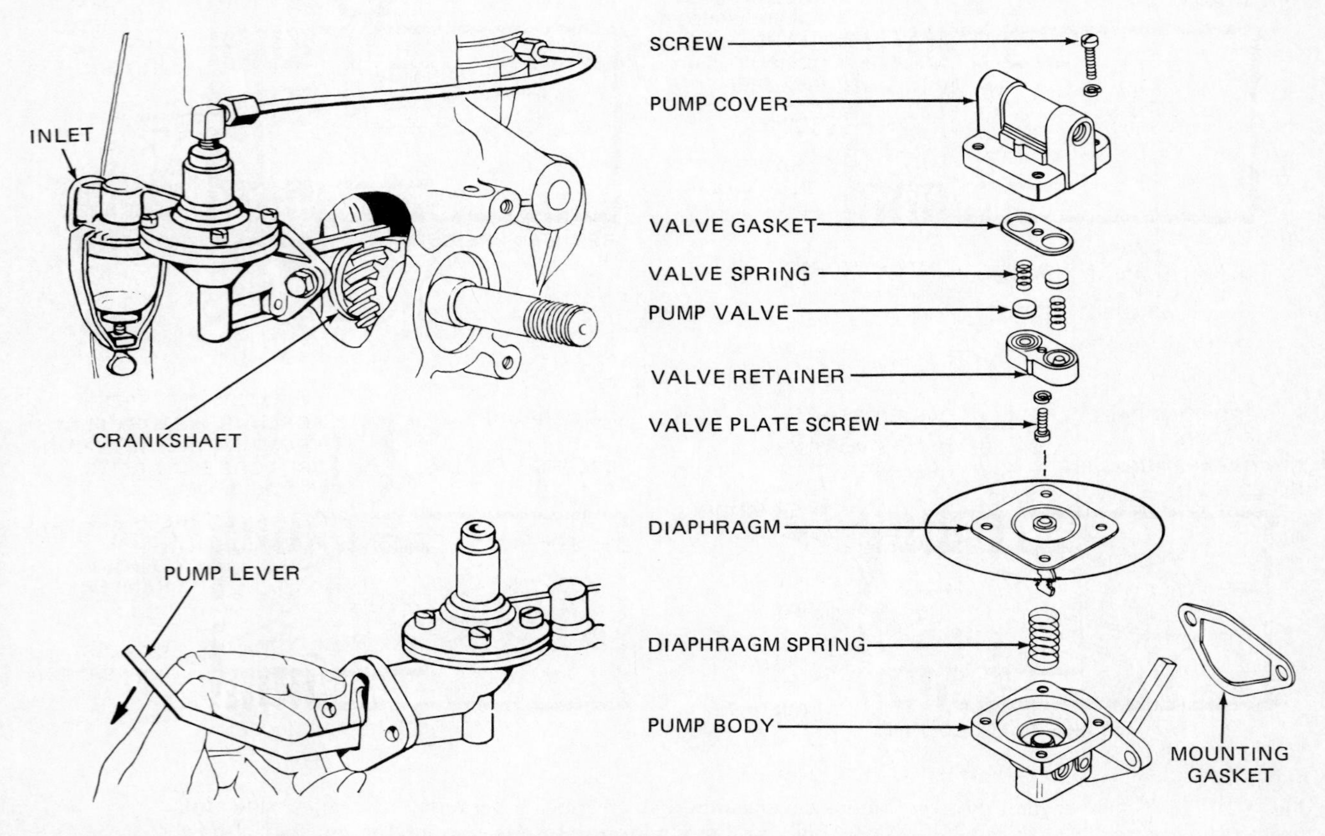


**Fig. 3-23.** Mechanical fuel pump. The diaphragm is made to flex up and down by the camshaft or crankshaft acting on the lever. (*Briggs & Stratton Corp. and Kohler Co.*)

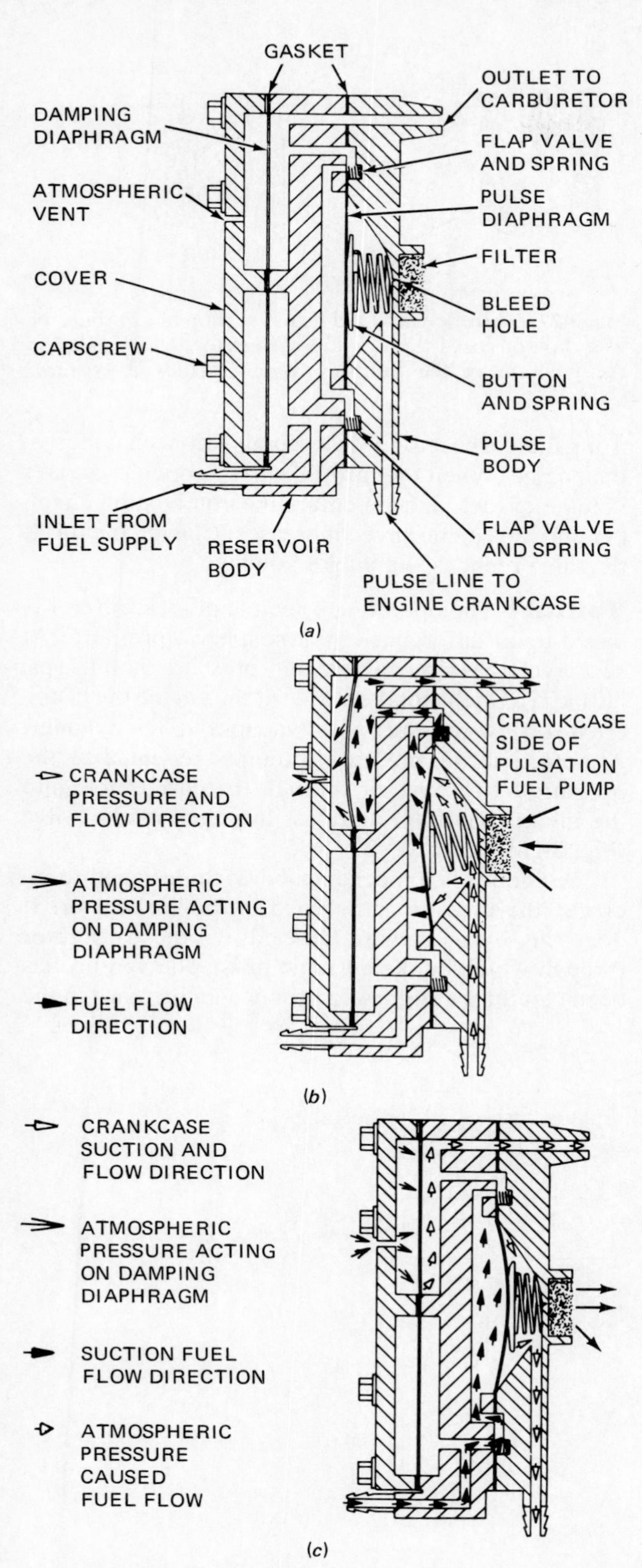


**Fig. 3-24.** Pulsation fuel pump. (*a*) Pulsations in the crankcase cause the pulse diaphragm to flex against spring pressure. (*b*) Crankcase pressure enters the pulse body, expanding the diaphragm. (*c*) Vacuum in the crankcase pulls the diaphragm toward the pulse body. Fuel is pumped by a series of flap valves and springs. (*Tecumseh Products Co.*)

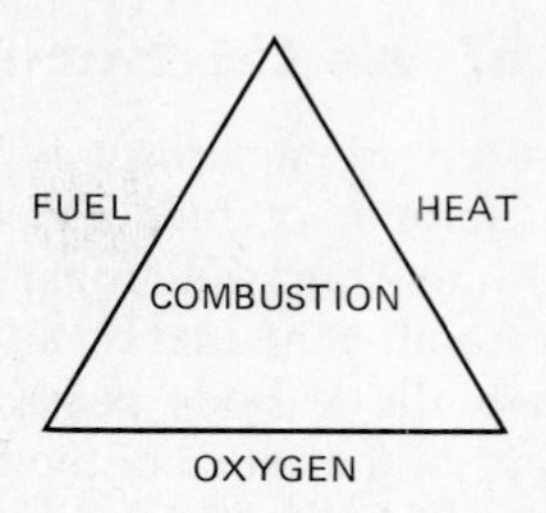


**Fig. 3-25.** The three essential elements for combustion.

cycle engines. This type of pump is especially popular on two-cycle engines because it can be built into the carburetor.

## FUEL INDUCTION SYSTEM

Fuel will not burn unless the requirements for combustion are satisfied. Figure 3-25 illustrates that combustion requires three elements—a fuel, oxygen, and heat. If one of the sides of the triangle is missing, combustion cannot take place.

To burn, a liquid fuel such as gasoline or diesel fuel must be mixed with the proper amount of air to obtain the necessary oxygen. It must then be heated to the kindling temperature. The *carburetor* is used in a spark-ignition engine to mix air and fuel. A compression-ignition (diesel) engine utilizes an *injector nozzle* to mix the fuel with the highly compressed air in the cylinder. In each situation the fuel must be mixed with air in the proper proportions. Also, the fuel must be broken up so finely (atomized) that it is in the form of a vapor (similar to the water vapor which is discharged as steam from a boiling pot of water) to obtain complete combustion.

### Air-Fuel Ratio

The relationship of mixing air and fuel is identified by the ratio of air to fuel. Table 3-1 identifies the air-fuel ratio requirements that must be provided by the carburetor. This ratio is based on the weight of air and fuel. On a volume basis (gallons) a ratio of 15 pounds (lb) of air to 1 lb of fuel [6.8 kilograms (kg) to 0.45 kg] would amount to approximately 9000 gallons (gal) of air to 1 gal of fuel [34,065 L to 3.785 L].

**TABLE 3-1 Air-Fuel Mixture Ratio Requirement, Gasoline Engine**

| | RATIO | |
|---|---|---|
| **Engine Operating Condition** | **Air, lb** | **Fuel, lb** |
| Starting, atmospheric temperature, 0°F | 0.4 | 1.0 |
| Slow idle, 400–600 r/min | 11.0 | 1.0 |
| Maximum power | 11.5–13.5 | 1.0 |
| Maximum fuel economy | 13.5–15.0 | 1.0 |

## Functions of the Carburetor

In addition to the primary responsibility of metering the correct amount of fuel, the carburetor has three additional important functions. First, a carburetor is used to atomize the fuel by adding air to it as it moves through the various passages. A second function is to control the speed of the engine by regulating the amount of fuel and air mixture taken into the cylinder on the intake stroke. The third function is to mix the correct proportion of air and fuel at varying engine speeds and loads.

## Parts of a Carburetor

A carburetor must be able to handle a liquid fuel such as gasoline. The basic parts consist of a container, or *bowl,* to confine a small amount of fuel and a nozzle to dispense fuel into the stream of air. The action is similar to the action of an airbrush. Figure 3-26*a* illustrates how the airbrush discharges a high-velocity stream of air past a nozzle. A vacuum is created at the nozzle which causes atmospheric pressure to push liquid into the nozzle where it is discharged with the air in a fine mist. A carburetor functions in a similar manner; however, certain other parts, as illustrated in Fig. 3-26*b*, are needed to make it operate an engine.

**Throttle Valve** The throat section of the carburetor is that portion of the structure which attaches to the intake manifold of the engine. It houses the *throttle* (butterfly) *valve*. The throttle valve controls engine speed (Fig. 3-27) by regulating the discharge from the carburetor to the combustion chamber of the engine. When the throttle valve is open, a greater volume of fuel mixture enters the combustion chamber and the engine speed increases. Closing the throttle valve reduces the intake volume.

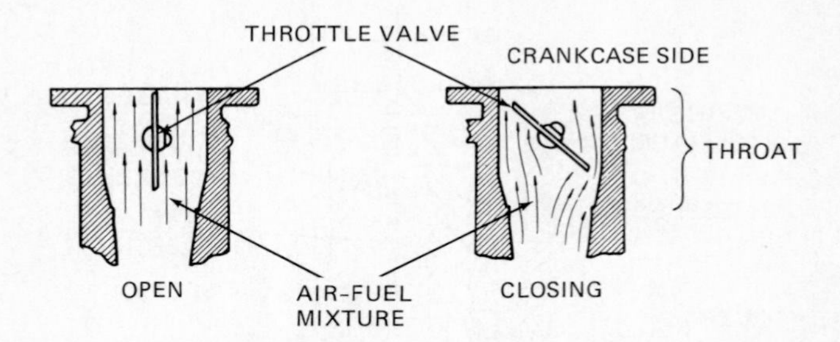


**Fig. 3-27.** Throttle valve and throat section of a carburetor. The flow of gases through the carburetor is controlled by the position of the throttle valve. (*Briggs & Stratton Corp.*)

**Venturi** The operating principle of a carburetor is based upon differences in atmospheric pressure. At sea level the atmospheric air pressure is 14.7 psi [101 kPa]. On the intake stroke of the engine the piston creates a low-pressure area (vacuum) in the cylinder. Atmospheric air pressure attempts to equalize the pressure by forcing air through the carburetor, into the intake manifold, and past the open intake valve (Fig. 3-28).

The venturi is a mechanical device designed to increase the pressure difference in the carburetor. It does this by narrowing the body of the carburetor through which intake air must pass. The venturi has been carefully designed to produce an increase in the

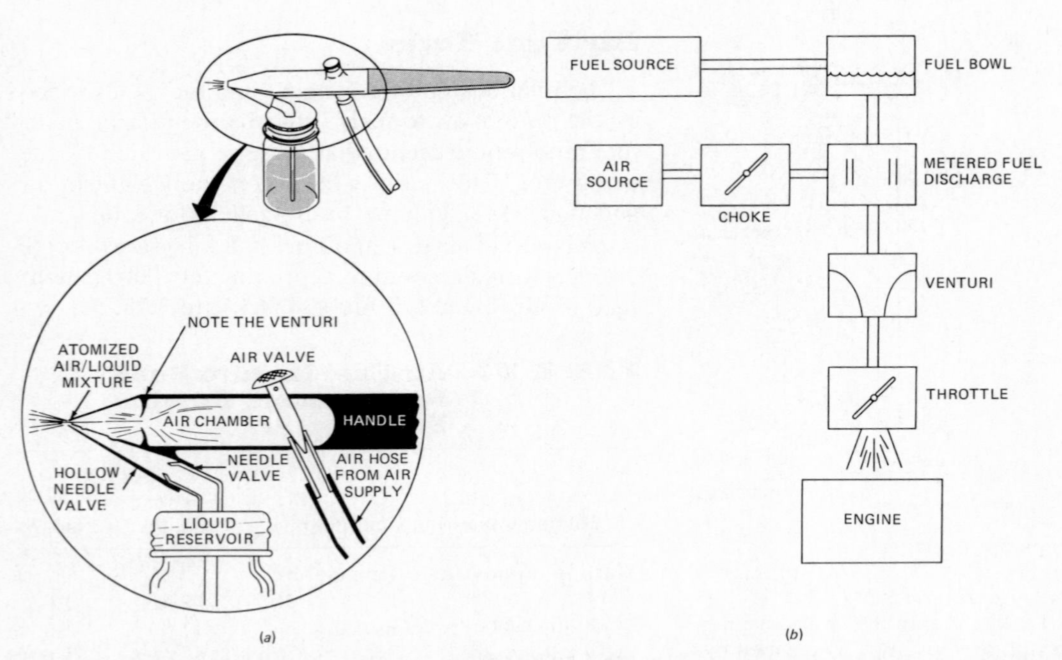


**Fig. 3-26.** (*a*) Airbrush operates on same principle as a carburetor. Air steam creates a vacuum at the nozzle causing liquid to rise and be discharged with air stream. (*b*) Essential parts of a carburetor.

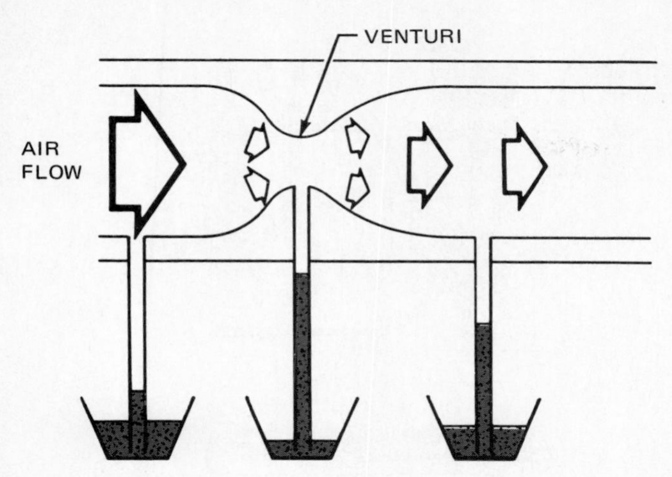


**Fig. 3-28.** Effect of a venturi on a vacuum through the carburetor. (*a*) Low vacuum in the air horn ahead of the venturi. (*b*) High vacuum. (*c*) Medium vacuum in the engine manifold. (*Crouse, Automotive Mechanics, 8th ed. Copyright © 1980 by McGraw-Hill, Inc. Reproduced with permission.*)

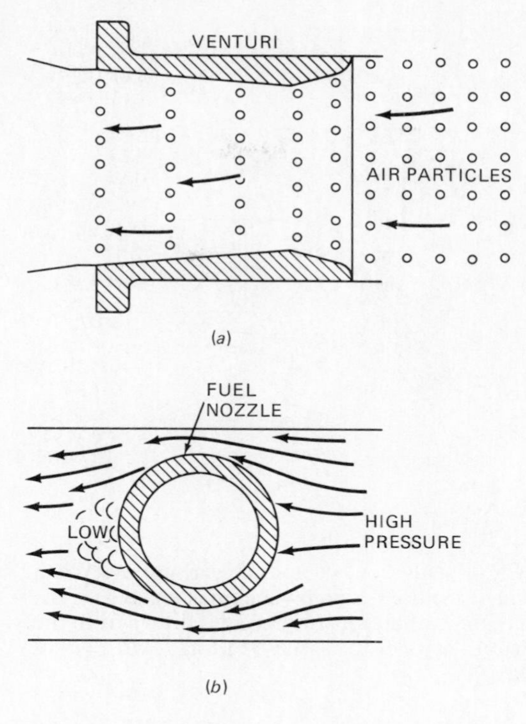


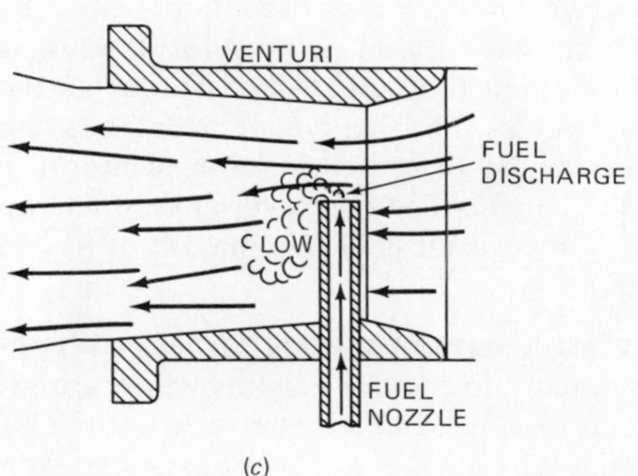


**Fig. 3-29.** Principles of the venturi and air foil. (*a*) Air particles are sped up as they move through a venturi by stretching the distance between them. (*b*) Air moving past the fuel nozzle creates an airfoil, where the lee side of the nozzle tube has an area of low pressure. (*c*) When combined, the venturi and airfoil work together to draw fuel through the fuel nozzle.

velocity (speed) of the air without decreasing its volume. As air picks up speed, the venturi also lowers the air pressure. A nozzle discharges fuel into the airstream. It is located in the venturi where the velocity of the air is the greatest and the pressure the least. The nozzle becomes an *airfoil* (Fig. 3-29). An airfoil has a rounded shape, like the nozzle, which also creates a low pressure. The venturi and nozzle work together to deliver fuel to the engine accurately.

**Choke Valve** The *choke* valve is used when starting the engine. It is similar in construction to the throttle valve. However, the choke valve may be larger in diameter and is located in the carburetor air intake ahead of the venturi. The choke can be closed to create a strong vacuum of 8 to 10 pressures less than atmospheric on the fuel supply system of the carburetor. The result is more fuel being drawn from the bowl, creating a richer mixture. The choke valve will have a small hole (auxiliary air valve) in one section to allow air to pass (Fig. 3-30). The opening is used to give the engine both fuel and some air to provide a combustible mixture. As soon as the engine starts, the choke valve is opened sufficiently to maintain operation. The choke valve is opened fully when the engine's temperature becomes warm enough and the fuel vaporizes more thoroughly.

**Float and Float Bowl** A carburetor requires a constant supply of liquid fuel. In the conventional float-type carburetor, the *bowl* (Fig. 3-31*a*) is used to store a small quantity of fuel which is immediately available for use. A float acting upon a needle valve at the fuel inlet controls the level. When the correct level of fuel enters the bowl from the supply, the float lifts the needle valve against its seat, which shuts off the supply by blocking the entry. During engine operation, fuel is drawn from the bowl. The float responds by allowing more fuel to enter through the open valve. Correct fuel level in the bowl is necessary in order to maintain a proper operating mixture.

Some small engines use *suction*-type carburetors. A relatively shallow fuel tank is attached directly to the bottom of this type of carburetor. The fuel pipe (Fig. 3-31*b*) draws fuel by vacuum directly from the fuel tank, which serves as a bowl. The level of the fuel in the tank varies with engine use. Another type

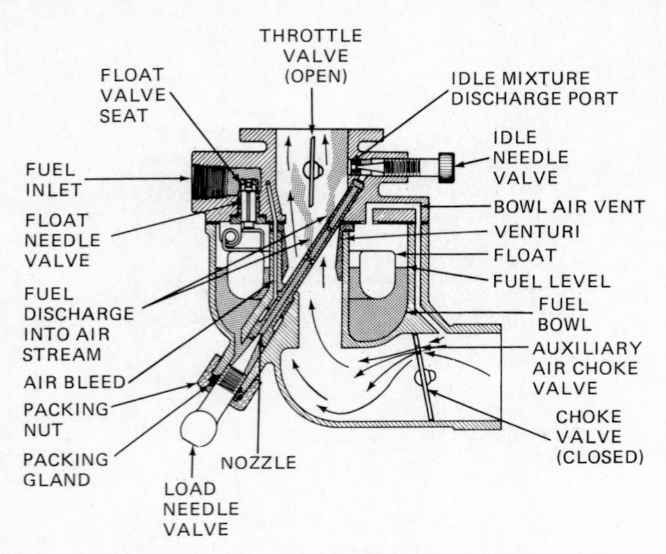


**Fig. 3-30.** Choke valve on an updraft carburetor. When the choke valve is closed, a high vacuum is created on the fuel system of the carburetor. Extra fuel is drawn from the main discharge nozzle to assist starting. (*Briggs & Stratton Corp.*)

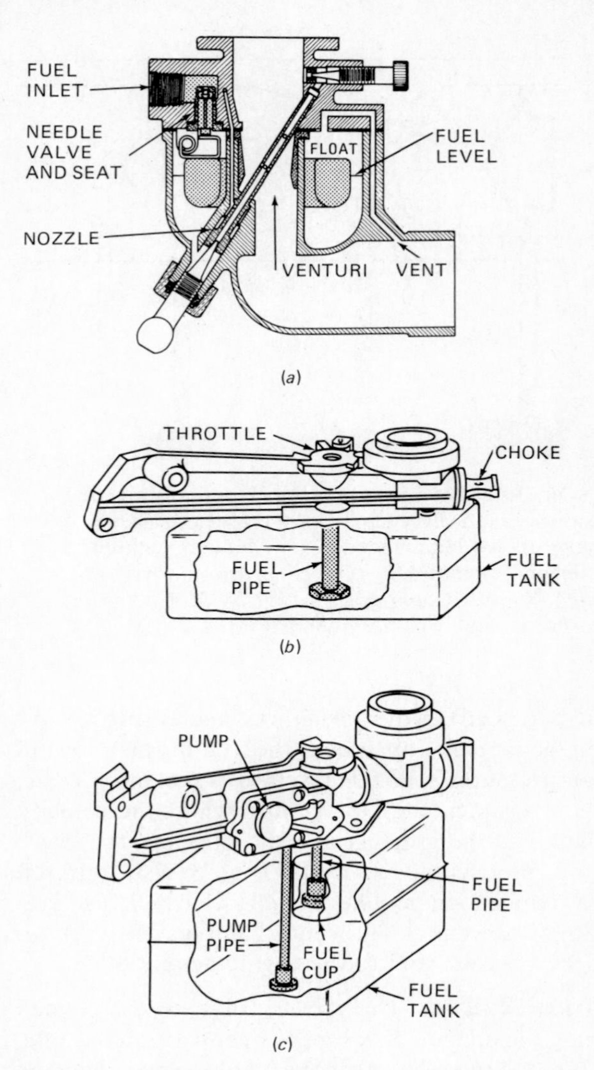


**Fig. 3-31.** Float and float bowl. (*a*) Conventional float-type carburetor. (*b*) Suction-type carburetor. (*c*) Pump-type carburetor. (*Briggs & Stratton Corp.*)

of suction carburetor has a larger fuel tank below the body of the carburetor. In this carburetor, the fuel is pumped into a small fuel cup, which serves as the fuel bowl (Fig. 3-31*c*). The fuel pipe then delivers fuel to the venturi region of the carburetor, where it is drawn into the engine. The fuel level in the fuel cup remains constant regardless of the amount of fuel in the supply tank.

**Float Carburetor Systems** A float carburetor is used on many four-cycle engines which are required to operate at varying engine speeds (revolutions per minute) and loads. For example, the operator of a small tractor with lawn mower attached stops to open a gate. The engine idles at low speed [1200 to 1800 revolutions per minute (r/min)] for a short time. To mow the lawn, the operator engages the mower drive, selects the correct forward speed, and moves the speed control to "high." The engine responds and accelerates to 3000 to 3500 r/min while driving the mower and moving the tractor.

The float carburetor includes an idle and high-speed load system to provide proper air-fuel mixtures for these conditions. The idle system (Fig. 3-32*a*) operates when the throttle is nearly closed. A small volume of air passes around the edges of the throttle valve, which places a vacuum on the idle discharge slot. A mixture of fuel and air is discharged on the intake manifold side of the throttle valve. The air-fuel mixture is adjusted by the idle valve screw to obtain the proper mixture for idle.

The high-speed load system (Fig. 3-32*b*) functions when the throttle valve is partially to fully open. The proper air-fuel mixture is controlled by the setting of the load mixture valve. The nozzle discharges the proper amount of fuel into the airstream at the venturi. When the throttle is fully open, the idle system does not operate.

## Diesel Injection Systems

Diesel engines operate on the ignition of an air-fuel mixture by compression. Fuel is sprayed into the cylinder near the end of the compression stroke. The air, heated by the compression, causes the fuel to burn. The mechanical system required to inject fuel to obtain an air-fuel ratio of approximately 0.04 lb of fuel [18 grams (g)] per 1 lb [453.6 g] of air must be very precise. The schematic diagram in Fig. 3-33 illustrates the parts of the diesel injection system.

**Injection Pump** The *injection pump* is designed to accomplish two important operations: (1) to de-

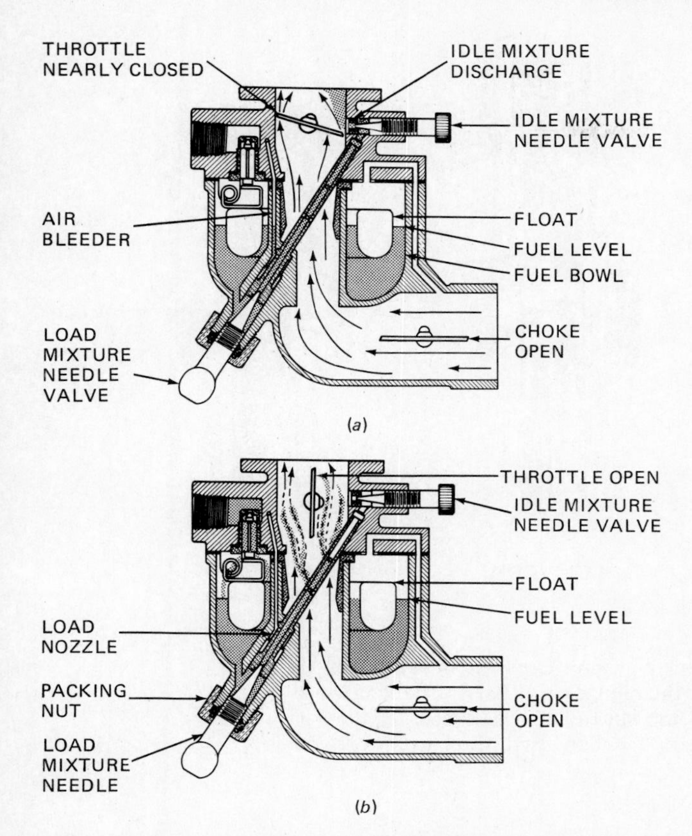


**Fig. 3-32.** Float carburetor systems. (*a*) Idle system. (*b*) High-speed load system. (*Briggs & Stratton Corp.*)

velop a high fuel pressure of over 2600 psi [17,924 kPa] in the injection system and (2) to regulate the amount of fuel discharge in order to control engine speed. Small one-cylinder diesel engines use a pump which is actuated by the engine camshaft. It consists of a cylinder element and a piston (Fig. 3-34), which is fitted with a bypass groove (helix) machined into the working end (Fig. 3-35). The rate of discharge from the pump is controlled by the effective length of the piston's stroke before the helix permits fuel to bypass. The piston is positioned by a geared sleeve which connects to the bottom of the plunger. The sleeve is rotated by the rack fitting the pinion gear

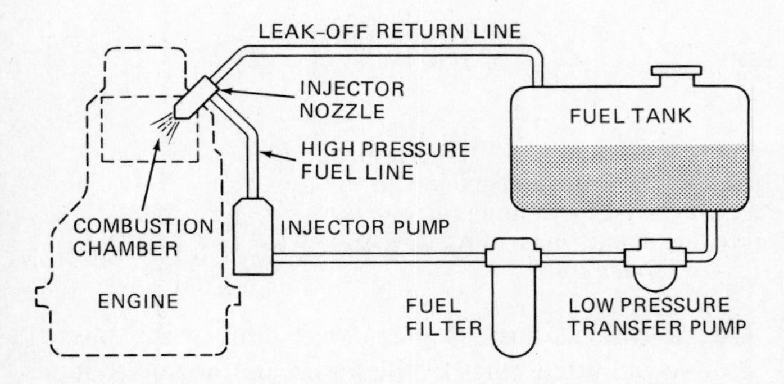


**Fig. 3-33.** Schematic diagram of a typical diesel injection system.

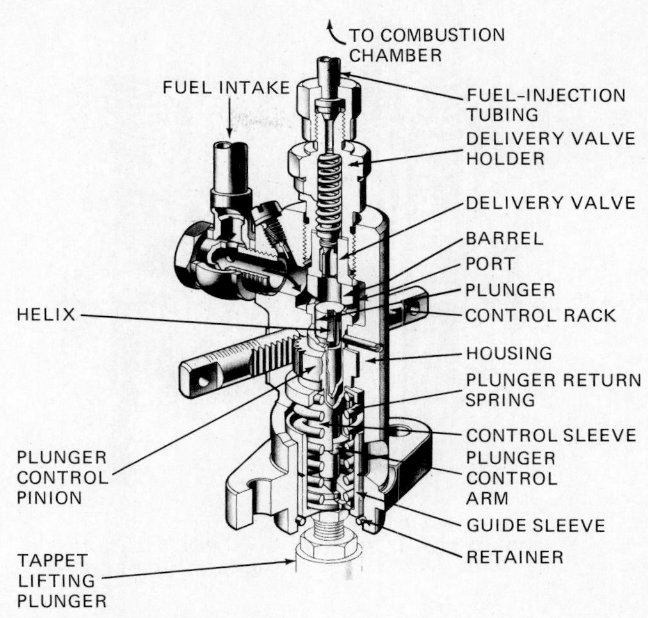

**Fig. 3-34.** Fuel injection pump for single-cylinder engines. The engine camshaft lifts the tappet. Fuel is pumped and metered to the injection nozzle by the upward movement of the plunger. (*Robert Bosch Sales Corp.*)

(Fig. 3-36). The helix portion of the plunger adjusts to produce the proper discharge. The length of time the fuel is sprayed into the cylinder determines the amount of power the engine will develop, as well as its speed in revolutions per minute (rpm). To stop the engine, the rack is moved to the cutoff position by the manual control. This permits the piston to bypass all fuel.

**Injector** The functions of the *injector* are (1) to start the injection of fuel into the cylinder when the fuel pressure reaches the specified pressure, approximately 2500 psi [17,237 kPa], (2) to atomize the fuel and direct it into the combustion chamber, and (3) to stop the spray when pressure from the pump drops below the injection pressure. In the cross section of the air-cooled diesel cylinders illustrated in Fig. 3-37, the nozzle discharges into a precombustion chamber within the cylinder head. In other systems the nozzle may discharge into a cavity in the piston. The latter is called *direct* injection. The systems are used to obtain greater atomization of fuel.

An injector consists of a nozzle body, needle valve, and pressure spring (Fig. 3-38). High-pressure fuel is pumped to the inlet of the holder, lifting the needle valve off its seat. Fuel is then sprayed into the cylinder. The design of the nozzle determines the type of discharge. A *pintle* nozzle (Fig. 3-39*a*) is designed to discharge an atomized stream of fuel into a precombustion chamber. A *hole*-type nozzle (Fig. 3-39*b*) is designed for direct fuel injection into the cylinder.

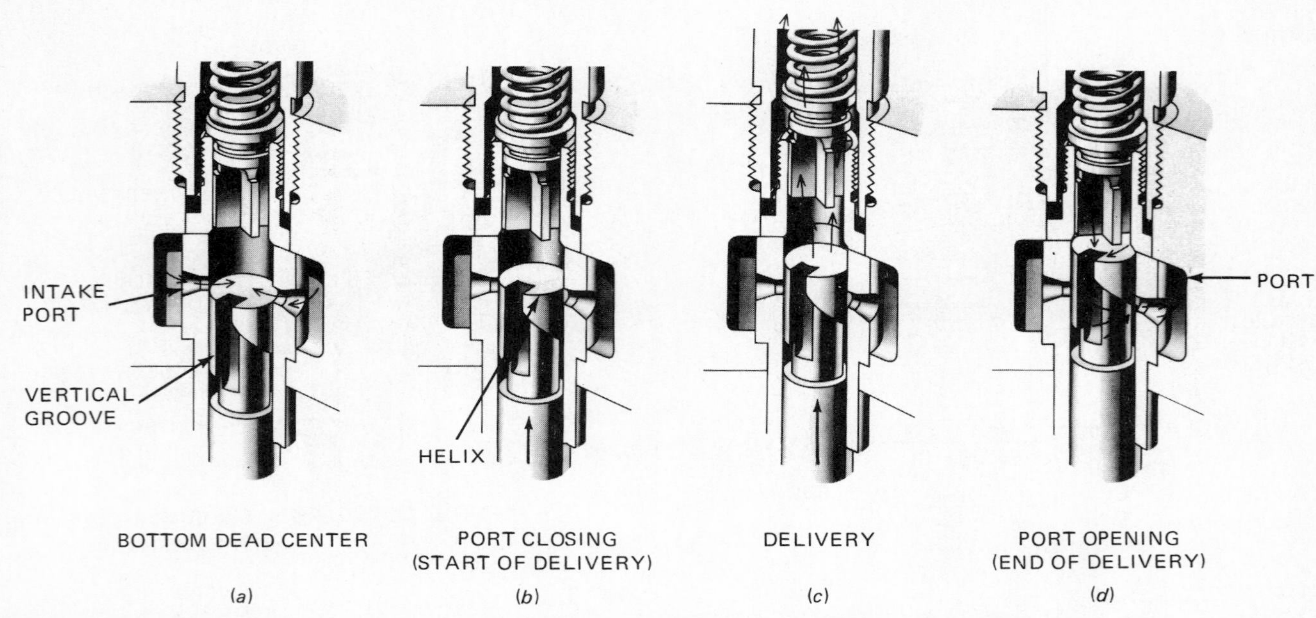


**Fig. 3-35.** The upward thrust of the plunger traps fuel in the barrel, resulting in the development of very high pressure. (*a*) Fuel enters the barrel of the pump when the plunger is at BDC; (*b*) the upward movement of the plunger closes the inlet ports in the barrel—the start of fuel delivery; (*c*) the upward movement of the plunger continues delivery until the helix (*d*) aligns with the barrel port and relieves pump pressure. (*Robert Bosch Sales Corp.*)

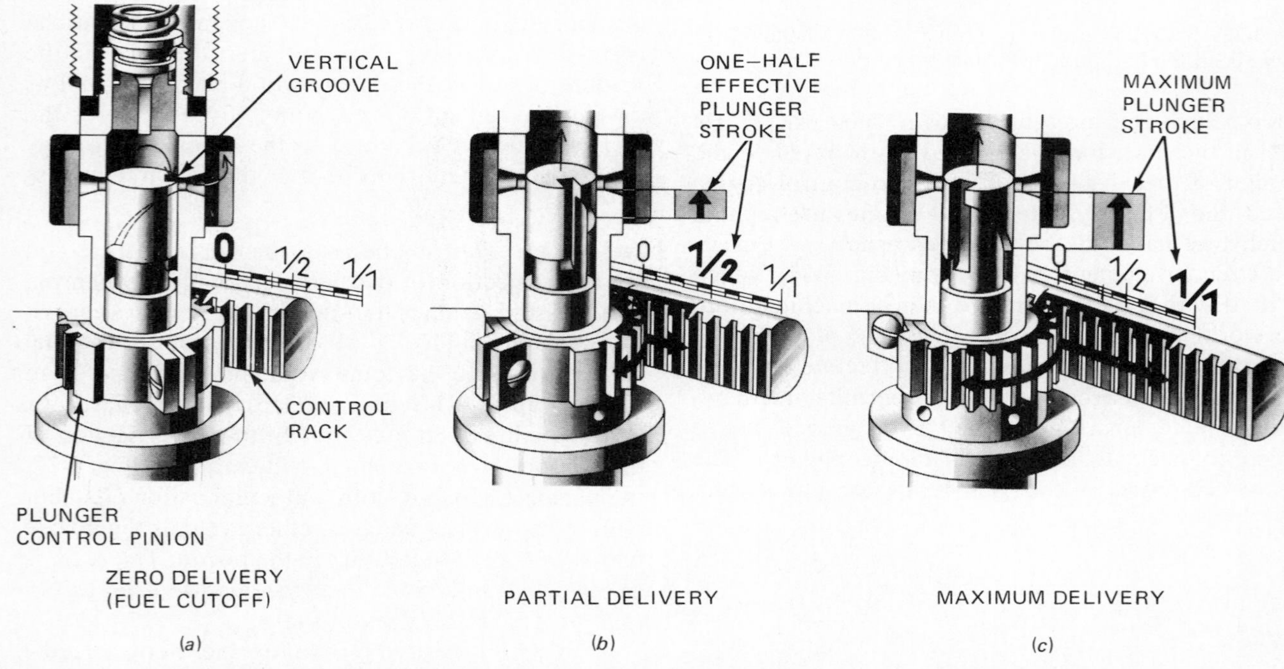


**Fig. 3-36.** Controlling the amount of diesel fuel discharge in a plunger-type pump. In (*a*), the control rack has moved the plunger to the zero delivery position with the vertical groove aligning with the port of the barrel. Partial fuel delivery (*b*) of one-half maximum is achieved by rotating the plunger with the rack and pinion; maximum delivery (*c*) is obtained because the plunger must move upward a greater distance before the helix aligns with the port. (*Robert Bosch Sales Corp.*)

Proper performance of the nozzle body seat and needle is necessary to control the time when injection starts and ceases. A nozzle which leaks small quantities of fuel past the valve seat is defective and must be serviced or replaced.

**Leak-off Line** Some diesel fuel must pass around the fitted parts of the pump and nozzle to lubricate the moving surfaces. Both the injection pump and nozzle return the leakoff fuel to the inlet side of the injector pump or to the fuel tank.

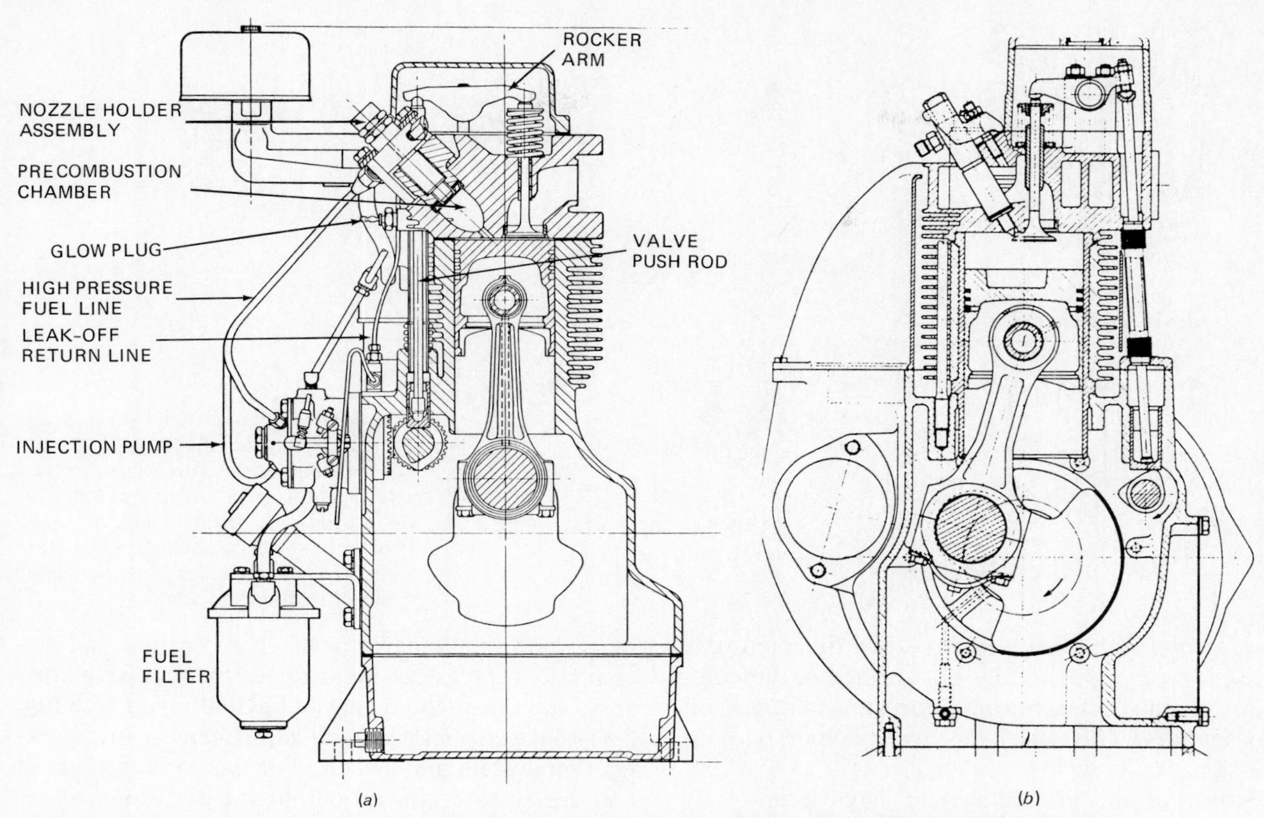


**Fig. 3-37.** Cross-section-views of two types of fuel injection. The nozzle sprays fuel (*a*) into the precombustion chamber and (*b*) directly into a cavity in the piston. (*Onan Corp.*)

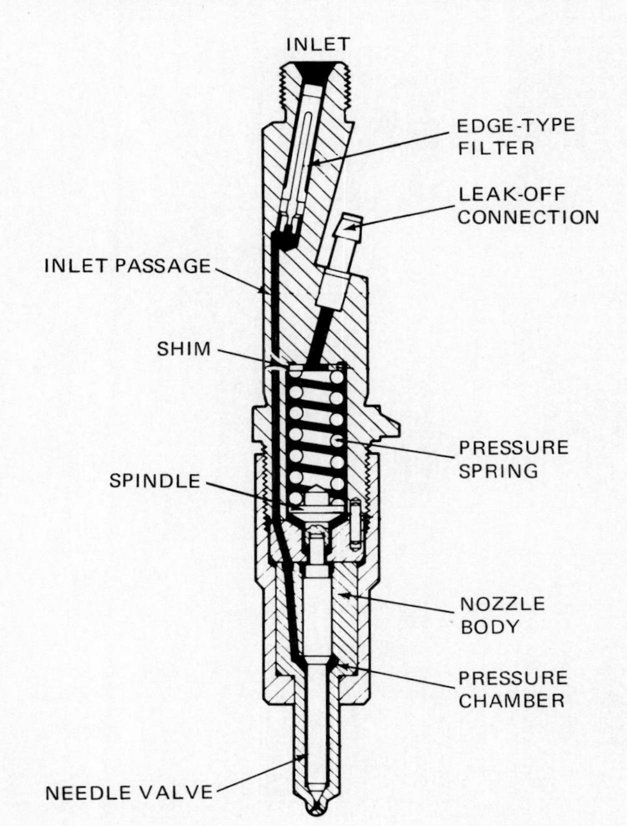


**Fig. 3-38** Cross section of a diesel fuel injector and identification of parts. (*Robert Bosch Sales Corp.*)

## IGNITION OF COMPRESSED GASES

As previously discussed, the two basic engines are the spark- and the compression-ignition types. The purpose of spark ignition is to provide a source of heat (the spark) which will raise the temperature of the air-fuel mixture to start the combustion process.

In the diesel engine the high-compression pressure raises the temperature of the compressed air to over 1000°F [538°C]. The heated air is capable of causing the diesel fuel to ignite when it is injected.

## THE COMBUSTION PROCESS

As illustrated in Fig. 3-40*a*, normal combustion of fuel-air mixture in the combustion chamber of a spark-ignition engine is a burning rather than an explosion. After ignition by an electric spark, a flame front burns evenly across the combustion chamber. The burning takes approximately three-thousandths of a second. The result is an even pressure rise of burning gases acting against the piston.

When the fuel is not the proper quality for the engine, *detonation* may occur, as illustrated in Fig. 3-40*b*. Detonation occurs following normal ignition, but as combustion pressures increase, the fuel becomes unstable and decomposes instantly, resulting in an explosion. A ''knocking'' sound is produced.

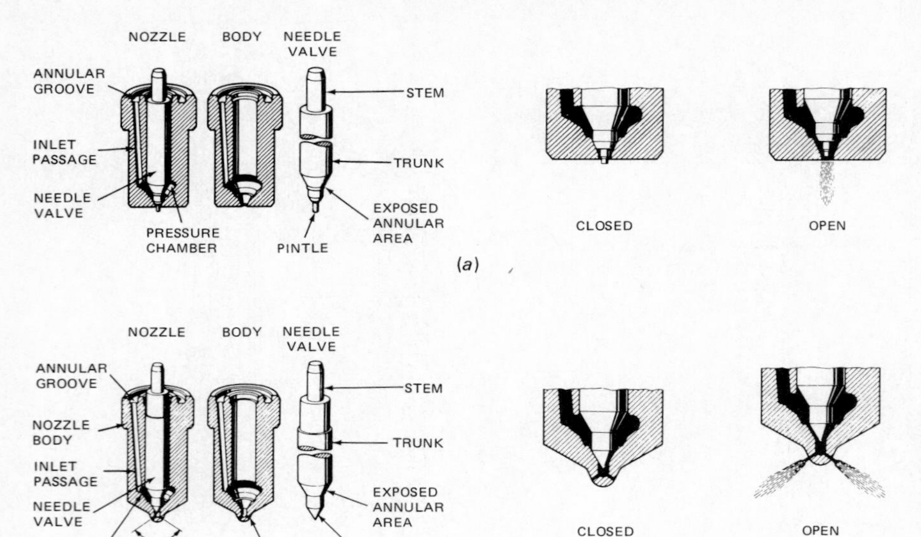


**Fig. 3-39.** Types of injection nozzles. (*a*) Pintle nozzle injects one stream. (*b*) Hole nozzle produces several streams at a specific spray angle. *Pintle* nozzles are used in engines with precombustion chambers, while *hole* nozzles are designed for engines using a direct injection system.

*Preignition* (Fig. 3-40*c*) occurs when the mixture is ignited by some abnormally hot surface in the combustion chamber. The burning starts ahead of normal spark ignition. The two flame fronts collide with an explosion.

Detonation and preignition can have harmful effects upon small-engine parts. The bearings and mechanical surfaces can be distorted by excessive pressure, and the piston damaged beyond repair (see Fig. 3-41). Proper operation and adjustment of the spark-ignition system and selection of the correct grade of fuel are essential for proper engine performance.

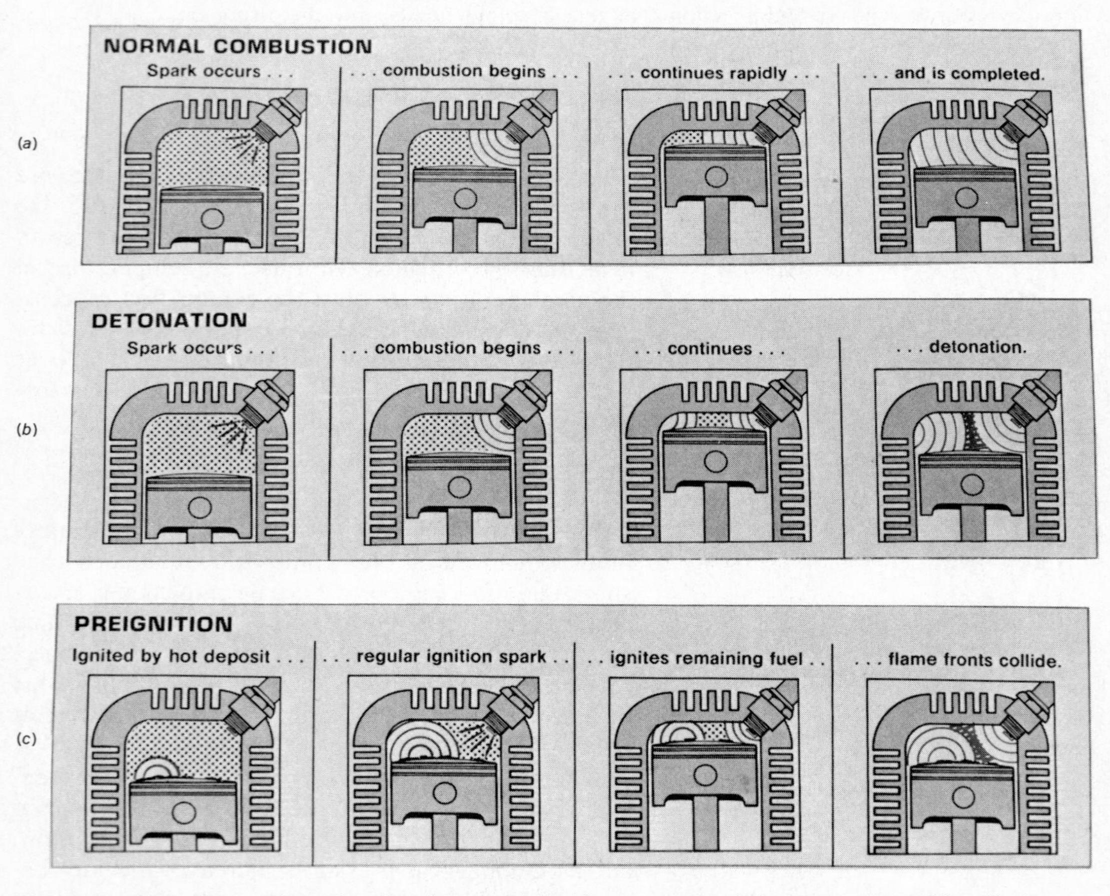


**Fig. 3-40.** Three types of combustion in a spark-ignition system. (*Champion Spark Plug Co.*)

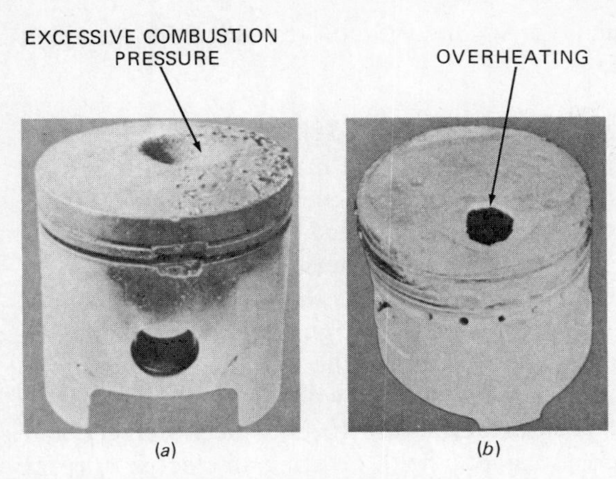


**Fig. 3-41.** Examples of pistons damaged by (*a*) detonation and by (*b*) preignition. (*Champion Spark Plug Co.*)

## Principles of Electric Spark Ignition

To achieve electric spark ignition it is necessary to have a source of electric energy. Gasoline engines generally utilize the *wet cell battery* or the *magneto* to operate the ignition system. The battery system produces electric energy from the chemical reactions which occur within the battery. The magneto develops electric energy from permanent magnets driven by the engine's rotation. The battery system is used in small engines when other sources of electric energy (for starting and lighting) are needed. The advantage of the magneto system is that it is a simple, reliable, and self-contained unit that requires no outside source of electricity.

**Electromagnetic Induction** Other than the difference in the source of electric energy, both systems achieve the development of a high-voltage spark of approximately 10,000 volts (V) to jump the spark plug gap. To convert the very low voltage from the battery or that developed by the magneto to very high voltage requires the use of a transformer called an ignition *coil* (Fig. 3-42).

The operating principle of the coil is based upon the principles of electromagnetic induction. This means that electric current in one winding generates a magnetic field which causes current to flow in a second winding. The coil used in small engines is composed of two sets of insulated wire wrapped about a soft iron core (Fig. 3-43). First, approximately 200 turns, or about 35 feet (ft), of No. 18 wire are wound about the core. This is the *primary* winding. When the primary circuit is complete, that is, the wires are not broken, an electric current will flow with a pressure of approximately 170 V. The flow of current will cause the iron core to become magnetized, and a very strong magnetic field will be developed around the coil.

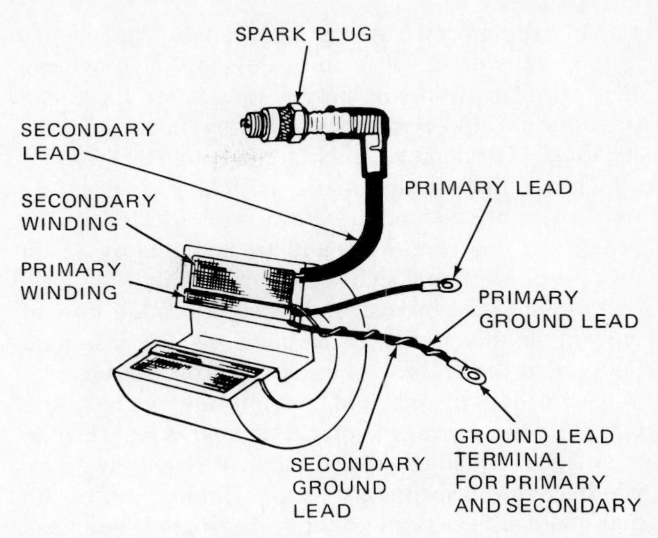


**Fig. 3-42.** Typical ignition coil used on a magneto-ignition system. (*Tecumseh Products Co.*)

Approximately 12,000 turns, or about 5000 ft, of a very fine (No. 32) insulated wire are then wound about the primary winding. This winding is referred to as the *secondary*. The primary and secondary windings are completely separate (insulated) from one another except that they share a common ground (Fig. 3-42). The opposite end of the secondary winding is connected to the insulated conductor that carries the high voltage to the spark plug center electrode. The secondary winding serves as the transforming mechanism for the voltage, which is proportional to the ratio of turns of wire used. The ratio is approximately 1 turn of primary winding to 60

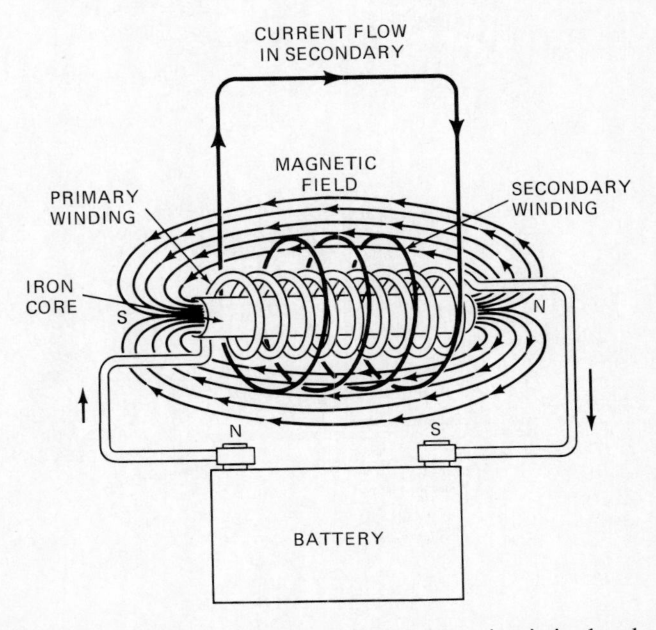


**Fig. 3-43.** Current flow in the secondary circuit is developed when the magnetic field, generated by current flow from the battery, is cut by the secondary winding.

turns of secondary winding. This causes a secondary voltage of over 10,000 V to be developed. However, the transformation will happen only when there is a disruption of the electric current flow in the primary winding. To achieve this action *breaker points,* which act like an automatic switching device, are used. The breaker points are always located in the primary circuit (Fig. 3-44) and are actuated by a cam located on the crankshaft or the camshaft.

As long as the breaker points are closed, a flow of current in the primary coil develops the magnetic field about the entire coil assembly from the electromagnetic action. When the points are opened, the current flow ceases; the magnetic field is lost. But as the magnetic field collapses, it cuts through the many windings of the secondary winding and generates the high voltage necessary to jump the spark plug gap.

To prevent arcing of the breaker points as they start to open, a "shock absorber" called a *condenser,* or *capacitor,* is connected across the points (Fig. 3-44). The condenser allows the points to break without an energy-absorbing spark. A clean break increases the speed at which the magnetic field cuts through the secondary coil to develop the high voltage.

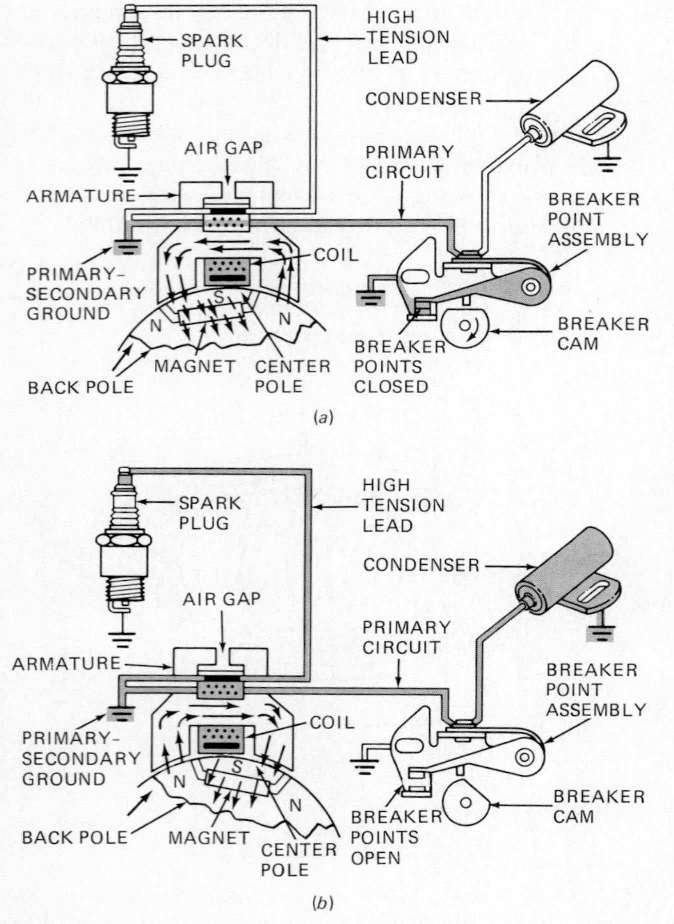


**Fig. 3-44.** Schematic of magneto operation.

**Flywheel Magneto Ignition** The electric energy needed to operate a magneto-ignition system comes from permanent magnets. In a flywheel magneto, the magnets are located in the outside edge of the flywheel. To transfer energy from the magnets to the ignition coil a specially shaped iron core called an *armature* is used. The ignition coil surrounds a part of the armature. As the flywheel rotates, two poles of the magnet align with the armature poles. This causes the armature to become a magnet and produce a magnetic field, creating an electric current flow in the primary circuit (Fig. 3-44*a*). At this point the breaker points are closed, completing the circuit to ground. The magnetic field is intercepted by the ignition coil. When the flywheel is rotated a few degrees, the poles of the magnet change position, which reverses the polarity of the armature poles (Fig. 3-44*b*). When the reverse flow is the fastest, the breaker points are opened by the rotating cam, which breaks the primary circuit. The reverse current flow into the condenser and primary windings of the coil cause the secondary windings to generate a high-voltage spark which jumps the gap at the spark plug.

**Battery Ignition** The primary difference between the magneto- and battery-ignition systems is the source of electric power. In the battery-ignition system a battery provides the power source. The diagram of a battery-ignition system is illustrated in Fig. 3-45. When the ignition switch is turned on, electric current flows from the battery through the primary winding of the ignition coil to the ground and returns to the battery through the closed breaker points (Fig. 3-45*a*). A strong magnetic field is created in the coil. As the crankshaft is rotated, it drives the breaker cam, causing the points to open at the correct time in the engine cycle. The opening breaks the primary circuit and current flow in the coil. The condenser absorbs the current flow to prevent an arc at the breaker points. The rapid collapse of the magnetic field cuts through the secondary coil winding, generating a high-voltage spark at the spark plug as in the magneto system (Fig. 3-45*b*).

**Solid-State Ignition** *Solid-state-ignition* systems utilize electronic components to serve the function of breaker points. There are no moving parts except the magnets. These switching devices are triggered, causing the ignition coil to develop a spark at the correct time. A wiring diagram of a solid-state magneto-ignition system is illustrated in Fig. 3-46.

The breakerless- (solid-state-) ignition systems include four major parts: (1) a low-voltage alternating

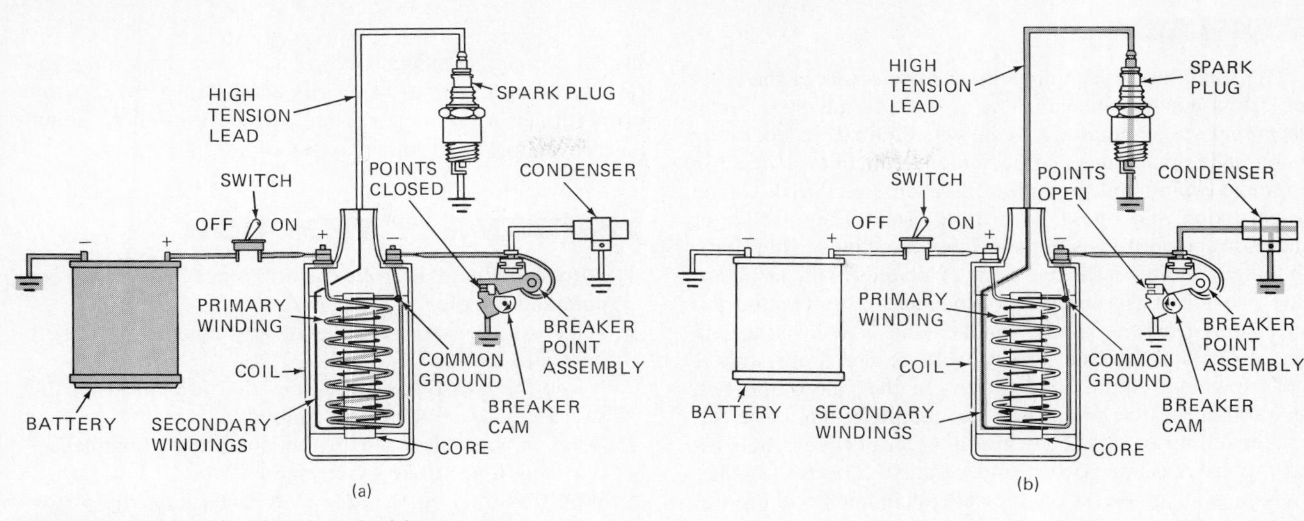


**Fig. 3-45.** Schematic of battery-ignition system.

current generator; (2) a silicon-controlled rectifier (SCR), which serves as a transistorized solid-state switching device; (3) an ignition coil assembly; and (4) a triggering device. The triggering device includes a trigger coil and a resistor to provide a very low switching current to activate the SCR solid-state switch. The function of the switch is to release the energy stored in the capacitor (condenser) to produce a low-voltage current flow into the primary windings of the ignition coil, where it is transformed to high voltage by a secondary winding. Spark timing is determined by the position of the flywheel magnets triggering the release of stored electric energy in the condenser. The process of generating high voltage in the ignition coil is identical with other systems. The steps in the operation of a solid-state system are illustrated in Fig. 3-46. In Fig. 3-46*a* the movement of the flywheel magnets develops a magnetic field in the low-voltage generating coil. This causes a current to flow in the low-voltage wires through the diode rectifier where it is stored in the condenser. In Fig. 3-46*b* slight movement of the flywheel reverses the magnetic field through the armature. The result is the development of alternating current in the low-voltage wire which is rectified by the diodes. Additional current is developed to charge the condenser to its full capacity. In Fig. 3-46*c* the flywheel magnets continue to rotate to the position of the trigger coil. The very low current developed in this coil is enough to trigger the transistorized switch to turn "on." With the transistor in the "on" position, the current stored in the condenser discharges into the primary windings of the coil. The magnetic field created by the primary windings is transformed into high voltage by the secondary windings, resulting in a spark at the spark plug.

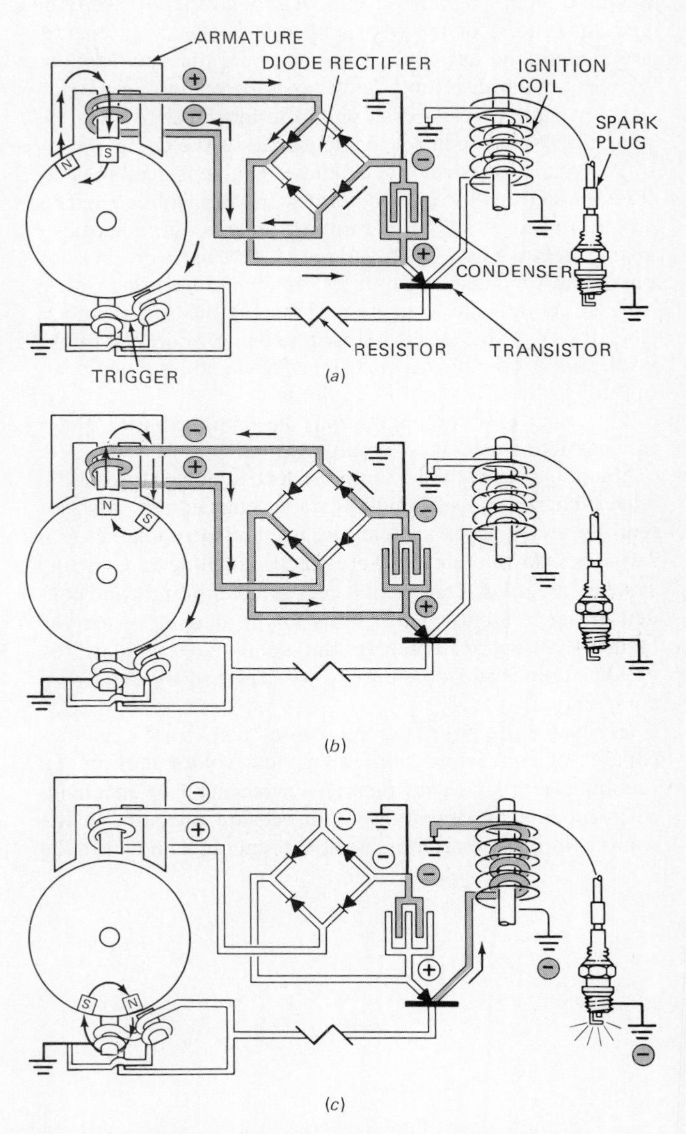


**Fig. 3-46.** Wiring diagram of a solid-state ignition system.

## SUMMARY

The small internal-combustion engine is one of the most important machines used in the agriculture industry. Small engines produce much of the power required for operating many of the machines and portable handheld tools used in groundskeeping and production horticulture. Small-engine construction and operation are patterned after the larger and more powerful gasoline and diesel engines. Therefore, an understanding of the operating principles of the small engine provides basic information needed to understand the operation of larger engines. Small engines may be either the two-stroke- or the four-stroke-cycle design. Four-stroke-cycle engines require four strokes of the piston and two revolutions of the crankshaft to complete the basic sequence of four events. They are, in order of occurrence, intake, compression, power, and exhaust. The two-stroke-cycle engine completes the four events in one revolution of the crankshaft. For this to happen, the intake and exhaust events occur as the piston completes the power stroke and begins the compression stroke. Although the construction and basic parts of the two- and four-stroke-cycle engines are similar, the external appearance and the valve system are very much different. The two-cycle engine uses cylinder ports and reed valves, while the four-cycle engine has poppet valves to control the sequence of the cycle.

The small four-stroke-cycle diesel engine is used in small tractors and small irrigation pumps. In the gasoline engine a spark is used to start the combustion process. The diesel engine relies upon heat generated by high-compression pressures to cause ignition as the fuel injection system sprays fuel into the heated air of the combustion chamber. This differs from the spark-ignition engine where the intake stroke allows the proper mixture of fuel and air from a carburetor to be taken into the cylinder.

The small gasoline engine may be equipped with either the magneto- or battery-ignition system.

Magnets provide the source of electric power for the flywheel magneto. Most small portable engines use this system. Battery-ignition systems are used on most engines that serve as a power source where electric lighting and starting systems are used. The ignition coil, breaker points, and condenser serve identical functions in the magneto- and the battery-ignition systems. In the solid-state system, the breaker points and condensers are replaced by electronic components.

In conclusion, internal combustion engines are machines capable of converting heat energy into rotary motion. To accomplish this difficult task an engine must be mechanically capable of developing compression pressure in the combustion chamber, must be able to mix fuel and air in the proper proportion, and must have a properly operating ignition system which will ignite the compressed gases at the proper time. When the three requirements are met, the engine is capable of producing power.

## THINKING IT THROUGH

1. How would you describe the difference between an engine and a motor?
2. What are the four events that are required to convert heat into motion?
3. Name the two types of engines that are based on the two methods of achieving ignition?
4. What is the difference in the operating principles of two- and four-stroke-cycle engines?
5. What information is required to determine the piston displacement of an engine?
6. When an engine has a compression ratio of 8:1, what is the piston displacement volume?
7. How does the valve operation of a two-cycle engine differ from that of a four-cycle?
8. Name four parts of the poppet valve system used on four-cycle engines and identify the purposes of each.
9. What are three types of valves that may be utilized in two-cycle engines?
10. What does it mean when engine manufacturers state that the exhaust valve should open 50° BBDC?
11. Compare the method of inducting fuel into the combustion chamber of the spark-ignition engine with the method used in the compression-ignition engine.
12. List the five basic parts of the carburetor and the purpose of each.
13. What part of the diesel injection system controls the amount of fuel taken into the combustion chamber?
14. What is the difference between the source of electric energy in a magneto-ignition system and that in a battery-ignition system?
15. What is the purpose of each of the following parts of an ignition system:
    (a) Ignition coil
    (b) Breaker points
    (c) Condenser
16. In the sequence of events required to fire a spark plug, explain what happens when the breaker points open in a magneto-ignition system.
17. What are the terms which describe the process of transferring electric energy from the primary to the secondary windings of an ignition coil?
18. What is a solid-state-ignition system?

# CHAPTER 4 SERVICING THE SMALL ENGINE

## CHAPTER GOALS

In this chapter your goals are to:

- Identify the factors which contribute to successful engine operation and trouble-free performance
- Service and maintain eight major operational "systems" of small engines
- Perform troubleshooting operations based upon a systematic procedure
- Prepare an engine properly for storage

When an athlete is said to be in "good condition", we immediately think of the person's physical well-being and the ability to perform at maximum effort. To run a 4-minute mile a track star must have the ability and stamina to go the distance at top speed and effort. An athlete achieves this ability through training and being in top physical condition.

Small engines are designed to operate at or near top speed. The conditions under which they operate often require full-power performance. Unlike larger engines which have several cylinders to produce power, a small engine must rely on one or two cylinders to do the work. For this reason, a small engine must be in "top condition" to perform effectively.

Service and maintenance specifications which manufacturers publish in owner-operator manuals provide the essential information to be applied by owner, operator, or service person to give lasting engine performance. Engine troubles are generally due to lack of service, improper adjustments, or poor maintenance. Small engines are usually subjected to some of the worst possible operating conditions. These include high temperatures, extremely dusty or dirty conditions, and the abuse of overspeeding and overloads.

The opportunities for the service person in power and machinery to enter into a rewarding career in small-engine sales, distribution, service, and maintenance are unlimited. Trained people who are technically competent to maintain such engine-powered equipment as chain saws, tillers, tractor mowers, trimmers, blowers, and pumps are very employable. One of the attributes which successful individuals who work with small-engine-powered equipment must have is the ability to troubleshoot and correct operational problems. The knowledge gained and the competencies developed working with small engines will prove invaluable in the understanding of larger, more powerful engines, should the opportunity develop.

## SERVICING THE LUBRICATION SYSTEM

The major function of lubricating oil is to reduce friction between the moving parts of an engine. Without lubrication the moving parts would come into contact with each other, and the engine would soon be destroyed from the heat of friction. To prevent metal-to-metal contact, a film of oil is used to separate the parts, thereby reducing friction (Fig. 4-1). The oil film also serves as a seal to the sliding surfaces between the rings and cylinder wall.

Small air-cooled engines rely greatly upon the "cooling ability" of oil. The oil that comes into contact with internal engine parts absorbs heat and transmits it to the interior surface of the engine block. The excess heat is then conducted to the exterior, where air can carry away the heat.

Lubricating oil "cleans" the moving parts by a washing action. Modern engine lubricants contain special chemicals, called additives, which wash the internal parts in a bath of oil. *Antirust* and *corrosion-resistant* chemicals reduce the effect of rust and acid attack on engine parts.

### Lubricating Systems

The lubrication system of small four-cycle engines consists of an oil supply reservoir in the crankcase called a *wet sump*. Oil from the sump is distributed to the moving parts under pressure from a pump, by splash, or by a combination of both methods. In the

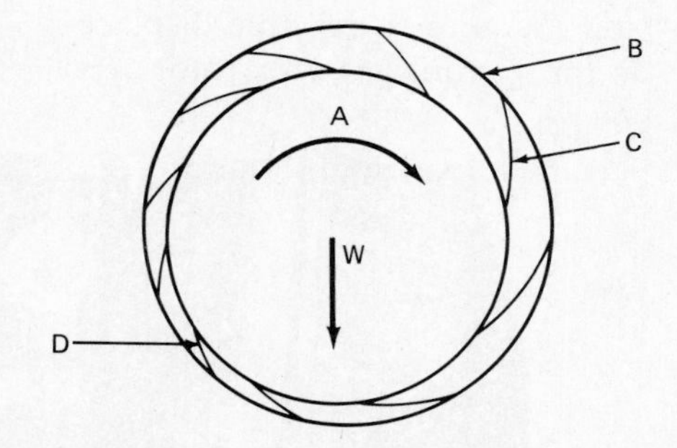


**Fig. 4-1.** The diagram illustrates the concept of friction-bearing lubrication. Shaft A rotates in the bearing journal B on a film of oil C. Layers of oil are dragged around with the shaft, wedging the two metal surfaces apart. The oil film at D is thin because W of shaft is downward. Weight is thereby supported on a film of oil.

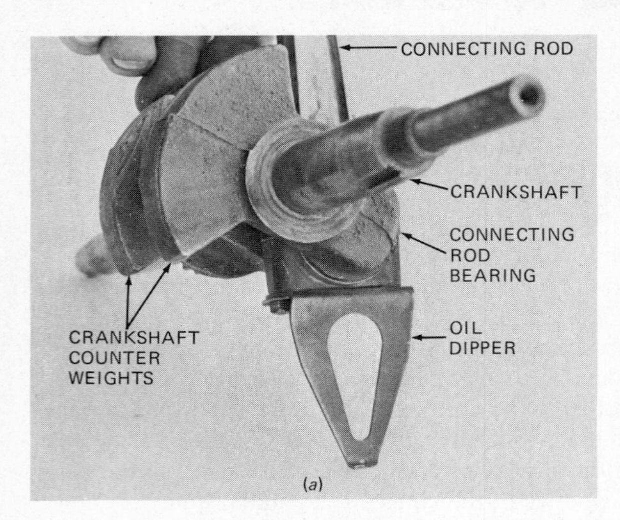


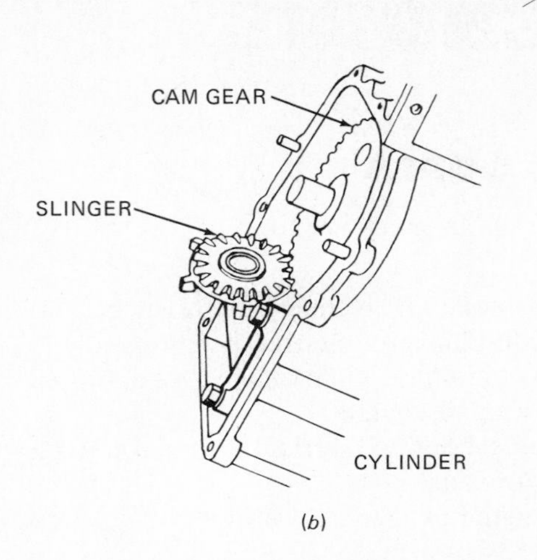


**Fig. 4-2.** The splash system of lubricating engine parts. On horizontal-shaft engines, the dipper (*a*) splashes oil on the engine parts. The slinger (*b*) serves the same purpose on vertical-shaft engines. (*Briggs & Stratton Corp.*)

splash system (Fig. 4-2*a*) of horizontal-shaft engines a dipper on the rotating connecting rod splashes engine oil from the sump in a mist of very fine droplets. The splashing is extremely intense, so that oil finds its way into drilled holes or grooves to bearings. On vertical-shaft engines the splashing action may be caused by a rotating slinger, operated from the camshaft gear (Fig. 4-2*b*). The oil in the sump of a splash system must be maintained to the proper level. If it is too low, the dipper may not touch the oil when the engine is operated at a severe angle. Some manufacturers have compensated for unusual operating positions by using a modified splash system (Fig. 4-3). In this system an oil pump lifts oil from the sump and fills a trough. The connecting rod dipper swings through the trough, splashing the oil. The pump maintains a constant level in the oil trough.

The pressure system requires a positive displacement pump such as the piston, gear, or rotor type (Fig. 4-4, A). The pump distributes oil under pressure to bearings through drilled passages in the block. A drilled passageway in the crankshaft permits oil pressure to reach the connecting rod bearing. Pressure systems are common on diesel and larger types of gasoline engines. Oil filters are used (Fig. 4-4, C) on pressure-lubricated engines so that the oil that is pumped to bearing surfaces will be free of harmful contaminants.

**Two-Cycle Engine Lubrication** The crankcase of the small spark-ignition two-cycle engine does not contain a reservoir for oil. It is classified as a *dry-sump* engine. The crankcase must serve as a pump to transfer the air-fuel mixture to the combustion chamber. The lubrication for the internal parts depends upon the addition of the correct amount of lubricating oil to the gasoline.

When the mixture of fuel and oil passes through the carburetor, the gasoline vaporizes, leaving very small droplets of oil to lubricate the internal parts. The ratio of the mixture of gasoline and oil varies according to manufacturers' specifications. The mixture may be as much as 16:1 to as little as 50:1 parts, respectively, of gasoline and lubricating oil. The

**Fig. 4-3.** Modified splash lubrication system. Oil is pumped into the trough and then splashed by the dipper on the crankshaft connecting rod. (*Wisconsin Motor Corp.*)

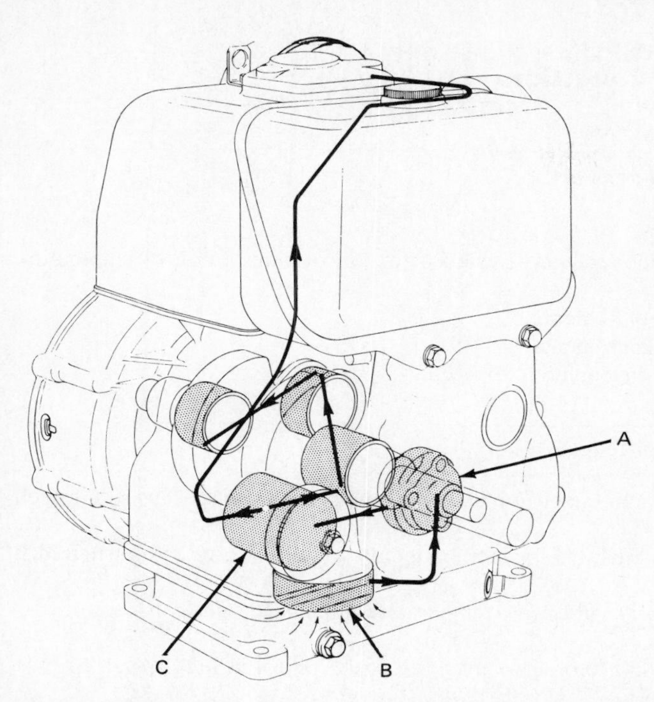


**Fig. 4-4.** Pressure lubrication system. The pump A lifts oil from the sump through the screen B, forcing the oil through the oil filter C and then to all bearing surfaces of the diesel engine.

| Season or Temperature | Grade of Oil |
|---|---|
| Spring, Summer, or Autumn<br>+120 to +40°F<br>[+50 to +5°C] | SAE 30 |
| Winter<br>+40°F to +15°F<br>[+5 to −10°C] | SAE 20 |
| Below +15°F [−10°C] | SAE 10W-30 |
| Crankcase capacity | 1.25 pt [0.6 L] |
| Use oils classified as Service SD or SE | |

**Fig. 4-5.** Typical oil grade and service classification chart. Always use the engine manufacturer's recommendation for grade and service classification of engine oil.

variation in oil requirement depends upon the design of the engine and the ability of oil to lubricate without burning in the piston ring area of the engine.

## Selecting Engine Oil

The operating condition of a small air-cooled engine is classified as "severe" when it comes to selecting an engine oil. The classification is based on the numerous opportunities for the oil to be contaminated with raw fuel, dirt, and combustion by-products. Small engines in equipment such as lawn mowers and tillers often operate near ground level. This location exposes them to severe dust conditions. Also, most small air-cooled engines do not have an oil filter. For these reasons, the proper oil must be selected according to the manufacturers' recommendations and the proper drain period must be observed.

Lubricating oils for spark-ignition and diesel engines are available in a variety of service classifications and viscosities.

**Oil Classification** The owner's manual and service specifications for small engines (see Fig. 4-5 for example) provide recommendations for engine oil lubricants. The information lists (1) service classification—often referred to as the "type" of oil—and (2) viscosity—often called the "grade" of oil.

Service classification (type of oil) is a rating standard of the American Petroleum Institute (API). The standard identifies the formulation of oils. Various chemicals are added to engine oils to improve their ability to withstand and resist bearing corrosion and oxidization. Detergent dispersants are used to pick up and hold dirt and sludge in suspension. The formulations are identified by letters such as SA, SF, CD, or others as listed in Table 4-1. The first letter in the pair indicates the type of engine for which the oil is recommended, for example, S is for spark-ignition engines, and C is for compression-ignition (diesel) engines. The second letter of the code signifies its place in the sequential development of the oils to meet new, and stricter service requirements.

Engine oil is also selected by viscosity grade. Viscosity is a rating of the oil's resistance to flow according to the Society of Automotive Engineers' (SAE) number index, such as SAE 30. The higher the number, the more viscous (thicker) the oil. Table 4-2 gives the viscosity-grade recommendation for six manufacturers of four-cycle engines.

When adding oil, use the service classification grade and SAE viscosity type recommended by the manufacturer. Choice of the correct SAE number oil will depend on the expected atmospheric air temperature (Fig. 4-6). A multiviscosity oil grade such as 10W-30 will serve the same purpose as SAE 10W oil when the temperature is 0 degrees Fahrenheit (°F) [−18 degrees Celsius (°C)] or SAE 30 oil when the temperature is above 30°F [−1°C].

## Mixing and Changing Oil

Most manufacturers of two-cycle engines recommend special two-cycle oil of their own formulation in their own ratio for mixing with gasoline. This ratio varies according to the kind of oil used and the construction of the engine. For example, one company recommends a ratio of 16 parts of gasoline to 1 part of oil when using SAE 30 equivalent oil and 32 to 1 when SAE 40 is used (Fig. 4-7). Since the two-cycle en-

**TABLE 4-1. American Petroleum Institute Oil Classifications and Recommended Operating Conditions**

| | | |
|---|---|---|
| | | **Spark Ignition** |
| SA | Light duty | Straight mineral oil. No performance requirements. |
| SB | ↓ | Medium-duty nondetergent. |
| SC | ↓ | Meets 1964–1967 automotive warranty requirements. Designed to control high- and low-temperature deposits. |
| SD | ↓ | Meets 1968–1971 automotive warranty requirements. |
| SE | ↓ | Meets 1972–1979 automotive warranty requirements. |
| SF | Severe duty | Meets automotive warranty requirements beginning in 1980. |
| | | **Compression Ignition** |
| CA | Light duty | Meets 1954 diesel warranty requirements. Intended for light duty with high-quality low sulfur fuel. |
| CB | ↓ | Meets 1958 diesel warranty requirements. Intended for moderate duty and offers protection when high sulfur fuels are used. |
| CC | ↓ | Meets 1964 diesel warranty requirements. Provides low-temperature antisludge and antirust performance in lightly turbocharged engines. |
| CD | Severe duty | These oils protect supercharged engines that require wear and deposit control in high-speed, high-output service. |

**TABLE 4-2. Recommended Oil Viscosity Varies with Temperature. Always Check the Operator's Manual for Specific Model Recommendations.**

| Manufacturer | Above 40°F [4.4°C] | Above 32°F [0°C] | Below 5°F [−15°C] | Below 0°F [−17.8°C] | Below −10°F [−23.3°C] |
|---|---|---|---|---|---|
| Briggs & Stratton | SAE 30 or 10W-30 | 5W-20 or 5W-30 | | SAE 10W diluted with 10 percent kerosene | |
| Clinton | SAE 30 | | 10W | | 5W |
| Honda | SAE 10W-40 | | | | |
| Kohler | SAE 30 | | SAE 10W | 5W or 5W-20 | |
| Tecumseh | SAE 30<br>10W-30 | 10W-40 | 5W-20<br>5W-30 | 10W diluted with 10 percent kerosene | |
| Wisconsin | SAE 30 | SAE 20 or SAE 20W | 10W | | |

gines are lubricated by gasoline and oil, the mixture enters the cylinder and is burned with the fuel. The mixing oil must not leave a residue that will foul the spark plugs or collect in the exhaust ports. Two-cycle oil conforms to the API classification, SA. Always mix oil and fuel in clean containers.

**Checking the Oil Level** The oil in the crankcase of four-cycle small engines must be kept in the safe operating range to assure proper lubricating action. Most manufacturers recommend an oil level check each time fuel is added. Check the level by first locating the filler plug on the crankcase. Then

| Air Temperature | Oil Viscosity | Oil Type |
|---|---|---|
| Above 30°F [−1°C] | SAE 30 | API Service SC* |
| 30° to 0°F [−1 to −18°C] | SAE 10W-30 | API Service SC* |
| Below 0°F [−18°C] | SAE 5W-20 | API Service SC* |

* SD-SF class oils may also be used.

**Fig. 4-6.** Grades of oil for four-cycle air-cooled engines, recommended on the basis of atmospheric air temperature.

clean the dirt or dust from around the cap before loosening. Make sure the engine is nearly level when checking the oil level. This will avoid taking a false reading. Withdraw the dipstick, wipe it with a clean waste rag, and then reinsert it. Be sure the plug holding the bayonet-type dipstick is pushed or screwed into the hole in order to obtain a true reading (Fig. 4-8). If there is no dipstick, the oil level should be to the top of or run from the filler hole. If the oil is low, add only enough of the correct type and grade of engine oil to bring it to the full mark.

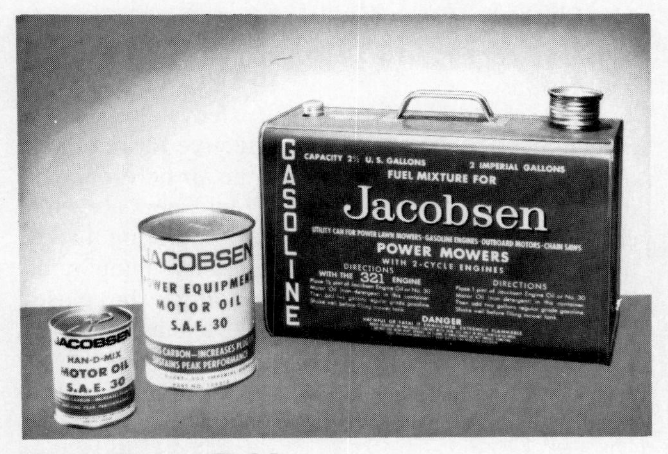


**Fuel Mixing Table**

| Gallons of Regular Grade Gasoline | Amount (SAE 40) 32:1 Oil | Amount (SAE 30) or Other 2-Cycle Motor Oil (16:1 Mix) |
|---|---|---|
| 1 | 4 oz [118 mL] | 8 oz [237 mL] |
| 2 | 8 oz [237 mL] | 16 oz [473 mL] |
| 3 | 12 oz [355 mL] | 24 oz [709 mL] |
| 4 | 16 oz [473 mL] | 32 oz [946 mL] |

**Equivalent Values**

| Ounces | 4 | 5 | 6 | 7 | 8 | 9 | 10 | 11 | 12 | 13 | 14 | 15 | 16 | 20 |
|---|---|---|---|---|---|---|---|---|---|---|---|---|---|---|
| U.S. Pint | 1/4 | 1/3 | 3/8 | 7/16 | 1/2 | 9/16 | 5/8 | 11/16 | 3/4 | 13/16 | 7/8 | 15/16 | Full Pint | — |
| Imperial Pint | 1/5 | 1/4 | 3/10 | 7/20 | 2/5 | 9/20 | 1/2 | 11/20 | 3/5 | 13/20 | 7/10 | 3/4 | — | Full Pint |

**Fig. 4-7.** Manufacturer's recommendations for mixing special two-cycle engine oil with gasoline. Always use clean containers and follow the mixing instructions. (*Jacobsen Div., Textron, Inc.*)

When the proper amount of oil is in the crankcase, return the plug and tighten it snugly so that it will not loosen from vibration, causing a loss of oil. When a gear reduction unit is attached to the engine (Fig. 4-9), check the oil level in the unit at least every 50 hours (h) of engine operation.

**Changing Oil** The purpose of changing engine oil is to remove contaminants from the crankcase. Small-engine manufacturers generally recommend that the oil be changed in new engines after the first 5 h of operation. This is to assure that metallic particles are flushed from the engine before damage can occur. The regular drain period that is recommended may vary between 10 and 50 h of operation. Most manufacturers recommend approximately 25 h, with words of caution that if the engine is operated in extremely dusty or dirty conditions, the time be reduced by one-half.

The proper time to drain the oil is after the engine has operated a sufficient length of time to heat the oil thoroughly. The heated oil will drain more completely and allow more of the contaminants to be removed due to the stirring action of engine operation. When the oil is completely drained, replace the drain plug snugly and refill the crankcase with the recommended service type and viscosity grade of oil for the season's air temperature conditions (Fig. 4-6).

## SERVICING THE AIR CLEANING SYSTEM

The air which enters the combustion chamber of an engine must be free of abrasive materials. When dirt and dust particles are mixed with the oil through carelessness, an extremely effective grinding compound is formed. The piston rings, cylinder wall, valve faces, valve seats, and valve guides will be badly worn in a very few hours of operation (Fig. 4-10).

### Types of Filters

When properly serviced, an air filter will remove an extremely high percentage of the small dust particles from the intake air. There are three types of air cleaners in use on small engines, depending upon the

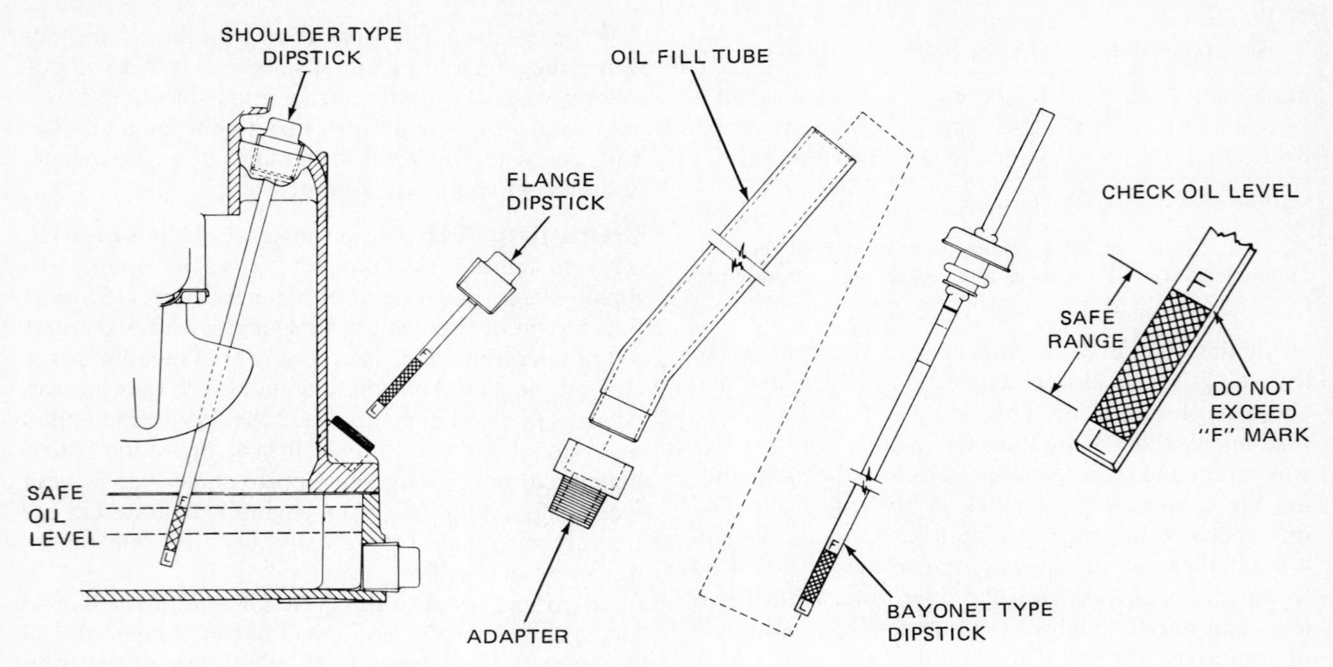


**Fig. 4-8.** Dipsticks and oil fill tube arrangements. (*Kohler Co.*)

manufacturer's fitting: the dry type, the oil-bath type, and the oil-saturated-foam type.

**Dry Type** *Dry-type* air cleaners (Fig. 4-11) filter air through a porous paper element which removes a very high percentage of the finest dust particles. The cleaners are serviced by replacing the element after a specified period of operation. In addition, some may be cleaned after each 50 h of normal operation by removing the element and tapping it lightly in your hand to loosen the dirt. If tapped on a hard surface, the element can be bent out of shape enough to cause improper sealing when reinstalled. Some elements can be washed in a solution of warm water and nonsudsing detergent. Always follow the manufacturer's instructions.

A heavy-duty plastic-foam sleeve precleaner placed over the dry element is effective for removing much of the abrasive material in the air before the air flows through the dry element. Precleaners are especially recommended for engines operating in extremely dusty or dirty conditions. Some manufacturers recommend washing and re-oiling the plastic foam every 25 h of operation or at 3-month intervals, whichever occurs first. *Do not oil the dry element!* Be sure to follow directions in the operator's manual regarding service of this type of precleaner.

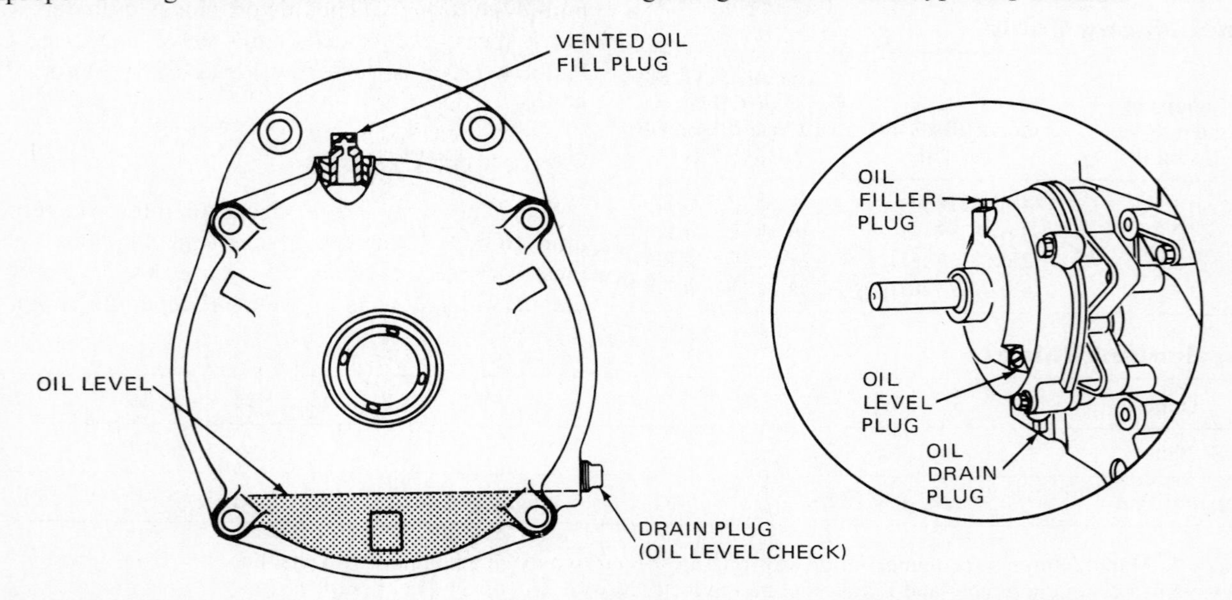


**Fig. 4-9.** Fill and drain locations for a reduction gear oil housing. (*Kohler Co. and Briggs & Stratton Corp.*)

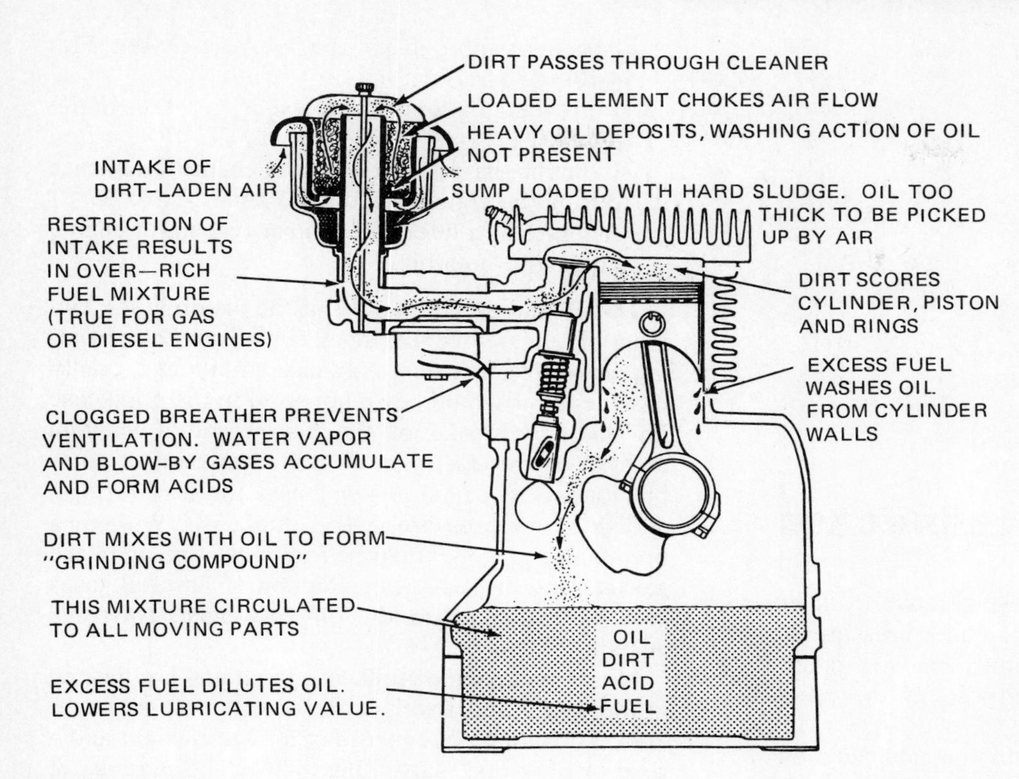


**Fig. 4-10.** The consequence of failing to service the air cleaner and crankcase breather. (*Onan Corp.*)

**Oil-Bath Type** The *oil-bath* air cleaner is used on some small diesel engines. This type of air cleaner filters the air by the action of a film of oil on a porous mesh. As the air passes the oil film, dirt and dust particles are held in suspension on the oil. The oil film develops as the air strikes the oil in the reservoir, which causes it to splash on the mesh (Fig. 4-12). The oil drains back into the reservoir, carrying some of the dust particles with it. Eventually the dust particles settle out of the oil into the bottom of the reservoir.

The washing action of the oil makes the upper element self-cleaning, providing the oil is changed often (8 to 10 h of operation in very dusty conditions) and the filter element is washed in solvent. When used on diesel engines, the solvent must be thoroughly drained and evaporated from the element. Otherwise the engine would accidentally run on the solvent vapor drawn into the cylinder on the intake stroke. As a result, the engine could overspeed, causing severe damage to internal parts. The oil-bath air cleaner should not be washed in gasoline because of the danger of fire and the poor washing characteristics of the fuel.

**Oil-Foam Type** *Oil-foam-type* air cleaners (Fig. 4-13) consist of a specially designed plastic-foam element that is saturated in engine oil. This filter is very efficient on small gasoline engines. It can be cleaned by washing it in hot water and a detergent or solvent, drying it, and saturating it with new engine oil. The element is then squeezed gently to remove excess oil. *Do not wash the element in gasoline.*

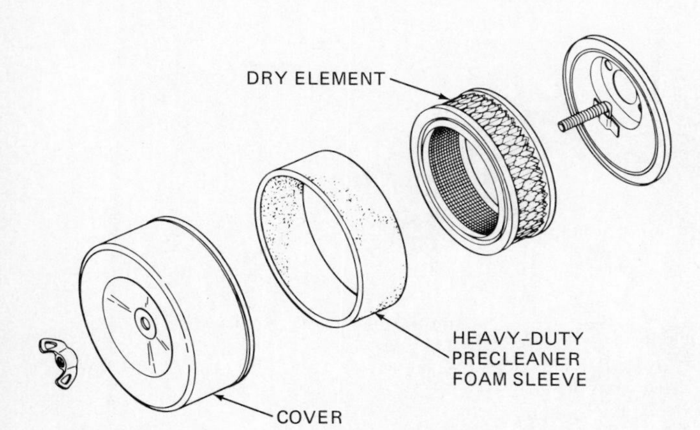


**Fig. 4-11.** Typical dry air cleaner with heavy-duty foam precleaner. (*Kohler Corp.*)

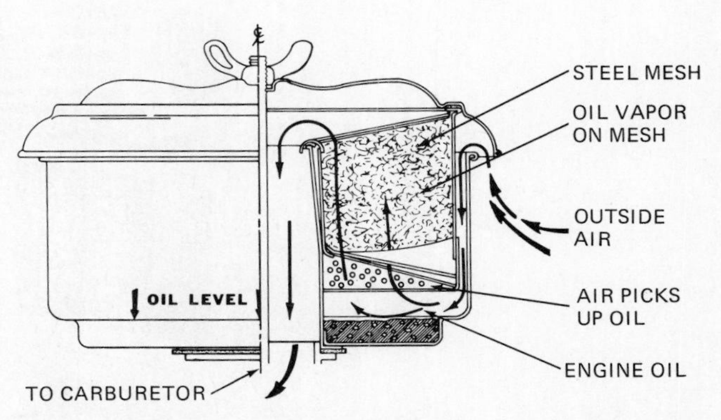


**Fig. 4-12.** Cutaway view showing air flow through a typical oil-bath air cleaner. (*Kohler Co.*)

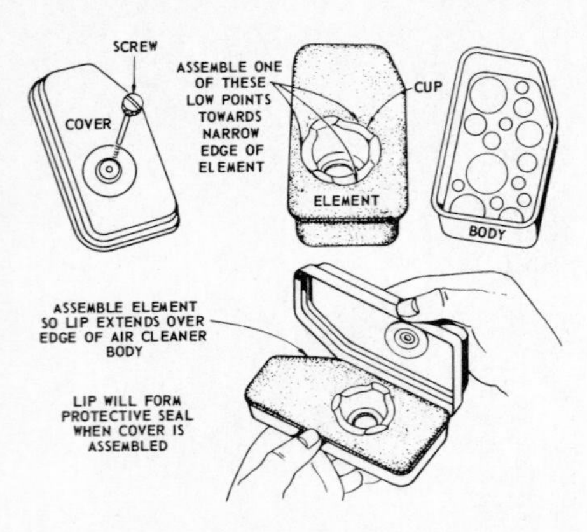


**Fig. 4-13.** Oil-foam air cleaner. (*Briggs & Stratton Corp.*)

## SERVICING THE CRANKCASE BREATHER

In contrast to the crankcase of a two-cycle engine, which is sealed and operates under pressure, a four-cycle engine's crankcase must be vented for it to "breathe." This is the purpose of the *crankcase breather*.

A breather serves several important functions:

1. It removes water vapor, unburned gases, and harmful vapors from the crankcase (see Fig. 4-10).
2. It prevents buildup of excessive pressures in the crankcase.
3. It maintains a slight vacuum in the crankcase during the compression and exhaust strokes.
4. It serves as a filter and control system for engine crankcase breathing.

The up-and-down motions of the piston in the cylinder of a single-cylinder small engine create a pulsating action of air in the crankcase. Without a crankcase vent, air would be compressed in the crankcase on the downstroke of the piston and a vacuum created on the upstroke. A limited amount of combustion gases, called *blowby*, pass the piston rings, adding to the pressure in the crankcase. Without a breathing vent, oil would be forced past many of the gaskets and oil seals. A collection of harmful gases would accumulate in the oil and cause corrosive acids to develop.

Most small-engine crankcase breathers are located in the valve tappet well or in the valve tappet cover area. The assembly consists of a filter element and a reed or fiber disk valve (Fig. 4-14). The purpose of the filter is to remove dust and dirt that would enter

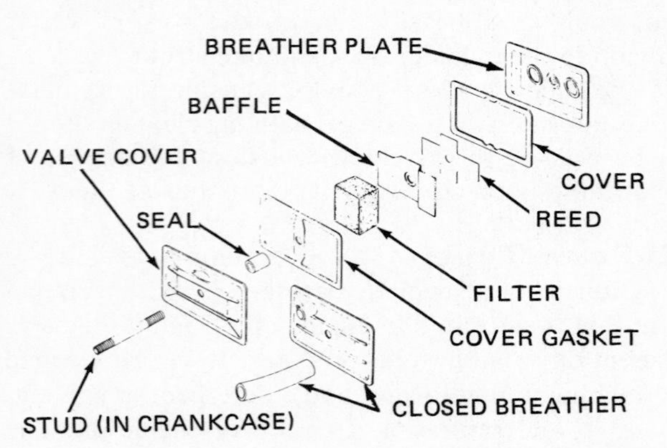


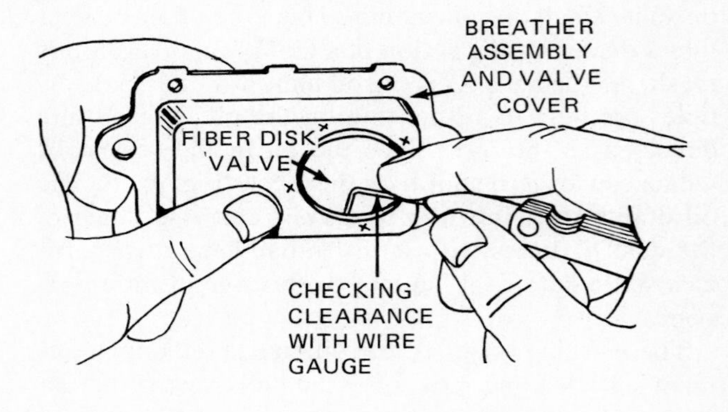


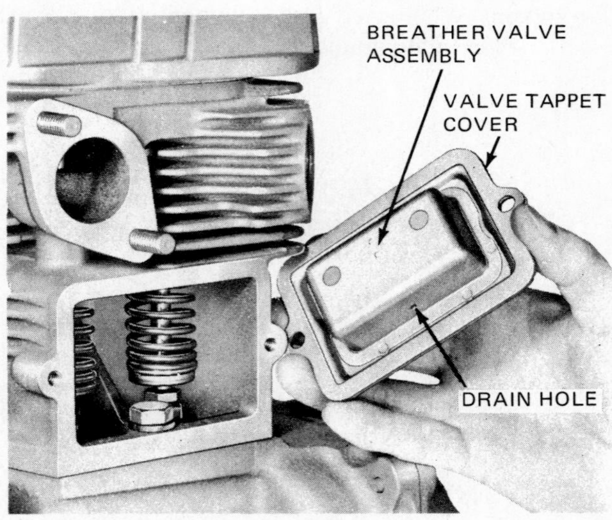


**Fig. 4-14.** Types of crankcase breathers located in the valve cover. (*Kohler Co., Teledyne Wisconsin Motor, and Briggs & Stratton Corp.*)

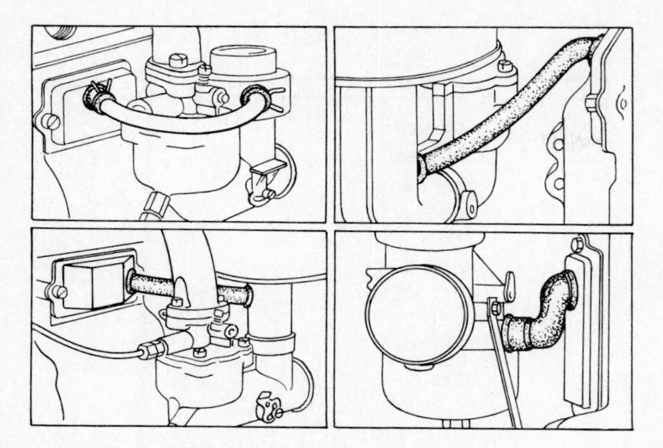

**Fig. 4-15.** Crankcase breather connections to the air intake system. The breather provides positive crankcase ventilation and creates vacuum in the crankcase. (*Briggs & Stratton Corp.*)

the crankcase. The valve controls the amount of vacuum maintained in the crankcase to prevent leakage of oil past the crankcase seals.

The valve cover is physically connected to the air induction system by a metal or rubber tube (Fig. 4-15). The connection is made at a point in the intake piping after the air cleaner but ahead of the carburetor. Thus, a slight vacuum is created by the intake system. Only clean air will enter the crankcase area.

Servicing the breather consists of removing and cleaning the breather filter element every 100 to 500 h of operation, depending upon the manufacturer's recommendation. The proper operation of the valve is checked by performing a vacuum test on the crankcase when the engine is at operating temperature and speed. A vacuum reading which is not within the manufacturer's specifications indicates faulty breather valve operation.

## SERVICING THE COOLING SYSTEM

The fuel burning in the combustion chamber reaches a temperature of approximately 3600°F [1982°C]. Only one-third of this heat is converted into power; of the remaining thermal energy an additional one-third is lost to the exhaust system and another third through the cooling system. Unless the cooling system of an engine functions properly, the metal parts of the piston and cylinder will fail.

The principle of the cooling system of an air-cooled engine is based upon moving a large amount of air over and around the hot cylinder and cylinder head. Air movement is developed by a fan which is a part of the flywheel (Fig. 4-16*b*). A housing, called a *shroud,* and *baffles* (Fig. 4-16*a*) direct the cooling air so that it is forced to flow in and around the cooling fins on the cylinder and cylinder head. The fins increase the surface area of the metal and allow the air to carry off more of the excess heat. Any foreign material such as trash, grass clippings, or dust collected on oily surfaces that will restrict the air flow around the cylinder can be a cause of both overspeeding and

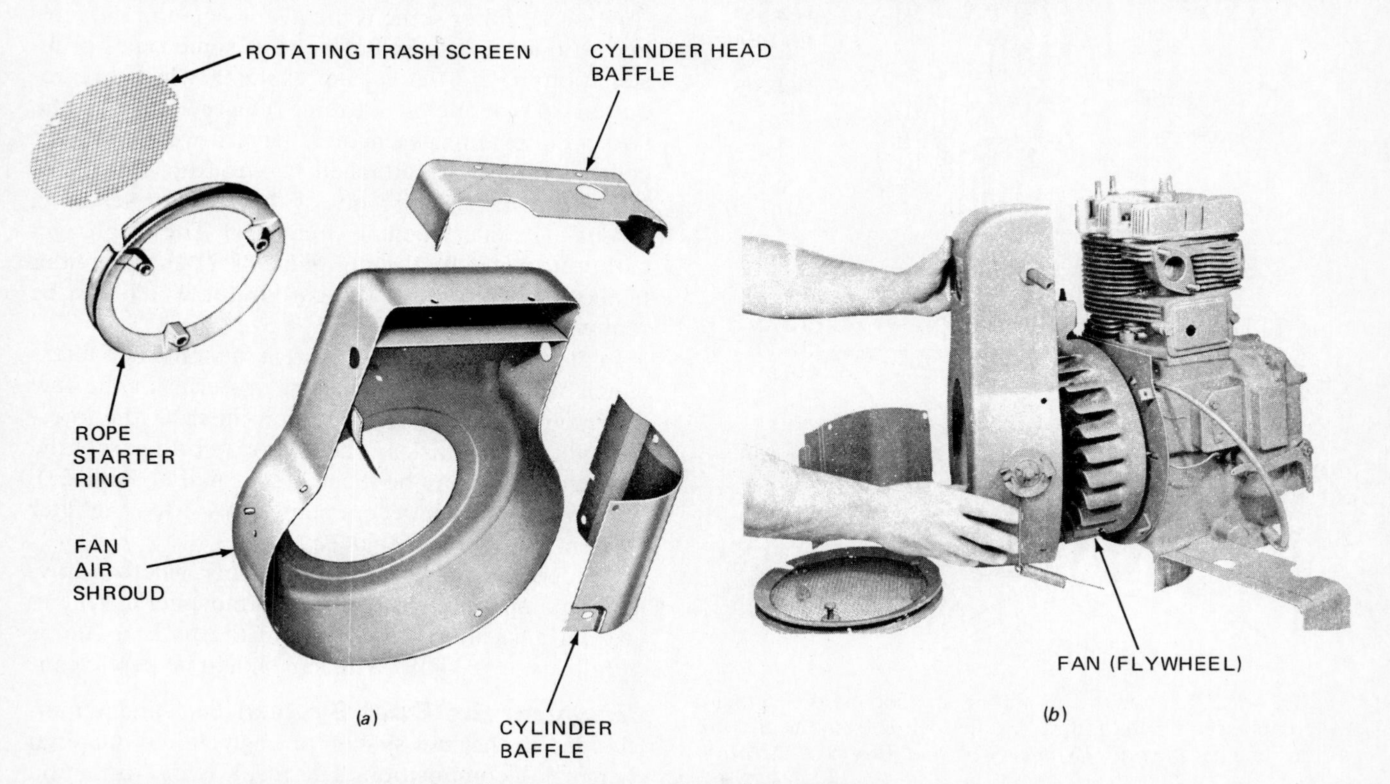


**Fig. 4-16.** Shrouds and baffles in the cooling system of an air-cooled engine. Do not operate an engine with the shrouds removed. (*Teledyne Wisconsin Motor*)

overheating. Excessive temperatures may cause the oil to burn on the cylinder walls and form damaging carbon deposits in the piston ring grooves. When the engine has been operating in severe dirty conditions, it will be necessary to remove the blower housing and baffles and remove all the accumulations (Fig. 4-17). Remove any oil-caked dust by scraping and using a degreasing solvent.

Be sure to replace all the baffles in the proper position before operating the engine. Without a baffle, the air flow cannot be directed to essential areas for cooling, and it will not be able to control the air-vane governor (see Fig. 4-51).

## SERVICING THE FUEL SYSTEM

The fuel system of both gasoline and diesel engines must receive periodic service to assure that the flow of fuel to the engine will not be affected. Most manufacturers recommend that periodic attention and service be given to gasoline engines every 100 h and diesel engines every 500 to 1000 h of operation.

### Fuel Filters and Screens

The fuel system of all gasoline engines should contain a fuel strainer which is located between the fuel tank and the carburetor. The strainer may be located in the fuel tank (Fig. 4-18) or in the fuel tubes which extend into the fuel tank of suction-type carburetors (Fig. 4-19). Filter screens are also located in the fuel-line strainer and sediment bowl of some types of installations (Fig. 4-20). Fuel systems which are required to operate at extreme angles, such as the two-cycle chain-saw engines, have a weighted filter called a *fuel finder* attached to a pickup hose. This filter must be "fished" out of the tank for servicing through the filler opening (Fig. 4-21). The diaphragm carburetor and fuel pump (Fig. 4-22) share a filter in the inlet portion of the carburetor which can be removed for cleaning.

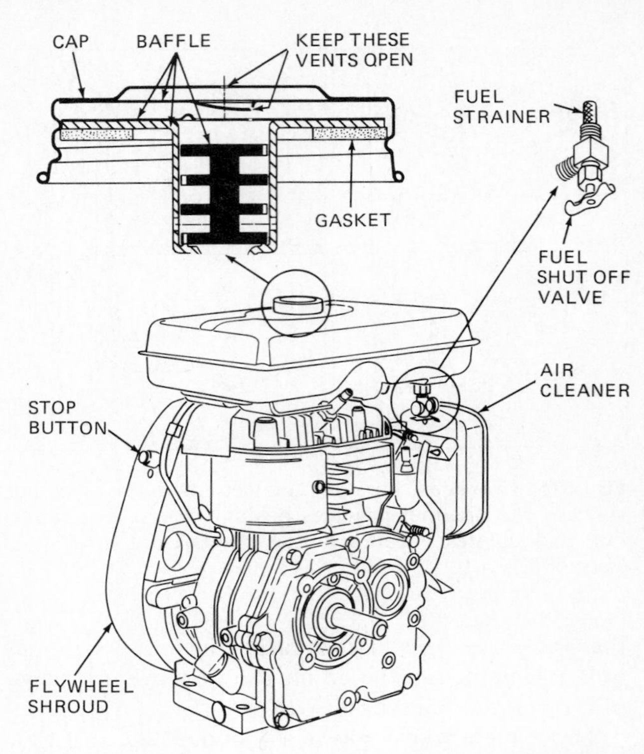


**Fig. 4-18.** Fuel strainer located within the fuel tank. It may be removed for cleaning. The vent hole in the cap must be kept open. (*Teledyne Wisconsin Motor.*)

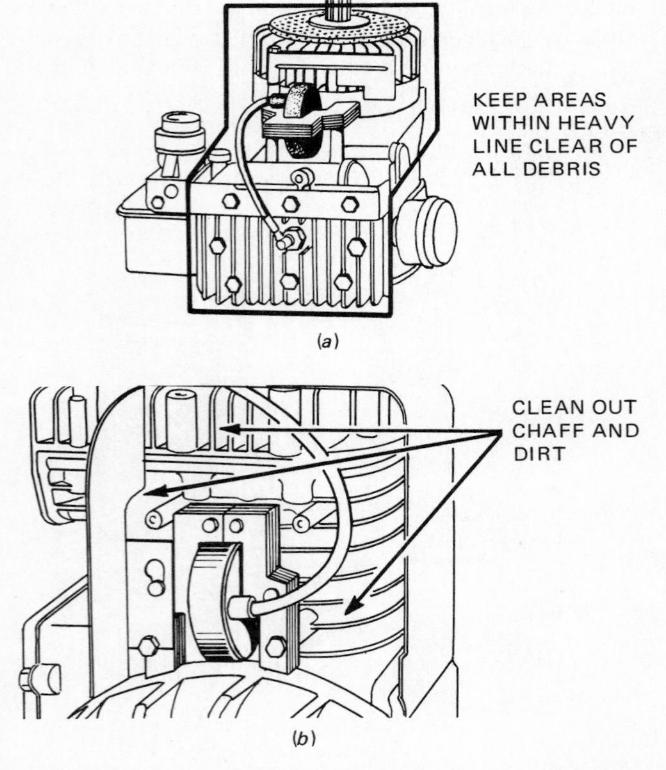


**Fig. 4-17.** Remove the air shroud and baffles to clean with compressed air chaff, dirt, or other debris from the areas of the engine shown. (*Kohler Co. and Briggs & Stratton Corp.*)

In diesel engines, a fuel system must include filters which will remove any foreign material in the fuel that would damage the precision finish of the injection pump and nozzles. The filter may be a part of the fuel tank, or it may be located externally (Fig. 4-23). Each manufacturer recommends how often the filter elements should be replaced.

The fuel tank cap of all fuel systems, whether gravity or pump type, contains an atmospheric vent to prevent a vacuum developing in the tank as fuel is withdrawn (Fig. 4-18). The cap should be kept clean.

**Cleaning the Fuel System** Gasoline left unattended in the fuel system of engines may undergo chemical decomposition. The result is the formation

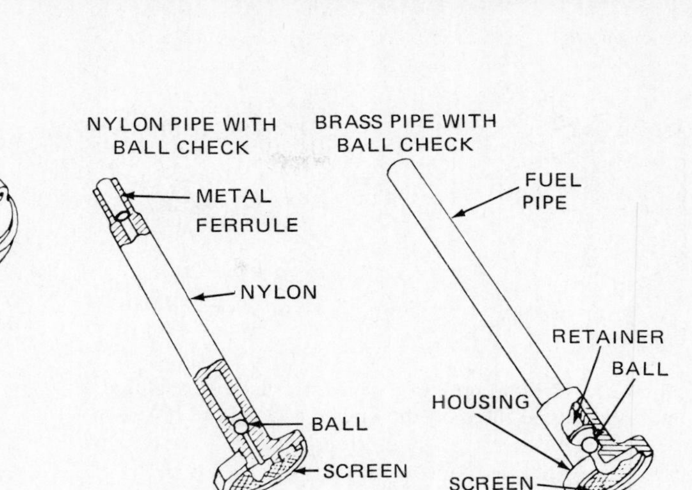


**Fig. 4-19.** Fuel screens located on the end of fuel pipes of suction-type carburetors. (*Briggs & Stratton Corp.*)

of "stale" gasoline, which forms a gum and varnish deposit on the internal surfaces of the system. The action may occur in a time period as short as 3 months. Stale gasoline may be identified by its foul odor. Before operating the engine, the stale fuel must be drained from the system and discarded. In some instances it may be necessary to soak and wash the fuel screens and lines in parts-cleaning solvent.

# SERVICING THE SPARK PLUG

Modern one- and two-cylinder small gasoline engines contain highly reliable ignition systems. Without a spark being delivered to the spark plug at the proper "time," power loss will be excessive or the engine will fail to operate altogether. The professional will be expected to provide ignition system maintenance information and service.

## The Spark Plug

Of all the various components that make up the ignition system, the *spark plug* has the most difficult role to perform. It must be able to conduct an electric spark to the combustion chamber with high reliability under severe heat, turbulence, and varying air-fuel mixture conditions.

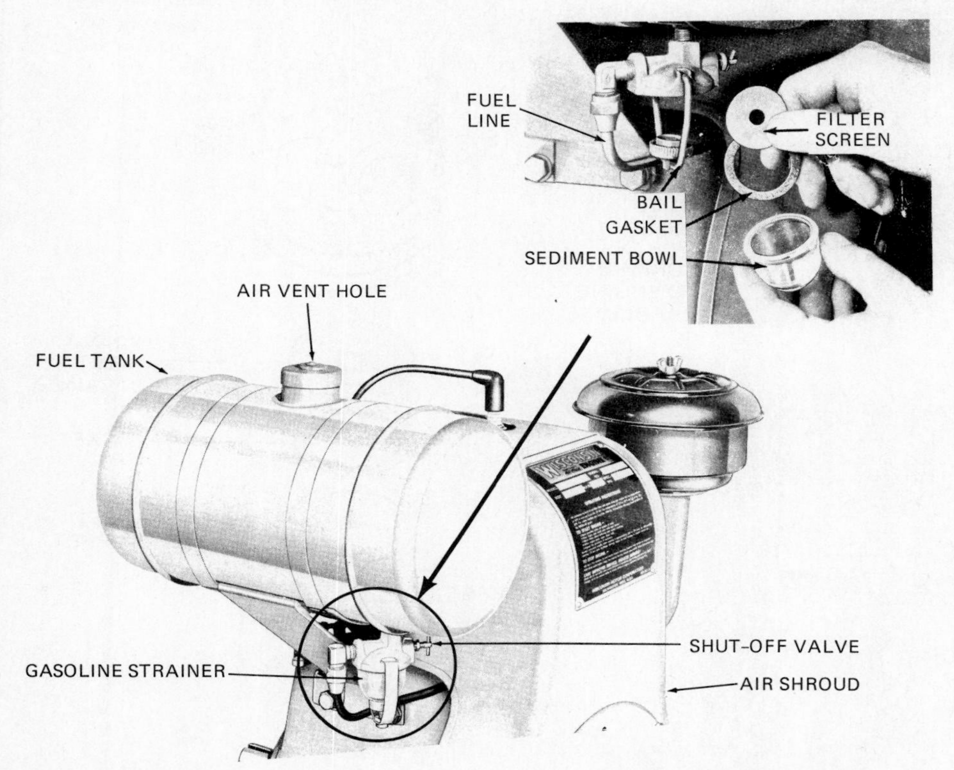


**Fig. 4-20.** External fuel strainer with screen and sediment bowl. (*Teledyne Wisconsin Motor*)

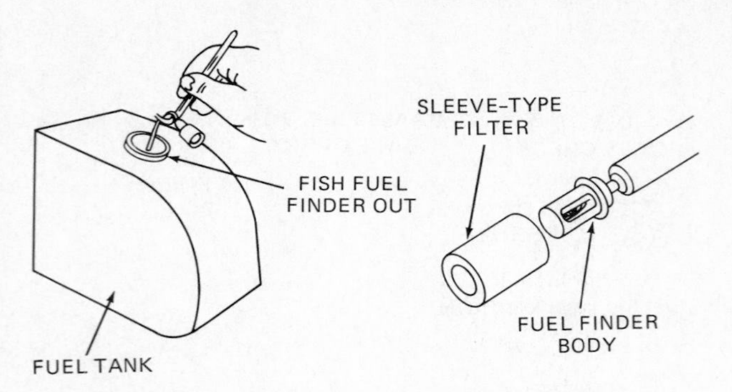


**Fig. 4-21.** The fuel tank of a chain saw uses a "fuel finder" and sleeve-type filter on the end of a weighted hose.

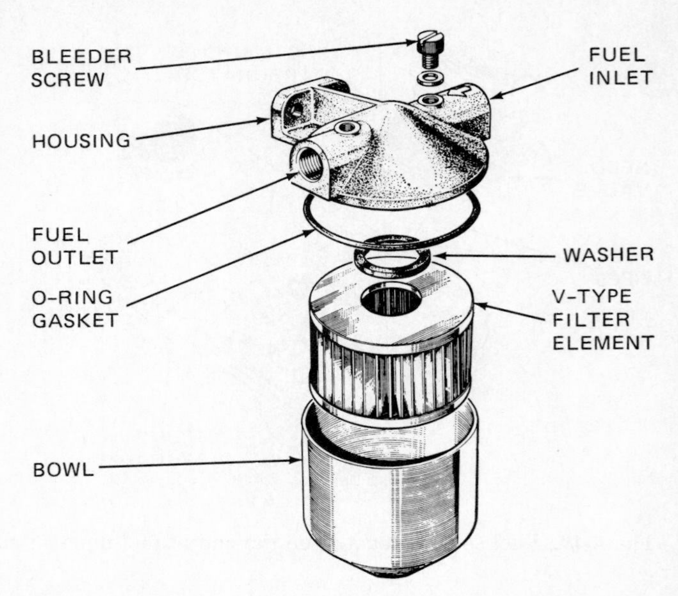


**Fig. 4-23.** Single stage external fuel filter for small diesel engine. The filter element is replacable. (*Petter Ltd.*)

**Parts of the Spark Plug** Spark plugs in general consist of five basic parts—(1) a steel shell, (2) an insulator core, (3) a center electrode, (4) a ground electrode, and (5) seals (Fig. 4-24).

The center *insulator core* is a porcelain material that is designed to resist very high combustion temperatures and insulate the center electrode to prevent high-voltage from leaking to the steel shell. The length of the cumbustion chamber end of the insulator identifies the operating *heat range* of the spark plug. When the insulator is long (Fig. 4-25), the spark plug is classified as a *hot* plug; a short insulator is a *cold* plug. The heat-range classification is based upon the distance the heat must travel in order to conduct heat to the cylinder head for cooling. A long heat path results in a hot plug, while a short heat path produces a cold plug. If the insulator operates at too low a temperature, it will collect deposits which "foul" the electrodes, causing misfiring. When the insulator operates too "hot," the electrodes will erode. Also the ceramic insulator may break from excessive heat or from detonation or preignition (see Fig. 4-26*e* and *f*).

An inspection of the color of the firing end of the insulator helps the service person determine how the spark plug is performing. For example, a tan or gray color (Fig. 4-26*a*) signifies normal engine operation. The carbon- or oil-fouled plugs in Fig. 4-26*b* and *c* indicate combustion or mechanical problems.

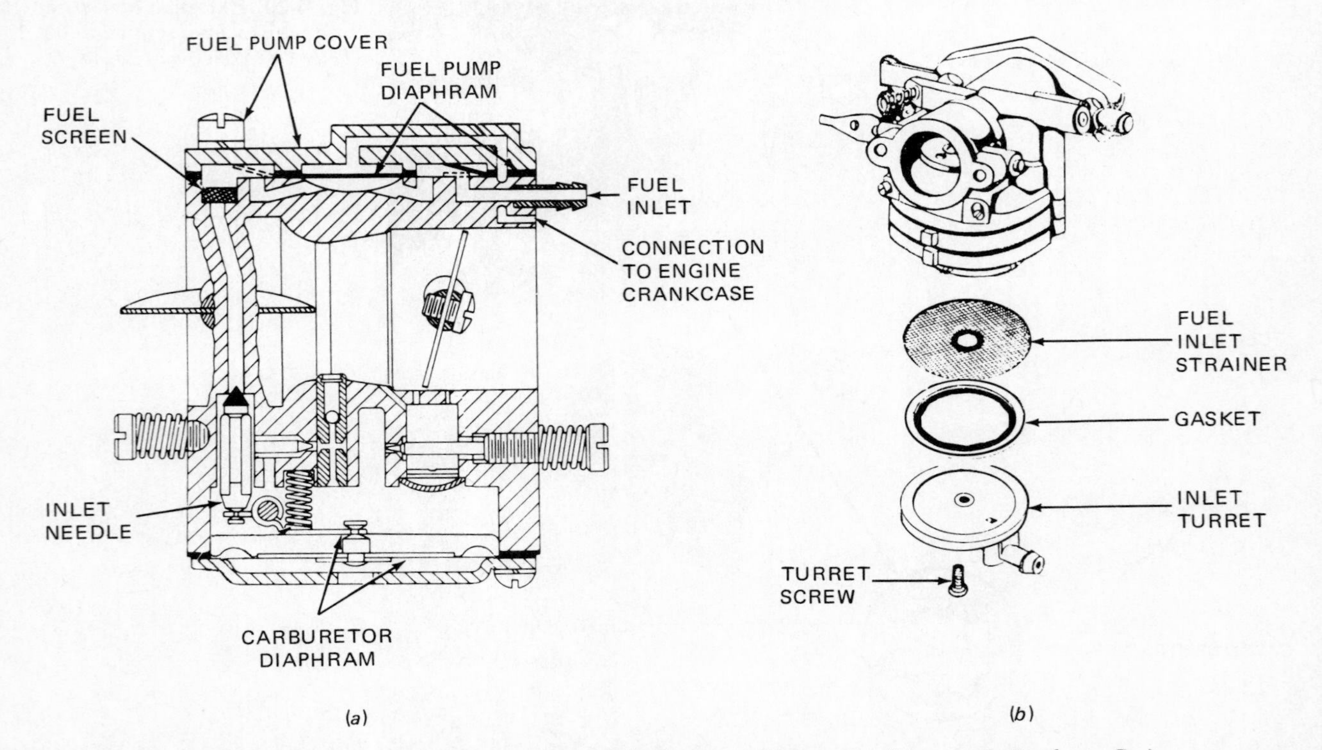


**Fig. 4-22.** Fuel screen in the fuel inlets of two types of diaphragm carburetors. (*Tecumseh Products Co.*)

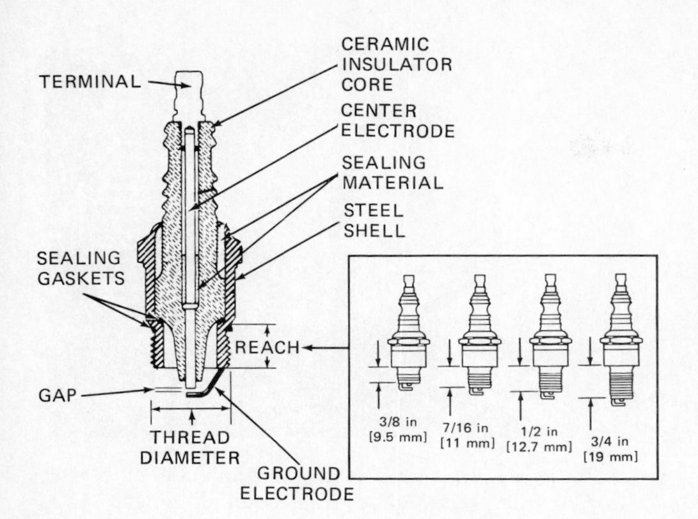


**Fig. 4-24.** Cross section, parts, and length of reach of a typical spark plug.

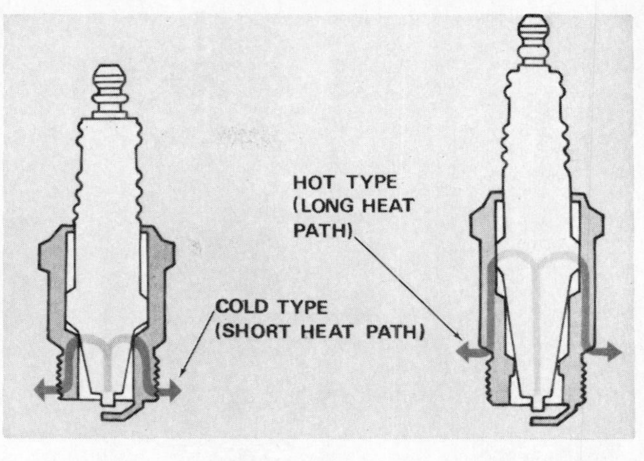


**Fig. 4-25.** Heat range. Heat must travel a longer distance in the "hot" spark plug from right to left. (*Champion Spark Plug Co. and Onan Co.*)

The *ground* and *center electrodes* are made of a corrosion-resistant metal alloy. The ground is welded to the steel shell. An air gap is developed at the point where the two electrodes meet by setting the ground so that it forms a gap of a specific distance. The spark jumps the air gap and ignites the fuel in the combustion chamber.

The ground electrode used in spark plugs for two-cycle engines is shortened slightly. This type of design reduces the tendency for loose particles of carbon to bridge the gap and cause the plug to short-circuit (Fig. 4-27).

**Selecting Spark Plugs** Each manufacturer of spark plugs identifies the various sizes and types by a letter, number, or letter-number combination. The code markings provide an index of thread diameter, reach, and heat range. Engine manufacturers' specifications give recommendations by the code index. The service person must base the selection of a replacement plug upon the manufacturer's recommendation and an evaluation of the firing chamber end for any adjustment needed for proper heat range.

**Servicing Spark Plugs** Hard starting and misfiring when the engine is under a heavy load are indications that the spark plug needs servicing. In ad-

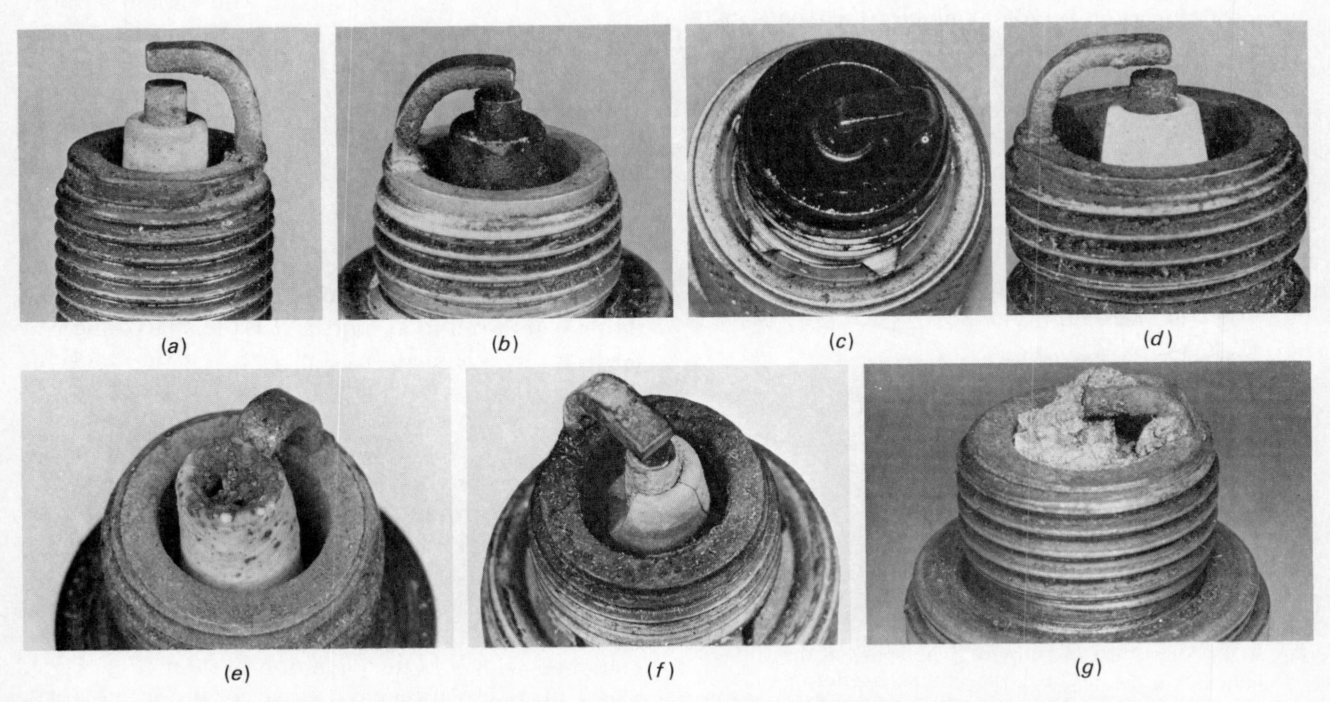


**Fig. 4-26.** Identifying engine operation and mechanical conditions from an inspection of firing chamber end of spark plugs. (*a*) Normal. (*b*) Oil-fouled (four-cycle). (*c*) Oil-fouled (two-cycle). (*d*) Overheating. (*e*) Preignition. (*f*) Detonation. (*g*) Aluminum throwoff. (*Champion Spark Plug Co.*)

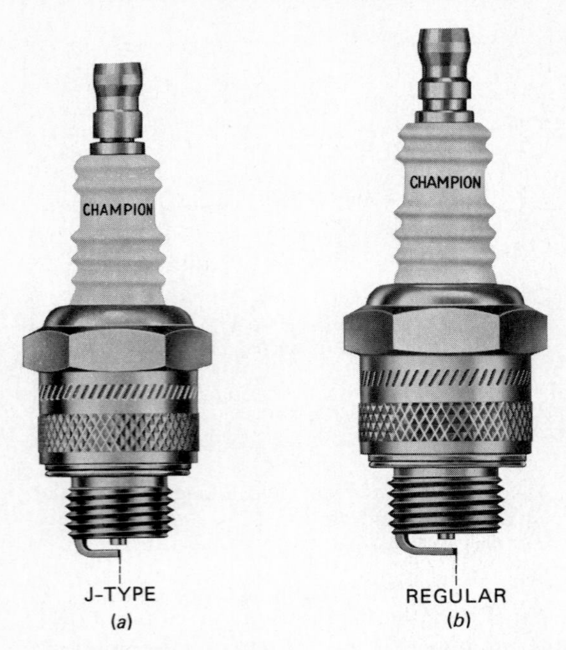


**Fig. 4-27.** Special length of the spark plug ground electrode for two-cycle engines (*a*) J-type. (*b*) Regular. (*Champion Spark Plug Co.*)

**Fig. 4-29.** Testing for spark. Disconnect and hold plug wire $^3/_{16}$ in from plug terminal A while cranking engine. If a spark occurs, the system is functional. If there is no spark, hold the wire $^3/_{16}$ in from plug shell B while cranking. A spark at B indicates a defective plug. (*Jacobsen Div. of Textron Corp.*)

dition, most manufacturers recommend removing, inspecting, cleaning, and regapping a spark plug every 100 h of operation whether or not the above problems are observed. One manufacturer provides a spark testing instrument (Fig. 4-28) to check the ignition system. The tester is clipped to a spark plug cable and grounded to the spark plug. If the spark jumps the 0.166-inch (in) tester gap when cranking, the ignition system is operating properly. When a tester of this type is not available, the spark plug wire is held approximately $^3/_{16}$ of an inch from the tip of the plug at point A (Fig. 4-29), and the starter rope is pulled rapidly. A spark should jump the gap. If it doesn't, position the wire so that it is approximately $^3/_{16}$ of an inch from the steel shell of the plug, point B, and pull the starter rope again. If a spark jumps at the base but not at the tip of the plug,

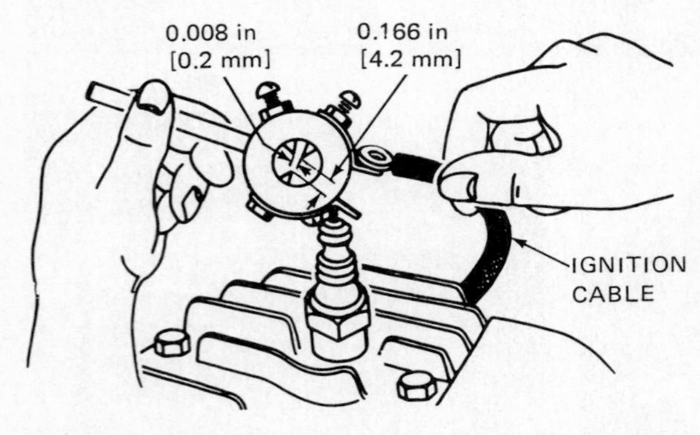


**Fig. 4-28.** Spark test instrument. The spark at the gap of the tester indicates that the magneto-ignition system is functioning properly. (*Briggs & Stratton Corp.*)

it indicates the plug is failing under compression and should be replaced. If the spark does not occur at either position, the ignition system must be checked. This practice is not recommended by all manufacturers and should not be used on solid-state- or electronic-ignition systems.

**Removing the spark plug** To remove a spark plug properly, disconnect the spark plug wire, and then use a spark plug socket wrench of the correct size to loosen the plug one turn. This will loosen dirt from around the base so that an air blast can remove the accumulations and prevent them from falling into the cylinder. If the plug turns hard, you should suspect that the threads are *seized* or *corroded* to the aluminum head threads. To prevent stripping the threads from the head, use a penetrating solvent and slowly rotate the plug back and forth until it works free.

**Inspecting the spark plug** Examine the condition of the electrode and the center insulator. If it appears that the spark plug is "carbon-grounded" (Fig. 4-30*d*), there will be a direct path of electric energy from the center electrode to the metal shell. This condition is referred to as *fouled,* and the spark plug will need to be cleaned or replaced. *Flashover* (Fig. 4-30*b*) may occur when the insulator becomes extremely dirty. A cracked insulator also may cause ignition failure. This condition can occur from failure to use the correct tool when installing or removing a plug, from overtightening during installation, or from some other type of external force.

Inspect the condition of the ground and center electrodes. If they both are rounded excessively, the plug will have to be replaced. If the small engine is equipped with a battery-ignition system (Fig. 4-31), check the connections at the coil for proper polarity. If the negative terminal of the battery is grounded,

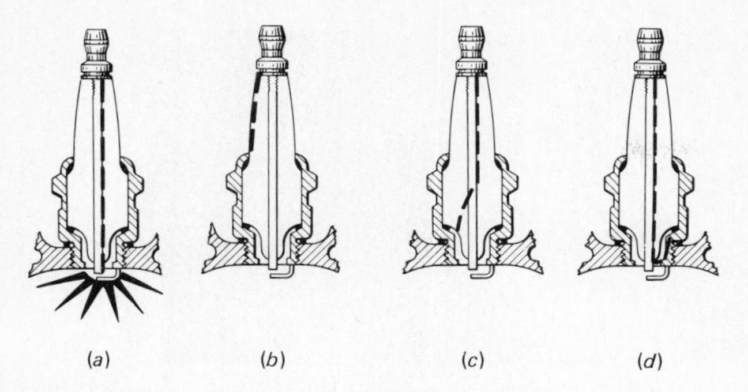


**Fig. 4-30.** Normal sparking does not occur when plugs are in poor condition. (*a*) Normal sparking. (*b*) Flashover. (*c*) Cracked insulator. (*d*) Fouled plug.

the wire connecting the coil to the breaker points should be connected to the negative terminal (−) of the coil.

**Cleaning the spark plug** A plug that has a light-tan or gray deposit and a small amount of center electrode wear may be cleaned and regapped. If the plug is serviceable, the insulator, shell, and firing chamber should be washed in solvent (Fig. 4-32). The use of an abrasive cleaner is not recommended since abrasives may remain on the plug and drop into the engine upon installation. (Occasionally, careful scraping of the excess accumulations of combustion deposits from the center insulator and shell with a penknife may be needed.) A spark plug gapping tool is then used to bend the ground electrode outward a sufficient amount to allow the insertion of a point file. Squaring the center electrode edges by filing will reduce the voltage required to fire the plug. Lastly, the ground electrode is bent back to the proper gap setting and tested with a wire gauge.

When installing a spark plug, run the plug into the threaded hole until finger-tight, and then tighten it a quarter turn; if a new plug is installed, tighten it a half turn, because the new plug will have a new, uncompressed gasket. When a *torque wrench* is available, use the values listed in Table 4-3 to install spark plugs to the proper tightness. The table gives the torque in pound-feet [newton-meters] according to cast-iron or aluminum head construction.

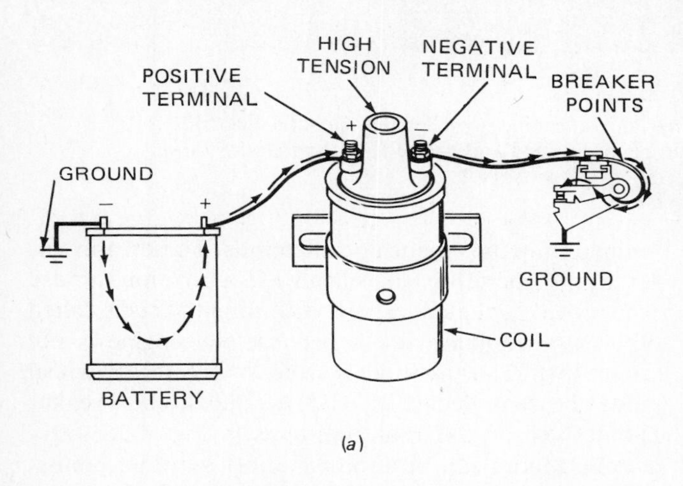

**Fig. 4-31.** Polarity of battery-ignition system. (*a*) Plus (+) terminal of the battery must be connected to the plus terminal of the coil. Breaker points are connected to the negative terminal of the coil. When reversed, excessive wear occurs on the ground electrode of the spark plug (*b*). (*Jacobsen Div. of Textron Corp. and Kohler Co.*)

## SERVICING THE IGNITION SYSTEM

When a small engine is difficult to start, the spark plug is assumed to be at fault. This is not always the case, and simply replacing the spark plug is not always the best practice. Spark plug deposits are often caused by a weak voltage output from the ignition system. The problem may lie in the spark generation system rather than in the spark plug. The small-engine mechanic must be able to analyze the ignition system components.

### Servicing the Flywheel Magneto

The principles of the small engine flywheel magneto operation were discussed in Chap. 3. In summary, the development and timing of the spark at the spark plug depend on the proper operation and adjustment of the system.

The spark is developed by a permanent magnet which generates an electric force in the ignition system strong enough to jump the gap at the spark plug.

**Adjusting Armature Air Gap** In order for the magnets to transfer a maximum amount of force to the coil, the armature must run close to the rim of the outside of the flywheel. The clearance between the armature and the flywheel is referred to as the *air gap* (Fig. 4-33). The armature air gap is adjusted by inserting a flat thickness gauge between the armature poles adjacent to the flywheel and then tightening the retaining screws. The gauge is then removed.

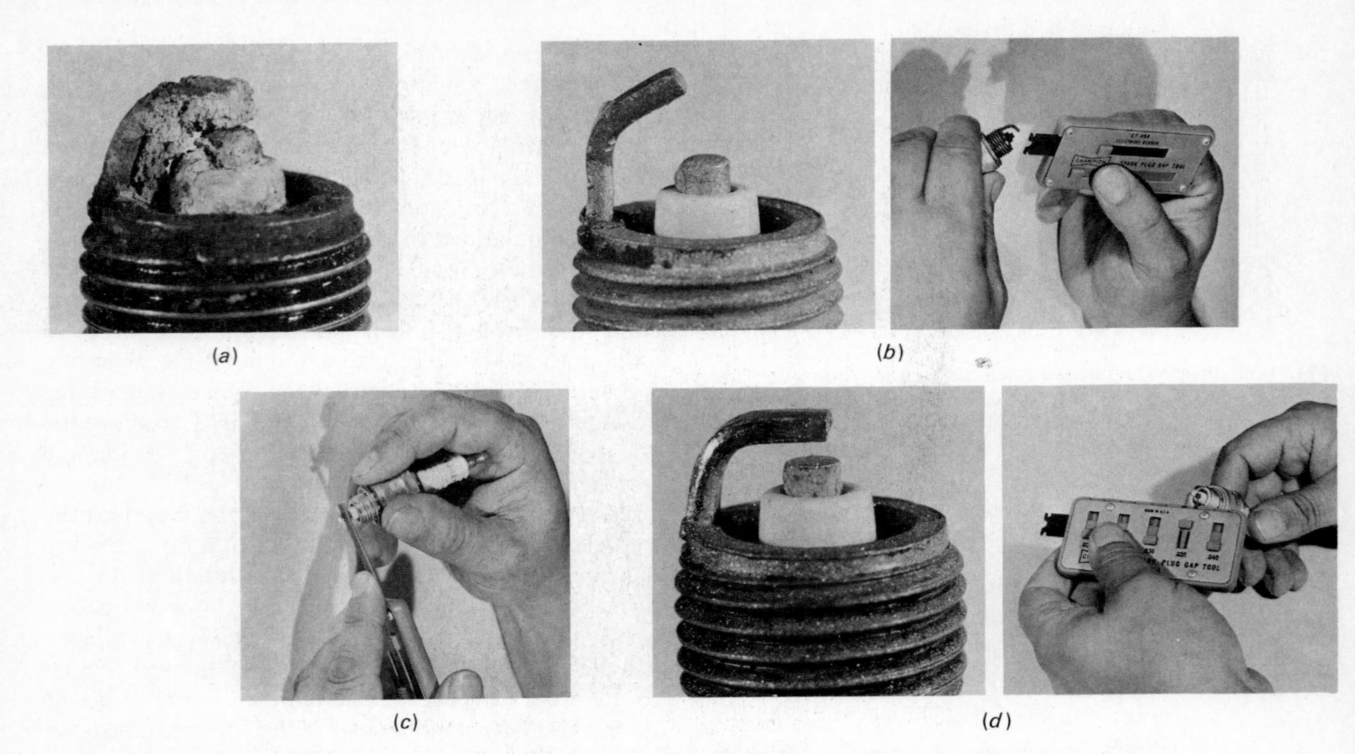


**Fig. 4-32.** Cleaning, filing, and setting proper electrode gap, (*a*) Deposit fouled. (*b*) Cleaned. Ground electrode bent outward. (*c*) Center electrode filed square. (*d*) Electrode properly filed and gapped. (*Champion Spark Plug Co.*)

**Timing the Magneto** For the best engine performance, ignition of the fuel must take place at the correct position of the crankshaft for the piston to exert the greatest force and to obtain the greatest benefit from the fuel. In an ideal situation the fuel should be burning and developing the greatest pressure just as the crankshaft passes top dead center. For this to take place the spark must occur approximately 20° before top dead center (BTDC) (Fig. 4-34). The timing of the spark must be very exact.

The spark occurs at the spark plug the moment the breaker points open. Thus, spark timing is determined by adjusting the breaker points to open at the desired time. In a magneto system timing is controlled by adjusting the breaker points to the correct gap.

The *point gap* is adjusted by moving the stationary contact (Fig. 4-35) until the correct gap is obtained, using a feeler gauge of the correct thickness as determined from the engine specifications. To perform this task, the operating cam must be in a position to raise the movable contact arm the full height. This is called the *static* timing process because the engine is not operating. This method is used where the flywheel must be removed (Fig. 4-35) to adjust the breaker points and on external points, as in Fig. 4-36*a*. *Dynamic* timing can be used to adjust external points. This is accomplished by using a timing light and adjusting the points to align the correct marks on the flywheel with a stationary indicator (Fig. 4-36*b*).

The flywheel magneto must be timed within itself to operate properly. In many instances this is accomplished by positioning the flywheel correctly on the crankshaft with a key. If the key is designed to shear, it is important to check the condition of the key and replace it if there is any indication of shear marks. (Fig. 4-37).

**TABLE 4-3. Torque Recommended for Small Engine Spark Plugs**

| SPARK PLUG THREAD SIZES, mm | With Torque Wrench | | Without Torque Wrench, Turns with Wrench (Beyond Finger Tightness) For New Plugs |
|---|---|---|---|
| | Cast-Iron Head | Aluminum Head | Engine Cool |
| 10 | 12–15 lb · ft [16–20 N · m] | 10–11 lb · ft [13–15 N · m] | 3/4 to 1 |
| 14 | 25–30 lb · ft [34–41 N · m] | 22–27 lb · ft [36–37 N · m] | 1/2 |
| 18 | 30–40 lb · ft [41–54 N · m] | 25–35 lb · ft [34–47 N · m] | 1/2 to 3/4 |

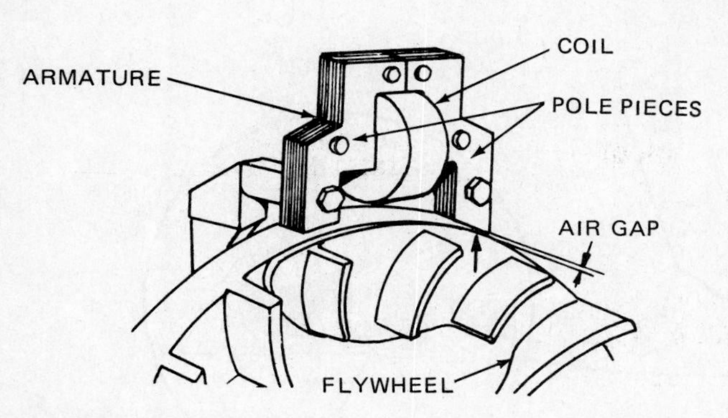


**Fig. 4-33.** The armature of the magneto must operate close to the flywheel with the proper air gap. (*Briggs & Stratton Corp.*)

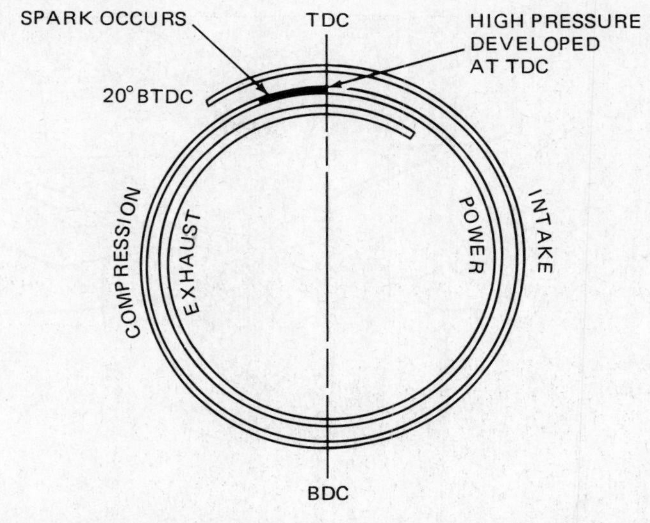


**Fig. 4-34.** Spark ignition is timed to occur before top dead center. A delay of the burning process causes the power stroke to start at top dead center.

**Checking the Condenser** The condenser acts as a safety valve for the primary circuit. It is connected across the breaker points to prevent an energy-destroying arc from forming when the points open. The condenser in an ignition circuit can be likened to placing a rubber bag over one end of a short pipe attached to the main source of water (Fig. 4-38). When the gate valve is suddenly closed (similar to opening the breaker points), water is discharged out the smaller pipe at high velocity (corresponding to the high voltage in the secondary). If the gate should rupture (equivalent to arcing of the points), there would be little discharge from the smaller pipe. To prevent rupture of the gate, the rubber bag expands and absorbs the shock. The bag must have sufficient capacity to absorb the shock.

The condenser consists of a series of aluminum foil

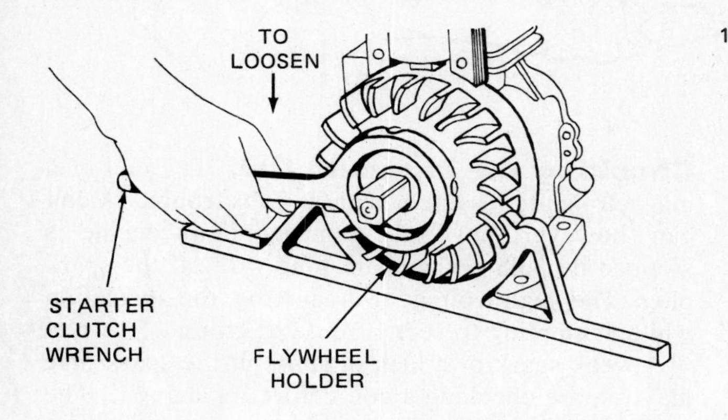


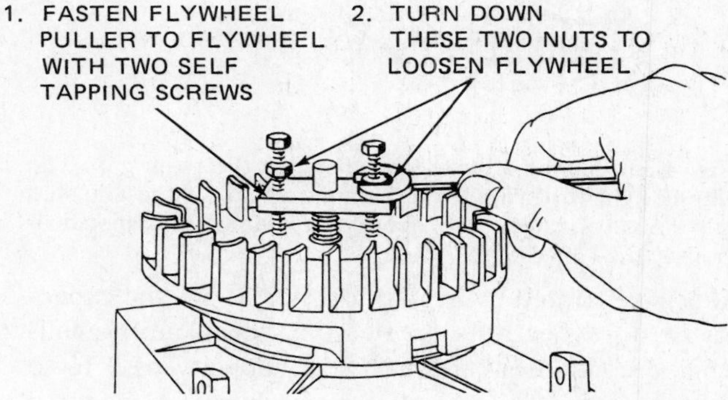


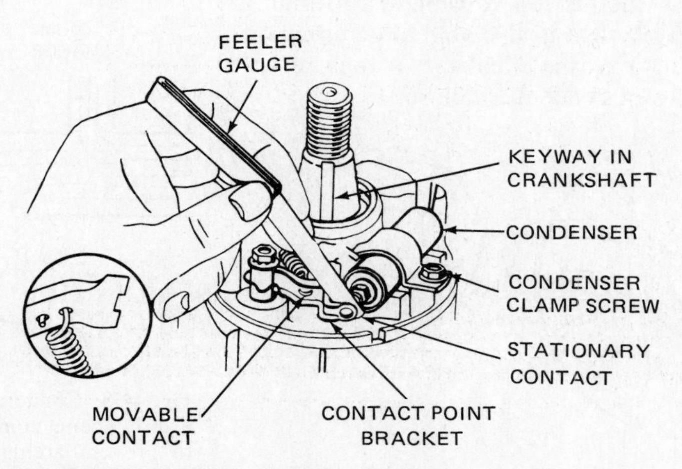


**Fig. 4-35.** Setting "static" timing by adjusting the gap of the internal breaker points. This process requires removal of the flywheel. (*Briggs & Stratton Corp.*)

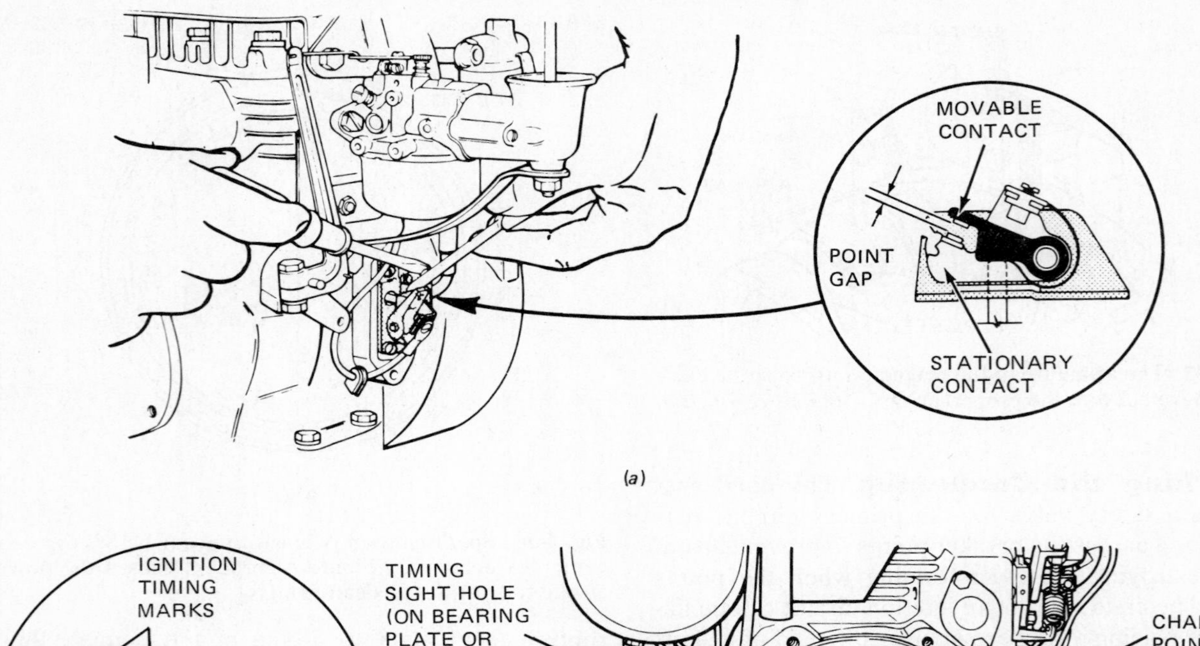


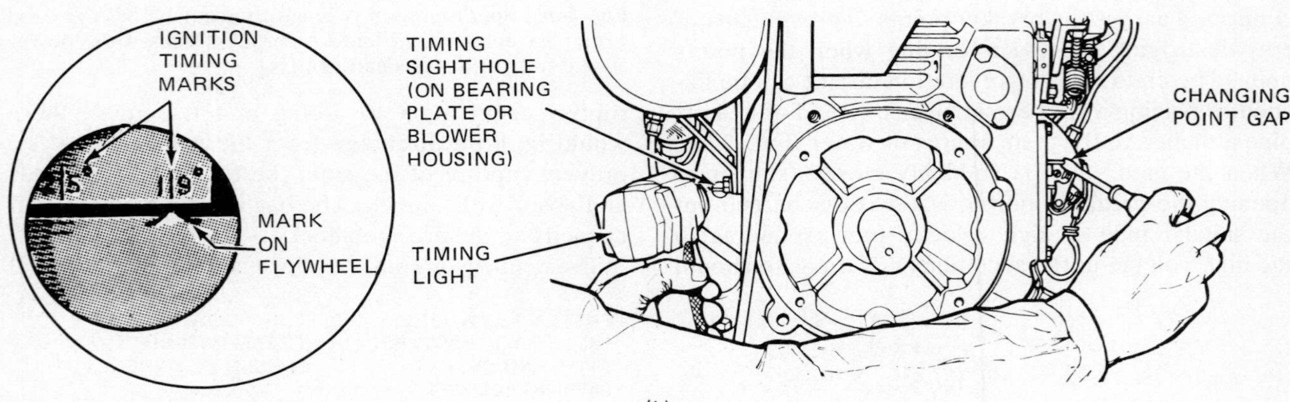


**Fig. 4-36.** Timing adjusted by changing the point gap. The timing light (*b*) provides a dynamic check of the adjusted gap. Always check the operator's manual for specifications. (*Kohler Co.*)

sheets separated by insulator paper. It has the capacity to store electrons. In electrical terms most small-engine condensers must have a capacity of 0.16 to 0.24 microfarads ($\mu$F) of storage ability. A magneto analyzing instrument is used to measure its capacity and to determine whether it is capable of performing properly. The instrument is also able to determine if the condenser has excessive leakage or high resistance, both causes of condenser failure.

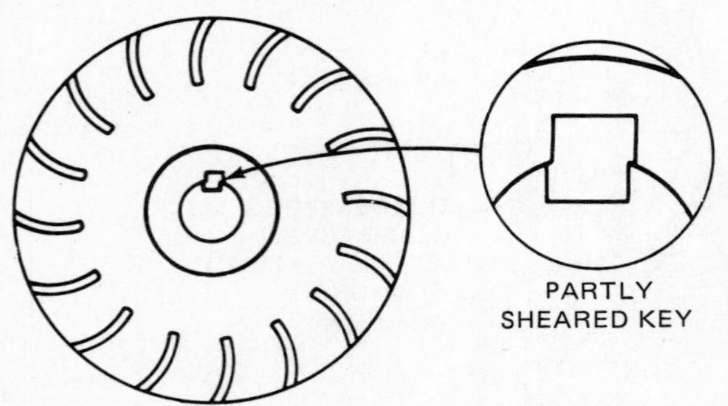


**Fig. 4-37.** A partly sheared flywheel key will throw the magneto out of time. (*Briggs & Stratton Corp.*)

**Checking the Magneto Coil** The coil of a magneto-ignition system seldom gives trouble. A coil can, however, be easily damaged if an engine is stopped by pulling the spark plug wire off the spark plug. The high voltage will destroy the insulation while attempting to seek a route to ground.

A weak spark or a lack of spark at the spark plug may require checking a coil before replacing it. The analyzer illustrated in Fig. 4-39 is used to evaluate a

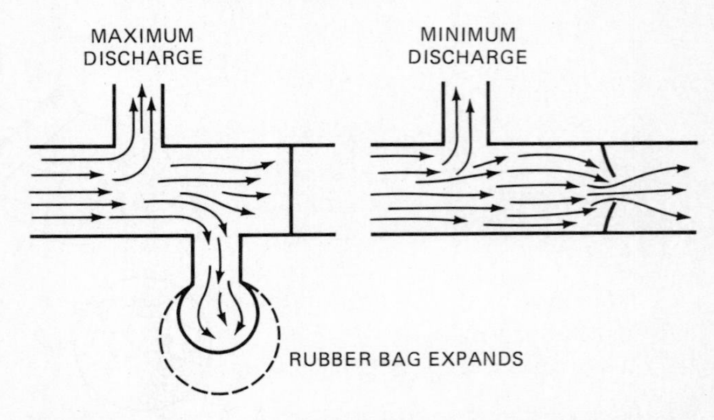


**Fig. 4-38.** Condenser action is similar to a shock absorber, a spring being compressed, or a rubber bag expanding (left) to prevent arcing (right) of points (bursting valve); the small pipe represents the discharge at the secondary system of the coil. (*Briggs & Stratton Corp.*)

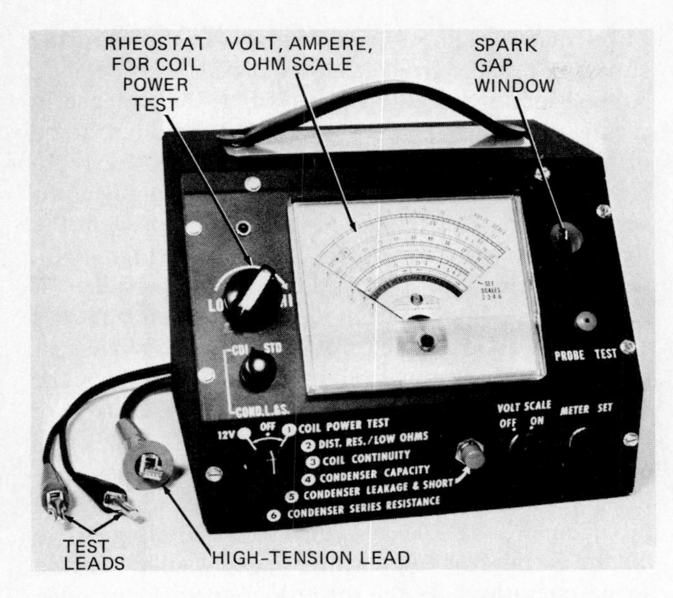


Fig. 4-39. Coil testing instruments will check the condenser and coil by simulating operating conditions. (*Mercotronic Instrument Corp.*)

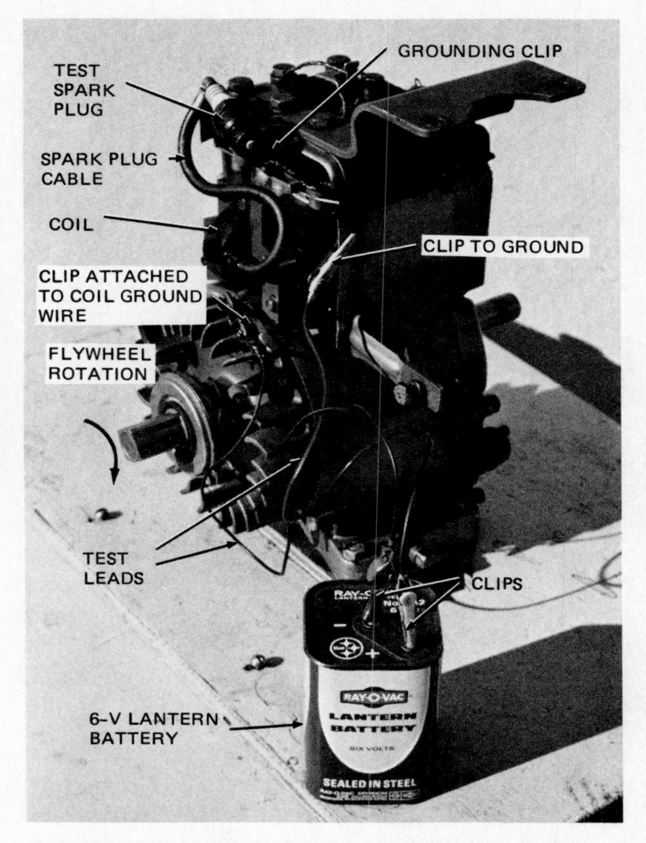


**Fig. 4-40.** Using a 6-V lantern battery to energize a magneto-ignition system to determine if components are operating properly.

coil for performance. The coil is tested for the ability to produce a spark when energized by the tester.

A 6-volt (V) lantern battery can be used to provide electric energy to the primary circuit of the ignition system. With battery power, the breaker points can be opened manually; if the components are functioning properly, a spark will occur at the spark plug when the points are opened (Fig. 4-40).

## Servicing the Battery-Ignition System

The battery-ignition system differs from the magneto system in two ways. First, the battery is a source of electric energy, and second, a switch is used to complete the circuit to the battery. When the switch is on or closed, the circuit is completed. The switch in a magneto system is open when the engine is running and closed to ground the primary windings.

## Servicing the Battery

Since the battery supplies the electric power, its state of charge must be maintained. Most small-engine batteries are the 12-V lead-acid type. One of the most common problems with this type of battery is self-discharge. This occurs because a light coating of corrosive acid develops across the top of the battery. To maintain a battery, it is necessary to perform the following service functions regularly.

1. Check the level of electrolyte—add distilled water as necessary to maintain the level above the plates—do not overfill, as this can cause poor performance or early failure due to loss of electrolyte. Do not add acid to a battery.

**⚠ WARNING Battery acid is extremely corrosive and caustic. Protect your eyes, skin, and clothing.**

2. Keep terminals and the top of the battery clean. Wash carefully with a commercial cleaner or baking soda and rinse with clear water. Do not allow the cleaning solution to enter the cells, as this will destroy the electrolyte.
3. Make sure that the battery holddowns are secure —if loose, vibration will cause premature failure. Be careful not to damage the battery case by overtightening the holddowns.
4. Clean the cable clamps and terminals occasionally with a wire brush. When reinstalling the clamps, press firmly into position—do not pound with a hammer.
5. An undercharged battery may freeze when unused during cold weather—keep the battery charged up or store it in a warm area.

**⚠ WARNING Adequate ventilation must be provided when batteries are being recharged. Sparks, open flames, or smoking should also be avoided since hydrogen gas is produced, which if ignited, could cause an internal explosion that could shatter the case of the battery. When charging, the cell vent caps should be removed to**

**prevent the gases generated from building up and exploding.**

## Servicing Battery-Ignition Components

All the remaining components of the battery-ignition system, such as the condenser, points, coil, and spark plug function as those in the magneto system and are serviced in a similar manner.

# CHECKING AND ADJUSTING THE CARBURETOR

The professional in power and machinery who operates a small engine service center or who services groundskeeping equipment such as small tractors, mowers and snow blowers, tillers, trimmers, chain saws, and many other types of equipment must employ people who can check and adjust carburetors. The carburetor is often ignored or improperly serviced.

The carburetor performs a number of functions, of which the most important are:

1. To provide the correct ratio of fuel and air to achieve the most desirable combustible mixture.
2. To break up the fuel into an extremely fine mist which will form a gaseous mixture with the air.
3. To control the amount of gaseous mixture entering the combustion chamber, regulating the ability of the engine to operate at a variety of speeds and power settings.

For the carburetor to perform the above tasks, it must be properly serviced and adjusted. Operating an engine with the carburetor out of adjustment can cause loss of power, excessive engine wear, and high fuel consumption. It is therefore important that service and maintenance people understand the various systems of the carburetor and their functions.

## Fuel Metering System

The carburetors used on most small heavy-duty four-cycle spark-ignition engines can be adjusted to provide the correct air-fuel mixture over a wide range of operating conditions and engine-loads. Carburetors of this type require two systems, the idle and the high-speed system.

**Idle System** When an engine is ''idling,'' it is running at a reduced speed of 1200 to 1800 revolutions per minute (r/min) and is not connected to a load. The throttle valve will be nearly closed, and the engine will be obtaining an air-fuel mixture from the idle system.

At idle speeds, the air velocity through the venturi is not great enough to draw a reliable supply of fuel from the nozzle. The idle system consists of a series of passages leading to openings in the throat of the carburetor on the crankcase side of the throttle valve. An idle-mixture needle valve (Fig. 4-41) is placed in the discharge area to permit the proper mixture to be made. An adjusting screw is also provided for regulating the idle speed. The adjustment is actually a setscrew which serves as a throttle stop to prevent the throttle valve from closing completely (Fig. 4-46). Adjusting the idle-speed adjusting screw makes it possible to adjust the amount of air which passes around the edges of the throttle valve. The bypass air mixes with the air-fuel mixture discharged from the idle ports. The ports may differ in design. One type of carburetor uses a long narrow slot; another uses two or more drilled openings. The function is the same—to serve primary and secondary idle conditions. In this manner, the idle system also makes it possible for the carburetor to provide for intermediate engine speed operation. As the throttle is opened, the edge exposes more of the slot opening or the secondary opening. This allows more mixture to be discharged, and the engine runs faster.

In some carburetors the idle-mixture adjustment screw will control air only. On other types it controls fuel only while others control a mixture of fuel and air. The maintenance person must identify the process used when making any adjustment. In all systems, the idle mixture and the idle speed settings must be coordinated in order to obtain the desired idle operating speed.

**High-Speed System** The basic parts of the high-speed system of most carburetors are the discharge nozzle, the regulating jet, and the compensation section. The discharge nozzle (Fig. 4-42) serves

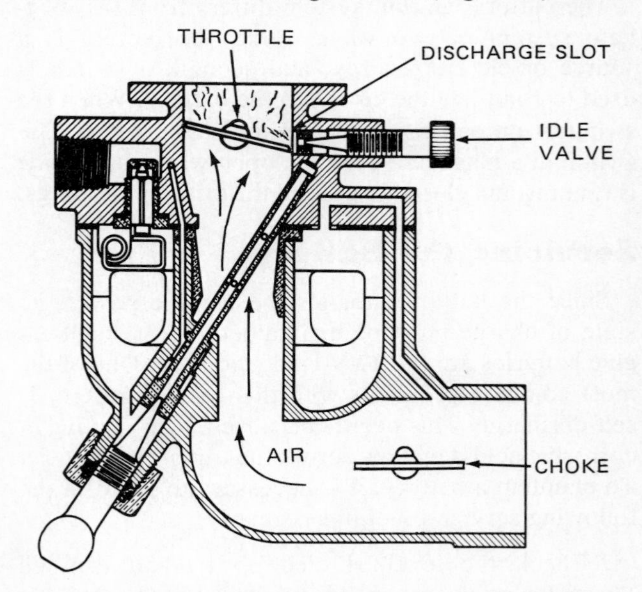


**Fig. 4-41.** Idle system. An air-fuel mixture is discharged by the idle jet into the intake manifold of the engine. Throttle stop screw is hidden by the position of the carb. (*Briggs & Stratton Corp.*)

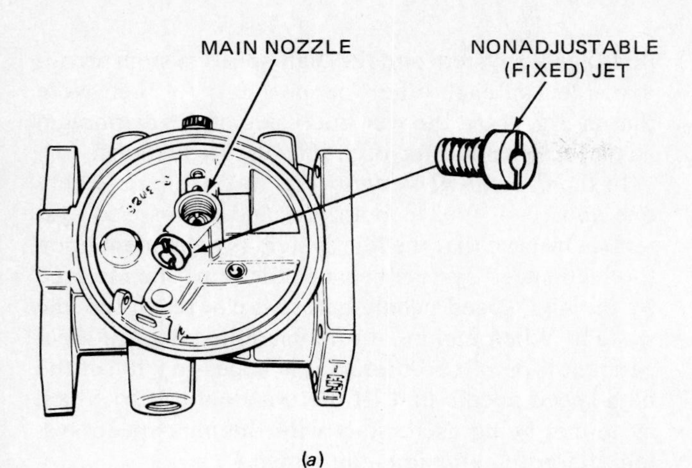


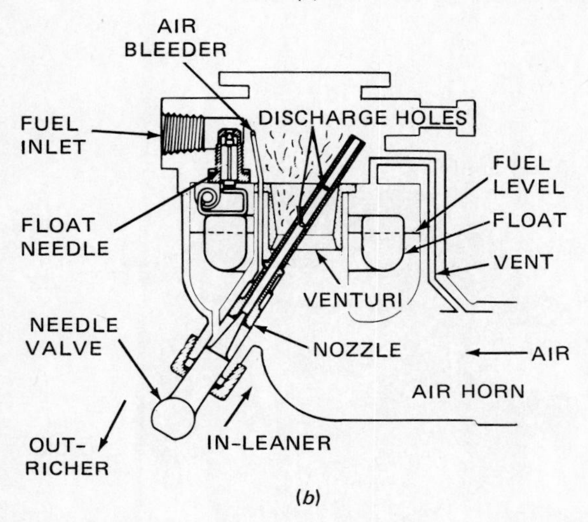


**Fig. 4-42.** High-speed system. (*a*) Fixed discharge (nonadjustable). (*b*) Adjustable needle valve for regulating fuel flow from the bowl supply. (*Briggs & Stratton Corp.*)

to discharge the fuel that has been metered by the regulating jet. The regulating jet is either a fixed (Fig. 4-42*a*) or an adjustable orifice (Fig. 4-42*b*). An adjustable orifice will include a needle valve which can be moved toward or away from its seat. Adjusting the needle in or out, respectively, leans or enriches the mixture (Fig. 4-42*b*).

**Compensating System** The compensating system of the carburetor is designed to prevent the fuel mixture from becoming increasingly richer as the throttle opening widens and engine speed increases. This action occurs because the vacuum on the high-speed system increases as the air velocity through the carburetor increases. The increase in velocity causes an increase in friction which decreases the density or weight of air. This results in the fuel-air ratio becoming richer as engine speed increases. The adjustable needle valve provides a means to achieve the proper air-fuel ratio when the throttle is wide open. The fixed jet system is used by some manufacturers to accomplish the same objective.

An *air-bleed* compensation system is used on

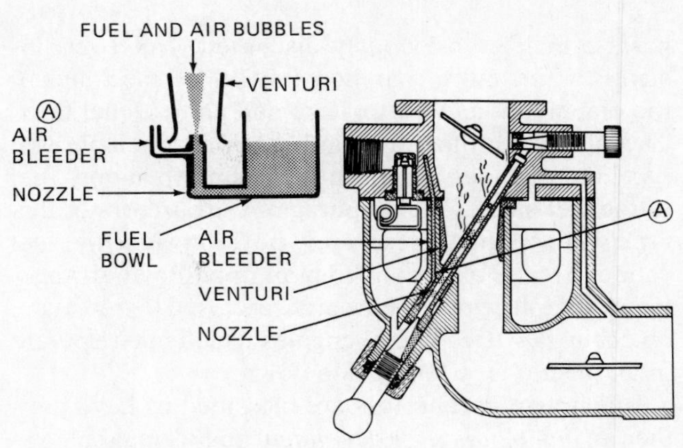


**Fig. 4-43.** Air-bleed compensation system. The air enters orifice A and mixes with fuel in the nozzle to "lean" the mixture at part-throttle speeds. (*Briggs & Stratton Corp.*)

many small-engine carburetors (Fig. 4-43). The system is used to control the air-fuel ratio automatically at less than full-power requirements. The air-bleed is an air passage which is calibrated to meter a small quantity of air into the fuel entering the nozzle (Fig. 4-43). As a result, a mixture of air and fuel is drawn out the nozzle. This overcomes the increase in richness with speed.

An atmospheric vent, or *balancing channel* (Fig. 4-44), is provided on most carburetors. It is used to balance air pressures on the fuel in the bowl and air flowing through the carburetor. Venting will permit an equilibrium to be established and prevents a pressure imbalance within the various systems of the carburetor. For this reason it is important that sealing gaskets are in good condition and that the mating surfaces are not distorted.

**Diaphragm-Control Fuel System** The diaphragm type of carburetor operates on the same

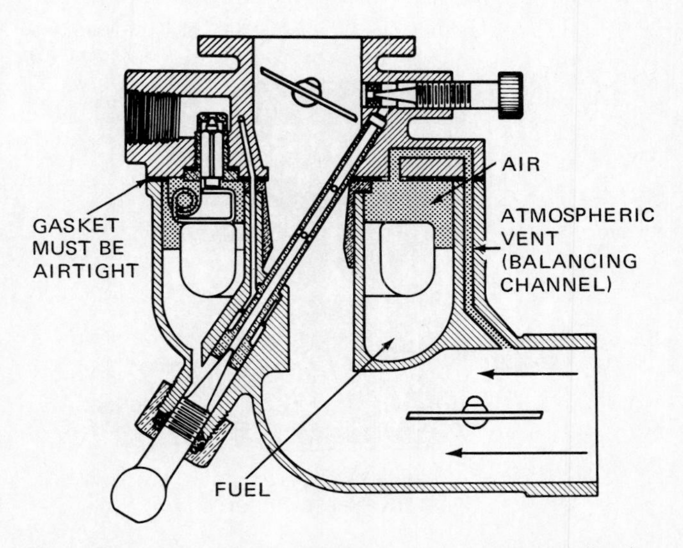


**Fig. 4-44.** An atmospheric vent is provided to equalize internal pressures. (*Briggs & Stratton Corp.*)

basic principles previously discussed. However, instead of the more common bowl-float mechanism, the diaphragm is used to store and control fuel flow. The carburetor illustrated in Fig. 4-45 is a diaphragm type and includes a built-in fuel pump to supply fuel to the wet side of the diaphragm. Carburetors of this type will operate in nearly any position since the fuel cannot escape as it can from an open-bowl arrangement. Diaphragm carburetors are used extensively on chain-saw (two-cycle) engines which must operate in all positions, even upside down.

Diaphragm carburetors are classified as having either an *independent* or *dependent* fuel discharge system. An *independent* carburetor is constructed so that the idle system and the high-speed system are independent of each other for the source of fuel. Note that in Fig. 4-45*c* the wet (fuel) side of the diaphragm has a fuel channel for both idle and high speed.

In the *dependent* system (Fig. 4-45*b*) there is only one source of fuel to both idle and high-speed systems. Observe that the idle system is dependent upon the high-speed system because fuel must be metered by the high-speed needle before it can reach the idle system. When making a running adjustment of a dependent type of carburetor, it is necessary to set the high-speed needle first. If there is doubt as to which system is being used, start with the high-speed system before making idle adjustments.

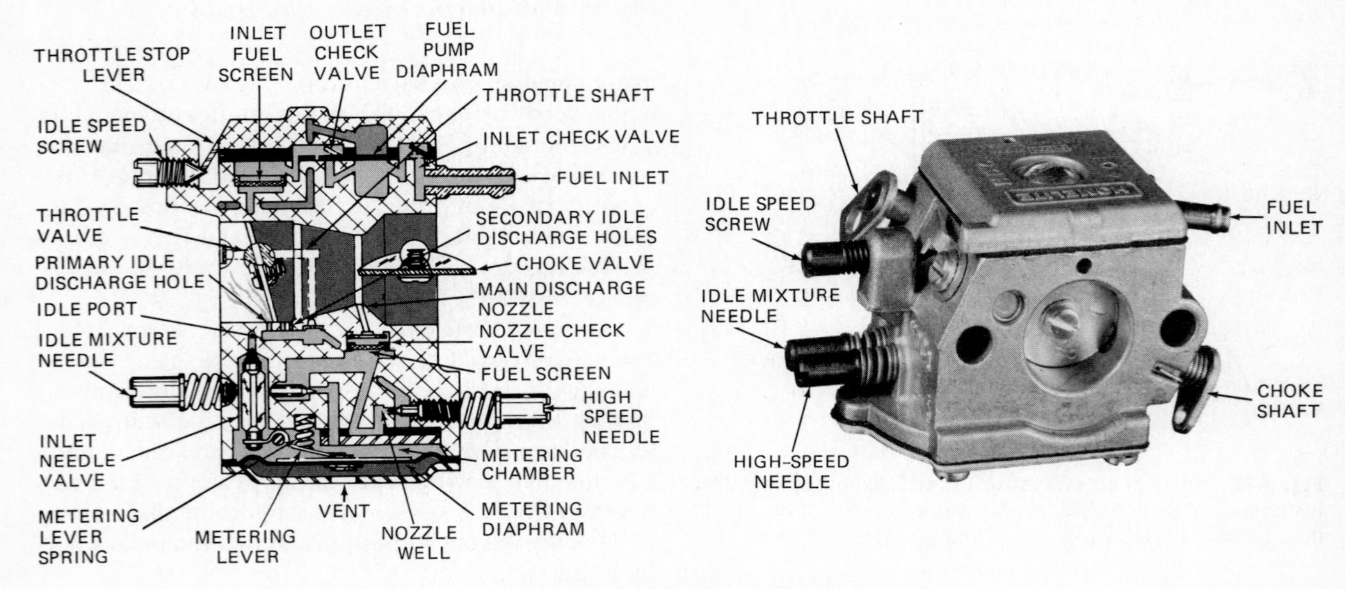


(*a*)



(*b*) (*c*)

**Fig. 4-45.** Cross section of a diaphragm-type carburetor and its air-fuel operating circuits. (*Walbro Corp.*)

## Adjusting the Carburetor

Small-engine carburetors can usually be adjusted for (1) idle speed, (2) idle mixture, and (3) high-speed load. These adjustments make it possible to control the fuel-air mixture for a variety of operating conditions involving engine speed and load. Most manufacturers state that carburetors are adjusted at the factory and should not be reset. Owners' and service manuals suggest, however, that if black smoke comes from the exhaust, the spark plug fouls, or the engine overheats or runs roughly with little power, the carburetor may need minor readjustment. Each manufacturer provides *static* adjustment procedures for each engine. This means that adjustment is made with the engine stopped or the carburetor removed from the engine, as follows:

1. Turn the load (main) and idle fuel-adjustment needle valve screw in a clockwise direction until it "bottoms" (touches) lightly on its seat. The needle valve must not be forced, or the needle and seat will be damaged.
2. Turn the main (load) screw out (counterclockwise) the specified number of turns. This usually is between 1½ to 2 full turns [be sure to follow the exact specifications (Fig. 4-46)].
3. Turn the idle fuel needle valve out (counterclockwise) ¾ to 1¼ turns following the exact specifications (Fig. 4-46).
4. Hold the idle-speed adjusting screw against its stop. At the same time turn the screw outward (counterclockwise) until there is a very small gap between the end of the screw and the stop; then turn the screw in (clockwise) 1 turn (Fig. 4-46).

For most operating situations, the static adjustment should provide satisfactory engine operation. However, because of altitude, air temperature, type of fuel being used, or load conditions, the carburetor may have to be adjusted when the engine is operating. Most manufacturers recommend this operation be performed when the engine is operating under a load. A small-engine dynamometer provides the safest method of placing a controlled load on an engine. When a dynamometer is not available, the engine can be operated at maximum governed throttle setting. For example, the best way to set the carburetor of a chain saw is to operate the engine under actual sawing conditions.

To adjust the carburetor when the engine is operating, proceed as follows:

1. Start the engine and allow it to warm up. This usually takes 3 to 5 minutes (min).
2. Operate the engine at maximum governed speed (Fig. 4-47), usually 3000 to 3600 r/min for four-cycle engines and as high as 12,000 r/min for a two-cycle chain-saw engine. *Be sure to use adequate hearing protection.*
3. Carefully turn the load (main) adjustment needle valve in (clockwise) until the engine slows down. This indicates the fuel mixture is becoming too lean for good power performance. Proceed slowly and allow enough time for the engine to adjust to the new setting. Note the position of the screw slot in the needle.
4. Slowly back out (counterclockwise) the load needle valve until the engine slows again from too rich a mixture. Note the position of the screw slot.

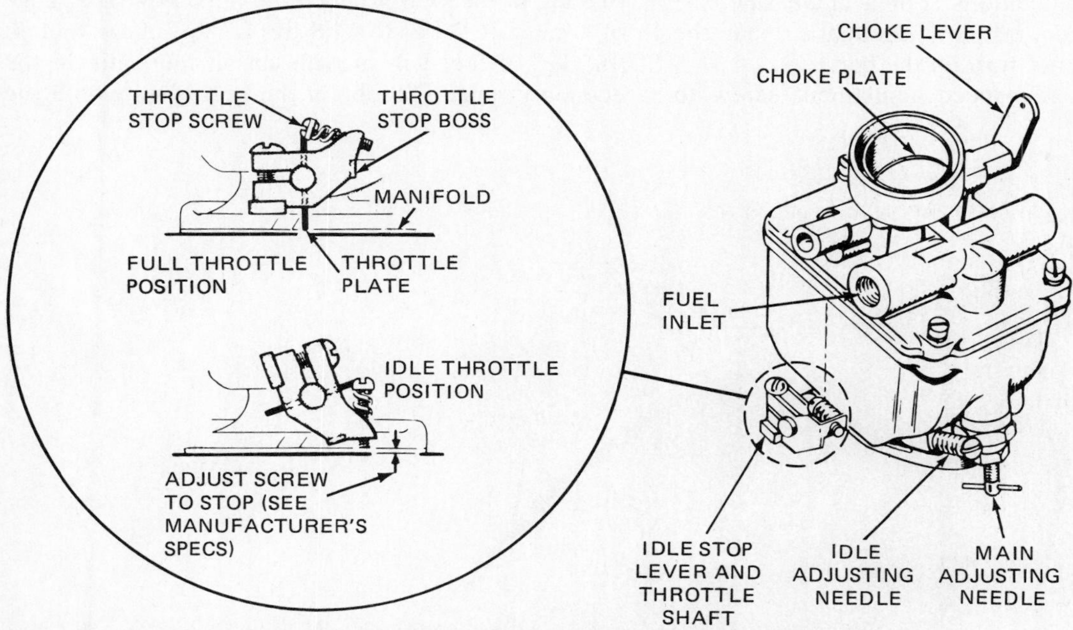


**Fig. 4-46.** Adjusting idle speed screw, idle mixture, and high-speed needle valves on a float-type carburetor. (*Onan Corp.*)

(a)

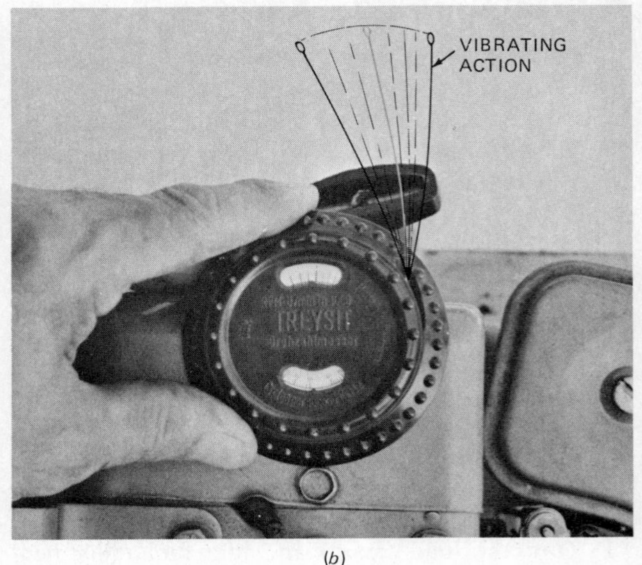


(b)

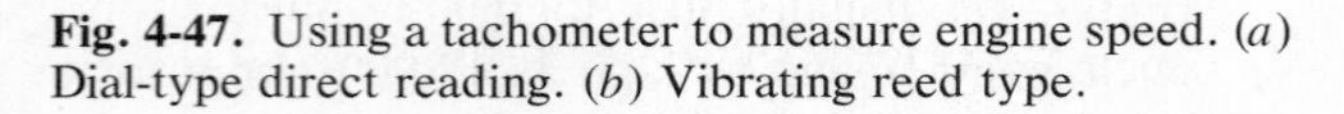

**Fig. 4-47.** Using a tachometer to measure engine speed. (*a*) Dial-type direct reading. (*b*) Vibrating reed type.

5. Turn the screw inward (clockwise) until the screw slot is halfway between the step 3 and 4 positions.
6. Remove the load from the engine and proceed to set the idle speed and mixture valve in the following manner:
   (a) Hold the idle-speed adjusting screw against its stop, or release all tension on throttle control linkage.
   (b) Set the idle speed of 1200 to 1800 r/min (Fig. 4-47) with a tachometer depending upon engine specifications. For a chain saw, the speed should be slow enough so that the chain does not turn on the bar.
   (c) Rotate the idle-speed adjustment screw to obtain the proper idle speed. Screw in to increase and out to decrease engine speed.
   (d) Continue to hold the idle-speed screw against its stop and move the idle-mixture adjusting screw until the tachometer shows the highest rpm and smoothest engine operation (Fig. 4-47).
   (e) Reset the idle-speed control to specifications.

When an exhaust gas analyzer (Fig. 4-48) is available, set the load (main) adjusting needle to obtain the desired air-fuel ratio. The information in Fig. 4-49 indicates that an air-fuel ratio of 11.5 pounds (lb) [5.2 kilograms (kg)] to 13.5 lb [6.2 kg] of air per 1 lb [0.5 kg] of fuel will produce the most power; a range of 13.5 lb [6.2 kg] to 14.8 lb [6.7 kg] of air to 1 lb [0.5 kg] of fuel will provide an air-fuel ratio in the economy range. To obtain the best results with the

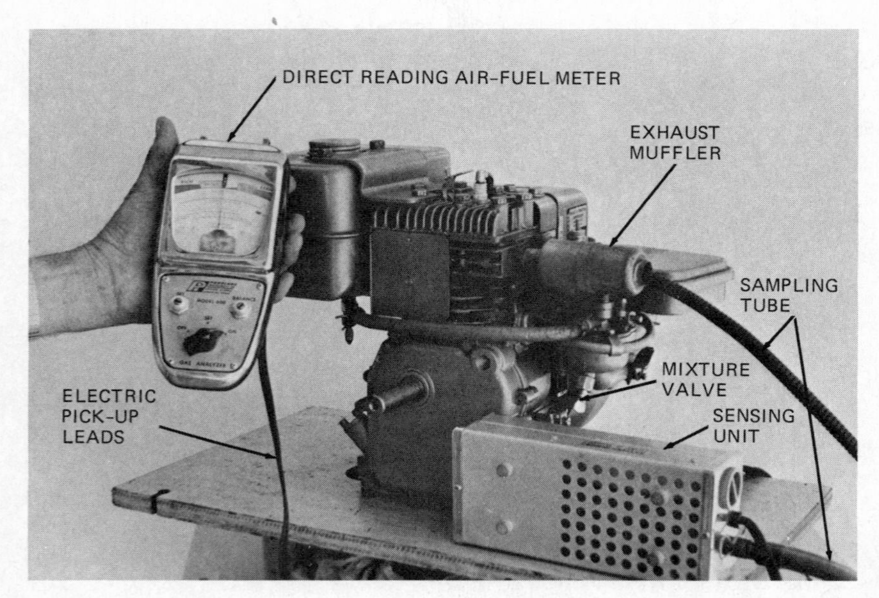


**Fig. 4-48.** Exhaust gas analyzer.

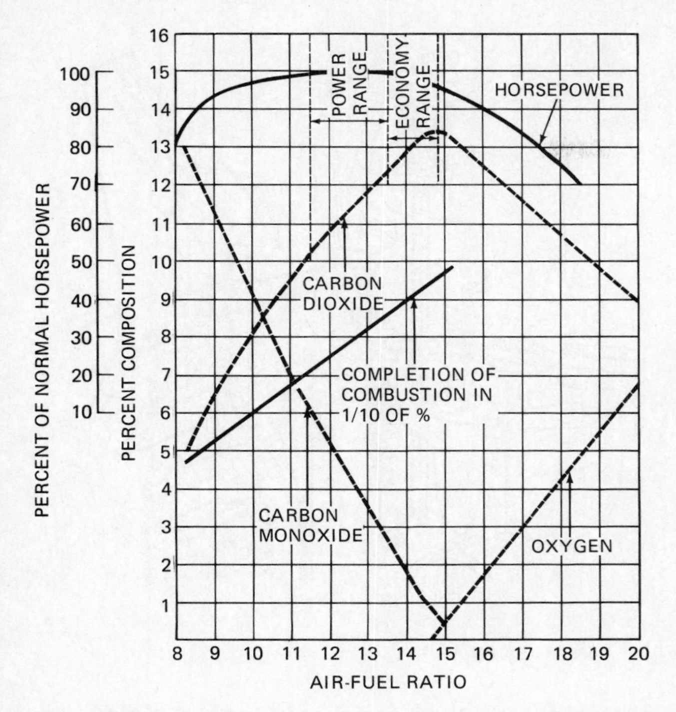


**Fig. 4-49.** The effect of the air-fuel ratio on both engine horsepower and the completeness of combustion.

exhaust gas analyzer, operate the engine at maximum governed rpm and near full load.

## ADJUSTING THE GOVERNOR

Small air-cooled gasoline and diesel engines have a built-in device called a *governor*. The governor controls the engine so that it will drive a piece of machinery at a designated operating speed. The purpose of the governor is to maintain a constant rpm by automatically adjusting the power output of the engine when the driven load requirement varies. In a gasoline engine the governor regulates the opening of the throttle valve in the carburetor. A governor will sense the change in power requirement, closing the throttle valve when the load is lightened and opening it when the demand is for more power. The governor of a small air-cooled diesel engine functions in much the same manner except that it controls the discharge rate of the injection pump. When the power requirement is low, the governor adjusts the amount of diesel fuel injected so that it is less than when a higher power output is needed.

### Types of Governors

There are two basic types of governors used on small gasoline engines. These are the *pneumatic,* or air-vane type and the *centrifugal,* or mechanical type. Diesel engines use only the centrifugal governor system. In both systems, any change in operating speed is sensed by the reaction of the governor spring (Fig. 4-50) to pneumatic or centrifugal force. The spring is accurately calibrated to fit the particular system and condition. A manual control is attached to the spring so that tension on the spring can be set by the engine operator. When the pull of the spring and the pneumatic or centrifugal force are equal, the engine is said to be operating at "governed" speed. When the engine speed drops because of increased load, the force working against the spring will be reduced and the spring will open the throttle and engine speed will increase.

**Pneumatic (Air-Vane) Governor** The pneumatic governor (Fig. 4-51) is operated by the force of the cooling air developed by the flywheel fan as it rotates. Air from the fan pushes against a vane. A mechanical linkage connects the air vane to the throttle valve of the carburetor. If the air flow pressure becomes greater because the fan is rotating faster, the vane overcomes spring tension and closes the throttle valve. If the operator wishes to increase the governed speed, the throttle control is moved to increase spring tension.

**Centrifugal (Mechanical) Governor** A centrifugal governor (Fig. 4-52) is generally a gear-driven device which is contained within the crankcase of the engine. The gears rotate a set of weights which expand or contract, depending upon the spring tension and centrifugal force developed by the speed of rotation. If the engine rpm slows because of increased load, the centrifugal force on the weights will decrease. This will allow the governor spring to contract the weights, opening the throttle valve so that the engine will develop more power to recover the loss of operating speed.

Reducing the load allows the rpm to increase, which will cause the centrifugal force to expand the weights against the spring's tension. This stretches the spring, and the linkage closes the throttle valve, reducing power output.

The operating principle of the centrifugal governor

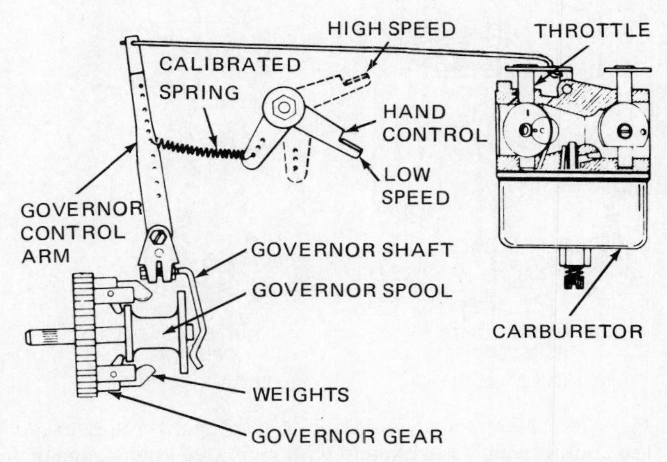

**Fig. 4-50.** The governor reacts to engine speed and setting of the hand control lever through the calibrated spring. (*Tecumseh Products Co.*)

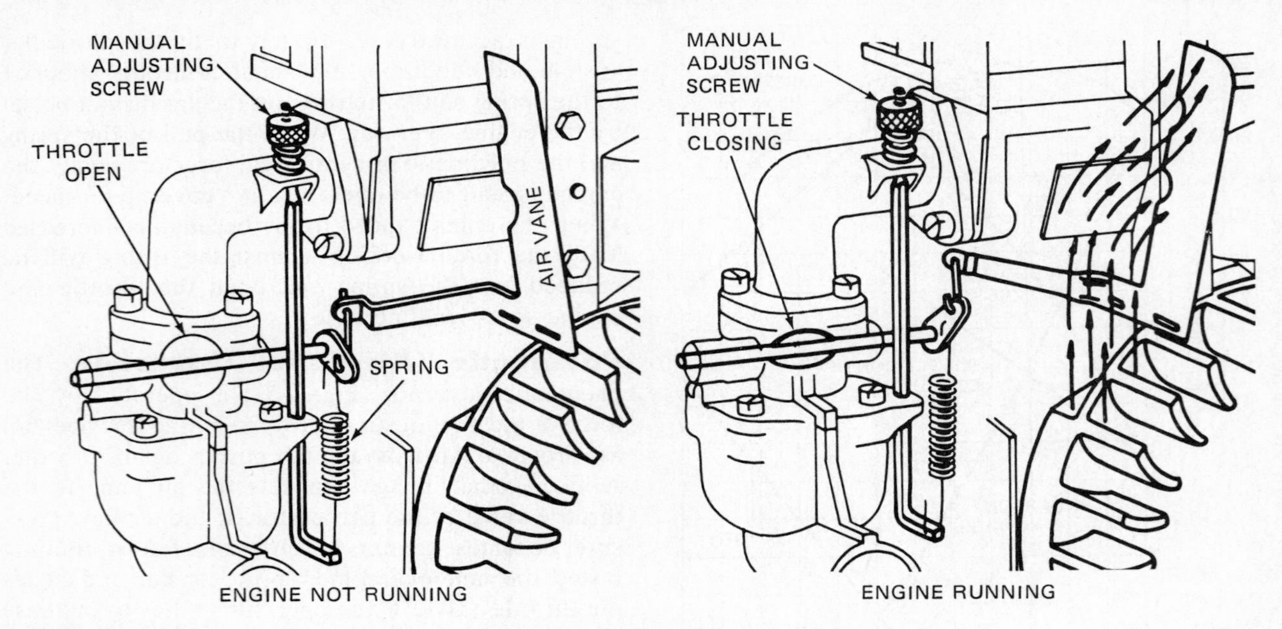


**Fig. 4-51.** Pneumatic governor. The force of air from the flywheel moves the air vane against the spring tension. (*Briggs & Stratton Corp.*)

of a diesel engine is identical to that of the gasoline engine. However, instead of regulating the opening of the throttle valve, the diesel engine is controlled by the movement of the injection pump plunger which controls the fuel discharge rate.

## Checking the Governor

When properly operating, a governor will respond quickly to any change in load. Trouble should be suspected with a governor system, however, when one of the following occurs: (1) sluggish action (a large variation in operating rpm), (2) hunting (surging) action with large and rapid movements of the throttle valve, and (3) poor reaction to change in the control lever position.

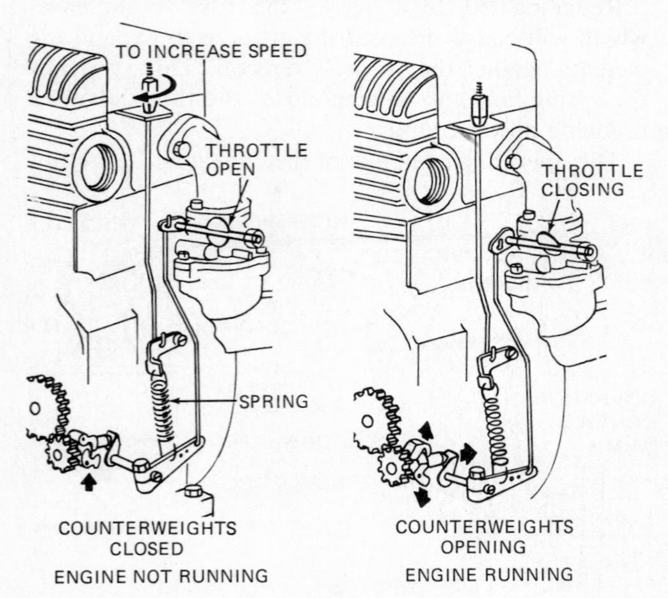


**Fig. 4-52.** Mechanical governor. Centrifugal force causes the counterweights to expand with increased engine speed; the weights react against the spring tension to close the throttle. (*Briggs & Stratton Corp.*)

To check a governor system, look for evidence of improper service. Pneumatic governors require unobstructed air flow around the engine. If dirt or debris has accumulated under the shroud or if the blower housing or baffles have been removed, the pneumatic governor will not work properly. You should also inspect the governor linkage for bent control rods and arms and for worn connections. Inspect the linkage and compare it with a service manual illustration showing correct setup and installation. Check to see that the governor spring is positioned in the correct holes in the control arms and linkage. Check the freedom of movement, and make sure that the throttle valve is fully open when the engine is stopped. Lubricate the adjusting screw or lever with penetrating oil to achieve freedom of movement.

# SERVICING THE VALVE SYSTEM

In four-cycle engines, the principal job that the service center will be required to perform is to check and adjust the valve tappet clearance. The recommended maintenance schedule for checking valve adjustment will vary with the engine manufacturer and the type of valve arrangement. For example, an L-head engine (Fig. 4-53*b*) may need to be checked only every 500 h of operation while an I-head (Fig. 4-53*a*) may require checking the valve clearance every 100 h.

The maintenance of the valve system required for two-cycle engines includes testing the reed valve and checking the cleanliness of the exhaust ports and muffler.

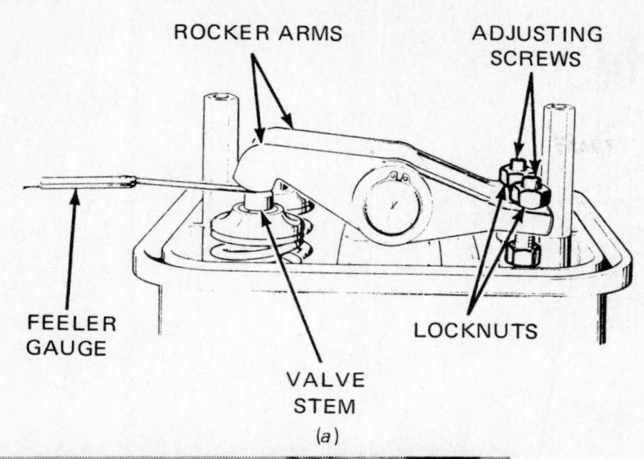


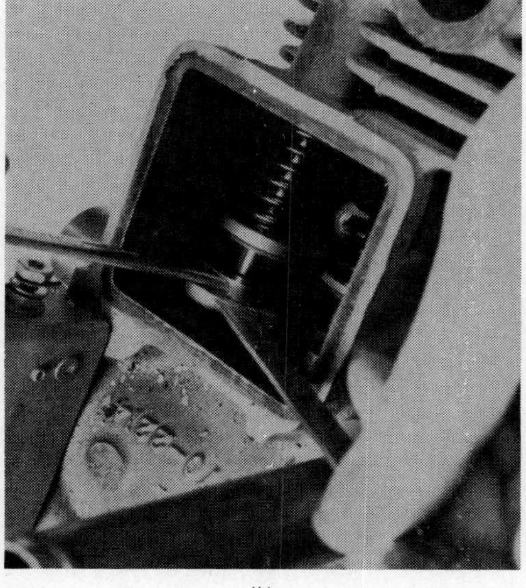


**Fig. 4-53.** Setting the valve clearances on (*a*) a valve-in-head engine and (*b*) an L-head engine. The piston must be at top dead center with both valves closed.

## Checking Four-Cycle Valve Clearance

The primary purpose of checking the valve tappet clearance of a four-cycle engine is to assure that the clearance is sufficient so that the valves will close completely and with the proper timing.

To perform the operation of checking and adjusting valve clearance, obtain the clearance specifications from the service manual. Also, determine whether the settings listed are for a cold or hot engine. To check the clearance and make necessary adjustments, be sure to position the piston at top dead center (TDC) of the compression stroke. This position will insure that the lifting cams are at their lowest point and that the subsequent valve clearance will be the greatest. Use the appropriate thickness feeler gauge for both the intake and exhaust valve (Fig. 4-53).

On a valve-in-head engine use a box end wrench to loosen the locknut on the rocker arm just enough so that the adjusting screw can be turned to give the correct clearance between the rocker arm face and the end of the valve stem. There should be a light drag on the feeler gauge. Hold the screwdriver steady and tighten the locknut snugly. On an L-head engine equipped with adjustable lifters, two tappet-adjusting wrenches are needed. One wrench is used to hold the lifter while the adjusting screw is turned in the proper direction to obtain the necessary clearance.

## Checking the Two-Cycle Engine Valves

To test the two-cycle engine for leaking reed valves, remove the air cleaner element from the carburetor. With the engine running, hold a clean piece of paper approximately 1 in from the carburetor throat (Fig. 4-54). If there is evidence of fuel spotting on the paper, the reeds are not seating. When this condition is observed, the reed seats may have to be cleaned, or major service repair may be needed. An accumulation of carbon deposits in the exhaust port or muffler will reduce power output. This condition is evidenced by slow acceleration and a quiet-sounding exhaust. If the exhaust ports are more than one-third

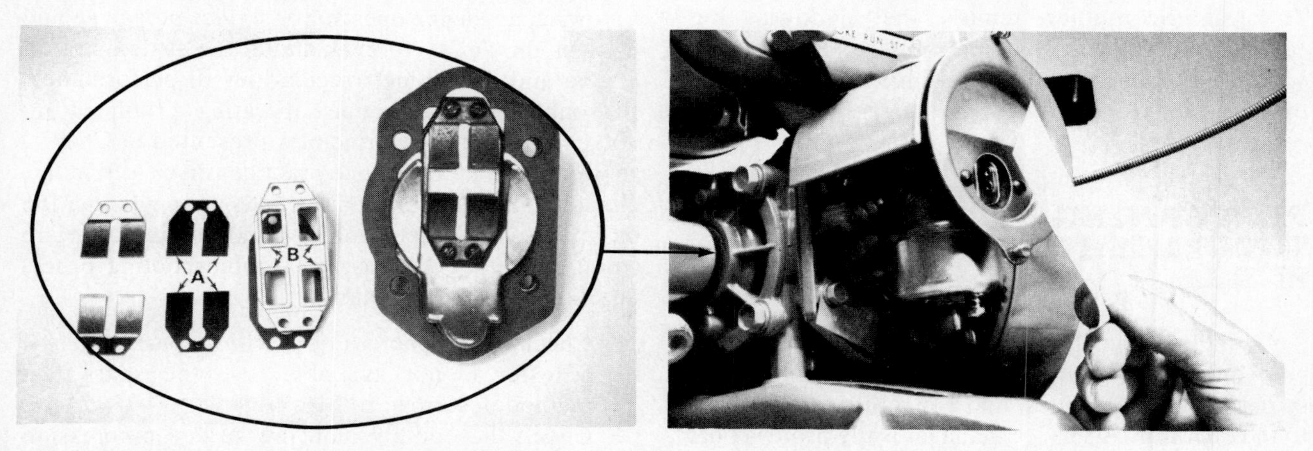


**Fig. 4-54.** Using a paper strip to check the operation of the reed valves on a two-cycle engine with the engine running. Reeds A may not be making good contact on seat B if fuel spotting is observed.

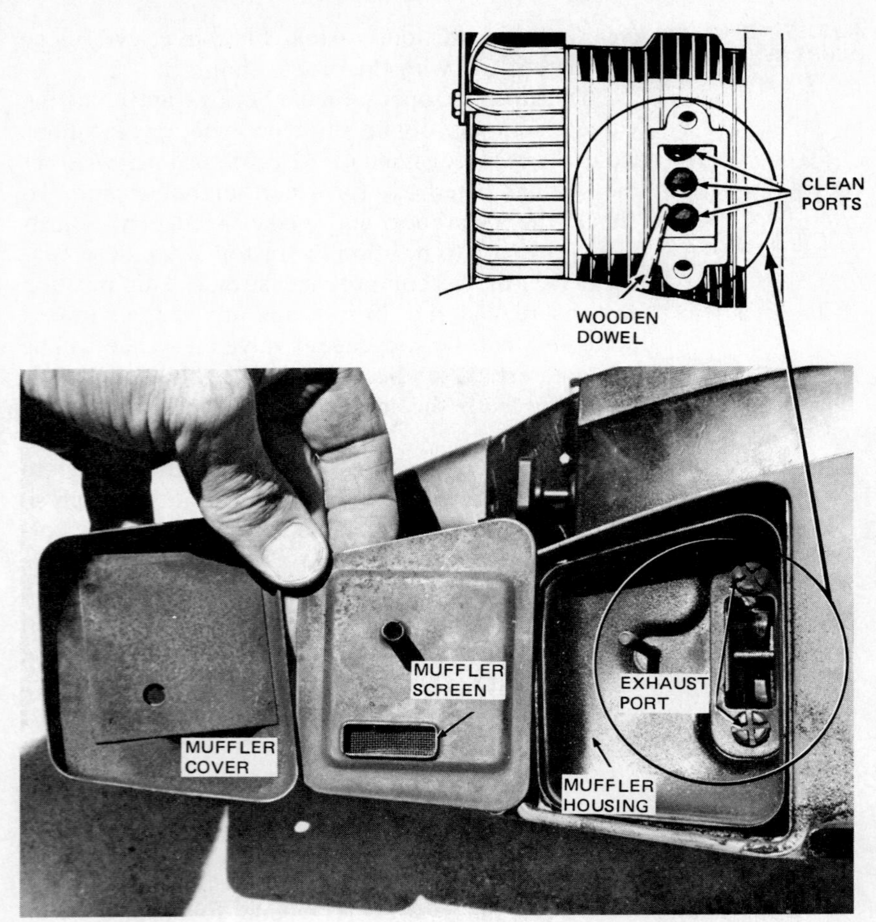


**Fig. 4-55.** Clean carbon accumulations from the muffler and exhaust ports of a two-cycle engine. (*Jacobsen Div. of Textron Corp.*)

clogged, use a wooden dowel or wood scraper to dislodge the carbon (Fig. 4-55). Be sure the piston is at bottom dead center (BDC) when scraping the ports, to prevent scratching the piston.

**⚠ WARNING It is necessary to remove dislodged carbon particles from the cylinder to prevent damage to the piston and cylinder by using compressed air or an oil-soaked cloth.**

To clean the muffler, remove the spark-arresting screen from the inside of the muffler. Replace the screen with a new one if one-third or more of the openings are clogged.

## PERFORMING TROUBLESHOOTING SKILLS

An engine which does not operate or perform properly can result in costly delays, disrupt the timeliness of production schedules, and constitute a safety risk to the operator. A service manager is the professional in power and machinery who is able to diagnose and plan the correction of malfunctions with a minimum of delay, thus performing a much needed service.

A person who has the ability to troubleshoot engine problems has much in common with a baseball coach who has the ability to analyze the problem of a player in a batting slump. The coach understands the principles of proper timing, coordination of impulses, and knows how to analyze the swing for proper use of the bat. The coach also possesses the ability to identify and correct a malfunction or make a minor adjustment in the player's stance or reaction timing. Likewise, an engine operator or service person should possess the ability to evaluate all the systems of an engine and determine irregularities in performance.

Troubleshooting engine operating problems involves the operating principles presented in Chap. 3 as well as the maintenance practices discussed in this chapter. Engine operation depends upon whether the basic factors of fuel, ignition, and compression are being provided. A four-step troubleshooting procedure is suggested as follows:

1. Check the magneto output with a spark tester. If a tester is not available, use the spark test method described in Fig. 4-29.
2. Check the compression pressure. Compression is tested by pulling the starter rope and noting whether resistance to turning can be felt when the piston enters the compression stroke. If little

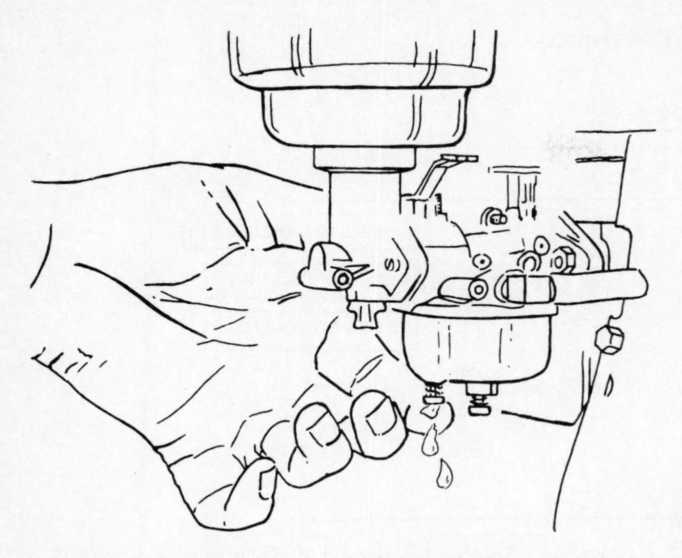

**Fig. 4-56.** Release of fuel from the carburetor bowl through the valve. Accumulations of water will show up as drops. If nothing flows from the valve, fuel lines and screens may be clogged. Check the quantity of fuel in tank.

or no resistance can be felt when pulling on the starter rope, serious mechanical troubles may be anticipated.

3. Check the supply of fuel to the carburetor. If the carburetor is equipped with a carburetor bowl drain, press this valve and let a small amount of fuel out of the carburetor (Fig. 4-56). If no fuel runs out, there is an obstruction to the flow from the tank to the carburetor. Check the fuel which comes out of the bowl for puddles of water or other foreign materials, which indicate the need to clean the tank, filters, lines, and carburetor bowl.
4. Check the spark plug condition. (Refer to the section in this chapter on inspecting the spark plug.) A spark plug which has heavy carbon deposits, burned electrodes, or cracked insulation must be replaced.

The *flowchart* in Table 4-4 presents a systemic method of checking the three operational requirements. The chart identifies 40 or more problems, any of which may prevent proper operation. Elimination of some malfunctions as identified in the chart may require major engine repair. These conditions are discussed in Chap. 5.

## PREPARING ENGINES FOR STORAGE

Small engines are subject to some of the most difficult operating conditions and lack of attention. They are expected to perform when called upon but seldom receive the ''rubdown'' a trainer gives a fine horse following a strenuous workout. Even with the best of daily care and maintenance, failure to take the necessary steps to protect an engine when in storage for an extended period of time may result in expensive damage to internal parts.

## CLEANING THE ENGINE

A dirty engine will collect moisture. Allow the engine to cool, then remove shrouding, and apply solvent or degreasing compound to the engine. Remove all trash from the screens and cooling vanes. Wash the engine with water or clean it with compressed air and reassemble the parts. Do not allow high pressure water to be forced under the flywheel or around the ignition system. Restart the engine and run 5–7 min to bring it up to operating temperatures. This will evaporate any moisture which may have accumulated.

### Protecting Internal Engine Parts

Engines that are to be stored must be protected from the damaging effects of rust and corrosion on internal engine parts. Bearing surfaces, piston rings, valve stems, faces, and seats are subject to damage.

Always drain the crankcase on four-cycle engines and fill with new engine oil of the recommended grade and classification. After refilling, operate the engine for a short time to distribute the oil.

### Draining Fuel

Over a period of time (depending upon the temperature), gasoline will become chemically unstable and undergo decomposition. When storing an engine or an engine-driven machine, it is a good practice to drain the fuel tank and carburetor to prevent the formation of gum and varnish deposits from ''stale'' gasoline. Stale gasoline can be identified by the foul odor. Tools such as chain saws are used at irregular time intervals throughout the year. Oftentimes chainsaw engines are stored with fuel in the tank for convenience. The manufacturer may recommend that a commercial gasoline stabilizer be added to the gasoline of these engines to prevent deterioration and to prevent carburetor diaphragms from drying out.

When four-cycle engines are to be stored for the season, drain as much fuel as possible from the tank. Then operate the engine to burn all remaining fuel from the tank and carburetor. Remove the spark plug and pour 1 tablespoon of new oil into each cylinder. Turn the crankshaft two or three revolutions by pulling the starter rope. Place the piston at top dead center position of the compression stroke. Both valves will be closed, sealing the combustion chamber. Replace the spark plug. Also place a tablespoon of oil into the fuel tank and tilt the engine to distribute the oil. This will prevent rust from forming in the metal tank. (This step is not necessary if a plastic tank is used.) Complete the preparation by storing the engine in a dry place.

**TABLE 4-4 Small Engine Troubleshooting Flowchart**

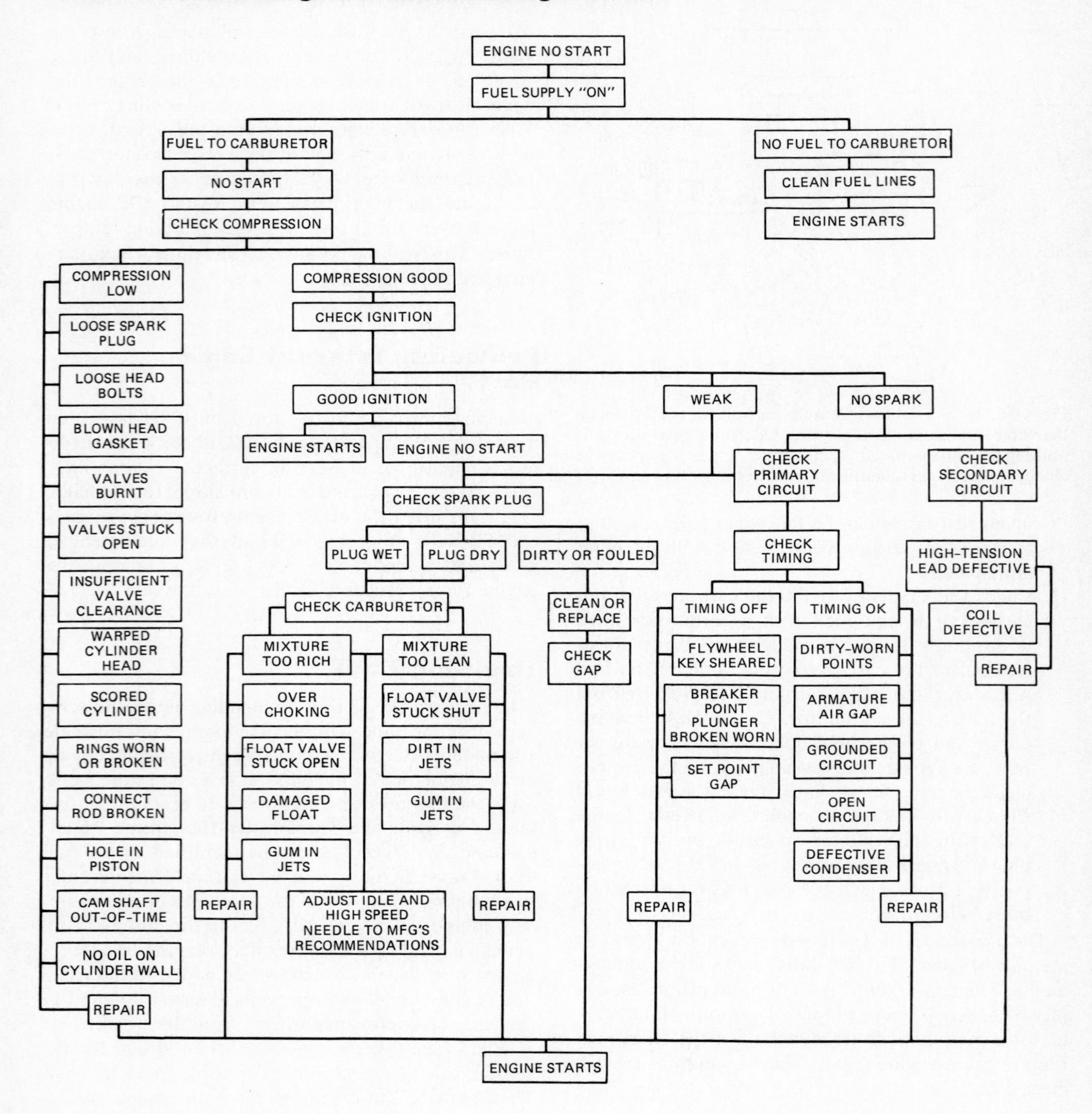

## SUMMARY

Small engines produce power for a large part of a highly specialized segment of the agricultural industry. Maintaining the operating efficiency and extending the useful life of these engines will contribute to conservation of fuel energy and human effort.

The service person in agricultural power and machinery has the knowledge and skill to service and maintain small engines and assist owners and operators to achieve the best results from engines.

Small-engine service and maintenance operations are not difficult to perform; the ability to make adjustments and to identify and correct faulty performance requires experience and knowledge. The information presented in this chapter is the basis for performing preventative maintenance operations on small engines.

## THINKING IT THROUGH

1. What four important functions does lubricating oil perform in an engine?
2. Describe the three systems of lubrication which are used in small engines.
3. Why must oil be mixed with the fuel in a two-cycle engine?
4. What is identified by the API-service classification of engine oils?
5. What is the meaning of the viscosity-grade index of engine oils?
6. What is the ratio by parts of fuel to oil recommended for two-cycle engines?
7. Why should the crankcase be hot when preparing to change oil?
8. How often should oil be changed in a small engine?
9. What are the three types of air cleaners used in small engines?
10. What is a heavy-duty dry-type precleaner? How is it serviced?
11. What are four important functions of a crankcase breather?
12. Where are crankcase breathers located?
13. How much heat energy must the cooling system be capable of removing from a small engine?
14. What are three important parts of the cooling system of an air-cooled engine?
15. List at least four places where fuel filters or screens might be found in an engine.
16. What is meant by a "cold" spark plug? A "hot" spark plug?
17. How can inspection of the firing chamber of a spark plug determine its operating condition?
18. What test is used for determining whether the spark plug or the ignition system is giving trouble?
19. What precautions must be observed when removing a spark plug from an aluminum head?
20. Why is an abrasive spark plug cleaner not recommended when servicing plugs in a small gasoline engine?
21. What is the purpose of adjusting the magneto flywheel air gap?
22. How does the setting of the breaker points affect ignition timing?
23. What may happen if an engine is stopped by pulling the ignition wire from the spark plug?
24. What are the three systems found on float-type carburetors?
25. What is the desirable air-fuel ratio for obtaining the maximum engine power?
26. What are the two types of small-engine governors?
27. What is the purpose of the governor spring?
28. What is the difference between checking the valves in a four-cycle engine and checking those in a two-cycle engine?
29. What are the three basic principles of engine operation that are involved when troubleshooting?
30. List four practices to follow in preparing an engine for storage.

# CHAPTER 5 REPAIRING THE SMALL ENGINE

## CHAPTER GOALS

In this chapter your goals are to:

- Evaluate mechanical problems requiring major engine repair
- Identify the purpose and use of measuring and indicating tools
- Perform specialized repair operations on the carburetor
- Inspect and repair engine valves and valve train
- Inspect and recondition pistons and cylinders
- Inspect and recondition main and crankshaft bearings and journals
- Assemble and operate the repaired engine

The knowledge and skill required to repair small one- and two-cylinder gasoline and diesel engines is equal to that which is needed for working on large agricultural tractors and machinery. Since the power that is produced by a small engine must be generated by only one or two cylinders, any abnormal condition which would weaken its power output nearly destroys the performance of the machine it is operating.

Many pieces of groundskeeping equipment such as those used for landscape, golf course, and nursery operation and maintenance are powered by small engines. Jobs such as mowing golf greens, fairways, and lawns and preparing seedbeds for cultivation must be performed on a definite schedule with properly operating equipment. Most city, county, and state agencies, as well as landscape maintenance contracting firms and privately owned businesses, have specialized service centers that employ specialists in small engine service, maintenance, and repair. Many equipment dealerships and private small-engine repair centers rely on personnel who can do more than replace parts. They require people who have dedication and are interested in keeping the equipment operating. They are the workers in agricultural power and machinery who not only are important members of the team, but have the training and ability to produce a quality product. For example, when an operator brings a power mower for golf greens to the service center and complains that the engine lacks power, is using too much oil, is leaking oil on the greens, is smoking, or just doesn't sound right, the service person is able to diagnose the problem quickly and make adjustments, thus minimizing the loss of productive time or the expense of major repair. When major repairs are required, the service person is qualified to correct malfunctions and return the engine to service with the least amount of time lost or expenses incurred.

## EVALUATING ENGINE PROBLEMS

As identified in Chap. 3, an engine should meet three basic requirements if it is to operate and develop the power it was designed to produce. These requirements are:

1. To receive a combustible mixture of air and fuel in an enclosed cylinder.
2. To compress the air-fuel mixture.
3. To ignite the compressed gases to obtain combustion.

When one of the three conditions is not met, an engine will not operate.

Correcting engine ignition or fuel malfunction is most often referred to as a *tuneup*. Problems associated with compression are generally classified as *top overhaul* operations. Mechanical repair operations which require the engine to be partially or completely disassembled may involve *major overhaul*. Whenever the valves, piston, cylinder, camshaft, crankshaft, and bearings need to be reconditioned or replaced, the service person must evaluate the costs of parts and labor to justify the service work. When mechanical problems occur causing the loss of compression or excessive oil consumption without visible external leaks, the troubles can usually be traced to the engine valves, piston, piston rings, or cylinder. Major repair problems which are caused by bearings and journals are identified by unusual knocking noises, broken connecting rod(s), and oil leaking from the bearing(s).

An experienced maintenance person can tell much about the condition of an engine by its general appearance. A neglected engine may be missing screws and bolts or even parts such as the muffler, shroudings, and air filter. The amount of dirt, an accumulation of caked and hardened oil, and the cleanliness of the cooling system are indications of the service and care the engine may have received. A turn of the crankshaft will allow the mechanic to feel or listen for broken internal parts and to tell whether there is excessive looseness, indicating wear.

One of the first evaluations of engine condition is to test cylinder compression pressure, using a compression gauge (Fig. 5-1). This test is especially necessary for small two-cylinder engines to determine if

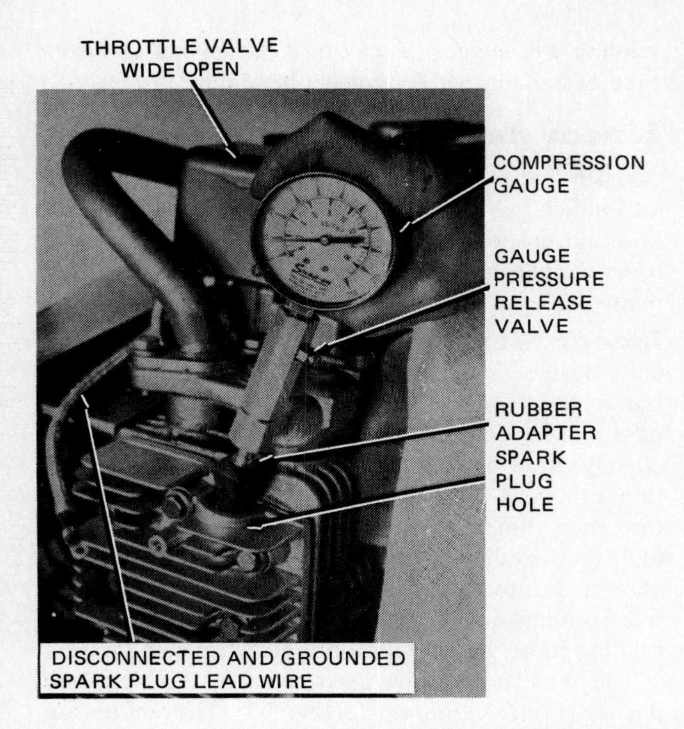


**Fig. 5-1.** A cylinder compression test is conducted to evaluate the ability of an internal combustion engine to develop rated power output. The test steps are to remove the spark plug(s), open the throttle, insert the compression gauge in the cylinder spark plug hole, crank the engine at normal cranking speed until the gauge needle ceases to rise, and read the compression pressure. *(Onan Corp.)*

the compression of each cylinder is even. Compare your readings to the specifications in the manufacturer's repair manual.

Several manufacturers of small single-cylinder engines do not recommend checking the cylinder compression if the engine is a compression-reducing type. Engines of this type are equipped with compression-reduction devices to reduce the amount of cranking effort on the starter. As a result, a compression gauge will not give an accurate reading of compression compared to that which is developed when the engine is operating at normal operating speed. It is recommended that the compression of these engines be checked by quickly spinning the flywheel backwards by hand against the compression stroke and letting go. If the flywheel rebounds sharply, the compression is considered normal.

The small-engine service and repair center personnel must be capable of making sound economic decisions for customers and business firms. The cost of repair parts, tools, and labor must be evaluated when determining whether to repair or to replace an engine with a new one. Generally, the cost of reconditioning or replacing engine valves and piston rings is not expensive. However, cost comparisons must be made when it is determined that parts such as the crankshaft or the cylinder block must be replaced. Most manufacturers list a *short block* as an item in their inventory of repair parts (Fig. 5-2). This unit consists of a new cylinder block fitted with crankshaft, connecting rod, piston, piston rings, camshaft, and valves. The short block is an alternative to re-

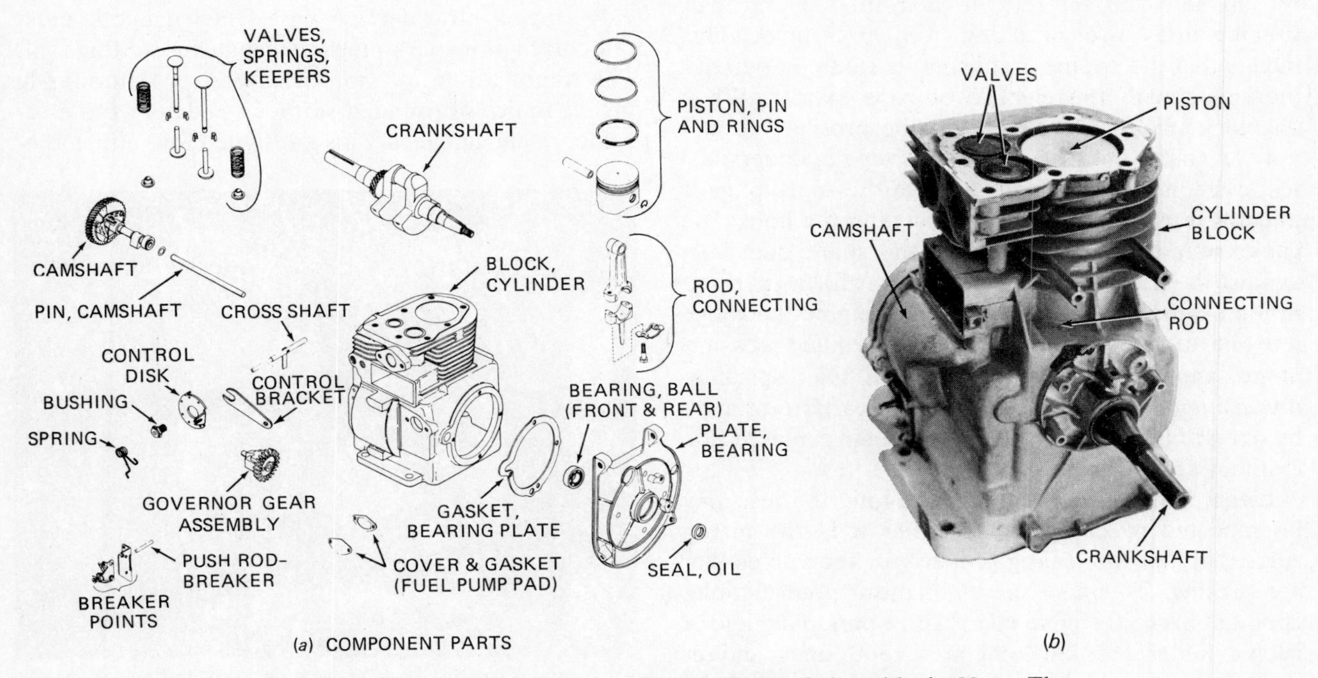


**Fig. 5-2.** (*a*) Components of a typical short block assembly. (*b*) Photo of short block. *Note:* The camshaft and connecting rod are not visible in the photo but are inside the assembly.

pair when the engine requires the replacement of expensive parts or when time is a major factor.

## MEASURING AND INDICATING TOOLS

To evaluate the condition of the many closely fitted parts in a small engine, the professional in agricultural power and machinery must be able to take accurate measurements and interpret the findings. In fact, the measurements which must be taken during engine repair are so numerous and of such importance that the ability to use special measuring and indicating tools is basic to working with small engines. Measuring tools are used to check cylinder pressure, to take linear measurements, and to determine the proper torque of nuts and screws.

### Measuring Cylinder Pressure

The cylinder compression gauge is one of the first diagnostic tools used by the small-engine mechanic (Fig. 5-1). The purpose of the tool is to measure the amount of pressure that is developed in the cylinder during the compression stroke of the piston. The amount of pressure that is developed depends upon the compression ratio of the engine and whether the valves and piston rings are forming a good seal. Normal compression pressures for small spark-ignition engines are usually 85 to 125 pounds per square inch (psi) [585 to 860 kilopascals (kPa)], while a small diesel engine may be 450 psi [3103 kPa].

The compression data values which are provided by engine manufacturers in their tuneup or repair specifications are cranking rpm pressures. This means that the engine crankshaft is rotated at cranking speed with the electric or rope starter. When checking two-cylinder engines, the pressure should not vary more than 10 percent between cylinders. To test cylinder compression, remove the spark plug or injector nozzle and insert the gauge in the hole. The gauge may be held by hand in the spark plug hole against the neophrene sealing adapter for a spark-ignition engine. However, in a diesel engine the gauge must be attached to the engine because hand pressure alone cannot provide a sufficient seal. On a spark-ignition engine the throttle valve at the carburetor must be open fully so that a full charge of air can enter the cylinder and be compressed.

If the compression readings are low, the test may be repeated to determine whether it is the piston rings that are not sealing properly or the valves that are leaking. To make the evaluation, pour 1 tablespoon of SAE 30 engine oil into the spark plug hole of each cylinder, crank the engine several times, and repeat the test. A higher reading will indicate that the piston rings are making a better seal and the cylinder, piston, or rings may be worn. Little change in the reading will suggest a leaking head gasket or an intake or exhaust valve that is not sealing properly.

### Linear and Torque Measurement

A service person must be able to use measuring and indicating tools to evaluate the engine's condition and properly assemble its internal parts. These tools include linear and torque measurement equipment.

**Linear Measurement** A number of specifically designed tools are used in engine repair to obtain exact dimensions of diameter, length, and thickness and to identify the space or clearance between closely fitted parts. Most engine repair and evaluation practices require taking measurements within tolerance limits of one-thousandths of an inch (0.001 in) [0.025 millimeter (mm)]. These tools are classified into the following three categories: (1) fixed-dimension leaf or plug gauges, (2) measurement reading tools, and (3) measurement transfer tools.

The most commonly used tools of this type are *thickness* or *feeler gauges,* (Fig. 5-3). These tools are precision-ground steel strips or wires arranged to provide a set of sizes. A thickness gauge set may include several leaves varying from 0.0015 to 0.025 in [0.04 to 0.64 mm] (Fig. 5-3*c*). Thickness gauges, such as the one shown in Fig. 5-3*a*, are classified as *go no-go gauges* because a small portion of the tip of each leaf is 0.002 in [0.05 mm] less than the rest of the leaf. If the tip enters the gap but the thicker part of the leaf does not, the setting is correct. This provides an optimum setting such as when adjusting valve tappet clearance. A wire feeler gauge is most often used as a spark plug gap gauge and setting tool. The round wires are more accurate for determining the clearance of rounded surfaces such as worn electrodes. Leaf gauges are also designed for setting igni-

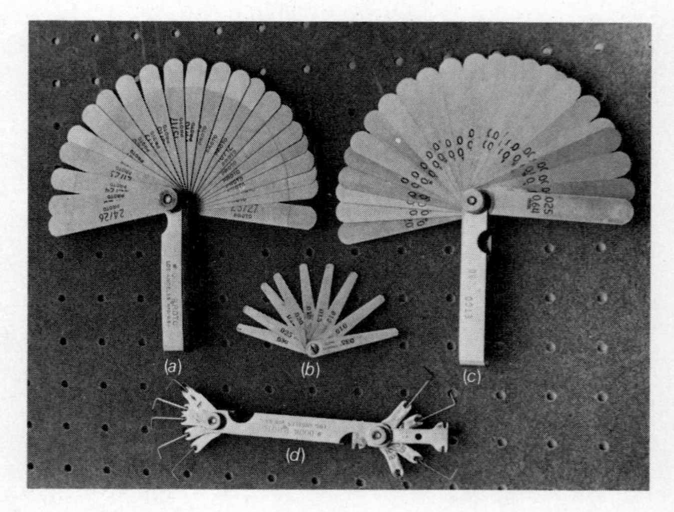


**Fig. 5-3.** Feeler, on leaf, gauges. (*a*) Go no-go gauge. (*b*) Ignition point gauge. (*c*) Standard 25-leaf feeler gauge. (*d*) Spark plug gap wire gauge and setting tool.

tion point gap and are provided in sets of common ignition gap settings.

Gauges are sometimes developed by the engine manufacturer for checking specific engine parts for excessive wear. Such gauges are classified as *plug gauges*. There are many forms and shapes of plug gauges because each gauge is constructed to check the wear of a specific engine. For example, if the plug gauge (Fig. 5-4*b*) will enter the breaker-point plunger hole ¼ in, the part is worn and must be repaired. Or if the plug gauge (Fig. 5-4*a*) enters the crankshaft bearing, it is worn excessively and must be reconditioned. A gauge to measure the wear of a piston ring groove (Fig. 5-4*c*) is constructed with a lip 0.006 in [0.152 mm] thicker than the standard piston ring groove width. When the ring groove is worn enough to allow the gauge to enter the opening, the top groove of the piston must be reconditioned or the piston must be replaced.

**Measurement Reading Tools** Several tools are capable of taking a linear measurement so that the dimensions can be read directly from the tool. Those tools that have special application to engine repair are (1) the dial indicator, (2) Plastigage, (3) the micrometer caliper, and (4) the slide caliper.

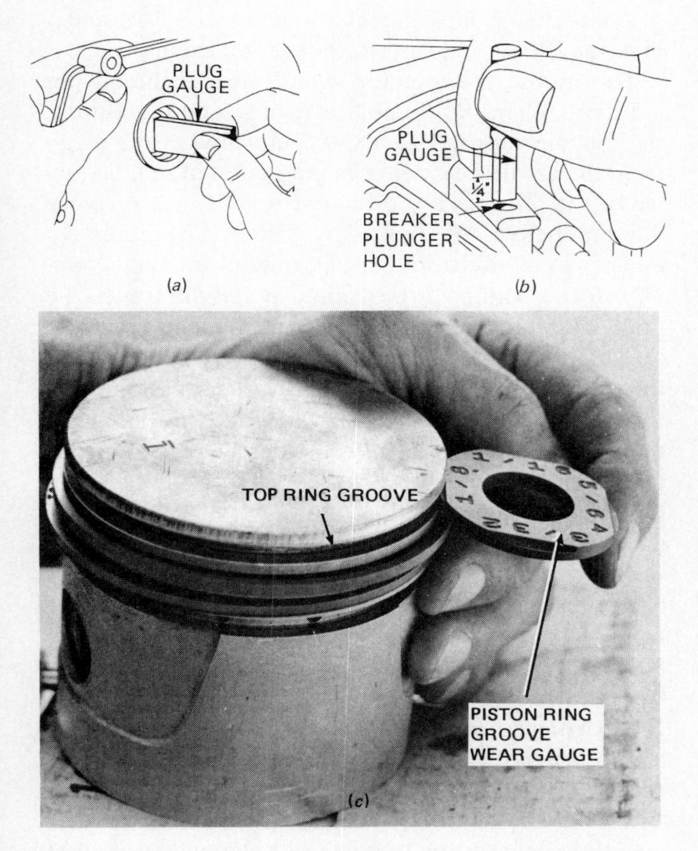


**Fig. 5-4.** Fixed-dimension plug gauges. (*a*) Checking for main bearing wear. (*Briggs & Stratton Corp.*) (*b*) Checking for breaking point plunger wear. (*Briggs & Stratton Corp.*) (*c*) Checking for piston ring groove wear.

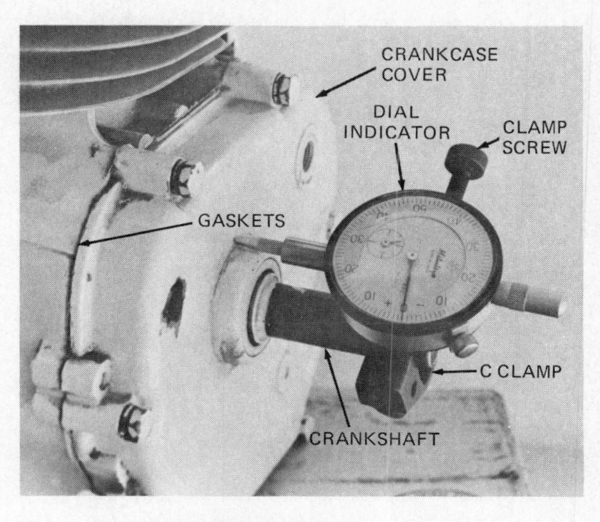


**Fig. 5-5.** A dial indicator being used to measure the exact amount of end play in the crankshaft.

The *dial indicator* is especially adapted to determining the amount of change, extent of movement, or wear in engine parts. For example, the exact amount of crankshaft end play can be measured in thousandths of an inch [millimeters] (Fig. 5-5). The dial indicator is attached to the power-takeoff end of the crankshaft by a C-clamp, and the shaft is moved in and out by hand. The sweep of the dial indicator needle will indicate the exact end play. The amount of runout (wobble) of a flywheel is easily checked with the dial indicator.

*Plastigage* is used to determine the exact amount of oil clearance in friction-type bearings such as those used on crankshaft main and connecting rod bearings. The procedure for using Plastigage is illustrated in Fig. 5-6. First, the bearing surfaces are wiped free of oil. Then a length of string is placed across the dry, clean bearing. The bearing half is returned to its proper position over the crankshaft and tightened to specifications. Care must be taken to prevent rotating the crankshaft or bearing surface during the tightening. When the bearing cap is removed, the string will have flattened to the thickness of the oil clearance. Width indicators, printed on the Plastigage envelope, are matched to the flattened string to read the oil clearance.

The *micrometer caliper* is also a direct-reading, precision measuring tool which has many useful applications to small-engine repair. Measurements can be taken and directly read to a tolerance of 0.001 in. The micrometer caliper consists of a threaded spindle which is rotated within a fixed nut by a cylinder called a *thimble* (Fig. 5-7). The part to be measured is placed between the measuring surfaces (anvil and spindle). The thimble is rotated until the two surfaces make contact, and the resulting dimension is read

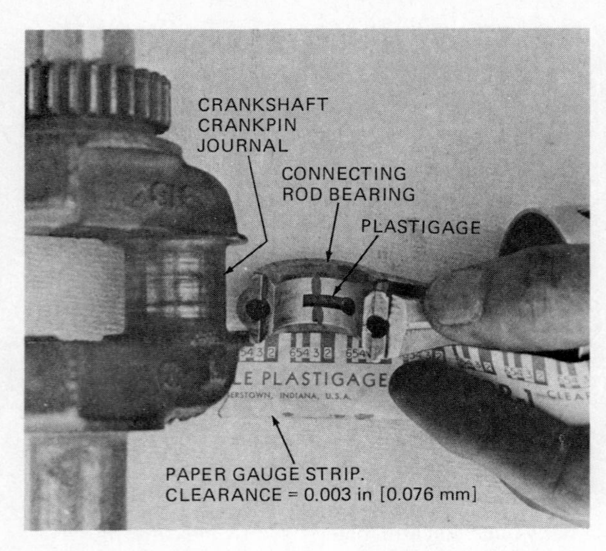


**Fig. 5-6.** Plastigage is a tool that measures clearance. The width of the flattened string indicates the oil clearance on the paper package used as a calibrating tool.

from the graduations found on the spindle sleeve and thimble. The maximum length of graduations on the sleeve is 1 in. The frame of the various sizes of micrometer calipers varies in 1-in ranges (Fig. 5-8).

The English micrometer has a spindle screw thread with 40 threads per inch. Therefore, when the thimble is rotated one turn, the measuring surfaces on the spindle advance or retract exactly $^1/_{40}$, or 0.025, in.

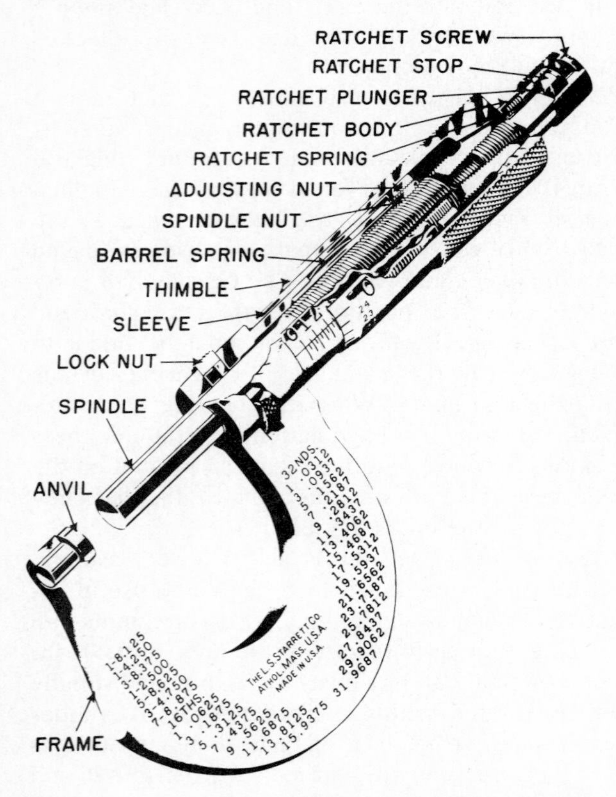


**Fig. 5-7.** Sectional view of an outside micrometer illustrating the operating parts. (*L. S. Starrett Co.*)

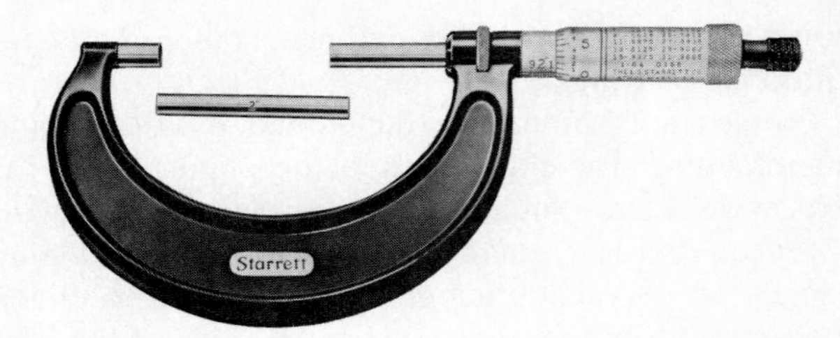


**Fig. 5-8.** Micrometer caliper with 2- to 3-in frame. Note the test bar for checking the adjustment for an accurate reading. Four micrometers, sizes 0–1, 1–2, 2–3, and 3–4, are required for small-engine repair work. (*L. S. Starrett Co.*)

To read the English micrometer, first identify the frame size of the tool being used. If the micrometer is a 1- to 2-in frame, write down 1.000 in. Next, add to this figure the number of tenths indicated by the largest number visible on the sleeve (each graduation on the sleeve represents 0.025 in and four of these equal 0.1 in). Assume the figure is 0.3; add 0.300 to 1.000. Now, count the number of lines easily visible following the tenths marking on the sleeve. Multiply the number of lines by 0.025 and add the product to the previous figures. Finally, add the number of thousandths of an inch (each graduation represents 0.001 in) from the *thimble* reading and sum the total of the digits. The examples shown in Fig. 5-9*a* and *b* illustrate how to read the English micrometer.

The metric micrometer spindle has 1 thread per 0.5 mm. When the thimble is turned one revolution, the spindle moves exactly 0.5 mm toward or away from the anvil. The sleeve is graduated in millimeters from 0 to 25 above the line with graduations every 0.5 mm below the line. The thimble is graduated equally in 50 divisions. Each graduation equals one-fiftieth of 0.5 mm, or 0.01 mm, in spindle travel. To read the metric micrometer, add the frame size and the total readings on the sleeve to the readings on the thimble, as in Fig. 5-9*c*. Note that the 5-mm graduation is visible on the sleeve with one additional 0.5 mm visible. The 28th line coincides with the longitudinal line on the sleeve for a reading of 0.28 mm (28 × 0.01 mm). Adding these numbers together equals a reading of 5.78 mm.

A *slide caliper,* as shown in Fig. 5-9*d*, is another direct-reading measuring instrument. This type of caliper is also called a *vernier caliper*. Some repair people prefer this tool when it is necessary to take fast, yet accurate measurements. To determine the English measurement obtained with a vernier caliper, locate the position of the zero line on the vernier slide. For example, the cylinder in Fig. 5-9*d* has a diameter in excess of 2 in because the zero on the vernier slide is between 2 and 3 in on the bar. Note that the zero is to the right of 0.5 in ($^5/_{10}$ of an inch). Each vertical line between the tenths mark represents 0.025 in; therefore the reading of two marks will re-

(FRAMESIZE = 0 to 1 in)

| | |
|---|---|
| FRAMESIZE: | 0.000 in |
| NUMBER OF TENTHS ON SLEEVE: | 0.300 in |
| LINES ON SLEEVE: | 0.050 in |
| THIMBLE READING: | 0.009 in |
| | 0.359 in |

(a)

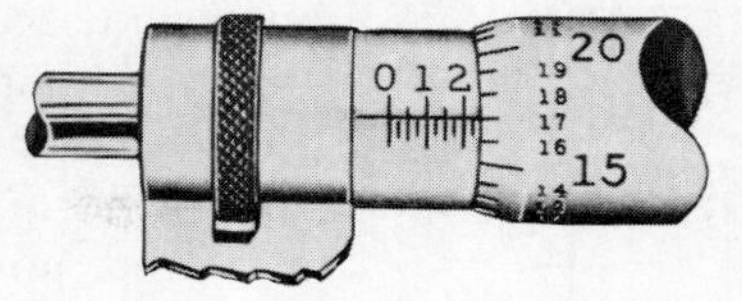

(FRAMESIZE = 2 to 3 in)

| | |
|---|---|
| FRAMESIZE: | 2.000 in |
| NUMBER OF TENTHS ON SLEEVE: | 0.200 in |
| LINES ON SLEEVE: | 0.025 in |
| THIMBLE READING: | 0.017 in |
| | 2.242 in |

(b)

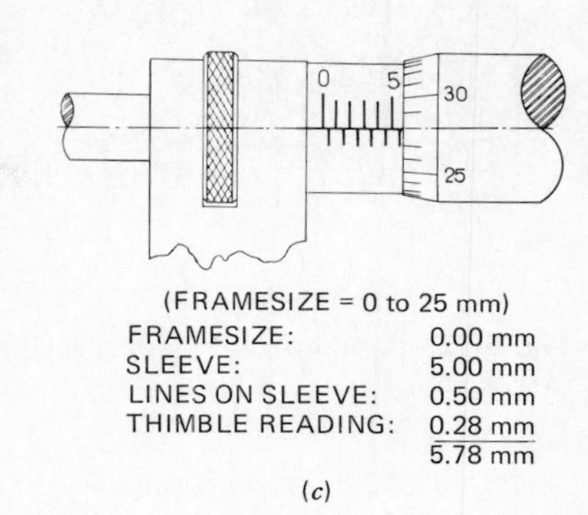


(FRAMESIZE = 0 to 25 mm)

| | |
|---|---|
| FRAMESIZE: | 0.00 mm |
| SLEEVE: | 5.00 mm |
| LINES ON SLEEVE: | 0.50 mm |
| THIMBLE READING: | 0.28 mm |
| | 5.78 mm |

(c)

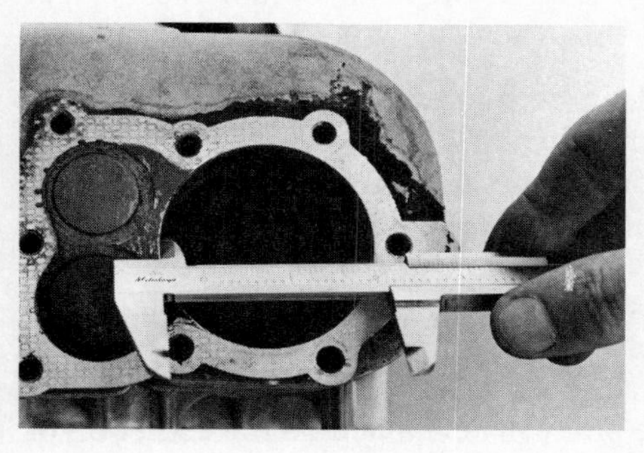

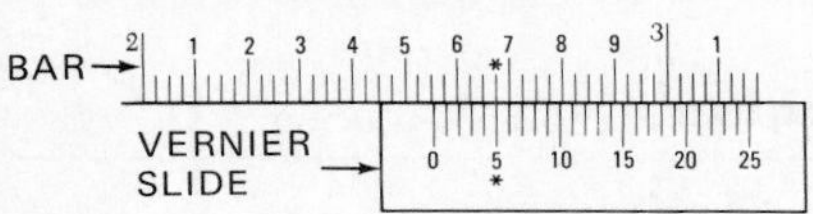


| | |
|---|---|
| BAR: | 2.000 in |
| | +0.500 in |
| | +0.050 in |
| VERNIER: | 0.005 in |
| | 2.555 in |

(d)

**Fig. 5-9.** (*a*) and (*b*) How to read the English micrometer. (*c*) How to read the metric micrometer. (*L. S. Starrett Co.*) (*d*) The inside diameter of a small-engine cylinder measured accurately with a vernier caliper, U.S. Customary measurement.

sult in an additional 0.050 in. Observe that the best alignment of the vernier scale marks, 0 to 25, is with line No. 5 on the bar scale. (Each line on the vernier scale represents 1/1000 in, or 0.001 in, therefore, $5 \times 0.001 = 0.005$ in.) This is the excess beyond the 0.050-in value, so that the total reading is 2.555 in in diameter.

**Measurement Transfer Tools** *Small-hole* and *telescoping* gauges are the two most common types of transfer tools used in small-engine repair. They are called *transfer* measuring tools because they cannot in themselves provide a direct reading of a measurement. The operating principle and purpose of the two gauges are identical; they transfer the internal measurement of an engine part to a micrometer caliper for reading (Fig. 5-10). A set of five telescoping gauges (Fig. 5-11) will transfer internal diameters ranging from ½ to 6 in [13 to 152 mm]. Small-hole gauges will transfer dimensions of holes (bores) 0.125 to 0.500 in [3.1 to 12.7 mm] (Fig. 5-11*b*).

**Torque Measuring Tools** The term *torque* is more completely discussed in Chap. 6. In brief, the term is used to describe the twisting force which causes a body to turn about its center of rotation. A torque wrench is used to determine the amount of

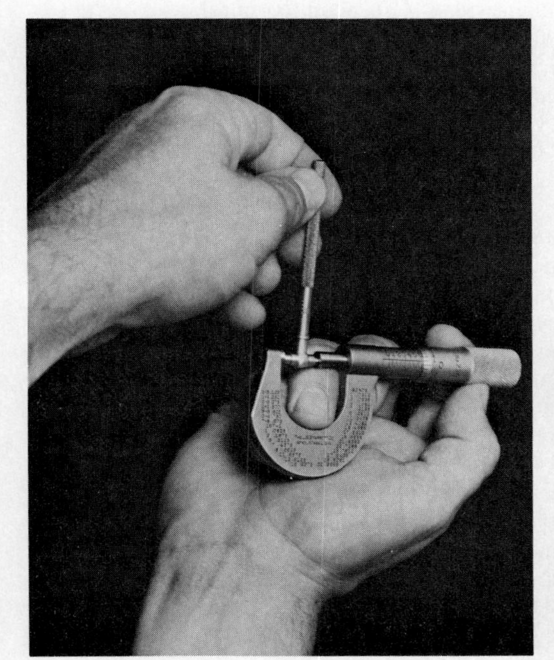

**Fig. 5-10.** After establishing hole size with a telescoping gauge, actual size is determined by measuring over the contacts with an outside micrometer.

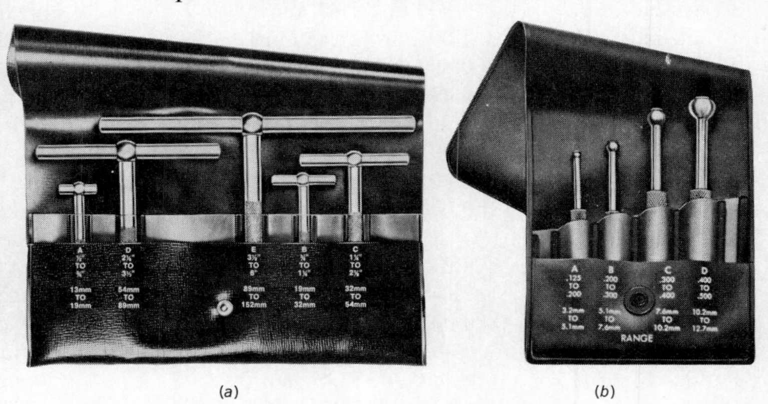

(a) (b)

**Fig. 5-11.** Transfer tools (*a*) Telescoping gauge set. (*b*) Small-hole gauge set.

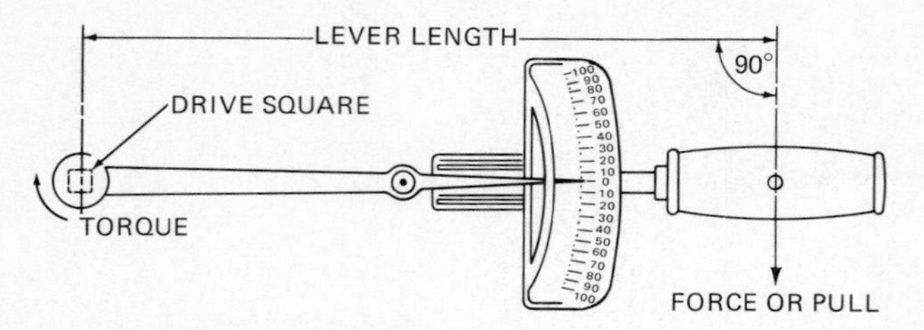


**Fig. 5-12.** Fundamental facts regarding the torque wrench. (*Sturtevant Co.*)

resistance to the twisting force that is exerted. A typical application is the tightening of a screw or nut. A torque wrench is used to measure the amount of effort being applied. The torque wrench is a measuring tool which can be compared to other accurate direct-reading measuring instruments.

Torque wrenches provide an indication of the force operating through the lever arm length (Fig. 5-12). The unit of measure may be in pound-inches (lb·in) or pound-feet (lb·ft), or newton-meters (N·m), depending upon the calibration of the indicator. To convert pound-inches or pound-feet into newton-meters, use conversion factors given in the Appendix. Engine manufacturers provide torque specifications similar to those illustrated in Table 5-1. Note that the torque specifications are listed in both pound-feet and pound-inches and newton-meters. Since there are 12 in per foot, a measurement from a torque wrench calibrated to read in pound-feet is easily converted to pound-inches by multiplying the reading by 12. A measurement given in pound-inches is converted to pound-feet by dividing the reading by 12.

When using a torque wrench, you must remember to do the following in order to obtain the correct torque value:

1. Always clean both internal and external threads before tightening. Cleanliness and condition of the threads will affect the amount of torque applied to the fastener. The resistance due to dirty or damaged threads will result in less effective torque in comparison to the value registered by the wrench.
2. Apply force to the handle of the torque wrench so that the pull will be applied on the center pivot

**TABLE 5-1 Specifications Typical for a 8.0 Cu In. Small Engine***

| Specifications | | Specifications | |
|---|---|---|---|
| Bore and stroke | 2⅜ × 1¾ in [60.3 × 44.4 mm] | Camshaft | |
| Displacement | 7.75 in³ [127 cm³] | Lobe height, min | 0.883 in [22.43 mm] |
| Cylinder diameter | | Journal diam, min | 0.4985 in [12.66 mm] |
| Standard | 2.374–2.375 in [60.25–60.3 mm] | Valve | |
| Max. oversize | 0.003 in [0.076 mm] | Stem diam, std. | 0.247 in [6.27 mm] |
| Max. out-of-round | 0.0025 in [0.064 mm] | Guide diam, std. | 0.252 in [6.4 mm] |
| Cylinder compression | 100–120 psi [690–827 kPa] | Guide diam, max. | 0.260 in [6.6 mm] |
| Crankshaft bearing journals | | Margin, min. width | 1/64 in [0.397 mm] |
| Main journals, std. | 0.875 in [22.2 mm] | Seat width | 3/64–1/16 in [1.19–1.59 mm] |
| Main journals, reject | 0.8726 in [22.16 mm] | Angle, valve | 45° |
| Con. rod journal, std. | 0.9963 in [25.31 mm] | Angle, seat | 44.5° |
| Main bearings | | Tappet clearance, intake | 0.005–0.007 in [0.127–0.179 mm] |
| Std. | 0.876 in [22.25 mm] | Tappet clearance, exhaust | 0.009–0.011 in [0.23–0.28 mm] |
| Reject | 0.878 in [22.28 mm] | Torque specifications | |
| Connecting rod bearings | | Flywheel nut | 57 lb·ft [77.3 N·m] |
| Crankshaft, std. | 0.999 in [25.37 mm] | Cylinder head | 140 lb·in [15.8 N·m] |
| Crankshaft, reject | 1.0013 in [25.43 mm] | Connecting rod | 100 lb·in [11.3 N·m] |
| Piston pin, std. | 0.490 in [12.45 mm] | End plate | 40 lb·in [4.5 N·m] |
| Piston pin, reject | 0.492 in [12.50 mm] | Spark plug | 60 lb·in [6.8 N·m] |
| Running clearances | | Electrical | |
| Main bearings | 0.001–0.0025 in [0.025–0.064 mm] | Air gap | 0.006–0.010 in [0.15–0.25 mm] |
| Connecting rod | 0.001–0.0025 in [0.025–0.064 mm] | Spark plug gap | 0.030 in [0.76 mm] |
| Piston pin | 0.0005 in [0.013 mm] | Condenser capacity | 0.18–0.24 μF |
| Piston | | Breaker point gap | 0.020 in [0.51 mm] |
| End gap, comp. ring | 0.035 in max. [0.89 mm] | Plunger bore, std. | 0.187 in [4.75 mm] |
| End gap, oil ring | 0.045 in max. [1.14 mm] | Plunger bore, reject | 0.189 in [4.8 mm] |
| Side clearance | 0.007 in max. [0.178 mm] | Operating speed | |
| Piston pin bore, std. | 0.490 in [12.45 mm] | Maximum | 3600–4000 r/min |
| Pin diam, std. | 0.4895 in [12.43 mm] | Idle | 1750 r/min |
| Pin diam, reject | 0.489 in [12.42 mm] | | |
| Skirt diam | 2.368 [60.15 mm] | | |

* *Note:* Use the above items only as examples of items to be checked. Obtain specifications from a repair manual for each engine.

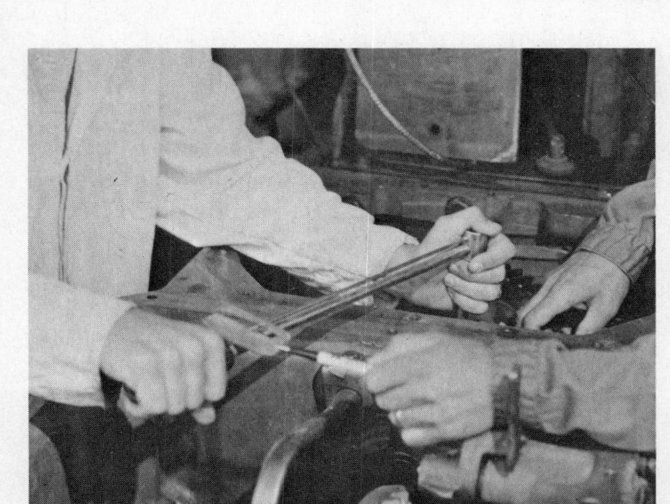

**Fig. 5-13.** The position of one's hands on the center pivot handle and the direction of pull affect the torque reading.

point of the handle and at right angles to the wrench body (Fig. 5-13).

3. When an extension is used to drive the socket, support the head of the wrench to prevent a cocking action on the fastener and a false reading.
4. Use the proper tightening sequence (Fig. 5-14). Do not torque the screws or nuts to full torque value the first time, but work up to full torque following the recommended sequence. For example, if the recommended torque is 75 lb·ft for a cylinder head, start by torquing all bolts in the proper sequence to 25 lb·ft. Continue by torquing to 50 and finally to 75 lb·ft and then recheck. The four steps are the minimum recommended sequence.

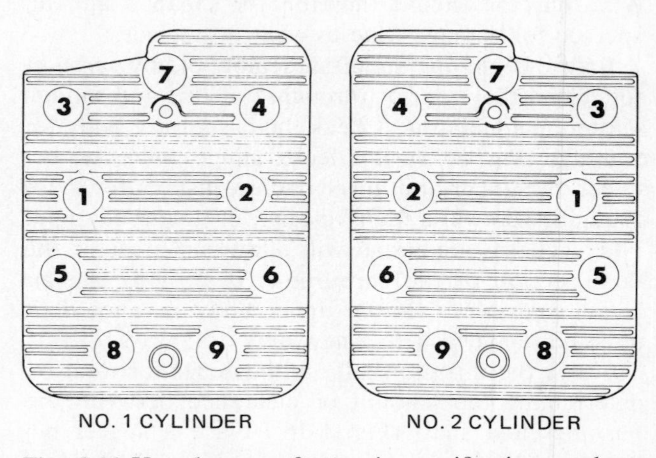


**Fig. 5-14** Use the manufacturer's specifications and sequence when torquing head bolts on an engine. *(Onan Corp.)*

## PREPARING THE ENGINE FOR REPAIR

The service center professional will refer to the job of repairing small engines as doing a *top overhaul* or doing a *major overhaul*. The difference between the two is in the extent of disassembly. A *top* overhaul indicates that only the cylinder head and valves are removed from the engine. A *major* overhaul, or major repair, requires a more complete engine disassembly.

Engines requiring repairs have usually been used under harsh conditions. As a result, the exterior is often covered with dirt and debris, the air cleaner probably needs servicing, and the carburetor should be inspected.

### Cleaning the Engine

It is a good practice to clean the exterior of the engine thoroughly after examining the engine but before starting to remove any parts, whether for top overhaul or major repairs. Make a note on the work order if there are excessive oil, coolant, or fuel leaks. A clean engine is easier to work on, and there is less chance of contamination when reassembly is started. Either a degreasing agent and high-pressure water spray or the steam-vapor method of cleaning is effective. A second cleaning may be needed after the air shrouding has been removed. Additional cleaning will be necessary to inspect any internal parts following disassembly.

### Cleaning the Air Cleaner

Improper maintenance of the air cleaner may be the cause of engine failure. Always inspect and clean the air filter as one of the first operations in top overhaul or major repair. Follow the procedure outlined in Chap. 4.

### Disassembling and Inspecting the Carburetor

A faulty carburetor will not provide the proper air-fuel ratio. An engine running on too lean a mixture may run too hot, while a rich mixture causes the oil

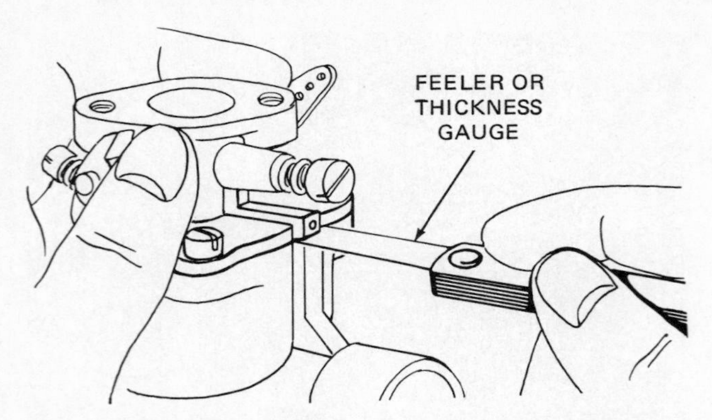


**Fig. 5-15.** Checking the upper body of the carburetor for warpage with a feeler gauge. (*Briggs & Stratton Corp.*)

to be washed from the cylinder wall(s) by excess fuel. A carburetor should therefore be cleaned and inspected for damages due to wear or neglect.

Before beginning the dissassembly of any carburetor, clean the exterior throughly of dirt and foreign matter accumulation. Check the carburetor body for warpage (Fig. 5-15). If a feeler gauge will enter between the upper and lower bodies, inspect for loose retaining screws, a faulty gasket, or a warped upper body. These conditions will allow air to enter the float section of the carburetor, destroying the air balancing system. If the upper body is warped, it must be replaced with a new unit.

Check the throttle shaft and bushings for wear by determining the amount of clearance between the bushings and shaft (Fig. 5-16). Place a spacer between the body of the carburetor and the shaft. Hold the shaft toward the spacer and measure the clearance with a feeler gauge. Remove the pressure and insert the feeler to measure the amount of wear. If the wear is more than the manufacturer's specification, the shaft and bushings should be replaced. To accomplish the job, it will be necessary to remove the throttle shaft (Fig. 5-17*a*) and the old bushings (Fig. 5-17*b*). First, remove the throttle valve from the shaft

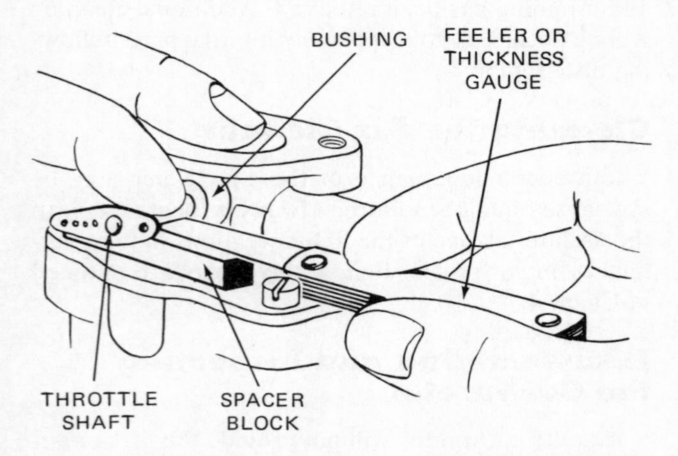


**Fig. 5-16.** Checking wear of the throttle shaft and bushings in the carburetor. (*Briggs & Stratton Corp.*)

and withdraw the shaft from the housing. Insert a tap of the proper size in each bushing and pull from the housing. New bushings are then pressed into place. The shaft may also need to be replaced.

Disassemble the carburetor by removing both idle and load needle valves from the body of the carburetor. Inspect each needle for damage (Fig. 5-18). Next, determine whether it is recommended to remove the main nozzle from the body of the carburetor. Some designs require a replacement nozzle if the original is removed. Check the specific repair manual *before* removing. If the nozzle projects diagonally into a recess in the upper body, it may be damaged when trying to separate the upper and lower bodies. Remove the float and float needle by pulling the axle shaft from the float with needle-nose pliers. Lift off the float and needle. Inspect the float for leakage by shaking it. A float that has fuel inside it should be discarded. Inspect the float needle for wear (Fig. 5-19). Remove all gaskets, seats, and the diaphragm. A carburetor repair kit includes new replacement parts, gaskets, and seals. (Fig. 5-20)

The carburetor parts should be cleaned by soaking in an approved cleaning solvent. Use caution in selecting a cleaner since some will dissolve metal parts. *Do not* use solvent on plastic or nylon parts. When all parts have been inspected and cleaned, replace the float needle valve and the float. Check the float for proper setting (Fig. 5-21). Set the float level to the manufacturer's specifications. To change the float height, use needle-nose pliers to bend the *float tang* (Fig. 5-21). Reassemble the carburetor and follow the manufacturer's specifications for static or bench setting of the needle valves.

## INSPECTING AND RECONDITIONING VALVES AND CAMSHAFT

The valves of a four-cycle engine must operate under very adverse conditions. For example, the exhaust valve may operate at near red heat and be exposed to the discharge of burning gases passing out the exhaust. The valves must seal the cylinder to retain high combustion pressures of 500 to 1000 psi [3447 to 6894 kPa]. If the compression test indicates valves are not sealing, repair of the valves and valve seats may be necessary. To repair the valves, remove the cylinder head and clean the carbon deposits from the combustion chamber and the exposed heads of the valves.

Remove the valve cover and the retainer (Fig. 5-22*a*) from the end of the valve using a spring compressor (Fig. 5-22*b*). Be sure to keep all valve components together that have been removed from each valve. Remove the valve from the guide and clean carbon deposits from the stem, face, and seating surface of the valve on a wire brush wheel. Use a small

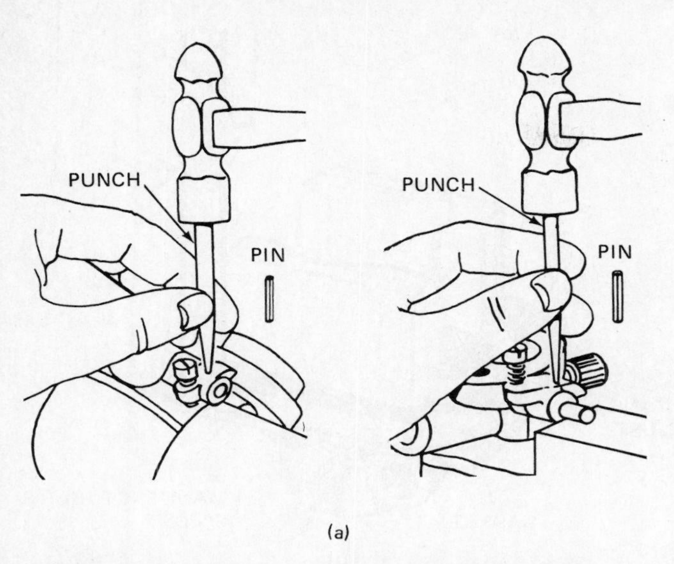


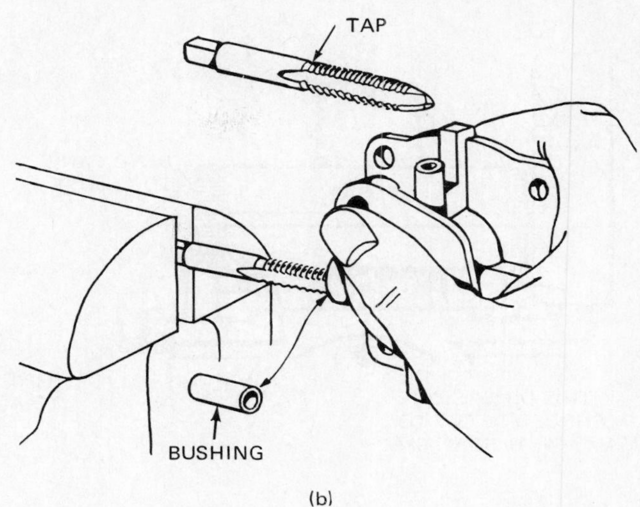


**Fig. 5-17.** When the throttle shaft and bushing wear are greater than specifications, (*a*) remove the throttle shaft and (*b*) the shaft housing bushings and replace. (*Briggs & Stratton Corp.*)

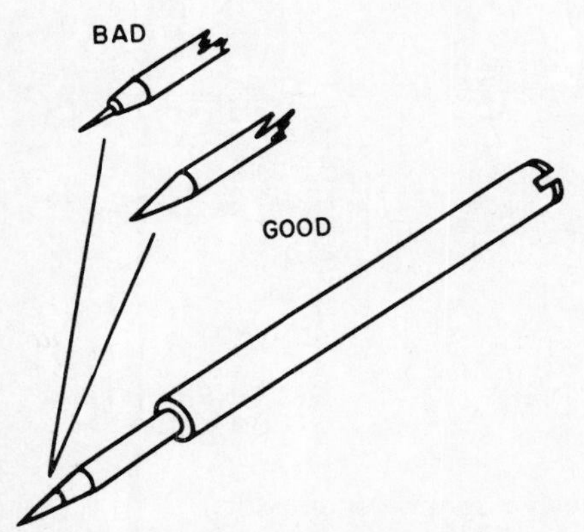


**Fig. 5-18.** When disassembling the carburetor, check the needle valves for damage. (*Tecumseh Products Co.*)

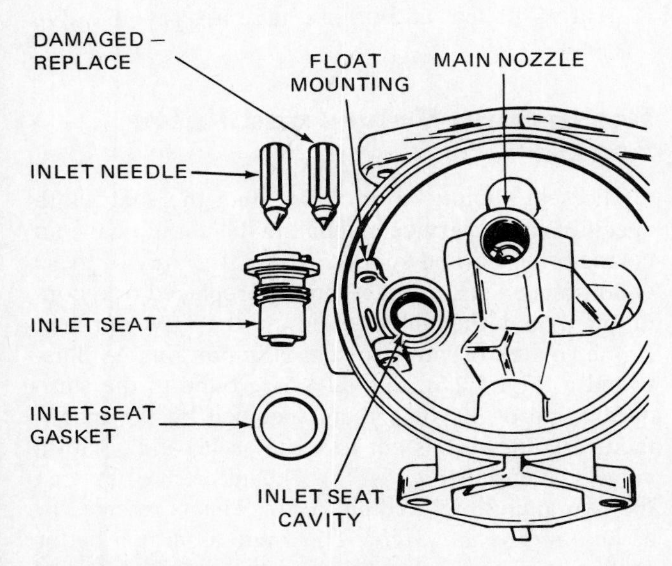


**Fig. 5-19.** Replace worn or defective float valves and inlet seat. (*Tecumseh Products Co.*)

wire brush in a ¼-in [6.4 mm] electric drill to clean the valve seats. Check the intake and exhaust valve spring for acid etching, squareness, free length, compressed length, and compression when specifications are given by the manufacturer.

## Evaluating Valve Parts and Specifications

When all the carbon deposits have been removed, do the following:

1. Check the valve guides for excessive wear. Worn valve guides contribute to sloppy valve action by changing the valve timing. Also, worn guides permit oil to be drawn into the combustion chamber, contributing to excessive oil consumption (Fig. 5-23). Clean the bore of the guide, using a guide cleaner. Check the bore for wear. One manufacturer recommends using a special plug gauge to determine the degree of wear. A small-hole gauge can be used to measure the guide and compare the dimension with the engine's repair specifications. The worn guides of some engines may be repaired by installing a special bushing

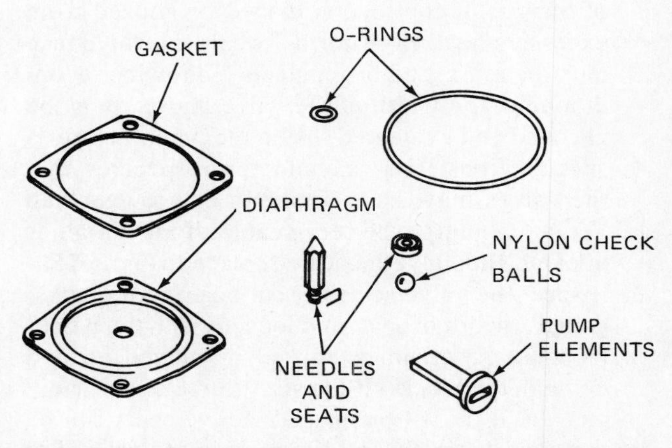


**Fig. 5-20.** Carburetor repair kits contain replacement parts, gaskets, and seals. (*Tecumseh Products Co.*)

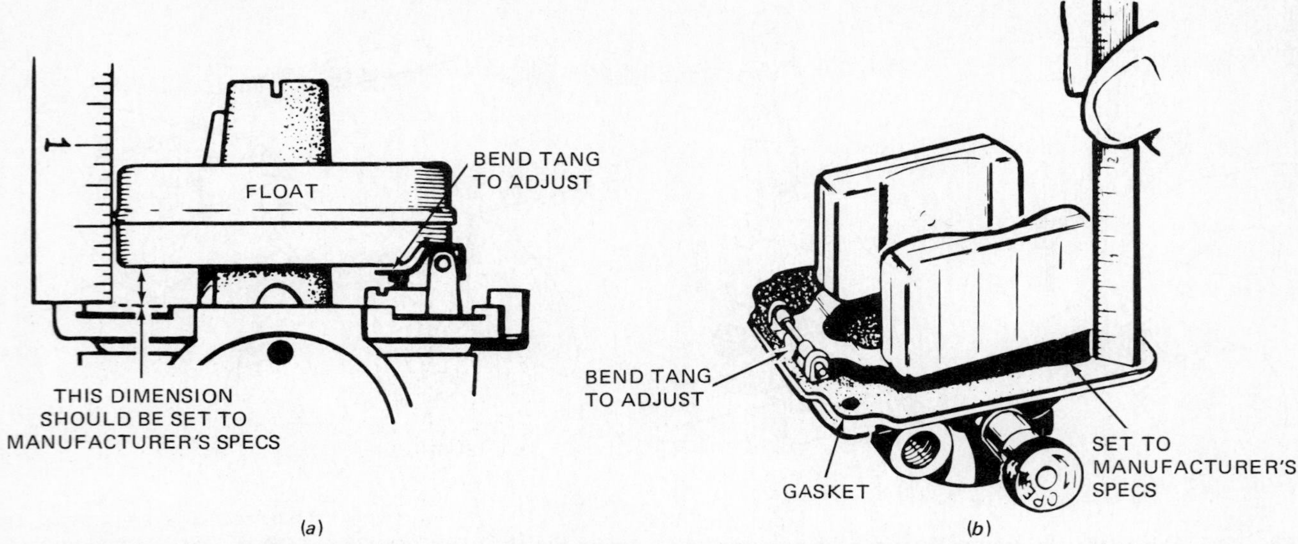


**Fig. 5-21.** Checking and setting the height of float level. (*a*) Round float. (*b*) Twin "kidney" float. (*Onan Corp.*)

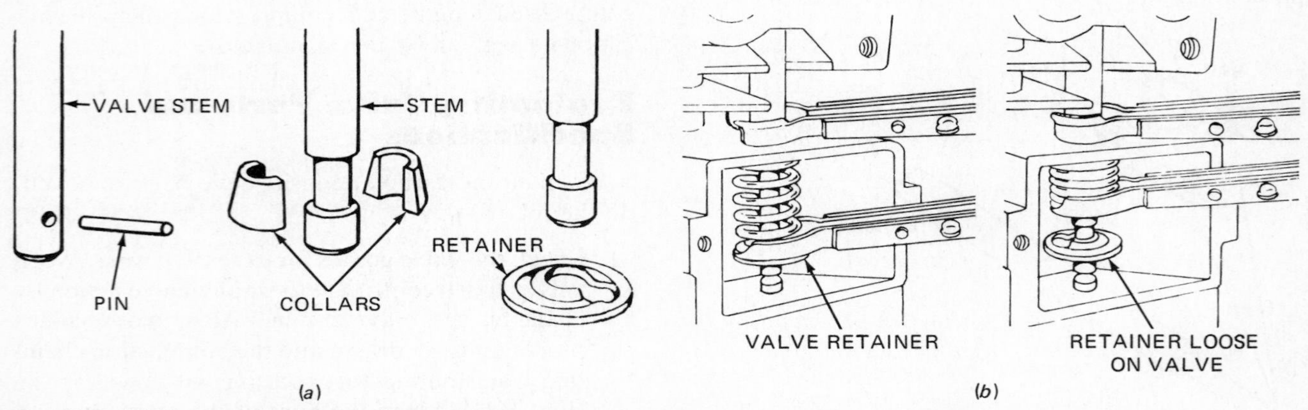


**Fig. 5-22.** (*a*) Types of valve retainers. (*b*) Removing the valve spring with a valve spring compressor. (*Briggs & Stratton Corp.*)

(Fig. 5-24). Engines may also be equipped with replaceable valve guides. Other manufacturers provide valve guide reamers to enlarge the bore. Then, a new valve with an oversized stem is installed.

2. Check the condition of the valve face and width of margin. If the face is dished or gouged from excessive heating or normal wear, the valve face must be refinished or replaced. Following reconditioning, the width of the valve margin must be checked and evaluated, using the manufacturer's specifications. One manufacturer specifies that the valves must have a margin of not less than $^1/_{64}$ in [0.4 mm] to be serviceable. If the margin is uneven, the valve must be replaced (Fig. 5-25).
3. Inspect the valve seats or seat inserts for cracks, gouges, width of seat, and looseness in the block. (A loose seat can be turned or moved up and down in the block.) If the seat is cracked or badly gouged, it must be replaced. Valve seats can be removed from the block using a special puller (Fig. 5-26*a*). The new seat can be installed with a driver (Fig. 5-26*b*) and peened around the edge (Fig. 5-26*c*) to hold it in place. Some engines use a shrink-fit seat and others have a screw-in valve seat.

## Refinishing Valves and Valve Seats

The seat should be no wider than the seat width specified in the service manual, and it should have no corrosion or burned spots (Fig. 5-25). If any of these conditions exists, the seat must be replaced or reconditioned and the valve fitted to the new surface. Proper match of valve to seat is important. As illustrated in Fig. 5-27*a*, the valve is ground to the standard angle of 30° or 45°, as specified by the manufacturer. The seat is cut to a 31° or 46° angle with a valve seat cutter (Fig. 5-27*b*). The difference between the two angles, 45° and 46°, is 1°. This is referred to as an *interference angle*. The angle assures a better sealing surface. Reconditioning the valve seat may cause it to become excessively wide (Fig. 5-28*a*). It

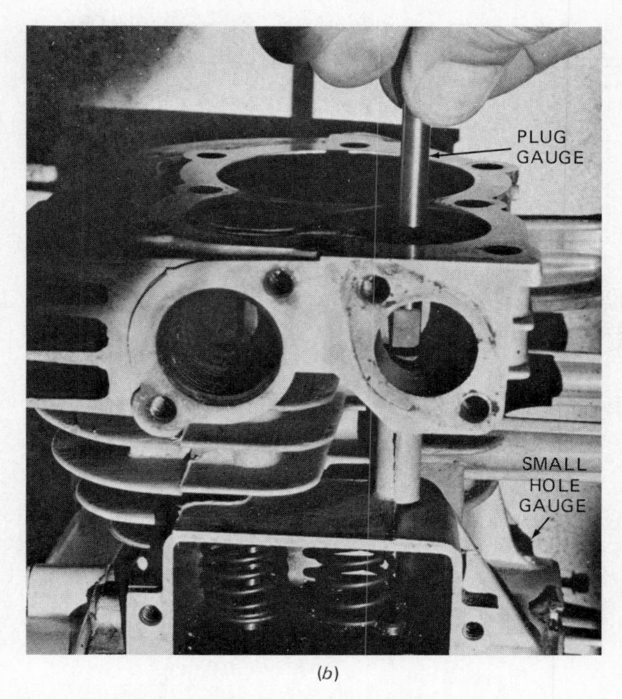


**Fig. 5-23.** Check the valve guide for excessive wear. A worn guide causes oil consumption. (*a*) Use a small-hole gauge or (*b*) special plug gauge. (*Briggs & Stratton Corp.*)

must be narrow so that the valve face will contact the seat near the center of the valve. This is accomplished by using the seat narrowing cutter (Fig. 5-28*b*).

## Inspecting the Camshaft

The design of the *lobes* on a camshaft is very exacting. When the shape of the cam is altered by wear, the operation of the engine will be affected. It is seldom possible to check the camshaft for wear when performing top overhaul repair since it is usually necessary to disassemble the engine to measure the lobes. However, in major overhaul operations, the camshaft lobes and bearings are checked with a micrometer (Fig. 5-29), and compared to the manufacturer's specifications. See Table 5-1.

## Installing Valves

In major overhaul, the engine valves are one of the last parts installed in rebuilding the block. However, the procedure is the same whether performing a top overhaul or major repairs.

The first operation in installing valves is to set the

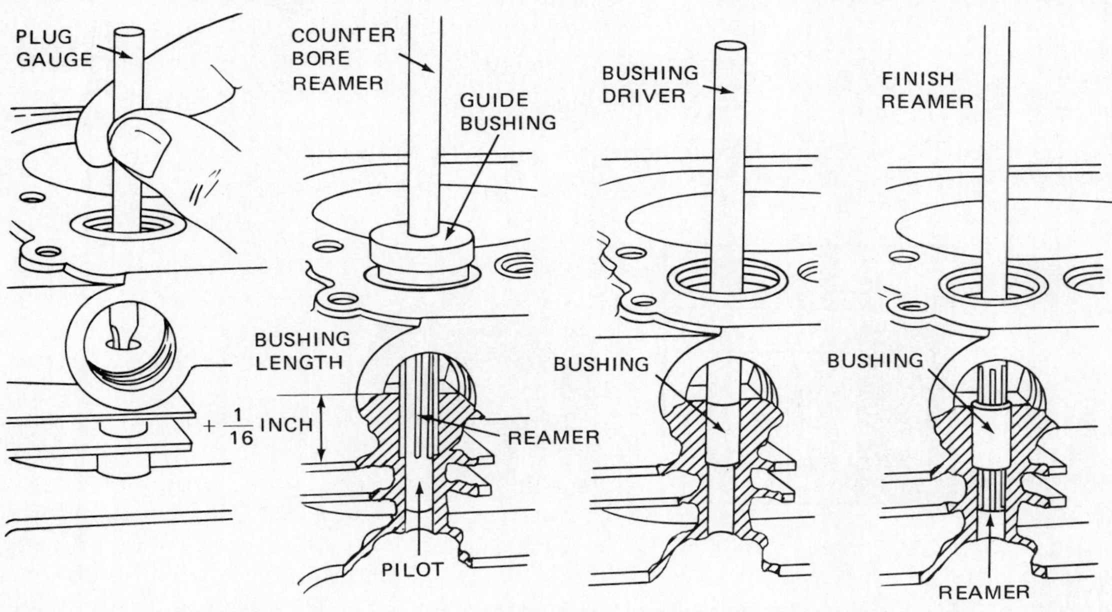


**Fig. 5-24.** Reconditioning a worn valve guide by reaming and installing a guide bushing. (*Crouse/Anglin, Small Engine Mechanics, 2nd ed. Copyright © 1980 by McGraw-Hill, Inc. Reproduced with permission.*)

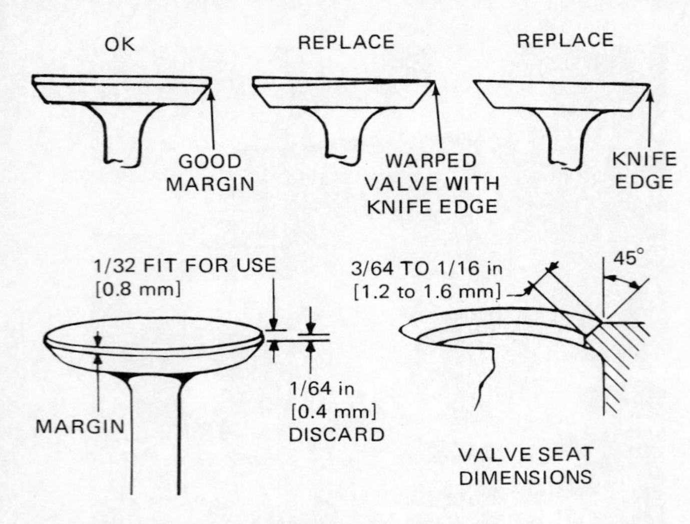


**Fig. 5-25.** Valve and seat dimensions. Always check specifications in the service manual. *(Briggs & Stratton Corp.)*

valve tappet clearance or valve lash. As previously indicated, the camshaft and tappets must be in place. If the crankcase cover has been removed, it must be replaced, since it usually supports one end of the camshaft. Each valve is then inserted into its respective guide. The crankshaft is rotated until the piston is at top dead center and both valves are closed. While holding down firmly on the valve with one hand, the clearance is checked with the proper-size feeler gauge according to engine specification for intake or exhaust (Table 5-1).

Manufacturers have several procedures for adjusting valve clearance or lash. One uses a set of rotator-adjusters which attach to the valve stem. The clearance is determined and the correct rotator-adjuster is selected. Another manufacturer uses an adjustable tappet (Fig. 5-30). The adjusting screw can be rotated to achieve a specific clearance. A third system re-

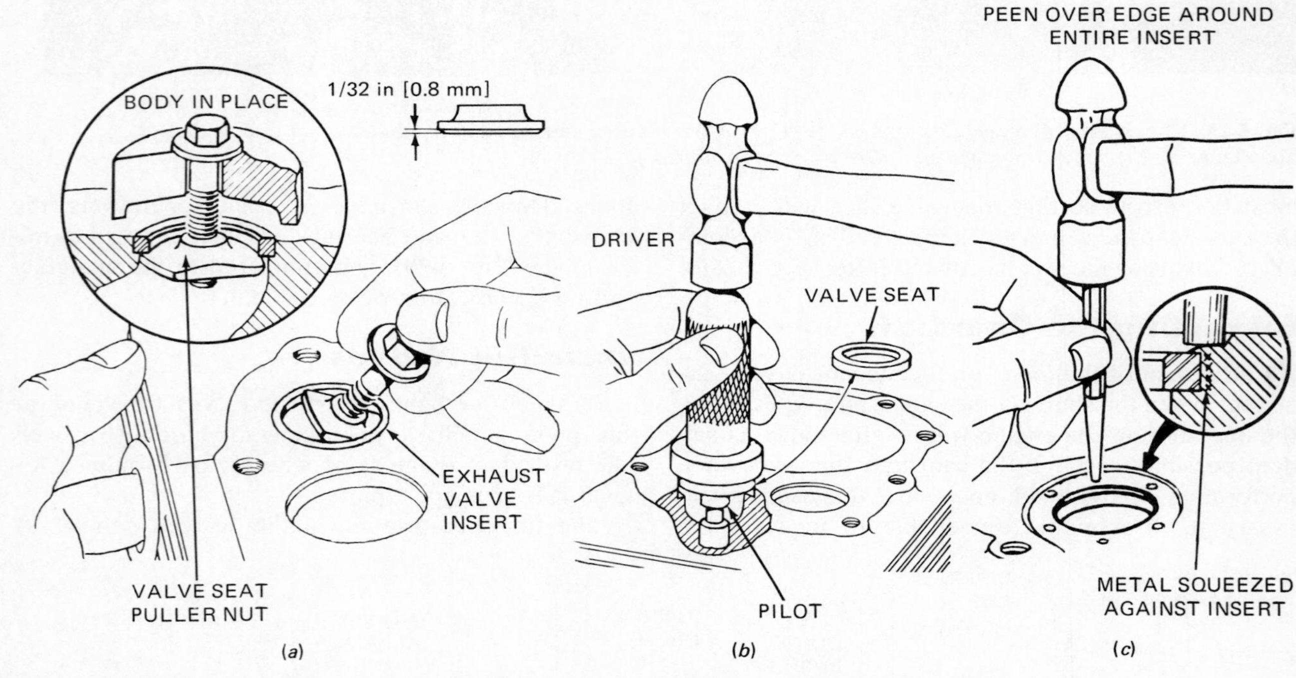


**Fig. 5-26.** (*a*) Removing defective valve seats. (*b*) and (*c*) Installing a new valve seat insert. *(Briggs & Stratton Corp.)*

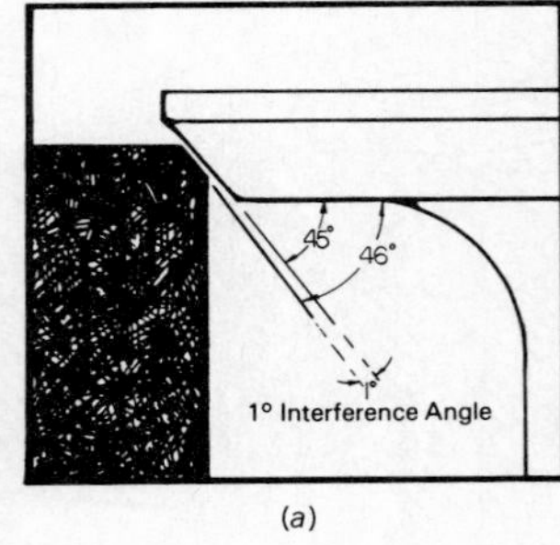


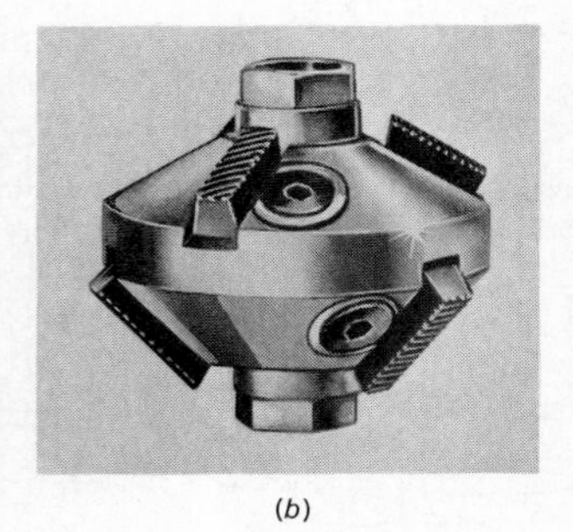


**Fig. 5-27.** (*a*) The valve face is resurfaced to an angle of 45°; the seat is cut to a 46° angle using the tungsten carbide blade cutter (*b*) to produce a 1° interference angle. *(Newway, Inc.)*

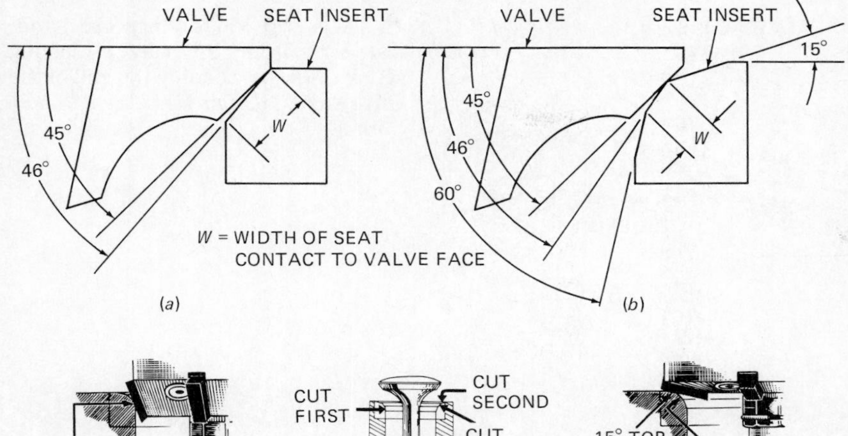


60° BOTTOM CORRECTION (FIRST CUT)
CUT FIRST
CUT SECOND
CUT SEAT LAST
15° TOP CORRECTION (SECOND CUT)
VALVE SEAT (THIRD CUT)

**Fig. 5-28.** (*a*) Refacing the valve seat causes the seat to become wider. (*b*) The width of the seat can be reduced by using a narrowing cutter that has a 15° top angle and a 60° bottom angle.

moves metal from the end of the valve stem. The filing or grinding must be done with extreme care to keep the end of the stem flat and square. A special valve-grinding machine is best adapted to this operation. The procedure is repeated for the other valve(s).

To install a valve, compress the spring and retainer until the keeper can be installed or until the retainer can be locked upon the end of the valve stem (Figure 5-31).

## INSPECTING AND RECONDITIONING PISTON, PISTON RINGS, AND CYLINDER

Excessive oil consumption will indicate to service center personnel that the piston rings may allow excess oil to be drawn into the combustion chamber. It may also indicate that major repairs requiring complete disassembly of the engine are necessary.

Before the piston can be removed from the cylinder, the deposits of combustion which have accumulated at the top of the cylinder must be removed. Generally the deposits can be scraped away with a stiff back scraper, using care not to cut into the part of the cylinder where the ring must form a seat. In addition to the carbon and fuel deposits, the upper part of the cylinder may be worn so that a "ring ridge" has formed (Fig. 5-32). Should the rings catch on the ridge when the piston is being removed, the piston ring and piston "lands" may be broken. A special ring ridge cutting tool may be required to remove the ridge before the piston can be taken from the cylinder.

The connecting rod bearing cap must be removed to release the connecting rod from the crankshaft. Carefully inspect how the connecting rod assembly fits together before removal. If the engine is a multi-

FUEL PUMP ECCENTRIC
*MEASURE THESE POINTS AND COMPARE TO ENGINE SPECIFICATIONS
CAMSHAFT

**Fig. 5-29.** Inspect the condition of the camshaft lobes and measure height of the lobes and diameter of the journals. (*Onan Corp. and Tecumseh Products Co.*)

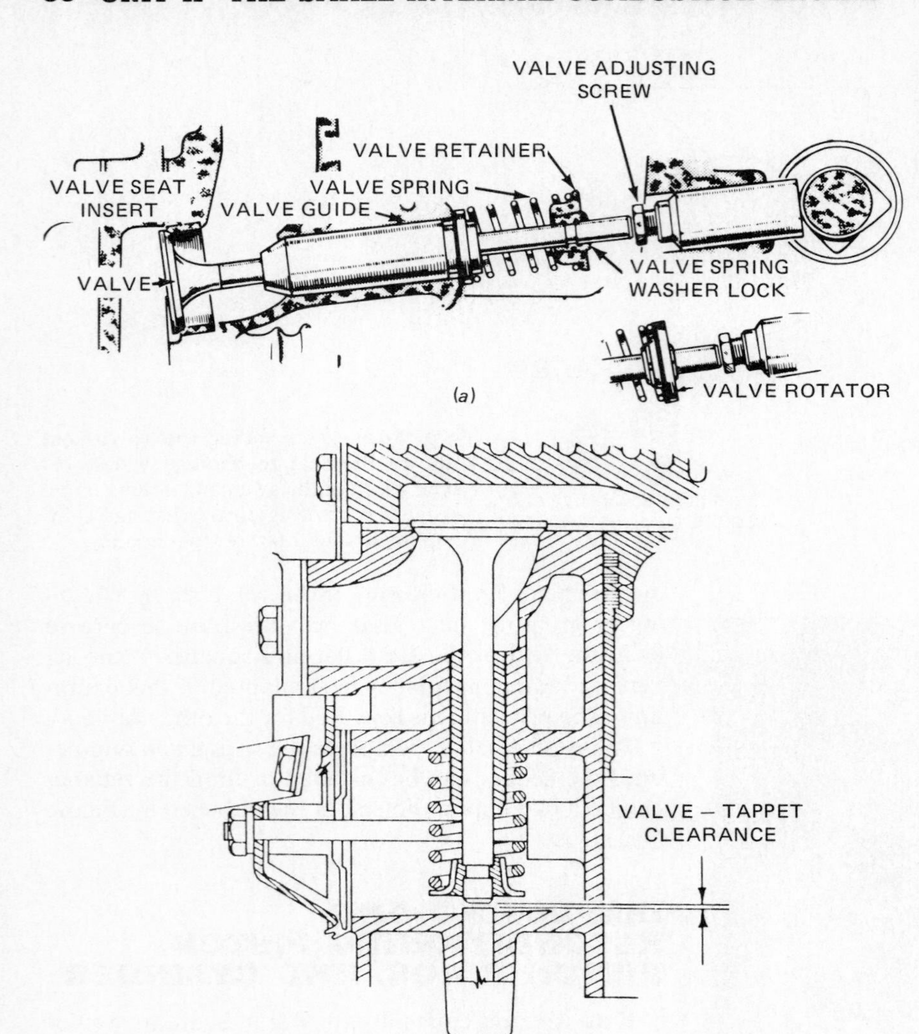


**Fig. 5-30.** (*a*) Valve clearance is adjusted by using the valve adjusting screw or (*b*) by grinding the end of the valve stem. (*Onan Corp. and Kohler Corp.*)

cylinder, be sure the connecting rod and cap are numbered. The professional will make a sketch of the arrangement on a work sheet to serve as a reminder when reassembly is started.

## Inspecting Piston and Cylinder

After the piston and connecting rod assembly have been removed from the open end of the cylinder, reattach the bearing cap to prevent loss of parts and damage to the bearing. Make a thorough visual check of the piston and cylinder for scored surfaces, vertical grooving, and scuff marks. If the surfaces of the piston and the cast-iron cylinder are scored, the piston may need to be replaced and the cylinder reconditioned. An aluminum cylinder (Fig. 5-33) with a scored cylinder bore tends to heal and the score will not severely affect performance.

Use a telescoping gauge and micrometer and measure the cylinder at six places (Fig. 5-34). The measurements should be taken in the cylinder at points at right angles to and parallel with the crankshaft. More cylinder wear can be expected in the upper one-fourth of the piston travel because combustion pressures are highest when the piston is in that position in the cylinder. Measuring the cylinder at six places permits the service person to determine the amount of cylinder wear, out-of-roundness (not perfectly round), and cylinder taper. For example, Table 5-1 indicates that if the cylinder is 0.003 in [0.076 mm] oversize or 0.0025 in [0.064 mm] out-of-round from the standard cylinder dimensions listed in Table 5-1, the cylinder must be resized.

From the measurements obtained and the visual inspection of the cylinder, decisions can be made on how to prepare the cylinder to receive piston rings. If the surface finish of a cast-iron cylinder is smooth and the amount of taper is within limits, the cylinder is honed with a flexible deglazing hone (Fig. 5-35*a*) to

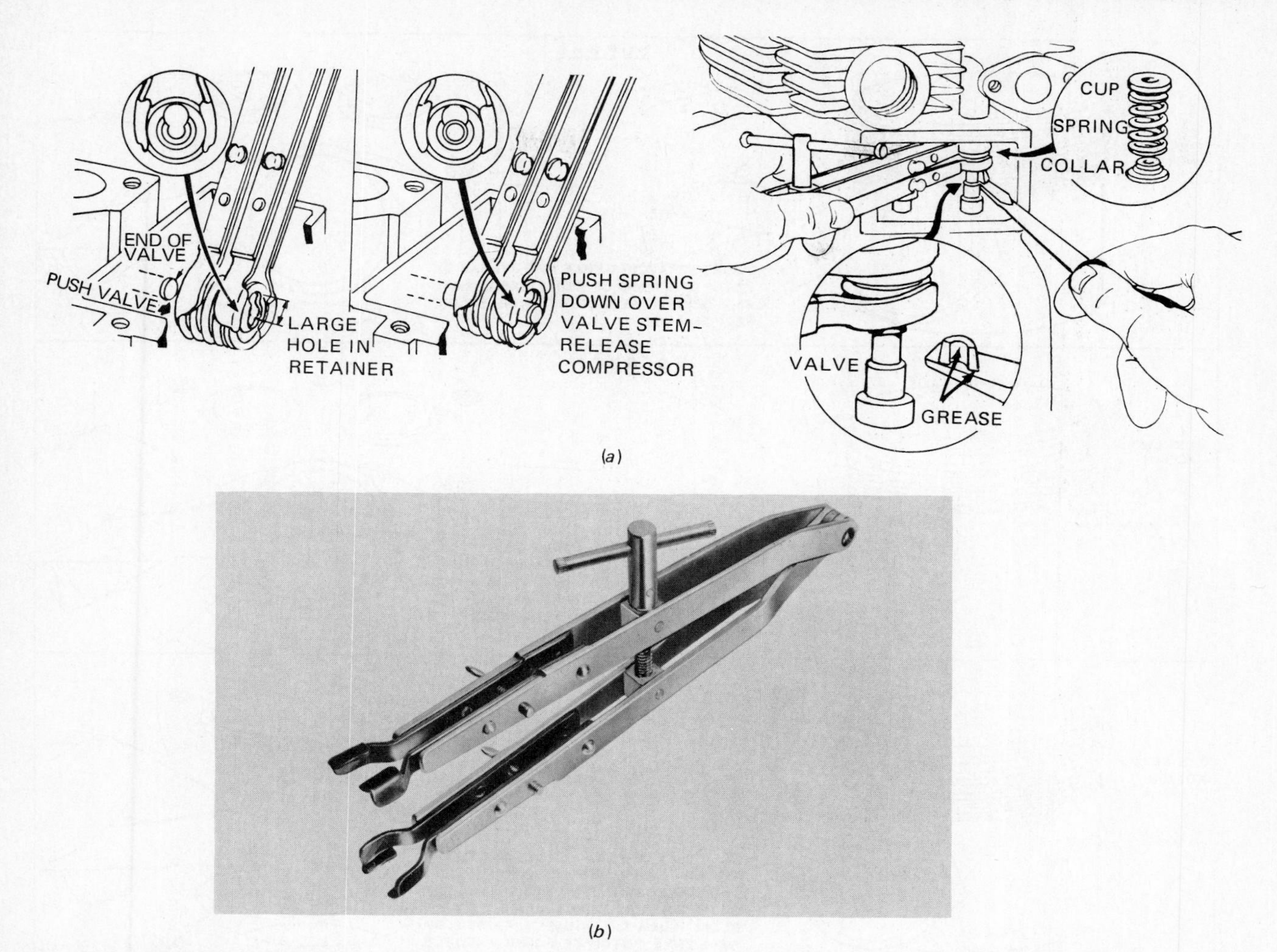


**Fig. 5-31.** (*a*) Installing valves by using a valve spring compressor. (*Briggs & Stratton Corp.*) (*b*) Valve spring compressor. (*Ammco Tools Inc.*)

make the cylinder ready to receive the piston and new rings (Fig. 5-35*b*). Occasionally during the honing operation the cylinder should be thoroughly cleaned and the piston should be checked. A deglazing hone should not be used on aluminum cylinders.

If the cylinder bore is scored badly or the amount of wear exceeds the maximum limits, the cylinder is bored oversize, using a cylinder grinder, and then honed (Fig. 5-36*a*). Thoroughly clean the cylinder with hot, soapy water. Scrub with a stiff bristle brush and rinse in hot water. It is *essential* that a good cleaning operation be performed. Any abrasive material remaining in the cylinder will cause rapid wear of both the cylinder and the rings. The cylinder should be swabbed with light oil and a clean shop towel and then wiped with a clean, dry cloth. *Cylinders should not be washed with gasoline.* An oversized piston and rings are fitted to the resurfaced cylinder. Oversize pistons are usually available in 0.010, 0.020, and 0.030 in [0.25, 0.50, and 0.76 mm] oversize. The desirable surface finish of a newly ground and honed cylinder will have a crosshatched surface (Fig. 5-35*c*).

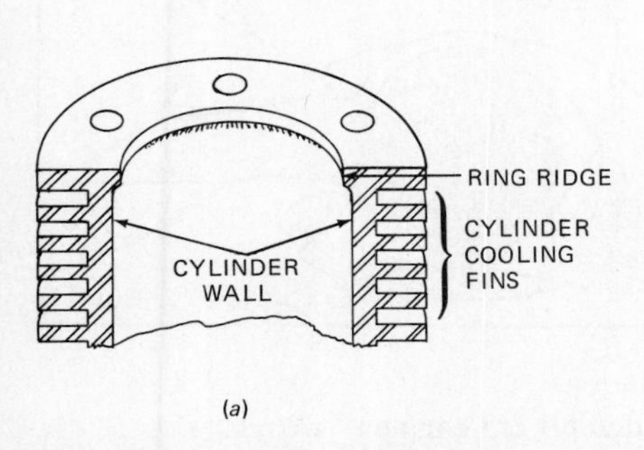


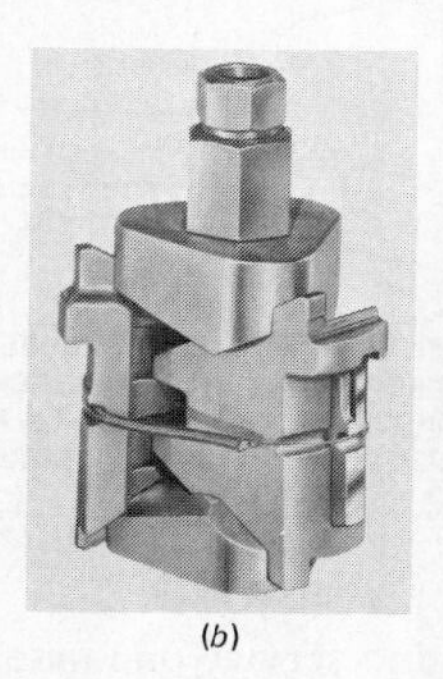


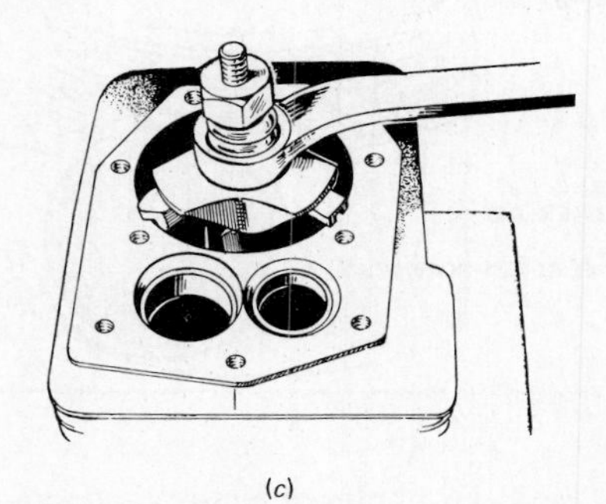


**Fig. 5-32.** The ring ridge (*a*) must be removed (*c*) with a ridge reamer (*b*) before the piston can be removed from the cylinder. (*Onan Corp. and Ammco Tools Inc.*)

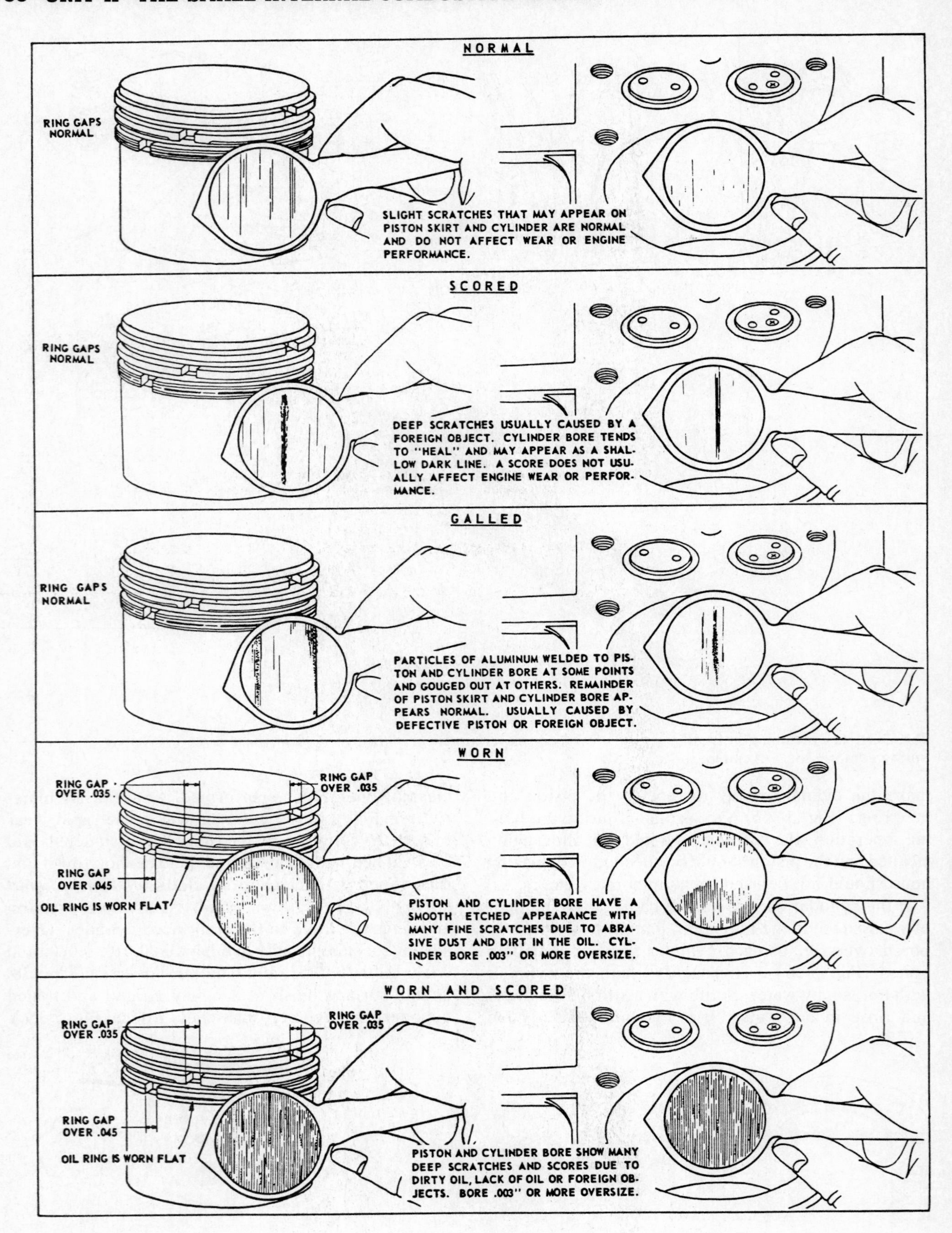


**Fig. 5-33.** Examples of pistons and cylinders found in service on model 6B.H and 6B.HS engines. (*Briggs & Stratton Corp.*)

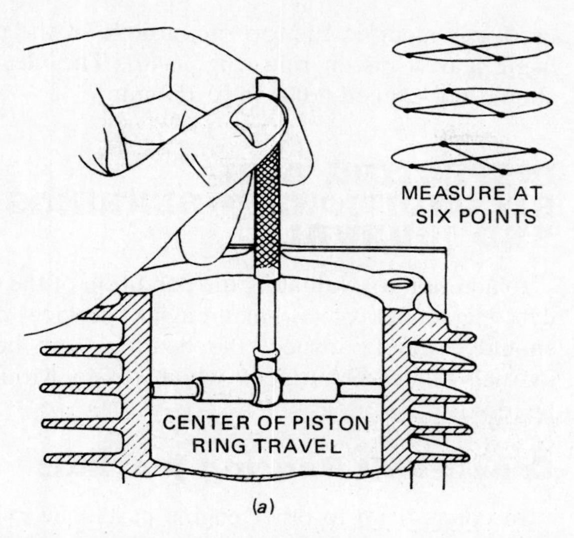


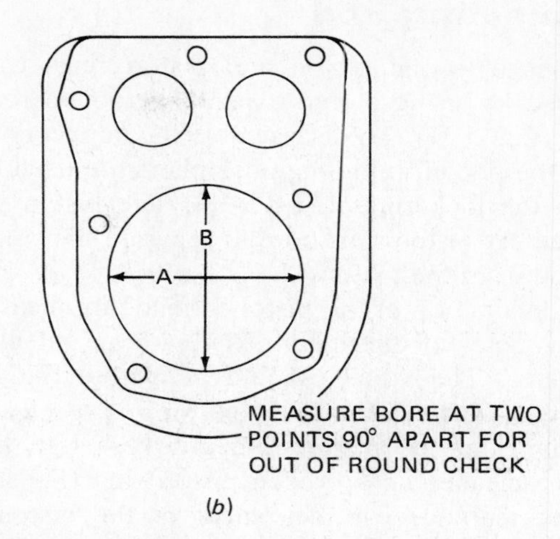


**Fig. 5-34.** Measure the cylinder at six points in two directions to determine the amount of taper and out-of-roundness. (*a*) Cross-sectional view. (*b*) Top view. (*Briggs & Stratton Corp. and Kohler Co.*)

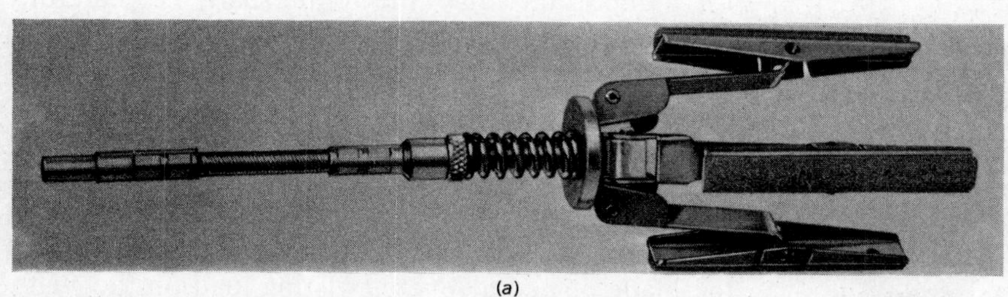
(a)

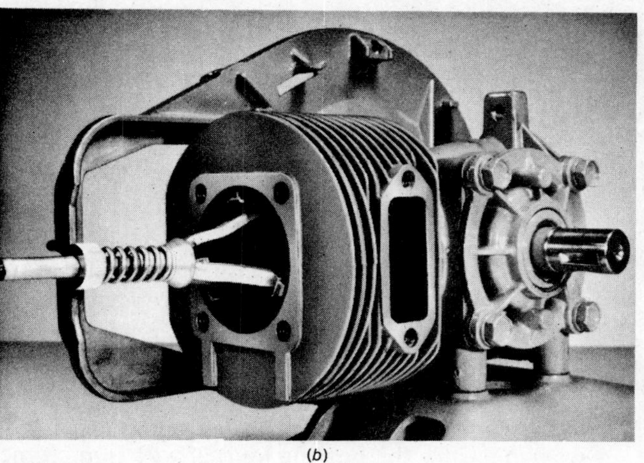
(b)

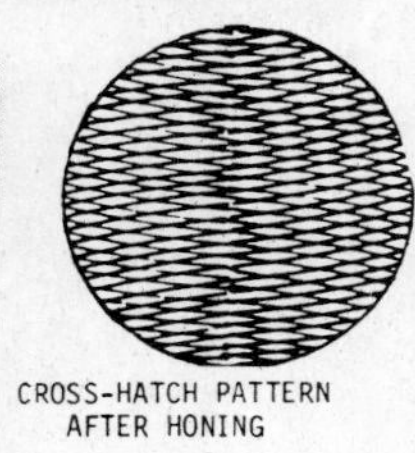

(c)

**Fig. 5-35.** Conditioning a cylinder bore. (*a*) A glaze-removing hone. (*Ammco*) (*b*) Conditioning a cylinder with hone to receive rings. (*Jacobsen Div. of Textron Corp.*) (*c*) Crosshatch pattern after honing.

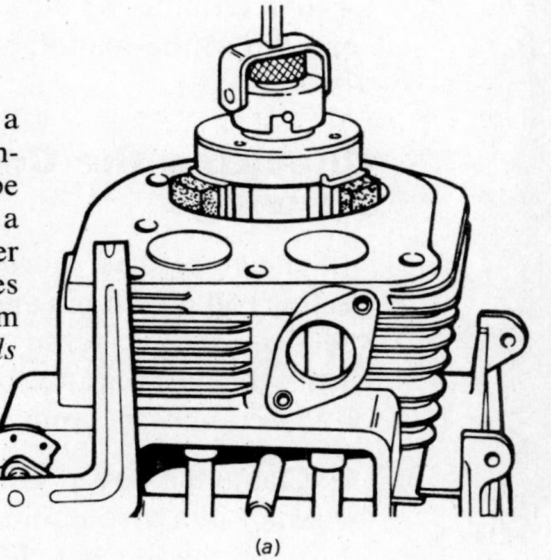
(a)

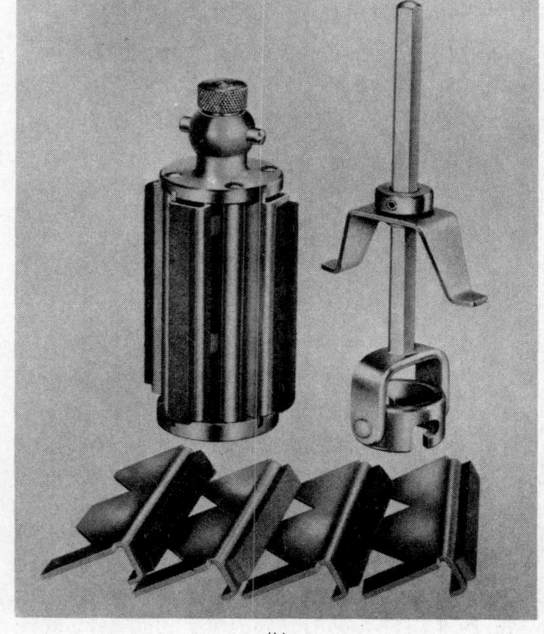
(b)

**Fig. 5-36.** (*a*) Reconditioning a cylinder. (*Onan Corp*) Cylinders of small engines can be ground to oversize to produce a cylindrical surface. (*b*) Cylinder hone showing abrasive stones which remove material from cylinder wall. (*Ammco Tools Inc.*)

### Remove Piston Rings and Connecting Rod

Damage to the piston and piston rings can be avoided by using a ring expanding tool to remove them (Fig. 5-37). To disassemble the connecting rod from the piston, thin long-nose pliers are used to remove the lock rings (Fig. 5-38). The piston pin is pushed out of the skirt bore to separate the connecting rod from the piston.

An evaluation of the piston is made by measuring the diameter of the piston skirt with an outside micrometer. The values should be within the limits specified (Table 5-1). The next step is to clean the carbon from the piston ring grooves. Use a ring groove cleaner or a broken piston ring (Fig. 5-39). "Side clearance" measurement of the piston ring groove is made in the top ring groove of the piston using a new piston ring (Fig. 5-40). The clearance should not exceed 0.006 in [0.152 mm].

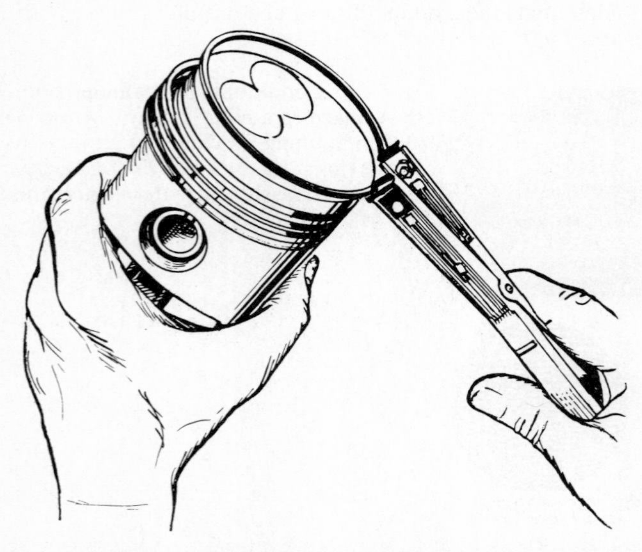

**Fig. 5-37.** To prevent damage, a ring expanding tool is used to remove or install a piston ring. (*Onan Corp.*)

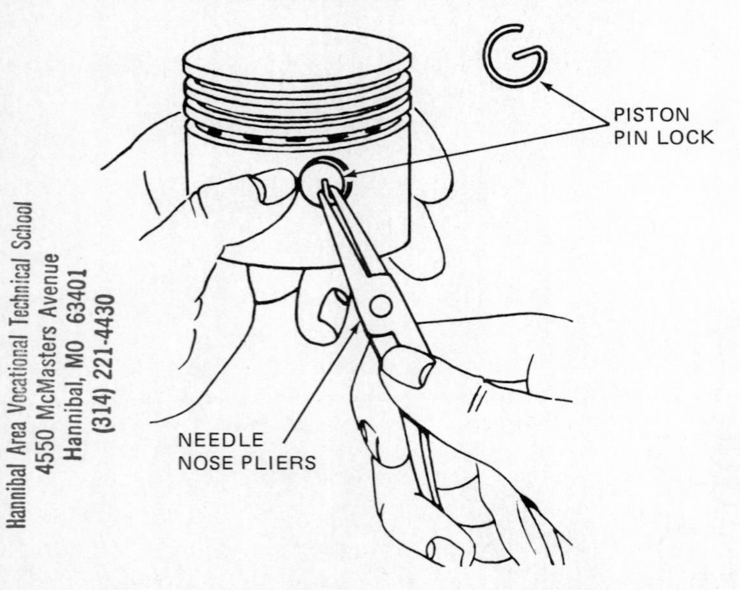


**Fig. 5-38.** The lock ring must be removed before the piston pin can be removed. (*Briggs & Stratton Corp.*)



## INSPECTING AND RECONDITIONING BEARINGS AND JOURNALS

In addition to evaluating the condition of the cylinder, personnel of a small-engine service center should carefully inspect the bearings and bearing journals (finished surfaces which rotate within the bearing).

### Crankshaft Bearing Journals

In comparison to other engine parts, the cylinder block and the crankshaft are the two most expensive replacement items in small-engine repair. It is therefore important that the crankshaft be carefully checked to determine its straightness, the condition of the journal surfaces, and whether there has been damage to the keyway(s).

The crankshafts of small engines driving rotary lawn mowers are subject to severe shock loads when the blade strikes a solid object. Very often the crankshaft is bent because a sudden stoppage of the blade placed a severe twisting load on the shaft. Always check the crankshaft of a rotary lawn mower for straightness with a dial indicator. The crankshaft bearing journals must be inspected for smoothness and wear. If badly grooved, or if the specification measurements cannot be met, the crankshaft will need to be replaced or in some cases the journal surfaces reground to undersize. Because it is seldom possible or practical to resurface the crankshaft of engines in the 3- to 6-horsepower (hp) [2.3- to 4.5-kilowatt (kW)] range, a defective crankshaft must be replaced. If the surface of the connecting rod journal is reground, it should be reduced to the size of the bearing to be fitted. Generally the bearings are available in 0.010, 0.020, or 0.030 in [0.25, 0.50, or 0.76 mm] undersize. Always measure the bearing journal with a micrometer to determine the out-of-roundness (Fig. 5-41), and compare to the manufacturer's specifications (Table 5-1).

### Checking the Connecting Rod Bearings

If the crankshaft journals are serviceable, the connecting rod piston pin and crankshaft bearings are inspected and measured to evaluate their condition. Check the oil clearance of the connecting rod bearing on the crankshaft connecting rod journal as follows:

1. Clamp the power takeoff (PTO) end of the crankshaft in a vise equipped with soft jaws.
2. Wipe the journal clean of oil.

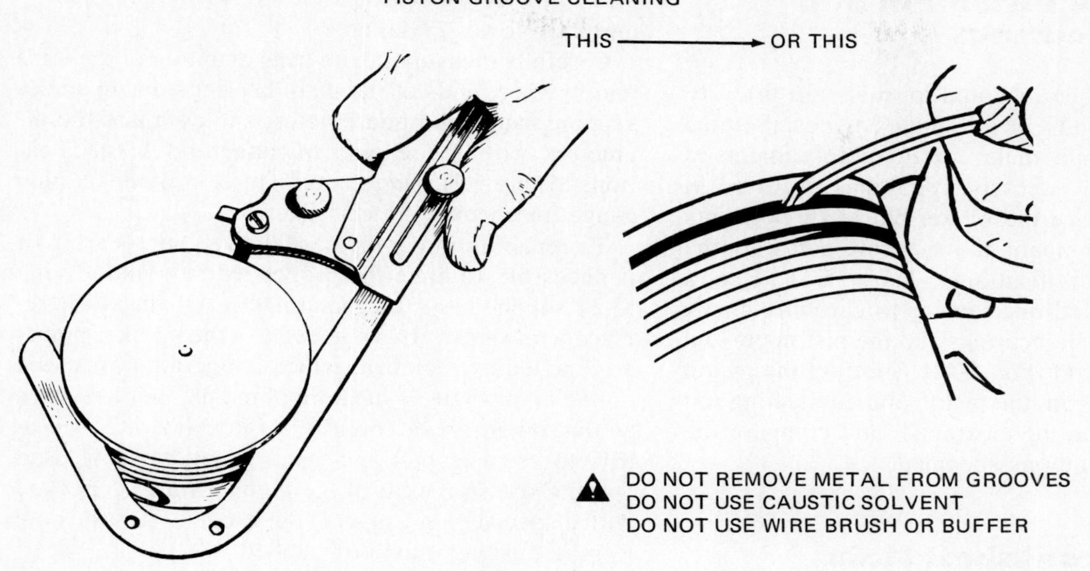


**Fig. 5-39.** Carefully clean the piston ring grooves with a special groove cleaner or a broken piston ring. (*Onan Corp.*)

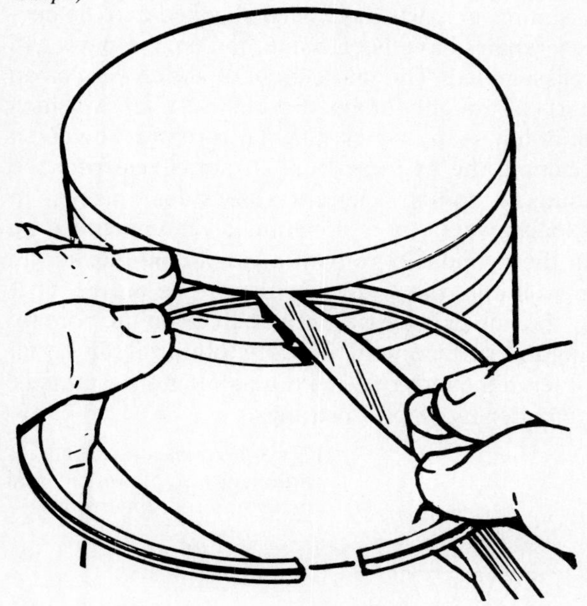

**Fig. 5-40.** Measuring top groove side clearance with a new ring.

3. Place a small strip of Plastigage on the crankshaft journal.
4. Attach the connecting rod to the crankshaft and tighten the cap screws to the torque specifications (Table 5-1). *Do not allow the connecting rod to rotate during tightening.*
5. Remove the cap and read the clearance from the flattened Plastigage. If the oil clearance is more than the specifications (Table 5-1), the connecting rod bearing and the crankshaft journal dimensions should be rechecked.

The bearing surface of most small-engine connecting rods is cast integral with the connecting rod bore. To replace the bearing it is necessary to replace the connecting rod. Most manufacturers make replacement connecting rods that will fit crankshaft journals which have been resurfaced.

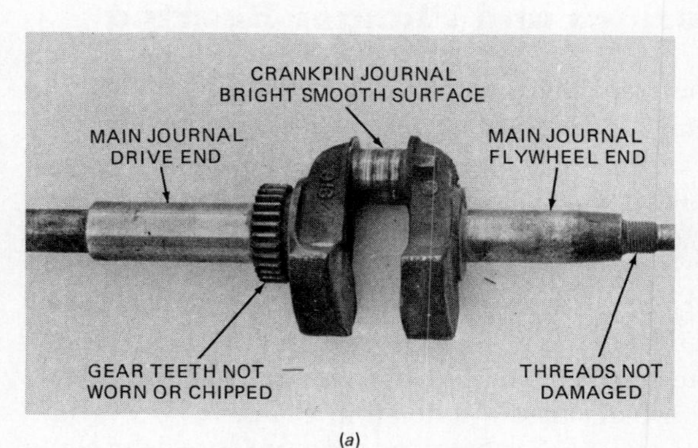


(*a*)

(*b*)

**Fig. 5-41.** (*a*) Checking the condition of the crankshaft by measuring the journals. (*b*) Visually inspecting the surfaces.

## Checking the Piston Pin and Piston Pin Bearings

The piston pin bearings and journals are fitted to close tolerance limits in all small engines. Service center personnel must make careful measurements to determine whether excessive wear has occurred in these parts. Measure the piston pin at three points (Fig. 5-42) and compare the measurements to the manufacturer's specifications (Table 5-1). Use a small-hole gauge and micrometer to carefully check each of the piston pin bearings and the piston pin end of the connecting rod (Fig. 5-42). Subtract the piston pin measurement from the piston and connecting rod values to determine the clearance, and compare the answer to the limitations specified.

## Checking Crankshaft Main Bearings and Plunger Bearing

The crankshaft main bearing of small engines may be one of two basic types of bearing: (1) plain bearings or bushings and (2) antifriction. *Plain bearings* are sleeve bearings which may be either an integral part of the bearing housing or a precision insert type (Fig. 5-43*a*). *Antifriction bearings* are either ball, roller, or needle. Some manufacturers use a ball bearing on the PTO end of the crankshaft and a sleeve bearing on the flywheel end. Bearings and seals are discussed in detail in Chap. 7.

If tapered roller bearings are used, one is used on each end of the crankshaft (Fig. 5-43*b*). This type of bearing is generally used on small engines that are classified "heavy duty" or "commercial."

Both types of bearings give long life; however, engines with sleeve bearings generally run more quietly and have lower initial cost.

Carefully measure the internal diameter of the PTO and flywheel ends of the main bearings, using a telescoping gauge and micrometer, and compare the diameters with the engine manufacturer's specifications (Table 5-1). One manufacturer provides a plug gauge for checking reject size.

To repair an integrally cast sleeve main bearing, it is necessary to have the proper reamers and driving tools supplied by the manufacturer to install a replacement sleeve. In some engines the breaker points are opened by a plunger which is operated by a lobe on the crankshaft. If the bore of the plunger has worn so that oil enters the breakpoint housing, it is necessary to use a reamer and install a new bushing (Fig. 5-44). Excessive wear of the plunger hole is checked with a special plug gauge. If the gauge enters the hole ¼ in, a bushing must be installed.

# ASSEMBLING THE ENGINE

Assuming all parts have been checked and the necessary repairs have been made, the engine parts can be reassembled. The first course of action is to clean all parts thoroughly of any dirt and debris by washing them in hot soapy water. Special care must be given to cleaning the cylinder bore of abrasives from the grinding or honing operations. Besides washing in hot soapy water, it is sometimes recommended to wash the cylinder bore with a clean cloth soaked in engine oil until the surface is clean. The oil will float away the fine particles. Continue to clean the bore by wiping the surface with a clean cloth until all the oil has been removed from the bore and interior parts of the cylinder block and bearings.

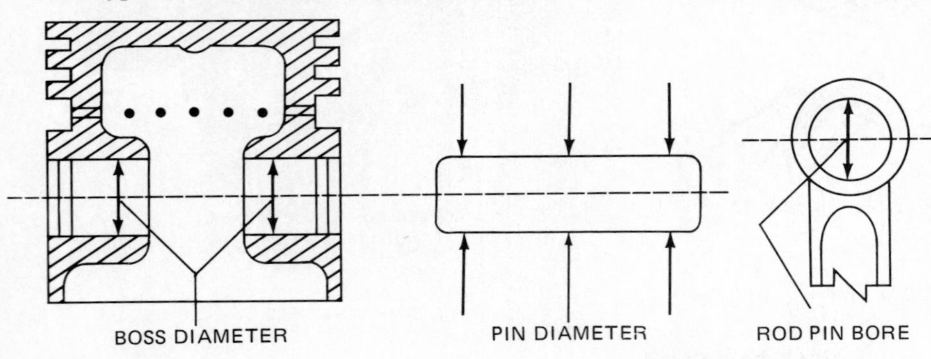


PISTON PIN

| | MFG. SPECS | AS MEASURED |
|---|---|---|
| BOSS DIAMETER | | |
| PIN DIAMETER | | |
| CLEARANCE | | |

CONNECTING ROD

| | MFG. SPECS | AS MEASURED |
|---|---|---|
| ROD PIN BORE | | |
| PIN DIAMETER | | |
| CLEARANCE | | |

**Fig. 5-42.** Measuring and recording piston, piston pin, and connecting rod dimensions.

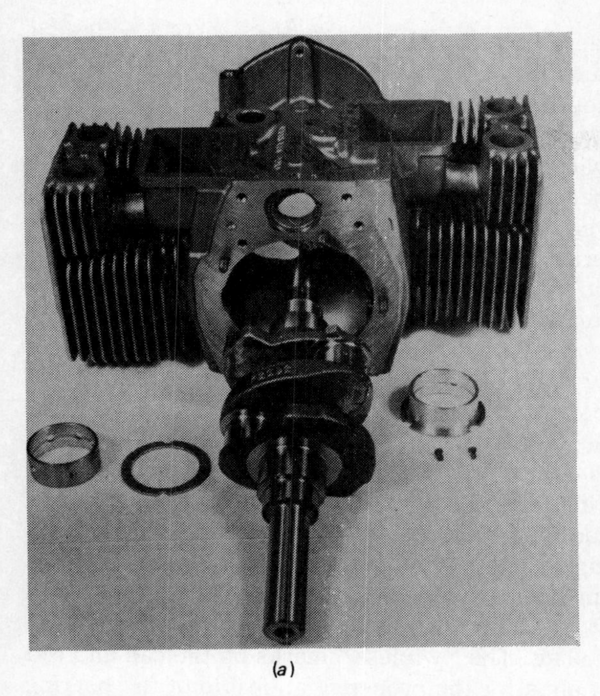

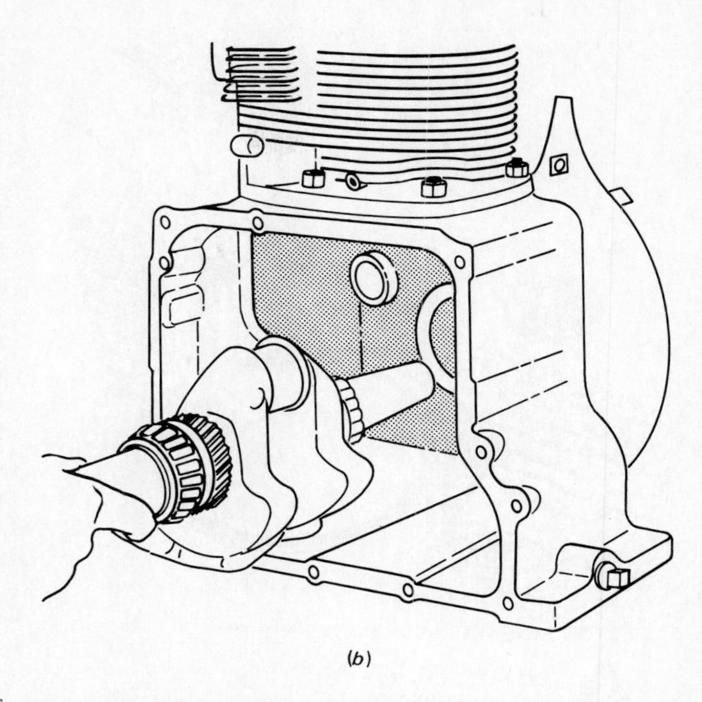

(*a*) (*b*)

**Fig. 5-43.** Crankshaft main bearings. (*a*) The friction types are the shell insert types. (*Onan Corp.*) (*b*) The antifriction types are tapered roller bearings on each end of the crank. (*Teledyne Wisconsin Corp.*)

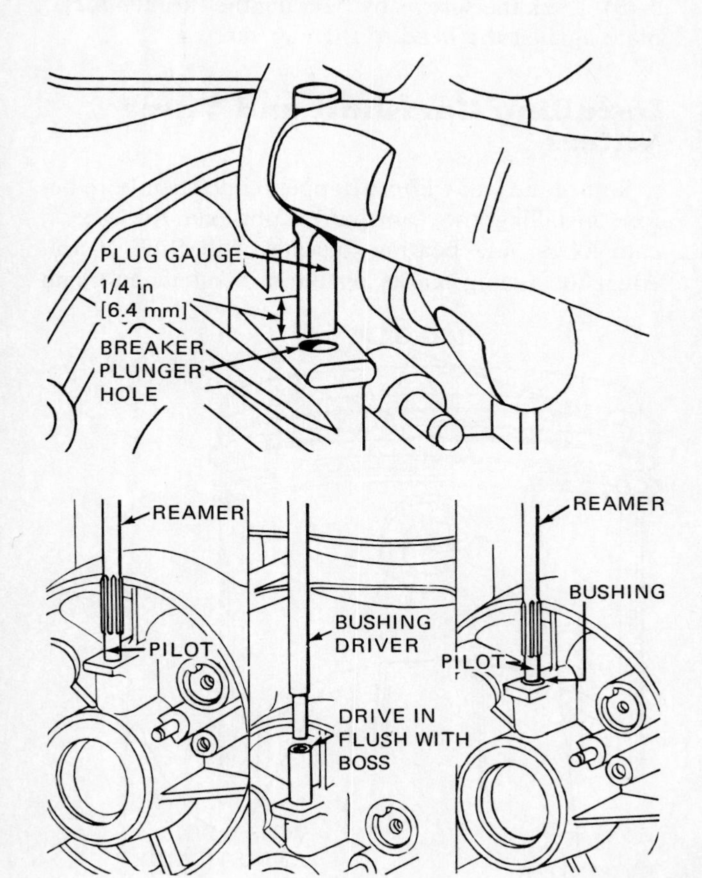


**Fig. 5-44.** Repairing a breaker point plunger hole with a replacement bushing. Always follow the manufacturer's recommendations. (*Briggs & Stratton Corp.*)

## Installing the Crankshaft

To install the magneto end of the crankshaft into the cylinder block, apply a small quantity of SAE 50 engine oil on the journal end. Then carefully slide the crankshaft into the main bearing.

## Installing Piston Pin and Rings

Before the piston rings are installed in the grooves of the piston, they must be checked for end-gap clearance in the cylinder. Place a ring into the cylinder with finger pressure. Then use the piston to push it approximately halfway down in the cylinder. Use a feeler gauge and measure the amount of end gap (Fig. 5-45). The gap must be within the limits specified for the engine (Table 5-1). If too great a gap occurs, an oversize set of rings must be used to produce the correct gap.

Attach the connecting rod to the piston by inserting the piston pin. Orient the connecting rod in the piston. Then lubricate the pin with engine oil and push it into the piston and connecting rod bearing with thumb pressure. Reinstall the lock rings so that they are properly placed. Install the rings in the proper piston grooves with the correct side up (Fig. 5-46).

Turn the rings on the piston so that none of the end gaps align, at the same time making sure that each gap is staggered approximately 120°.

## Installing the Piston and Connecting Rod

Lubricate the ring grooves thoroughly with SAE 50 engine oil (Fig. 5-47*a*). Attach a ring compressor to

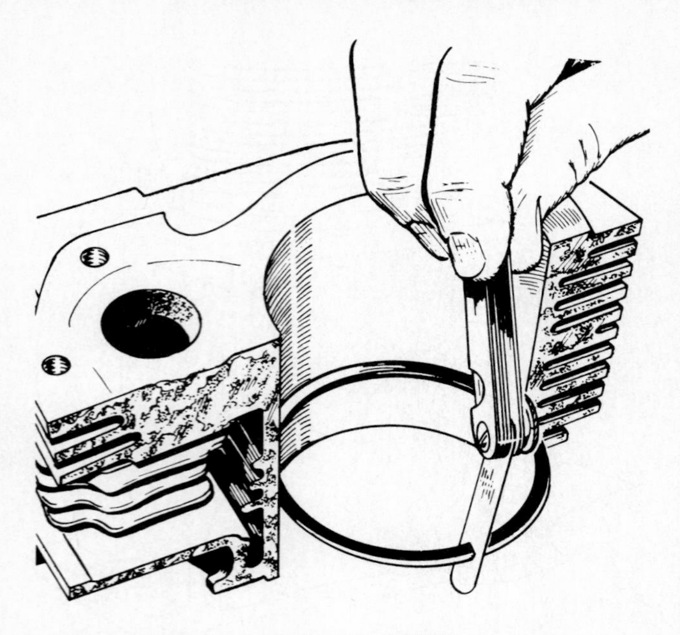

**Fig. 5-45.** Measuring the end gap in a piston ring. (*Onan Corp.*)

the piston and compress the rings so that the piston can be inserted into the cylinder (Fig. 5-47*b*). A short piece of 1/4-in [6.4-mm] hose should be placed over the connecting rod bolts to protect the journals and the bolt thread during installation. Place the piston so that the connecting rod throw is aligned with the crankshaft journal before tapping the piston into the bore with the wooden handle of a hammer. The crank throw should be as far from the connecting rod as possible while inserting the piston. This will reduce the possibility of the connecting rod damaging the journal.

Plastigage should be used to determine the oil clearance between the crankshaft journal and the connecting rod. Wipe the oil from the journal and the rod cap. Place a short piece of the plastic gauging material across the journal and install the rod cap. Lightly oil the connecting rod bolts and torque to the manufacturer's specifications. *Note: Do not rotate the crankshaft during this measurement.* Remove the rod cap and determine the oil clearance by comparing the width of the flattened Plastigage to the graduations on the envelope. The number indicates the clearance in thousandths of an inch or in millimeters. Check this clearance with the recommendations.

For final assembly, lubricate the journal with SAE 50 engine oil. Attach the connecting rod to the crankshaft by replacing the bearing cap, oil slinger, and lock plate. The "witness" marks on the cap and rod must align for the proper relationship of the bearing surfaces. The marks are used to identify the correct alignment for assembly. Use a torque wrench to tighten the cap screws or nuts to specifications (Fig. 5-48). Lock the screws by bending the "french lock" plate against the head of the cap screw.

## Installing Camshaft and Valve Lifters

Return the valve lifters (tappets) into their bore before installing the camshaft. Lubricate the lifters, cam lobes, and bearing journals with SAE 50 oil. Align the timing marks on the crankshaft and cam

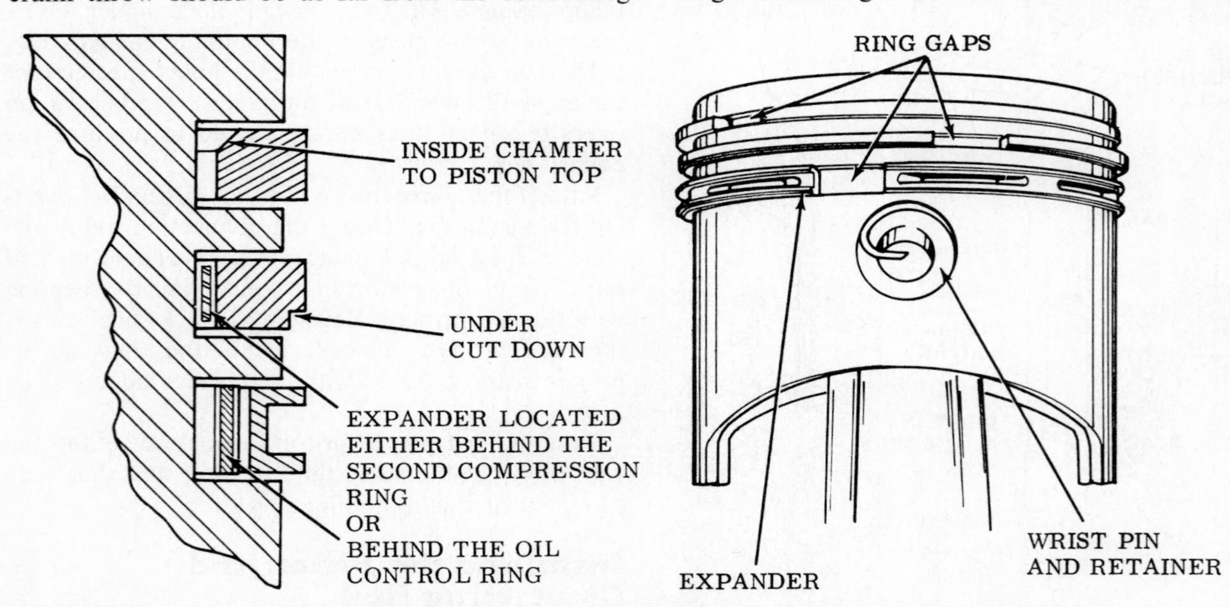


**Fig. 5-46.** Install the piston rings in the proper groove of the piston with proper orientation of the first and second compression ring. *Note:* stagger the end gaps 120° but do not align with the piston pin. Always check the manufacturer's service manual for specifications. (*Tecumseh Products Co.*)

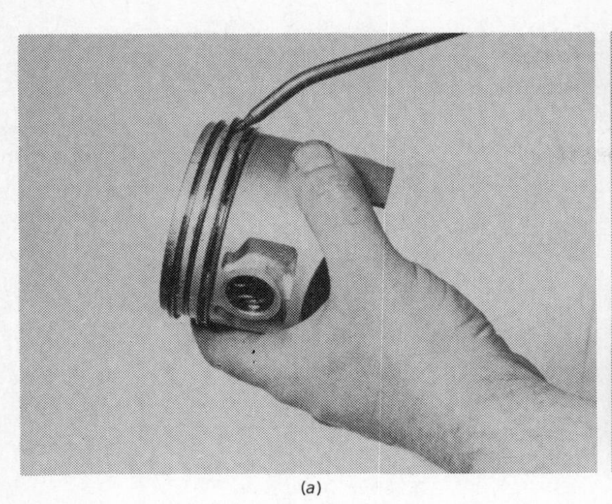


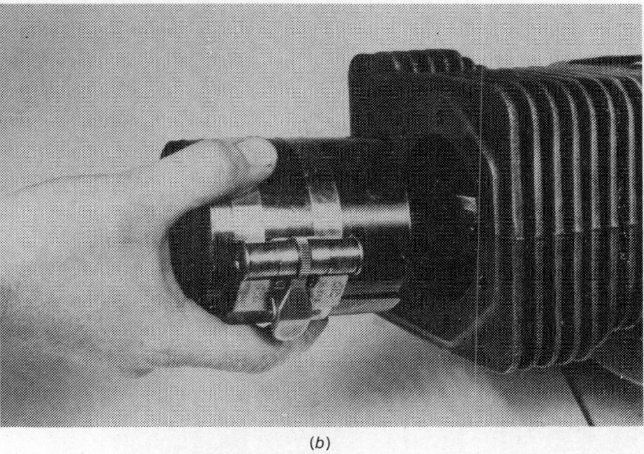


**Fig. 5-47.** (*a*) Oil the rings and ring grooves liberally with SAE 50 engine oil. (*b*) Compress the rings with a ring compressing tool before inserting the assembly in the cylinder. (*Onan Corp.*)

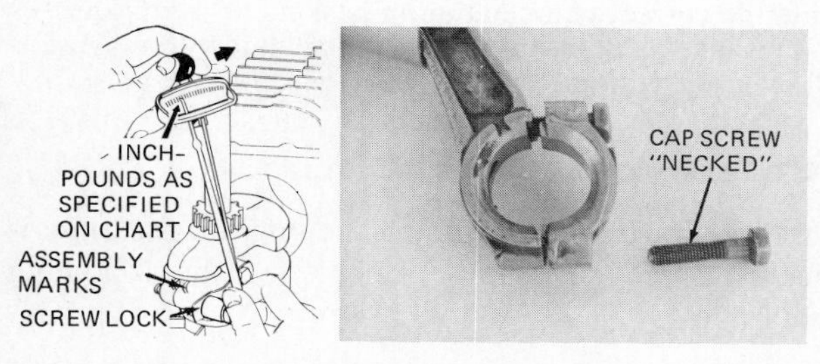


**Fig. 5-48.** Assemble and torque the bearing cap screws. Note the cap screw that is elongated (necked) as a result of excessive tightening. Always follow the manufacturer's specifications.

gear as the camshaft is fitted into place (Fig. 5-49). If the engine has an oil pump, check to see that it is properly installed.

## Adjusting Crankshaft End Play

Return the crankcase cover plate to the cylinder block. Use a sleeve over the end of the crankshaft to prevent any damage to the oil seal (Fig. 5-50) as the cover is installed.

Be sure to install a new cover plate gasket of the same thickness as the one removed. Tighten all the cover plate cap screws with a torque wrench set to the specification of the manufacturer (Fig. 5-51). Attach a dial indicator to the crankshaft (Fig. 5-5) and measure the end play of the crankshaft. The amount of end play must be within the limits established by the manufacturer (Table 5-1). If the end play is too great, a thinner cover plate gasket or a thrust washer must be used. Conversely, if there is not enough end play, a thicker (or several) gasket(s) must be used Some manufacturers specify the minimum gasket thickness.

## Installing Engine Valves and Cylinder Head

The procedure for installing the valves has been previously discussed in this chapter. When the valve

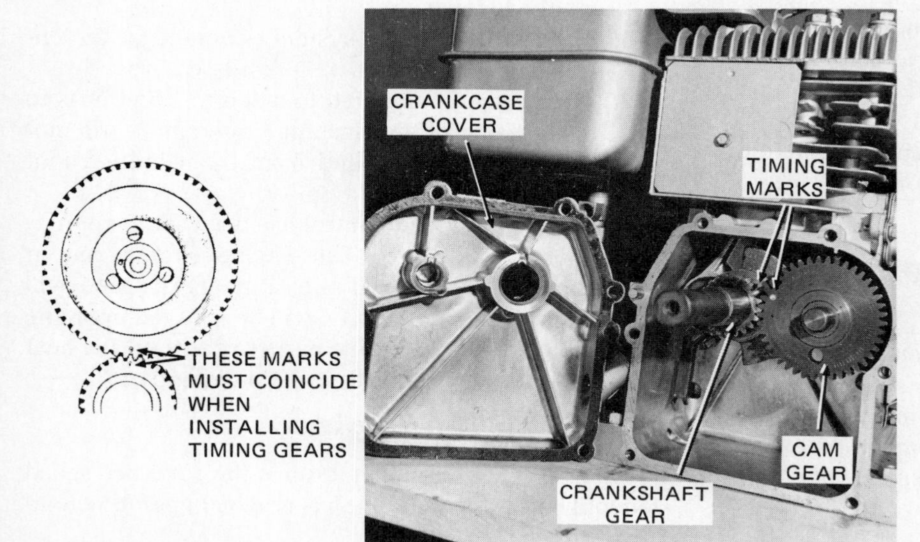


**Fig. 5-49.** When installing the camshaft, make sure the timing marks are properly aligned. (*Onan Corp.*)

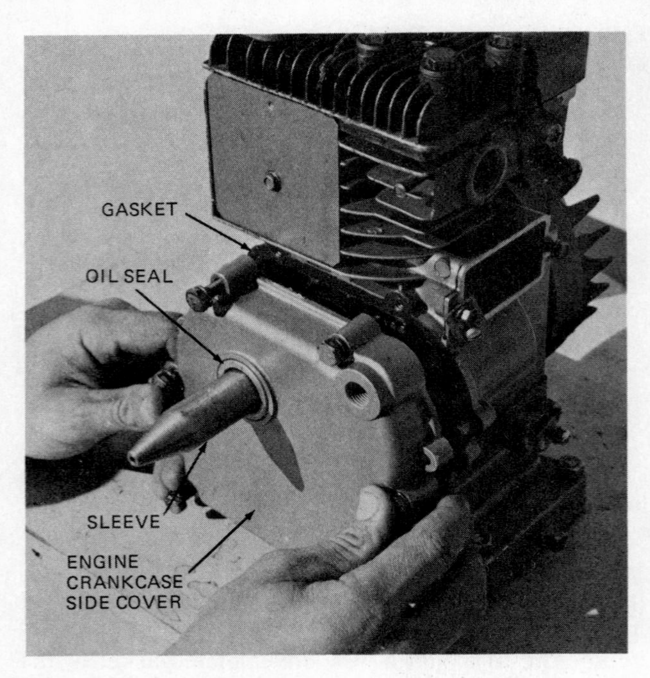


**Fig. 5-50.** Use the proper-size sleeve to protect the oil seal from damage when installing the crankcase cover plate or bearing housings.

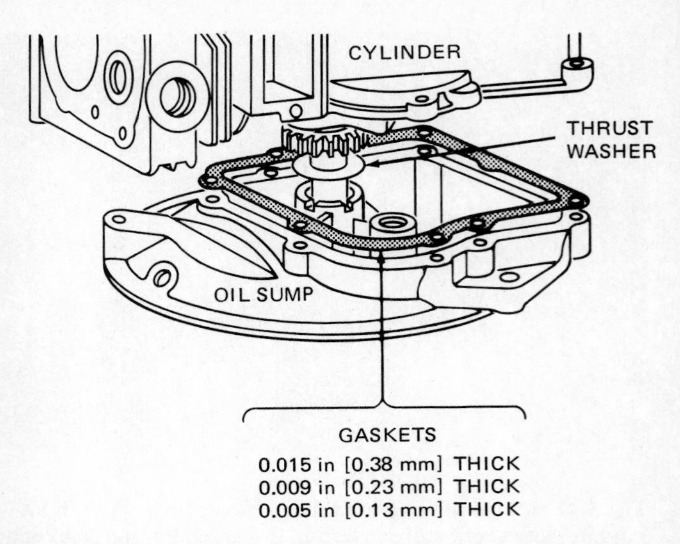


**Fig. 5-51.** Adjust the crankshaft to the proper end play by using the correct combination of gaskets. The thicknesses of the gaskets vary with the engine manufacturer. Always follow specifications. (*Briggs & Stratton Corp.*)

tappet clearance has been properly set, the valve cover plate can be replaced. The cylinder head and a new gasket are attached to the block. *Note:* Do not use sealing compound on the head gasket. Follow the tightening sequence and torque recommended for the cylinder head cap screws according to that specified by the manufacturer (Fig. 5-52 and Table 5-1).

### Installing the Magneto and Flywheel

Set the breaker points to the proper gap spacing. Clean the points by drawing a small strip of lintless paper between them. Replace the breaker point cover. When applicable, install a new flywheel shear key and install the flywheel. Use an appropriate flywheel holder and flywheel nut wrench to torque the flywheel nut to the recommended value. (See Table 5-1.) Complete the overhaul by replacing the blower housing, shrouds, and baffles.

## OPERATING THE REPAIRED ENGINE

Before attempting to start the small air-cooled gasoline or diesel engine, recheck to see that all the air shrouds are in place. Cooling air must flow over the cylinder to prevent the development of excessive heat on the new rings and the possibility of damaging the reconditioned cylinder.

### Starting the Engine

Before starting and running the engine, the experienced service person will check for the following items:

1. Engine oil of the proper SAE grade and quantity has been added to the crankcase of four-cycle engines. In two-cycle engines, the proper amount of oil has been mixed with the gasoline.
2. Static ignition or injection timing has been set.
3. The needle valves of the carburetor have been adjusted to the recommended initial setting and the governor linkage to the carburetor moves freely without binding or sticking through its entire range of movement.
4. The air cleaner has been cleaned and properly installed.
5. There are no unusual noises when the engine is cranked slowly.
6. The engine exhaust system is vented or the engine is operated in a well-ventilated area.
7. The engine is mounted to a run-in stand or is in the chassis of the machine so that it will not ''walk'' (move) around from the vibration that develops.
8. The engine speed control has been set to approximately midrange. The engine should run at about 2500 r/min at the start. This speed will assure that sufficient oil will be sprayed upon the various internal engine parts to carry off the heat of friction and lubricate the moving parts.

### Running the Engine

Start the engine and let it run at the governed speed identified in step 8 until it has reached operating tem-

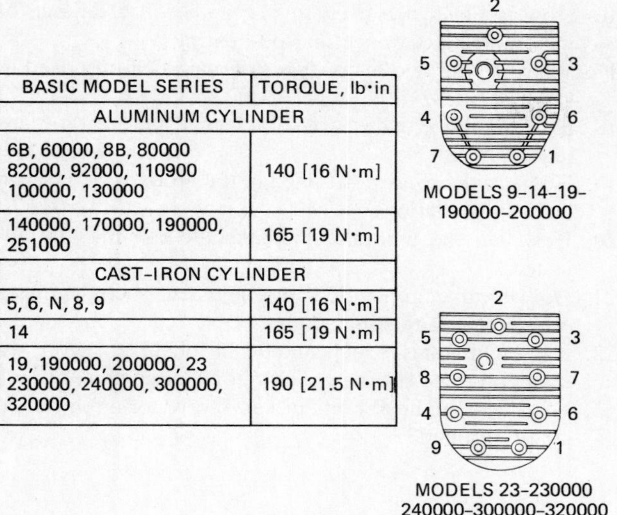

| BASIC MODEL SERIES | TORQUE, lb·in |
|---|---|
| ALUMINUM CYLINDER | |
| 6B, 60000, 8B, 80000 82000, 92000, 110900 100000, 130000 | 140 [16 N·m] |
| 140000, 170000, 190000, 251000 | 165 [19 N·m] |
| CAST-IRON CYLINDER | |
| 5, 6, N, 8, 9 | 140 [16 N·m] |
| 14 | 165 [19 N·m] |
| 19, 190000, 200000, 23 230000, 240000, 300000, 320000 | 190 [21.5 N·m] |



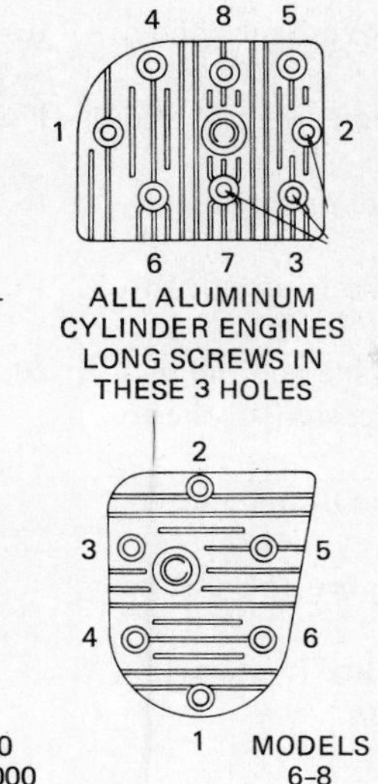


**Fig. 5-52.** Use the manufacturer's torque specifications and tightening sequence when installing head bolts. Given here are the specifications for Briggs & Stratton engines. Use the proper service manual for other manufacturers' specifications. (*Briggs & Stratton Corp.*)

perature. If the carburetor is equipped with an adjustable main needle valve, it may be necessary to make slight adjustments as described in Chap. 4 to get the engine to run smoothly. During the time that it is warming up, be alert for oil leaks or for unusual noises. When satisfied that the engine is in good operating condition, the engine can be stopped. When the engine is cool, the cylinder head cap screws are retorqued to specifications following the torque sequence as described earlier.

## Refurbishing the Engine

A small-engine service center not only will turn out a quality repair job, but will restore the engine to original appearance by painting it the proper color and applying a set of decals. This is an essential procedure in engine repair.

It may first be necessary to restore the appearance of air shrouds and other sheet metal parts of the engine so that when it is painted it will look like a working piece of equipment. Remove dents and grooves by carefully working the damaged spots with a rubber- or plastic-faced mallet and suitable backup blocks. The metal surface is smoothed with a power grinder-sander and then with metal finishing paper. Bare metal must be treated with a primer before application of the finishing coat. When dry, the entire engine is spray-painted to the desired color. When the paint is completely dry, the decals should be applied.

## Operating the Repaired Engine

A newly reconditioned engine should be "broken in," following the procedure used for new engines. Operate the engine on a reduced power setting of not more than ¾ full power for 3 to 4 hours (h). The load should be increased and decreased alternately so that a load requiring lugging power is limited. This type of service will permit the rings to seat to the cylinder walls properly.

After the first 5 h of operation, drain the engine oil and replace it with new oil of the grade and classification recommended by the manufacturer for temperature or season.

# SUMMARY

The repair of small engines requires the attention of skilled people. Although the repair operations are not difficult, special tools and dedicated people are required to perform all the repair operations successfully.

The small engine is the primary source of power of the commercial nursery, landscape, and groundskeeping businesses. Service and repair personnel who can analyze and economically evaluate the best engine repair procedures provide an important role for the business by permitting engines to maintain proper production schedules.

# THINKING IT THROUGH

1. What are the three essential requirements that must be satisfied before an engine will operate satisfactorily?
2. Which of the above requirements is most affected by the mechanical condition of the engine?
3. Why is a compression gauge not recommended to take the cylinder pressure of an "easy start" small engine?
4. List three types of tools that can be used to determine linear measurements and give an example of each.
5. What is the best method of determining the inside diameter of a crankshaft bearing?
6. List four factors which may affect the accuracy of torquing an engine cap screw to proper specifications.
7. If the torque wrench is calibrated in pound-inches, what should the reading be to obtain a torque of 30 lb·ft?

8. What is the difference between a "top overhaul" and a "major" engine repair?
9. What parts of the carburetor are inspected before and after disassembly?
10. What effects do worn valve guides have on engine operation?
11. How should the valve and valve seat be resurfaced to obtain a 1° "interference" angle?
12. Why must the crankcase cover be in place and the piston be at top dead center of the compression stroke to set the valve clearance?
13. What condition would prevent the piston from being removed from the top end of the engine cylinder?
14. List two conditions which would require the engine cylinder to be replaced or resurfaced.
15. Why must the new piston rings be checked in the cylinder and on the piston before installation?
16. How is Plastigage used to check the oil clearance of the crankshaft connecting rod bearing?
17. What are the two major types of main bearings used in small engines?
18. How should engine parts be cleaned before they are reassembled?
19. How are the rings positioned in the grooves of a piston before installation?
20. How can the amount of crankshaft end play be adjusted?
21. How is valve tappet clearance adjusted when the lifters do not have adjusting screws?
22. List eight items which should be checked before attempting to start a newly rebuilt small engine.
23. When are the head bolts or cap screws of a rebuilt engine retorqued?

# UNIT III POWER AND POWER TRANSMISSION COMPONENTS

| Competencies | Agricultural Machinery Service Manager | Agricultural Machinery Parts Person/ Manager | Agricultural Tractor and Machinery Mechanic | Agricultural Machinery Setup and Delivery Mechanic | Agricultural Tractor and Machinery Mechanic's Helper | Small-Engine Mechanic | Farmer, Grower, and Rancher |
|---|---|---|---|---|---|---|---|
| Apply and interpret specifications to power transmission components | Essential | Essential | Essential | Desirable | Desirable | Essential | Desirable |
| Maintain belts, chains, and U-joint drive | Desirable | Not necessary | Essential | Essential | Essential | Essential | Desirable |
| Install and adjust bearing and seals | Desirable | Not necessary | Essential | Essential | Essential | Essential | Desirable |
| Maintain gear transmiss-ions | Desirable | Not necessary | Essential | Essential | Essential | Desirable | Desirable |
| Service hydrostatic transmiss-ions | Desirable | Not necessary | Essential | Essential | Desirable | Desirable | Not necessary |
| Maintain hydraulic systems components | Desirable | Not necessary | Essential | Essential | Essential | Desirable | Desirable |

Essential    Desirable    Not necessary

Almost everyone that works with or uses agricultural machinery has a direct contact with power and the various forms in which it is transmitted. The mechanical methods used in power transmission applications are vital to machine operation. In baseball, the bat is a tool which permits the player to increase the strength of his body. The force developed when hitting the ball multiplies muscle power many times. Likewise, the power takeoff (PTO) shaft, which transmits power from the tractor engine to a hay baler, makes it possible to use the tractor as a power unit for the baler.

It is estimated that the mechanical power used in the United States' agricultural operations is approximately equivalent to all other forms of production industry. To gain an idea of what this represents in machinery, the current agricultural statistics reported 5,270,000 tractors and 3,179,000 trucks on farms in the United States. Each one of these machines is powered by an engine. In addition, thousands of other farm machines perform important services in agricultural production and processing operations.

The power requirements of these machines vary from those that are pulled or pushed by another force to those that are self-propelled. Self-propelled equipment requires specialized power transmission devices in order for the machine to be propelled in the field at speeds selected by the operator. In addition, a variety of power transmission systems are used to operate the cutting, elevating, and processing components of harvesting machinery. The operator also has controls which are power-assisted to guide, stop, and change ground travel speeds. For example, the operator mowing a golf green must have complete control of the forward travel of the machine, the speed of the cutting reel, and the height of cut if the proper shape and contour of the grass are to be obtained. These mechanical operations as well as others are accomplished by gears, belts, chains, rotating shafts, hydraulic pumps, motors, and cylinders. Bearings, clutches, brakes, and fluid piping may be an integral part of each system. The professional in power and machinery occupations must understand the relationships of machinery components to make effective use of the power that is developed.

# CHAPTER 6 BASIC PRINCIPLES OF POWER

## CHAPTER GOALS

In this chapter your goals are to:

- Identify the terms which are used to describe power and be able to explain them
- Describe the mechanical advantage effect of a lever and its influence upon torque
- Identify the relation which exists between developing torque and power
- Describe the method by which an engine develops torque and power
- Identify three terms used in agricultural power and machinery that describe power
- Describe the two ways by which agricultural tractors are rated for power and how each can be measured
- Interpret engine performance test data in accordance with power and torque requirements

A person who is preparing for a career in an agricultural industry—or any career for that matter—uses technical terms to describe accepted ideas and specific understandings. The terms permit the communication of specific meanings. When properly used, technical terms become a word picture of an idea, process, or action that may occur. As a result, when the written or spoken term is properly interpreted, accuracy of understanding will save the loss of valuable time and expense or will prevent the use of the wrong product. For example, a pharmacist must be able to fill a prescription accurately, or an individual's life may be endangered. A skilled welder needs to understand the classification of structural steel in order to apply the correct welding process. Likewise, workers in agricultural power and machinery use technical terms to describe and compare machines, to make the correct decisions, and to perform the correct operations. The parts manager of an agricultural machinery center must be able to interpret specifications to provide the correct replacement parts. The mechanic is involved with making decisions, taking measurements, making adjustments, or assembling parts in order to make corrections in the way a machine performs. Terms such as *force, torque, horsepower,* and *kilowatt* have specific meanings. These terms are tools, and like a socket wrench that is properly applied, they increase the precision of communication, and reduce the chances of making incorrect decisions or providing improper information. This chapter contains a number of technical terms that are universally used to describe power and its use. These terms will become a part of your everyday language as a professional in agricultural power and machinery.

## MECHANICAL POWER

As a worker in service and maintenance occupations in agricultural mechanics, you will be working with machines that produce power or require power to operate (Fig. 6-1). To be able to understand power and how it applies to agricultural machinery requires a knowledge of the terms used to define power. As a salesperson or service manager, you will be assisting

(a)

(b)

(c)

(d)

**Fig. 6-1.** Methods of power transmission: (*a*) pulling, (*b*) self-propelled, (*c*) hydraulic, and (*d*) power takeoff (PTO).

in the selection of the proper-size machine for the job. For this type of work you will need to interpret manufacturers' specifications and performance data. The literature you will be confronted with will present terms used to indicate measurements in both the United States system of units (U.S. Customary System) and in the International System (SI) of units (the modernized metric system). The U.S. system is being changed to the SI system for the purpose of unifying the measurement system internationally. In this text units of measure are given first in U.S. Customary units followed by SI units in brackets. Although it may be confusing to include two different systems of measurement, both are necessary during the transition if there is to be common understanding and communication.

## Force

For power to be developed there must be a force that exerts energy. Wind blowing a sailboat is an example of the action of force (Fig. 6-2). *Force* is energy that is exerted by one source (the wind) upon another (the boat sail) with a tendency to change the action of the one being acted upon.

The drawing in Fig. 6-3 illustrates an engine mechanic and helper lifting the cylinder head of an engine to place it on the workbench. The weight (called *mass*) of the head is 100 pounds (lb) [45.36 kilograms (kg)]. Weight expresses the mass which must be held by the workers to overcome the action of gravity acting upon the material.

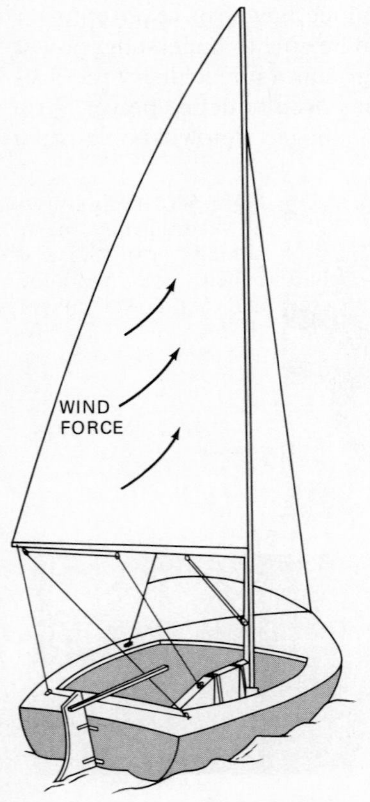


**Fig. 6-2.** Wind striking the sail creates a "force" to move the boat.

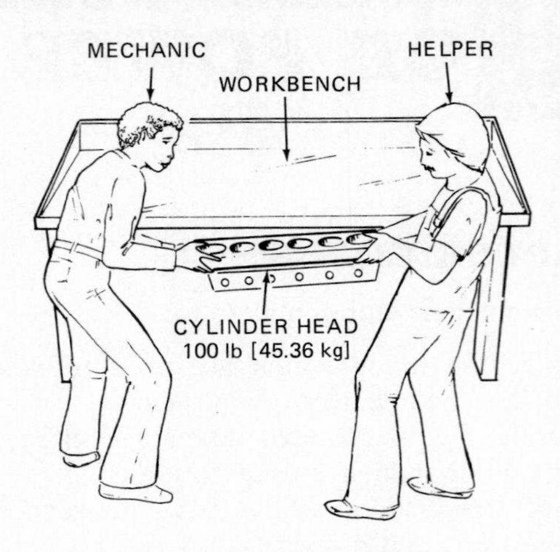


**Fig. 6-3.** The mechanic and helper lift an engine cylinder head onto a workbench. The workers must exert a force of 100 lb [45.36 kg] to overcome the weight caused by gravity acting upon the iron mass.

The force required by the workers to hold the cylinder head is expressed in the SI system as newtons (N). One newton is equal to 9.80665, or 9.81, kilograms of force. The force required to support the cylinder head is 45.36 kg times 9.81, or 445 N. There is *no motion* involved in the expression of force.

## Work

If the workers raise or lower the cylinder head, work is performed. *Work,* then, involves the action of force to move a body at rest. Force must actually move the object through a distance if work is to be performed. If the cylinder head is placed on the workbench or lowered slowly to the floor, the mass has been placed in motion, which results in work.

Work involves force and distance, both of which can be measured. In the U.S. system, work (energy) is measured as distance, in feet, times force, in pounds. The SI unit of measure for work is *joules* (J). One joule is equal to one newton-meter (N·m).

As an example, in the U.S. system, if the workers carry the 100-lb cylinder head 32.8 ft, they have performed the following work:

$$\begin{aligned}\text{Work} &= \text{distance (ft)} \times \text{force (lb)}\\ &= 32.8 \text{ ft} \times 100 \text{ lb}\\ &= 3280.0 \text{ foot-pound-force (ft·lbf)}\end{aligned}$$

In the SI system, work energy is expressed as

$$\text{Work (J)} = \text{force (N)} \times \text{distance (m)}$$

In the cylinder head problem, force is given in kilograms (45.36 kg), but the equation for work requires that force be expressed in newtons. To convert from kilograms to newtons, the equation for force is used:

$$\text{Force (N)} = 45.36 \text{ kg} \times 9.81 \text{ kilogram-force (kgf)} = 445 \text{ N}$$

Now the problem can be solved:

Work (J) = 445 N × 10.0 m
= 4450.0 J, or 4450.0 N·m

Force and work differ in that force is developed when pressure or stress is applied to a mass. The force of wind on the sail acts upon the boat, or the force of a bat hitting a ball develops a pressure or stress on the ball.

Work is accomplished when the force of the wind moves the boat through the water. The amount of work accomplished is the product of the force times the distance the boat is moved or the ball travels.

## Torque

The term *torque* is used to describe a rotating (twisting) force. It requires the use of one of our most basic machines, the lever. A common example of a lever is a wrench. The handle of the wrench extends the length of the worker's arm when force is applied to the tool. While work is used to describe the distance through which the force moves, torque identifies the amount of twisting (turning) effort that is applied by using a lever to achieve rotation. Torque does not produce work unless movement (rotation) is achieved. For example, if you wish to tighten a nut using a wrench, you must exert enough force on the wrench to turn the nut. The action results in work being performed. If you pull on the wrench without moving the nut, only torque is created (Fig. 6-4*a*).

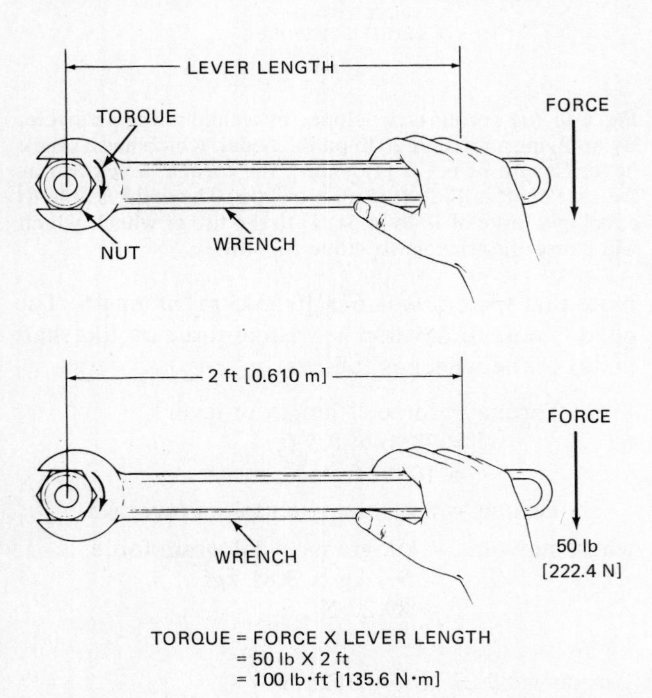


**Fig. 6-4.** (*a*) Torque is developed when a twisting action occurs. The wrench serves as a lever to increase the ability of the person to tighten the nut. (*b*) The amount of torque which is developed depends upon the force applied and the length of the lever.

Torque is equal to the force on the lever times the length of the lever:

Torque = force × length of lever

Figure 6-4*b* illustrates the use of units of force and lever length applied to the nut and wrench. If the wrench has an effective length of 2 ft [0.610 m] and the worker exerts a force of 50 lb [222.4 N] on the wrench, the amount of torque generated on turning the nut is

U.S.: Torque = force (lb) × length of lever (ft)
= 50 pound-force (lbf) × 2 ft
= 100 pound-force-feet (lbf·ft)

SI: Torque = 222.5 N × 0.610 m
= 135.7 N·m*

You will note that torque and work are calculated using the same units of measure: pounds [newtons] of force and length or distance in feet [meters]. The terms differ in that *work* expresses the distance through which the force moves, while *torque* identifies the importance of the length of the lever and the force it can produce. Notice that for both terms, the force in pounds [kilograms] is multiplied by distance in feet [meters]. In order to differentiate the two terms, work is written as foot-pound-force (ft·lbf), while torque is listed as pound-force-foot (lbf·ft). In the SI system, torque is expressed in newton-meters, while work is measured in joules.

Torque is influenced by the length of lever. Assume the mechanic and helper working on the engine cylinder head had access to an engine hoist. As illustrated in Fig. 6-5, force is applied to the handle of the wrench which has a crank lever (handle) length of 1 ft [0.305 m]. The lever transmits a twisting action on the drum, which has a radius of 3 inches (in) [0.0762 m]. The mechanic would need to exert a turning force on the crank handle to raise the 100-lb [45.4-kg] cylinder head. The amount of force required is as follows:

$$\text{U.S.: Force at lever} = \frac{\text{100 lb} \times \text{0.25-ft (3-in) drum}}{\text{1-ft crank lever}}$$
$$= \text{25 lb}$$

$$\text{SI: Force at lever} = \frac{\text{445 N} \times \text{0.0763-m drum}}{\text{0.305-m crank lever}}$$
$$= \frac{\text{111.32 (N)}}{\text{9.81 kgf/N}}$$
$$= \text{11.35 kg}$$

By using a mechanical device (the hoist), it is necessary for the mechanic to exert only 25 lb [11.35 kg] on the handle of the wrench to raise the 100-lb [45.4-kg] cylinder head. The force of 25 lb was multiplied four times by the mechanical advantage provided by

* *Note:* N·m = 1.357 × lbf·ft.

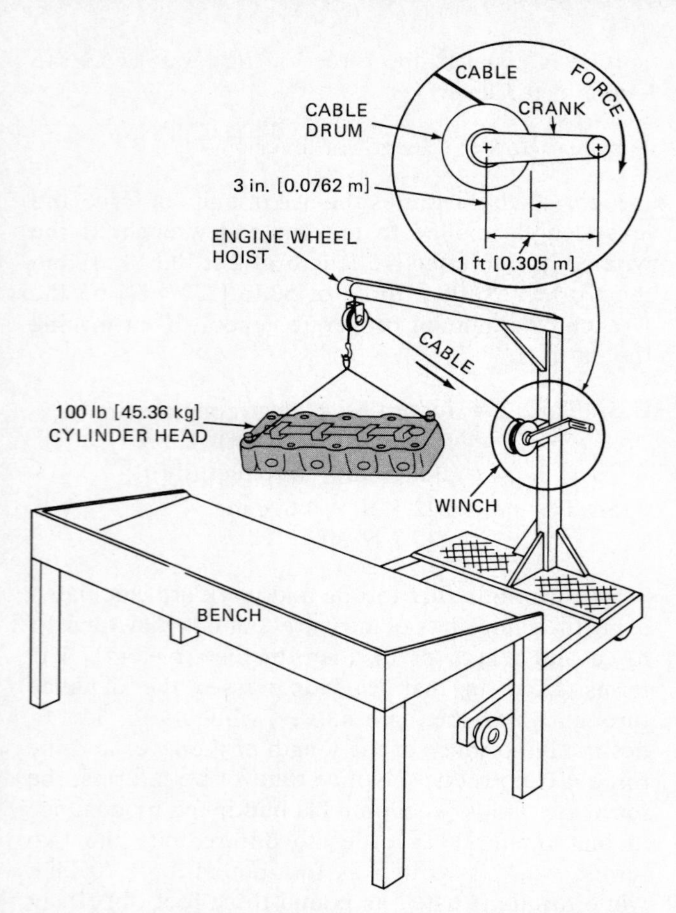


**Fig. 6-5.** Force acting against the crank handle of the winch attached to the engine hoist generates torque on the cable to lift the 100-lb [45.36-kg] cylinder head. The cable drum has a radius length of 3 in [76.2 cm].

the 1-ft crank lever and the 0.25-ft drum lever as follows:

Ratio of levers = 1-ft crank : 0.25-ft drum
= 4 : 1

Thus, the force required to lift the cylinder head is

$$\text{Force on handle} = \frac{100 \text{ lb } [45.36 \text{ kg}]}{4} = 25 \text{ lb } [11.34 \text{ kg}]$$

## Engine Torque

The previous examples illustrate the application of force used to develop torque and produce work. Much of the force which is used to operate agricultural machines is developed by the internal-combustion engine. The operating principle of this type of engine involves high pressure developed by the expansion of burning gases acting upon a piston contained within a closed cylinder. The force of the piston is transmitted to a crank (lever), thus developing a twisting motion (torque) on the crankshaft.

To illustrate the principles involved, assume a child riding a tricycle (Fig. 6-6*a*) is capable of applying 20 lb [9.1 kg] of force to one pedal of the tricycle.

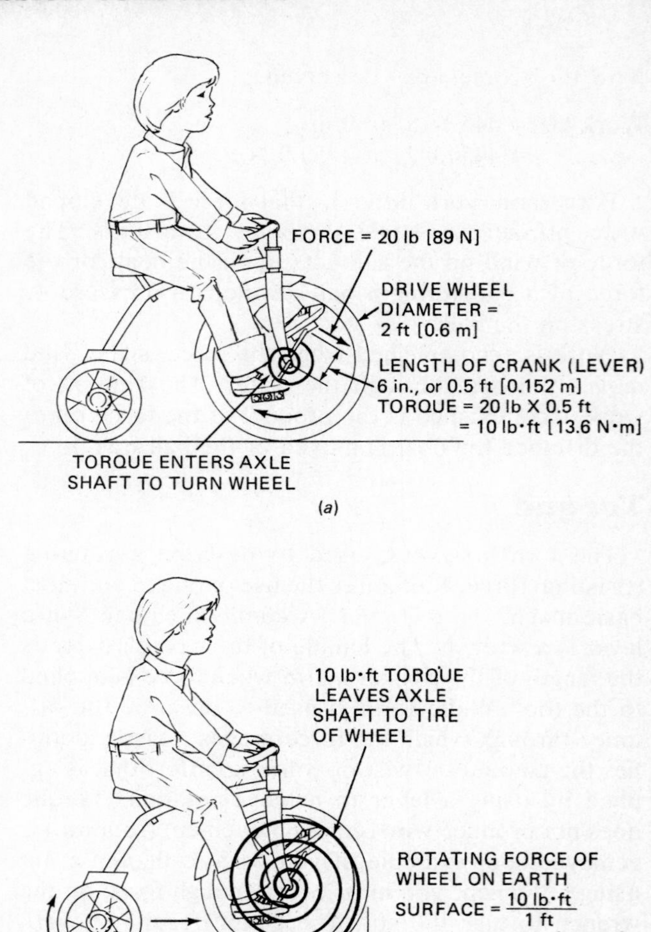


**Fig. 6-6.** (*a*) Torque is developed by a child riding a tricycle. By applying a force of 20 lb on the pedal, which has a crank (lever) length of 0.5 ft [152 mm], the torque developed at the axle shaft is 10 lb·ft [13.6 N·m]. (*b*) The child is able to develop a force of 10 lb [4.54 N] to the tire of wheel, which will cause the tricycle to move forward.

Note that the crank is 6 in [0.1525 m] in length. The child is able to develop a twisting force on the shaft (axle) of the wheel as follows:

U.S.: Torque = force × length of lever
= 20 lb × 0.5 ft
= 10 lbf·ft

SI: Torque = newtons × length of lever (m)

where newtons = kilograms × kilogram-force
= 9.1 kg × 9.81 kgf
= 89.27 N

Then

Torque = 89.27 N × 0.1525 m
= 13.6 N·m

In order for the child to make the tricycle move, the force from the pedal must be transmitted to the tire of the large drive wheel. For this to occur, the 10 pounds-feet (lb·ft) [13.6 N·m] of torque that have been transmitted from the pedal to the axle shaft of the wheel must now go from the axle shaft to the tire of the wheel. Because the wheel has a radius of 1 ft [0.305 m], it has a lever length of 1 ft (Fig. 6-6*b*). The amount of force the tire exerts against the surface upon which it rolls is as follows:

$$\text{Force} = \frac{\text{torque}}{\text{radius of wheel}} = \frac{10\ \text{lb·ft}}{1\ \text{ft}} = 10\ \text{lb}\ [4.54\ \text{kg}]$$

The piston, connecting rod, and crankshaft of an internal-combustion engine can be compared to the child riding the tricycle. Force from the child's leg moves the pedal attached to a crank (lever) which rotates the wheel shaft. In the engine, a connecting rod transmits the force from the piston to the crankshaft where torque is utilized (Fig. 6-7). Note that the piston exerts force on the throw of the crankshaft (the lever arm). Assume the stroke of the piston is 4 in [102 millimeters (mm)]. The length of the crankshaft throw is therefore 4 in [102 mm]/2 = 2 in [51 mm]. If the piston exerts 100 lb [45.4 kg] of force on the crankshaft, the torque developed is as follows:

$$\text{U.S.: Torque} = 100\ \text{lb} \times 0.167\ \text{ft (2 in)} = 16.7\ \text{lb·ft [200 pound-inches (lb·in)]}$$

$$\text{SI: Torque} = 45.4\ \text{kg} \times 9.81\ \text{kgf} \times 0.051\ \text{m} = 22.7\ \text{N·m}$$

Note that in the U.S. system, torque may be expressed as pound-feet or pound-inches. To convert pound-feet to pound-inches, multiply by 12. In the above example, a torque of 16.7 lb·ft would be converted to pounds-inches as follows:

$$16.7\ \text{lb·}\cancel{\text{ft}} \times \frac{12\ \text{in}}{1\ \cancel{\text{ft}}} = 200\ \text{lb·in}$$

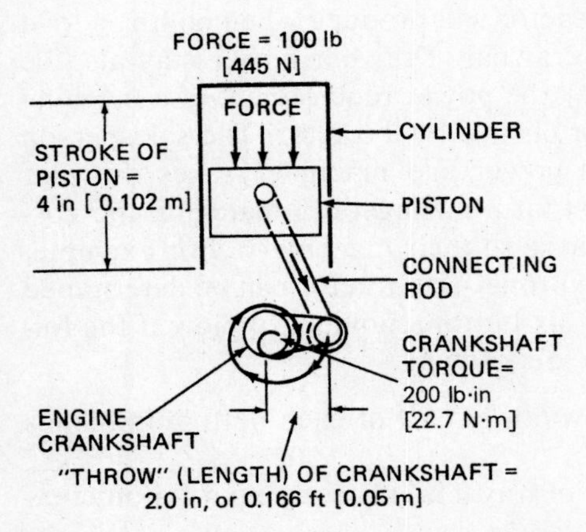


**Fig. 6-7.** The force of the piston, 100 lb [445 N], acts upon a 2-in [51 mm] lever (crankshaft), which produces a torque of 200 lb·in [22.7 N·m] at the engine's crankshaft.

The end product of the force developed by the combustion of gases in the engine's cylinder produced 16.7 lb·ft [22.7 N·m] of torque at the crankshaft. In this example the torque would be equivalent to that developed by a small lawn and garden tractor.

## Power

The term *power* is used to express the rate at which work is being performed. Therefore, power identifies the amount of work being accomplished in a given period of "time." The formula for expressing power is as follows:

$$\text{Power} = \frac{\text{work}}{\text{time}} \quad \text{or} \quad \frac{\text{force} \times \text{distance}}{\text{time}}$$

In the U.S. system, power is measured in foot-pounds per minute (ft·lb/min); that is,

$$\text{U.S.: Power} = \frac{\text{force (lb)} \times \text{distance (ft)}}{\text{time (min)}} = \frac{\text{ft·lb}}{\text{min}} \quad \text{or ft·lb/min}$$

In the International System, power is measured in watts. The term *watt* (W), named for the Scottish inventor, James Watt, is a very small unit of power in which work is done at the rate of one newton-meter per second, or:

$$1\ \text{w} = \frac{1\ \text{N (force)} \times 1\ \text{m (distance)}}{1\ \text{second (s) (time)}}$$

In the International System, power is expressed as follows:

$$\text{SI: Power (W)} = \frac{\text{force (N)} \times \text{distance (m)}}{\text{time (s)}}$$

For example, Fig. 6-8 illustrates a small tractor pulling a one-bottom moldboard plow. A force gauge is connected between the tractor and the plow. The plow creates a load on the tractor which is illustrated by a gauge reading of 1000 lb [454 kg]. The forward speed of the tractor and plow is 4 miles per hour (mi/h) [6.4 kilometers per hour (km/h)]. The amount

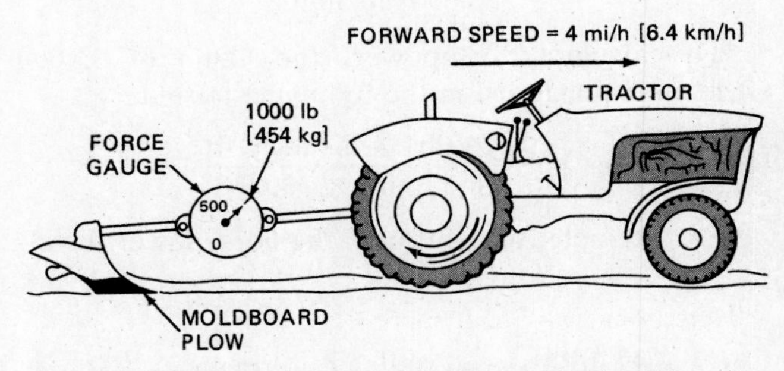


**Fig. 6-8.** A small tractor pulling a moldboard plow at a forward speed of 4 mi/h [6.4 km/h] develops a pull of 1000 lb [454 kg].

of power required for this operation is determined as follows:

$$\text{U.S.: Power} = \frac{\text{force (lb)} \times \text{distance (ft)}}{\text{time (min)}} = \frac{1000 \text{ lb} \times 352^* \text{ ft}}{1 \text{ min}} = 352{,}000 \text{ ft·lb/min}$$

$$\text{SI: Power (W)} = \frac{\text{force (N)} \times \text{distance (m)}}{\text{time (s)}} = \frac{454 \times 9.81 \times 1.79^* \text{ m}}{60 \text{ s}} = 7957 \text{ W}$$

If in the above problem the speed of plowing was reduced to 2 mi/hr [3.2 km/h], the amount of power required would be reduced by half as follows:

$$\text{U.S.: Power} = \frac{1000 \text{ lb} \times 352 \text{ ft}}{2 \text{ min}} = 176{,}000 \text{ ft·lbf/min [3980 W]}$$

Thus, the number of foot-pounds-force of power required at 2 mi/h is ½ that at 4 mi/h because it took twice as long (2 min) to do the same amount of work (force × distance).

**Horsepower** The term *horsepower* is a description of the amount of work that an average horse can produce in a specific time period. During the time that James Watt was developing the steam engine, he was concerned with rating the power which the engine developed. Because horses were a common draft animal, he developed a standard of measurement to describe the power of a horse—hence the term *horsepower*. The term is used only in the U.S. Customary system to designate power. Watt observed what he considered an average horse doing work and determined that it was able to move 330 lb a distance of 100 ft in 1 min. Thus

$$1 \text{ horsepower} = \frac{\text{force (330 lb)} \times \text{distance (100 ft)}}{\text{time (1 min)}} = \frac{33{,}000 \text{ ft·lb}}{1 \text{ min}} = 33{,}000 \text{ ft·lb/min}$$

To calculate horsepower, the value of 33,000 ft·lb/min is included in the following formula:

$$\text{Horsepower} = \frac{\text{force (lb)} \times \text{distance (ft)}}{\text{time (min)} \times 33{,}000}$$

For example, to determine the horsepower developed by the small tractor (Fig. 6-8) the observed data are inserted into the formula as follows:

$$\text{U.S.: Horsepower} = \frac{1000 \text{ lb} \times 352 \text{ ft}}{1 \text{ min} \times 33{,}000} = \frac{352{,}000 \text{ ft·lb}}{33{,}000 \text{ ft·lb/min}} = 10.67 \text{ hp}$$

The small tractor was able to develop 10.67 hp to pull the plow through the soil. The same problem can be solved in SI units, but first *horsepower* must be converted to an equivalent unit of measure in the SI system. This can be done using the following equation:

$$1 \text{ hp} = 0.746 \text{ kilowatts (kW)}$$

To determine the power in kilowatts that the small tractor developed, horsepower is multiplied by 0.746; thus:

$$\text{Power (W)} = \text{hp} \times 0.746 = 10.67 \times 0.746 = 7.96 \text{ kW}$$

**Types of Power** The professional in agricultural power and machinery not only must be able to understand how power is determined, but must be able to interpret technical data that identify the various types of power. In most cases, the term *power* will be modified by the specific type of power or power rating. The three types used to explain agricultural power are

1. Drawbar power
2. Power takeoff (PTO) power
3. Brake power

These terms may be expressed in horsepower (U.S. system) or kilowatts (SI system).

**Drawbar Power** *Drawbar power* refers to the power that a tractor will produce when pulling a load attached to a drawbar. Drawbar power may also be used to identify the power requirement of a machine being towed or pushed by a tractor. The salesperson in agricultural power and machinery uses drawbar power to assist farm managers in matching the correct power source to their machinery. For example, in Fig. 6-9, the proper-size tractor can be determined for pulling the six-bottom moldboard plow if the following factors are known:

1. The size (width of cut) of each bottom in inches [meters]
2. The speed of travel in miles per hour [kilometers per hour]
3. The depth of plowing in inches [millimeters]
4. The classification of soil (gumbo, clay, loam, etc.).

* Note: 4 mi/h = 352 ft/min

$$\frac{5280 \text{ ft/mi}}{60 \text{ min}} \times 4 = \frac{88 \text{ ft} \times 4}{1 \text{ min}} = 352 \text{ ft/min}$$

[107.3 m/min, or 1.79 m/sec]

**Fig. 6-9.** *Drawbar power* is the term used to express the power required to pull the six-bottom moldboard plow connected to the tractor's drawbar. (*International Harvester Co.*)

We will assume that correct plow speed is 4.5 mi/h [7.2 km/h], or 396 ft/min [121 m/min]. At a plowing depth of 8 in [203 mm], each bottom requires 120 lb [54.43 kg] of pull per 1 in [25.4 mm] of cut when working in loam soil. If each of the bottoms cut 14 in [356 mm], the total pull required for the six bottoms is

$$\text{Pull} = \text{total width of cut} \times \text{pull in lb/in}$$
$$= \frac{14 \text{ in}}{1 \text{ bottom}} \times 6 \text{ bottoms} \times \frac{120 \text{ lb pull}}{1 \text{ in of cut}}$$
$$= 10{,}080 \text{ lb } [4572 \text{ kg}]$$

The size tractor required to furnish the power requirements for the plowing operation is as follows:

$$\text{Power} = \frac{\text{force} \times \text{distance}}{\text{time}}$$

$$\text{U.S.: Horsepower} = \frac{\text{force (lb)} \times \text{distance (ft)}}{\text{time (min)} \times 33{,}000 \text{ ft}\cdot\text{lb/min}}$$

$$\text{Drawbar horsepower} = \frac{10{,}080 \text{ lb} \times 396 \text{ ft}}{1 \text{ min} \times 33{,}000 \text{ ft}\cdot\text{lb/min}}$$
$$= 120.96 \text{ dhp}$$

$$\text{SI: Drawbar power} = \frac{\text{force (N)} \times \text{distance (m)}}{\text{time (s)} \times 1000 \text{ W/kW}}$$
$$= \frac{4572 \text{ kg} \times 9.81 \text{ kgf} \times 121 \text{ m/min}}{60 \text{ s/min} \times 1000 \text{ W/kW}}$$
$$= \frac{44{,}891 \text{ N} \times 121 \text{ m/min}}{60 \times 1000}$$
$$= 90.5 \text{ kW}$$

**PTO Power** PTO power identifies the power a tractor develops measured at the power takeoff shaft (Fig. 6-10). A power takeoff shaft transmits power from the tractor PTO shaft to provide machinery with a rotating source of power.

The rotating speed of PTO drive shafts is standardized at 540 and 1000 r/min. Therefore, the power which is developed at the PTO is the result of measuring the torque transmitted from the engine through the PTO shaft when it is rotating at one of the standard speeds.

**Brake Power** Brake power is the power which an engine or motor delivers when measured at the output shaft. In a tractor, brake and PTO power are generally considered the same even though a very slight loss of power occurs when power is transmitted from the engine through the PTO shaft. Drawbar power is generally 15 percent less than brake or PTO power because of the losses involved with overcoming the rolling resistance of the wheels and transmission gears due to friction.

**Measuring Power** Both brake and PTO power are measured by an instrument called a *dynamometer*. The word *dynamometer* comes from *dynamo*, which means power, and *meter*, which is a measuring device. A dynamometer is capable of placing a load on a rotating output shaft in a controlled manner similar to applying the wheel brakes of an automobile—hence the term *brake* power. The instrument is capable of measuring the amount of torque which an engine or motor develops as follows:

$$\text{Torque} = \text{force} \times \text{length of lever}$$

A *prony brake* (Fig. 6-11) is one of the simplest forms of dynamometer. The instrument consists of a friction wheel which is attached to the drive shaft of

**Fig. 6-10.** The tractor engine provides the power to drive the mower through the PTO shaft. (*New Holland*)

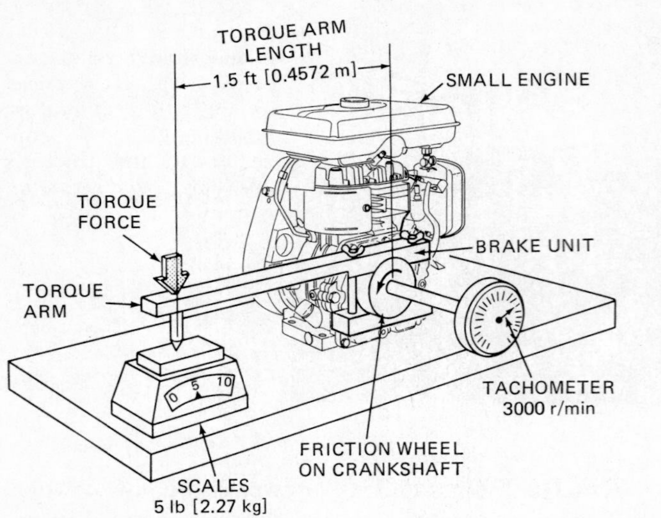


**Fig. 6-11.** A prony brake dynamometer attached to the crankshaft power output shaft (PTO).

the engine or motor. A braking unit contacts the friction wheel, which causes a force to be transmitted to a fixed-length torque arm. As power from the source rotates the friction wheel, the brake transmits the force through the braking unit to the torque arm. A scale measures the force produced and converts it into weight. In Fig. 6-11, the small engine converts heat energy into rotating force. The prony brake converts this force into torque measured in pound-force-feet or newton-meters, based upon the length of the torque arm in feet [meters] and the weight of the force in pounds [kilograms] as registered by the scales. The number of revolutions per minute which the crankshaft and friction wheel rotate is measured by a tool called a *tachometer*. Brake power developed by the engine can then be determined as follows:

$$\text{U.S.: Brake power (bhp)} = \frac{\text{force} \times \text{distance}}{33{,}000}$$

where

$$\text{Distance} = 2\pi L^* \times \text{revolutions per minute (r/min)}$$
$$= 2 \times 3.1416 \times L \times \text{r/min}$$

$$\text{Brake power (bhp)} = \frac{F \times 2 \times 3.1416 \times L \times \text{r/min}}{33{,}000}$$

or

$$= \frac{F \times L \times \text{r/min}}{5252}$$
$$= \frac{5 \text{ lb} \times 1.5 \text{ ft} \times 3000 \text{ r/min}}{5252}$$
$$= 4.3 \text{ bhp}$$

* If free to rotate, the torque arm would cover a distance equal to the circumference of a circle. Thus, $2\pi L$ is used to determine the distance in feet (meters) per revolution when $L$, the length of the torque arm, is the radius of the circle circumscribed.

$$\text{SI: Brake power (kW)} = \frac{F(\text{N}) \times \text{distance}}{\text{s/min} \times 1000}$$

where

Force = N

$$\text{Distance} = 2\pi L \times \text{r/min}$$
$$= 2 \times 3.1416 \times L(\text{m}) \times \text{r/min}$$

Brake power (kW)

$$= \frac{\text{force (N)} \times 2 \times 3.1416 \times L\text{ (m)} \times \text{r/min}}{60 \times 1000}$$
$$= \frac{F(\text{N}) \times L\text{ (m)} \times \text{r/min}}{9549}$$
$$= \frac{2.27 \text{ kg} \times 9.81 \text{ kgf} \times 0.4575 \text{ m} \times 3000 \text{ r/min}}{9549}$$
$$= 3.2 \text{ kW}$$

A simple prony brake dynamometer of the type illustrated in Fig. 6-11 has limited use in small engines and electric motors because there is no effective way to reduce the heat buildup in the friction wheel and brake. For this reason mechanics who service small engines and do tuneups will use hydraulic-type dynamometers (Fig. 6-12). This type of dynamometer incorporates the principles of the prony brake by measuring torque, but it can be operated for longer periods and has greater accuracy in measuring brake power.

The PTO dynamometer (Fig. 6-13) is one of the most useful tools found in agricultural power and machinery service centers. As a technician you will need to be skilled in the use of this device and have an understanding of how it operates and measures power. By applying proper testing procedures, the performance characteristics of a tractor can be evaluated, and operational problems can be identified. From this information the technician will be able to make decisions to correct inefficient or improper engine operation. For example, one of the service practices in a modern agricultural machinery dealership is to check the performance of a tractor following service, maintenance, or repair operations. The professional in power and machinery will be able not only to perform these tests but to evaluate the results.

As previously stated, the PTO dynamometer is especially adapted to measuring the engine power of agricultural tractors. This is because the engine can transmit power to the PTO shaft without having to propel the wheels. Thus, the power output at the PTO shaft of the tractor is representative of the power developed by the engine. For this reason agricultural tractor manufacturers list PTO power (in horsepower and kilowatts) as an *indicator of engine performance,* and they list drawbar power (in horsepower and kilowatts) as the *pulling performance* of the tractor.

It is necessary for a technician preparing to use a

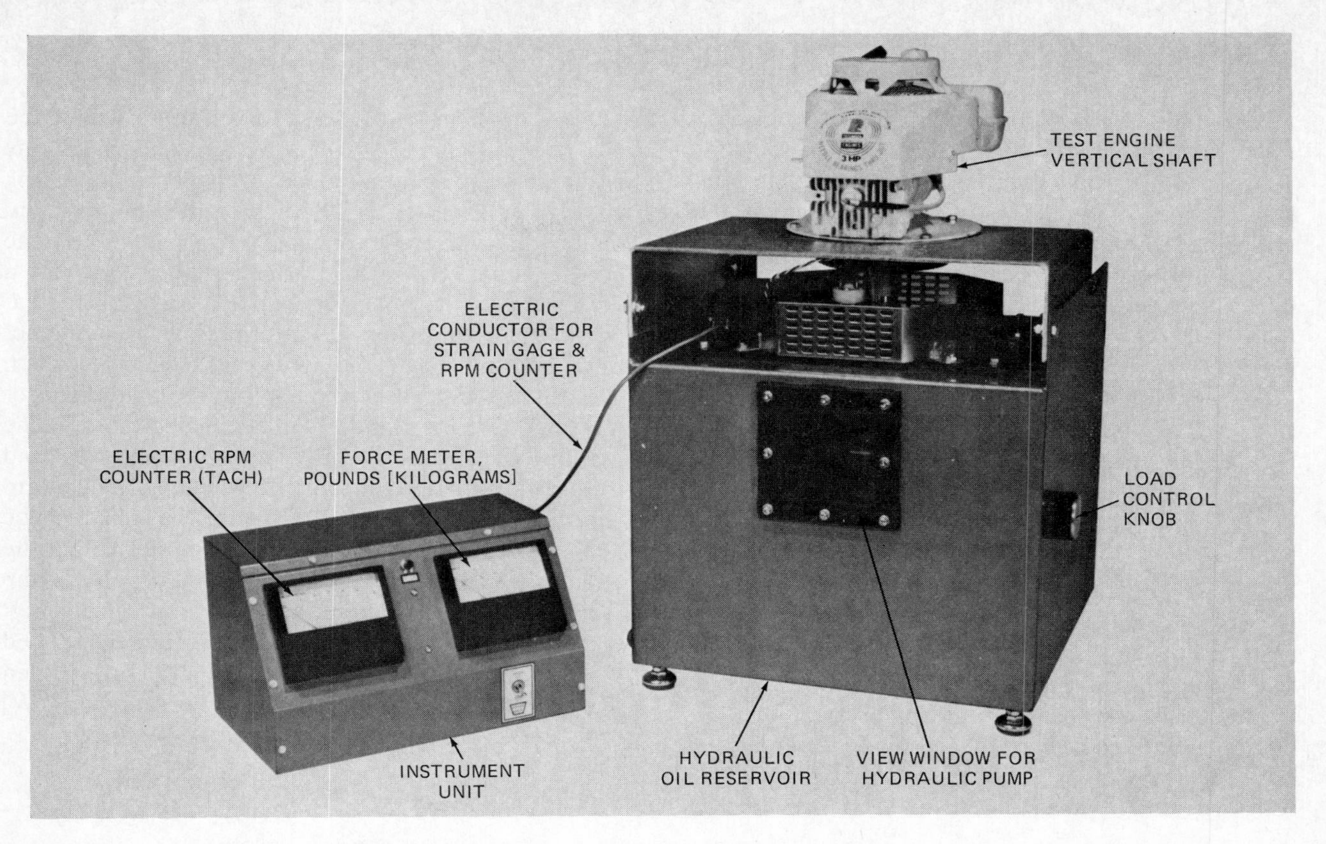


**Fig. 6-12.** A hydraulic dynamometer is used to measure the torque developed by a small engine when operating at a specified speed (rpm).

**Fig. 6-13.** The PTO dynamometer attaches to the PTO shaft of an agricultural tractor. This tool permits the technician to adjust the amount of load placed on the engine and to measure the power developed and transmitted to the PTO. (*M & W Gear*)

PTO dynamometer to follow carefully the manufacturer's instructions to ensure its proper use. Chapter 12 will provide you with information on how to utilize the PTO dynamometer to make engine adjustments when maintaining or repairing tractor engines. Chapter 26 shows how to use dynamometer tests when comparing new tractors.

## INTERPRETATION OF DATA

As a professional in power and machinery, you will find it necessary at various times to interpret the literature provided by tractor and engine manufacturers. In one instance, a salesperson must interpret information on engines when assisting a customer in the selection of the proper size and type of engine as a replacement engine for a PTO-driven hay baler. In another, the technician will operate a reconditioned engine on a dynamometer to determine if it performs according to the manufacturer's specifications. For example, the sales literature provides performance data for a model CCKA engine (Fig. 6-14) with the following design specifications:

Type: spark-ignition
No. of cylinders: 2, horizontally opposed
Cycles: 4
Cylinder bore: 3.25 in [82.44 mm]
Piston stroke: 3.0 in [76.2 mm]
Piston displacement: 49.8 in$^3$ [816.2 cm$^3$]
Compression ratio: 7.0 : 1
Brake power, maximum rated rpm: 16.5 bhp [12.30 kW] at 3600 r/min

These data are used to give information regarding the size of the engine. Total piston displacement and the brake power output are closely related in identifying the size and power capabilities of the engine. Generally, the greater the displacement, the greater the power output. However, these data do not provide a very graphic picture of the engine's performance capabilities. To provide this information the manufacturer gives torque and power curves (Fig. 6-15) which illustrate tests of performance that the manufacturer has conducted on this particular engine model. The graph and curves identify maximum power and torque developed by the engine when operated at engine speeds of 1800 to 4200 r/min. Note that in the above specifications the manufacturer stated that the engine would produce a maximum of 16.5 bhp [12.3 kW] when operating at 3600 r/min. From Fig. 6-16*a* it can be determined that the torque developed was approximately 24.0 lbf·ft [32.5 N·m] when operating at 3600 r/min. The brake horsepower [kW] produced can be determined as follows:

$$\text{Brake horsepower} = \frac{\text{torque, lbf·ft} \times \text{r/min}}{5252}$$

$$= \frac{24.0 \times 3600}{5252}$$

$$= 16.5 \text{ bhp}$$

$$\text{Brake power (kW)} = 16.5 \text{ bhp} \times 0.746$$

$$= 12.3 \text{ kW}$$

The brake power curve is therefore developed to express the product of the torque times the rpm divided by 5252. When the rpm is reduced to 2000

**Fig. 6-14.** CCKA model, two-cylinder engine. (*Onan Corp.*)

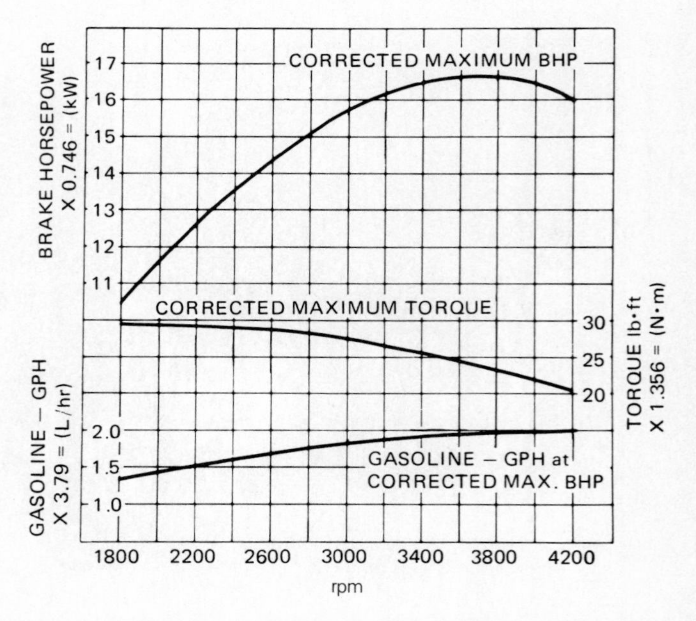


**Fig. 6-15.** Performance test data provided by the engine manufacturer include brake power, engine torque, and fuel consumption at varying rpm. (*Onan Corp.*)

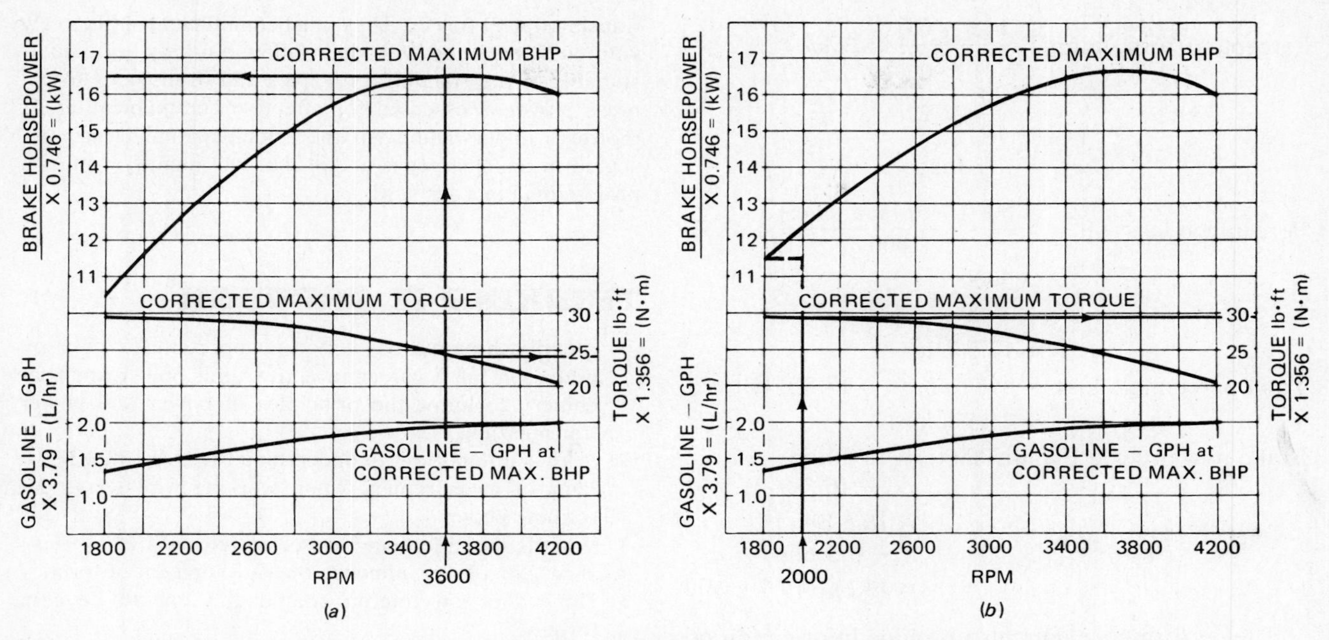


**Fig. 6-16.** Engine performance curves for (*a*) brake horsepower [kilowatts] and (*b*) torque in pound-force feet [Newton meters]. Maximum brake horsepower [kilowatts] is obtained at 3600 r/min, while maximum torque is produced at 1800 to 2000 r/min. Note that as the engine speed increases, the brake horsepower increases and torque decreases. At 3800 to 4200 r/min, an increase in friction with a decrease in torque causes the brake horsepower [kW] to decline.

(Fig. 6-16*b*), the engine will produce maximum torque (29.5 lbf·ft) but only 11.5 bhp [8.6 kW]. Most manufacturers specify that an engine should not operate at more than 75 percent of the maximum brake power for continuous duty. Therefore, this engine should not be used to operate equipment requiring more than 12.0 bhp [9.0 kW] (Fig. 6-17*a*).

The temperature and altitude at which the engine is to operate affect the power the engine will develop. Engine power decreases approximately 2 percent for each 10 degrees Fahrenheit (°F) [5.5 degrees Celsius (°C)] above 60°F [16°C] and 3.5 percent for each 1000 ft [305 m] above sea level. For example, if the engine is to operate in conditions of 80°F [26.7°C] and 5000 ft [1525 m] elevation, the loss of maximum brake power at 3600 r/min would be

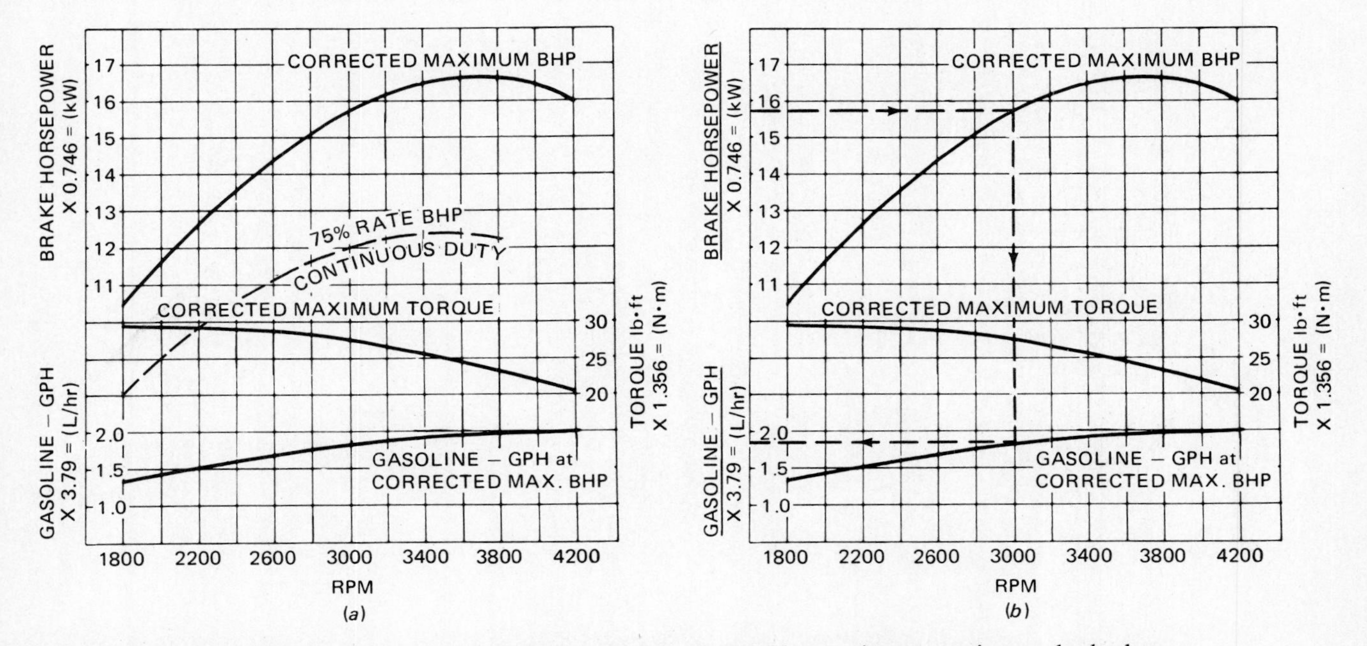


**Fig. 6-17.** Engine performance curves (continued from Fig. 6-16). The maximum continuous brake horsepower curve (*a*) is based on 75 percent of the maximum rate brake horsepower. Consumption of gasoline (*b*) is given in gallons per hour [liters per hour] at maximum brake horsepower from 1800 to 4200 r/min.

$$\text{Percentage loss, temp} = \frac{80° - 60°}{10°} \times 2\%$$

$$= \frac{20°}{10°} \times 2\%$$

$$= 4\% \text{ loss}$$

$$\text{Percentage loss, alt} = \frac{5000 - 0 \text{ (sea level)}}{1000} \times 3.5\%$$

$$= \frac{5000}{1000} \times 3.5\% = 17.5$$

$$= 17.5\% \text{ loss}$$

$$\text{Total percentage loss} = 4\% \text{ (temp)} + 17.5\% \text{ (alt)}$$

$$= 21.5\% \text{ loss}$$

$$\text{Brake power @ 80°F and 5000 ft} = 16.5 \text{ bhp} - (16.5 \times 21.5\%)$$

$$= 16.5 \text{ bhp} - 3.5 \text{ bhp}$$

$$= 13 \text{ bhp [9.7 kW]}$$

The performance data also provide information on the expected fuel consumption of the engine when it is operating under maximum loaded conditions. For example, when the engine is developing approximately 15.8 bhp [11.8 kW] running at 3000 r/min, the consumption of gasoline will be approximately 1.7 gallons per hour (gal/h) [6.4 liters per hour (L/h)] (Fig. 6-17*b*). This information is useful in estimating fuel cost per hour of operation.

## SUMMARY

Service and salespeople in agricultural power and machinery are constantly involved with the development and transmission of power. They must communicate effectively with customers and with their fellow workers. An understanding of the standard principles and terms used to express power is essential for effective communication. A thorough understanding of the principles and terms presented in this chapter is a skill that will identify you as a professional in your work.

## THINKING IT THROUGH

1. Identify three reasons why it is important for a person preparing for a career in agricultural power and machinery to know the principles of power and power transmission.
2. Why is it important to understand the U.S. and SI systems of measurement when working with terms describing power?
3. What is the difference between *force* and *work?*
4. How can a lever influence the development of torque?
5. How does an internal-combustion engine develop torque?
6. What is the difference between the terms *torque* and *power?*
7. What is the meaning of each of the following forms of power?
   (a) Drawbar
   (b) PTO
   (c) Brake
8. Determine the drawbar power of a tractor that pulls 2000 lb [907 kg] and travels 4 mi/h [6.44 km/h].
9. A tractor develops a torque of 200 lbf·ft [2.71 N·m] on a PTO dynamometer. If the PTO shaft turns at 1000 r/min, how much power is the tractor producing?
10. Why is it important to be able to interpret manufacturers' performance test data?

CHAPTER 7

# CLUTCHES, TRANSMISSIONS, AND FINAL DRIVES

## CHAPTER GOALS

In this chapter your goals are to:

- Describe the components of the power train
- Identify the types and purposes of gears
- Identify the types and purposes of bearings and seals
- Identify the types and characteristics of clutches
- Identify the types and characteristics of transmissions
- Identify the types and characteristics of final drives

Throughout history machines were developed to increase the capacity of people to do work. Modern agricultural machinery is an extension of the early machines that were used to till the soil and harvest the food and fiber-producing crops. Some of the greatest advances in machines have been in the development of efficient methods of transmitting power. For example, in the early tractors as much as one-half of the power developed by the engine was lost in transmitting it to the drive wheels. In comparison, the modern tractor is capable of transmitting nearly 85 percent of the engine's power to useful work. These advances in efficiency have been made possible by refinements and new developments in methods of transmitting power.

*Clutch, transmission,* and *final drive* are terms which describe various components in transmitting power. They are a necessary part of many machines but are especially descriptive of the power transmission units found in tractors and other self-propelled equipment. When these units are combined, they form the system for transmitting power. Because of the complexity of the equipment, skilled workers are

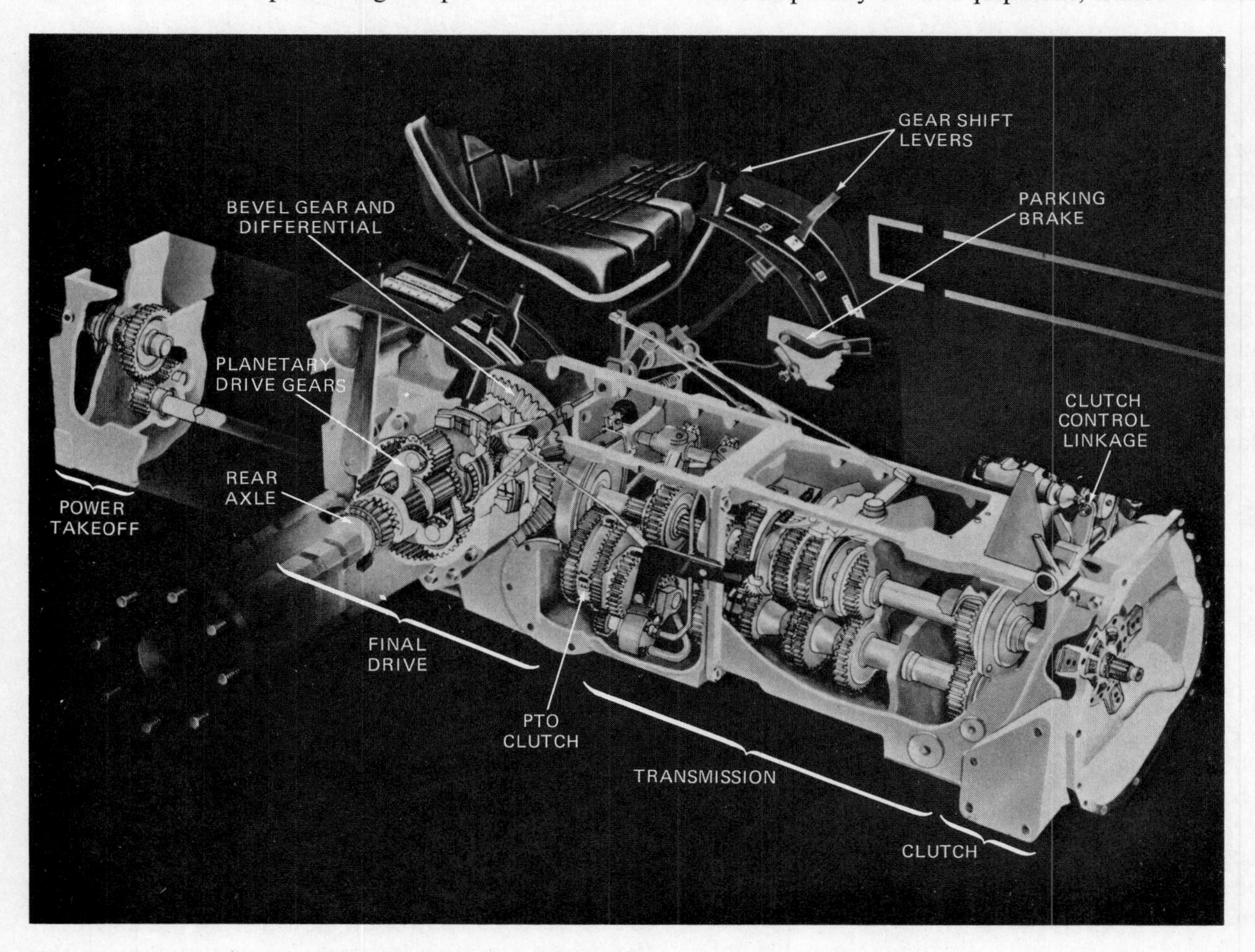


**Fig. 7-1.** Power train of a farm tractor. (*International Harvester Co.*)

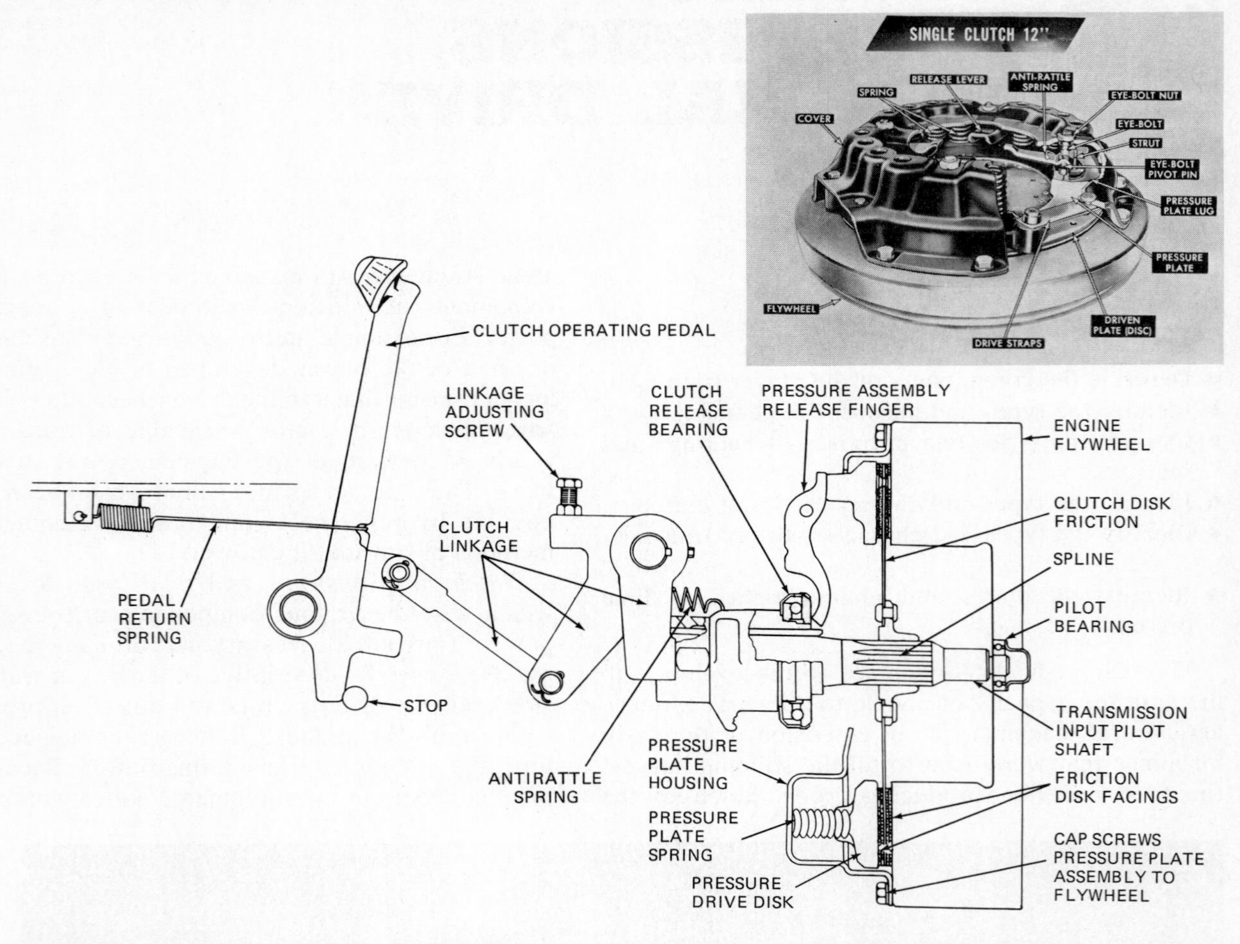


**Fig. 7-2.** (*a*) Parts of a typical single dry-plate disk clutch. (*b*) Linkage arrangement when the clutch is in the engaged position. (*Ford Tractor Operations and Deere & Co.*)

required to service and maintain the components of most machinery.

Upon completion of this unit you should understand the principles of clutches, transmissions, and final drives and the characteristics of servicing and maintaining the power train system.

## THE POWER TRAIN

Power from the engine is transmitted to the drive wheels or output shaft of a machine by means of the power train. In order for the transfer of power to take place properly, the operator uses the following parts of the power train to control the machine's functions:

1. A *clutch* is used to connect or disconnect power when starting or stopping the movement of drive wheels or power shaft.
2. A *transmission* provides the means for selecting the direction of travel, the speed ratio for operating speed, and the torque output to the drive wheels or power shaft.
3. A *final drive assembly* distributes equal power flow to the drive wheels when turning and provides the ideal speed ratio and torque transfer from the transmission to the drive wheels.
4. A *power takeoff* (PTO) shaft transfers power to operate attached or towed equipment.

The relationship of these functions is illustrated in the power train of a farm tractor (Fig. 7-1 on page 113). First, the crankshaft of the engine is separated from the transmission by the clutch. When the operator needs to connect or disconnect the engine power from the rest of the power train, the clutch is operated. The clutch is designed to permit smooth and positive release or engagement of engine power to the gears of the transmission. A shifting lever for the transmission allows the operator to select the desired gear for forward or reverse travel. The drive shaft of the transmission is connected to the final drive gear assemblies to drive the rear wheels through the axles. The PTO shaft is housed within the transmission and final drive.

### Clutches

Two basic kinds of clutches are used in agricultural machinery: the power transmission clutch and the safety clutch. The *power transmission clutch* allows the operator to control the power from the engine to the driven machine by engaging or disengaging the clutch. *Safety clutches* are used to protect parts of the machine from damage due to excessive loads or operating conditions. The power transmission clutches discussed in this section are the disk, electric, belt, and centrifugal types, and the safety clutches are the overrunning, disk, and shear-pin types.

**Disk Clutches** The disk-type clutch is used on farm tractors and trucks as well as other types of self-propelled agricultural machinery which are equipped with shift-type transmissions. The disk clutch is made up of four basic parts (Fig. 7-2). These parts are the engine flywheel, the friction disk, the pressure plate, and the release bearing and linkage. The flywheel is attached directly to the engine crankshaft. It provides the surface from which the clutch can receive torque to drive the transmission. The disk has specially constructed friction faces attached to both surfaces. Some clutch designs use two or more friction disks to transmit power. The pressure plate is bolted to and rotates with the flywheel. In operation, the friction disk is tightly clamped between the machined facings of the flywheel and the pressure plate by spring tension. The friction disk is spline-fitted to the transmission power shaft.

When the clutch pedal is depressed (Fig. 7-3), the clutch release bearing is forced against the operating levers. This action causes the levers to apply lifting force to the pressure plate and compress the clutch springs. As a result, the clamping force on the friction disk is released and the flow of power is no longer transferred from the flywheel and pressure plate to the friction disk.

To engage the clutch, the clutch pedal is slowly released. This moves the release bearing rearward and relieves pressure on the operating levers. The compressed springs force the pressure plate toward the flywheel and clamp the friction disk between the machined surfaces of the pressure plate and the flywheel. If the operator moves the pedal slowly, the power is transferred to the friction disk smoothly. To prevent the release bearing from constantly running against the release levers when engaged, the linkage is adjusted to provide a specified amount of *free pedal* travel (Fig. 7-3). As the clutch disk facings wear from long service, the disk becomes thinner and the pressure plate moves closer to the flywheel. The result is a reduced amount of free pedal travel. If the travel is reduced to the point where the clutch release bearing contacts the clutch release fingers, the disk may begin to slip because plate pressure on the friction disk will be reduced.

Unless the friction disk is designed to run in oil, lubricants *must not* reach the friction surface of the clutch disk. For this reason, the pilot bearing and the clutch release bearing are generally sealed units

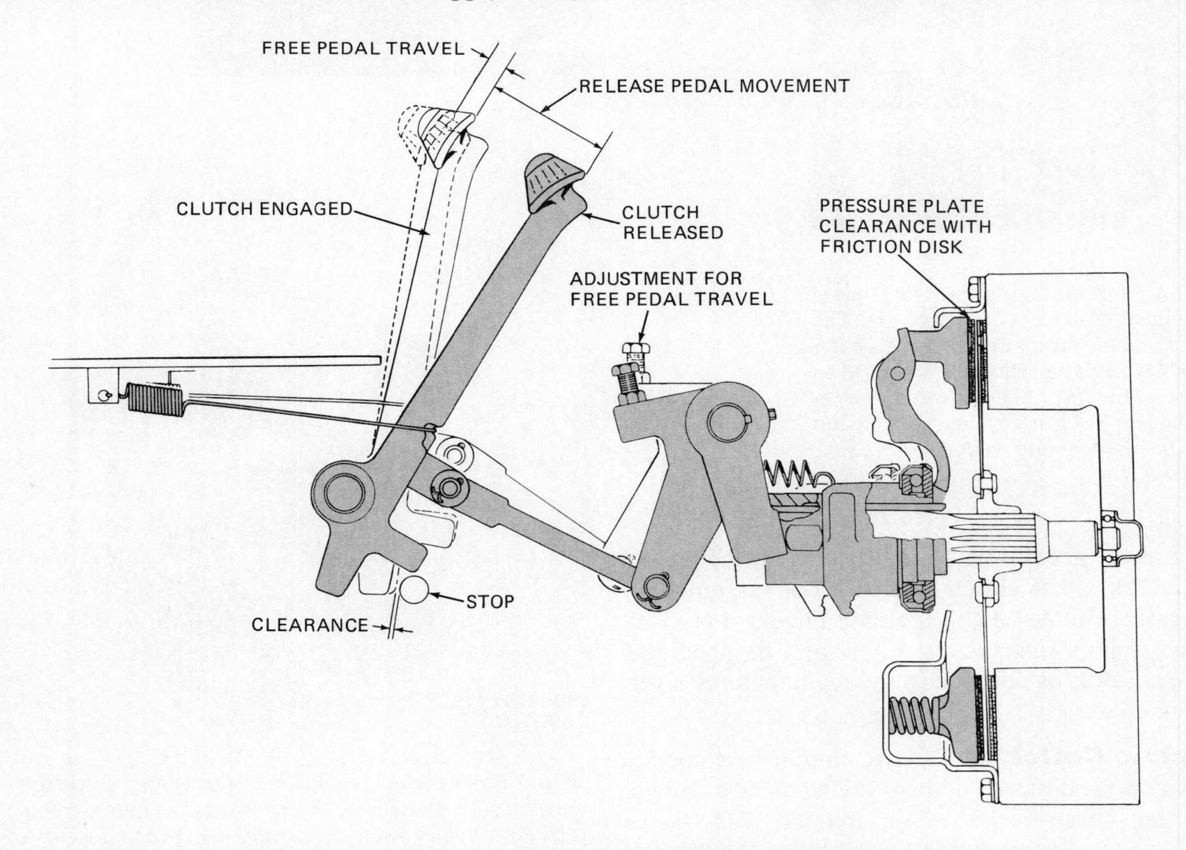


**Fig. 7-3.** Sectional view of a disk clutch in the disengaged position. Note that a disk clutch must be provided with the correct free pedal travel to prevent excessive wear to the friction disk and release bearing.

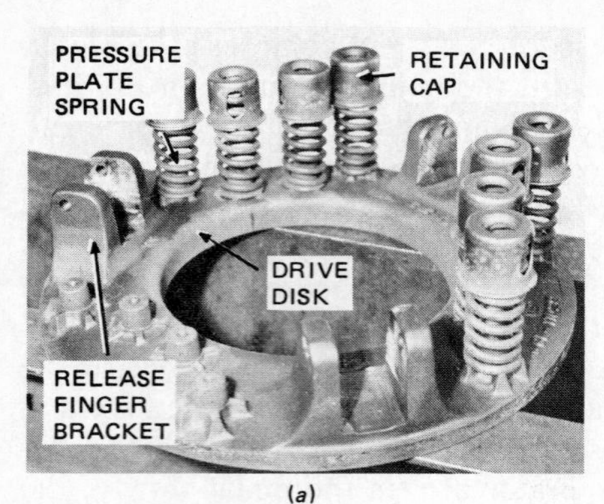


(a)

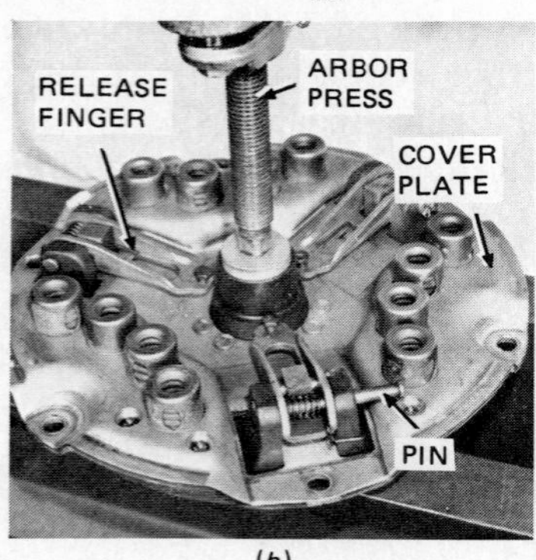


(b)

(c)

**Fig. 7-4.** Assembling a pressure plate and adjusting release fingers. (*a*) The pressure plate springs are installed on the drive disk. (*b*) The arbor press is used to compress the cover plate to install the release fingers. (*c*) Finger height is adjusted with the appropriate gauge. (*Massey-Ferguson, Inc.*)

which never need to be lubricated. The principles of overhauling a disk clutch involve proper assembly of the pressure plate and the release fingers (Fig. 7-4). Always follow instructions involving the specifications and tools as outlined in the manufacturer's service manuals.

**Electric Clutches** Electric clutches are used in special applications to control the transfer of power from the source to the driven unit. A small tractor may have an electric clutch connected to the drive of the power takeoff shaft (Fig. 7-5). An electric clutch

(a)

(b)

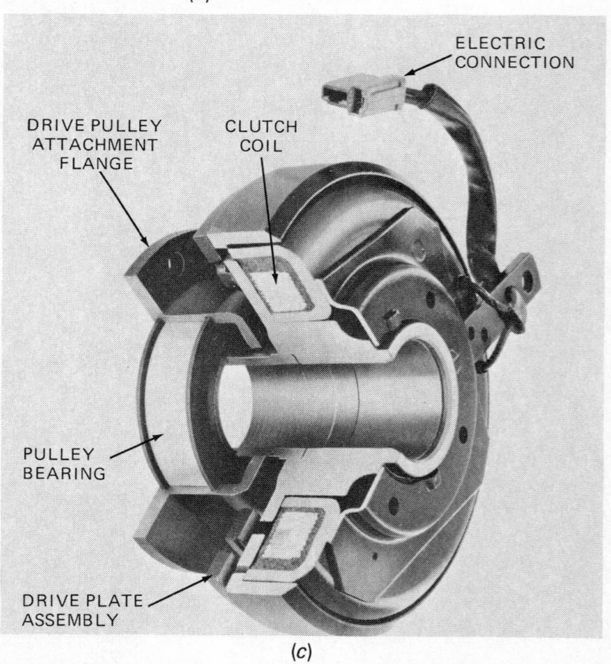


(c)

**Fig. 7-5.** The electric clutch controls the belt drive of a small tractor to operate the PTO-driven equipment. (*a*) Belt drive on small tractor. (*Ariens Co.*) (*b*) Closeup of belt drive. (*Ariens Co.*) (*c*) Sections of actual clutch. (*Warner Electric Brake and Clutch Co.*)

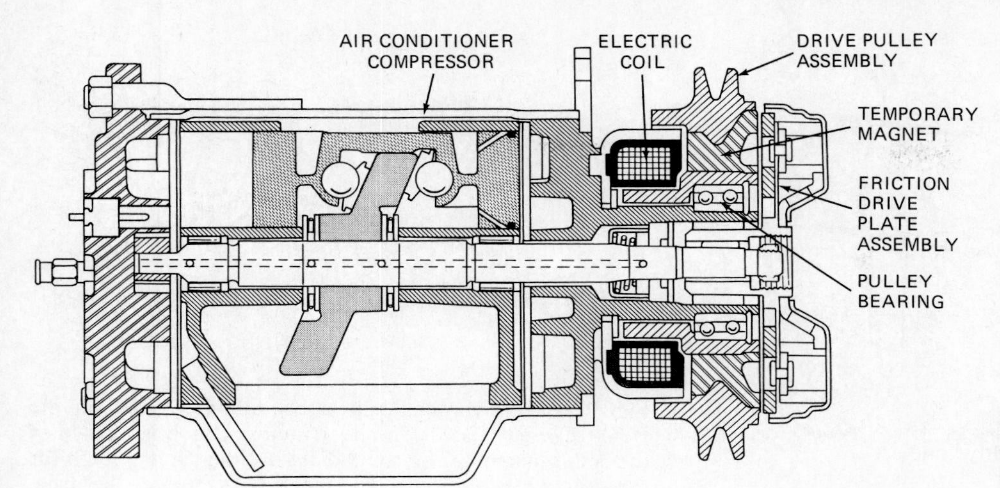


**Fig. 7-6.** The electric clutch is used on air-conditioning compressors. (*Deere & Co.*)

is used on air-conditioning compressor drives (Fig. 7-6). In this type of clutch, the engagement is very positive when the operating switch is activated. Closing the switch causes an electric current to flow through the field coils within the clutch assembly. The current sets up a magnetic field which pulls the clutch shoe assembly against the facing on the drive pulley assembly (Fig. 7-7). Power is then transmitted to the drive shaft. When the switch is opened (off), the magnetic field is broken, which allows the pulley assembly to turn freely on the hub bearing.

**Belt Clutches** Many agricultural machines use belt clutches to engage and disengage the power source. Generally, belt clutches are used in low-horsepower applications. The belt clutch is operated by tightening or loosening the tension on the V belt which transmits the torque through friction on the pulleys. The small tractor clutch illustrated in Fig. 7-8 uses a belt clutch. The idler and belt tightener pulley is operated by the foot pedal. When the foot pedal is depressed, it raises the pulley which reduces the belt tension and relieves the drive friction. When the clutch pedal is raised, the tensioning spring forces the idler and belt tightener pulley to tighten the belt so that power is transmitted to the transmission input pulley. Belt guides are necessary to maintain the belt position and relieve belt friction when the belt tension is removed.

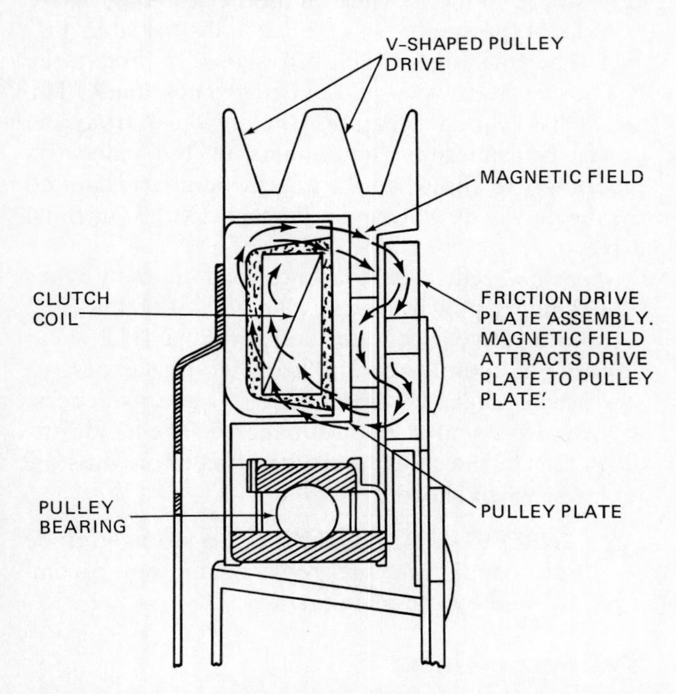


**Fig. 7-7.** The electric clutch operates from magnetic attraction created by the field coil. The friction drive clutch is attracted to the drive plate and held in contact by magnetic attraction.

A belt clutch is also used for controlling the implement drive in Fig. 7-9. When the operating lever is moved to the position illustrated it tightens the belt. Continued forward movement locks the lever into an overcenter position. This results in the compression spring maintaining constant tension on the belt. Moving the lever to the rear releases the spring tension. This also reduces the belt-pulley friction, which stops the power transfer to the implement.

**Centrifugal Clutches** The chain saw (Fig. 7-10) and the small engine are two examples of machines which use a centrifugal clutch. This type of clutch is automatic in operation. It is usually mounted directly on the crankshaft of the engine.

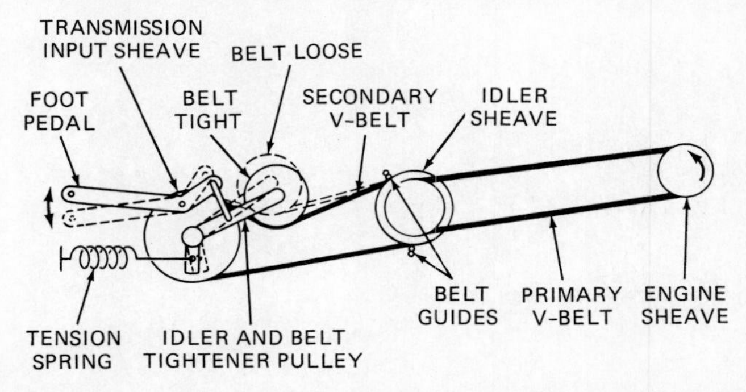


**Fig. 7-8.** The belt clutch of a small tractor. Tightening or loosening the belt causes clutch action.

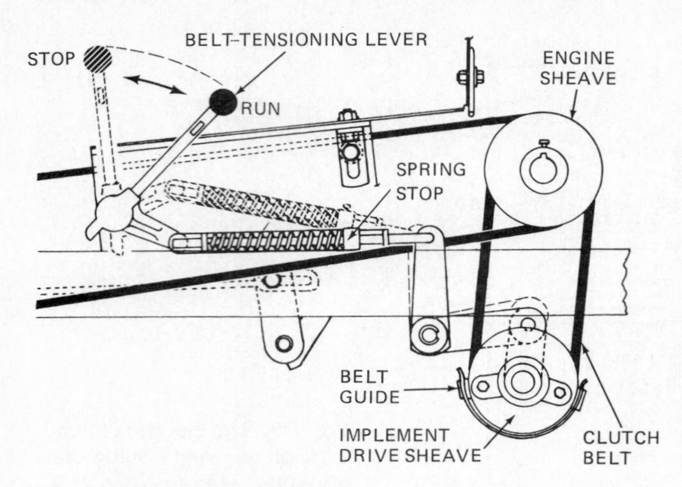


**Fig. 7-9.** The belt clutch of an implement drive and power train. (*Massey-Ferguson, Inc.*)

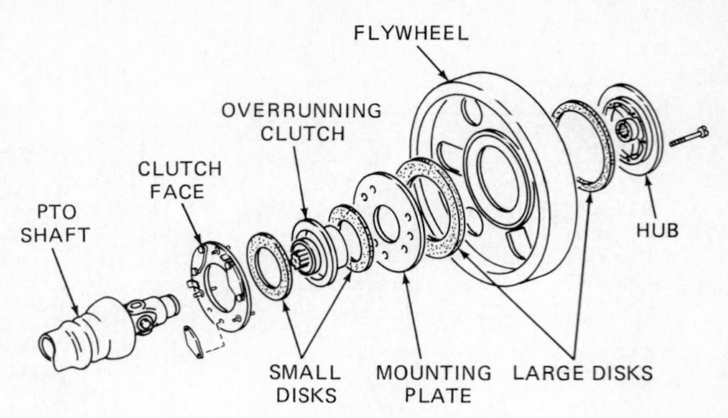


**Fig. 7-11.** Overrunning and safety clutches used on the PTO-drive hay baler. The overrunning clutch is located in the roller housing. The small disks are the safety clutch for the power takeoff shaft. The large disks protect the drive from excessive torque energy stored in the flywheel.

When the engine speed is increased, the clutch friction shoe weight is "thrown" outward against a drum due to the increased centrifugal force. The friction of the clutch shoes cause the drum to be rotated. The greater the engine speed, the greater the force of the friction clutch shoes against the drum. When engine speed is decreased to normal idle, a set of springs overcome centrifugal force and pull the shoes away from the drum. The output to the driven part of the machine is then stopped.

**Overrunning and Safety Clutches** Overrunning and safety clutches have a common purpose. They are used to prevent damage to machine parts. The overrunning clutch does this by allowing the drive shaft to transmit torque in one direction but disengage from the driven shaft in the overrunning or stopped position. The PTO baler drive illustrated in Fig. 7-11 includes an overrunning clutch. The clutch in the baler drive permits the tractor's PTO drive shaft to stop by disengaging from the machine while the energy stored in the flywheel continues to rotate the baler. This action prevents accidental damage to the PTO drive of the tractor. The operation of an overrunning roller clutch is illustrated in Fig. 7-12. When power is developed in the drive line, it forces the rollers in the clutch housing outward. The rollers wedge themselves in the housing because of the inclined surface, and both shafts are locked together. When the drive line power is reduced, the rollers loosen their friction by rolling down the incline. This allows the driven machine to coast freely from energy stored in the flywheel and other rotating parts.

A disk-type safety clutch is also illustrated in Fig. 7-11. The small disk clutch is designed to protect the PTO drive from excessive twisting shock loads. The large disk clutch separates the flywheel from the power transmission components of the baler. Its function is to slip when the energy being transmitted by the flywheel will cause damage to the machine drive.

*Shear bolt* safety clutches are used on many types of machinery power drives. This type of clutch (Fig. 7-13) is positive, accurate, and low-cost. The shear bolt or pin operates on the principle that excessive stress will break the pin or bolt. When this occurs, the two drive shafts are disconnected. The major disadvantage of the clutch is that the bolt or pin must be replaced when sheared.

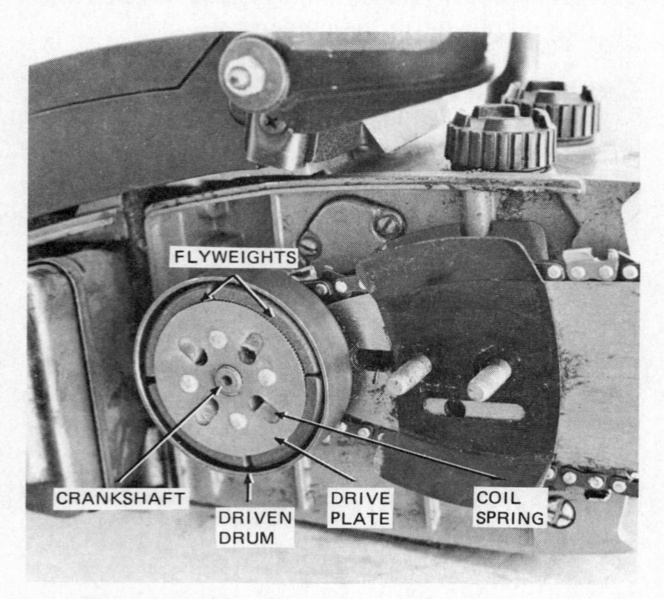


**Fig. 7-10.** Centrifugal clutch for a chain saw drive.

**CAREFUL** Do not substitute a stronger-grade bolt or pin for the replacement pin recommended by the manufacturer.

## Transmissions

The transmission section of a power train is a necessary part of most agricultural machinery for the following reasons:

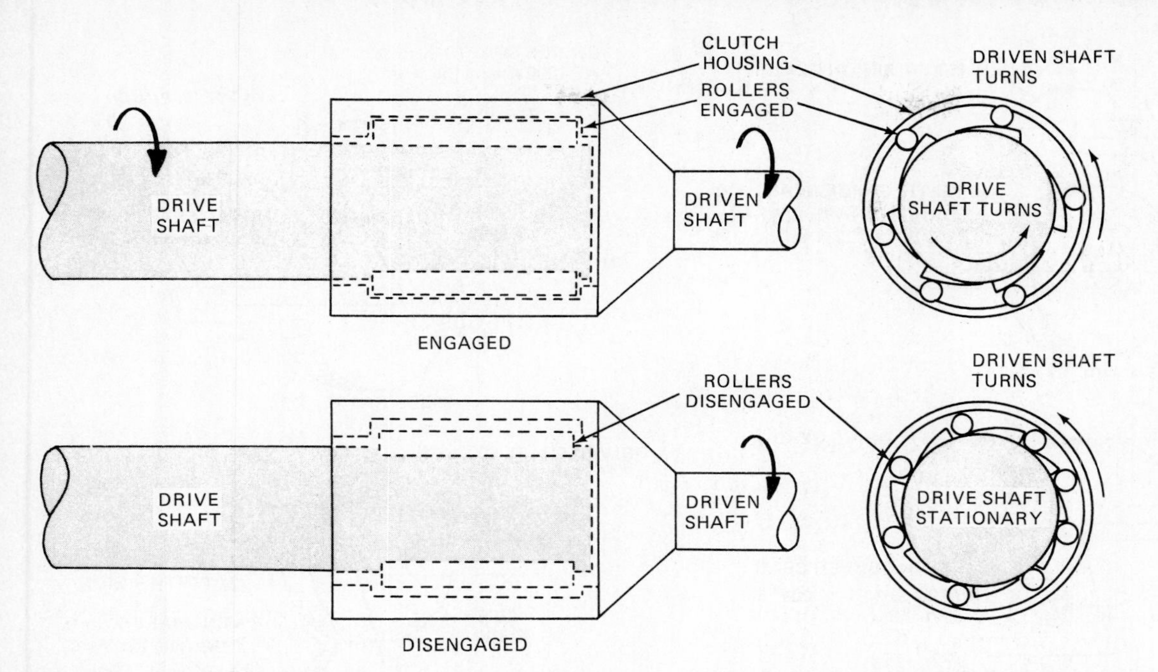


**Fig. 7-12.** Overrunning roller clutch. The roller wedge themselves against the roller housing to lock both drive shafts together. The clutch disengages when the drive line power is reduced, and the driven shaft freely coasts to a stop. (*Deere & Co.*)

1. The transmission is able to produce changes in the speed of rotation or movement of the machine or machine part.
2. It can change the torque output from the source of power.
3. It can produce changes in the direction of the force exerted.

For example, the gear transmission of a farm tractor (Fig. 7-1) illustrates the application of all the reasons listed above for transmitting the power produced by the engine to the drive wheels and the power takeoff. By selecting the correct gear combinations the operator is able to control the movement of the tractor and affect the speed of travel in the forward and reverse directions.

The principle of accomplishing this is based upon the mechanical effect produced when a gear of one diameter is meshed with a gear of another diameter. This effect is referred to as a *gear set*. Thus, a gear set will produce a *gear ratio* and a specific *speed ratio*. The relationship of these two terms is very close. Gear ratio describes the ratio of the diameter or number of teeth of one gear when in mesh with another. In Fig. 7-14 the larger gear has 48 teeth and the smaller (pinion) gear has 24 teeth. The expression of gear ratio depends upon the direction of power flow. When a rotating force is transmitted from the 48-tooth to the 24-tooth gear pinion, the ratio is 48:24, or 2:1; if the power flow is from the pinion to the larger gear, the ratio is 24:48, or 1:2. Gear ratio can

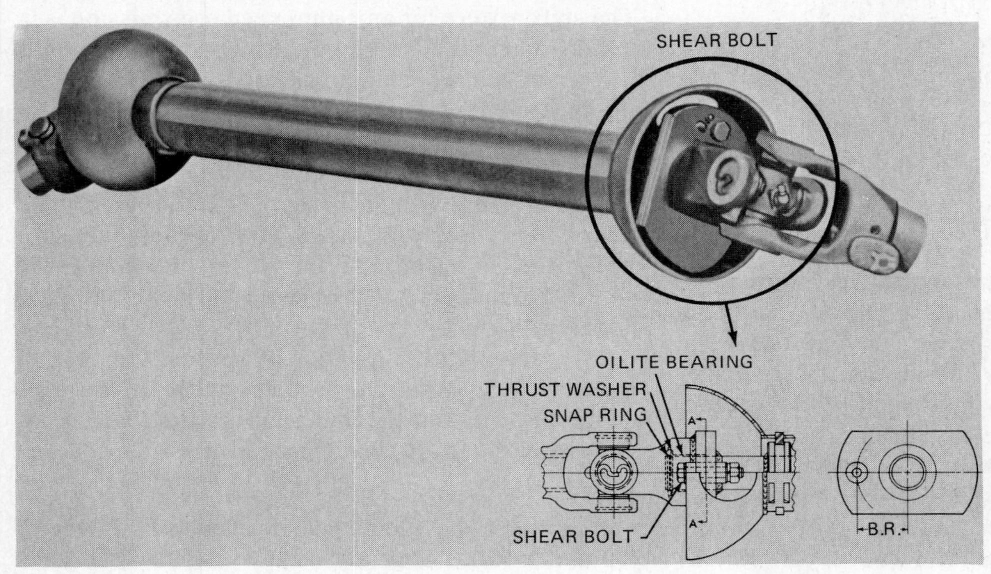


**Fig. 7-13.** Shear-bolt safety clutch. When the torque being transmitted by the PTO exceeds the power transmission components limits of the machine, the bolts are sheared (cut off), disconnecting the driven shaft from the power source. (*Rexnord*)

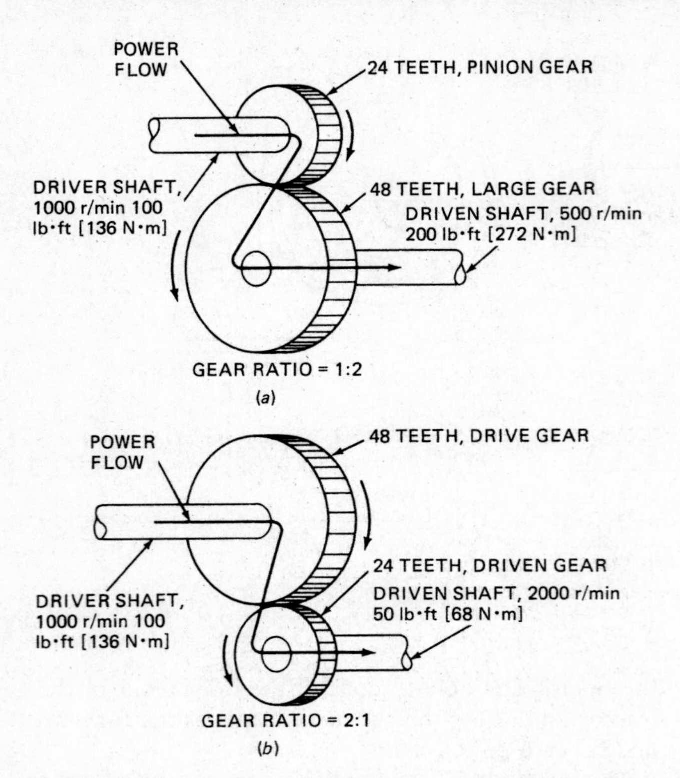


**Fig. 7-14.** Gear ratio is the relation between the size of the gears in a gear set and the direction of power flow.

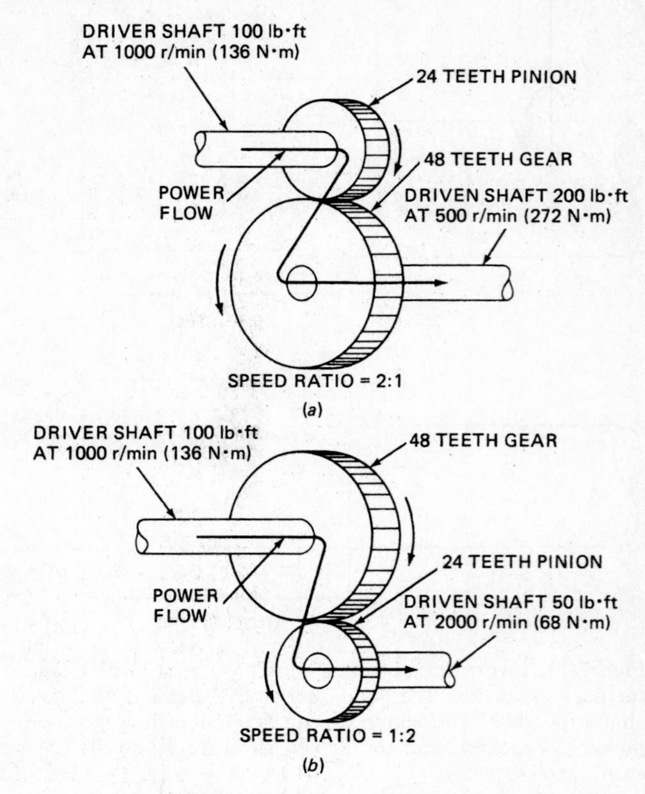


**Fig. 7-15.** Speed ratio is the relation between the size of the gears in a gear set and the operating speed of the driver and driven shaft, measured in rpm.

be found easily by counting the number of teeth of each gear in a set and then determining the direction power is flowing through the set.

Speed ratio describes the relationship which exists between the speed of the driving (drive) and the driven shaft to which the gears are attached. In Fig. 7-15, the 24-tooth pinion gear drive gear is rotating clockwise at 1000 r/min and is in mesh with the 48-tooth gear. The larger gear will rotate at 500 r/min in a counterclockwise direction. The pinion gear will rotate two turns for each turn of the larger gear and produce a speed ratio of 2:1, or

$$\frac{24 \text{ teeth}}{48 \text{ teeth}} = \frac{1000 \text{ r/min}}{500 \text{ r/min}} = \frac{2}{1} \quad \text{or} \quad 2:1$$

If the flow of power is from the larger gear to the pinion gear, the speed ratio will be 1:2, or

$$\frac{48 \text{ teeth}}{24 \text{ teeth}} = \frac{1000 \text{ r/min}}{2000 \text{ r/min}} = \frac{1}{2} \quad \text{or} \quad 1:2$$

Speed ratio and gear ratio provide information which identifies changes that take place in rpm and torque. In Fig. 7-15*a* the power flow results in a "torque increase"; that is, driven shaft rpm is reduced one turn for every two turns of the driver while torque is doubled. Thus we have the basis for the relationship that torque is inversely related to speed (rpm). This means that when gears are used to provide a decrease in the speed ratio (2:1), the torque will increase by the inverse ratio of 1:2.

**Mechanical Transmission** Shift levers are provided on a *mechanical* type of transmission to permit the operator to shift gears manually. Many farm tractors and self-propelled machines are equipped with this type of transmission because it is an efficient and low-cost unit.

Mechanical transmissions allow the operator to select a gear ratio giving three or more forward speeds, one or more reverse speeds, and a neutral setting. The shift pattern for an eight-speed transmission is illustrated in Fig. 7-16. When the high-low shift lever is in low and the gear shift lever is in the No. 2 position, the forward travel is 1.0 mi/h [1.6 km/h] while running the engine at 1000 r/min and 2 mi/h [3.2 km/h] at 2000 engine r/min. When the high-low lever is shifted to high, the transmission is in the No. 6 position. This setting provides a gear ratio giving a forward travel of 3.6 mi/h [5.8 km/h] at an engine speed of 1000 r/min and 7.2 mi/h [11.6 km/h] at 2000 r/min.

In addition to the number of speeds, the type of mechanical transmission is determined by the type and method of shifting that is used. The three types of transmissions are *sliding gear, constant mesh,* and *synchromesh.*

**Sliding gear** The sliding gear transmission is one of the oldest forms of gear-change systems. Spur gears

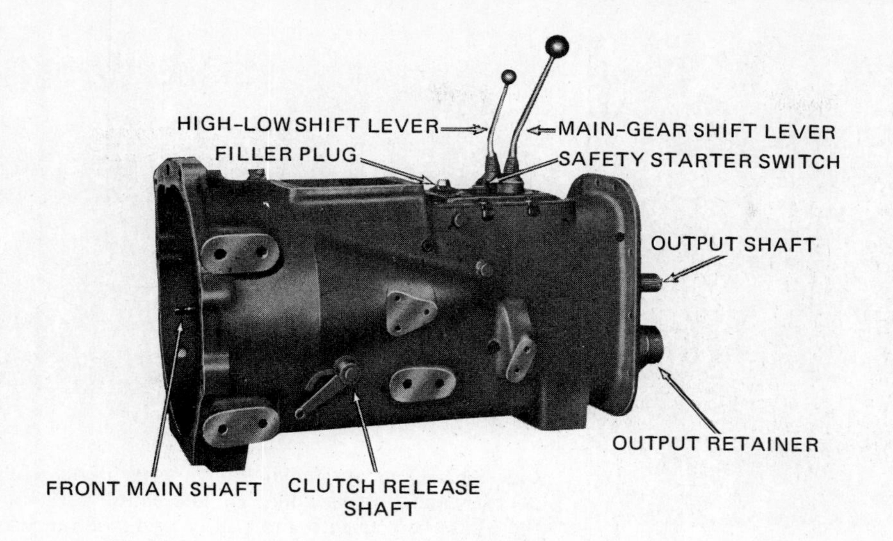


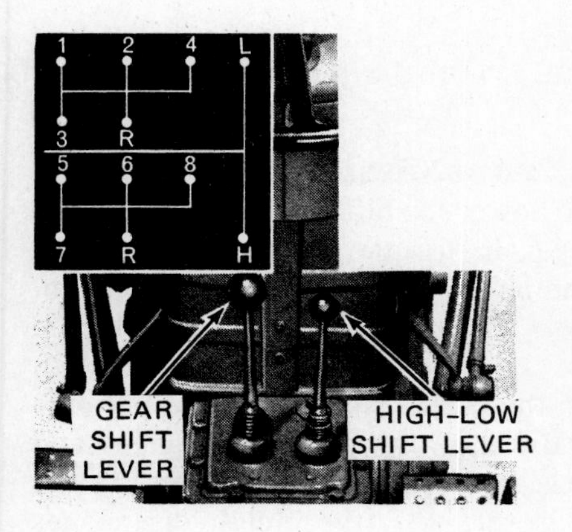


(a)

TRAVEL SPEEDS, with 12.4–28 tires, mi/h [km/h]

| Gear | Eight-speed transmission 1000 r/min | 2000 r/min |
|---|---|---|
| 1st | 0.8 [1.3] | 1.6 [2.6] |
| 2nd | 1.0 [1.6] | 2.0 [3.2] |
| 3rd | 1.8 [2.9] | 3.6 [5.8] |
| 4th | 2.4 [3.9] | 4.8 [7.7] |
| 5th | 2.9 [4.7] | 5.8 [9.3] |
| 6th | 3.6 [5.8] | 7.2 [11.6] |
| 7th | 6.4 [10.3] | 12.8 [20.6] |
| 8th | 8.7 [14.0] | 17.4 [28.0] |
| Reverse low | 1.2 [1.9] | 2.4 [3.9] |
| Reverse high | 4.2 [6.8] | 8.4 [13.5] |

(b)

**Fig. 7-16.** An eight-speed tractor transmission. Two shift levers (*a*) allow the operator to select one of eight speed ratios (*b*) for forward travel and two for reverse through a four-speed transmission and a high-low gear assembly. (*Ford Tractor Operations*)

(Fig. 7-17) are used in this type of transmission because the gears can be held in mesh by the "shifter forks." Spur gears do not create side thrust when power is being transmitted, and thus the shifter forks can hold them in mesh. A cutaway view of a typical sliding gear transmission is shown in Fig. 7-17. Note that the shifter forks normally operate in the shifter collars to move the sliding gears.

The operating principles of a sliding gear transmission are illustrated in the schematic diagram in Fig. 7-18. The three gears that are shifted to change the speed or direction of the splined output shaft are D, E, and F. As illustrated, F on the output shaft has been slid into mesh with C on the input shaft. This places the transmission in the third (high) gear range because the output shaft will rotate faster than the input shaft in any of the additional combinations.

When D is shifted to mesh with the A gear and F is moved out of mesh with C the transmission is in first (low) gear. In a similar manner, if E is meshed with the B gear and D and F are not in mesh, the transmission is in second (intermediate) gear.

To place the transmission in reverse, the reverse idler gears G and H are used. Note that H is in permanent mesh with A on the input shaft and G is driven by H on a common shaft. By sliding D in mesh with G the direction of the output shaft is reversed.

The transmission is placed in neutral by shifting the D, E, and F gears on the output shaft so they do not engage any gear on the input or the reverse shaft.

**Constant mesh** As the name suggests, all the gears in a constant mesh transmission remain in mesh with one another whether they are transmitting power or not. This type of transmission may use either the spur or the helical-type gear (Fig. 7-19). To transmit power, a shifter collar is slid across a splined gear and engages the teeth of the shifter gear portion of the driven gear. This locks the shifter gear and the driven gear together. Two shifter collars, A and B, are used in the constant-mesh transmission (Fig. 7-20). In first gear, shifter collar A is centered, locking the countershaft into a solid unit. Shifter collar B is moved to the right, locking to first gear. Input

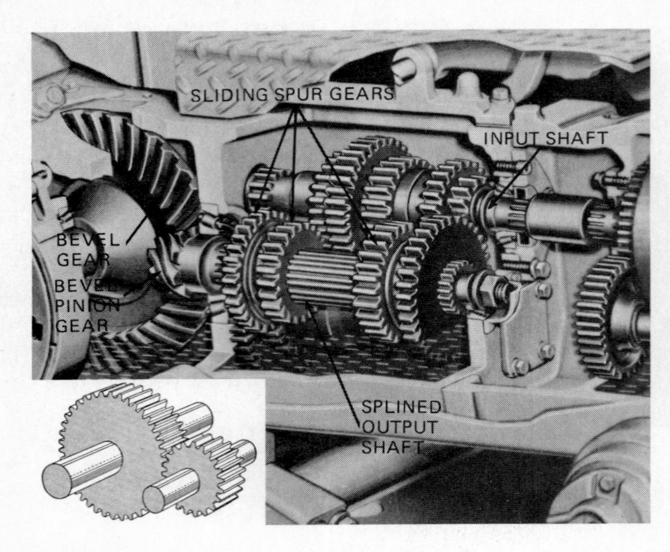


**Fig. 7-17.** Three-speed sliding gear transmission. The shift lever is in the neutral position. To engage, the spur gears (see inset) are slid on the spline shaft by shifter forks to engage the spur gears on the input shaft. (*Deere & Co.*)

torque is then transferred from the input shaft through the first reduction gear to the countershaft drive gear, through the countershaft cluster gears and the first gear, to the output shaft. Shifting of the two collars A and B allows the proper connections to obtain power flow through second, third, and fourth gears.

**Synchromesh** One of the problems with the sliding gear and the sliding collar transmission systems is the difficulty of shifting gears when the tractor or machine is in motion. Unless the forward motion of the equipment is stopped, the sliding gear or collar is difficult to engage with another gear without the grinding or grating of gears. A synchromesh transmission is a constant mesh to which synchronizer sleeves (Fig. 7-21) have been added to the shifting collars. The purpose of the synchronizers is to bring the speed of adjacent gears to identical rotating speeds so that the shifting collar will align with the driven gear's shifter teeth. This is accomplished by a blocking ring which acts as a friction clutch, speeding up or

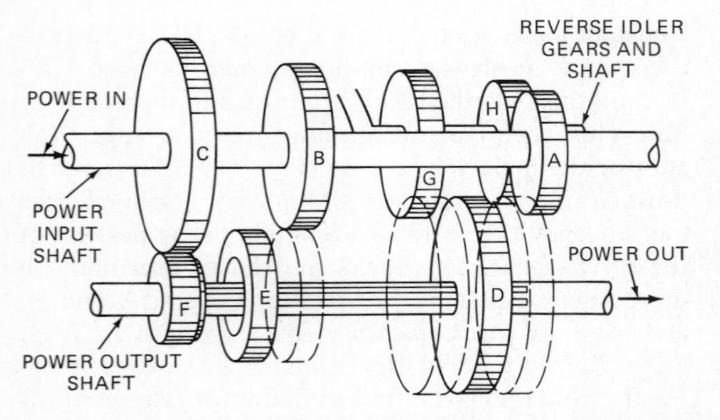


**Fig. 7-18.** Schematic diagram of a sliding gear transmission.

**Fig. 7-19.** Constant mesh transmission parts. Helical gears A and shifting collar B. The shifter fork moves the shifter collar into the shifter teeth of the driven gear, locking the shifter gear and driven gear together. (*Deere & Co.*)

slowing down the gear to be joined by the shifter collar.

**Power Shift Transmission** Some farm tractors use hydraulic power to shift the transmission and to provide changes in torque and speed ratio. In many tractors, the power shift is a part of a constant mesh transmission and uses a hydraulic clutch in connection with a countershaft and overriding clutch (Fig. 7-22). When the clutch is engaged by hydraulic pressure, the overriding low clutch disengages and the transmission is in the high range. When the hydraulic pressure is released, the clutch disengages and the overriding clutch on the countershaft engages, driving the shaft at low range. Shifting can be done without stopping forward travel. The high-low ranges increase an eight-speed transmission to sixteen forward speeds.

Another type of power shift transmission uses hydraulic clutches and planetary gearing. Planetary gears are extensively used in power train systems for the following reasons.

1. They are simple in design.
2. They provide a change in speed and torque.
3. They act as a solid shaft.
4. The direction of rotation can be reversed.
5. They provide a neutral (disconnected driver and driven shaft).

A planetary gear set consists of four parts—the sun gear, the planet gears (two or three), the carrier, and the ring gear (Fig. 7-23). The action of a planetary gear set depends upon the attitude of the carrier. Referring to Fig. 7-24*a*, when the carrier is the input

**Fig. 7-20.** Constant mesh transmission. Sliding collars A and B are used to connect the proper sequence of gears. The helical gears cannot be shifted because they create side thrust and would slide out of mesh. (*Ford Tractor Operations*) (See facing page 123.)

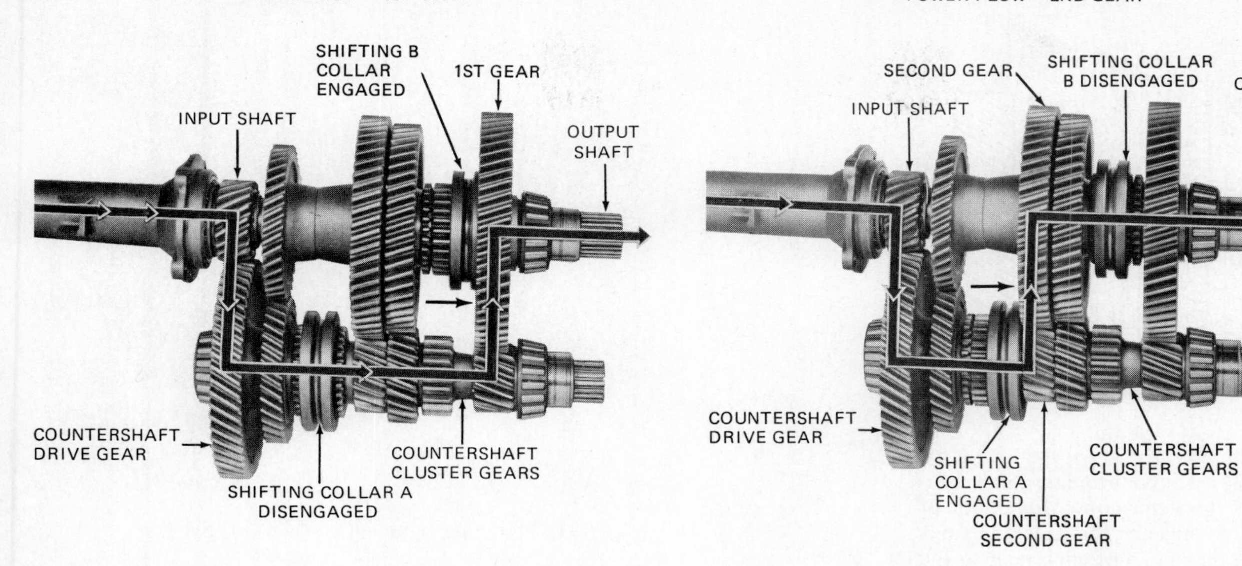


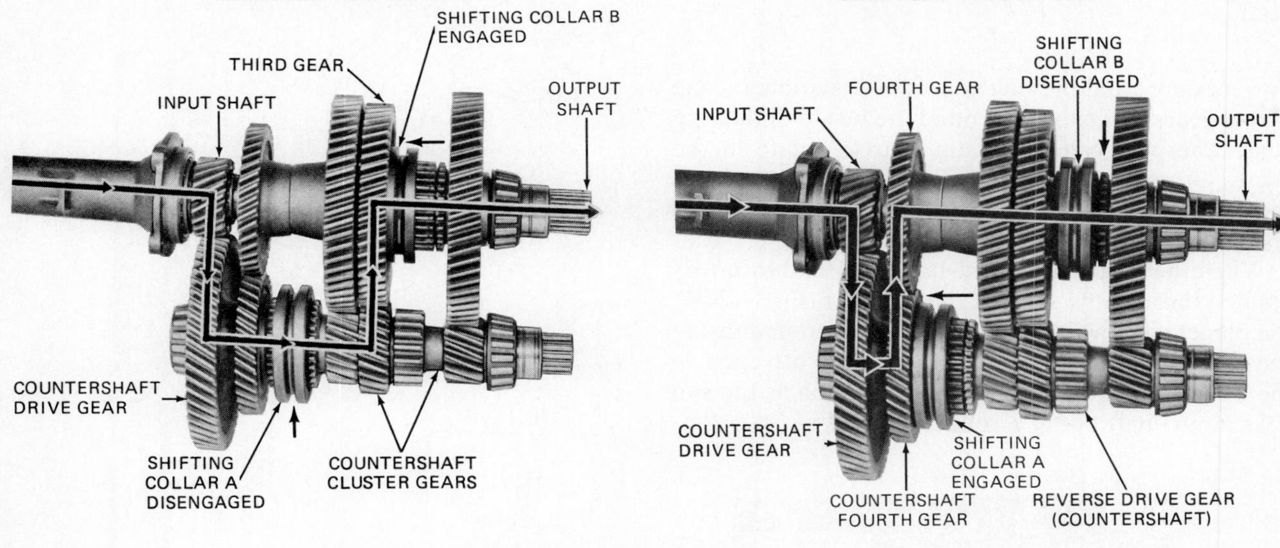


POWER FLOW – REVERSE

SHIFTING COLLAR B ENGAGED WITH 1ST GEAR
SECOND GEAR
OUTPUT SHAFT
INPUT SHAFT
COUNTERSHAFT
REVERSE GEAR ENGAGED WITH REVERSE DRIVE GEAR ON COUNTERSHAFT ASSEMBLY
COUNTERSHAFT DRIVE GEAR
REVERSE IDLER
SHIFTING COLLAR A ENGAGED WITH COUNTERSHAFT
REVERSE DRIVEN GEAR

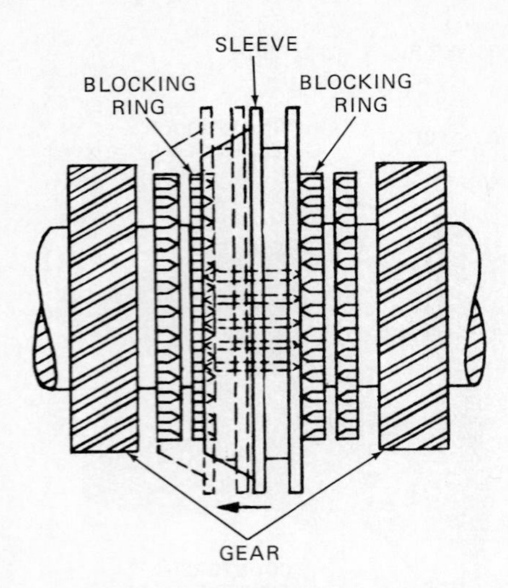


**Fig. 7-21.** A synchronizer of a syncromesh transmission. Movement of the sleeve engages the blocking rings to bring the speed of each gear to the same rpm. The shifting collar will then align with the shift teeth of the driven gear.

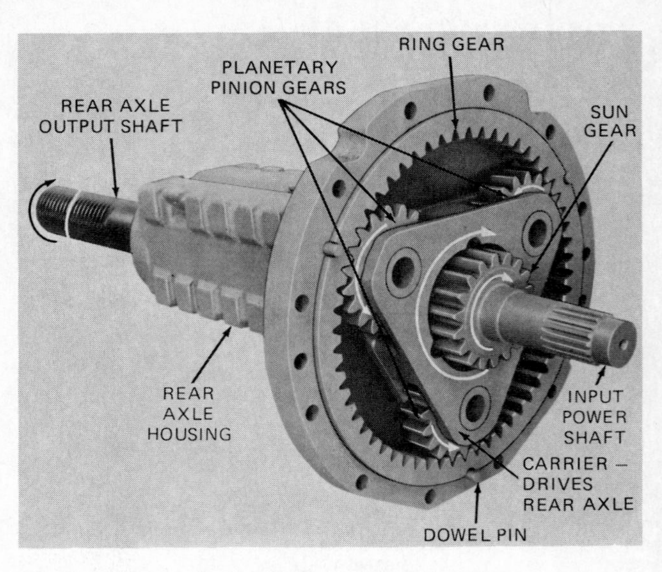


**Fig. 7-23.** Planetary gear set. (*Deere & Co.*)

power source and the ring gear is held stationary, the planet gears will revolve around the inside of the ring gear. This will cause the sun gear to rotate in the same direction as the carrier. Torque output is through the sun gear with an increase in output speed.

When the ring gear is held stationary and torque is applied through the shaft of the sun gear (Fig. 7-24*b*), the planet gears will be forced to rotate around inside the ring gear. Since the planet gears are attached to the carrier, it rotates in the same direction as the sun gear input shaft but at a reduced speed.

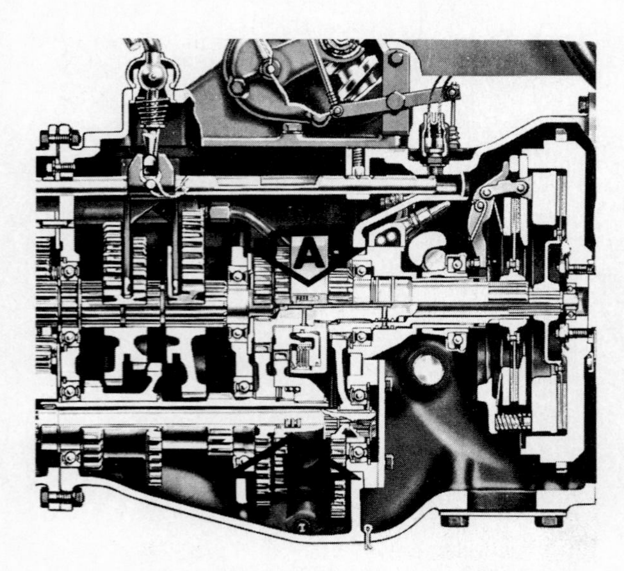

**Fig. 7-22.** The transmission input shaft speed is changed by using a hydraulically actuated clutch (*a*) and an overriding clutch (*b*). (*Massey-Ferguson, Inc.*)

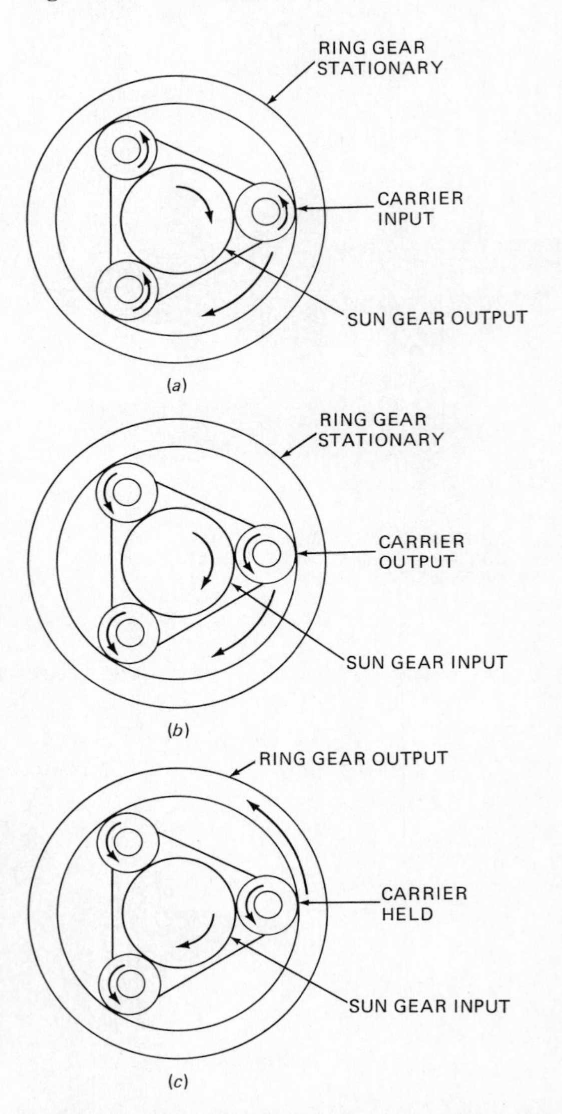


**Fig. 7-24.** Actions of the planetary gear set: (*a*) when the carrier is torque input; (*b*) when the carrier is torque output; (*c*) when the carrier is held.

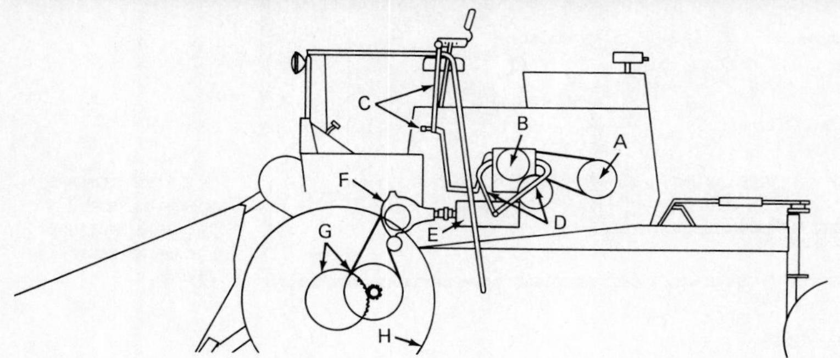


Fig. 7-25. Windrower operated by variable speed hydrostatic drive. (*International Harvester/New Holland*)

In the last condition (Fig. 7-24*c*), the carrier is held. The planet gears rotate when torque is applied to the sun gear shaft. The ring gear will then rotate in the opposite direction to the input shaft. Thus when the carrier is held, the ring gear will rotate in the opposite direction, reversing the output shaft.

To obtain direct drive through a planetary set, it is necessary to hold (lock) any two of the three units.

When the planetary gear set is coupled with power (hydraulic) clutches, high and low speed and neutral operation of the output shaft can be achieved without clutch operation.

**Hydrostatic Transmission** Hydrostatic transmissions are used on many types of harvesting machinery and some tractors. The hydrostatic transmission is made up of hydraulic power components which utilize high fluid pressure and low fluid flow. Engine power drives a pump which transfers high fluid pressure to a hydraulic motor. The windrowing machine, shown in Fig. 7-25, illustrates the principle of hydrostatic drive. Engine power (A) is transmitted by belt to the hydraulic pump (B), which uses pistons to develop high fluid pressure. The operator controls the forward speed of the machine by moving the speed control lever (C). This determines the amount of fluid the pump will discharge through the hydraulic lines (D) to the motor (E). The amount of fluid supplied by the pump determines the rate at which the motor transmits power to the differential (F) and final drive (G) and then to the drive wheels (H). The result is an infinite variation in forward and reverse travel and a neutral setting.

In order to achieve this type of speed and power response, the pump must be capable of developing high pressure and a variable discharge rate. Nearly all hydrostatic systems use the *axial-type* piston pump and motor. Axial-type pumps have pistons which are mounted parallel to the drive shaft. The pump consists of a rotating cylinder block into which are inserted a number of pistons. When the drive shaft rotates the cylinder block, the pistons follow the rotation, producing a pumping action by an in and out movement in the cylinder block according to the angle of the swash plate (Fig. 7-26). The swash plate of the pump and motor is an inclined plate against which the pistons exert their force.

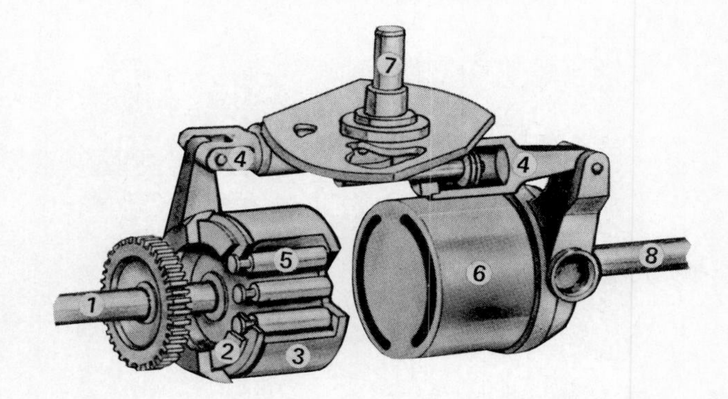


Fig. 7-26. The drive shaft (1) of a hydrostatic drive transmission rotates the cylinder block (3) and oil pumping pistons (5). The swash plate (2) is tilted by the cylinders (4) in response to the control level (7). Oil under high pressure is transferred to the motor (6), which drives the output shaft (8). (*International Harvester Co.*)

The swash plate of the pump is designed so that it can be tilted (Fig. 7-27) to affect the amount of pumping action of the pistons. The motor swash plate may be held at a fixed angle, or in some designs it can also be tilted. A movable swash plate is often called a variable-displacement unit, while a fixed swash plate

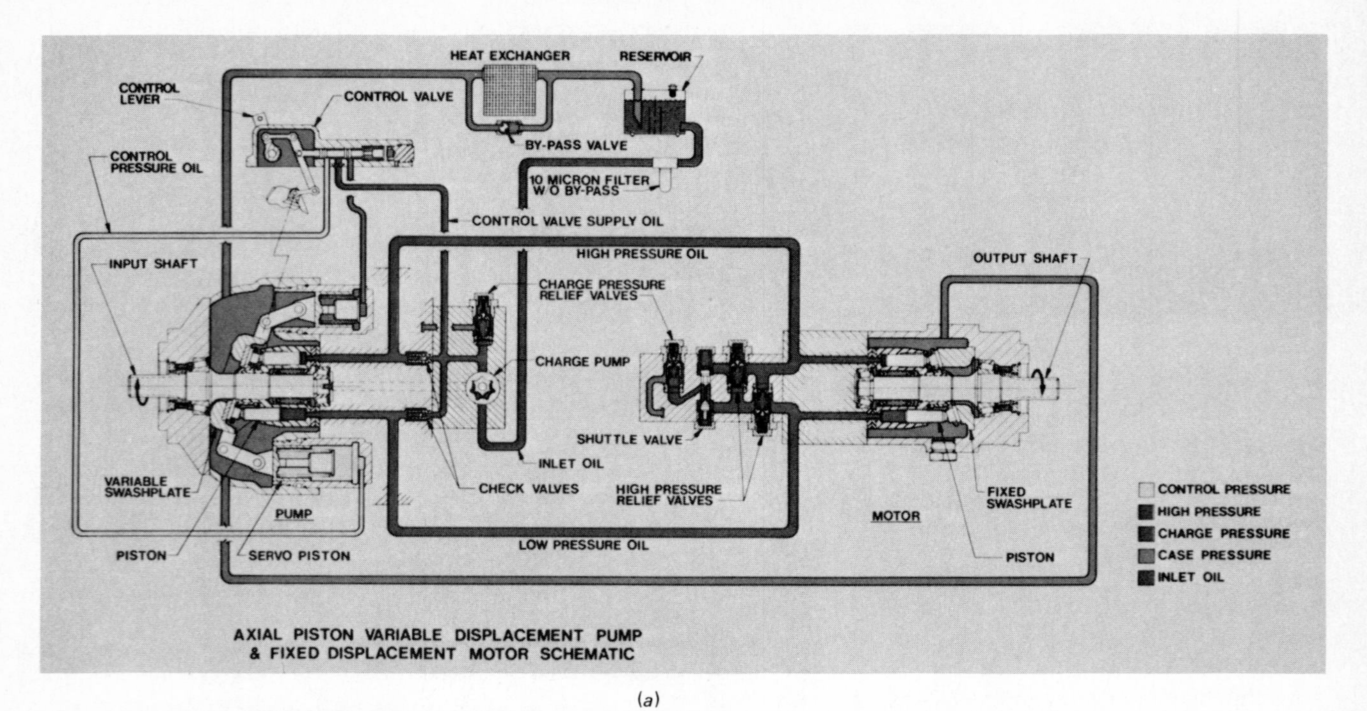


(*a*)

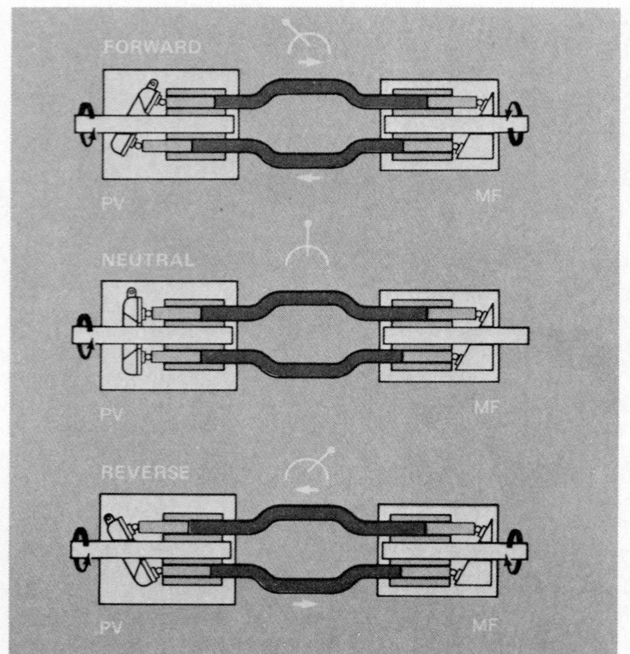


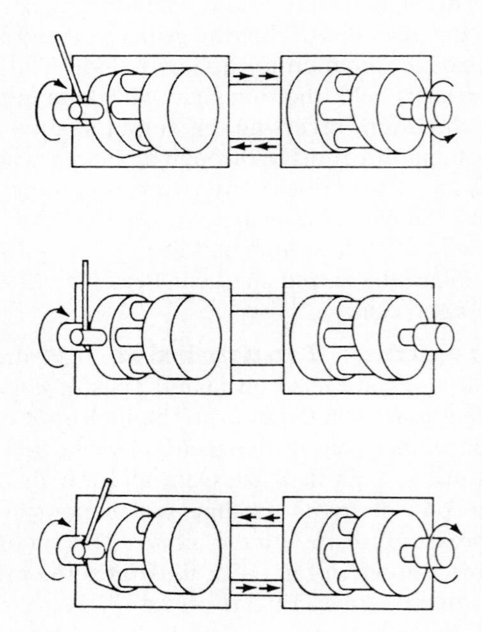

(*b*)

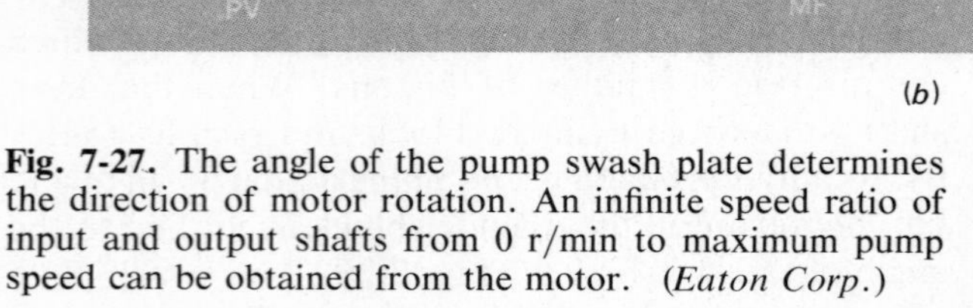

**Fig. 7-27.** The angle of the pump swash plate determines the direction of motor rotation. An infinite speed ratio of input and output shafts from 0 r/min to maximum pump speed can be obtained from the motor. (*Eaton Corp.*)

is referred to as a fixed-displacement unit. Only the cylinder blocks rotate and the ends of the pistons react against the swash plate. Forward, neutral, and reverse direction of motor rotation can be achieved by the position of the swash plate and the resulting effect it has on the pumping action of the piston. When the control lever is in neutral, the pump does not deliver oil because the swash plate is also in the neutral position and the pistons are not stroking. Three factors control the hydrostatic transmission: pressure of the fluid produces power, direction of the flow determines direction of travel, and rate of flow determines the speed of travel.

Hydrostatic transmissions may be an integral unit, as illustrated in Fig. 7-28*a*, or the components of the pump and motor may be separated a considerable distance and connected by oil transfer lines, as shown in Fig. 7-28*b*. Cleanliness of oil is vital to the life of hydrostatic transmissions. An oil filter having a rating of 10-micrometer ($\mu$m) filtering capacity is used in most hydrostatic systems to protect the finely

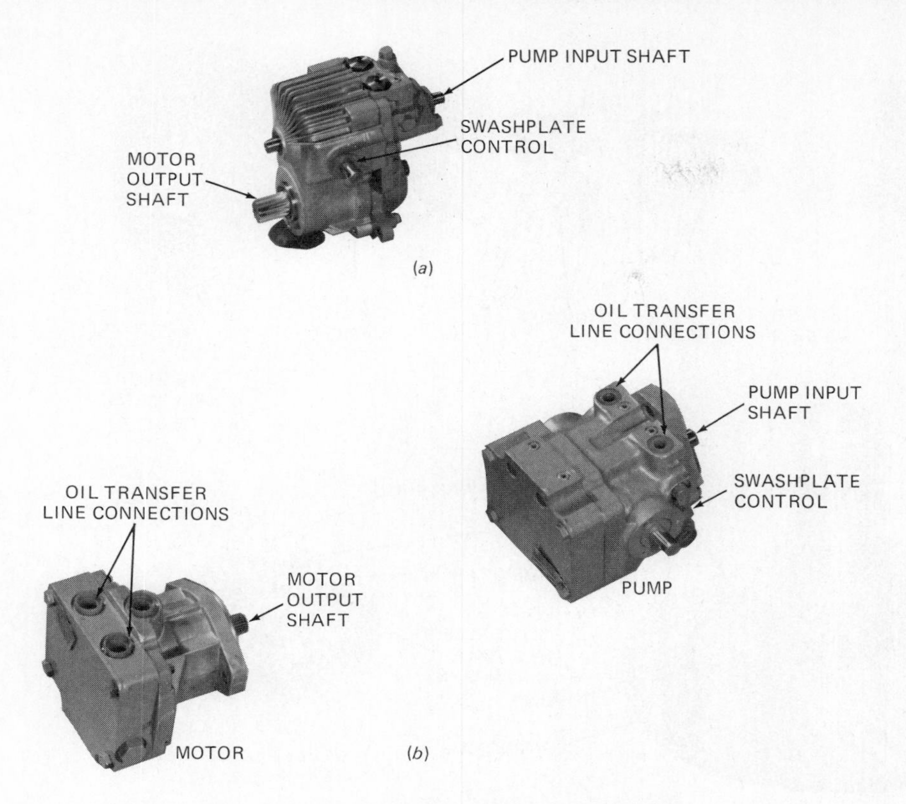


**Fig. 7-28.** A self-contained hydrostatic drive transmission (*a*) requires no external oil reservoir. When the motor and pump are separate (*b*), oil transfer lines carry the high-pressure oil between the two units.

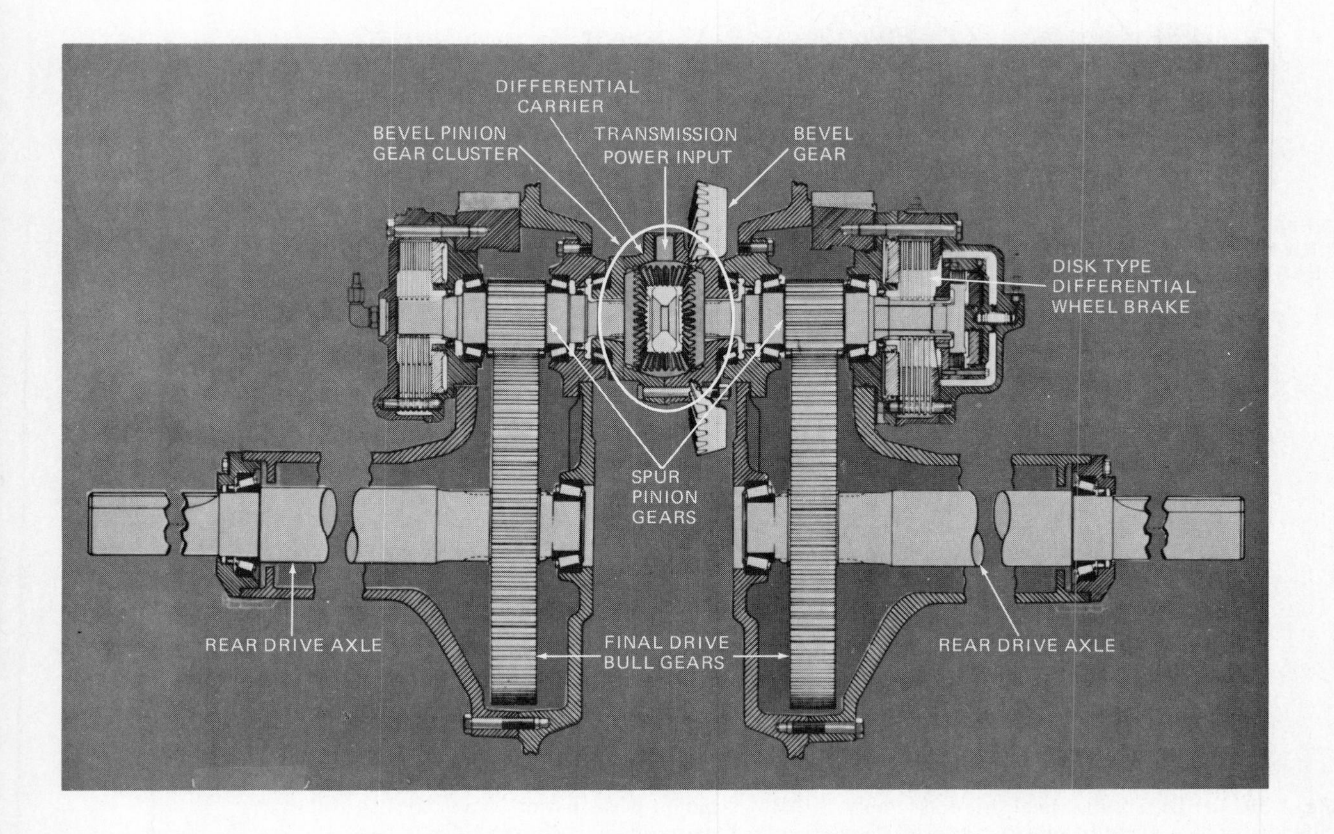


**Fig. 7-29.** Final drive components of a farm tractor. (*International Harvester Co.*)

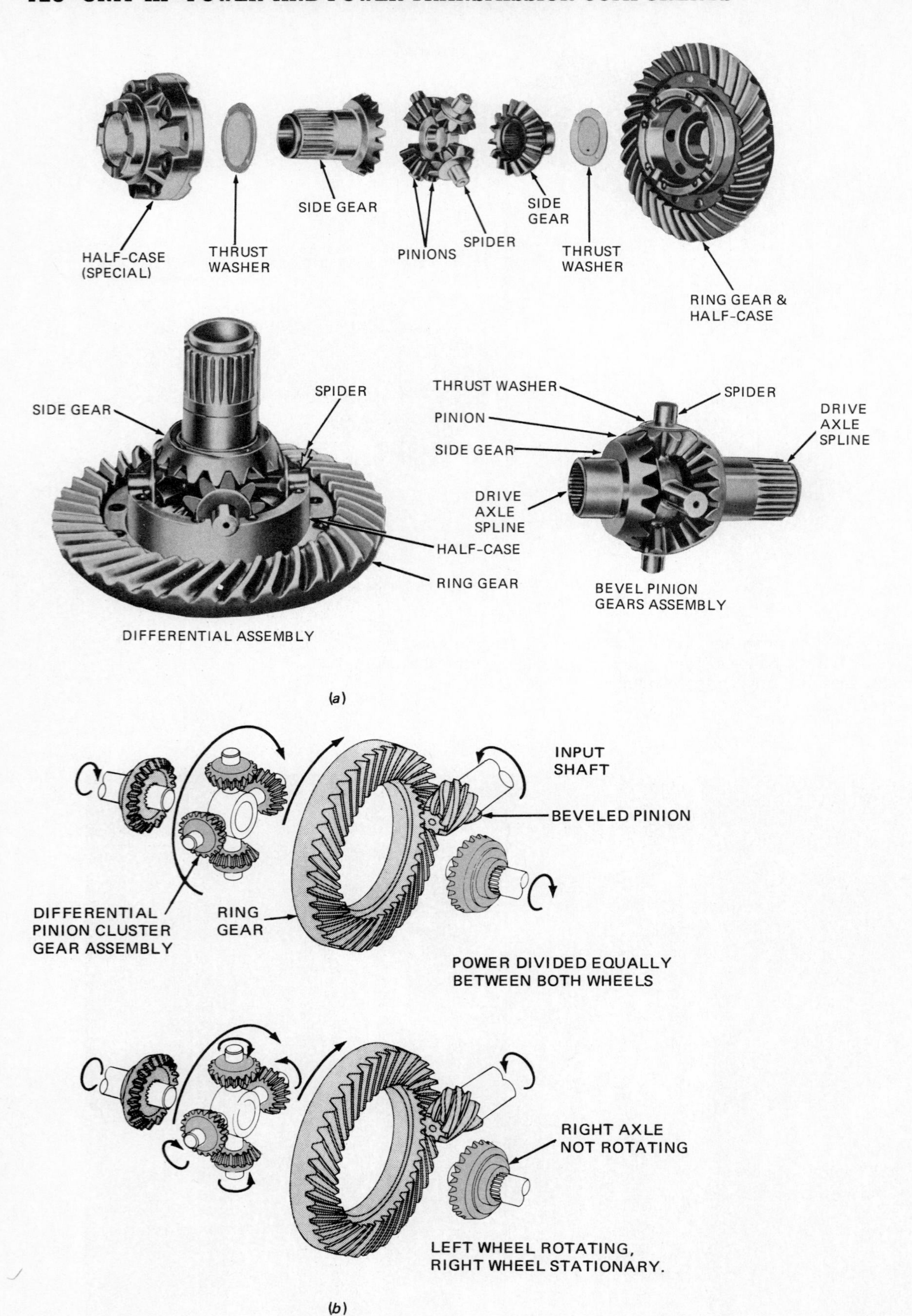


**Fig. 7-30.** (*a*) Differential final drive assembly. (*b*) Bevel pinion gears permit drive axles to rotate at different speeds when turning a corner. (*Ford Tractor Operations and Deere & Co.*)

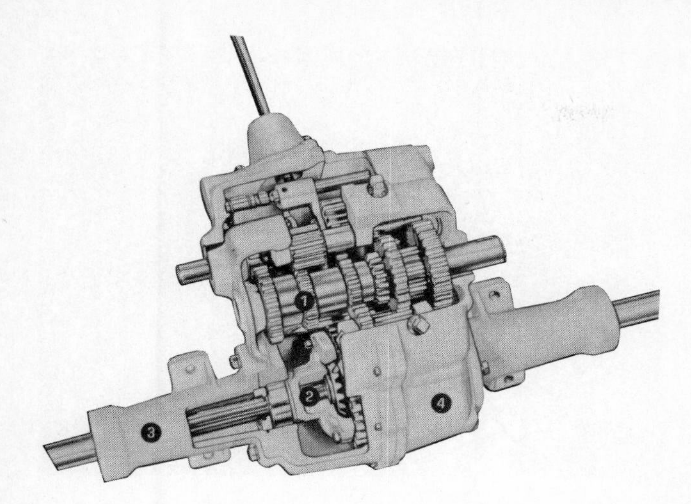

Fig. 7-31. Cutaway view of a transaxle. The spur gears (1) provide four forward speeds, neutral, and reverse. The final drive consists of a spur reduction gear and differential (2) and a drive axle (3). A cast-iron housing (4) provides support and a reservoir for the lubricant. (*Tecumseh Products Co.*)

machined parts from damage due to foreign materials.

## Final Drives

The *final drive* of the power train includes the differential gear, bull gears or planetary gear sets, wheel brakes, and drive axles (Fig. 7-29 on page 127).

Power is directed from the transmission to the spiral bevel pinion gear. This gear meshes with the larger spiral bevel gear to achieve an increase in torque, a decrease in speed of the output shaft, and a 90° change in the direction of power flow from transmission to drive axles. The bevel gear (Fig. 7-30 on page 128) is attached to the differential carrier. The purpose of the differential is to separate the two drive axles so that power can be transferred equally to each drive wheel. As the machine turns, the outside wheel must complete more revolutions than the inside wheel. The differential permits the two drive axles to operate independently by utilizing the bevel pinion cluster gear assembly contained in the differ-

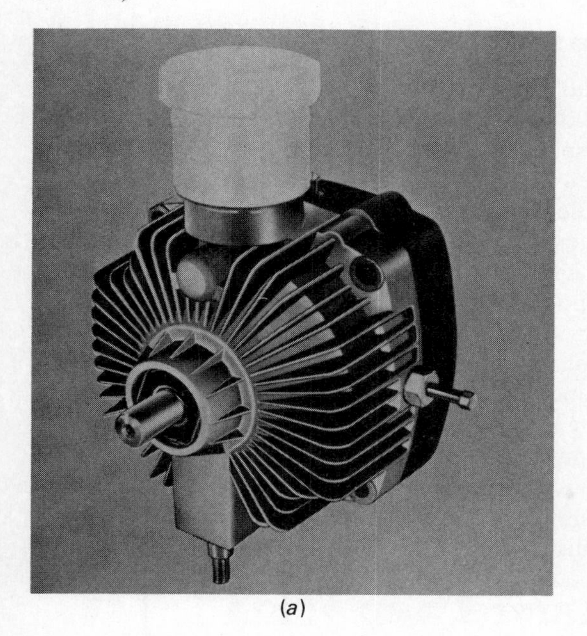

(*a*)

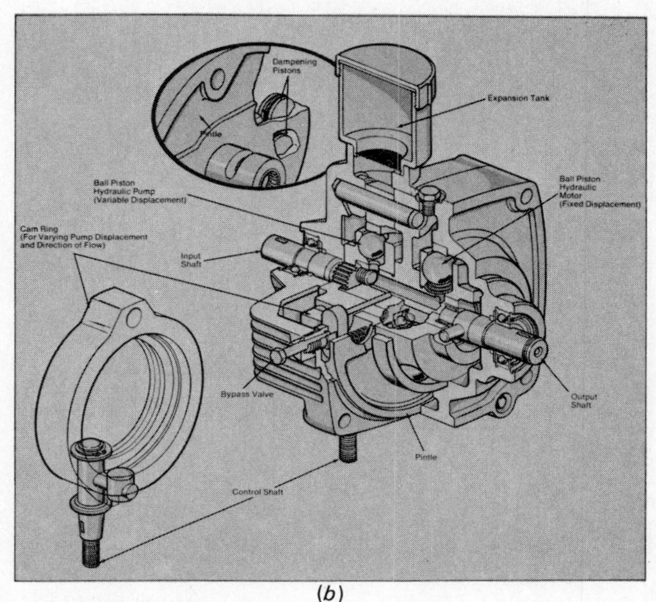


(*b*)

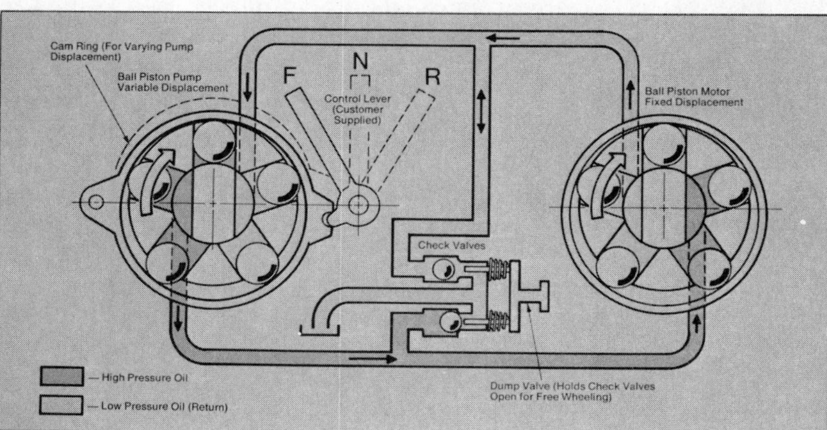


(*c*)

Fig. 7-32. (*a*) Ball-piston-type hydrostatic transmission for powering small tractors and lawn and garden machinery. (*b*) and (*c*) The pump output and the direction of rotation of the motor are controlled by varying the displacement of the ball piston pump with the cam ring. (*Eaton Corporation*)

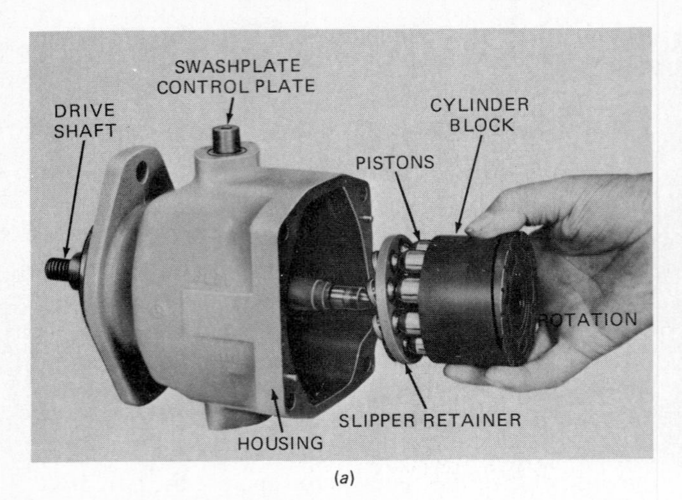


(a)

(b)

**Fig. 7-33.** Piston-cylinder-type hydrostatic transmission for direct attachment to the transaxle of small tractors and lawn and garden equipment. The speed of the motor output shaft is controlled by varying the stroke of the pistons through the swashplate control shaft. (*Deere & Co. and Sunstrand Corp.*)

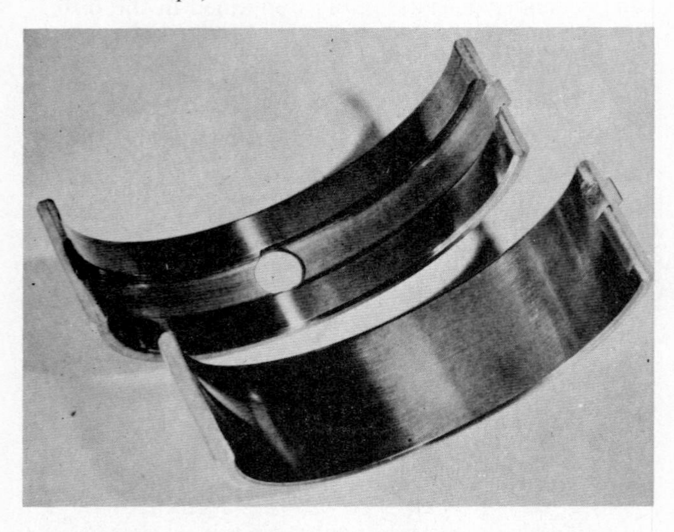

**Fig. 7-34.** Plain bearing used as a precision insert for an engine connecting rod. (*The Texas Co.*)

ential housing. A differential lock control may be used on some tractors. The control "locks" the bevel pinion cluster gear assembly causing the two axles to be locked together as one. The machine should *not* be turned while the differential is locked.

Final reduction gears and differential brakes are illustrated in Fig. 7-29. Final reduction gears may be utilized to rotate the drive axles. They may be spur or planetary gears. The purpose of the final drive gears is to reduce the speed and increase the torque transmitted to the axles. Spur gear drives employ a large bull gear fixed to each axle. These gears mesh with the small spur pinion at each end of the drive axle.

In planetary gear final drives (Fig. 7-23), the ring gear is held stationary. Power is applied to the sun gear, and the carrier drives the axle at a reduced speed. The individual brakes are designed to assist in turning as well as stopping the forward motion of the tractor or machine. Brakes may be located on each end of the differential shaft or on the final drive axle shafts.

### Transaxles

Small tractors and other groundskeeping equipment utilize a power train which is very compact. For these specific applications, the *transaxle* has been developed (Fig. 7-31 on page 129). The transaxle contains the transmission, differential, and final drive assemblies within one complete package. The cutaway illustration of a transaxle in Fig. 7-31 includes (1) a four-speed forward, a reverse, and a neutral gear selection, (2) a differential gear assembly, (3) rear-axle drive shafts, and (4) an iron casting, to house all the components. Hydrostatic drives (Figs. 7-32, page 129, and Fig. 7-33) are also used in compact tractors and machinery. The units provide infinitely variable speed control of the output shaft. This is accomplished by controlling the discharge rate of the pump. (See the previous section on hydrostatic transmissions.)

## BEARINGS AND SEALS

*Bearings* and *seals* are an integral part of power transmission systems. Bearings serve the power train by allowing gears and rotating shafts to turn freely when heavy forces are applied. Seals prevent the loss of lubricant from bearings and gears and prevent abrasive dirt from entering the working parts. To a large degree the amount of energy loss in a power train depends upon the amount of friction in the bearings which support the rotating gears, pulleys, and drive shafts. Bearings also serve the power transmission system by maintaining the alignment and spacing of gears and absorbing the *thrust* which is developed when belt pulleys and gears transmit power.

One of the common activities of workers in agricultural mechanics operations will be to evaluate the condition of bearings and seals and to remove and replace defective units.

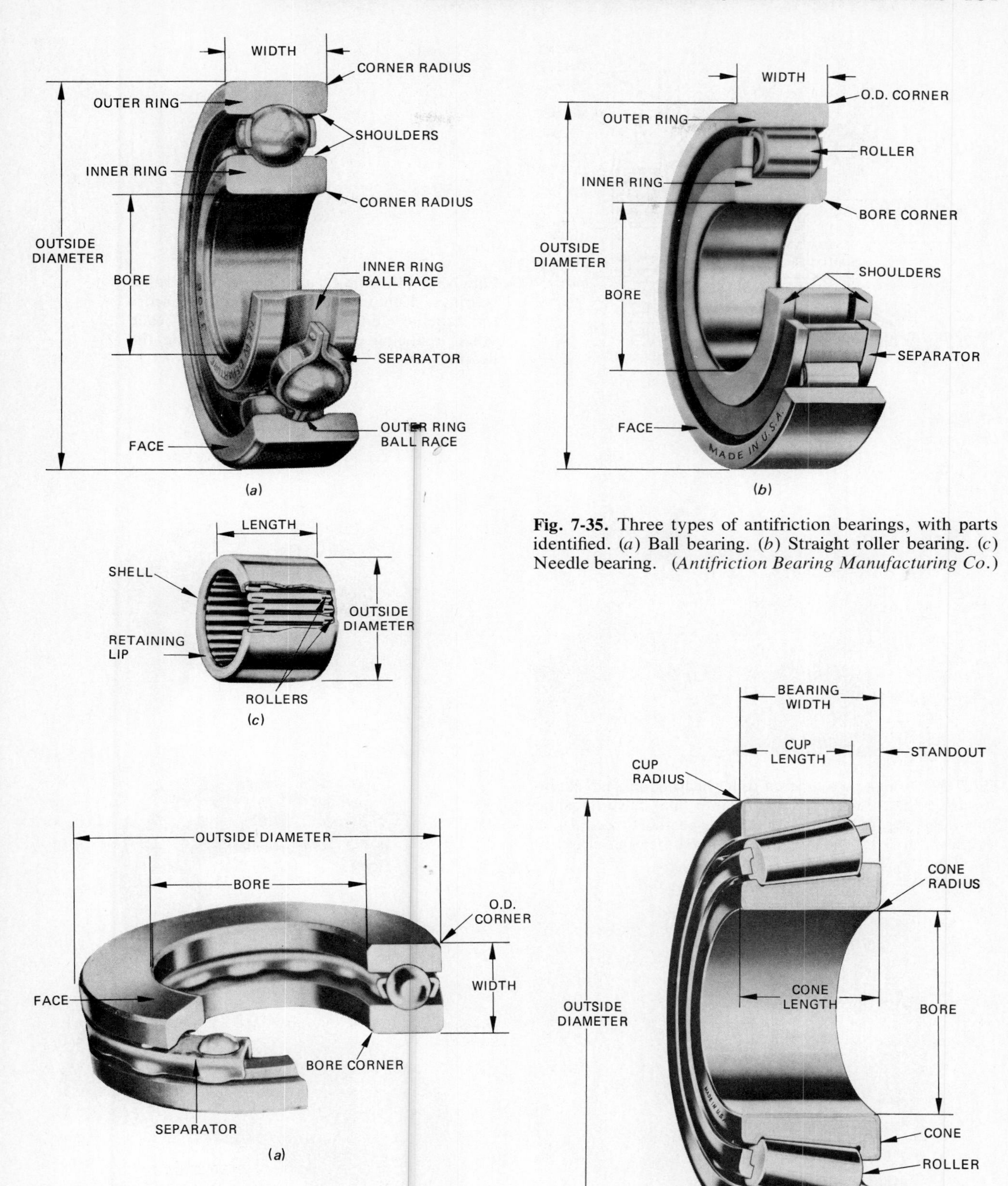


**Fig. 7-35.** Three types of antifriction bearings, with parts identified. (*a*) Ball bearing. (*b*) Straight roller bearing. (*c*) Needle bearing. (*Antifriction Bearing Manufacturing Co.*)

**Fig. 7-36.** Bearings designed to carry thrust loads. Ball thrust (*a*) is commonly used as a clutch release bearing. Tapered roller (*b*) is used extensively in agricultural equipment. (*Antifriction Bearing Manufacturing Co.*)

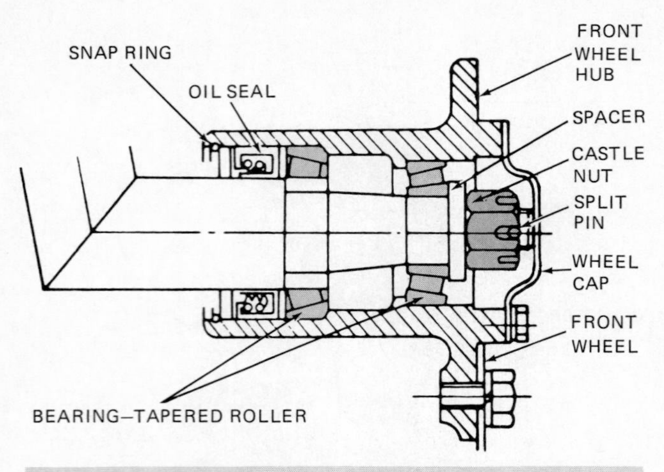


**Fig. 7-37.** Principle of adjusting tapered roller bearings. Tightening the castle nut (6) forces the tapered cone of bearings toward each other, reducing end play and increasing the bearing preload. (*Ford Tractor Operations.*)

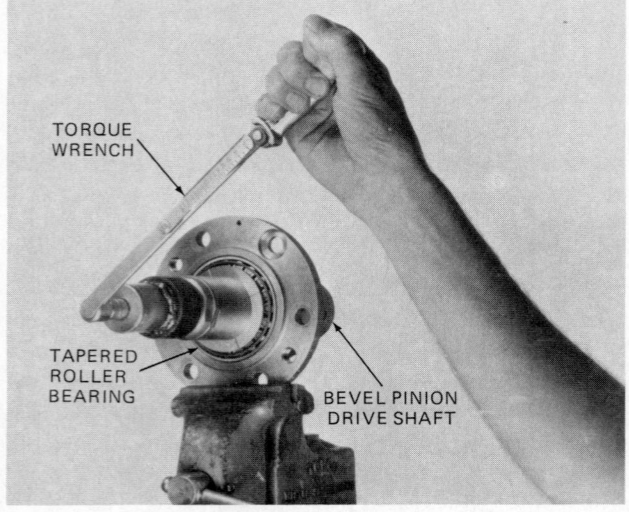


**Fig. 7-38.** Adjusting a tapered roller bearing on a bevel pinion drive shaft before installation into final drive housing. The adjusting nut is tightened to specifications with a torque wrench to achieve the desired bearing preload. (*Massey-Ferguson, Inc.*)

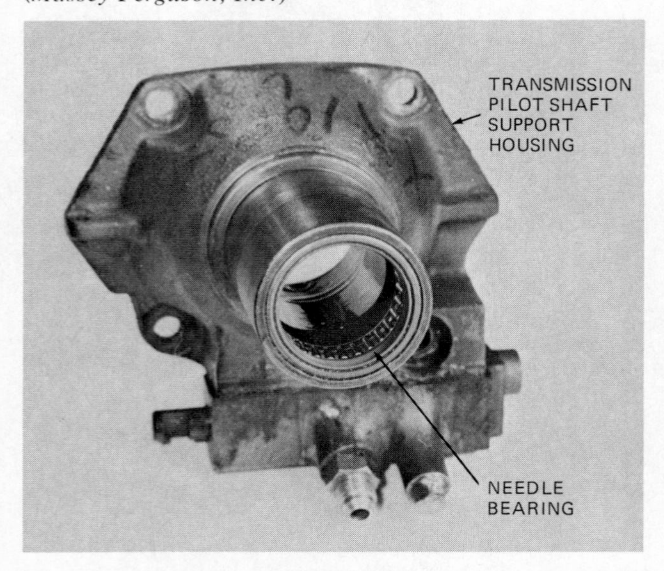


**Fig. 7-39.** Needle bearings are used where space is limited and the pressure is great on shaft and bearing supports. (*Massey-Ferguson, Inc.*)

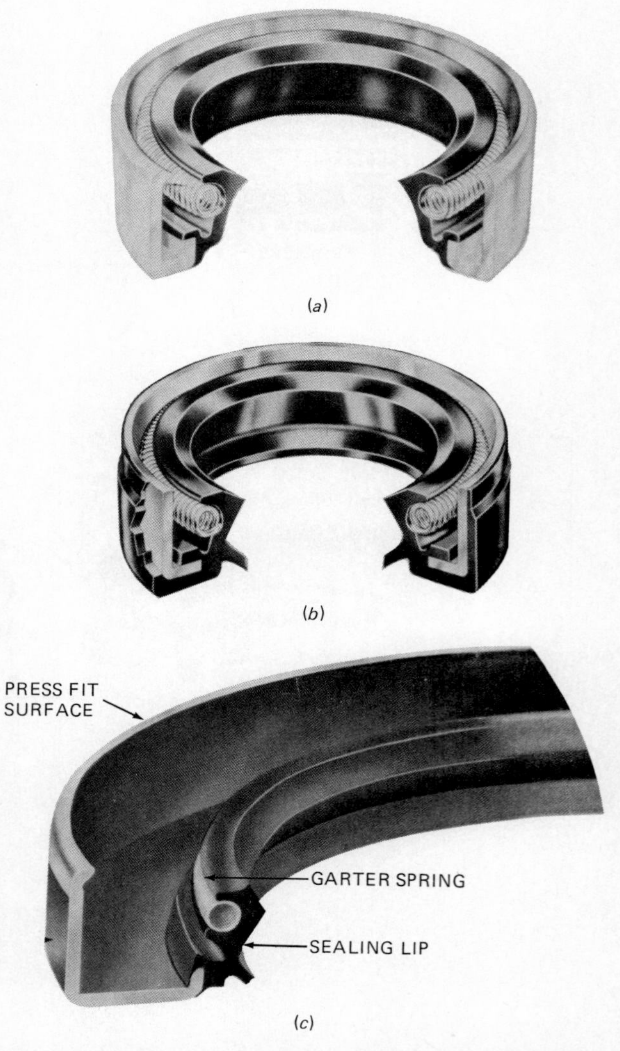


**Fig. 7-40.** Oil seals. (*a*) Single- and (*b*) double-lip type seal. (*c*) Cross-section shows the *garter spring,* which keeps the oil sealing lip in contact with the shaft. (*Perfect Circle*)

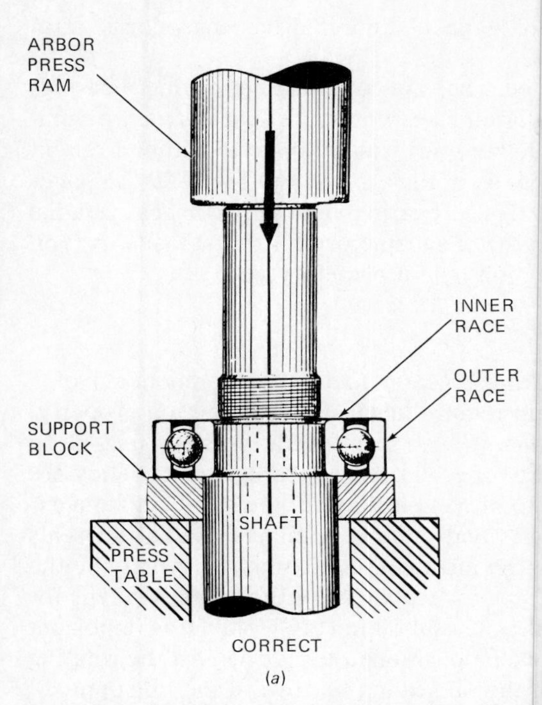


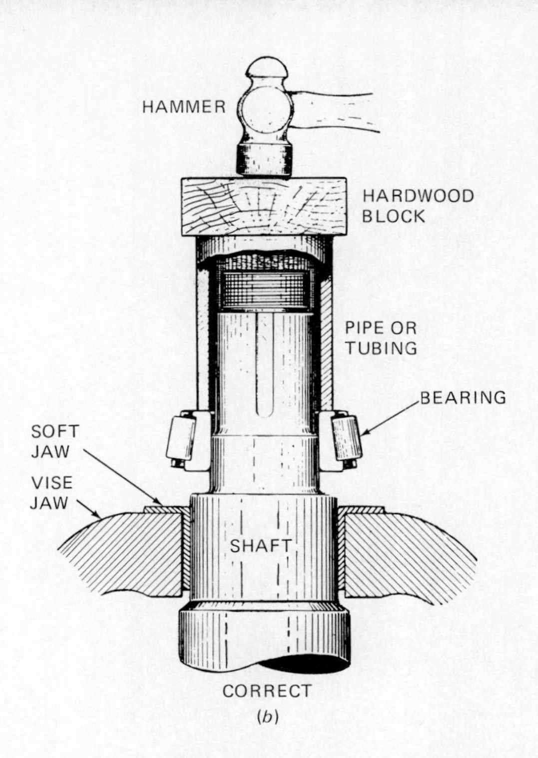


**Fig. 7-41.** Removing and installing a bearing. Never hammer on a ball bearing to remove it. (*a*) When using an arbor press, support the inner race with blocks which should touch the shaft. (*b*) Use a sleeve and wood block to install a bearing. Do not use a hammer directly on the bearing when installing it. (*Anti-Friction Bearing Manufacturer's Association*)

## Bearings

The bearings used in agricultural machinery are classified as plain bearings or bushings and antifriction types. *Plain* bearings are generally used in applications which are classified as low-speed. Most plain bearings are lubricated by hand, using the conventional oil can or grease gun. Specially designed plain (insert) bearings are used in engines for connecting rod, camshaft, and crankshaft bearings (Fig. 7-34 on page 130).

The three types of *antifriction* bearings (Fig. 7-35 *a–c*) used in power trains are ball, roller, and needle.

*Ball* bearings are used in a great many applications because they can carry both radial and thrust loads. Radial loads in a bearing occur as the force is applied at right angles to the shaft. Thrust loads, sometimes called axial loads, are applied parallel to the drive shaft. A special ball thrust bearing is used as the clutch release bearing (Fig. 7-36*a*).

*Roller* bearings are used in the transmission and final drive assemblies of most agricultural machinery. The most common is the tapered roller. This type of bearing is designed to handle heavy thrust load conditions. It consists of four parts: cup, rollers, separator, and cone (Fig. 7-36*b*). Because of its construction the bearing can be used to make adjustments in a

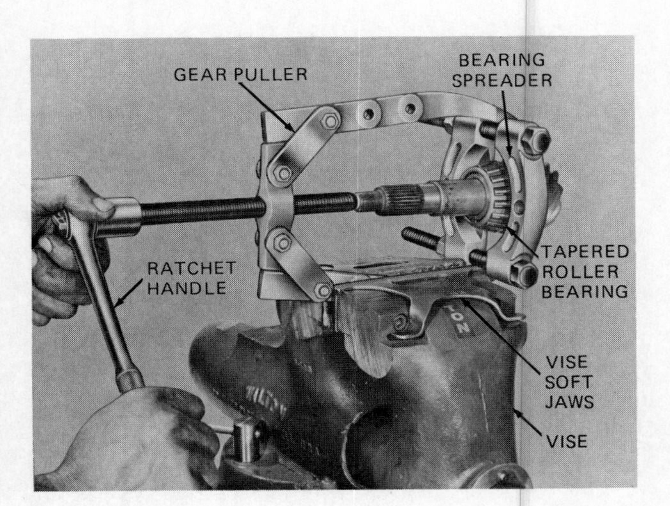


**Fig. 7-42.** Removing a tapered roller bearing from a pinion gear shaft with a special bearing attachment. (*Owatonna Tool & Co.*)

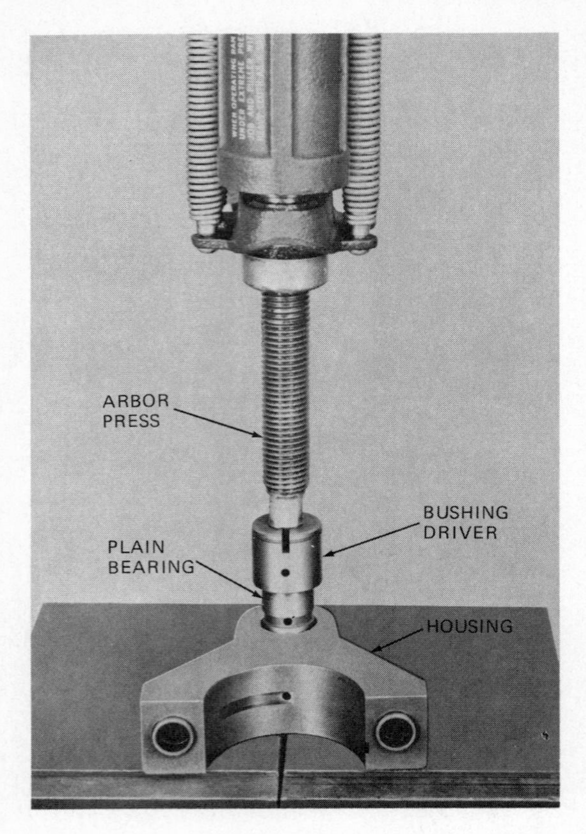


**Fig. 7-43.** An arbor press is an important tool for "pressing" bearings and shafts. In this illustration, the bushing driver is used to press the plain bearing into the housing. (*Massey-Ferguson, Inc.*)

gear train. For example, the tapered-roller-bearing front-wheel assembly (Fig. 7-37 is adjusted by tightening the axle castle nut to reduce the amount of end play in the wheel hub. Increasing the tension on the nut will place additional tension on the bearings, causing a "drag" in rotation. This is referred to as *preloading* the bearing and is specified by the equipment manufacturer. Figure 7-38 illustrates the procedure for adjusting the preload on the tapered roller bearings of a bevel pinion drive shaft for a farm tractor.

*Needle* bearings are small-diameter roller bearings which are often used where the space is very confining. Needle bearings usually operate within a retaining housing, as in Figs. 7-35*c* and Fig. 7-39 (on pages 131–132). These bearings can support heavy radial loads but are *not* satisfactory for thrust loads. A typical application is in a planetary gear set.

## Seals

All bearings, gears, and other components of a power train require lubrication to function properly. Seals confine the lubricant to transmission final drive cases and to special bearing arrangements. They are also used to prevent leakage of the lubricant from exposed shafts and control openings. Radial lip seals are one of the most common kinds. They may be the single-lip type for sealing oil or the double-lip type for keeping dirt out and oil in (Fig. 7-40). It is important that the seal be positioned to ride against the rotating shaft with the oil sealing lip toward the oil supply.

## Tool Requirements

Special tools are required to remove and install bearings and seals. These pullers, drivers, and sleeves are necessary to prevent distortion or damage (Fig. 7-41). The pulling attachment illustrated in Fig. 7-42 is being used to remove a tapered roller bearing from a final drive pinion gear shaft.

Care must be exercised when installing plain bearings to prevent damage to the thin surface. An arbor press is often used (Fig. 7-43) when the bearing is installed with a *press-fit*. When a bushing is installed into a housing with the force of a hammer, a special *bushing* driver fitted with the proper-size adapter is used.

Seals must be installed using driving equipment, as shown in Fig. 7-44. This equipment prevents damage

**Fig. 7-44.** An oil seal installed in the final drive housing. Note the special driver for installing the seal without distortion. (*Massey-Ferguson, Inc.*)

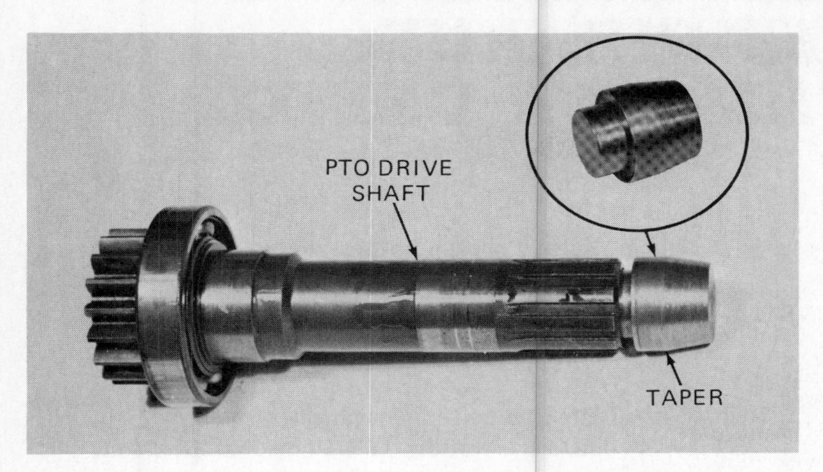


**Fig. 7-45.** A taper is used to protect seals when installing them on shafts with sharp edges on the spline end. (*Massey-Ferguson, Inc.*)

to the seal and assists in performing quality workmanship. Protection must be provided to the sealing lip when pushing the seal over shafts. Sleeves and protective tapers prevent damage from keyways, threads, burrs, and sharp edges (Fig. 7-45).

## SUMMARY

This chapter has described many of the principles of power transmission that are a part of engaging, disengaging, and changing the speed ratio and torque to control engine power. These operations are accomplished by various types of clutches, transmissions, and final drive assemblies that are used in modern agricultural machinery. Service people who will repair and make adjustments to these units must rely upon specific service and repair manuals provided by manufacturers of machinery to obtain specifications and detailed instructions. The data in the manuals provide procedures and specifications to make the proper decisions. Also, special tools are listed and often illustrated to assist in repair operations and to prevent unnecessary damage to bearings and precision-fitted parts.

## THINKING IT THROUGH

1. Identify the four major components of a power train and list the function of each.
2. What is the purpose of each of the following parts of a disk clutch?
   (a) Flywheel
   (b) Pressure plate
   (c) Friction disk
   (d) Clutch release bearing
3. Why is it important to have free pedal travel when making linkage adjustments of a disk clutch?
4. List the five types of clutches used in agricultural equipment and identify one feature of each.
5. What is the difference between a constant mesh and a synchromesh transmission?
6. Why are planetary gears used in many power shift transmissions?
7. Why does a hydrostatic transmission have an infinite range of speed from the low to high settings of the control lever?
8. What controls the rate of oil discharge from the pump of a hydrostatic unit?
9. What is the purpose of each of the following parts of a final drive assembly?
   (a) Bevel pinion and bevel (ring) gear
   (b) Differential assembly
   (c) Individual wheel or differential brakes
   (d) Bull gears
10. Why is the size of the final drive axle shaft used to turn the drive wheels so much larger in diameter than the transmission input shaft of a farm tractor?
11. What is a transaxle?
12. Why are transaxles used on lawn and garden tractors?
13. Why are tapered roller bearings often used on the bevel pinion ring gear assembly?

# CHAPTER 8 POWER TRANSMISSION BY BELTS, CHAINS, AND POWER TAKEOFF DRIVES

## CHAPTER GOALS

In this chapter your goals are to:

- Identify the types of belt drives
- Identify the types of chain drives
- Describe the characteristics of PTO drives
- Service and maintain belt, chain, and PTO drives
- Observe safe practices when working with belts, chains, and PTO drives

Historically, manufacturers of agricultural machinery have made extensive use of belts and chains to supply power. For many years leather belts made from tanned steer hides were used. Leather combined the characteristics of strength, flexibility, and friction desired in a belting material. Chains were used when the slippage associated with belts was not desirable. Many early agricultural machines used a drive shaft which included universal joints to transfer power.

Although many refinements have been made in construction, manufacturers of modern agricultural machinery still make extensive use of belts, chains, and the power takeoff (PTO) to serve as the power transmission system. As an operator or machinery mechanic, you will be required to perform maintenance operations on machinery equipped with these methods of transmitting power.

## POWER TRANMISSION BY BELTS

The most common form of transmitting power to various operational units of farm machinery is the *belt*. Belts are friction drives; that is, the belt makes contact with a specially shaped wheel called a *sheave* (or *pulley*). A portion of the belt makes contact with the sheave (or pulley) and establishes a frictional force. When the driver sheave (or pulley) rotates, it exerts a force on the belt, which in turn transfers the driver force to the driven sheave (or pulley) (Fig. 8-1). *Note:* In most agricultural applications, the term "pulley" refers to a flat-faced wheel used for flat belts and also for idlers and tighteners for V belts.

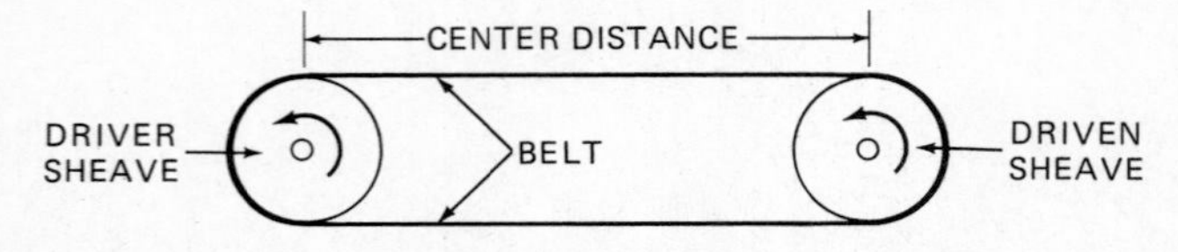


**Fig. 8-1.** In a belt drive, the driver sheave (pulley) transfers power to the driven sheave (pulley) by a continuous belt.

The term "sheave" refers to a grooved wheel designed for V belts.

Belts have particular advantages compared to other methods of power transmission:

- Belts are economical and relatively trouble-free.
- Belts require no lubrication.
- Belt drives produce little noise in operation.
- Belts will operate with some misalignment of sheaves or drive shafts.
- Belts can transmit power with very little loss in efficiency.
- Belt drives will absorb shock loads, thus reducing the effect of irregular forces on engines, drive trains, and bearings.
- Belts are quite resistant to the abrasive effect of operating in dusty conditions commonly associated with farm machinery.

When compared to chains, belt drives have the disadvantage of being more easily damaged or affected by oil, grease, sunlight, and extreme heat.

### V-Belt Drives

*V belts* are the most common form of belt drives for transmitting power in modern agricultural machinery. The different types are conventional, double-angle, adjustable-speed, multiple-strand banded, and ribbed belts.

V belts are constructed by using a tough fabric and rubber material which in cross section are formed to a specific trapezoidal shape (Fig. 8-2). The internal construction of a single V belt is composed of three basic parts (Fig. 8-3) as follows:

1. The *tensile member* is a cordlike material running endlessly around the belt. The tensile member transmits the power from one sheave to the next and keeps the belt from pulling apart when under load.
2. The *undercord* absorbs the power forces of the driver sheave and transmits it to the driven sheave.
3. The *band* is a cover wrapping which provides correct frictional characteristics and protects the internal parts.

V belts are sized according to an industry standard. Figure 8-3 illustrates the single, double-angle, and adjustable-speed V belts used in agricultural industry. A belt marked with the letters and numbers HC485 describes a conventional agricultural-type V

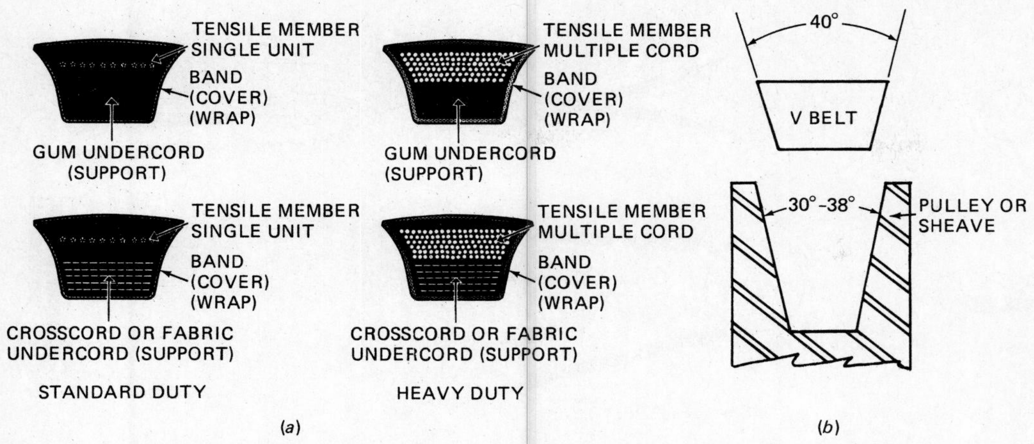

Fig. 8-2. Cross section of V belts, showing the three basic parts (*a*) and the shape of the belt and sheave (pulley) (*b*). (*Gates Rubber Co.*)

belt having a nominal width of ⅞ inches (in) [22.2 millimeters (mm)]. The number 485 means that the belt is 48.5 in [1232 mm] in length.

Single V belts are normally used to transmit rotation so that one shaft rotates in the same direction as the other. A crossed single V-belt drive (Fig. 8-4) will cause a reversal in the direction of rotation of the driven shaft from the drive shaft. A quarter-turn drive allows the shafts of the driver and driven to be 90° out of parallel and rotate in the same direction.

Double-angle V belts are used in installations where both the top and bottom of the belt must transmit power, such as the serpentine drive illustrated in Fig. 8-5.

It is common practice to use single V belts in sets of two or more. However, most agricultural machinery utilizes the banded V belt or the V-ribbed belt. The banded V belt (Fig. 8-6) consists of multiple single-angle V belts which have a tie band permanently attached to the back of each belt. This type of belt eliminates the problem on multiple-strand drives of single strands turning over or jumping off the sheave. It also eliminates the chance of obtaining belts of different lengths during replacement. V-ribbed belts are a combination V belt and flat belt (Fig. 8-7). The V-ribbed belt has the strength and simplicity of a flat belt and retains the positive positioning of the V belt. V-ribbed belts do not rely upon the wedging action of the V belt to transmit power. They must operate with the belt tight between the sheaves to achieve friction between the sheave and belt.

**Variable Speed** Adjustable-speed V belts are designed for use on variable-speed belt drives. Variable-speed drives make it possible to adjust the rotation of the driven sheave through a wide range of operating speeds. This drive is popular for combine harvester cylinder drives and some ground wheel drive applications. The principle of operation involves the use of sheaves which are constructed so that the effective diameter can be changed by moving one or both sheave halves closer or farther apart (Fig. 8-8). The change in diameters can be controlled manually to achieve the desired change in speed ratio. Figure 8-9 illustrates the operation of the variable-speed belt drive for the cylinder of a combine harvester. The hand crank (1) changes the width of sheave (4), which drives both the forward cylinder, through variable-width sheave (3), and the rear beater (5). Lever (2) locks the hand wheel. The sheave width in modern combines is changed by using a hydraulic control in the operator's cab.

## Determining Speed Ratio

As identified in Chap. 7, speed ratio expresses the difference in rpm of power transmission components. Figure 8-10*a* shows the driver shaft turning a belt sheave at 1000 r/min. The belt drives the driven sheave at the same speed because both sheaves are

| DETAILS | CROSS SECTION | SIZE DESIGNATOR | in. | | mm | |
|---|---|---|---|---|---|---|
| | | | WIDTH | DEPTH | WIDTH | DEPTH |
| | WIDTH / DEPTH | HA | 1/2 | 5/16 | 12.7 | 7.9 |
| | | HB | 21/32 | 13/32 | 16.7 | 10.3 |
| | | HC | 7/8 | 17/32 | 22.2 | 13.5 |
| | | HD | 1-1/4 | 3/4 | 31.8 | 19.0 |
| | | HE | 1-1/2 | 29/32 | 38.1 | 23.0 |
| | WIDTH / DEPTH | HAA | 1/2 | 13/32 | 12.7 | 10.3 |
| | | HBB | 21/32 | 17/32 | 16.7 | 13.5 |
| | | HCC | 7/8 | 11/16 | 22.2 | 17.5 |
| | | HDD | 1-1/4 | 1.0 | 31.8 | 25.4 |
| | WIDTH / DEPTH | HI | 1.0 | 1/2 | 25.4 | 12.7 |
| | | HJ | 1-1/4 | 19/32 | 31.8 | 15.0 |
| | | HK | 1-1/2 | 11/16 | 38.1 | 17.5 |
| | | HL | 1-3/4 | 25/32 | 44.4 | 19.8 |
| | | HM | 2.0 | 7/8 | 50.8 | 22.2 |

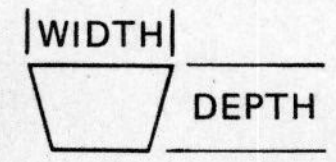

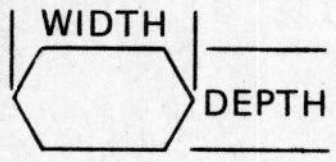

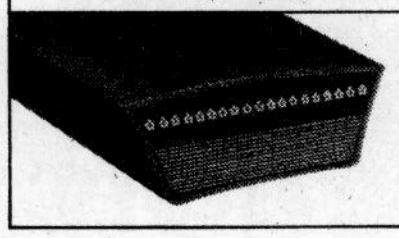

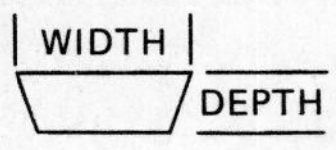

Fig. 8-3. Size and designator for (*a*) conventional, (*b*) double-angle, and (*c*) adjustable-speed V belts. (*Gates Rubber Co.*)

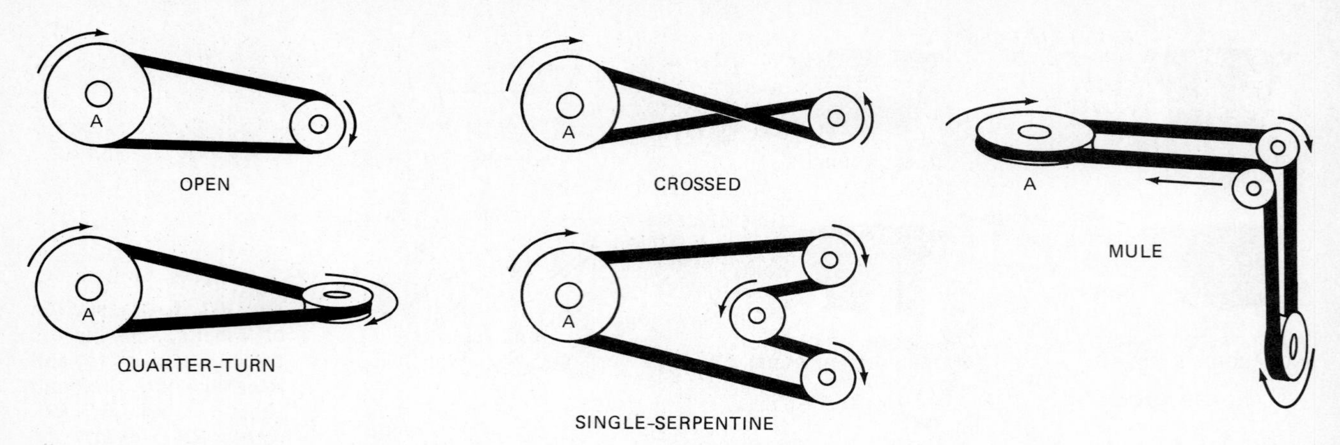


**Fig. 8-4.** Types of single V belt drives using conventional and double-angle V belts. Pulley (sheave) A is the driver in all drawings. (*Gates Rubber Co.*)

the same size. In Fig. 8-10*b* the driven sheave is 10 in [254 mm] in diameter, which is twice as large as the driver. As a result, the driver sheave, rotating at 1000 r/min, drives the driven shaft at 500 r/min. The rpm of the driven shaft can be determined by applying the following formula:

$$\frac{\text{rpm of driver}}{\text{rpm of driven}} = \frac{\text{diameter of driven}}{\text{diameter of driver}}$$

Thus,

$$\frac{1000 \text{ r/min}}{x \text{ r/min}} = \frac{10 \text{ in}}{5 \text{ in}}$$

Multiply both sides by $x$ and 5 in, and cancel. Thus,

$$10x = 5000$$

Divide both sides by 10:

$$\frac{\not{1}\not{0}x \text{ (r/min)}}{\not{1}\not{0}} = \frac{500\not{0}}{1\not{0}}$$

$$\text{Driven} \times \text{(r/min)} = 500$$

The expression of speed ratio identifies the difference in rotating speeds of the driver and driven shafts. Speed ratio is figured by dividing the rpm of the driver by the rpm of the driven:

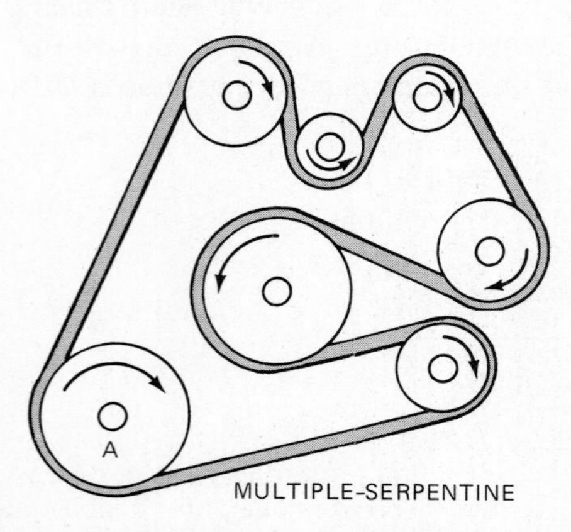


**Fig. 8-5.** A multiple serpentine-belt drive requires the use of a double V belt.

$$\text{Speed ratio} = \frac{1000 \text{ r/min}}{500 \text{ r/min}}$$

$$= \frac{1000}{500} = \frac{2}{1} \quad \text{or} \quad 2:1$$

The formula given previously to determine the rpm of the driven can also be used to solve an equation for an unknown sheave size. For example, suppose you were operating a grain elevator with an electric motor (Fig. 8-11). The speed of the electric motor is 1750 r/min and is equipped with a 3-in [76-mm] sheave. The grain elevator is designed to run at

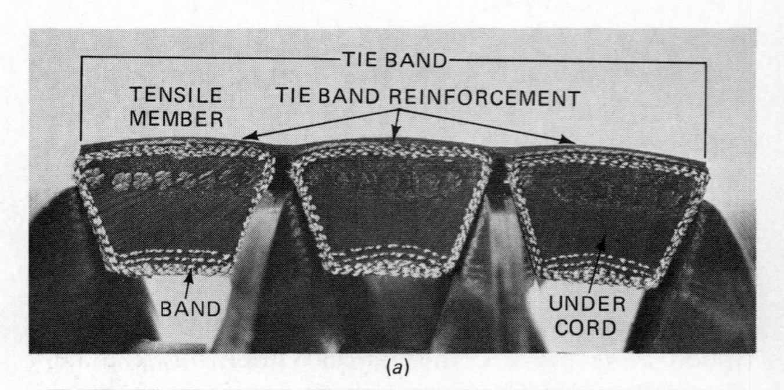


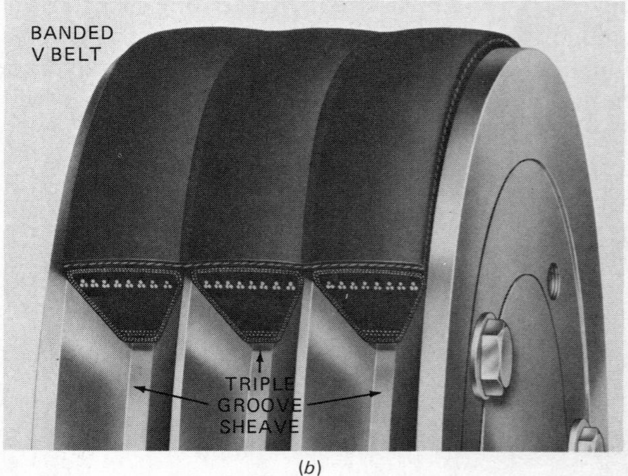


**Fig. 8-6.** Multiple strand, banded V belts are used to transmit power for heavy loads. (*Gates Rubber Co.*)

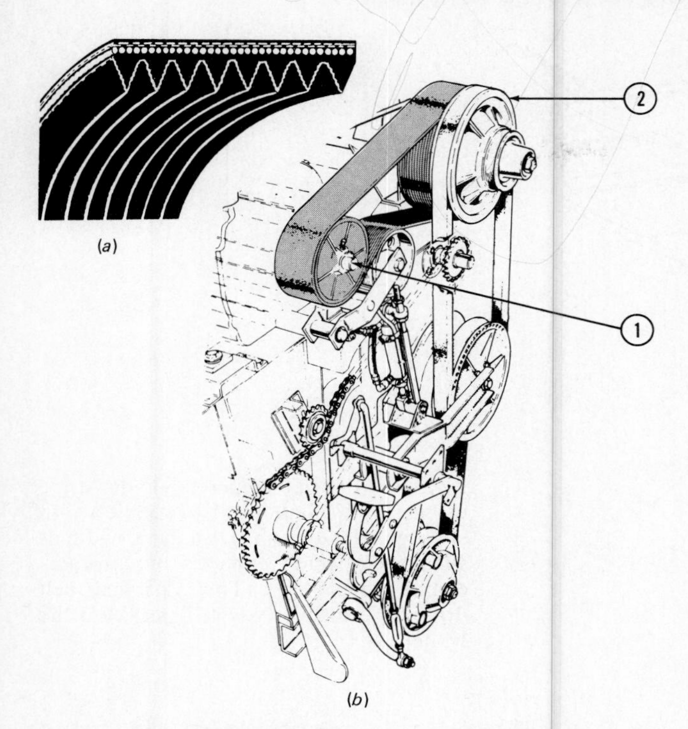


**Fig. 8-7.** (*a*) Cross section of a V-ribbed belt. (*b*) A V-ribbed belt transmitting engine power (1) to a combine harvester separator's pulley (2).

750 r/min. How large a sheave should be installed on the grain elevator? Solve by inserting into the formula as follows:

$$\frac{1750 \text{ r/min of driver}}{750 \text{ r/min of driven}} = \frac{\text{diameter of driven}}{\text{3-in [76-mm] driver}}$$

$$\text{Diameter of driven} = \frac{5250}{750}$$

Diameter of grain elevator sheave = 7 in [178 mm]

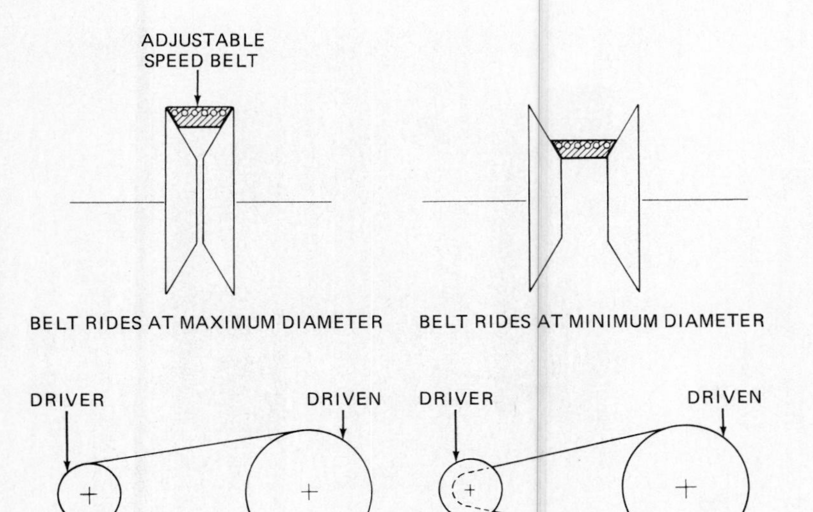


**Fig. 8-8.** Variable-speed V belt drive. (*Gates Rubber Co.*)

## Installing, Adjusting, and Maintaining V Belts

Although V-belt drives are capable of providing dependable service in adverse operating conditions, their useful life can be extended, thus reducing operating costs, by applying proper maintenance practices. A characteristic of V-belt drives is that they can generally be easily inspected during operation. Belts should be checked for misalignment of sheaves, slippage and the wear it can cause, and damage from oil, fuel, and grease.

A good operator and maintenance person will use a straightedge to check the alignment of the drive shafts and sheaves. When the shafts are not parallel, the sheaves will be misaligned (Fig. 8-12). Misalignment may also occur when the sheaves become loose on the shafts. Excessive misalignment will result in reduced belt life.

To install V belts on sheaves, loosen the belt tension idlers or the mounting bolts to slide the drive shafts together and reduce the center distance. Place the belt over the sheaves and adjust the tension. *Never attempt to pry a belt over the sheaves* (Fig. 8-13), as this will cause cords within the belt to break.

V belts must be operated at the proper tension. When belts are too loose, slippage will occur which will cause excessive band wear, belt overheating, or even burned spots. The best tension is the lowest amount which will prevent the belts from slipping under full load. Too much tension will cause belt stretching, belt overheating, and sheave damage. Excess tension will also place stress on the drive shaft bearings, shortening their life.

Information on the correct tension to place on V belts is provided in most operators' or service man-

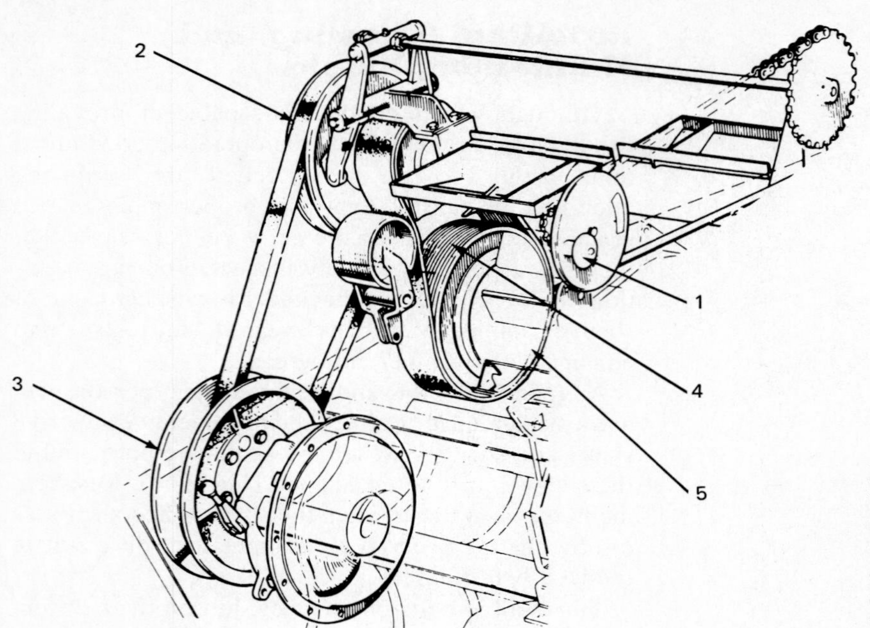


Fig. 8-9. A variable-speed cylinder drive using a V belt and variable-width sheaves. Handcrank (1) changes width of sheave (2) which expands or contracts engine sheave (3). The V-ribbed belt drive (4) transmits power from (2) to the beater (5).

uals. These data may be expressed by the amount of "deflection" the belt will develop (1) when measured at a specified number of pounds of pull or (2) when depressed at a point midway between the sheaves (Fig. 8-14).

During the run-in period, V belts become seated to the sheaves and some initial stretching takes place.

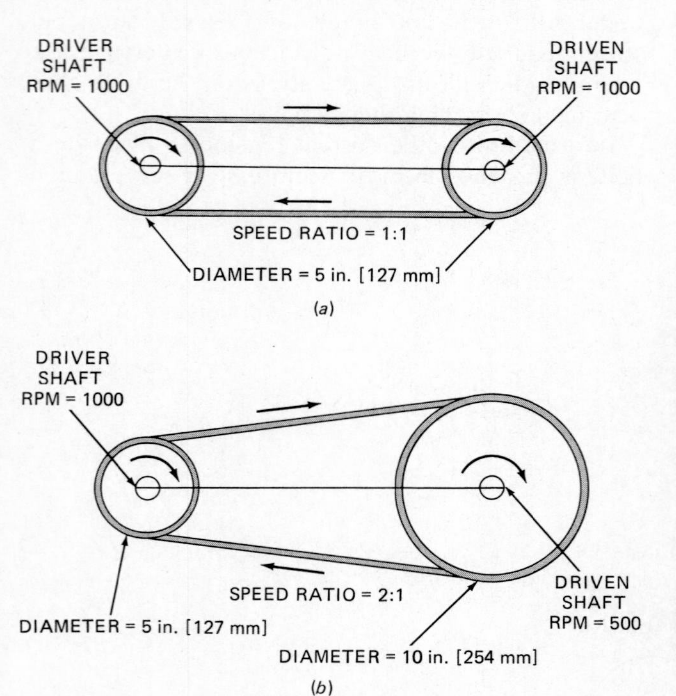


Fig. 8-10. Speed ratio illustrated. In (*a*) the rpm of the driver and driven shafts are equal for a speed ratio of 1:1. In (*b*), the driver shaft rotates at twice the speed of the driven (1000 r/min/500 r/min) for a speed ratio of 2:1.

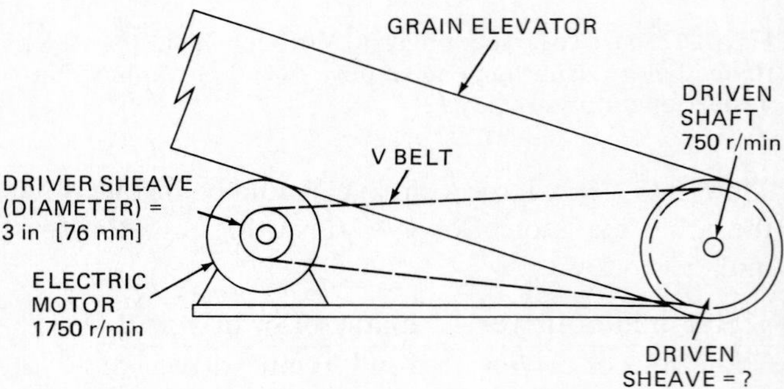


Fig. 8-11. Determining what size sheave is required by a grain elevator operated by an electric motor running at 1750 r/min with a 3-in (76-mm) driver sheave, if the driven shaft is designed to operate at 750 r/min.

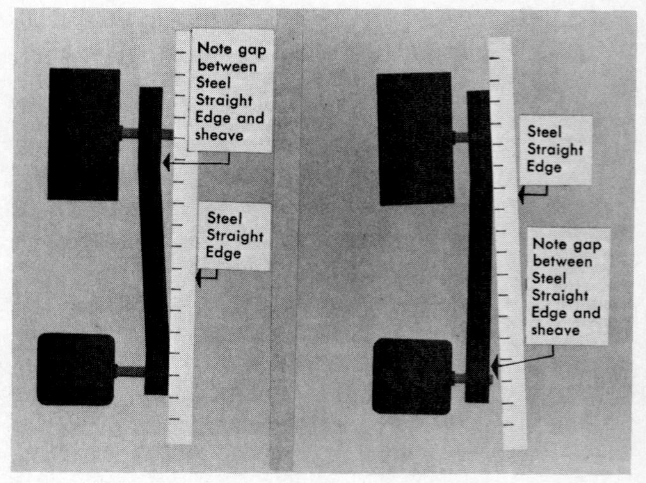


Fig. 8-12. Checking the alignment of the shafts and sheaves using a straightedge. (*a*) Nonparallel alignment. (*b*) Shaft angular misalignment. (*Gates Rubber Co.*)

(a) (b)

**Fig. 8-13.** (*a*) When installing or removing belts, loosen the mount or tighteners and provide belt slack. (*b*) Avoid prying belt(s) over the (sheaves). (*Gates Rubber Co.*)

Watch all new belts carefully, especially during the first hour of operation. It is a good plan to check tension periodically thereafter and make the necessary adjustments. The belt dressing used to prevent slippage for flat belts is not recommended for V belts. A screeching sound will develop if the belt is slipping. When this sound occurs, an immediate check of belt tension is necessary to prevent damage to the belt. If the V belts develop a squeaking sound, a light spray of silicone-base belt lubricant will relieve the condition.

Maintenance of belts also involves proper cleaning and storage. Greases and oils can be removed by using a waste cloth. Dampen the waste cloth with a detergent solution and then wipe dry with a clean cloth.

Belts should be loosened or removed from machines that are out of use for seasonal storage. It is good practice to clean belts as they are removed and then spray them very lightly with a silicone compound and store them in a cool, dry place out of direct sunlight. Matched belts used in double-or triple-groove sheaves must be stored as a set to prevent mixups. Try to maintain the original shape of the belts instead of coiling them tightly. This will help prevent distortion and damage to the cords.

The bright surfaces of the sheaves should be coated with an antirust compound when storing machines. In addition, the sliding parts of variable-speed drives should be given a protective coating of a rust- and corrosion-prevention agent.

When it is not convenient to remove a V belt from a machine, loosen the tensioning devices to prevent the belt from taking a "set" around idler pulleys and to prevent sharp turns in the belt.

## POWER TRANSMISSION BY CHAINS

Chains of various types have been used to transmit power in agricultural machinery for many years. Early machines were often powered by the friction of drive wheels in contact with the earth as they were being pulled by animals. Chains were used to transfer this rotating force to other parts of the machine. Chain drives function as gearing because they transmit power without slippage. As a result, chain drives maintain a positive speed ratio between the driver and driven shaft. Chain drives will transmit heavier loads and perform adequately using small-diameter "sprockets" and short center distances (Fig. 8-15). Chains are also impervious to deterioration due to oils, greases, and sunlight. For slow-speed drives, chains are more practical than belts. The principle disadvantages of chain drives are that (1) they are generally noisy, (2) they may require frequent lubrication, and (3) they can accept only slight misalignment.

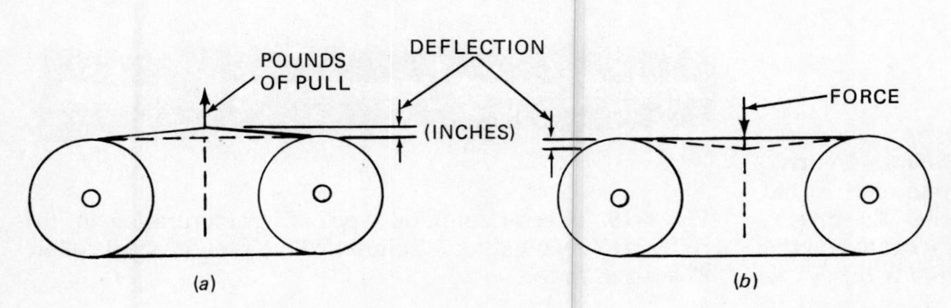


**Fig. 8-14.** The correct belt tension is obtained by applying a specified amount of outward pull (*a*) or inward force (*b*) while measuring the deflection at the center distance midpoint. (*Gates Rubber Co.*)

## Types of Chains

Many different types of chains are used in power transmission. The types most common to agricultural machinery applications are (1) the roller chain, (2) the detachable steel or cast-iron link chain, (2) the pintle chain, and (4) the conveyor steel-roller chain (Fig. 8-16).

**Roller Chain** Roller chain is constructed by assembling alternate roller and pin sets (Fig. 8-17). The roller links consist of two side bars, two bushings, and two rollers. Pin links have two side bars and two pins which are assembled by press-fit (pressed or driven together) into the side plates. Some roller chain pins are designed so that one side plate is held in place by small cotter pins through the ends of the pins. A split spring steel keeper is also used to hold side plates and pins in place on "disconnecting or master links." Roller chains with offset links (roller and pin link combined) are used to connect chains when an odd number of links are required (Fig. 8-17*e*). Since offset links wear faster than straight links, they are to be avoided when possible.

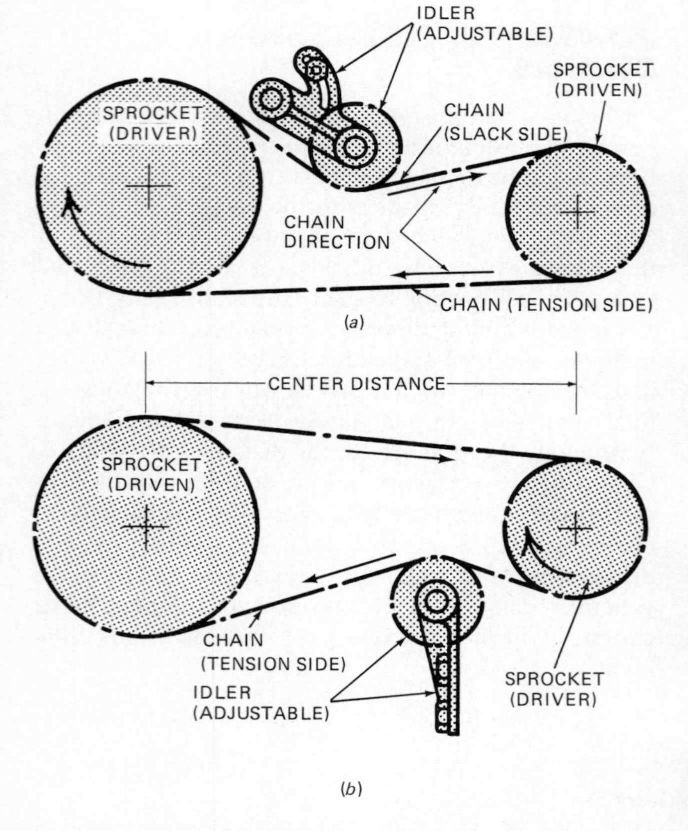


**Fig. 8-15.** Two types of chain drives: (*a*) The *step-up* drive with the large sprocket driving the small one and (*b*) the *step-down* drive, where the small sprocket is the driver. The chain tensioning idler sprocket is always located on the slack side of the chain drive. (*FMC Link Belt Corp.*)

Roller chain is sized according to a number system. The number refers to a specific standard for *pitch, width,* and *roller diameter*. The pitch of a roller chain is the distance between the center of its rollers. The standard sizes of single-pitch roller drive chain are listed in Fig. 8-18. Center distance identifies the spacing between the center lines of the driver and driven shafts.

Double-pitch drive chains, which have twice the distance between roller centers, are used only on light loads and slow speeds. (See the sizes listed in Fig. 8-19.)

**Detachable Link Chain** Detachable link chain is a simple low-cost chain of stamped steel or malleable cast iron (Fig. 8-20). One end of the link is a hook, and the opposite end is a bar. The direction of travel for a detachable link chain is important (Fig. 8-21). This type of chain must travel with the hook ends forward and the slot openings outward to prevent excessive wear of the bars during operation.

**Pintle Chain** Pintle chain is made up of a series of pressed-steel links joined together by pins (Fig. 8-22). This type of chain is used for slow-speed operations where abrasive materials and exposure to weather are critical. Pintle chains must travel with the barrel forward to reduce pin wear.

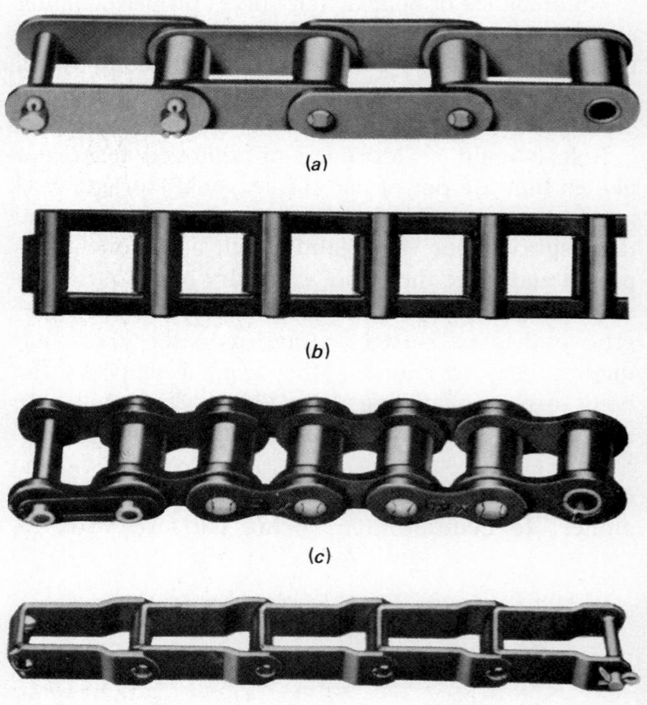


**Fig. 8-16.** Several common types of agricultural chain: (*a*) roller; (*b*) detachable; (*c*) pintle; (*d*) conveyor, steel roller. (*Rexnord, Inc.*)

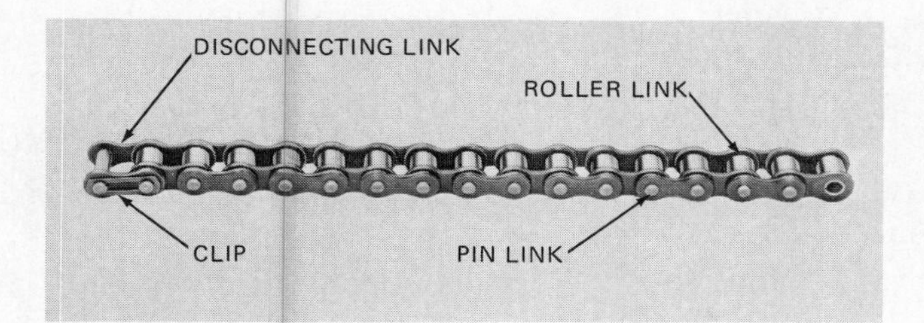


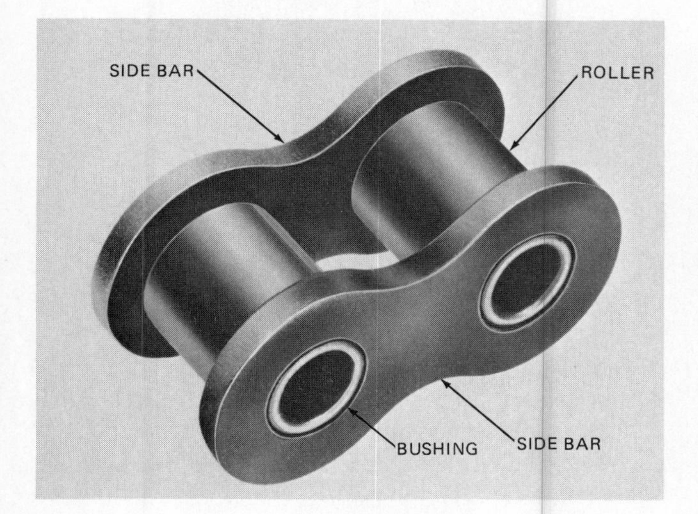


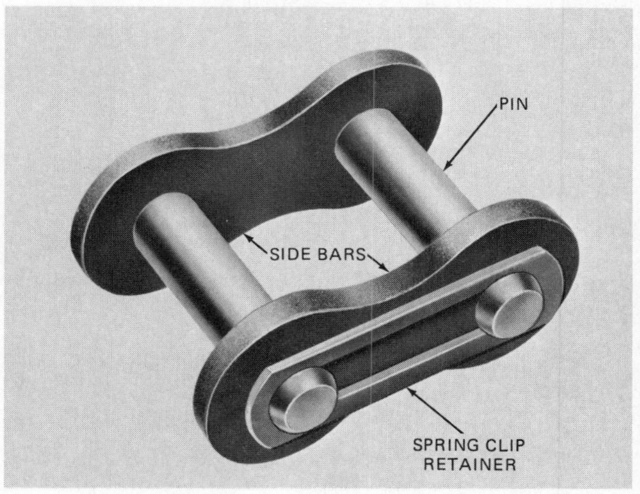


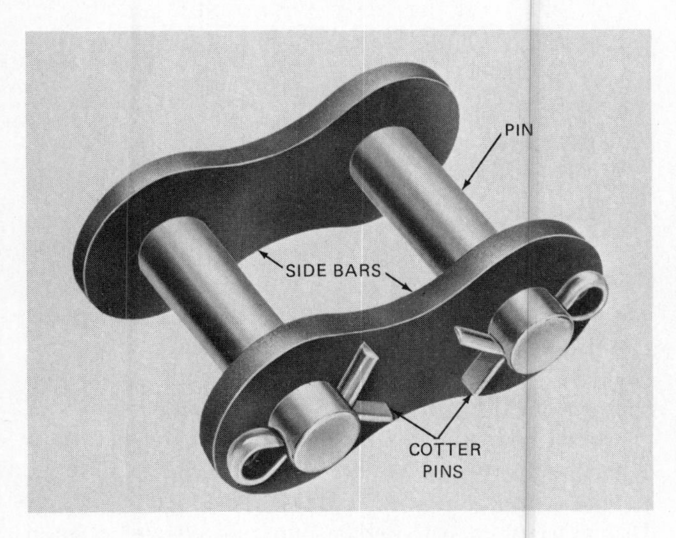


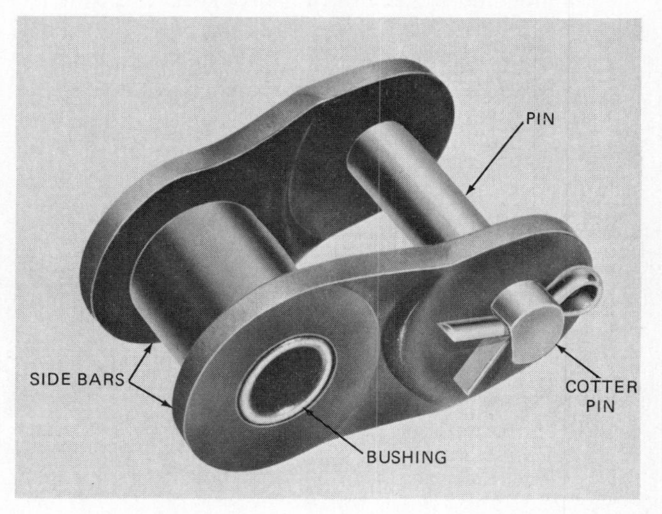


**Fig. 8-17.** Types and parts of roller chain links. (*a*) Roller chain assembly. (*b*) Roller link. (*c*) Connecting pin link, spring clip type. (*d*) Connecting pin link, cutter pin type. (*e*) Connecting offset link assembly. (*FMC Link Belt Corp.*)

**Conveyor Steel-Roller Chain** Straight sidebars of equal height on conveyor steel-roller chain provide a good sliding surface for increased wear resistance.

Conveyor chains are used as gathering chains on corn pickers and forage harvesters; as elevator chains on combines, tobacco harvesters, and corn pickers; and as raddle, apron conveyor, and feeder chains on combines and harvesters.

## Installing, Adjusting, and Maintaining Chain Drives

When installing chains, proper alignment of the sprocket is very important for maximum chain and

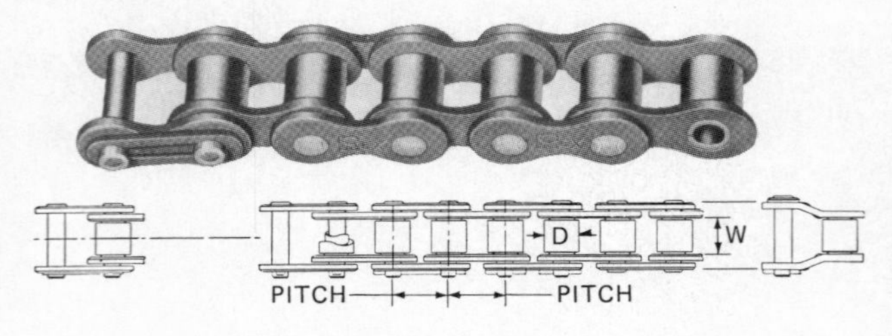


| CHAIN NO. | PITCH, in. [mm] | WIDTH (W), in. [mm] | ROLLER DIAMETER (D), in. [mm] |
|---|---|---|---|
| 40 | 1/2 [12.7] | 5/16 [7.9] | 5/16 [7.9] |
| 50 | 5/8 [15.9] | 3/8 [9.5] | 0.400 [10.2] |
| 60 | 3/4 [19.0] | 1/2 [12.7] | 15/32 [11.9] |
| 80 | 1 [25.4] | 5/8 [15.9] | 5/8 [15.9] |
| 100 | 1-1/4 [31.8] | 3/4 [19.0] | 3/4 [19.0] |
| 120 | 1-1/2 [38.1] | 1 [25.4] | 7/8 [22.2] |
| 140 | 1-3/4 [44.4] | 1 [25.4] | 1 [25.4] |

**Fig. 8-18.** Standard roller chain sizes. *(Rexnord)*

sprocket life (Fig. 8-23). The correct alignment involves the following operations:

1. Level the shafts so that they are running parallel.
2. Align the shafts on their center distances to see that they are parallel. At the same time, align the sprockets on their shafts using a straight edge on the finished surface of the sprockets. If the shafts have end play, align the sprockets with the shafts in their running position. Make sure that the sprockets will not slide on the shafts during operation by tightening any setscrews or collars to the shafts.

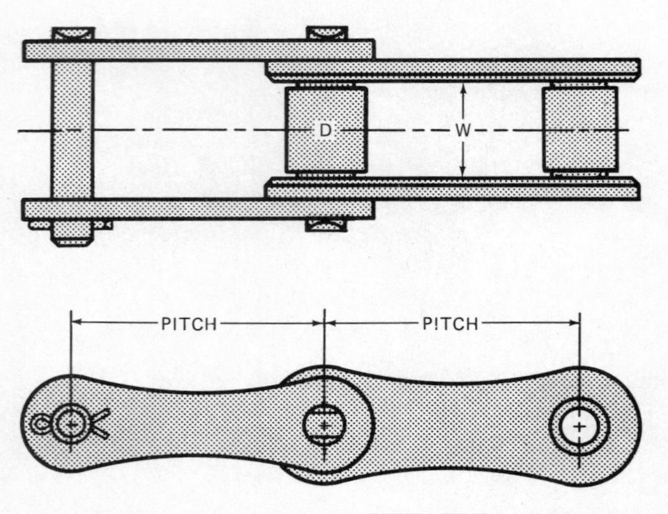


| CHAIN NO. | PITCH, in [mm] | WIDTH (W), in [mm] | ROLLER DIAMETER (D), in [mm] |
|---|---|---|---|
| 2040 | 1 [25.4] | 5/16 [7.9] | 5/16 [7.9] |
| 2050 | 1 1/4 [31.8] | 3/8 [9.5] | 0.400 [10.2] |
| 2060 | 1 1/2 [38.1] | 1/2 [12.7] | 15/32 [11.9] |

**Fig. 8-19.** Double-pitch roller drive chain. *(FMC Link Belt Corp.)*

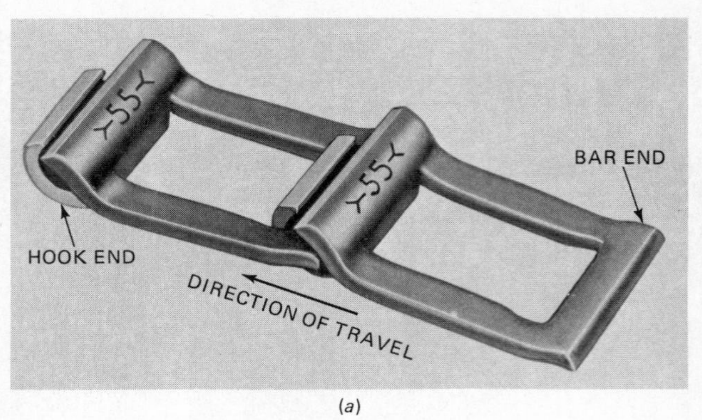


(*a*)

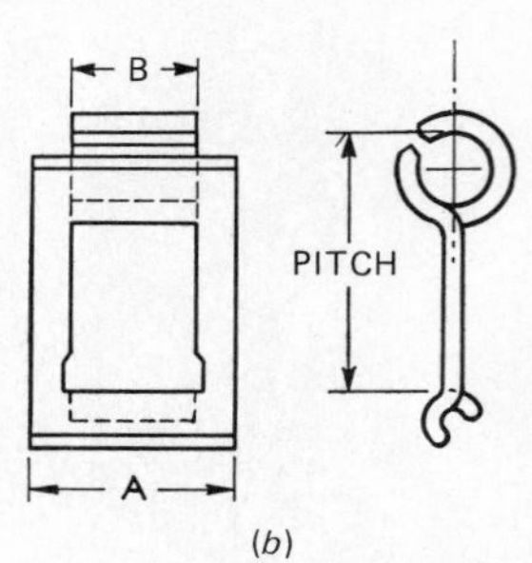


(*b*)

| CHAIN NUMBER | PITCH, in [mm] | DIMENSIONS, in [mm] HOOK (B) | BAR (A) |
|---|---|---|---|
| 25 | 0.904 [23] | 27/64 [10.7] | 45/64 [17.9] |
| 32 | 11.157 [29.4] | 19/32 [15.1] | 15/16 [23.8] |
| 42 | 1.375 [35] | 25/32 [19.8] | 1 1/16 [27.0] |
| 52 | 1.508 [38.3] | 27/32 [21.4] | 1 13/32 [35.7] |
| 55 | 1.630 [41.4] | 51/64 [20.2] | 1 9/32 [32.5] |
| 62 | 1.654 [42] | 63/64 [25.0] | 1 9/16 [39.7] |
| 67 | 2.313 [58.8] | 1 3/32 [27.8] | 1 7/8 [47.6] |

**Fig. 8-20.** Detachable link chain and standard sizes. *(FMC Link Belt Corp.)*

Installing chains on sprockets varies with the type of chain used. The easiest process for installing roller and pintle chain is to loosen the chain tighteners to obtain working slack. Chain tighteners are always located on the *slack side* of the chain. Loop the chain over the sprockets so that the open ends meet over the arc of one of the sprockets. The pin or link pin can then be installed and secured by cotter pins or a split keeper (Fig. 8-24). Readjust the chain tightener so that the chain will depress approximately ¼ in [6.4 mm] for each 12 in [305 mm] of distance between centers of the shaft (Fig. 8-25). For example, if the center distance is 24 in [610 mm], the chain should depress approximately ½ in [12.7 mm].

Care must be exercised when assembling or repairing chains to prevent the side bars from becoming distorted or bent. When the pin is a press-fit (requiring a force to assemble) in the side bar, be sure to support the chain on a flat surface so that it will be level and not distorted during handling and assembly.

Specially designed tools are required to detach press-fit pins from the side bar to prevent damage to the precision roller chain. One type of tool is illustrated in Fig. 8-26. The important rule of thumb is al-

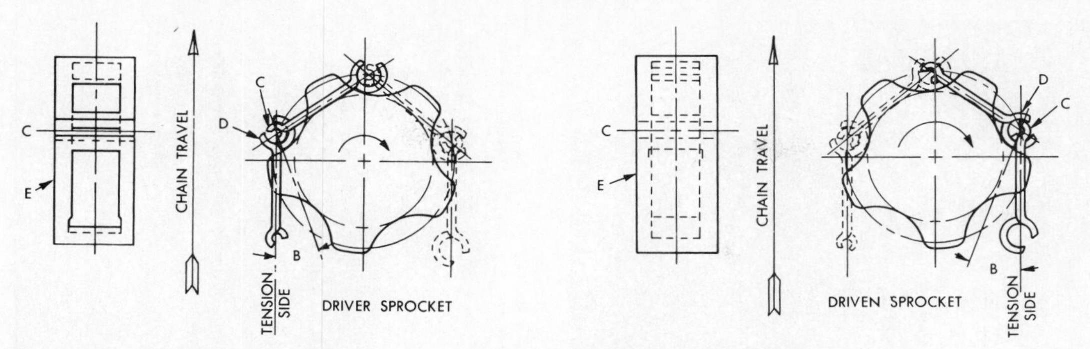


**Fig. 8-21.** The normal direction of travel for a detachable steel link chain is with the hook end (E) traveling forward, as in sketch (*a*). The least wear in the bar (C) and hook (D) occurs because tension has been removed as the chain articulates around the driver sprocket. Sketch (*b*) illustrates that wear occurs in the bar (C) and hook (D) at the driven. However, in step-down drives the diameter of the driven sprocket will be larger than the driver thus reducing the flexing in the hook-bar surface. (*Rexnord, Inc.*)

ways to support a chain when it is being detached! During assembly do not drive the connecting link too far downward on the pins, or it will bind, and therefore oil will not be able to reach the pins and rollers.

Steel or malleable hook-type detachable link chains are installed on or removed from sprockets by first obtaining sufficient slack in the chain to permit one of the links next to a sprocket to be rotated until the side of the link on the bar end is aligned with the opened end of the hook. The side of the link can then be struck lightly with a hammer to drive it out. A detaching tool can also be used. Place the chain in the slot and align the bar with the hook opening and strike the side of the link lightly with a hammer.

When installing link chain, make sure that the hook end is traveling forward and that the hook opening is outward (Fig. 8-21).

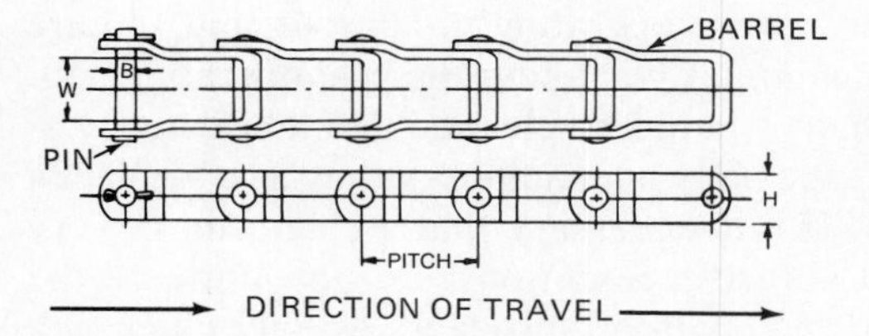


| CHAIN NUMBER | DIMENSIONS, in [mm] PITCH, in [mm] | PIN DIAMETER (B), in [mm] | WIDTH (D), in [mm] | HEIGHT (H), in [mm] |
|---|---|---|---|---|
| 667H | 2.5116 [58.7] | 5/16 [7.9] | 1 1/64 [25.8] | 7/8 [22.2] |
| 667J | 2 1/4 [57.2] | 3/8 [9.5] | 1 1/16 [27.0] | 15/16 [23.8] |
| 667X | 2 1/4 [57.2] | 7/16 [11.1] | 1 1/16 [27.0] | 15/16 [23.8] |

**Fig. 8-22.** Steel pintle chain and standard sizes.

(*a*)

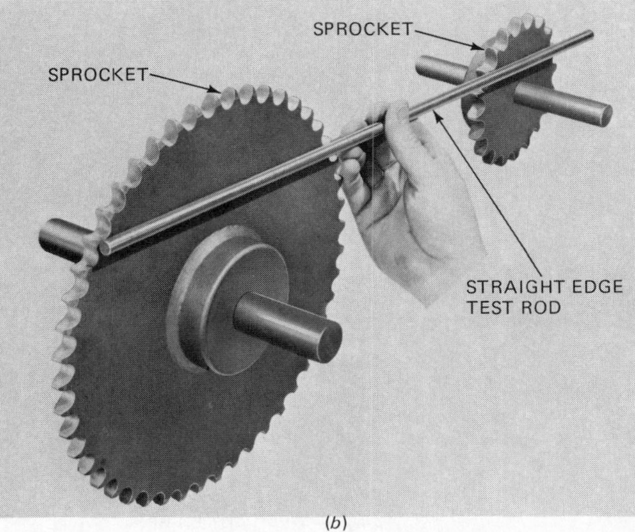


(*b*)

**Fig. 8-23.** (*a*) Align the shafts with a feeler bar to ensure they are parallel. Use a machinist's level on each shaft to check horizontal parallelism. (*b*) Move the sprockets on the shaft to obtain axial alignment. Test with a straightedge or heavy cord to determine accuracy. (*FMC Link Belt Corp.*)

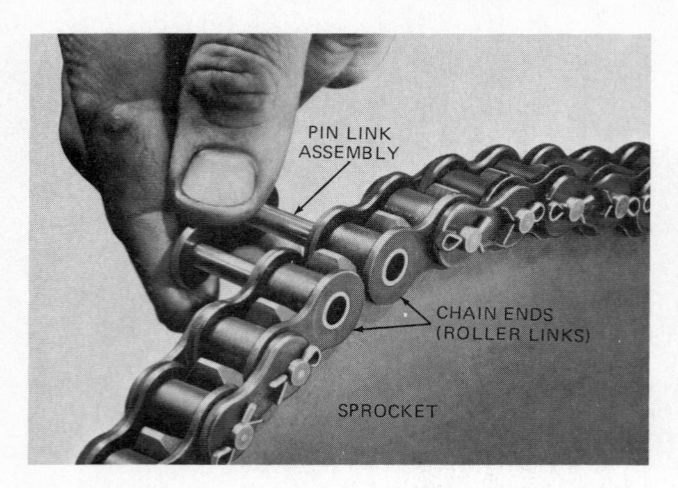


**Fig. 8-24.** Installing roller chain—place the ends of the chain over a sprocket and insert the pin link assembly. (*FMC Link Belt Corp.*)

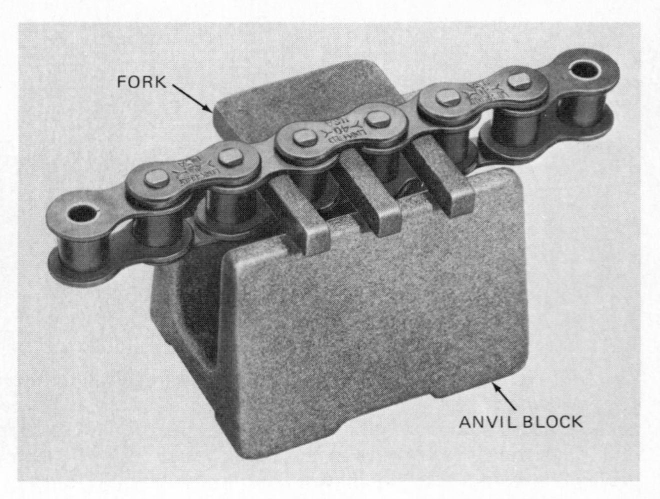


**Fig. 8-26.** The fork and anvil tools for detaching roller chain links are specific for each chain size and are ordered by the chain size number. A pin punch is used to remove the pins from the link bar. (*FMC Link Belt Corp.*)

Roller chains wear in the pin and bushing portion of the links. As this occurs, the chain becomes longer, or "stretches." The pin-bushing wear causes a change in the pitch length of each link. As the wear develops, the chain will no longer fit the sprocket. A badly worn chain may climb over the driven sprocket, breaking the chain.

The rate of chain wear is related closely to the amount of lubrication the pins receive. Therefore, keep chains lubricated with clean engine oil.

Many of the chain drives on farm machinery are exposed to severe operating conditions and require periodic cleaning. Even under the best operating conditions, chains should be cleaned and lubricated seasonally—especially on machinery that is to be placed in storage. If it is necessary that a chain run in very dusty and abrasive conditions, do not lubricate the exterior of the chain or sprocket. An oily chain will pick up dirt, which will accelerate wear. Always refer to the operator's manual for specific service requirements.

To clean chains, remove them from the sprockets and wash them in a suitable solvent such as diesel fuel. When all the exterior dirt accumulations are removed, hang the chains to drip-dry. Next, place them in a pan of heated oil and soak. A chain will take warm oil into the pin and link bushings. The chains should then be allowed to hang for several hours to drain off excess oil. If the machine is to be stored for an extended period of time, place the cleaned and oiled chains in an oil-soaked burlap bag and store them out of the weather. Be sure to apply a coating of grease to the finished surface of all sprockets.

A chain which cannot be removed from the sprockets because of its inaccessibility should be periodically operated by running the machine until the chain becomes warm. Lubricating oil should be manually applied to the chain. This will prevent "frozen" pin and bushing joints due to rust formation (Fig. 8-27).

**WARNING Do not attempt to lubricate the chain while the machine is operating.**

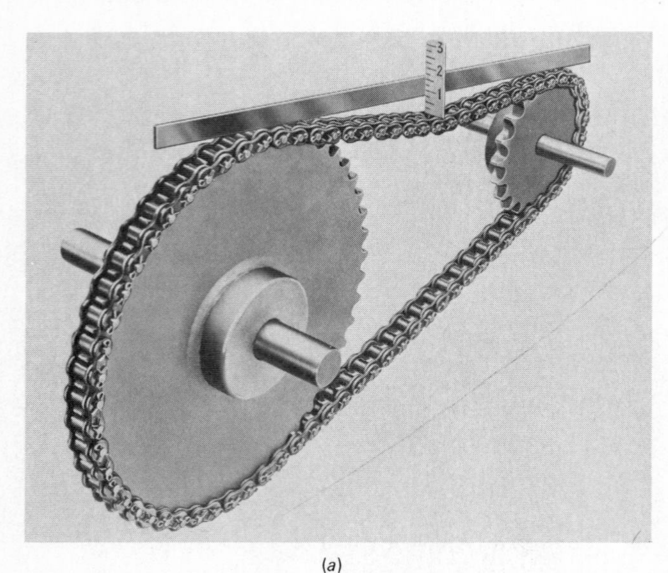


(*a*)

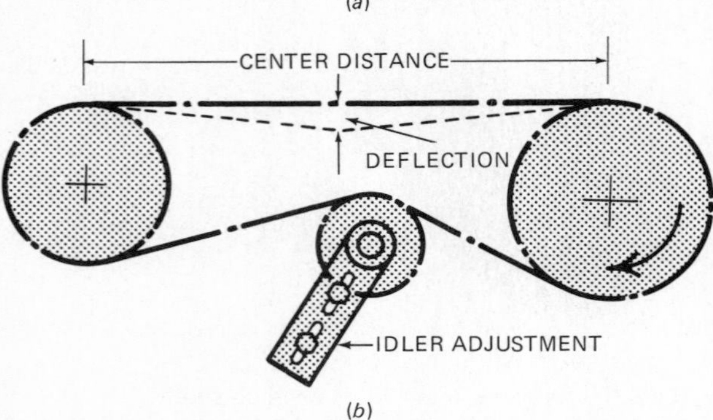


(*b*)

**Fig. 8-25.** Adjusted chain tension deflection should be limited to 1/4 in per foot of sprocket center distance. (*FMC Link Belt Corp.*)

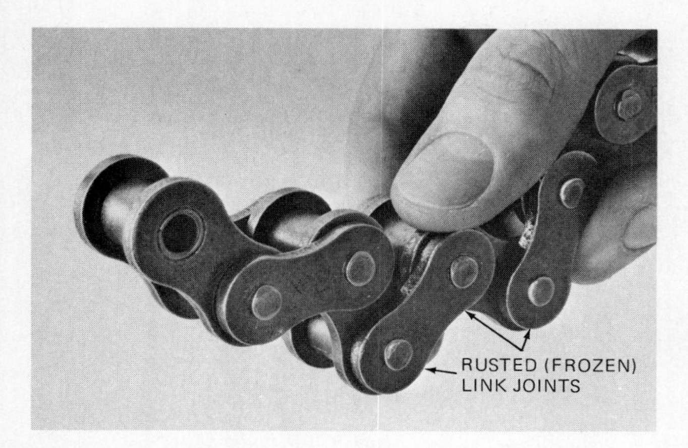


**Fig. 8-27.** Roller chain must be serviced regularly by cleaning and lubricating to prevent tight joints. The lubricant must penetrate between the bushing and pin, and the roller and side links. (*FMC Link Belt Corp.*)

There are several important points to consider in the installation and maintenance of chain drives, as follows:

1. It is never advisable to install a new link in a worn chain. The pitch of the new link will be shorter, and a shock load will occur each time the link engages the sprocket, destroying the chain.
2. When sprockets have a hooked appearance (Fig. 8-28), they are worn and must be replaced. A new chain on worn sprockets or a worn chain on new sprockets will result in a short life for both the sprockets and chain.
3. Maintain proper chain tension—a chain that is too tight will place excessive loads on bearings; if too loose, the chain will whip and may override the driven sprocket on the slack side.
4. After adjusting the chains, slowly turn the drive several times to make sure the tension is not too tight at any point. There should be slight slack at all times. Quite often the drive sprockets are not perfectly concentric with the shaft, and the chain becomes alternately loose and tight during operation. Tension that is proper in one position may be too tight in another.

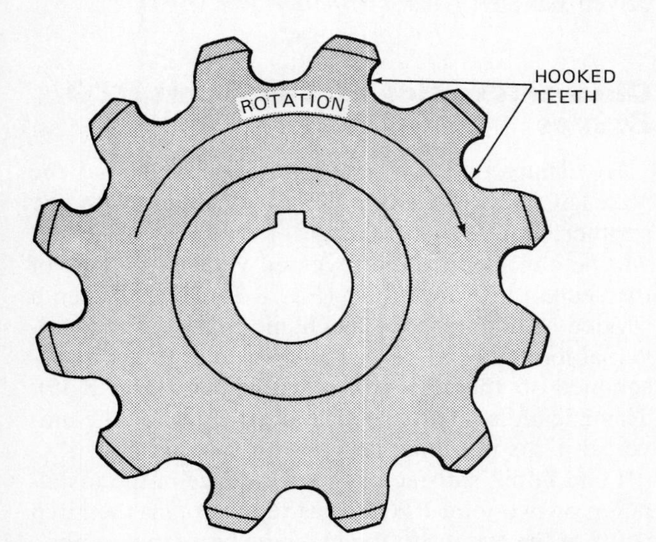


**Fig. 8-28.** A worn sprocket. (*FMC Link Belt Corp.*)

## POWER TRANSMISSION BY PTO DRIVES

PTO drives used in agricultural machinery are most often identified with the transmission of rotary power from a tractor to operate a machine at the same time that it is being pulled through the field. Occasionally, similar drives are used to transmit the rotating force developed by a drive wheel to another part of a machine. A very common application of this type drive is connecting the transmission of trucks and automobiles to the rear axle drives. PTO drives are a positive means of transmitting power from the source to the driven machine. A PTO drive consists of a drive shaft that transmits the source of power through a coupling made flexible by the use of one or more universal joints. This enables the PTO drive to transmit power efficiently and with flexibility when it is necessary to change angles of power flow.

### Characteristics of PTO Drives

The operating speeds of the tractor power takeoff shaft have been standardized by the American Society of Agricultural Engineers (ASAE) in cooperation with the Society of Automotive Engineers (SAE). Since 1964, PTO speeds of 540 and 1000 r/min are standard. Diameters of the shafts and the number of splines for each rpm have also been standardized (Fig. 8-29). The principle advantage which the faster shaft speed provides is that smaller-sized PTO components can be used to transmit the same power compared to that at 540 r/min (Fig. 8-30).

| | TYPE 1 | TYPE 2 | TYPE 3 |
|---|---|---|---|
| SHAPE OF SPLINE |  | | |
| r/min | 540 ± 10 | 1000 ± 25 | 1000 ± 25 |
| DIAMETER OF SHAFT | 1-3/8 in. [35 mm] | 1-3/8 in. [35 mm] | 1-3/4 in. [45 mm] |
| NO. OF SPLINES | 6 | 21 | 20 |

**Fig. 8-29.** ASAE/SAE standards for agricultural tractor PTO for shaft speed, shaft diameter, and number of splines.

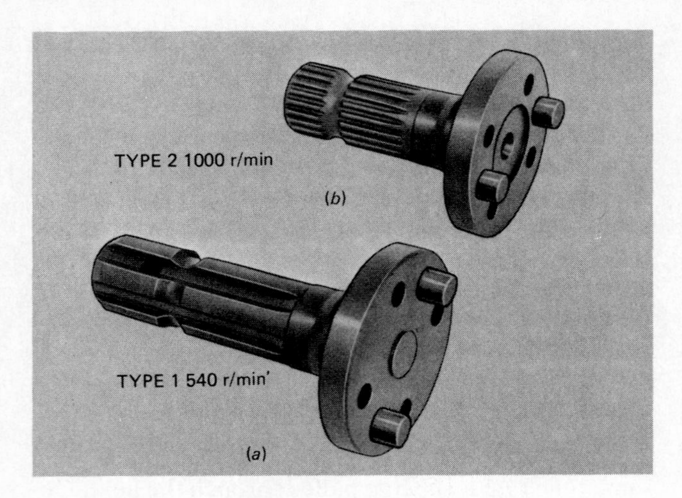


Fig. 8-30. Type 1 and Type 2 conversion PTO drive shaft adaptors for the modern farm tractor. (*a*) 540 r/min. (*b*) 1000 r/min. (*Deere & Co.*)

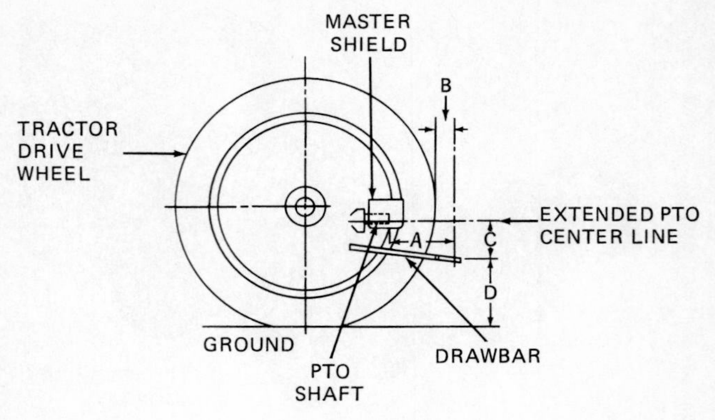


| LOCATION | DIMENSION, in. [mm] | |
|---|---|---|
| | TYPE 1 | TYPE 2 |
| A | 14 [356] | 16 [406] |
| B | 1 TO 5 [25 TO 127] | 1 TO 5 [25 TO 127] |
| C | 6 TO 12 [152 TO 305] | 6 TO 12 [152 TO 305] |
| D | 15 ± 2 [381 ± 50] | 15 ± 2 [381 ± 50] |

**Fig. 8-31.** Standardized locations for Type 1 and Type 2 PTO shafts and a drawbar hitch on agricultural tractors. The PTO shaft should be within ± 3 in [76 mm] of the tractor centerline, and the hitch point should be directly beneath the extended centerline of the PTO shaft.

In order to be able to hitch different tractors to the same farm equipment, certain hitching and shielding specifications have also been standardized. As illustrated in Fig. 8-31, the location of the PTO shaft in relation to the drawbar connection, the extension of the drawbar back of the drive wheels, and the height of the drawbar above the ground surface are identified.

## Characteristics of Universal Joints

Nearly all agricultural machinery PTO drives utilize the Cardan, or Hooke, type of universal joint. This device consists of two U-shaped yokes connected with a cross-shaped linkage which fastens the yokes together and allows them to pivot on journal surfaces and needle bearings (Fig. 8-32). The universal joint allows the connecting drive shafts to be located as much as 30° out of alignment. This type of universal joint is low-cost, easily serviced, and stronger than most other types of universal joints. However, if only one universal joint is used, a uniform flow of power can be obtained only if the center line of the driving and driven shafts lies in a straight line. If any angle exists between the shafts, a velocity change will develop which will increase in magnitude as the angle becomes more acute.

During the first 45° rotation of the driving yoke, the driven yoke will accelerate its rotation. This is referred to as lead. The amount of acceleration will be approximately 2° when the angle between the two shafts is 20°. At a 30° angle the acceleration will approach 4°; at 40° the lead will be 8° (Fig. 8-33).

When the driving yoke has rotated a full 90°, the driven yoke has decelerated until both shafts are turning at the same speed. However, continued rotation through the next 45° of rotation results in a completely opposite reaction—an equal deceleration of the driven yoke. This is called lag.

The significance of this mechanical action of constantly changing velocity in the output yoke of a single universal joint is directly related to proper service, installation, and maintenance. For example, assume that the driving yoke is connected to a tractor PTO, and the driven yoke is attached to a flywheel. If the assembly is operated at more than a 20° angle, the universal joint would be immediately destroyed. This is because the flywheel mass could not change velocity as it rotated to comply with the action of the driven yoke.

## Characteristics of Two-Joint PTO Drives

By using a two-universal-joint drive which has the yokes of the intermediate shaft in alignment with one another, the velocity change of one universal joint will be canceled by the reversed velocity change of the second universal joint (Fig. 8-34). Thus, when a service or maintenance mechanic repairs a two-universal-joint drive, it is essential that the shafts be assembled so that the yokes are aligned (Fig. 8-35). This alignment is often referred to as placing the universal joints in proper *phase* with each other.

In addition, satisfactory performance in the installation of two-joint PTO drives requires that the hitch point of the tractor to the machine be correctly positioned. In Fig. 8-36*a* the machine is attached to the

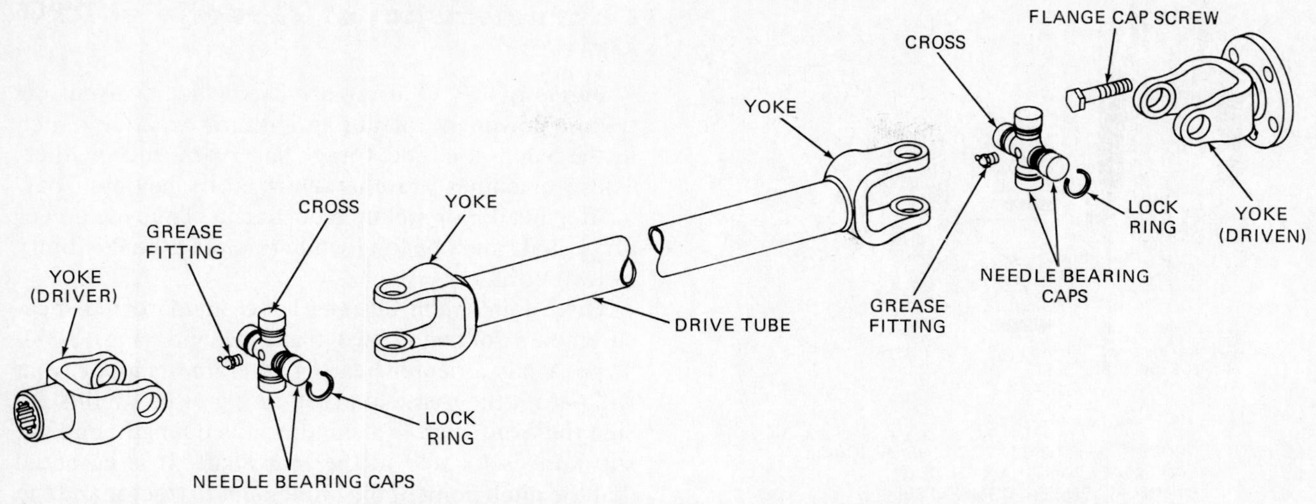


**Fig. 8-32.** Cardan, or Hooke-type, universal joints, commonly referred to as U-joints because of their U-shaped yokes. (*FMC Link Belt Corp.*)

tractor drawbar with the PTO drive positioned on the center line of the tractor PTO drive and implement gear box. In part *b* of the figure, the tractor is turned 90° to the center line of the implement. The PTO drive shaft is compressed in length through a *slip joint* to achieve the turn. The result is that the compressed length (LC) of the intermediate shaft (LC/2) is the same for the tractor joint and implement joint in relation to the implement-tractor hitch pin. This results in smooth operation and transmission of power from tractor to implement. The equidistant hitch point allows a short turning radius without excessive PTO shaft vibration or joint damage from unequal velocity change in the universal joints which would occur if the angles were different.

The slip joint is illustrated in Fig. 8-37*a* for a two-joint shaft. The slip joint may be covered by a protective shield, as shown in part *b*. In part *c* the standard measurement specifications to permit proper operation of two-joint shafts are identified. In both types of drive, the tractor drawbar *must be locked* in order to prevent swinging.

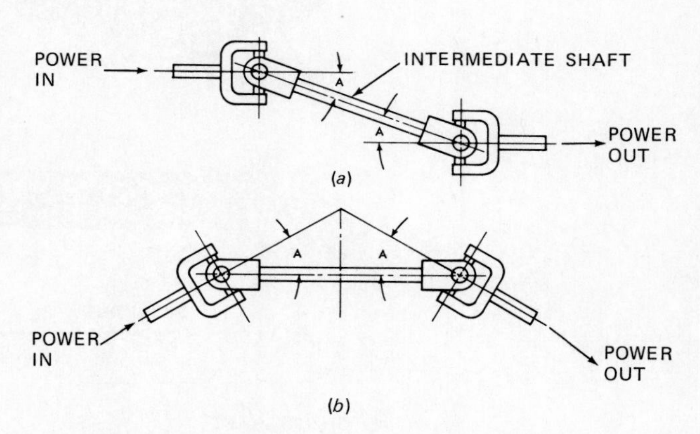


**Fig. 8-34.** The velocity change in one universal joint is offset by the opposite velocity change of the other when the amount of angle A is the same in both universals. (*a*) Power "out" is transmitted parallel to the power "in" shaft. (*b*) Power "in" is taken through angle A to the power-out shaft. (*Rexnord*)

**Fig. 8-33.** Lead-and-lag cycle of the driven shaft. During a 180° rotation of a single universal joint, lead lag varies as much as 8° when the driving and driven shafts are at a 40° angle to each other. In the first 90°, the driven shaft accelerates (as much as 8°), and in the second 90°, it decelerates (up to 8°).

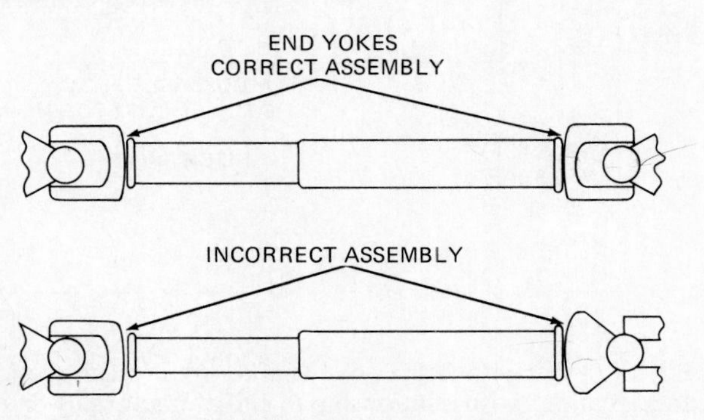


**Fig. 8-35.** Proper universal joint *phasing* is necessary to achieve desirable performance.

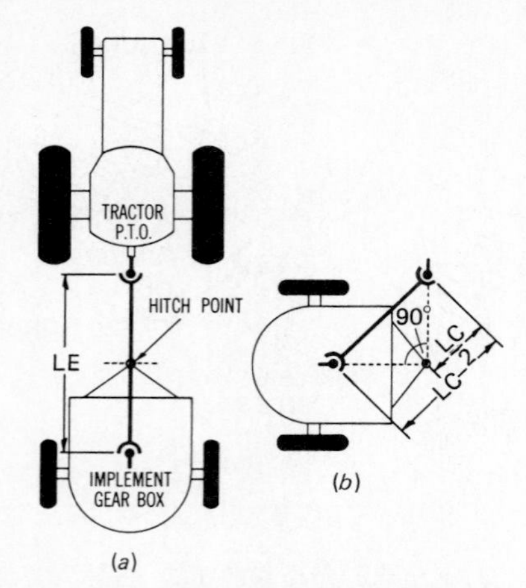


LENGTH: CENTER LINE TO CENTER LINE OF JOINTS
LE: MAXIMUM LENGTH, EXTENDED
LC: MINIMUM LENGTH, COMPRESSED
$\frac{LC}{2}$: ONE-HALF OF COMPRESSED LENGTH

**Fig. 8-36.** Proper location of hitch point permits universal joints to have equal angles when turning. (*a*) PTO straight. (*b*) Turning PTO 90° angle. (*Haynes-Dana, Inc. and Ag-Master Products*)

## Characteristics of Three-Joint PTO Drives

Three-joint PTO drives are used when transmitting tractor power to operate agricultural machines such as the baler, combine, forage harvester, and swather. These machines are characterized by having a harvesting header or pickup mechanism. Thus the power drive shaft must be long while retaining the flexibility to turn corners.

Three-joint hitch designs have standardized measurements following two basic designs (Fig. 8-38). Type A has a center bearing (A-frame) support that will permit the frame to pivot during turns. In this design the front shaft is a standard fixed length, and the slip joint is located in the rear shaft. It is essential that the hitch point of the implement to tractor and the hitch to pivot point be the standard 14 in [356 mm].

In type B the center bearing support cannot pivot because it is fixed to the implement hitch. The slip joint must therefore be located in the front shaft. The

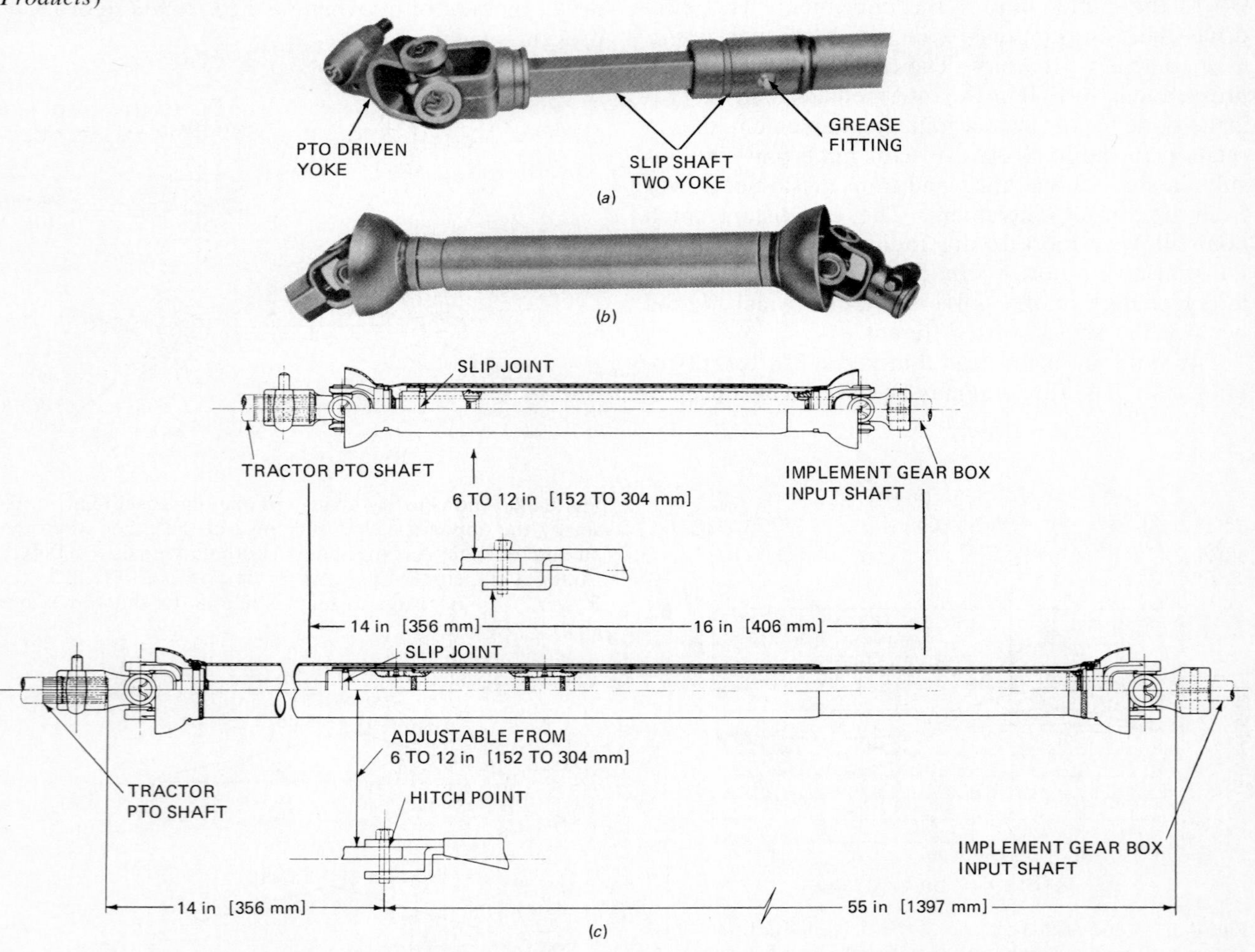


**Fig. 8-37.** (*a*) Two-joint PTO drive shafts. (*b*) Typical two-joint PTO shaft with protective shield. (*c*) Dimension specifications with relationship to end of tractor power shaft and hitch point and hitch point to input shaft on machine being driven. Dimensions are for short intermediate shafts with slip couplings and for two-joint slip joint shafts with a long intermediate shaft with a slip coupling. (*Haynes-Dana, Inc. and Ag-Master Products*)

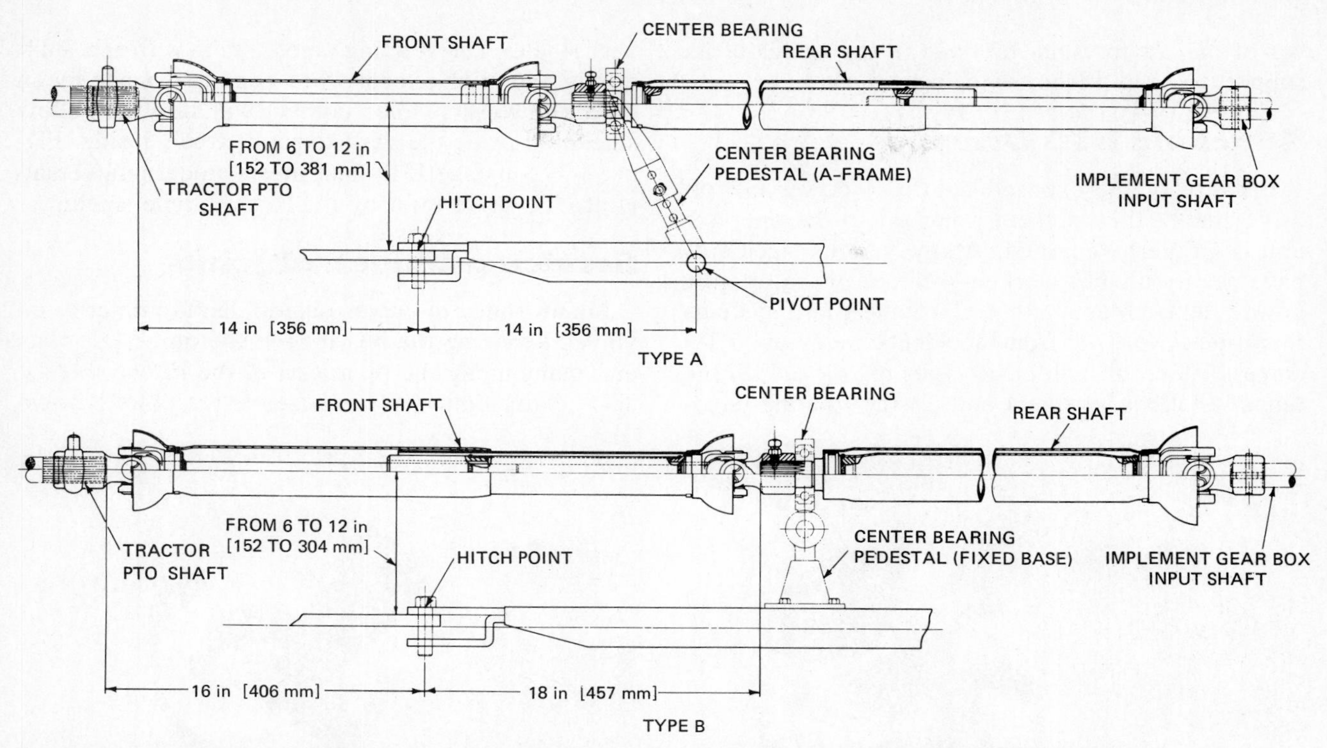


**Fig. 8-38.** Three-joint PTO drive shafts. Type A has a slip joint in the rear shaft; type B has a slip joint in the front shaft. Hitch point measurement specifications and center bearing support assemblies differ. *(Haynes-Dana, Inc. and Ag-Master Products)*

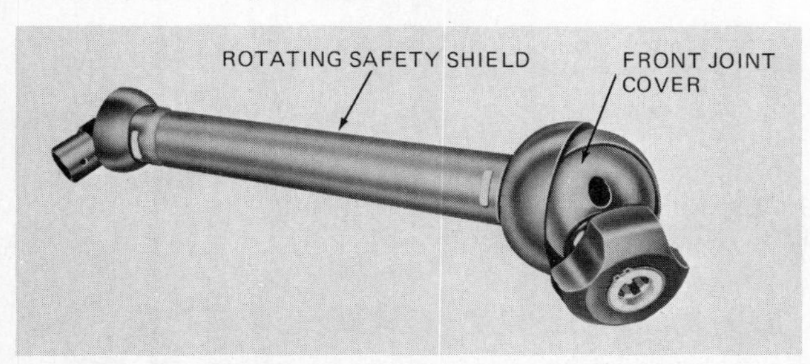


**Fig. 8-39.** Modern Z-joint PTO drive shaft with rotating safety shield and fully enclosed joint protection. *(Haynes-Dana, Inc.)*

tractor to implement hitch must be 16 in [406 mm] and the hitch point to the end of the rear shaft spline must be 18 in [457 mm] to achieve vibration-free power transmission.

It is also important that the center joint be adjusted parallel in a vertical plane (side view) to the first and third joints. This adjustment is provided at the center bearing pedestal. Alignment of the universal joints in the horizontal plane (top view) should be as straight as possible, with the front shaft directly over the hitch pin. As in the two-joint power shaft, the draw-

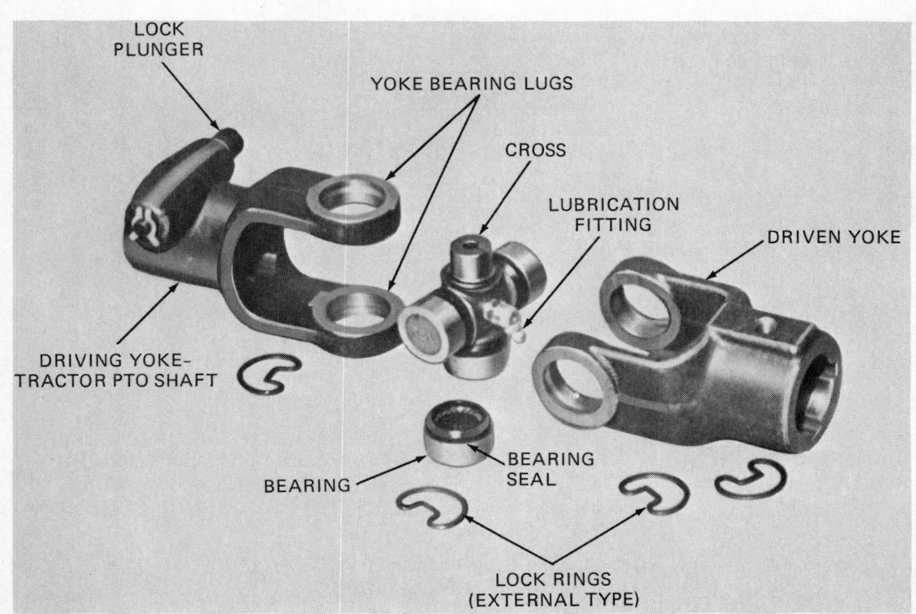


**Fig. 8-40.** Typical Hooke-type universal joint. The repair kit includes the cross, bearings with seals, and snap rings. *(Rexnord Inc.)*

bar of the tractor must be fixed in the center of its support to keep it from swinging.

## Attaching PTO Drives

Because of the exposed nature of power takeoff drive shafts, they present a hazard to the operator unless properly guarded. Many serious accidents have occurred when workers were caught in rotating power shafts. Manufacturers provide guard systems to protect workers from accidents involving PTO drives. There are two basic types of shields: (1) the tunnel (stationary) shield and (2) the rotating (spinner) shield. The rotating shield spins with the shaft during operation except when contact is made by an object or worker. Such contact will stop the hollow shield, allowing the shaft to rotate freely inside. Figure 8-39 (on page 151) illustrates a modern universal-joint PTO drive utilizing the rotating-type shield.

## Servicing Universal Joints

Maintenance of power takeoff shafts primarily involves servicing the bearings in the universal joints and maintaining the operation of the safety shields. The bearings may fail because of (1) lack of lubrica-

(*a*)

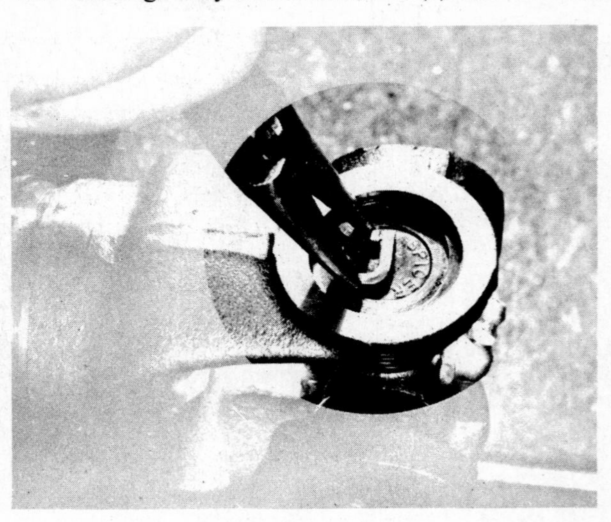

(*b*)

STEP 1. REMOVE SNAP RING. USE A SCREWDRIVER. TO REMOVE INTERNAL TYPE (*a*) BEARING RETAINERS OR PLIERS WITH EXTERNAL TYPE (*b*).

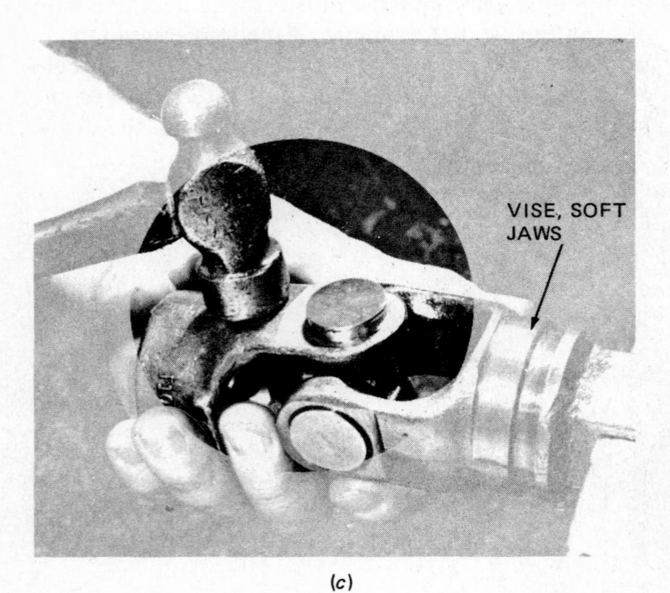


(*c*)

(*d*)

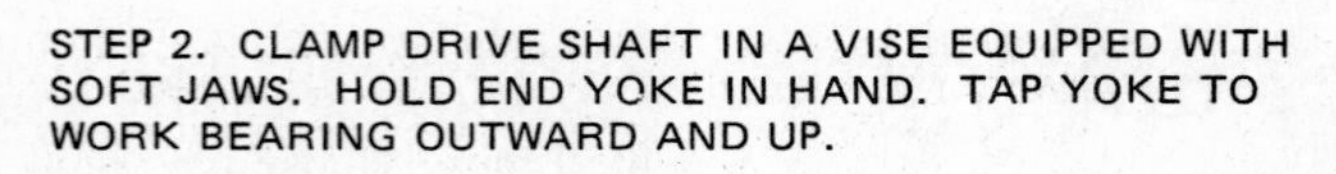

STEP 2. CLAMP DRIVE SHAFT IN A VISE EQUIPPED WITH SOFT JAWS. HOLD END YOKE IN HAND. TAP YOKE TO WORK BEARING OUTWARD AND UP.

STEP 3. CLAMP PROTRUDING BEARING IN VISE. TAP YOKE TO PULL BEARING FROM YOKE. TURN JOINT OVER AND TAP EXPOSED END OF JOURNAL CROSS TO REMOVE SECOND BEARING.

**Fig. 8-41.** Disassembly procedure for a universal joint. (*Haynes-Dana, Inc.*)

tion, (2) overloading, (3) improper phasing, or (4) misalignment of drive shafts. The machine operator or maintenance mechanic should periodically check the universal joints for (1) noisy operation, (2) looseness between the cross and yokes, (3) excessive vibration, and (4) freedom of telescoping shaft movement.

If the bearings in the universal joints are worn, replacement cross and bearing kits (Fig. 8-40) can be purchased and installed by the maintenance mechanic. There are two types of universal joints used on most agricultural machinery PTO drives. Part *a* of Fig. 8-41 shows the type in which the bearing cups are retained in the yokes by an internal snap ring, and part *b* shows the external snap ring type. After the snap rings are removed, the drive shaft is clamped in a vise equipped with soft jaws, and the yoke is tapped lightly with a soft-faced hammer. This will pull the bearing cup (Fig. 8-41*c*) out of the yoke. The cup can then be grasped in the vise jaw and the yoke tapped with a hammer (Fig. 8-41*d*) permitting disassembly.

When all the bearings have been removed, the cross is washed in cleaning solvent and inspected for

(*a*)

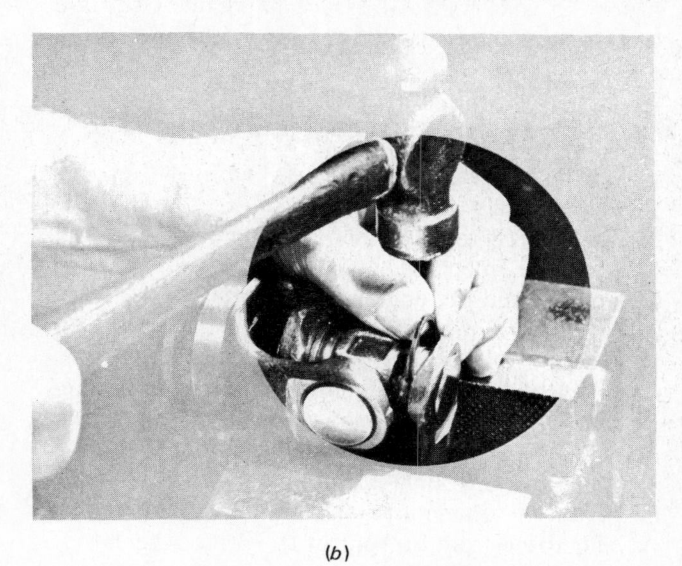

(*b*)

STEP 1. CLAMP END YOKE LIGHTLY IN VISE EQUIPPED WITH SOFT JAWS. LUBRICATION FITTING FACING AWAY FROM SHAFT. LIFT SHAFT TO LEVEL POSITION.

(*c*)

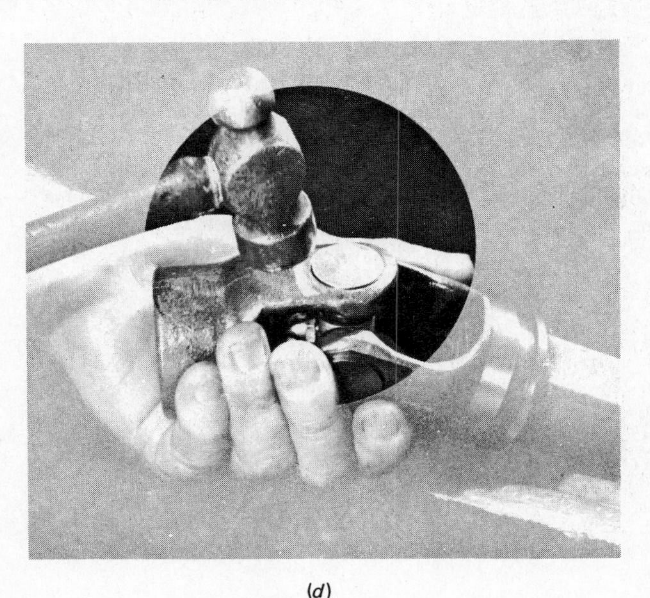

(*d*)

STEP 2. TAP BEARING DOWN (*a*) UNTIL SNAP RING (*b*) OR (*c*) CAN BE INSTALLED.

STEP 3. PROCEED TO TAP SECOND BEARING AND INSTALL SNAP RING. IF ASSEMBLED JOINT IS STIFF, SHARPLY STRIKE FORGED SURFACE OF YOKE LUGS (*d*).

**Fig. 8-42.** Assembly of a universal joint. (*Haynes-Dana, Inc.*)

wear grooves caused by the needle bearings. If pitted, a new cross and bearing kit must be installed. Inspect the needle bearing cups for contamination by removing needles and washing in solvent. After complete inspection and evaluation, return *all* the needles to the bearing cups. Use SAE 140 lubricant or Type 1 grease to service the bearings. Install the proper number of needles in each cup.

Assemble the universal joint following the procedure outlined in Fig. 8-42.

When universal joints are equipped with lubricating fittings, apply SAE 140 lubricant or Type 1 multipurpose grease with a pressure handgun. Use care not to apply excessive lubricant, or damage to the seals may occur.

### Safe Practices with PTO Drives

1. Always operate PTO drives with the proper shielding in place.
2. Service the machine or make adjustments only after the machine and the PTO drive have stopped and the engine is shut off.
3. Do not ever permit anyone to ride on the tractor except the operator.
4. Check the rotating shields for freedom of rotation.
5. Do not connect a machine designed for a 540-r/min to a 1000-r/min power source.
6. Mount or dismount a tractor equipped with a PTO drive from in front of the drive wheels.

## SUMMARY

Belts, chains, and PTO drives are a common part of many agricultural machines. They perform important roles of transmitting power from one machine part to another. Belt and chain drives are also used to change the speed ratio of rotating parts. Setup, predelivery of new machinery, and the many different types of service, maintenance, and repair operations are common service department jobs. Service personnel are required to install, align, and make adjustments of belts, chains, and PTO components.

This chapter has described the various type specifications, safety procedures, and operating principles for a number of common installations. The information can only serve as a basis for understanding specific applications and practices as outlined in the manufacturers' service bulletins.

## THINKING IT THROUGH

1. Identify one use for each of the following types of belt drives on agricultural machinery:
   (a) V belt
   (b) Variable-speed belt
2. What are the advantages of using V belts to provide power transmission in agricultural machinery?
3. Describe the operation of a variable-speed V-belt drive.
4. Why is it poor practice to force V belts on sheaves without relieving tension?
5. What are two methods recommended for determining proper belt tension?
6. How should V belts be stored if removed from a machine?
7. What are the principle advantages of using chain drives compared to belt drives on agricultural machinery?
8. How should roller chain be cleaned and lubricated?
9. Define the following terms:
   (a) Chain pitch
   (b) Center distance
10. What determines the proper direction to operate a chain?
11. What are the standard speeds of PTO operation in agricultural machinery?
12. What is meant by *lead, lag,* and *phasing* when referring to universal-joint mechanics?
13. How should PTO drives be inspected to determine operating condition?
14. What are the safety practices which must be observed when working with the following:
    (a) Belts
    (b) Chains
    (c) PTO drives

CHAPTER 9

# HYDRAULIC POWER TRANSMISSION

## CHAPTER GOALS

In this chapter your goals are to:

- Identify operating principles of hydrostatic power
- Examine the types of hydraulic components used in agricultural machinery
- Describe methods of controlling various hydraulically operated actuators
- Identify hydraulic system service and maintenance procedures
- Identify procedures for testing the performance of a hydraulic system
- Identify procedures for repair of hydraulic hose and fittings

Have you ever watched an automobile in a service station or garage as it was raised from the floor by one or more large-diameter tubes attached to a supporting frame? It seemed no problem at all to raise the 3500-pound (lb) [1587.6-kilogram (kg)] machine. Such car lifts are *hydraulic*. Unless you have had a chance to work as a helper on a construction site, on a farm, or in an automotive repair station, the chances are that the only direct contact you have had with hydraulic power is when operating an automobile or truck equipped with hydraulic brakes, power steering, or automatic transmission. Many vehicles are equipped with these forms of hydraulic power.

A great many agricultural machines are operated hydraulically. As illustrated in Fig. 9-1, hydraulic power is used to lift, control depth of tillage tools, and to provide rotating power for driving such parts as the ground wheel of a groundskeeping machine. Many modern agricultural tractors rely on hydraulic power for brakes, steering, and transmission speed control. Some tractors use hydraulic power to drive the transmission.

As a serviceperson in agricultural power and machinery, you will recognize that knowledge of the mechanical operation and working principles of hydraulic power is essential to be a successful worker. Once you know how hydraulic power can be utilized and transmitted, you will find it easier to develop the mechanical skills needed to service and maintain complex equipment.

## PRINCIPLES OF HYDROSTATIC POWER

The word *hydraulic* is taken from the Greek words *hydro*, meaning "water," and *aulis*, meaning "tube" or "pipe." Today the term *hydraulic power* is used to

(*a*)

(*b*)

(*c*)

**Fig. 9-1.** Hydraulic power is used in agricultural equipment as shown here. (*a*) The front loader uses hydraulic cylinders to *lift*. (*International Harvester*) (*b*) A cylinder is used to raise or lower tillage equipment. (*Sunflower Mfg. Co.*) (*c*) The hydraulic motor rotates the ground drive wheel of a groundskeeping mower. (*Toro*)

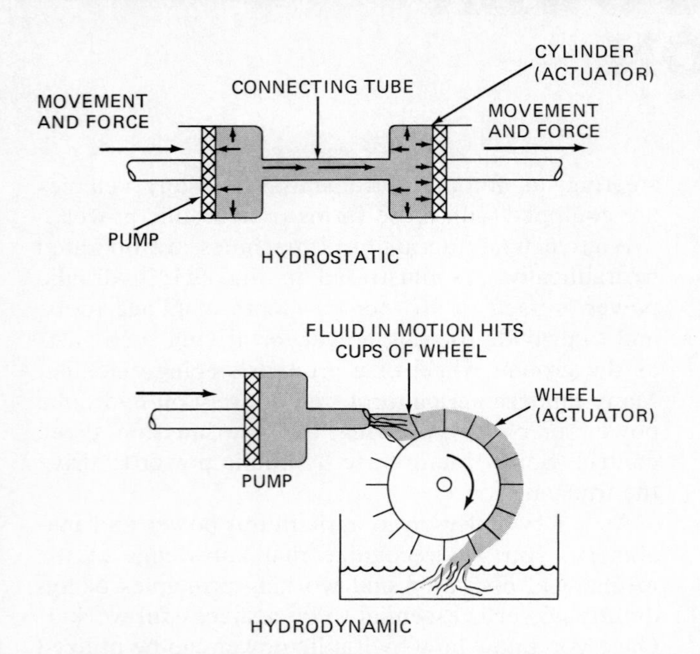


**Fig. 9-2.** In the hydrostatic system, a liquid under pressure causes movement. In the hydrodynamics system, a liquid under high velocity hits the fins of a wheel, causing it to move.

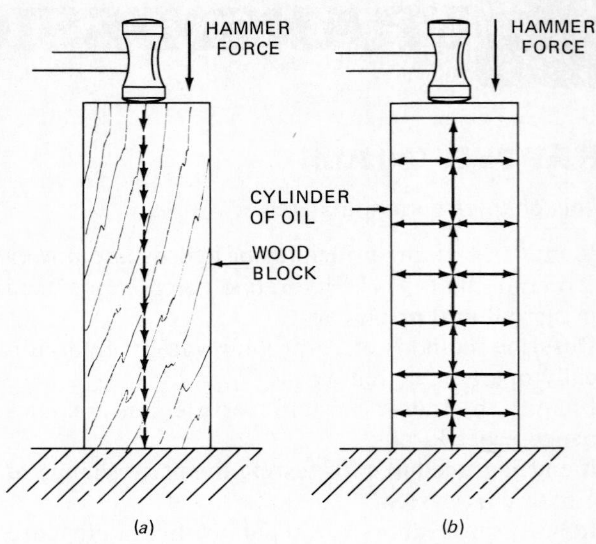


**Fig. 9-3.** Comparison of transmission of force. (*a*) The force of the hammer striking the wood block (a solid) is transmitted in a straight line. (*b*) In the liquid the force is exerted equally in all directions within the container.

describe the way in which oil, a liquid, is used under controlled pressure and flow to do work.

There are two forms of transmitting power hydraulically. They are classified under the broad headings of hydrostatic and hydrodynamic processes. The difference between the two processes is the way each uses the pressure and flow of the liquid oil to transmit power. The *hydrostatic system* uses high pressure and relatively slow movement of fluid to do work. The opposite is true for the *hydrodynamic system,* where the fluid is at a high speed but at a low pressure (Fig. 9-2). This chapter is limited to a discussion of the hydrostatic principle, since it is the basic system used in agricultural machinery.

Hydrostatic power is based upon a physical principle, or law, which was discovered by the Frenchman Blaise Pascal in 1653. *Pascal's law* states that when a pressure is developed on fluids it acts equally in all directions, regardless of the shape of the container. Figure 9-3 illustrates the relation of solids to liquids in transmitting a force. Striking a wood block with a hammer (Fig. 9-3*a*) causes the force to be applied in one direction only—directly in line with the blow. In Fig. 9-3*b* the force of the hammer is through the liquid but also equal force is exerted in all other directions. Unlike air or other gases, liquids create this action because they are fluid and cannot be compressed into a different volume. Fluids act as "flexible solids" (Fig. 9-4). When pressure is applied to a liquid by a weight resting on a piston in a cylinder, the pressure is transferred to the fluid. The amount of pressure that is transmitted to the fluid is in direct relation to the area of the piston and the force. A 1000-lb [454-kg] load on a piston with an area of 100 square inches (in²) [645 square centimeters (cm²)] develops a pressure of 10 pounds per square inch (psi) [69 kilopascals (kPa)]. Or it can be stated that the pressure of 10 psi will exert a force of 1000 lb on a piston having an area of 100 in² (Fig. 9-4*b*).

One of the advantages of transmitting power by hydraulic force is the flexibility of the system. The energy carried by a fluid under pressure is not limited to straight-line action; it can also go around corners (Fig. 9-5*a*). This makes it possible to use flexible hose to carry the fluid to various and multiple locations. By connecting a pump to the cylinder, as illustrated in Fig. 9-5*b*, and attaching a lever on the pump, a force of 100 lb [45.4 kg] acting on the area of the pump cylinder will develop a pressure of 10 psi [69 kPa] on the fluid. The 10-psi pressure will balance the 1000-lb weight resting on the lift position because it acts against 100 in² ($10 \text{ psi} \times 100 \text{ in}^2 = 1000 \text{ lb}$). Note, however, that the pump piston must be moved 5 in [127 millimeters (mm)] to lift the 1000-lb weight ½ in [12.7 mm]. This is because the *mechanical advantage* of 100 lb of force raising 1000 lb is 10 to 1.

If, as in Fig. 9-6*a*, we increase the force on the 10-in² [64.5-cm²] piston to 200 lb [90.7 kg], an internal oil pressure of 20 psi [138 kPa] ($^{200}/_{10} \text{ in}^2 = 20 \text{ psi}$) will be developed. When this pressure is exerted against the 100-in² [645-cm²] piston, a lift of 2000 lb [907.2 kg] will be created. Therefore,

$$\text{Pressure} \times \text{area of piston} = \text{load capacity}$$

If the lift piston area is reduced to 50 in² [323 cm²] (Fig.9-6*b*), the 20-psi [138-kPa] pressure will lift

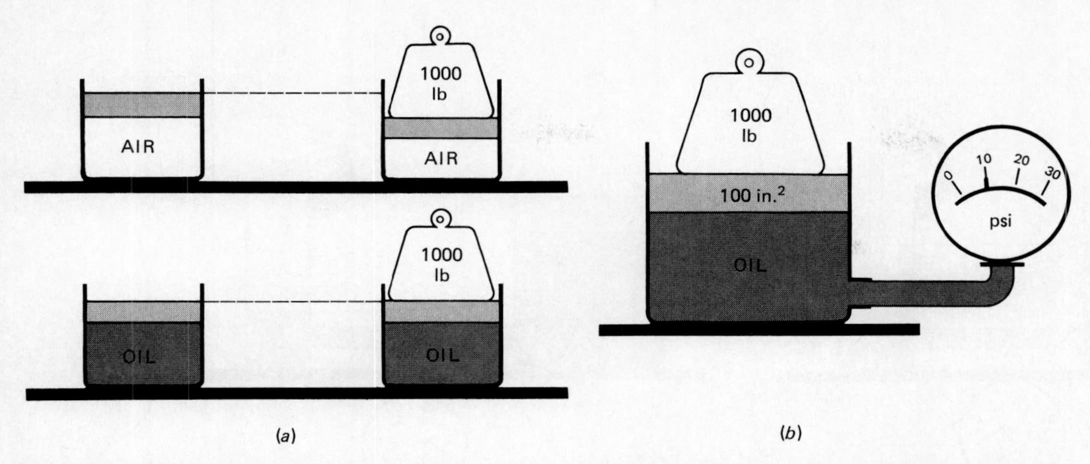


**Fig. 9-4.** (*a*) Unlike air, liquids cannot be compressed into a smaller volume. (*b*) An applied force, divided by the area of the cylinder, equals the pressure on the liquid.

1000 lb [454 kg]. The load, however, will be moved 1 in [25.4 mm], or twice as far. Therefore,

Volume of oil displaced × area of piston
= movement of lift piston

Hydraulic power also involves a second basic law, called the *law of conservation of energy*. This law states that energy can be neither created nor destroyed. That is, work done, or the output of energy, cannot exceed the input. For example, a mechanical lever jack is used to lift the right front wheel of a tractor in order to repair a flat tire (Fig. 9-7). The lift hook of the jack is placed under the wheel axle, and the handle of the jack is "pumped" up and down to raise the axle. The fulcrum point (the point where the handle of the jack pivots in its support) is 3 in [76.2 mm] from the center line of the jack. The lift handle (the lever to which force is applied to raise the axle) is 3 ft [0.914 meter (m)] in length. If the front wheel axle assembly weighs 720 lb [327 kg], a force of 60 lb [27.2 kg] applied to the lever would raise the wheel assembly as follows:

$$\text{Lever force} = \frac{720\text{ lb} \times 0.25\text{ ft}}{3.0\text{ ft}}$$
$$= \frac{180\text{ lb} \cdot \text{ft}}{3.0\text{ ft}}$$
$$= 60\text{ lb [27.2 kg]}$$

## Hydraulic Jack

Most service centers use a hydraulic *jack* to lift heavy objects such as a tractor. This type of jack provides a more efficient mechanism than the mechanical lever jack method and serves as a good example of the practical application of the principles of hydraulic power previously illustrated.

The piston (ram) of the hydraulic jack is placed under the axle of the tractor. The service worker pumps a short lever on the jack (Fig. 9-8), which

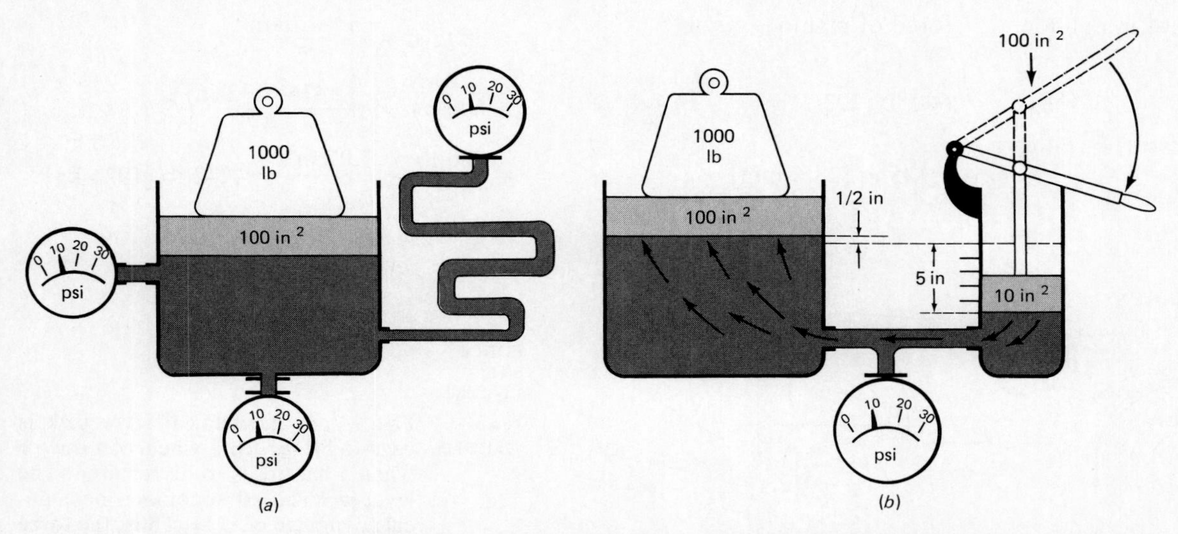


**Fig. 9-5.** (*a*) Liquids are classified as flexible solids. (*b*) A force of 100 lb divided by the pump cylinder volume of 10 in$^2$ equals a force of 10 psi on the fluid. Acting on 100 in$^2$, the pressure will support 1000 lb.

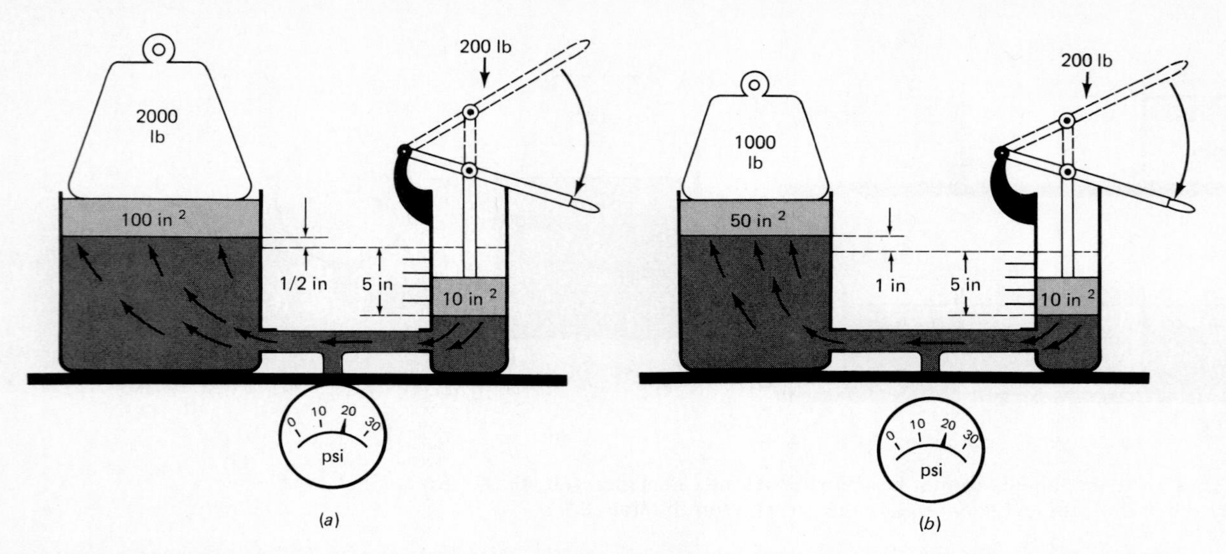


**Fig. 9-6.** Pressure multiplied by the area of the piston equals the load-carrying capacity. (*a*) A pressure of 20 psi acts against 100 in² and supports 2000 lb. (*b*) The same pressure acting against 50 in² supports 1000 lb, but lifts the weight twice as far (1/2 in to 1 in).

raises the axle. The mechanical construction of a hydraulic jack is illustrated in Fig. 9-9. Note that the downward force on the "pump" handle forces the piston of the pump downward in its cylinder. The action forces oil out of the pump cylinder opening past the ball check valve. The amount of oil displaced by the pump and the force exerted on the oil in the pump cylinder is as follows:

Mechanical advantage of pump handle $= \dfrac{\text{10-in handle}}{\text{1-in lever}}$

$= 10:1$

Force on pump piston = force on lever × mechanical advantage

$= 60 \text{ lb} \times 10$

$= 600 \text{ lb}$ [272.2 kg]

Oil pressure in cylinder = force of piston ÷ area

where

Force = 600 lb [272.2 kg]

Diameter (diam) of pump cylinder = 1.125 in [28.6 mm]

Area of pump cylinder $= \dfrac{\pi \times \text{diam}^2}{4}$

$= \dfrac{3.1416 \times (1.125 \text{ in})^2}{4}$

$= 1.0 \text{ in}^2$ [6.45 cm²]

Thus

Oil pressure $= 600 \text{ lb} \div 1.0 \text{ in}^2$

$= 600 \text{ psi}$ [4137 kPa]

With the pressure release valve in the closed position (needle valve seated), a pressure of 600 psi enters the line connecting the pump to the lift cylinder and distributes an equal pressure to all other parts of the system. The amount of force which the piston (ram) is able to exert is as follows:

Lift force = oil pressure (psi) × area cylinder (in²)

$= 600 \text{ psi} \times \dfrac{\pi \times \text{diam}^2}{4}$

$= 600 \text{ psi} \times \dfrac{3.1416 \times (3 \text{ in})^2}{4}$

$= \dfrac{600 \text{ lb}}{\text{in}^2} \times \dfrac{7.07 \text{ in}^2}{1} = 4242 \text{ lb}$ [1924 kg]

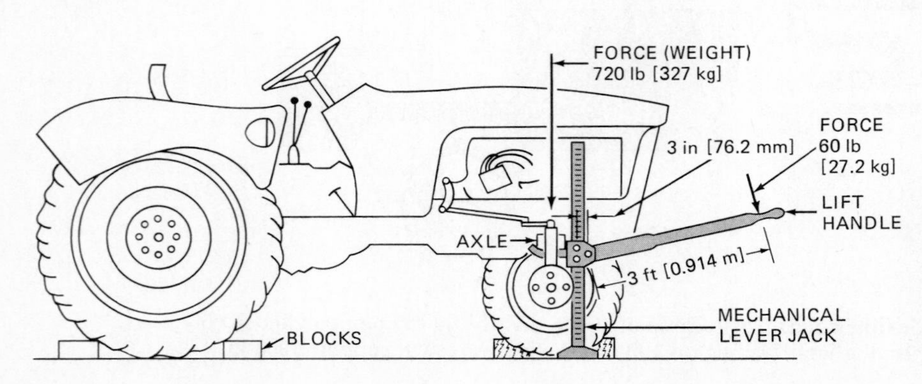


**Fig. 9-7.** A mechanical lever jack is used to lift the front wheel of a tractor when a flat tire is to be repaired. The lever jack is used to obtain a mechanical advantage of 12:1. Thus, the force required to lift the 720 lb [327 kg] weight of the axle is 60 lb [27.2 kg].

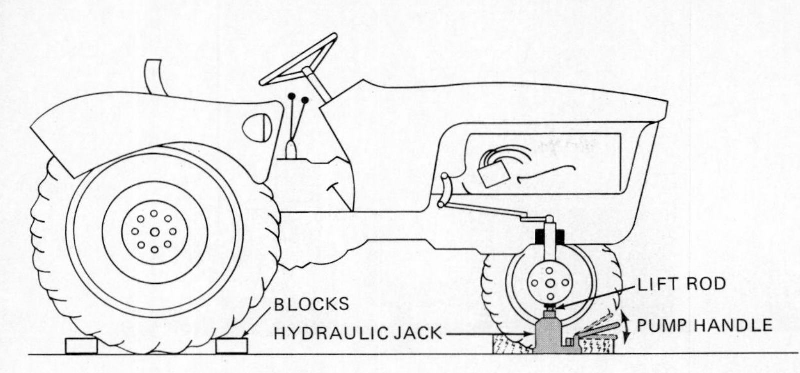


**Fig. 9-8.** A safer, more efficient method for raising the tractor wheel is to use a hydraulic jack.

By using the hydraulic jack the mechanical advantage obtained was 4242 lb of force output from 60 lb of force input, or approximately 71 : 1 ($^{4242}/_{60}$). Thus, the serviceperson would need to apply approximately 10.2 lb [4.63 kg] of force ($^{720}/_{71}$ lb) on the handle to lift the 720-lb [327-kg] tractor axle (Fig. 9-8).

## HYDROSTATIC POWER COMPONENTS

The hydraulic jack (Fig. 9-9) illustrates the basic parts required of a hydrostatic hydraulic system. These parts include (1) a reservoir to hold a supply of oil, (2) a pump capable of moving the fluid (oil) in the system against a resistance, (3) control valves to direct the flow of liquid, and (4) a device such as a hydraulic cylinder (ram) or hydraulic motor to be operated by the fluid.

### Hydraulic Pump

The heart of a hydrostatic power transmission system is the *positive-displacement* pump. A pump for a hydrostatic system will displace (move) a certain quantity of oil in a stated period of time within the working pressure of the system. For example, a positive-displacement pump may be rated in gallons per minute [liters per minute] at a pressure of 1000 to 5000 psi [6900 to 34,500 kPa]. It is important to remember that a pump does not itself develop pressure. A pump creates movement of the liquid (oil). The pressure created is the result of resistance to the flow of liquid from the pump.

To produce a steady movement of the hydrostatic power transmission system, the pump must be capable of producing a continuous flow of liquid. The piston pump illustrated in Fig. 9-9 provides "positive displacement"; that is, each stroke of the piston displaces a given volume of oil. However, the up-and-down movement of the piston would produce a pulsating flow of oil, and the lift ram would move in jerks. Modern hydraulic systems use valved positive-displacement pumps to provide a constant flow of oil and smooth piston action.

### Control System

A system of valves is used to control the flow of fluid for the hydrostatic system. Valves not only are used to control the amount and direction of the fluid,

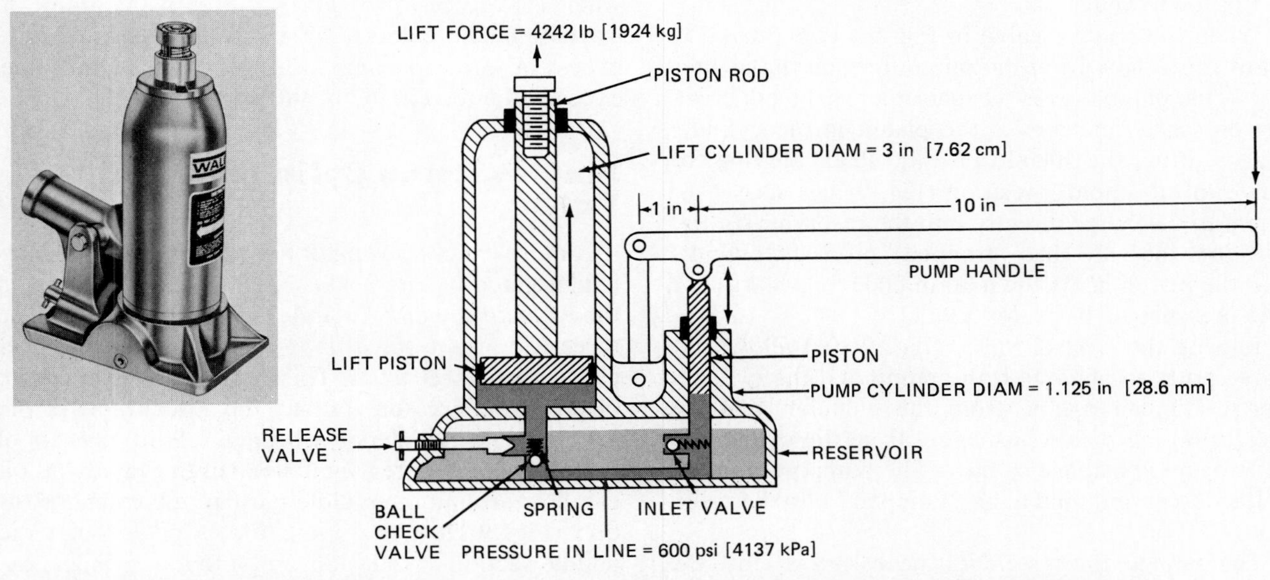


**Fig. 9-9.** Cross section of a hydraulic jack. A force of 60 lb [27.2 kg] on the pump handle is capable of developing a ram lift force of 4242 lb [1924 kg]. (*Walker Mfg.*)

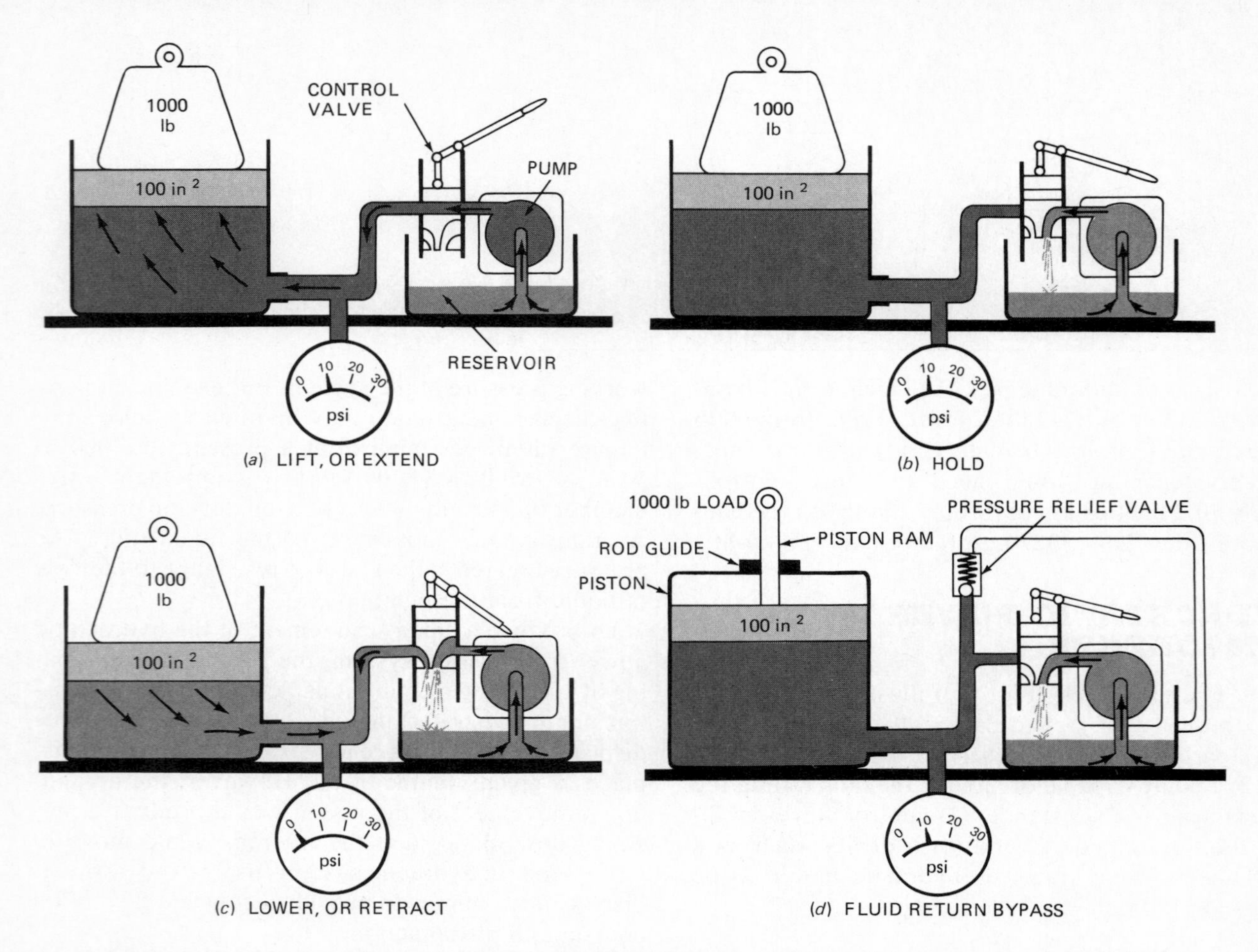


**Fig. 9-10.** Hydraulic systems designed for agricultural machinery require a valving system to control fluid flow and a pump which develops a flow of fluid to the lift cylinder, depending upon the position of the valve.

but are used to protect the system from excessive pressure. The illustrations in Fig. 9-10 show how a slide valve is utilized to control the action of the piston in the cylinder.

When the control valve in Fig. 9-10*a* is moved to allow oil to flow from the pump through the aligned ports, the pump moves oil against a resistance, creating pressure, and moves the piston in the cylinder that is lifting the 1000-lb [454-kg] load. Moving the valve to the "hold" position (Fig. 9-10*b*) allows the pump to discharge directly into the reservoir. In this position, the valve has "trapped" oil in the cylinder and the piston holds the load of 1000 lb [454 kg] at a line pressure of 10 psi [69 kPa].

Raising the control valve (Fig. 9-10*c*) allows the valve ports to align with the pump and the cylinder ports. Oil can escape from the cylinder, and the weight on the piston pushes oil from the cylinder at little or no resistance to flow. The pump port is also aligned to permit fluid to be "dumped" into the reservoir.

To complete the system it is necessary to close the open end of the cylinder, install a rod guide, attach a piston rod (ram) to the piston, and install a pressure relief valve (Fig. 9-10*d*). The pressure relief valve prevents pressures from developing in excess of the capacity of the equipment by allowing fluid to return to the reservoir. Excessive pressures may occur when (1) the piston reaches the outward limits of travel in the lift cylinder, (2) the load being lifted is in excess of safe capacities, or (3) when a sudden force exceeds the load limits of the system.

## Single-Acting Cylinder Spool Valve

A more exact drawing of a hydraulic control valve is illustrated in Fig. 9-11*a–c*. The system illustrates the *extending, holding,* and *retracting* positions of a three-way *spool* valve. It is called "three-way" because it has three ports. This type of valve is capable of providing pressure to only the extend port of the *single-acting* cylinder. The big ends of the valve spool fit snugly in the bore (round hole) of the valve, but oil can flow around the middle part. In the extend position (Fig. 9-11*a*), the pump forces oil through the middle section of the spool and into the cylinder, extending it.

When the spool is moved to the hold position (Fig. 9-11*b*), one spool blocks the opening to the cylinder,

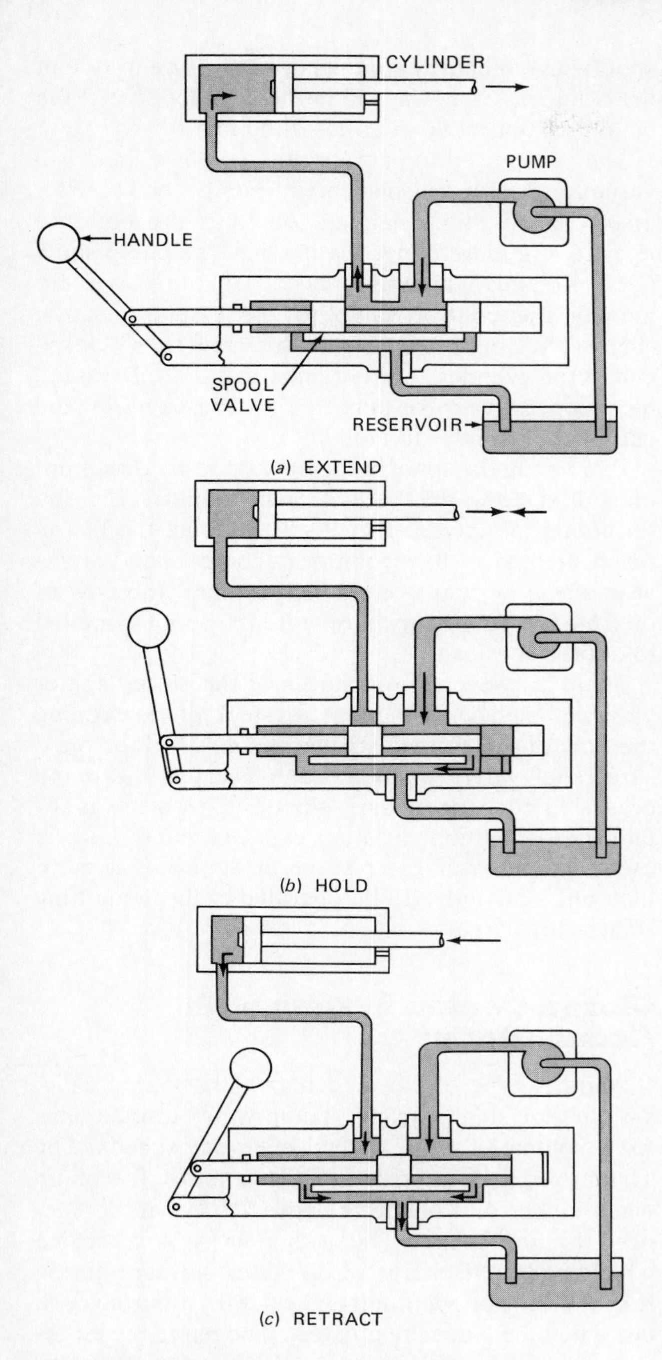


**Fig. 9-11.** The three-way spool valve is designed to control a single-acting cylinder.

trapping oil in the cylinder, while the second spool opens the dump port at one end of the valve body and pump oil flows to the reservoir.

To be able to retract the piston, the spools are moved completely forward by the control lever. In this position (Fig. 9-11*c*) the two discharge ports to the reservoir are opened as well as the port to the cylinder. This allows oil to drain from the cylinder to the reservoir and the pump to discharge directly into the reservoir. A load on the rod is required to force the piston back and move oil out of the cylinder. The weight of the equipment forces oil out of the cylinders during the lowering (retract) cycle.

## Double-Acting Cylinder Spool Valves

The hydraulic valves used on many agricultural machines are designed to control cylinders that can both push and pull. A typical application is found in the cylinders used to regulate the bucket position of a front loader (Fig. 9-1). When the control lever is moved forward, hydraulic pressure forces the cylinder rams to extend, dumping the bucket full of material into the spreader. When the bucket is empty, the operator moves the control lever in the opposite direction and returns the bucket to the loading position.

Systems of this type utilize *double-acting* cylinders and four-way spool valves to control their "push-pull" response. A four-way valve receives its name from the four connections on the valve body. The illustrations in Fig. 9-12 show how a four-way valve directs fluid to and away from the cylinder simultaneously.

When the valve spool is moved to the extend position (Fig. 9-12*a*), the spool ends cover two of the three oil return ports. This allows the pump to force oil around the spool, out the port into the cylinder, thereby pushing the piston outward and extending the rod. As the rod extends, oil on the opposite side of the piston is pushed back through the open valve port and to the reservoir through the uncovered escape port.

In the neutral, or hold, position (Fig. 9-12*b*), the pump dumps oil back into the reservoir through the open center while the spool ends block off the ports which connect with either end of the cylinder. Fluid is "locked" on each side of the piston so that it cannot move in either direction.

To retract the piston rod, the spool valve is moved to the extreme opposite end of the valve body. The spools seal two of the three discharge ports to the reservoir while the open port directs pump fluid to the rod end of the cylinder. The retracting piston forces oil out the open discharge ports and into the reservoir.

A typical example of a complete system utilizing a four-way, open-center valve is illustrated in Fig. 9-13. Hydraulic force can be applied to raise or lower the dozer blade. In this type of system, a relief valve is necessary to protect the equipment from excessive pressure. A relief valve is always installed in the line that connects the discharge side of the pump to the control valve. The relief valve will allow oil to bypass to the reservoir when system pressure in excess of normal is developed.

## Closed-Center Valving System

The open-center valving system is generally limited to operating one single-function system such as one single cylinder at a time. Many modern machines

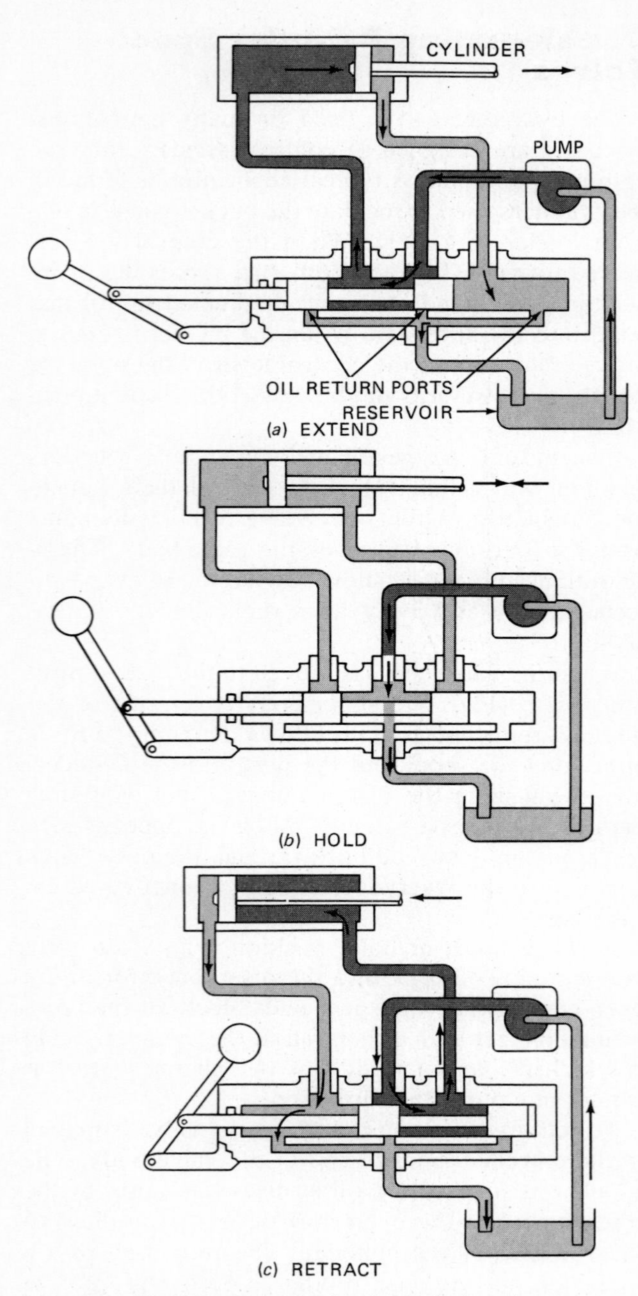


**Fig. 9-12.** The four-way spool valve is designed to control a double-acting cylinder.

require more hydraulic power than can be provided by the open-center system. A modern farm tractor may require hydraulic pressure for power steering, power brakes, remote cylinders, loaders, and other mounted equipment. Closed-center valving is often used to operate equipment where multiple functions are operated.

The closed-center control system differs from the open-center in the number and arrangement of spools in the control valve. Instead of allowing oil to flow through the center of the valve body when the valve is in neutral, spools in the valve block the inlet pump port in the body of the valve (Fig. 9-14). Also, the spool valve blocks the return of oil from each side of the cylinder. Oil is trapped in the cylinder so that the piston cannot move in either direction.

The pump used to operate this type of system is a variable, positive-displacement pump. A variable-displacement pump delivers oil until the pressure rises to a predetermined level. Then the pump shuts itself off and maintains pressure to the valve. By moving the control valve to the extend position (Fig. 9-15*a*), oil is routed from the pump to the closed end of the cylinder. At the same time, the valve spool has opened the valve port to allow oil from the rod end of the cylinder to flow back to the reservoir.

By moving the spool in the fully opposite direction, oil is directed to the rod end of the cylinder, thereby retracting the piston (Fig. 9-15*b*). Should the piston reach the end of its stroke or become loaded in excess of the system's operating pressure, the flow of oil from the pump must stop until the spool is moved to a different position.

There are several advantages of the closed-center variable-displacement pump system of controlling the flow of oil. Foremost, there is no need for pressure relief valves in the system because the pump ceases to function when "standby" pressure is attained. In addition, a larger-capacity pump can be used, thus providing a reserve of fluid flow if more than one function is to be operated at the same time (Fig. 9-16).

## Closed-Center System with Accumulator

Another type of closed-center valve system (Fig. 9-17) uses a small-capacity pump which discharges a fixed volume of oil at normal operating speeds. The system requires an accumulator to assist the pump when higher oil volume is needed than can be supplied by the pump. The accumulator is a device which stores the energy of oil under pressure during slack periods of operation and releases it during periods when the loads are greatest. The pump recharges the accumulator after each peak use. Accumulators are sometimes used as safety devices to protect against oil pressure failure. For example, hydraulic power brakes may be operated by accumulator pressure in the event that the pump has stopped and cannot supply enough pressure to operate the brakes to stop the machine. Most accumulators are of the bladder type (Fig. 9-18) with one side of the bladder charged with nitrogen gas. Oil under pressure compresses the bladder against the gas until pressure in the oil side of the accumulator is equal to the relief valve setting of the system. If the pressure in the system drops more than the pump can maintain, gas pressure forces fluid from the accumulator into the system to maintain maximum hydraulic pressure.

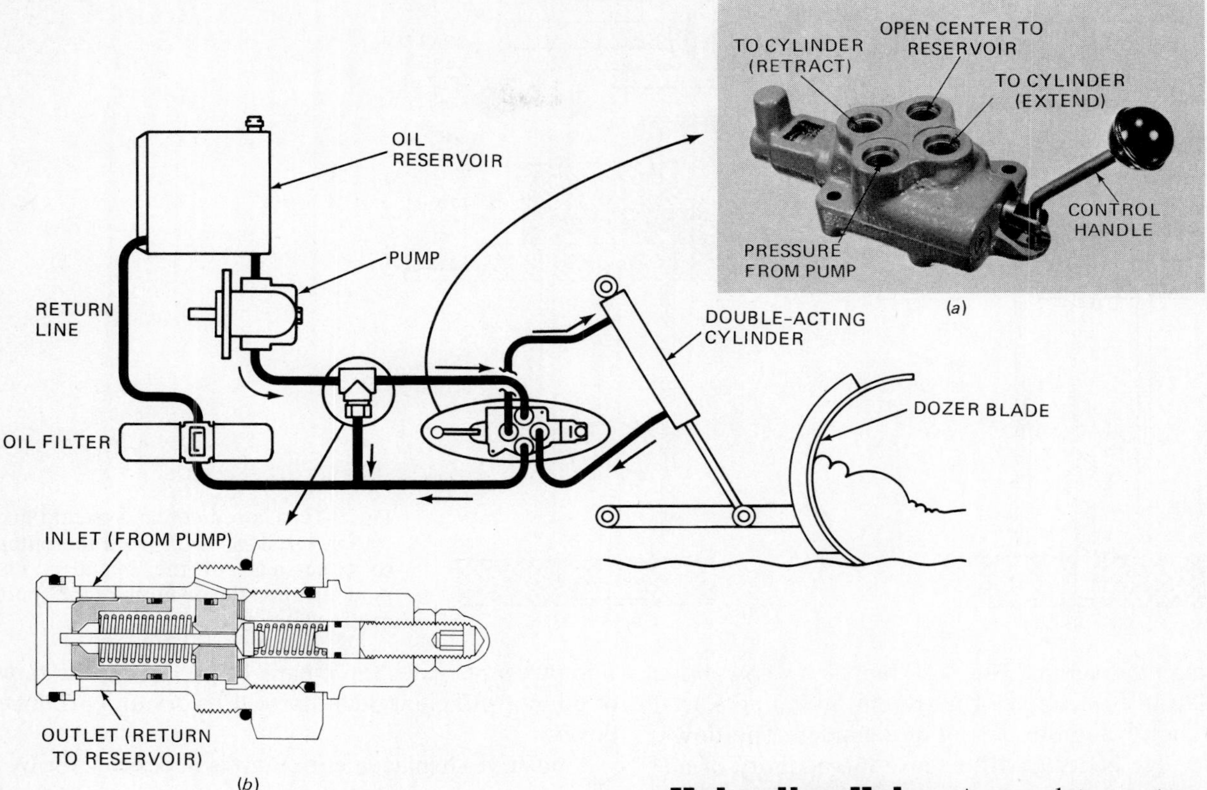


**Fig. 9-13.** Typical circuit diagram for a four-way, open-center system. A relief valve protects the system from excessive pressure. (*a*) Four-way control valve. (*b*) Pilot-operated relief valve. (*Eaton Corporation.*)

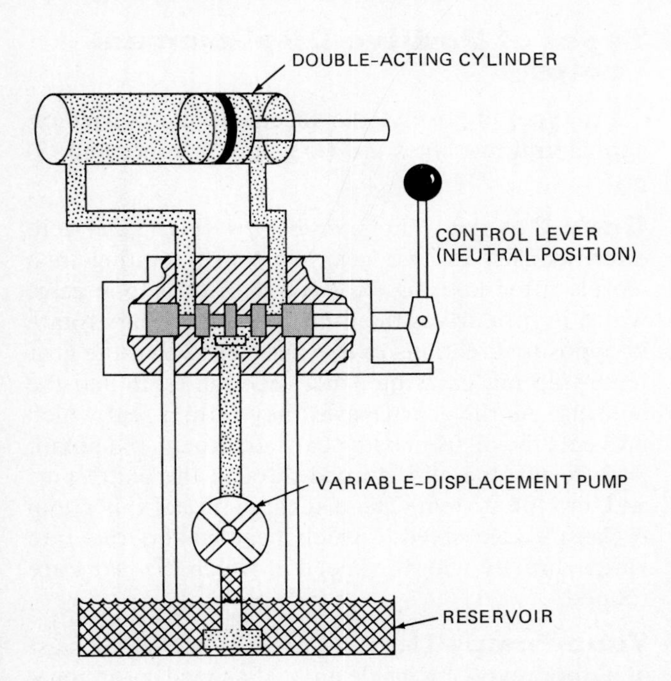


**Fig. 9-14.** A closed-center spool valve in the neutral position blocks the flow of oil from the pump to the cylinder and prevents the return of oil from each end of the cylinder to the reservoir. The ram of the cylinder is locked in the position shown.

**Unloading Valve** Accumulator systems of this type require an *unloading valve* to prevent excessive pressures from being developed by the pump. The unloading valve also serves as a pressure relief valve by reacting to system pressure through a sensing line (Fig. 9-19). The valve makes it possible for the pump to operate at a reduced power requirement by routing the pumped fluid back to the reservoir during periods of low demand.

**Check Valve** A *check valve* (Fig. 9-19) is also required for this system, to prevent the loss of pressure in the operating functions at the time the unloader is decreasing pressure on the pump.

**Flow-Control Valve** In Fig. 9-17 the three functions involve different-size cylinders. When these conditions exist, the volume of oil to the cylinders must be controlled to get equal response. For example, in Fig. 9-20, a cylinder and a hydraulic motor are operated from one system. In order to provide sufficient oil to maintain the motor's speed when the cylinder is operated, it is necessary to regulate the flow of oil to each. This is provided by a *flow-control* valve. The orifice (1) reduces the flow to the small cylinder, while the flow-control valve (2) allows oil to bypass and return to the reservoir when the pump pressure becomes too great in the supply system. The size of the orifice is calibrated to furnish the motor with the correct amount of oil to maintain proper operating speed.

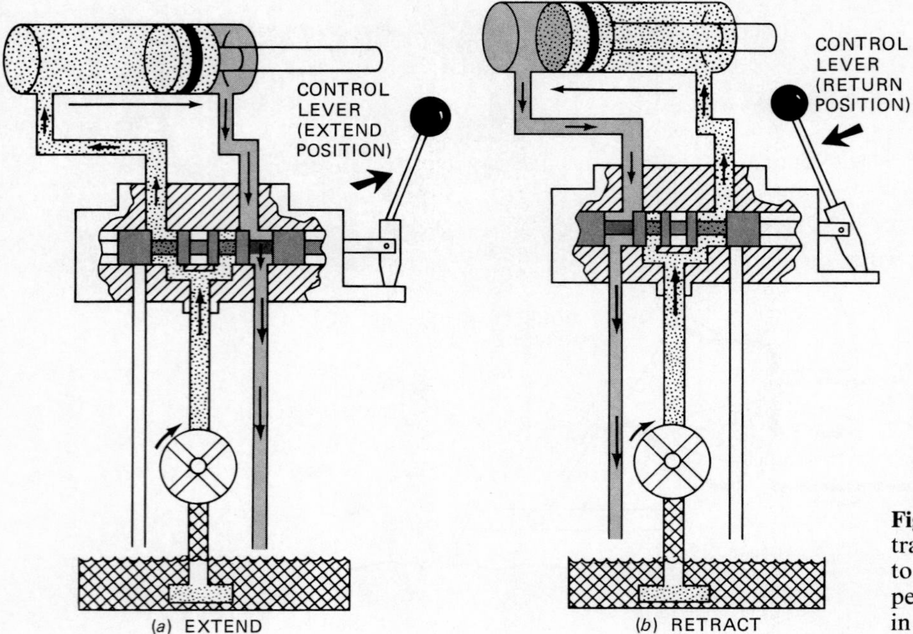


**Fig. 9-15.** Closed center system illustrating the flow of oil from the pump to either end of the cylinder, depending upon the position of the spool in the valve body.

The circuit diagram (Fig. 9-21) includes a flow-control valve in the circuit of a system operating a hydraulic motor and double-acting cylinder. The flow-control valve is designed to sense the quantity of oil required by the motor to maintain operating speed whenever the cylinder is put into operation by its control valve.

## PUMPS

Now that you have developed an understanding of how the flow of hydraulic fluid is controlled, you can see that a pump is the heart of the hydraulic system.

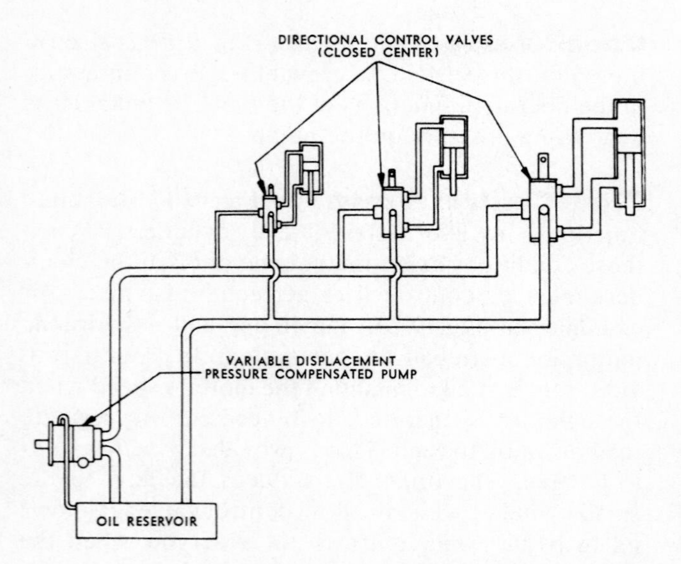


**Fig. 9-16.** Closed-center, variable-displacement pump system. Note that the system can operate several functions, each with a different oil volume requirement. *(Deere & Co.)*

The pump provides the means of producing the flow of oil in a sufficient quantity to do work and produce power.

A positive-displacement pump is necessary for hydraulic systems because a constant volume of oil must flow against a resistance. Fluid pressure is developed when there is resistance to the flow of oil. Hydraulic power is produced when a measured quantity of oil flows at specified pressure.

### Types of Positive-Displacement Pumps

The types of positive-displacement pumps used on agricultural machines are (1) gear, (2) vane, and (3) piston.

**Gear Pumps** The *gear pumps* (Fig. 9-22) generally consist of two gears in mesh rotating in a closely fitted housing. A drive shaft turns one gear, which in turn drives the other gear. The gears rotate in opposite directions to one another so that the gear teeth trap oil, carrying it between the teeth and the housing. As the gears mesh, they form a seal which prevents the oil from being carried through the pump. As a result, the oil is forced through the outlet port and into the system. The discharge rate of this pump is classified as ''fixed,'' which means that output rate varies directly with the speed at which the gears are rotated.

**Vane Pumps** The *vane pumps* (Fig. 9-23) consist of a rotor revolving inside an oval-shaped rotor housing. Metal vanes are fitted into the rotor slots and are free to move in and out. As the rotor turns, the vanes are thrown out against the inside surface of the pump housing by centrifugal force. A pumping action is

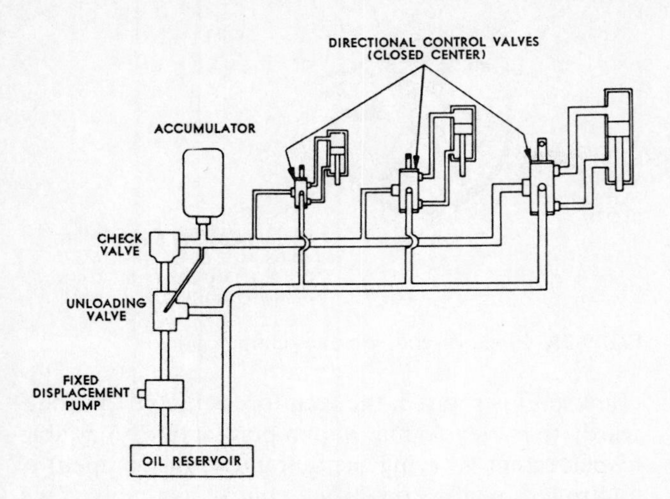


**Fig. 9-17.** Closed-center valve systems with an accumulator. The accumulator releases its stored energy (oil under pressure) when the demand is greater than the pump can meet. An unloading valve allows the pump to dump fluid directly back into the reservoir during slack periods. A check valve presents loss of pressure in the control and operating system. (*Deere & Co.*)

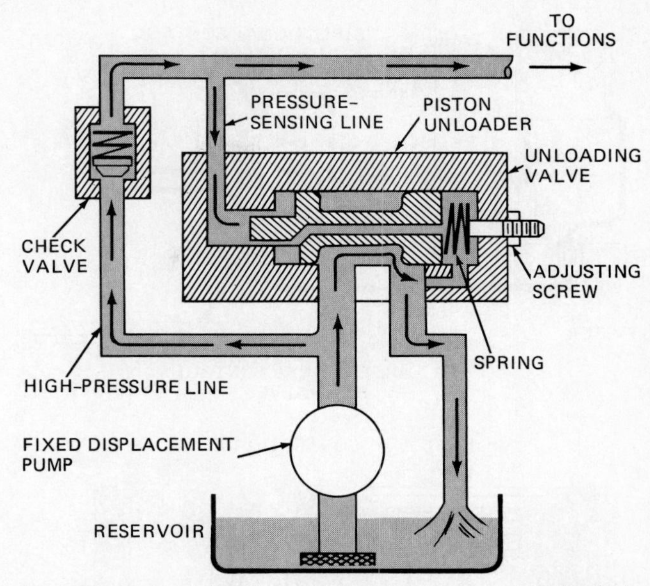


**Fig. 9-19.** The unloading valve senses pressure in the high-pressure line and then moves the unloader piston against a compression spring, opening the return (dump) port to the reservoir, allowing the pump to operate at a reduced power requirement. A check valve prevents any loss of pressure in the system when the unloading valve opens.

created as oil is carried around the inside of the housing by the vanes. Vane pumps are usually classified as fixed-discharge types.

**Piston Pumps** The *piston pumps* are capable of operating at high speeds and developing high pressures. They are the most common pump found in today's hydraulic systems. Piston pumps are further classified into two types: axial and radial.

**Axial piston pumps** The *axial piston pumps* used in agricultural machinery are the in-line type (Fig. 9-24). They may be fixed- or variable-discharge pumps, depending upon whether the swash plate (a plate which can be tilted) is fixed or movable. In operation, the drive shaft turns a rotating cylinder block in which six to nine pistons are fitted. These pistons move inward and outward in response to the angle of the swash plate and produce a pumping action.

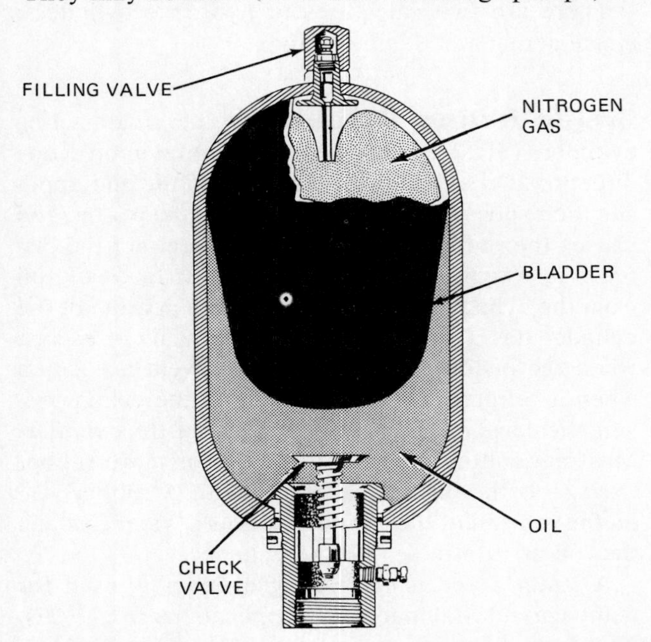


**Fig. 9-18.** Hydraulic system with a bladder-type accumulator. A check valve presents the bladder from being drawn into the oil port when the bladder expands.

**Radial piston pumps** The variable-displacement *radial piston pump* is a popular type of hydraulic pump for agricultural machinery. This pump consists of a series of small cylinder bores fitted with pistons arranged like spokes in a wheel. The pistons are hollow, contain a spring, and are actuated by a roller cam acting as a crankshaft (Fig. 9-25). When the

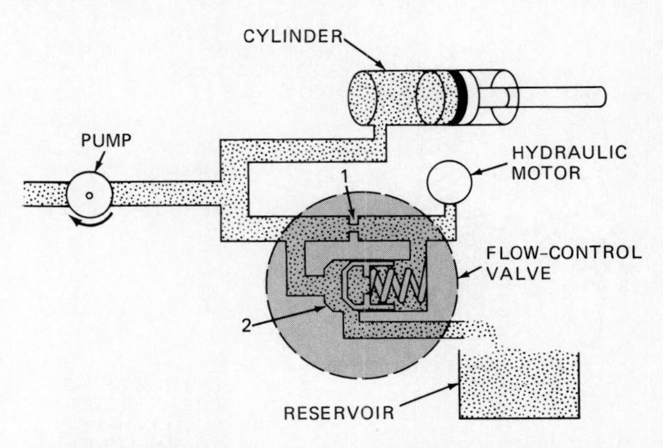


**Fig. 9-20.** A flow-control valve is used to regulate the volume of oil flow to two, separate functions. The orifice (1) restricts flow to the hydraulic motor, and the "dump" valve (2) reduces excessive pressure due to orifice restriction.

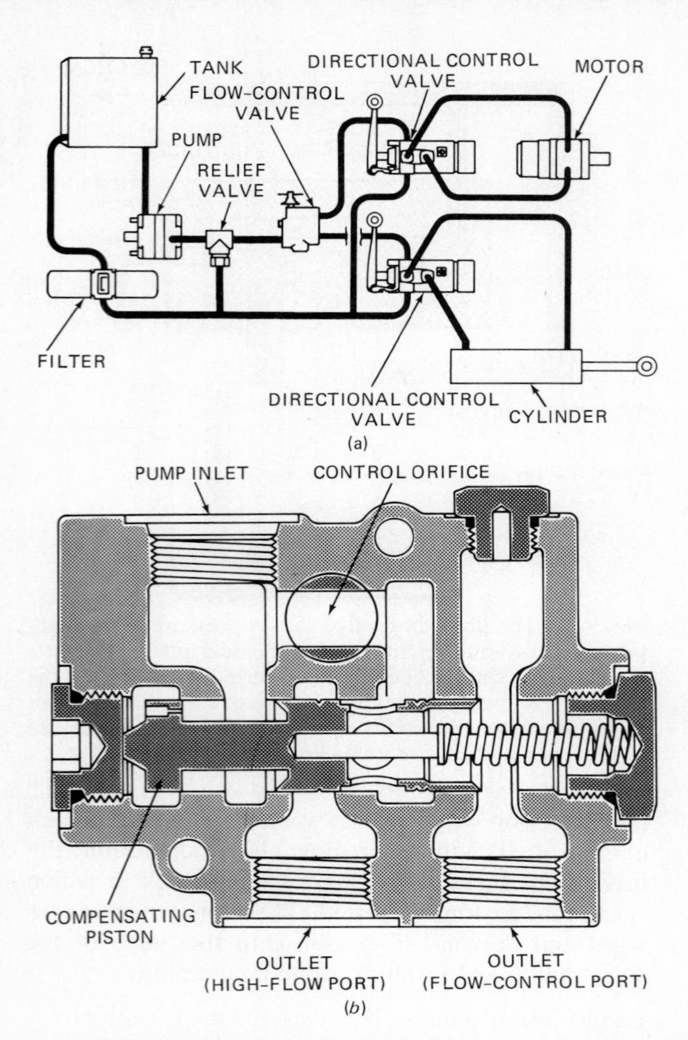


**Fig. 9-21.** Flow diagram of a hydraulic system utilizing a flow control valve to equalize the amount of oil required when operating a hydraulic motor and separate cylinder. Inset shows cross section of flow control valve. (*Eaton Corp.*)

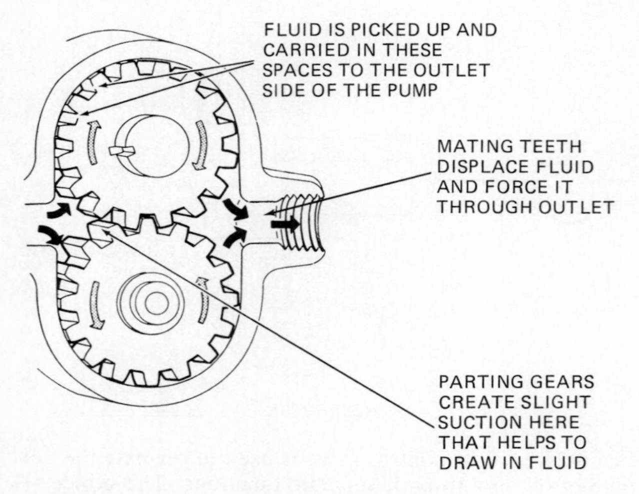


**Fig. 9-22.** Fixed discharge external gear pump. The top gear is driven by the power source. The bottom gear runs in mesh with the top gear.

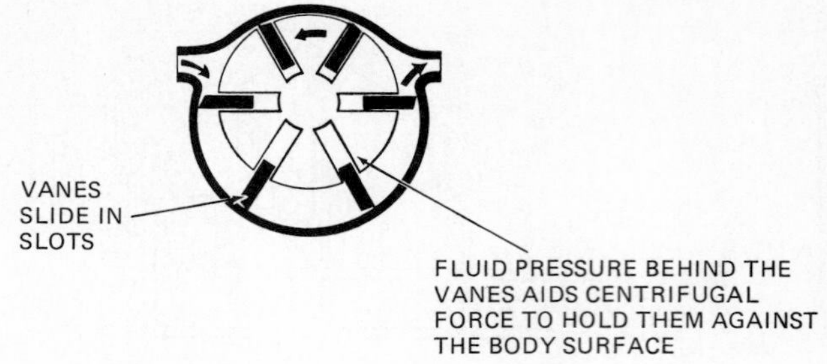


**Fig. 9-23.** Fixed-discharge vane pump.

crankshaft is rotated, the cam forces the pistons outward, thus developing a pumping action. Variable displacement (varying the amount of oil pumped) is controlled by a piston stroke control assembly. This assembly senses the quantity of oil the system requires, limiting the inward movement of the pistons toward the cam and the resulting stroke. This governs the amount of piston travel in the cylinder and the amount of oil pumped by each.

## ACTUATORS

In most hydraulic systems used on agricultural machinery, there are two types of *actuators*. These are better known as hydraulic *cylinders* and hydraulic *motors*. These parts of a hydraulic system are called actuators because they utilize the hydraulic force of the oil to create mechanical movement. Hydraulic cylinders give straight-line motion, and hydraulic motors produce rotating motion.

### Hydraulic Cylinders

There are two categories of hydraulic cylinders: single-acting and double-acting.

**Single-Acting Cylinders** The single-acting cylinders (Fig. 9-26) can produce a force in only one direction. Oil enters the cylinder from the end opposite the exposed piston rod, acts upon the piston, and causes the rod to extend. In order to retract the piston, the weight of the lifted load must force the oil from the cylinder to the reservoir. The dry end of the cylinder has a small air vent to allow air to escape when the piston extends, and to prevent a vacuum when it retracts. An oil seal fitted in the piston prevents leakage of oil into the dry side of the cylinder. Most agricultural cylinders use one or more rubber O-rings as the piston seal (Fig. 9-26). A rubber seal on the rod end of the cylinder acts as a wiper to clean the rod as it retracts into the cylinder.

A ''ram''-type, single-acting cylinder is used for many agricultural machinery applications (Fig. 9-27). This type of cylinder does not use a conventional piston. Instead, the end of the rod serves as a piston. The rod (ram) is slightly smaller than the inside of the

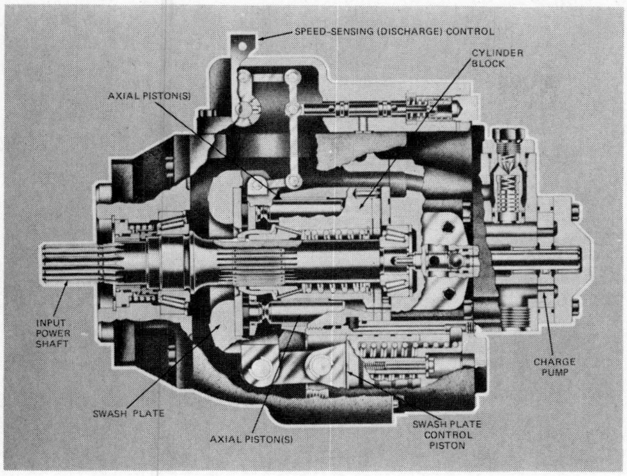

**Fig. 9-24.** Variable-discharge axial in-line piston pump. The position of the swash plate is controlled by the hydraulic control cylinder and connecting linkage. (*Eaton Corp.*)

cylinder. A lock ring prevents the rod from being pushed out of the cylinder. Ram-type cylinder construction has several advantages: (1) the large-diameter rod resists bending due to side loads, (2) no air vent is needed, (3) the rod does not make contact with the cylinder wall which eliminates scoring, and (4) the rod is easily removed from the cylinder to replace wiper and rod seals. This is accomplished by moving the ring from the normal groove to a deeper recess (Fig. 9-28).

**Double-Acting Cylinders** When it is necessary to provide a force in both directions, double-acting cylinders are used. In order to accomplish this action, fluid must be directed to both sides of the piston (Fig. 9-29). A rod seal must be used to prevent leakage of oil around the rod when pressure is applied to the rod end of the cylinder. A characteristic of double-acting cylinders is that the total force on the rod side of the piston is less than that on the closed or blank side of the piston. This is because the rod re-

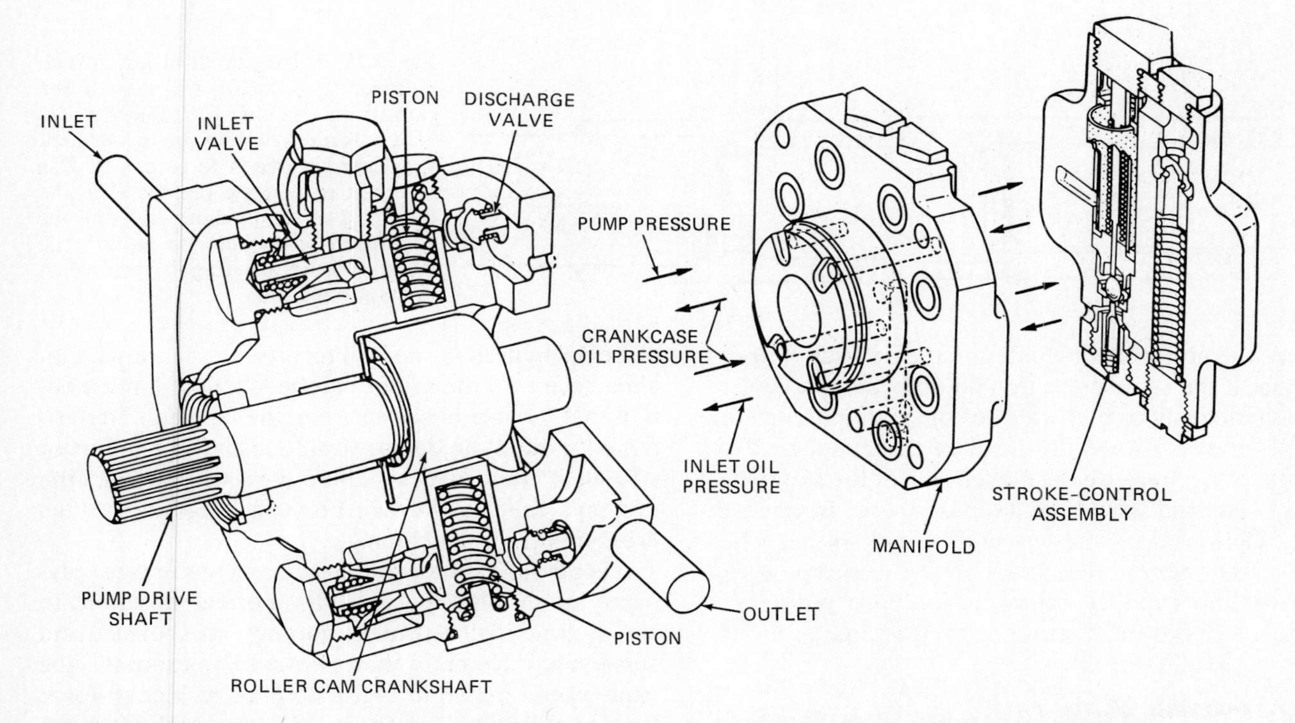


**Fig. 9-25.** Variable-discharge radial piston pump. The stroke control assembly limits the length of the piston movement. (*Deere & Co.*)

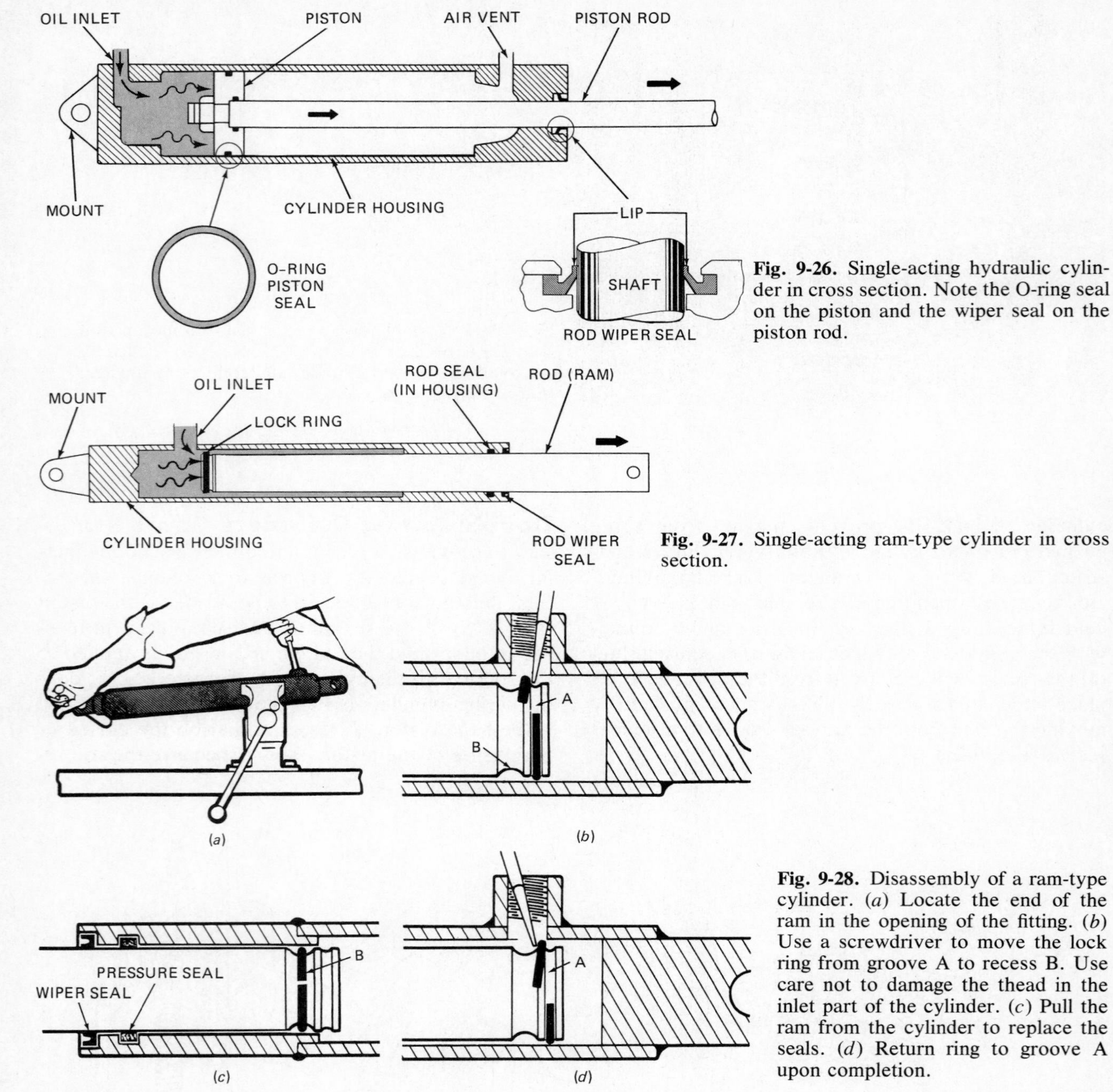


**Fig. 9-26.** Single-acting hydraulic cylinder in cross section. Note the O-ring seal on the piston and the wiper seal on the piston rod.

**Fig. 9-27.** Single-acting ram-type cylinder in cross section.

**Fig. 9-28.** Disassembly of a ram-type cylinder. (*a*) Locate the end of the ram in the opening of the fitting. (*b*) Use a screwdriver to move the lock ring from groove A to recess B. Use care not to damage the thead in the inlet part of the cylinder. (*c*) Pull the ram from the cylinder to replace the seals. (*d*) Return ring to groove A upon completion.

duces the area of the cylinder on the rod side. For example, if the same pressure and flow rate are applied simultaneously to both sides of the piston, the piston would move toward the head of the cylinder. This would occur because the closed side of the piston has more area and would create more force. Because of this relationship, cylinders are located so that when the load condition requires a slower, more powerful stroke, the cylinder extends. This also provides a faster, less powerful stroke for retracting.

## Hydraulic Motors

Hydraulic motors are actuators which produce rotating mechanical power. The important advantages of using hydraulic motors to drive agricultural machinery are (1) the motor can be located in any position that flexible hoses carrying the fluid can be positioned, and (2) no damage can occur to the motor or system if the driven machine part should stall; the system's relief valve will prevent dangerously high pressures from developing.

It is possible for a pump to run as a motor by applying a flow of oil opposite to the normal direction. In actual practice, however, motors are constructed somewhat differently than pumps. For example, the vane motor shown in Fig. 9-30 uses springs to force the vanes outward to form a positive seal when the motor is operating at slow speeds.

The internal gear motor, illustrated in Fig. 9-31, in-

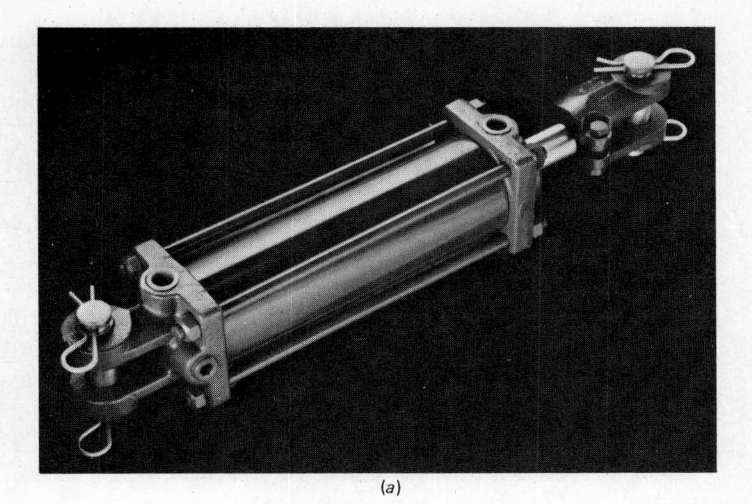

(a)

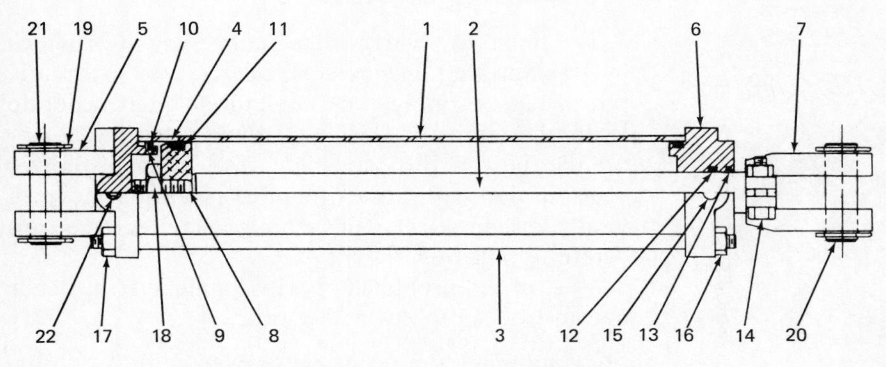


1. CYLINDER BARREL
2. PISTON ROD
3. TIE RODS
4. PISTON
5. BUTT END CLEVIS
6. ROD END CASTING
7. ROD CLEVIS
8. STATIC ROD O-RING
9. STATIC O-RINGS
10. STATIC BACKUP WASHERS
11. PISTON SEAL
12. U-CUP ROD SEAL
13. ROD WIPER
14. BOLT AND NUT
15. FLUID PORT (RETRACT)
16. TIE ROD NUTS
17. TIE ROD NUTS
18. PISTON NUT
19. HAIRPIN CLIPS
20. & 21. CLEVIS PINS
22. FLUID PORT (EXTEND)

(b)

**Fig. 9-29.** (*a*) A typical double-acting hydraulic cylinder. (*b*) A cross section of its parts. Cylinders of this type are classified as agricultural and meet ASAE standard size specifications for agricultural machinery. Operating pressure is 250 psi [1724 kPa], and standard bore sizes are 2 in [51 mm], 3 in [76.2 mm], $3^{1}/_{2}$ in [89 mm], and 4 in [102 mm] with a stroke of $10^{1}/_{4}$ in [260.3 mm]. (*Command Hydraulics, Inc.*)

volves a toothed rotor which is forced around the stator assembly by hydraulic pressure. As it rolls over each of the vanes, it turns the drive link through gear action. This kind of motor is commonly used as traction wheel drives for self-propelled harvest machinery.

The axial piston, shown in Fig. 9-32, is generally considered to be the most efficient of the hydraulic motors. It can be either fixed or variable-speed, depending upon the design of the swash plate. It is often used as the motor for driving self-propelled machinery or where variable speed is desired.

## HYDRAULIC SYSTEM SCHEMATIC

The diagram in Fig. 9-33 identifies the arrangement of hydraulic system components for a windrower designed to cut and condition hay and forage crops. The system includes hydrostatic pumps and wheel motors at each drive wheel to propel the machine forward or backward. There is also a pump and motor for operating the header section. A header lift pump provides oil under controlled flow through the header lift valve to the lift cylinders. Residual case pressure oil is returned to the oil cooler, where heat developed in utilizing the oil is removed before the oil returns to the reservoir for reuse. Return oil is also filtered before entering the reservoir. Diagrams such as this help the serviceperson in agricultural power and machinery visualize the relationship of all the hydraulic components of an agricultural machine.

### Observing Hydraulic Safety Rules

Before performing repair work on hydraulic systems, there are important safety rules that must be observed to avoid physical harm and mechanical damage to equipment. It is essential that the hydraulic service person remember to:

1. Wear eye and face protection when working on pressure systems.

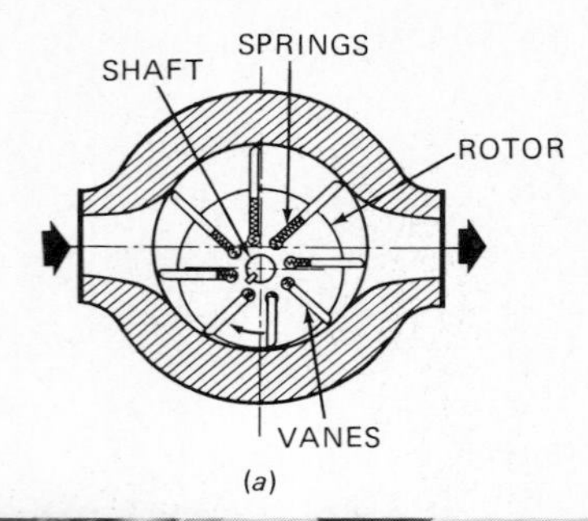


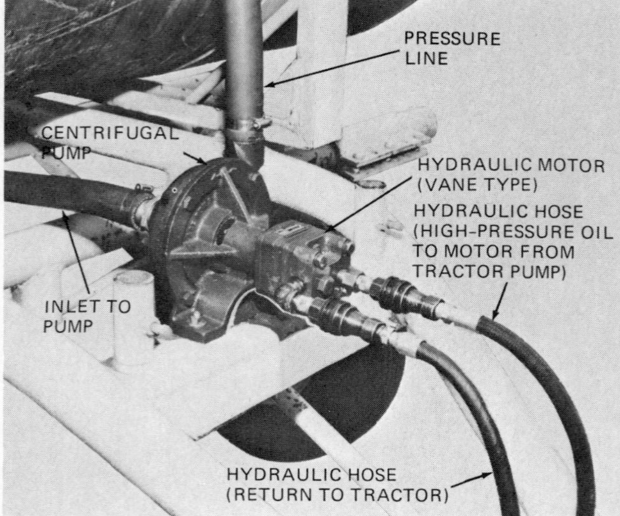


**Fig. 9-30.** Hydraulic vane motor. Unbalanced hydraulic forces work against the vanes to turn the rotor. (*a*) Springs behind the vanes are used to hold them against the casting when operating at slow speeds. (*b*) Vane motors are adapted to operating equipment such as the centrifugal pump for sprayers. (*Ace Pump Corp.*)

2. Avoid examining potential leaks or breaks when the system is under high pressure.
3. Block or lower hydraulic equipment before repair and during service.
4. Always tag and remove keys from equipment being repaired or serviced to avoid accidental starts.
5. Bleed air from the system before working around repaired equipment to avoid unexpected movement or dropping of heavy equipment supported by cylinders.
6. Relieve all pressure from a system before loosening system fittings to avoid high-pressure oil sprays.
7. Beware of oil spray, drips, or leaks on hot engine manifolds or exhausts.
8. Fasten hoses securely; a loose hose may whip, causing serious injury.
9. Use gloves to handle hot hydraulic equipment.
10. Route lines and hoses away from heat sources.
11. Keep all valves and equipment that might leak away from heat sources.
12. Wear heavy gloves when replacing hose couplings to avoid cuts from sharp reinforcing wire braid.
13. Avoid burns from hydraulic oil. Be very cautious when checking overheated equipment.
14. Clean up oil spills. Avoid creating slippery or flammable conditions.
15. Avoid using air pressure to disassemble hydraulic cylinders. Compressed air can turn the cylinder into a cannon.
16. Always read the service manual before repairing accumulators. They contain very high spring or gas pressures which may be released unexpectedly.

## Servicing and Maintaining the Hydraulic System

The flexibility which a hydraulic system provides in transforming mechanical power into hydraulic power has encouraged manufacturers to incorporate hydraulics into the design of their machinery. Hydraulic power provides safe and efficient power transmission with a maximum of performance, providing certain service and maintenance practices are carefully followed.

Most of the problems of a hydraulic system occur because of the following factors:

1. Contamination of the oil by water, dirt, and foreign matter
2. Failure to follow the recommended oil drain schedule
3. Insufficient oil in the reservoir
4. Use of the wrong type of filter
5. Use of the wrong type of oil
6. Air that has entered the system

**Cleanliness** Dirt, moisture, and other foreign materials are the No. 1 enemy of hydraulic systems. The closely fitted parts of pumps, motors, cylinders, and valves are susceptible to the scratches and erosion of the highly polished surfaces. The result is poor performance, leaks, and overheating of oil. Following improper service procedures often results in dirt entering hydraulic systems.

Many of the problems associated with dirt getting into the system can be avoided by taking special precautions, including the following:

1. Always clean away all visible dirt, moisture, and trash around the filter cap before removing it. Also be sure to clean around the dipstick opening before checking the oil level. Use a clean lint-free cloth to wipe the bayonet dipstick.
2. When adding oil to the reservoir, be sure to use clean oil. Hydraulic oil must be handled and stored in closed containers to prevent contamination. Always use oil that has been kept in sealed cans or dispensed in clean containers—avoid transferring oil from the original container

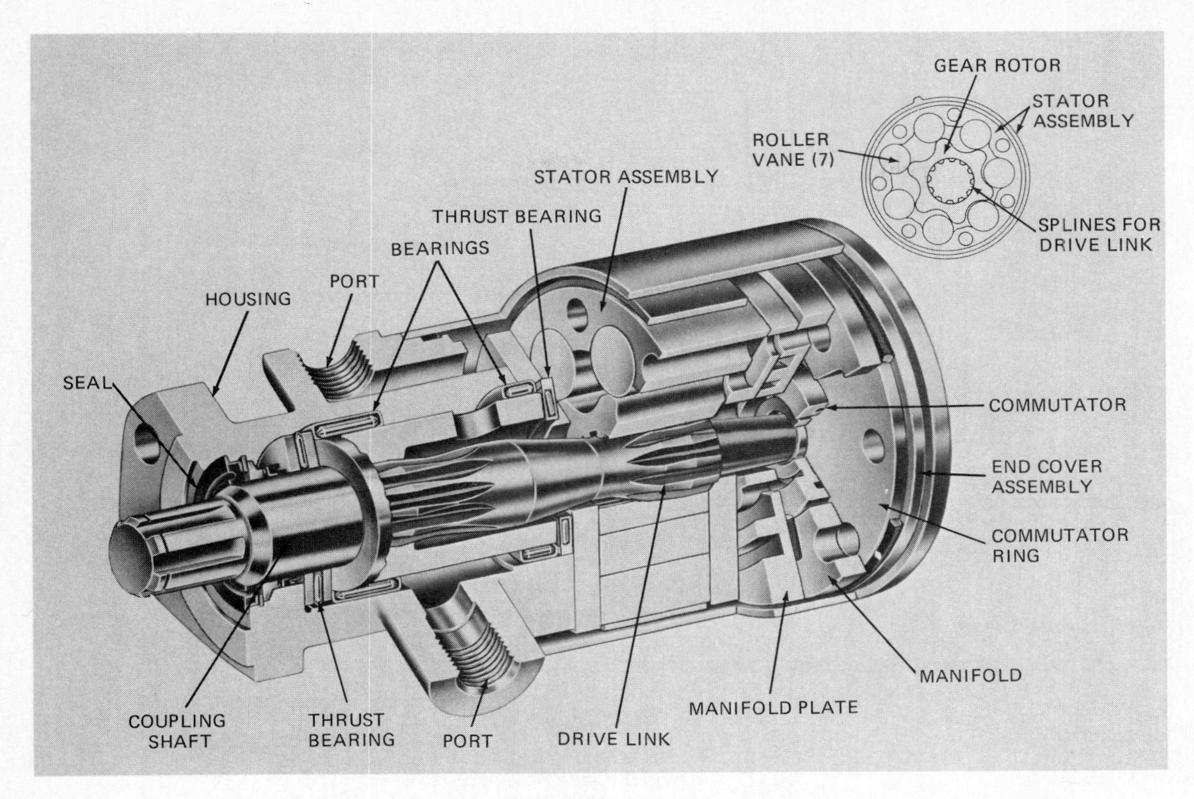


**Fig. 9-31.** Roller-gear-type hydraulic motor. The direction of rotation is controlled by changing the flow of oil through the system. (*TRW, Ross Gear Div.*)

as much as possible and seal the opening when sufficient oil has been withdrawn.

3. Avoid the use of funnels and other dispensing hardware that are not properly cleaned of all dust and dirt.
4. Clean all fast-coupling fittings before attaching remote cylinders or motor drive hoses and replace protective caps when the couplings are not used (Fig. 9-34).
5. When it is necessary to remove any parts from a

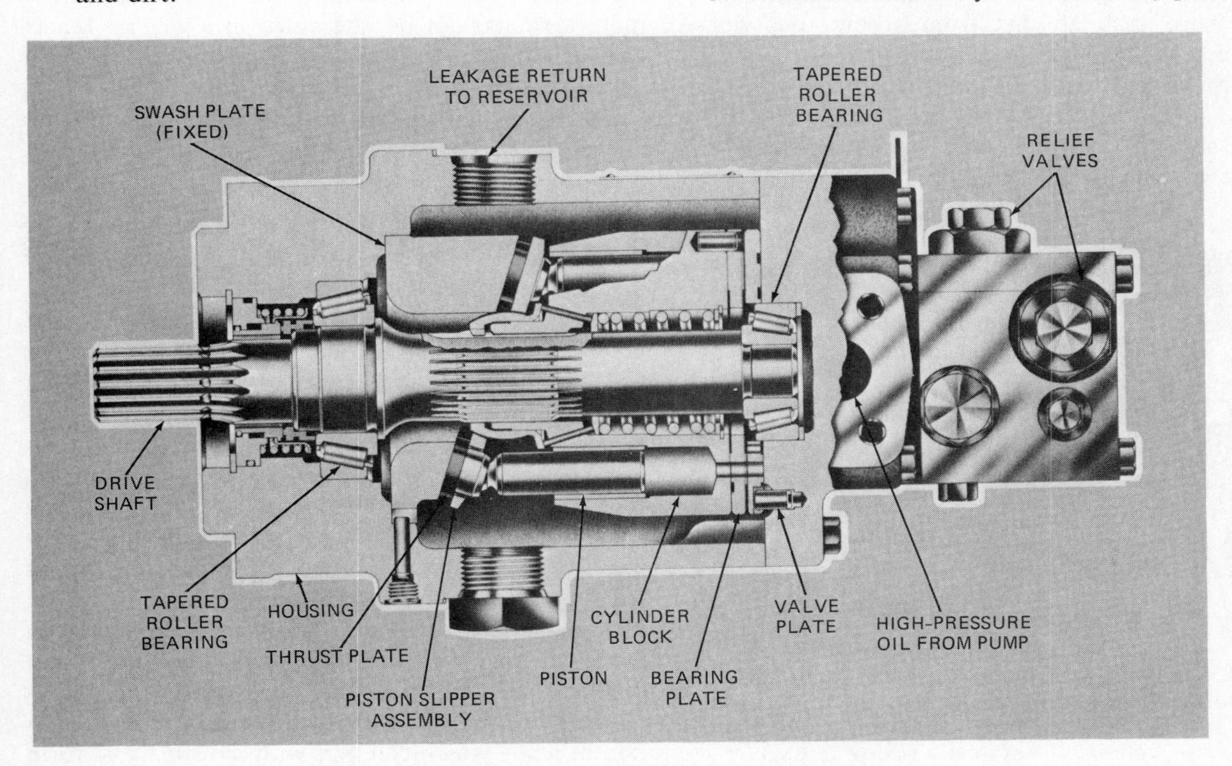


**Fig. 9-32.** Cross section of an in-line axial piston hydraulic motor. (*Deere & Co.*)

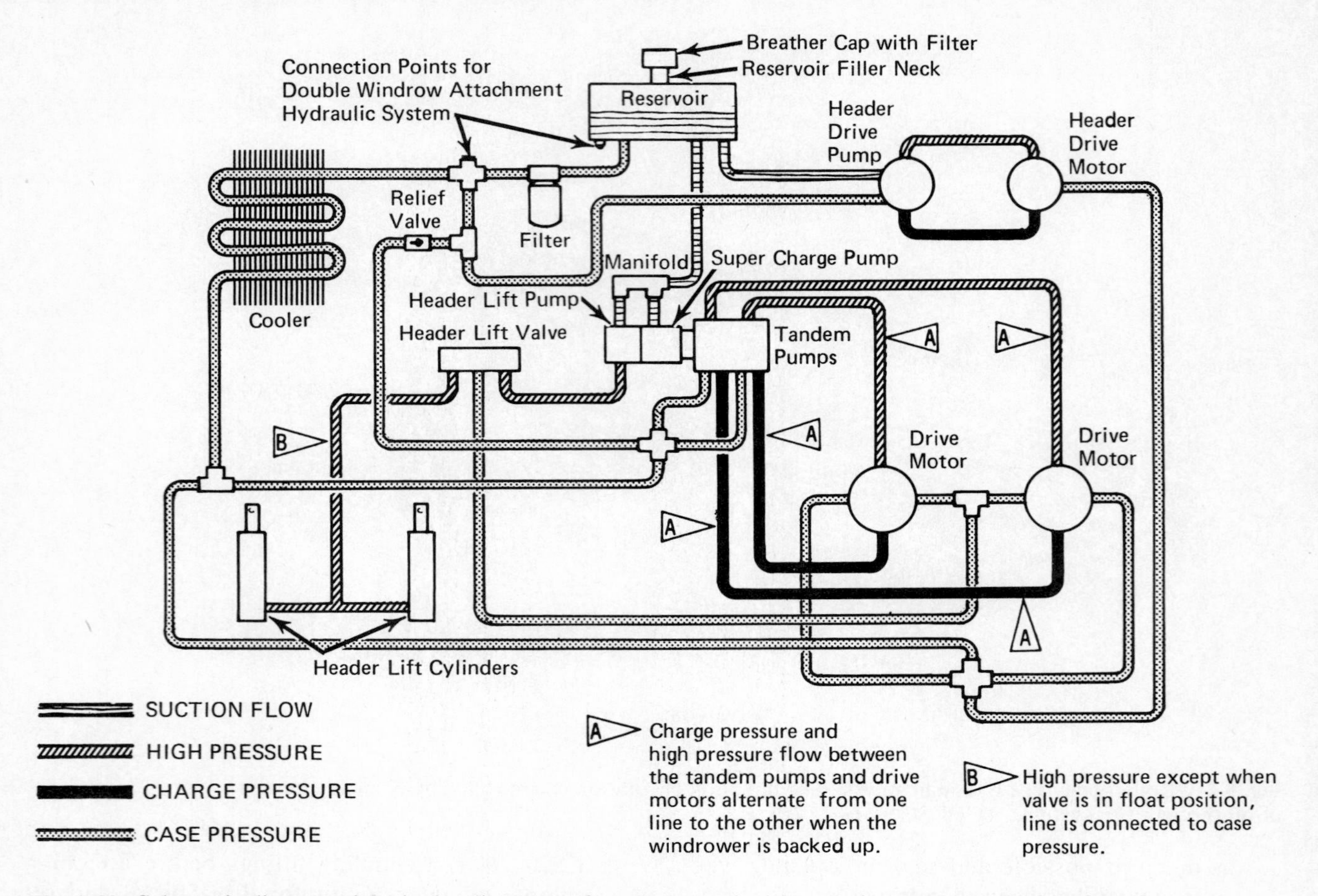


**Fig. 9-33.** Schematic diagram of the hydraulic system for a hydrostatic drive windrower. Note the location of the oil cooler and oil filter. (*Hesston Corp.*)

system for repair, always steam-clean or high-pressure-wash all dirt from around the work area.

6. During repair, wash all the parts in a clean, suitable solvent and store them in a plastic bag until reassembled. Be sure the work area is clean before starting and during the work period.

**Fig. 9-34.** Clean any dirt from the hydraulic connectors before removing the protecting plugs. After the fast-couplers have been removed, reinstall the protective plugs to prevent dirt from entering the system. (*Deere & Co.*)

Overhaul of hydrostatic piston pumps and drive motors should not be attempted in a general repair or field service area. Precision fits in the units require "white room care" during overhaul. These units should be sent to a hydraulic overhaul facility because of the clean environment and the precision fit which is required.

**Oil Drain Schedule** Many manufacturers recommend that the hydraulic oil be drained and new oil added after 50 to 100 hours (h) of operation and at regularly scheduled periods thereafter. Hydraulic oil contains many different additives to ensure proper operation over a long period of service. However, these additives will wear out and cannot provide continued protection from oxidation, corrosion, foaming, and moisture absorption. Draining and refilling with the recommended lubricant is good insurance protection. Hydraulic oil gets hot because it absorbs considerable energy from friction and the transfer of work energy. Many hydraulic systems are equipped with oil coolers to dissipate this heat. However, trash can block the air passing through the cooling fins and reduce cooler effectiveness. As a result, the oil becomes too hot and oxidizes. This causes the formation of acids which can mix with moisture and form sludge. The end result is sticking or slow-acting valves and excessive wear to moving parts.

**Maintaining Oil Reservoir Level** Always maintain the proper reservoir oil level to retain the cooling ability of the oil and reduce the chance of drawing air into the system.

**Replacing Hydraulic Oil Filter** Hydraulic systems use filters to remove harmful particles from the oil. During operation, small pieces of metal wear from the hydraulic system, and unless trapped in the filter, they can cause severe damage to valves, motors, and pumps.

Most manufacturers provide filters in the reservoir breather cap and for the main oil supply. Always replace these filters with the element approved by the manufacturer. Oil-filtering elements are designed to filter particles from the oil as small as 10 micrometers (10-millionths of a meter, or about one-tenth the size of a grain of sugar). Do not substitute an engine oil filter, because it will not provide the necessary filter protection.

**Selecting Hydraulic Oil** Hydrostatic drive and lift systems used to transfer power from the engine to the machine require special hydraulic oil. Always use an oil that is approved by the manufacturer. Other oils may be damaging to the seals, causing premature leaking or a reduction in the wear protection provided by a recommended oil.

**Keeping Air Out of System** Hydraulic systems use oil to transfer forces because of its ability to lubricate, because it can act as a flexible solid, and because it is noncompressible. Air can, however, be compressed; therefore it is a contaminant which promotes foaming of the oil and jerky-noisy operation.

**WARNING Air trapped in a lift cylinder could create a sudden and dangerous lowering of a lifted machine part or load.**

Air enters the system either through the reservoir or through loose pump intake fittings.

## Checking Hydraulic System Performance

The problems which develop in hydraulic systems are (1) visible external leaks, (2) poor performance due to malfunctions, and (3) complete failure of the system to respond to controls.

Most leaks develop at seals of actuators, pumps, and directional control valves. Proper tools and procedures are required to replace oil seals. Always refer to the service manual for the equipment when repairs are necessary.

Leaks which occur in fittings and quick connectors may only require the replacement of O-rings. Hoses which leak due to damage or deterioration from use will require replacement or new fittings.

Poor performance of a hydraulic system is associated with the capabilities of the components to develop the specified (1) operating pressure and (2) quantity of oil flow at a specified oil temperature and operating speed. Special hydraulic circuit testers are available for evaluating the performance of hydraulic system components (Fig. 9-35). The tester is attached to the hydraulic circuit by external hoses, so that the flow of hydraulic oil is routed through the tester. For

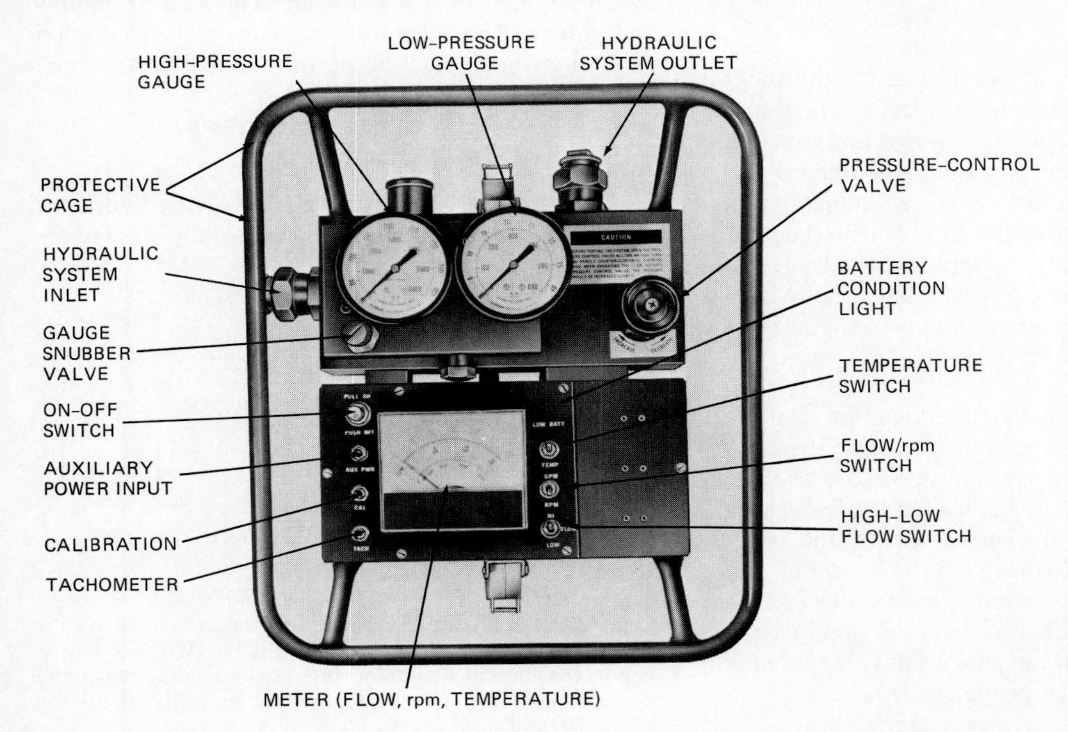


**Fig. 9-35.** A hydraulic system tester is used to evaluate hydraulic system pressure, oil flow rate, and pump rpm. (*Owatona Tool Co.*)

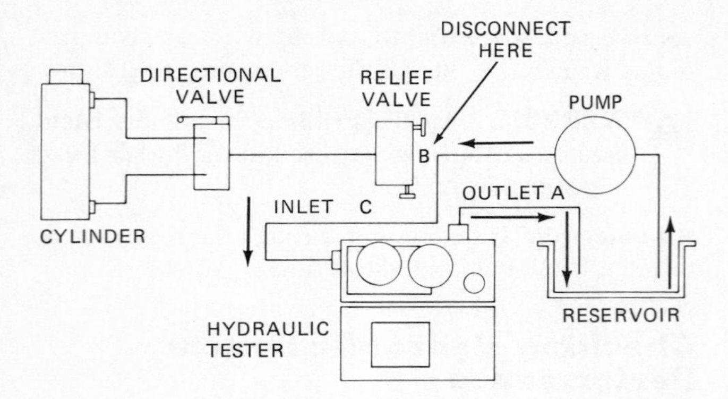


**Fig. 9-36.** Circuit diagram using the tester to evaluate the performance of the system's pump. The outlet of the pump is disconnected from the relief valve (B) and routed to the inlet of the tester. (*Owatona Tool Co.*)

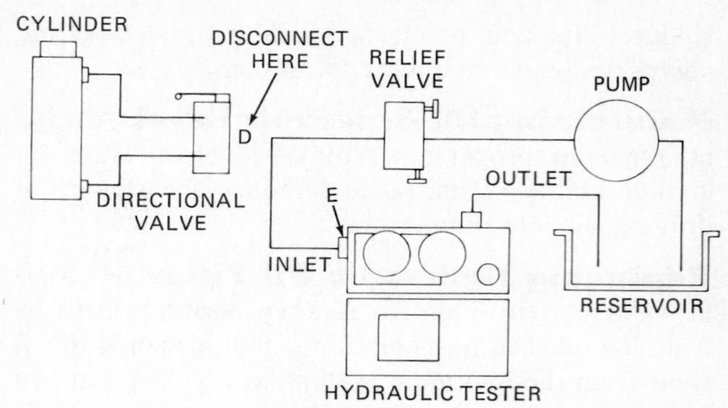


**Fig. 9-37.** Circuit diagram using the tester to check the relief valve operation. The hose at the inlet to the directional valve (D) is connected to the inlet of the tester (E). (*Owatona Tool Co.*)

example, there is a malfunction in the open-center system shown in Fig. 9-36. The piston in the double-acting cylinder operates normally at no-load. As the load is increased, the movement slows down. To locate the source of the problem the tester is connected to isolate each of the components in the following order: (1) pump, (2) relief valve, (3) directional valve, and (4) cylinder.

**Testing the Pump** Connect the outlet port side of the tester to the system reservoir, as shown in Fig. 9-36. This connection is used for all three tests. Disconnect the line (B) between the pump and the relief valve. Couple the test hose onto the discharge port of the pump and the inlet port of the tester. Determine the maximum relief pressure setting of the system from the manufacturer's specification. Never increase the pressure on the system by the tester control valve in excess of the relief valve setting, or serious damage to the pump may result.

Operate the pump at rated speed until the oil operating temperature is reached. Determine the volume of oil being pumped at both low and maximum system pressures at the flowmeter. Compare these values with the manufacturer's recommendations. When the flow rate is low, the trouble is in the pump, suction line, or filter. If the pump test results are normal, the trouble is in the relief valve, directional valve, or cylinder. Remember that closed-center systems do not have a relief valve.

**Testing Relief Valve** Attach the tester as shown in Fig. 9-37. Proceed as in testing the pump, increasing the pressure gradually. Record the pressure when the relief valve opens and the tester gauge falls to zero. If this pressure is below the recommended operating pressure, adjust the relief valve setting to achieve the manufacturer's specification. The flow rate at maximum pressure should be the same as at the pump. If not, the relief valve is leaking and must be repaired or replaced. *Note:* This test is applicable only to open-center systems.

**Testing Directional Valve** If the relief valve is functioning properly, attach the tester as shown in Fig. 9-38. Proceed as with the previous test, shifting the directional control valve to direct oil from the cylinder port (G) to the inlet port of the tester. If full volume flow and oil pressure cannot be obtained, the problem is in the directional valve. Always check the discharge from each outlet port of a control valve operating a double-acting cylinder. When conditions are normal, the problem is in the cylinder.

## Replacing Hydraulic Hoses and Fittings

When hydraulic hoses leak, are swollen or cracked, or have been accidently damaged, replacement is a necessary maintenance task. In a similar manner, a hose which has broken down internally will contaminate hydraulic oils and damage close-fitting pumps and valves. The selection, makeup, and installation of the new hose must be performed skillfully.

## Selecting Hydraulic Hoses, Fittings, and Couplings

Repair and service operations involving hydraulic systems on agricultural machinery usually involve re-

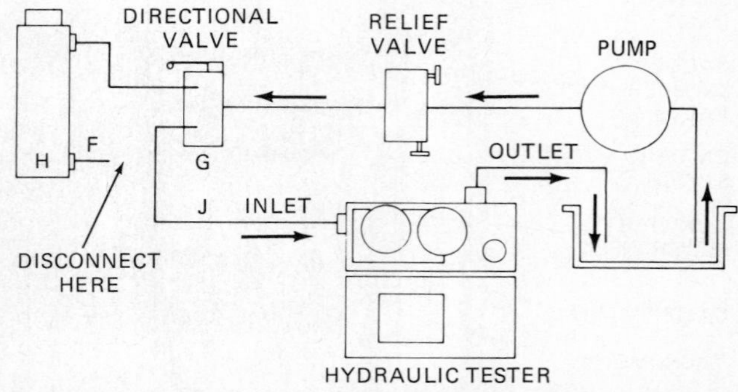


**Fig. 9-38.** Circuit diagram using the tester to evaluate the directional valve operation. The cylinder (H) hose is disconnected at point F, and the outlet to extend the ram at the directional control valve (G) is connected to the inlet of the tester (J). (*Owatona Tool Co.*)

placement of hoses, fittings, and couplings. Parts and service workers must recognize that selecting the right hose, fitting, and coupling for the job is the first consideration to performing the work properly. Hoses, fittings, and couplings are manufactured in many types and designs to meet the pressure requirements of the system, withstand the heat generated, and to be fully compatible with the specific fluid used.

**Selecting Hydraulic Hose** Hydraulic hose consists of three basic parts: an inner synthetic rubber tube, a braided or spiral wire reinforcement carried within this tube to prevent its rupture from high fluid pressure or vacuum, and an oil- and weather-resistant outer rubber cover. The outer cover is designed to prevent damage from normal abrasion and the elements to which the hose is exposed. The cover may carry the manufacturer's name, part number, hose size, SAE number, and date of manufacture. The Society of Automotive Engineers' (SAE) standard 100 R rating identifies the performance in pressure rating and construction for hydraulic hose. The cross-sectional construction of five types of hose is illustrated in Fig. 9-39. The corresponding pressure rating of each hose according to inside diameter is

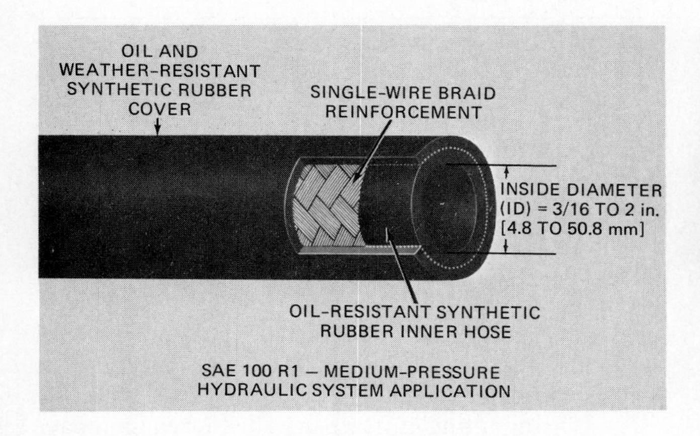


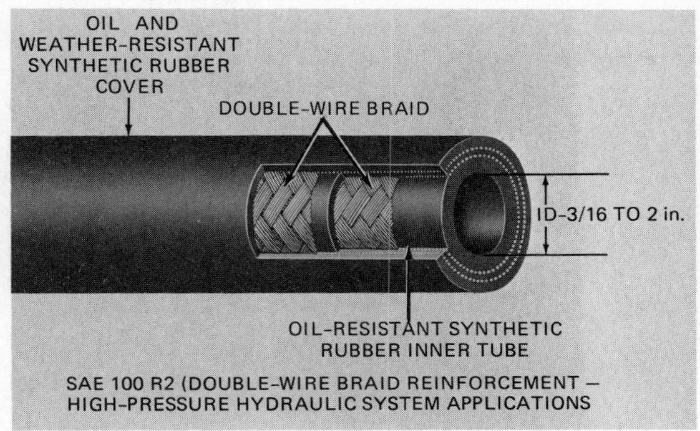


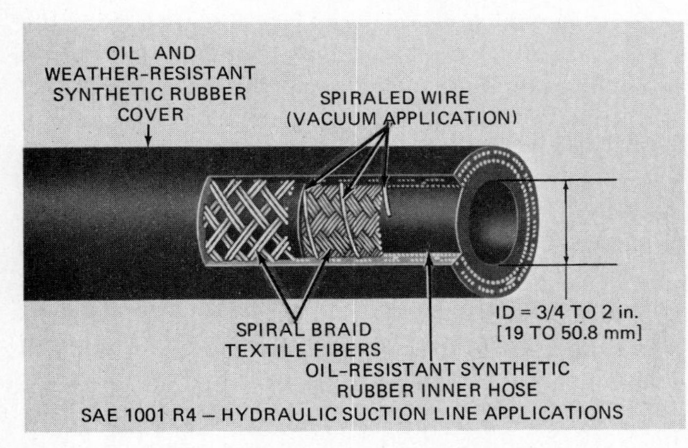


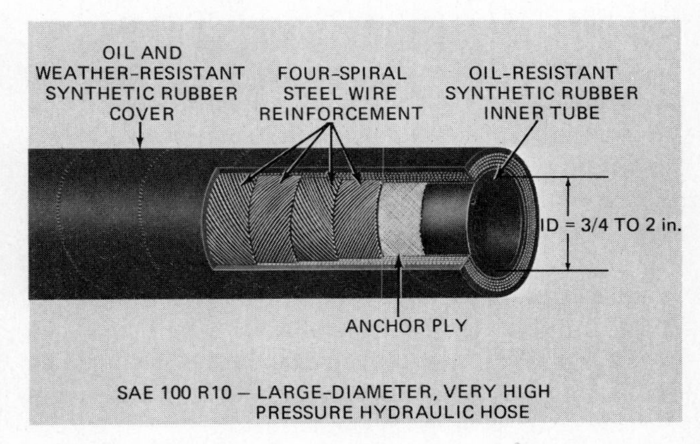


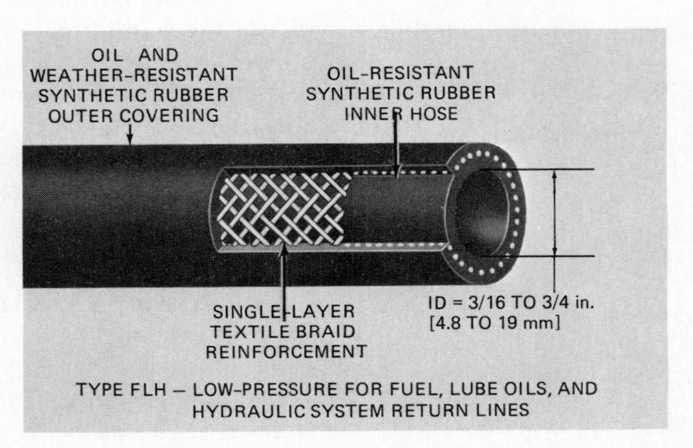


**Fig. 9-39.** Construction and cross sections of five types of hose used in agricultural hydraulic systems. Compare the differences in the reinforcing layers. (*Gates Rubber Co.*)

**TABLE 9-1. Maximum Operating Pressure of Hydraulic Hose by SAE Type and Inside Diameter***

| | Hose Inside Diameter (*ID*) Vs. Working Pressure, psi | | | | | | | | | | |
|---|---|---|---|---|---|---|---|---|---|---|---|
| **HOSE TYPE** | **3/16 in** | **1/4 in** | **5/16 in** | **3/8 in** | **1/2 in** | **5/8 in** | **3/4 in** | **1 in** | **1 1/4 in** | **1 1/2 in** | **2 in** |
| SAE 100R1 | 3000 | 2750 | 2500 | 2250 | 2000 | 1500 | 1250 | 1000 | 625 | 500 | 375 |
| SAE 100R2 | 5000 | 5000 | 4250 | 4000 | 3500 | 2750 | 2250 | 2000 | 1625 | 1250 | 1125 |
| SAE 100R4 | — | — | — | — | — | — | 300 | 250 | 200 | 150 | 100 |
| SAE 100R10 Wire spiral | — | — | — | — | — | — | 5000 | 4000 | 3000 | 2500 | 2500 |
| FLH Textile Braid | 500 | 400 | 400 | 400 | 400 | 350 | 300 | — | — | — | — |

* Inside diameter expressed in fractional size. Dash numbers may be used; –8 identities 8/16 or 1/2 in; –16 equals 16/16 or 1 in.
—Indicates no SAE specification available.

presented in Table 9-1. For example, a hydraulic hose rated SAE 100 R1, no. 1, has a single steel braid wire reinforcement (Fig. 9-39). The ¾ in [19 mm] size will withstand a maximum pressure of 1250 psi [8619 kPa]. An SAE 100 R10 hose of the same size has four steel spiral bands and will withstand 5000 psi [34,475 kPa].

The size of the hose may be expressed in *dash* numbers. In this system the ID converts to 1/16 of an inch [1.59 mm]. A dash number of −8 identifies an 8/16 in or 1/2 in [12.7 mm] ID hose. A dash − 16 size indicates a 16/16 in or 1 in [25.4 mm] ID hose.

**Selecting Fittings** Hydraulic fittings are manufactured to very close tolerances. Each is designed to be used in a specific application. Like hoses, fittings have different pressure capabilities. Six of the most common types of hydraulic fittings used on agricultural equipment are illustrated in Fig. 9-40. Fitting No. 1 is a dry-seal male pipe thread coupling. National pipe thread (NPT) and national pipe thread fuel (NPTF) fittings look very similar, but the difference is extremely important for hydraulic application. The NPTF, also known as dry-seal pipe threads, are made to provide nearly 100 percent thread contact and seal more effectively. Ordinary NPT fittings will leak under pressure. They are not suited to hydraulic systems.

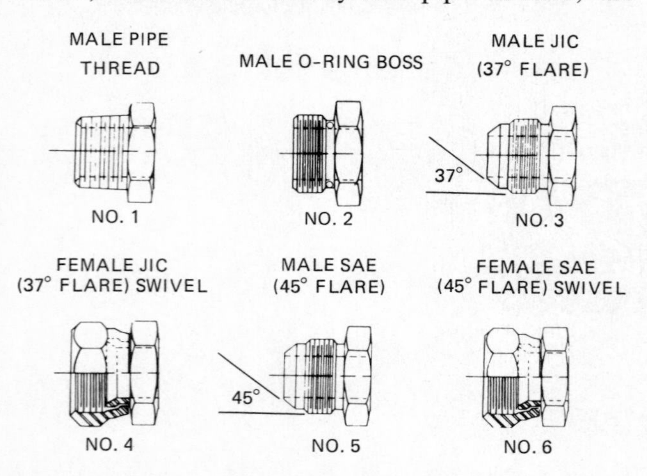


**Fig. 9-40.** Six types of hydraulic fittings used on agricultural equipment. (*Gates Rubber Co.*)

The male O-ring boss and the Joint Industry Council (JIC) 37° flare fittings (Nos. 2 and 3, respectively) are popular types of fittings for medium- and high-pressure application. The JIC 37° flare is available in male and female (No. 4) fittings. The latter has a swivel coupling.

The O-ring fitting must be installed into a companion fitting which is recessed slightly to accept the ring. The seal is positive, and when properly installed, there is no seepage. The JIC 37° flare fitting seals by clamping the matching angle of the female to the male tubing flare between the seals. Thread sealing compounds or tapes will not serve any useful function on flare fittings and should not be used.

The male and female SAE 45° flare fittings (Nos. 5 and 6, respectively) are used for low- and medium-pressure application. *Do not mix 37° and 45° flare fitting bodies and nuts.* The seal is imperfect and will leak when fluid is under pressure.

The information in Table 9-2 identifies the standard hose sizes and the size fitting appropriate for each of the six types of fittings. The size of the fitting is expressed in inches and thread pitch (number of threads per inch). These data are necessary when matching components with hose. To determine what hose and fittings to use, consult the parts and service manual for the piece of equipment. The data in the manual will give the part number and performance requirements necessary to make the correct selection.

**Assembling Fittings** Care must be taken when assembling fittings to avoid damage. When possible, use a tubing wrench to avoid distorting the nut. The data presented in Table 9-3 identify the proper-size wrench to use with each of the six types of fittings. Always lubricate the O-ring and the threads of all fittings before assembly. Do not overtighten

NPTF fittings which are inserted into a casting. The wedging action could crack the casting, requiring expensive repair.

Several of the common problems which cause leakage or failure of the hydraulic connections are:

1. Overtightening—This causes distortion of nut seats and flares, and can cause cracks in the castings.
2. Undertightening—The seats do not seal, and will probably loosen due to vibration.
3. Mismatched flares—The seats do not match for a proper seal.
4. Dirt and foreign material—The seats cannot seal.
5. Sealing tape or compound in the system—Compound or Teflon tape covering NPTF is squeezed into the system.

**Types of Hose Couplings** Threaded couplings must be secured to the ends of the hose so that they can be attached to and complete the fluid circuit between the functioning parts. These couplings are either the permanent or reusable types (Fig. 9-41). Permanent couplings consist of a grooved stem which is inserted in the hose. A ferrule section is then tightly crimped around the hose end, forming a leakproof seal. The screw-type reusable coupling consists of an internally threaded socket and a threaded nipple. The socket is screwed on the outside of the

**TABLE 9-2. Thread Sizes for Six Common Hydraulic Fittings in Nominal Size and Thread Pitch**

| | Dryseal Pipe Thread | | SAE 45° Flare Male and Female | | SAE 37° Flare Male and Female and Straight Thread O-Ring Boss | |
|---|---|---|---|---|---|---|
| Hose ID, in | in | Thread Pitch | in | Thread Pitch | in | Thread Pitch |
| 1/8 | 1/8 | 27 | — | — | — | — |
| 3/16 | — | — | — | — | 3/8 | 24 |
| 1/4 | 1/4 | 18 | 7/16 | 20 | 7/16 | 20 |
| 5/16 | — | — | 1/2 | 20 | 1/2 | 20 |
| 3/8 | 3/8 | 18 | 5/8 | 18 | 9/16 | 18 |
| 1/2 | 1/2 | 14 | 3/4 | 16 | 3/4 | 16 |
| 5/8 | — | — | 7/8 | 14 | 7/8 | 14 |
| 3/4 | 3/4 | 14 | 1 1/16 | 14 | 1 1/16 | 12 |
| 7/8 | — | — | — | — | 1 3/16 | 12 |
| 1 | 1 | 11 1/2 | — | — | 1 5/16 | 12 |
| 1 1/4 | 1 1/4 | 11 1/2 | — | — | 1 5/8 | 12 |
| 1 1/2 | 1 1/2 | 11 1/2 | — | — | 1 7/8 | 12 |
| 2 | 2 | 11 1/2 | — | — | 2 1/2 | 12 |

—Indicates no SAE specification available.

**TABLE 9-3. Six Common Hydraulic Fitting Nut Sizes (Minimum Hex Size, Inches Across Flats)**

| Hose ID, in | Straight Thread O-Ring Boss—Male, in | Dryseal Thread Pipe—Male, in | SAE 37° Flare—Male, in | SAE 37° Flare—Female, in | SAE 45° Flare—Male, in | SAE 45° Flare—Female, in |
|---|---|---|---|---|---|---|
| 1/8 | — | 7/16 | — | — | — | — |
| 3/16 | — | — | 7/16 | 9/16 | — | — |
| 1/4 | 9/16 | 9/16 | 7/16 | 9/16 | 7/16 | 9/16 |
| 5/16 | 5/8 | — | 1/2 | 5/8 | 1/2 | 5/8 |
| 3/8 | 11/16 | 11/16 | 9/16 | 11/16 | 5/8 | 3/4 |
| 1/2 | 7/8 | 7/8 | 3/4 | 7/8 | 3/4 | 7/8 |
| 5/8 | 1 | — | 7/8 | 1 | 7/8 | 1 |
| 3/4 | 1 1/4 | 1 1/16 | 1 1/16 | 1 1/4 | 1 1/16 | 1 1/4 |
| 7/8 | 1 3/8 | — | 1 3/16 | 1 3/8 | — | — |
| 1 | 1 1/2 | 1 3/8 | 1 5/16 | 1 1/2 | — | — |
| 1 1/4 | 1 7/8 | 1 11/16 | 1 5/8 | 2 | — | — |
| 1 1/2 | — | 2 | 2 | 2 1/4 | — | — |
| 2 | — | 2 7/16 | 2 1/2 | 2 7/8 | — | — |

—Indicates no SAE specification available.

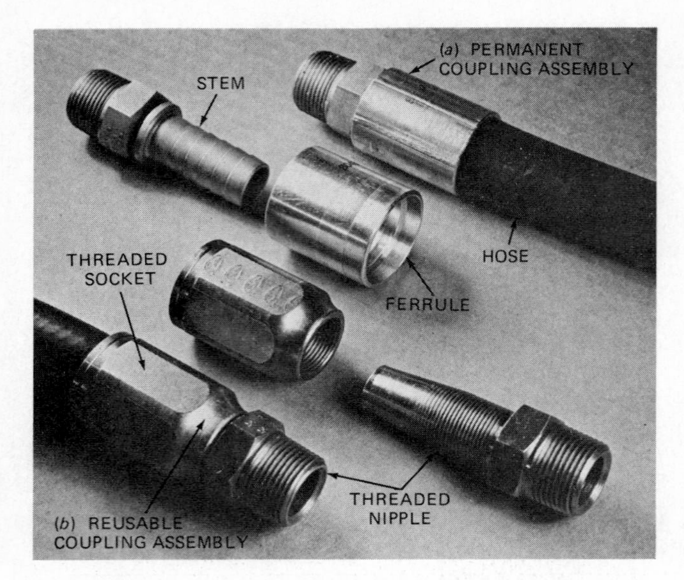


**Fig. 9-41.** (*a*) Permanent couplings cannot be removed following installation. (*b*) Reusable couplings can be removed from old hose and reinstalled on new hose. (*Gates Rubber Co.*)

hose, followed by threading the nipple into the socket and hose to form a tight seal. Reusable couplings cost approximately twice as much as permanent couplings, but require a minimum amount of equipment to fit the ends and can be used more than once.

**Installing Hose Couplings** The first step in attaching hose coupling to a new hose is to cut the hose to the proper length. A square end cut is necessary to obtain a good seal. Use a fine-tooth hand hacksaw and miter box, a cutting wheel, or an arm-mounted power saw as shown in Fig. 9-42.

**Fig. 9-42.** Cut hydraulic hose using a handsaw, power saw, or cutting wheel. Make sure the end is cut square. (*Gates Rubber Co.*)

To install the reusable coupling, proceed as follows:

1. Dip the end of the hose into hydraulic oil.
2. Clamp the threaded socket in a vise and turn the hose counterclockwise, feeding it into the socket.
3. Back the hose off one-half turn after the hose bottoms in the socket. This will provide a gap for the hose to expand.
4. Lubricate the end of the nipple and thread it into the socket and hose clockwise until the nipple bottoms against the socket (see Fig. 9-43).
5. Install protective caps on both hose ends to prevent foreign material from entering the hose. Carefully plan the routing of hydraulic hose to prevent sharp turns, twisting, or abrasion from rubbing and contact with hot surfaces.

Special equipment is required to install permanent couplings because the ferrule must be evenly crimped around the hose with the stem inserted. The procedure is as follows (Fig. 9-44*a* to *d*):

1. Insert the ferrule over the hose, lubricate the end of the stem with hydraulic fluid, and push the hose over the stem until the hose is solidly against the stem shoulder (Fig. 9-44*a*).
2. Insert the proper crimping dies for the size hose into the chamber of the crimping machine (Fig. 9-44*b*).
3. Adjust the machine for the size hose; then insert the hose and coupling assembly into the crimp cylinder. Start the hydraulic pump until the cylinder stops. Release the pressure and remove the finished assembly (Fig. 9-44*c*).

**Fig. 9-43.** Installing reusable couplings. (*Gates Rubber Co.*)

(*a*)

(*b*)

(*c*)

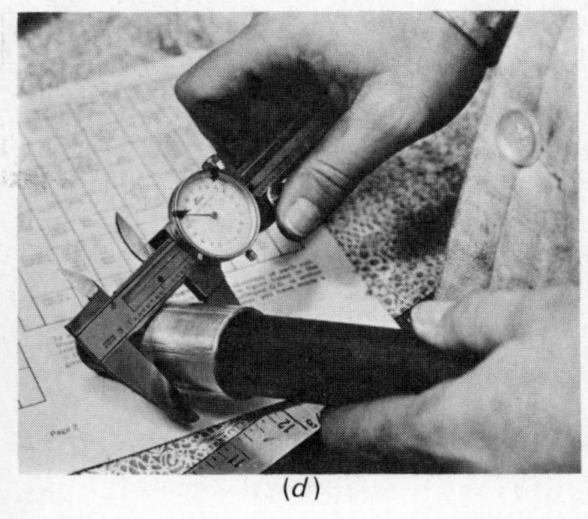

(*d*)

**Fig. 9-44.** Installing permanent couplings using a crimping machine. (*The Gates Rubber Co.*)

4. Measure the outside diameter with a vernier caliper and compare the reading obtained with the specification. The crimp must be within $\pm 0.005$ in [0.127 mm] of the stated specifications. If not, the die must be adjusted and the assembly remade (Fig. 9-44*d*).
5. Install protective caps on both hose ends to prevent foreign material from entering the hose.

## SUMMARY

The serviceperson in agricultural power and machinery takes pride in quality workmanship and attention to detail when servicing and maintaining hydraulic systems. By further study of manufacturers' literature and experience gained through on-the-job training, you can become qualified as a specialist in hydraulic systems.

## THINKING IT THROUGH

1. What is the principle difference between the two basic forms of transmitting hydraulic power?
2. Why do fluids act as flexible solids?
3. How much weight will a piston lift operating within a 10 in² [64.5 cm²] cylinder when the fluid pressure in the cylinder is 100 psi [68.9 kPa]?
4. Name four basic components of the hydraulic system of the hydraulic jack.
5. What is the primary purpose of the pump in a hydraulic system?
6. When is a pump capable of developing fluid pressure?
7. What are three type of positive-displacement pumps?
8. What type of valve is used to control a single-acting cylinder with a positive-displacement pump?
9. Describe the flow of the fluid from the pump of a three-way control valve for:
   (a) Extending (piston)
   (b) Holding (piston)
   (c) Retracting (piston)
10. What kind of pump is required when using a four-way (closed-center) valve?
11. Why do many modern hydraulic systems use closed-center valving?
12. What is the primary purpose of an accumulator? An unloading valve?
13. When is a flow-control valve required in a hydraulic system?
14. What are the desirable features of a "ram"-type cylinder?
15. List six service and maintenance procedures which are necessary to assure proper hydraulic equipment operation.
16. List six safety precautions which should be observed when servicing hydraulic systems.

# UNIT IV THE AGRICULTURAL TRACTOR

| Competencies | Agricultural Machinery Service Manager | Agricultural Machinery Parts Person/ Manager | Agricultural Machinery Mechanic | Agricultural Machinery Setup and Delivery Mechanic | Agricultural Tractor and Machinery Mechanic's Helper | Tractor Mechanic | Farmer, Grower, and Rancher |
|---|---|---|---|---|---|---|---|
| Set up a service schedule | Essential | Desirable | Desirable | Essential | Essential | Desirable | Essential |
| Perform daily preventive maintenance | Desirable | Not necessary | Essential | Essential | Essential | Essential | Essential |
| Perform major maintenance services | Desirable | Not necessary | Desirable | Desirable | Essential | Essential | Desirable |
| Perform minor repairs and service | Essential | Desirable | Essential | Desirable | Essential | Essential | Desirable |
| Use engine test equipment to diagnose problems and service engines | Essential | Desirable | Essential | Desirable | Desirable | Essential | Not necessary |
| Perform major engine repairs | Desirable | Not necessary | Essential | Not necessary | Desirable | Essential | Not necessary |
| Use precision measuring tools | Essential | Essential | Essential | Essential | Essential | Essential | Desirable |
| Paint tractors and other agricultural equipment | Desirable | Not necessary | Essential | Essential | Essential | Not necessary | Desirable |

**Essential**

**Desirable**

**Not necessary**

The internal-combustion engine is an almost universal tool in agriculture. In production agriculture, farmers, ranchers, and other agricultural producers depend on engines to provide the power to operate tractors and other machines they use to produce food and fiber. Engines are also used in the marketing, processing, and distribution industries that process agricultural products and deliver them to the consumer.

Most agricultural workers should understand how an engine operates. Many jobs require that a person be able to operate equipment powered by engines. A person will be a better operator if the basic principles of engine operation are understood. Also, many operators are required to perform maintenance recommended by the engine manufacturer.

There are employment opportunities for persons in sales of agricultural equipment. A person interested in a career in sales must be able to explain operating principles to a potential buyer. This person must be able to instruct the buyer in operation and maintenance procedures. He or she must also be able to determine the condition of agricultural equipment in order to estimate the value of used equipment.

When farmers, ranchers, or other agricultural producers buy agricultural equipment they expect service. All machines need service. Agricultural tractors, combines, forage harvesters, and other power equipment are no exception. Dealers must rely on trained service persons to provide this service. There are excellent employment opportunities for young men and women that understand engines and can perform the service needed. Manufacturers also employ service persons to work with dealers and local service persons.

To provide service, a person must have replacement parts. A parts person must understand engines and machinery in order to supply the parts needed. Customers often rely on parts people to make service recommendations. In order to provide this information, a parts person must understand the operation of the equipment sold by the dealership.

Farmers, growers, and ranchers, and other agricultural producers can increase the life of a machine and reduce operating costs if they follow the maintenance program suggested by the manufacturer. When they understand the operation of an engine and other tractor components, they should recognize the importance of proper service.

Upon completion of this unit, you should know how to perform routine and major service and repairs on an engine. You should have an understanding of the systems that are combined to produce power and the tools and equipment that are used in maintenance, service, and repair.

# CHAPTER 10 PREVENTIVE MAINTENANCE

## CHAPTER GOALS

In this chapter your goals are to:

- Set up a service schedule for an agricultural tractor
- Service the lubrication system of an agricultural tractor
- Service the cooling system of an agricultural tractor
- Service the electrical system of an agricultural tractor
- Service the intake and exhaust system of an agricultural tractor
- Service the fuel system of an agricultural tractor
- Check the fluid levels and change filters in other systems of the agricultural tractor

Modern agricultural equipment represents a major portion of the capital investment in agricultural production. A machine that is not operating properly is a liability to the owner and may result in delayed tilling, planting, or harvesting. Any of these delays can mean the difference between profit or loss for the producer.

Machines may fail because of a design defect, physical damage, or normal wear, but many times machines fail because of neglected service and the lack of properly scheduled maintenance. Manufacturers supply operator's manuals with all major equipment. The operator's manual provides important information on safety, operation, adjustments, lubrication, and owner-operator service. It is recommended that the operator's manual be read, studied, and understood before the machine is operated.

The performance of the scheduled routine maintenance services and checks outlined in the operator's manual is called *preventive maintenance*. Preventive maintenance will not totally prevent equipment failure, but it can reduce it to a minimum. The cost of preventive maintenance will be returned many times in the form of better performance, greater reliability, and longer equipment life, and less downtime during peak operating periods.

This chapter deals with preventive maintenance for the agricultural tractor, but the basic principles apply to all agricultural engines and equipment.

## USING THE OPERATOR'S MANUAL

Manufacturers invest considerable time and money to supply an operator's manual for new agricultural equipment (Fig. 10-1). If used equipment is purchased, it is recommended that the correct operator's manual be obtained from the previous owner, the equipment dealer, or the manufacturer.

The first portion of an operator's manual deals with safety. Then other subjects such as controls and instruments, operator's station, engine break-in, prestarting checks, procedures for starting and stopping the engine, and general operation and maintenance are discussed. As tractors become more sophisticated, some equipped with turbocharged engines and other advanced designs and options, it is essential that you study the manual to protect yourself as well as the tractor. Major mechanical damage can be caused by failure to follow the operational procedures outlined in the operator's manual.

Universal symbols are used to identify gauges, warning lights, adjustments, and controls (Fig. 10-2). Their placement and meanings should be completely understood before the tractor is operated.

### Maintenance Schedule

The operator's manual includes diagrams, pictures, and/or lubrication charts which show lubrication and recommended service for systems and component parts at predetermined intervals. These recommendations should be followed for typical conditions. If the tractor is operated in extremely dusty conditions, the manufacturer usually recommends that the hours between service be reduced. The types of lubricant are also specified by the manufacturer and should be used in order to maintain the warranty on new tractors.

**Fig. 10-1.** Manufacturers provide an operator's manual with new agricultural equipment. The manual should be read and understood before equipment is operated. (*Jane Hamilton-Merritt*)

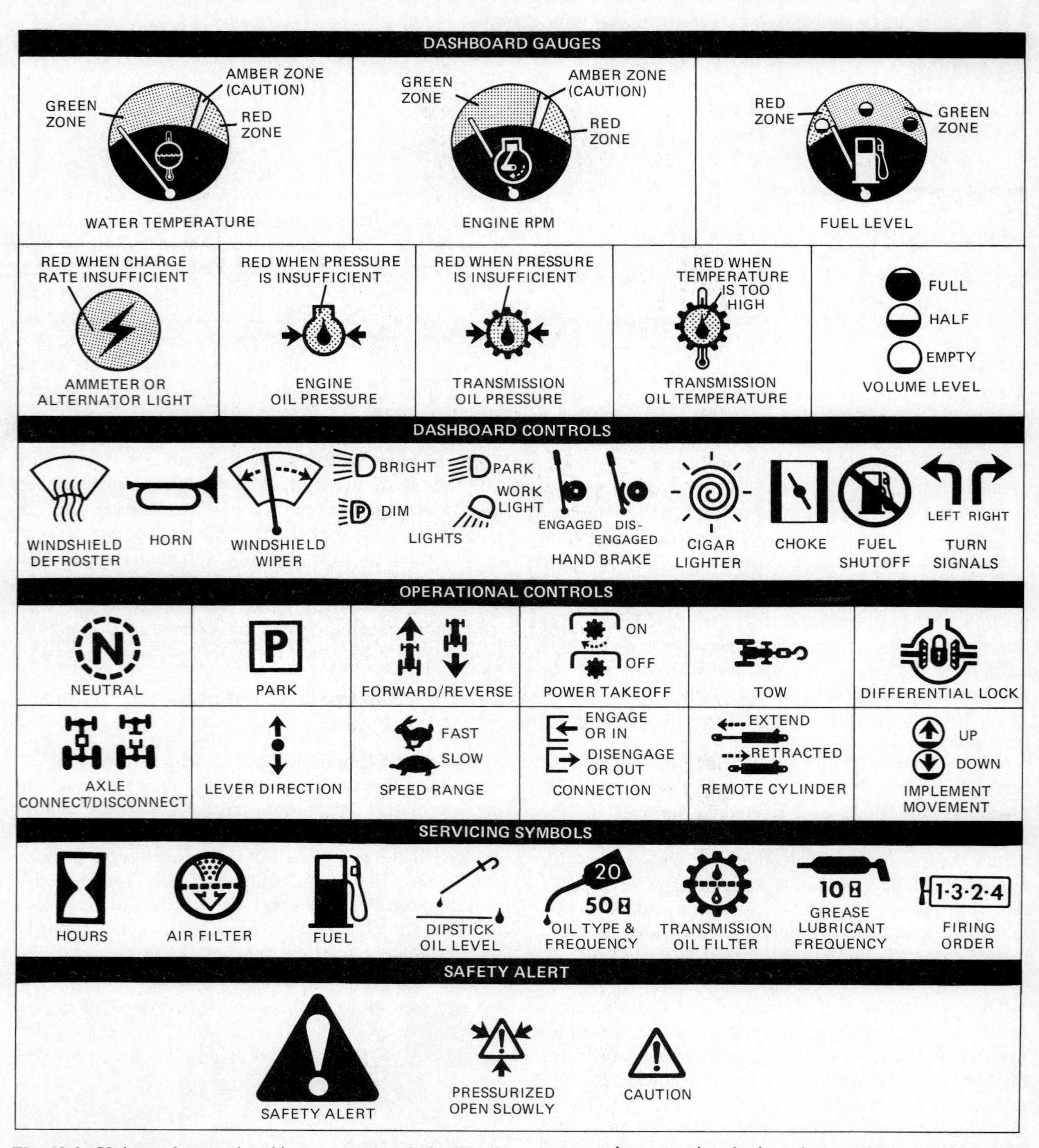


**Fig. 10-2.** Universal control and instrument symbols. Word captions shown are for identification only and are not part of the universal symbols. (*American Society of Agricultural Engineers.*)

## Service Records

Service records are essential when service is performed at recommended intervals. Records can be used to document that recommended service has been completed. This can be of great value if a tractor requires repairs during the warranty period. Accurately kept service records also increase the value of the tractor when it is sold or traded.

Service schedules are available at many dealers. Figure 10-3 shows a simple service schedule and record that can be duplicated and prepared for each tractor. A record of this type can be used to ensure that service is performed at recommended intervals even when different persons are operating the tractor. Service on a fleet of tractors and other equipment could be performed by a service mechanic because the date

SERVICE SCHEDULE AND RECORD
MACHINE IHC 1466-1975 No. 1
RECORD DATE AND HOUR—METER READING 6610 Hours
SIGN OR INITIAL

SERVICE PERIOD RECOMMENDED

| NAME OR INITIALS | DATE | 1. CRANKCASE OIL CHANGE 2. LUBE FRONT WHEEL BEARINGS 3. LUBE COMPENSATING RING BUSHING 100 HR. | 1. CRANKCASE OIL & FILTER CHANGE 2. LUBE TA HOUSING 3. LUBE BELL PEDAL BRAKE-CLUTCH 200 HR. | 1. CHANGE HYDRAULIC FILTER & LUBE TACH DRIVE 250 H.R. | 1. CHANGE TRANSMISSION-HYDRAULIC OIL Periodic | 1. PACK FRONT WHEEL HUBS Periodic | 1. CHANGE DIFFERENTIAL OIL Periodic | | | |
|---|---|---|---|---|---|---|---|---|---|---|
| GS | 1/04 | 5991 | 5991 | 5991 | 5991 | 5991 | | REPLACED HOSES AND BELTS | | |
| GS | 3/8 | 6108 | | | | | | | | |
| RS | 5/28 | 6212 | 6212 | | | | | | | |
| GS | 6/17 | 6296 | | 6296 | | | | | | |
| RS | 7/23 | 6403 | 6403 | | | | | | | |
| RS | 8/27 | 6507 | | 6507 | | 6507 | | REPLACED 12V BATTERIES | | |
| GS | 9/29 | 6610 | 6610 | | | | | | | |
| | | | | | | | | | | |
| | | | | | | | | | | |
| | | | | | | | | | | |
| | | | | | | | | | | |
| | | | | | | | | | | |
| | | | | | | | | | | |
| | | | | | | | | | | |

**Fig. 10-3.** The service schedule and record helps ensure that the recommended maintenance is performed. They are also very beneficial when selling the used machine.

and hour-meter reading are recorded at each service period.

**Using the Service Schedule and Record** The operator's manual is used to fill in the service period (hours of operation) section and to determine the recommended service at each interval. For convenience and to ensure that all recommended service is performed, it is good procedure to attach a list of services to be accomplished at each interval in the service schedule. Types of oil and other specific service information can be given to assure that the manufacturer's recommendations are followed.

## INTAKE AND EXHAUST SYSTEMS

Engines must "breathe." It takes about 15 parts of air to 1 part fuel (by weight) for gasoline engines to operate. The air-fuel ratio for diesel engines can go as high as 40:1. Therefore, air is the most abundant matter in the combustion of fuel to produce power.

### Types of Air Intake Systems

Three types of air intake systems are used on agricultural engines. Engines are commonly identified by the type of intake system used.

**Naturally Aspirated Engines** Naturally aspirated engines (Fig. 10-4) draw air into the combustion chamber on the intake stroke. The air is drawn in as the piston moves down creating a low-pressure area in the cylinder. This is referred to as natural aspiration because air pressures tend to equalize. Air moves from high- to low-pressure areas. Most gasoline tractors and smaller diesel tractors are naturally aspirated.

**Turbocharged Engines** Turbocharged engines (Fig. 10-5) use an exhaust-driven turbine to increase the power of the engine by forcing more air into the cylinder than is drawn in a naturally aspirated engine. This means that more fuel can also be injected into the cylinder. Thus more power can be developed.

**Turbocharged and Intercooled Engines** Turbocharged engines using an intercooler to get even more air into a cylinder are often used on large diesel tractors (Fig. 10-6). Turbochargers increase the temperature of the air, and as the tempera-

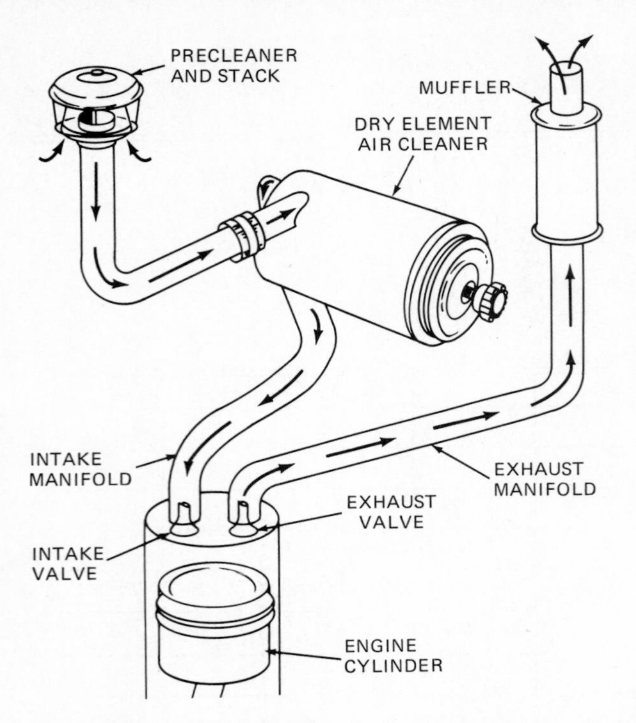


**Fig. 10-4.** Naturally aspirated diesel engine intake-exhaust system. (*Ford Tractor Operations.*)

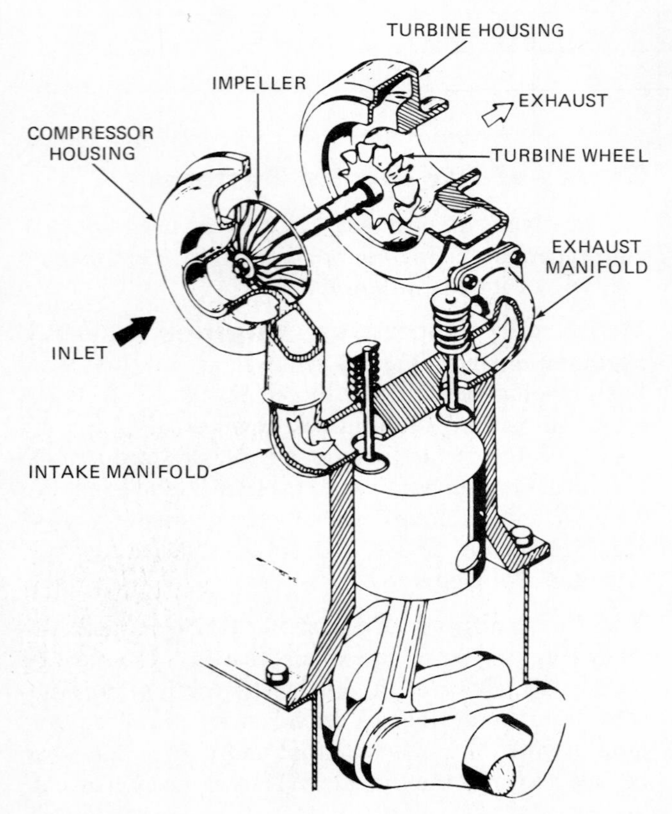


**Fig. 10-5.** Turbocharged intake-exhaust system used on many larger diesel agricultural engines. (*Allis-Chalmers Manufacturing Co.*)

1 TURBOCHARGER
2 INTERCOOLER
3 AIR CLEANER

(*a*)

1 AIR CLEANER TO TURBOCHARGER TUBES
2 ASPIRATOR TUBE
3 EXHAUST TUBE
4 AIR TO TURBOCHARGER
5 INTERCOOLER
6 AIR CLEANER

(*b*)

**Fig. 10-6.** Intercoolers using (*a*) engine coolant or (*b*) air reduce the temperature of the intake air on some larger diesel engines. (*Deere & Co.* and *Ford Tractor Operations.*)

ture rises, the space between the air molecules gets larger. The intercooler uses the coolant in the cooling system or cooler air to reduce the temperature of the intake air. Therefore the space between the air molecules becomes smaller, and more air can be packed into the cylinder. Thus even more fuel can be injected into the cylinder, and more power can be derived from the engine. Large diesel tractors are often equipped with this system.

## Air Cleaners

Engines are equipped with air cleaners. The function of the air cleaner is to remove abrasive materials from the intake air.

An engine takes in 12,000 to 15,000 gallons [45,420 to 56,775 L] of air (by volume) for each gallon [3.785 L] of fuel used. Any dirt or other solid particles that enter the engine through the intake system mix with the engine oil on the cylinder walls. This forms an abrasive solution that causes wear on the cylinders, piston rings, and other internal engine components.

Properly serviced air cleaners remove 99 percent or more of the dirt and other solid particles in the intake air. Tests have proven that engine wear is increased at the rate of 100:1 when an engine is operated under normal operating conditions without an air cleaner. Tractors and other agricultural equipment often operate in extremely dusty conditions. When an air cleaner fails, an engine may wear so rapidly that major repairs are needed after only a few hours of operation. The importance of proper air cleaner maintenance and service cannot be overemphasized.

**Dry-Element Air Cleaners** The dry-element air cleaner is standard equipment on most modern agricultural tractors. The final cleaning of the air entering the engine is done by a replaceable paper cartridge. The dry-element unit is used because it is efficient, relatively compact, may be mounted in any position, and restriction indicators can be installed to indicate when the unit needs servicing (Fig. 10-7).

**Oil-Bath Air Cleaners** Medium-duty oil-bath air cleaners were used on agricultural tractors for many years. They are still in limited use on some smaller models and on small engines. Oil-bath air cleaners are mounted vertically with an oil cup on the bottom (Fig. 10-8). Air is directed down through a center tube, and centrifugal force causes most of the dirt particles to be trapped in the oil in the cup as the direction of the air flow is reversed. Dirt not trapped in the oil and a small amount of the oil are carried into

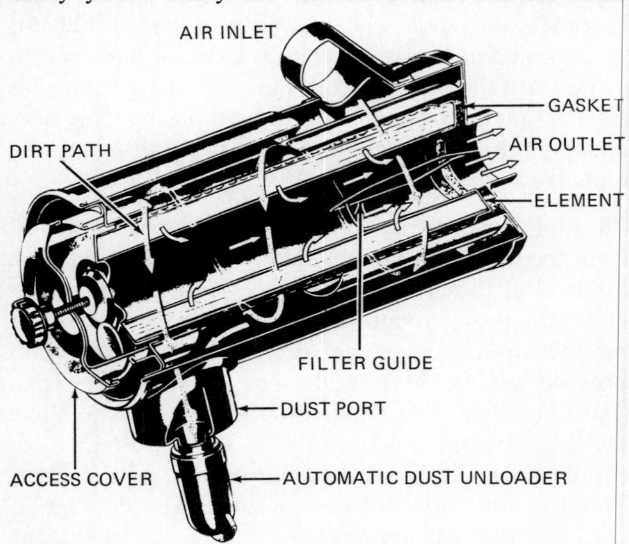


**Fig. 10-7.** Air flow in a dry-element air cleaner which is used on most agricultural tractor engines. (*Ford Tractor Operations.*)

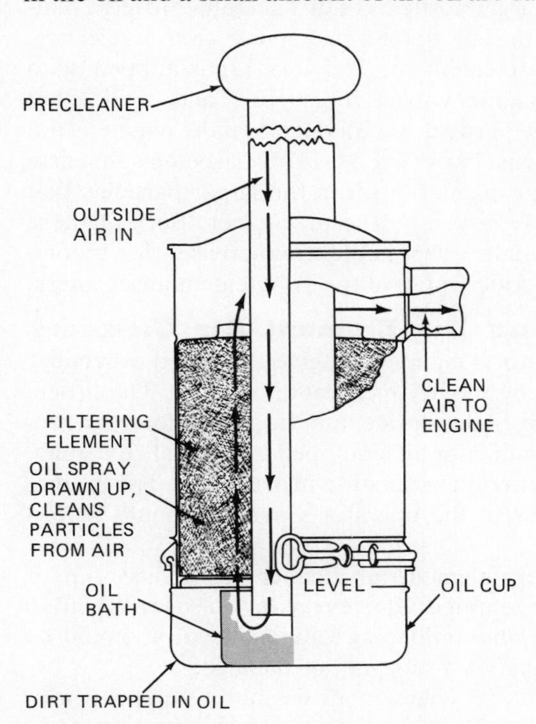


**Fig. 10-8.** Oil-bath air cleaner.

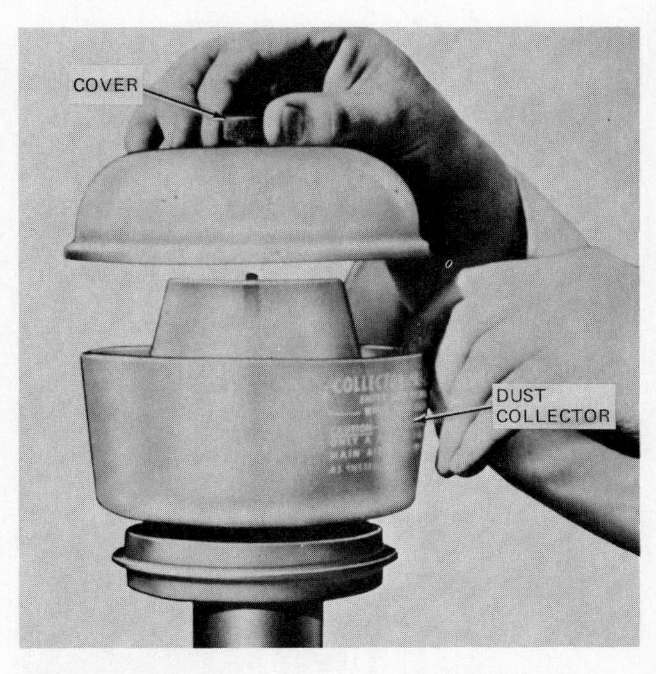


**Fig. 10-9.** The air cleaner precleaner is designed to remove larger particles of dirt and other foreign material from intake air before it enters the air cleaner. (*Ford Tractor Operations.*)

the cleaner element. The oil and dirt eventually flow back into the cup. Oil-bath air cleaners will allow dirt to enter the engine if they are not serviced properly at regular intervals.

**Precleaners** All air cleaners are equipped with some type of precleaner (Fig. 10-9). The precleaner removes larger particles of dirt and other foreign matter from the air before they enter the air cleaner. Some dry-element air cleaners are equipped with dust unloading valves (Figs. 10-7 and 10-10) that allow large particles to fall out when the engine is not running. Dust cups (Fig. 10-11) are used on some dry-element air cleaners to trap the larger particles that were not removed by the precleaner. Both of these systems make it possible to increase the recommended service interval for dry-element air cleaners.

## Servicing Dry-Element Air Cleaners

The operator's manual will give a detailed procedure for servicing the dry-element air cleaner. The procedure will include service interval, operation of the restriction indicator (if equipped), removal, cleaning, checking, and reinstallation instructions. The recommendations in the operator's manual should be followed.

If the dry-element air cleaner has a dust cup, it should be emptied after every day of use. When an automatic dust unloading valve is used, it should be checked daily for clogging or damage.

When the dry-element air cleaner is equipped with a restriction indicator, most manufacturers recommend that the filter element be serviced only when a restriction occurs. A restriction may activate a light on the instrument panel or cause a colored band to appear in a sight glass when the air cleaner element is clogged. When the system does not include a restriction indicator, the air filter element should be serviced at the regular intervals recommended in the operator's manual. Excessive smoke or a loss of power are signs of a restriction in the dry-element air cleaner.

**Fig. 10-11.** Many dry-type air cleaners are equipped with a primary filter, secondary filter, and dust cup. (*Allis-Chalmers Manufacturing Co.*)

**Fig. 10-10.** A dry-element air cleaner equipped with a dust unloading valve helps to extend filter element life. (*Deere & Co.*)

Many agricultural tractors are equipped with both a primary (outer) dry-element filter and a secondary (inner) dry-element filter (Fig. 10-11). The secondary filter serves as a safety filter in case the outside filter becomes damaged. The secondary filter should not be removed unless it is to be replaced or has dust on it. Dust on the inner filter means that the primary filter has failed, and both should be replaced. The secondary filter should never be cleaned and should be replaced once a year.

**Cleaning primary air filters** Dust can be removed from the primary (outer) filter by tapping it gently on the heel of the hand (Fig. 10-12). If this does not remove the dust, compressed air can be used. Usually the maximum pressure recommended for compressed air is 30 pounds per square inch (psi) [207 kPa]. The air is applied to the inside of the filter, moving up and down the pleats (Fig. 10-13). Air should *never* be applied to the outside of the filter. This can force the dirt into the filter and damage it.

Check the instructions on the filter, or check the operator's manual to see if the filter can be cleaned with water or with water and commercial filter element cleaner or nonsudsing detergent. When water

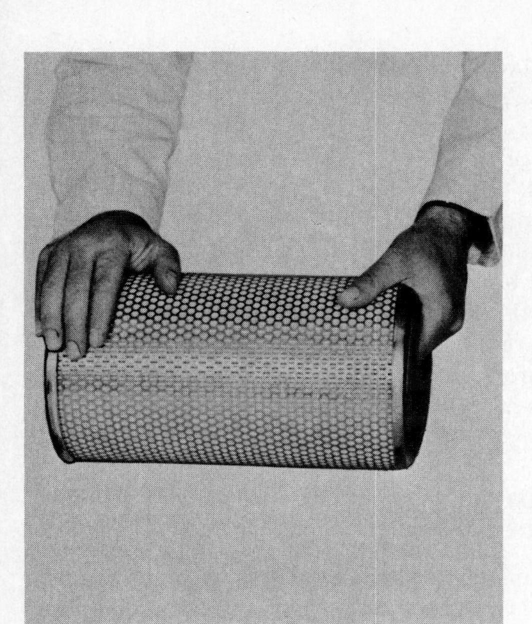

**Fig. 10-12.** Remove the dust from a primary dry-element air filter by tapping. *(Deere & Co.)*

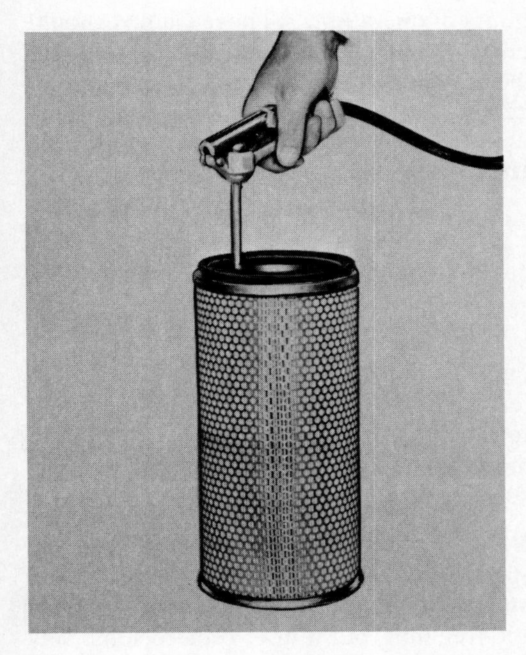

**Fig. 10-13.** Clean a primary dry-element air filter with compressed air, using care not to damage the element. *(Deere & Co.)*

and a cleaner are used, the filter should be soaked in a lukewarm mixture and agitated slowly. The filter should then be flushed with clean water to remove all the cleaning solution. (Fig. 10-14). (When rinsing, use a garden hose to rinse the filter from the inside out.) Excess water should be removed from the filter, and it should be allowed to air-dry for 24 hours (h) or more. For this reason, it is desirable to keep at least two filter elements for each engine.

**Fig. 10-14.** Many manufacturers recommend cleaning a primary dry-element air filter by using a commercial filter cleaner or nonsudsing detergent. Always check the service manual for specific recommendations. *(Deere & Co.)*

**Checking primary air filters** After the filter has completely dried, the paper element should be checked for damage. This is done by dropping a light into the filter and looking for bright spots in the paper from the outside of the filter (Fig. 10-15). If any damage is found, replace the filter. Check the sealing gasket for damage and the metal case for leaks. The filter should be replaced if it has been used for the recommended interval or has been cleaned the maximum recommended number of times.

Two important things to remember are that dry-element filters are never to be washed in fuel oil, solvent, or gasoline and that compressed air should not be used to dry a wet filter.

Before replacing the filter, the inside of the air cleaner case should be wiped with a clean damp cloth. The filter should be installed gasket end first. Replace the wing nut, outer gasket, and cover. Reset the restriction indicator if the tractor is equipped with one. Check the operator's manual for the exact procedure for each tractor.

## Servicing Oil-Bath Air Cleaners

Even though oil-bath air cleaners are in limited use on new agricultural tractors, many older tractors equipped with them are in use and will continue to be used for many years. The general procedure is to service oil-bath air cleaners daily or after every 10 h of operation. More frequent service is recommended when the tractor is operated in extremely dusty conditions.

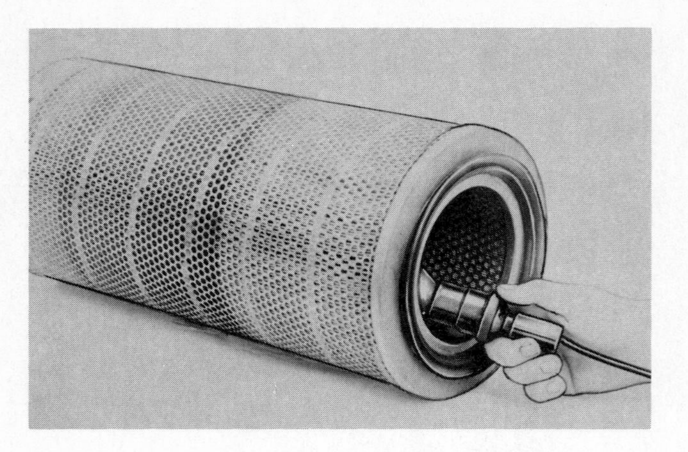

**Fig. 10-15.** Check the paper element in a primary dry-element air filter for damage by using a drop-light or a flashlight. (*Deere & Co.*)

The oil cup is loosened and removed from the filter for service. If there is ½ inch (in) [12.7 mm] of sediment in the bottom of the cup or the oil has thickened, the cup should be thoroughly cleaned and refilled with the proper grade oil to the fill line (Fig. 10-16). When high-detergent oil is used in the cup, dirt may never settle to the bottom and so it is important to look for thick oil. Nondetergent oil is recommended, but it is not always available. Some oil-bath air cleaners are equipped with a tray in the oil cup. When a tray is used, it should be cleaned.

Check the air intake pipe for dirt. Although it will not need cleaning often, it should be checked each time the oil cup is serviced. If it is dirty, clean it with a cloth soaked in solvent. The lower part of the filter element may eventually become clogged with dirt. It should be removed and cleaned at least once a year and more often under severe operating conditions.

**Fig. 10-16.** The cup on an oil-bath air cleaner should be cleaned and filled with the proper viscosity oil to the fill line at the interval recommended in the operator's manual. (*Ford Tractor Operations.*)

Reinstall the oil cup and tray (if used). Check the air hose for dirt, damage, or obstructions between the filter and carburetor on gasoline engines or between the filter and intake manifold on diesel tractors. Check the air inlet screen or precleaner for trash, and remove any accumulated material. For additional information on a particular filter, check the operator's manual and follow the specific recommendations.

## Checking Air Cleaner Hoses and Connections

Hoses and connections need to be checked at regular intervals on both the oil-bath and dry-element air cleaners. It may be necessary to remove the hood or cowl in order that hoses can be checked for cracks, pin holes, or other damage. Some manufacturers recommend that all hoses and connections be checked every 500 hours of operation. All hose clamps should be tightened. The filter case should be checked for damage. Check the operator's manual for the recommended interval.

## Servicing the Exhaust System

The exhaust system on a naturally aspirated tractor has three jobs. It removes heat, carries away burned and unburned gasses, and reduces engine noise. Little service is required other than replacing burned-out parts. Tractors equipped with vertical exhaust systems should be protected to prevent moisture from entering the engine through the exhaust system. Most new tractors are equipped with rain caps, and protection is not a problem as long as these caps close properly.

**Turbochargers** Many agricultural tractors are now equipped with turbochargers to increase the power output. The turbocharger is a reliable and relatively trouble-free unit, but it does require some service and special attention. It should be inspected for free rotation, bearing clearance, and cleaned at recommended intervals. This is done usually after each 1000 h of operation or annually. At 2000 h, the turbocharger should be disassembled, new seals and gaskets installed, and all bearing tolerances readjusted. This will usually be done by a technician.

The turbocharger operates at very high speeds of 40,000 to 100,000 r/min or more and special care should be taken to prevent damage due to improper operation. Due to the high temperature of the exhaust gases it is important that the turbocharger be allowed to cool down before stopping the engine. An engine that has been operating at full load should be allowed to idle at about 1000 r/min approximately 2 to 5 minutes (min). This will rapidly cool the turbocharger exhaust turbine and prevent excessive heat from reaching the bearing areas. Extreme heat will cake the oil which will restrict the oil passage causing inadequate lubrication to reach the bearings, resulting in premature bearing failure. If an engine is accidentally stopped while under load, it should be restarted immediately. Tractors equipped with turbochargers should be transported with a bonnet over the exhaust. The air movement can cause the turbine to turn. This can cause premature bearing failure because the bearings are not fully lubricated when the engine is not running.

**Starting an engine in storage** When an engine has not been run for several weeks, it is recommended to hold the engine fuel shut-off knob (the kill control on a diesel) out and crank the engine with the starter until engine oil pressure is attained. This technique forces lubrication into the turbocharger bearings and other vital bearing surfaces before the engine starts. The starter should not be operated for more than 30 seconds (s) at a time to prevent overheating.

It is also recommended that the engine be operated at approximately 1000 r/min for several minutes. An engine should not be accelerated or have a heavy load applied to it until it has reached operating temperature. This is good practice for all engines, but it is essential when operating turbocharged engines.

**Performing periodic checks** Periodic checks should be made to be sure that the mounting and connections of the turbocharger are secure and that there is no leakage of oil or air. The engine crankcase vent should be checked to be sure there is no restriction to the air flow. Unusual noise or vibration may indicate that bearings are about to fail or that there is improper clearance between the turbine wheel and the housing. In case of excessive noise or vibration, the turbocharger should be removed for inspection and service. Turbocharger maintenance is usually done by a service shop because of the special tools and equipment that are required.

## ENGINE LUBRICATION SYSTEM

Engines in agricultural tractors and equipment operate almost continuously under full loads, at near constant speeds, and for long periods of time. Engine oils that meet the special needs of agricultural engines and other off-road vehicles have been developed. The operator's manual will specify the type of oil that should be used (Table 10-1).

### Functions of Engine Oil

Engine oil performs functions that enable an agricultural engine to operate efficiently for several thousand hours. Engine oils reduce friction between moving parts and bearing surfaces in the engine and keep them clean. They help seal the rings to the cylinder walls and reduce cylinder and ring wear. Oils also help to cool moving parts. Many new tractors are equipped with oil coolers to help remove heat from the oil and improve its heat-carrying capacity.

**TABLE 10-1. The Operator's Manual Will Give the Manufacturer's Recommendations for Engine Oil**

| | Crankcase Oils | | |
|---|---|---|---|
| | | Other Oils* | |
| Air Temperature | Manufacturers' Special Oil | Single Viscosity Oil | Multi-Viscosity Oil |
| Above 32°F [0°C] | SAE 30 | SAE 30 | Not recommended |
| −10 to 32°F* [−23 to 0°C] | SAE 10W-20 | SAE 10W | 10W-20 |
| Below −10°F [−23°C] | SAE 5W-20 | SAE 5W | SAE 5W-20 |

* If air temperature is below 10°F [−12°C], use an engine heater. SAE 5W-20 oils also may be used to ensure optimum lubrication of engine and turbocharger.

*Note:* If oil other than manufacturer's special oil is used, it must conform to one of the following specifications.

*Single-viscosity oils:* API Service CD/SD
MIL-L-2104C

*Multiviscosity oils:* API Service CC/SE, CC/SE, CD/SD, or SF
MIL-L-46152

In order for engine oils to do these jobs, the oil must keep a protective film on moving parts and be resistant to high temperatures. The oils must resist corrosion and prevent acid and sludge formation. Oil must also prevent ring sticking, flow easily at low temperatures, and resist foaming and chemical breakdown after prolonged use.

### Classification of Oils

Oil is classified by the Society of Automotive Engineers (SAE) for its *viscosity*. This refers to the oil fluidity or its resistance to flow at a given temperature. For an example, water has a lower viscosity than syrup. The viscosity is normally referred to as the weight of the oil, and the operator's manual will recommend a given viscosity for the outside temperature range in which the engine will be operating (see Table 10-1). (For another discussion of oil classification see the section Selecting Engine Oil in Chap. 4.) The viscosity of oil is indicated by an SAE stamp printed on the oil can or barrel.

The American Petroleum Institute (API) has established a performance or service classification for engine oils. Table 4-1 explains the API classifications.

The operator's manual will specify the minimum grade of oil that will meet the requirements for the engine. For example, CD oil is currently the best diesel grade, and is recommended for use in late model turbocharged diesel agricultural tractor engines.

## Servicing the Engine Lubrication System

Specific instructions are given in most operator's manuals for the break-in period of new engines. These recommendations should always be followed.

The operator's manual will give the recommended change intervals for the crankcase oil and oil filter. To be sure that you are using the correct oil, follow the manufacturer's recommendations and purchase only a reputable brand of oil from a reliable supplier. Oil should be stored and handled in a manner that will keep it from being contaminated with dirt and water.

**Changing Engine Oil** When changing engine oil, the engine should be at operating temperature. This will permit the oil to drain more freely and more contaminants will be removed from the engine. Many agricultural tractors are equipped with magnetic drain plugs. These plugs should be checked for metal particles, and if any are found, they should be removed. Allow the oil to drain for several minutes or until all old oil is removed. Replace the drain plug, and tighten it securely but carefully to avoid stripping the threads in the oil pan.

Wipe any dirt from the area around the filter and then remove the filter. Some oil filters are the spin-on type, and some are filter elements housed in a bowl or cannister. If in doubt about the removal procedure, check the operator's manual. Remove the old filter and gasket and discard both. Clean the inside of the filter cannister and the engine block mating surface with a suitable solvent (not gasoline) and wipe it dry with a lint-free rag. Insert the new filter and install a new gasket. Tighten the retainer bolt. When spin-on filters are used, lubricate the gasket with engine oil. Turn the filter until the seal contacts the base, and then tighten no more than one-half turn, or as specified in the operator's manual. Recheck the drain plug to see that it is secure.

Fill the crankcase with oil that meets the manufacturer's specifications for SAE viscosity and API service. Check the operator's manual for capacity. It is important that the crankcase not be overfilled. Overfilling can cause excessive oil consumption and oil foaming. Be sure that the tops of oil cans, funnels, and other equipment are clean to avoid adding dirt to an engine.

Start the engine and allow it to idle. The oil pressure indicating light should go off in a few seconds, indicating that the pump has picked up oil and that the oil pressure has been attained. If the tractor is equipped with an oil-pressure gauge, check the gauge for operating pressure. Check for oil leaks at the drain plug and around the filter.

It should be remembered that detergent oil has the ability to carry contaminants in suspension, rather than depositing them in the engine. This is especially true of the CD oils recommended for use in diesel tractors. Due to the presence of these contaminants, the oil will become dark soon after changing. Thus, the color of the oil cannot be used to determine when it should be changed. The schedule recommended in the operator's manual should always be followed except when extreme operating conditions warrant a shorter interval.

# ENGINE COOLING SYSTEM

Agricultural tractors are powered by engines which are liquid-cooled or air-cooled. The air-cooled engine uses fins, shrouds, and blowers to dissipate the heat created by the combustion of fuels. The operator's manual should be followed when checking the operation of an air cooling system. Since most tractors used in agricultural production are liquid-cooled, this chapter will deal with the liquid cooling system.

The cooling system regulates engine operating temperature. The coolant, usually water and ethylene-glycol-based antifreeze, is circulated in a water jacket around the cylinders, valve ports, and other vital engine parts (Fig. 10-17). As the coolant circulates, it picks up the heat that is created by the combustion of fuel in the cylinders. When the coolant in the water jacket reaches a specified temperature, the thermostat opens and allows the coolant to circulate through the radiator.

**The Thermostat** The thermostat is simply a temperature control valve for the engine. It holds the coolant in the water jacket until the engine reaches operating temperature, usually 180 to 190° Fahrenheit (°F) [82 to 88° Celsius (°C)]. Then it opens and allows the coolant to circulate through the radiator where some of the heat is removed as the coolant circulates down through the cores or coolant passages in the radiator. The actual temperature drop depends on atmospheric temperature but it usually ranges from 2 to 5°F [1° to 3°C]. The fan aids in air circulation through the core of the radiator.

**The Water Pump** Modern tractor engines are equipped with water pumps to increase the flow of the coolant through the cooling system. Water pumps are equipped with a bypass to allow the coolant to circulate in the water jacket when the thermostat is closed during engine warmup. When the thermostat opens, the water pump aids in circulation throughout the cooling system.

**Pressure Cap and Hoses** The radiator pressure cap and upper and lower radiator water hoses complete the cooling system. The hoses prevent the normal vibration of the engine from damaging the ra-

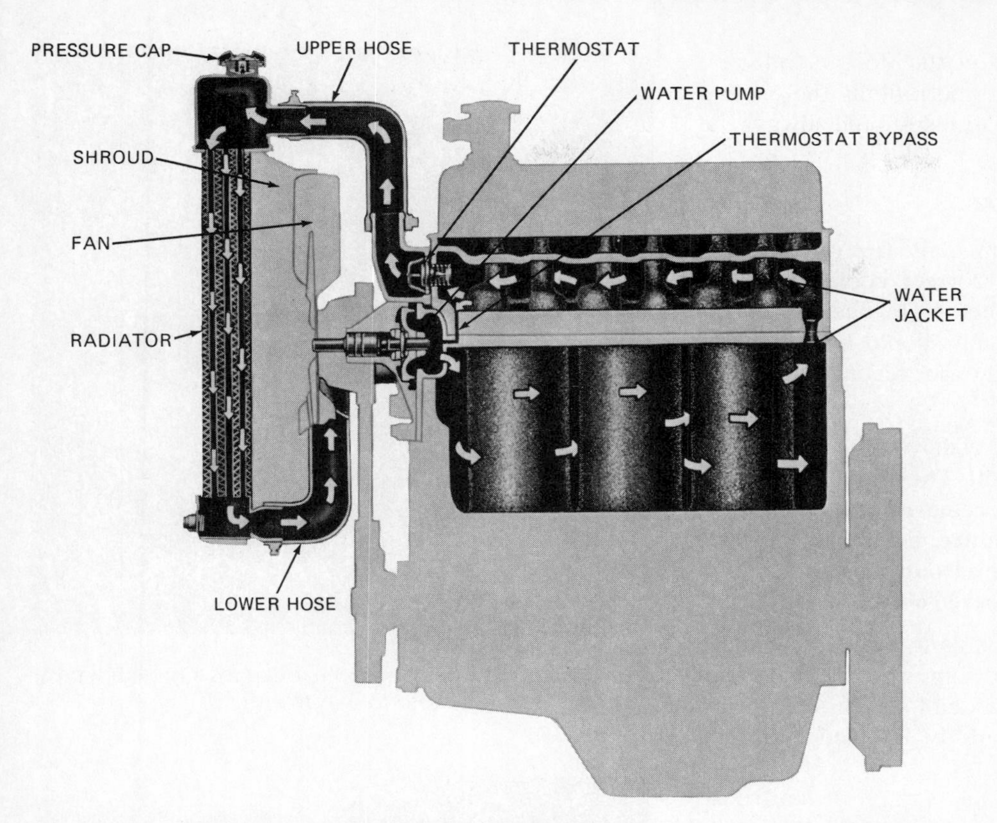


**Fig. 10-17.** Most larger agricultural engines are liquid cooled. Arrows show the coolant circulation in the cooling system. (*Ford Tractor Operations.*)

diator. The care of the hoses and the operation of the pressure cap will be discussed later in this chapter.

**Operating temperature** It is important that the cooling system operate correctly because overheating can result in extensive damage to engine parts. If an engine is operated below 165°F [74°C], sludge and acid will form in the crankcase. Since engine operating temperature is controlled by the thermostat, an engine should never be operated without one. Proper care of the cooling system is an important part of a complete preventive maintenance program on an agricultural tractor or engine.

## Engine Coolant

Water is a universal coolant but it has several physical characteristics which cause problems when used in an engine cooling system. Water freezes at 32°F [0°C] and boils at 212°F [100°C] at sea level. Ethylene-glycol-base antifreeze is added to water to lower the freezing point and raise the boiling point. Ethylene-glycol-base antifreeze boils at 223°F [106°C], so a 50-50 mixture of water and antifreeze boils at a considerably higher temperature than water alone. Antifreeze is added in the winter to prevent the coolant from freezing. The operator's manual will suggest the mixture ratio and will give recommendations for warm weather operation. Some manufacturers recommend that antifreeze be used year-round while others advise the use of summer coolant conditioners for warm weather operation. Modern antifreeze solutions combine both winter protection from freezing and summer protection against overheating as well as chemicals to retard rust formation. Water alone is not recommended because it will cause rust and corrosion to form in the cooling system.

Water with a high mineral content will cause scale to build up and reduce the efficiency of the cooling system. Dirt and other foreign materials will settle out in the water jacket and result in reduced coolant capacity and heat transfer (Fig. 10-18). It is recommended that only clean soft water be used in the cooling system. Distilled water or rain water is best.

Ethylene-glycol-base antifreeze may be referred to as permanent antifreeze. This does not mean that it should be left in the system and not changed. The term *permanent* refers to its boiling point which as stated earlier is higher than the boiling point of water.

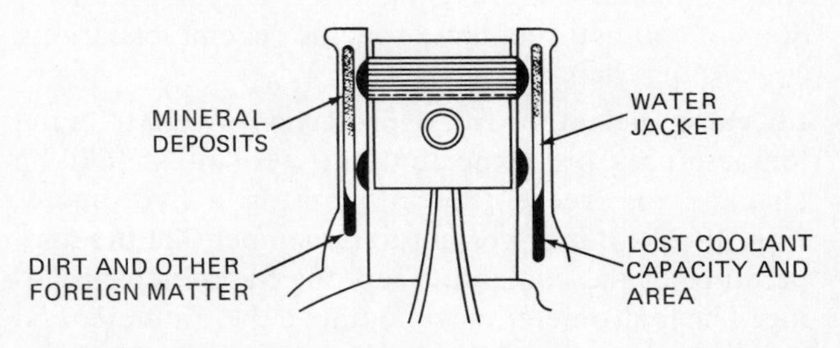


**Fig. 10-18.** Deposits in the cooling system reduce the heat transfer and cause hot spots in the engine.

The operator's manual will give the recommended interval for draining, flushing, and refilling the cooling system. These recommendations should always be followed.

## Checking the Coolant

Check the cooling system daily (every 10 h) for correct coolant level. The correct coolant level is usually about 2 in [51 mm] below the neck of the radiator. The operator's manual should be consulted for a more specific level and exact recommendations given by the manufacturer.

⚠ **WARNING DO NOT remove the radiator cap while the engine is hot! The radiator is pressurized and if the cap is removed when the engine is at operating temperature, hot steam and boiling liquid will be sprayed out. This will injure you and result in excessive coolant loss.**

When you are sure the coolant is well below the boiling point, the radiator cap may be removed safely. The cap should be turned to the first stop and any remaining pressure should be allowed to escape before removing the cap.

If the coolant level is low, add water or antifreeze and check for the cause of the coolant loss. *Do not* add coolant to a hot or overheated engine. This could crack the block or cylinder head. Always allow the engine to cool before adding additional coolant. Check the radiator cap and its sealing gasket. The cap is designed to maintain a specific pressure in the cooling system. If the cap sealing gasket is damaged or if it is not holding pressure due to failure, coolant will escape through the overflow pipe. The holding pressure of a radiator cap can be checked using a radiator cap tester (Fig. 10-19). A cap that holds less than specified pressure should be replaced.

Leaks can be the cause of low coolant levels. Check the radiator, the hoses, and the hose connections. Coolant can also leak into the crankcase. If this condition exists, the engine oil will have a milky color. This can usually be prevented by checking the cylinder head bolt torque at recommended intervals.

Stop-leak products may be used to stop small leaks at least temporarily. They will not successfully stop cylinder head gasket leaks because of the heat and pressures of combustion. You should check your operator's manual before using any stop-leak products. Some manufacturers have specific recommendations concerning their use.

**Checking coolant for freeze protection** When freezing temperatures are expected, the coolant should be checked for freeze protection using a hydrometer (Fig. 10-20). It is important to remember that the temperature of the coolant affects the hydrometer reading. The hydrometer must contain a thermometer and a temperature correction table in order to check the freeze protection accurately.

**Fig. 10-19.** Test radiator cap holding pressure with a pressure cap tester. (*Photo by Westen McCoy.*)

**Fig. 10-20.** Check coolant for freeze protection with a radiator hydrometer. (*Photo by Westen McCoy.*)

Normally when the hydrometer shows the protection level is not low enough, it is recommended that the system be drained and flushed and new coolant added. The procedure for draining, flushing, and refilling the cooling system is discussed later in this chapter. However, some coolant can be drained from the system and sufficient antifreeze added to increase the level of freeze protection.

After the antifreeze has been added, run the engine until it reaches operating temperature. If the tractor is equipped with a cab heater, it should be turned on. This will allow the solution to circulate through the entire system. Costly repairs such as cracked engine blocks and cylinder heads have been caused by failure to thoroughly circulate coolant.

### Flushing the Cooling System

The operator's manual will give the recommended interval for draining, flushing, and refilling the cooling system with the proper antifreeze and water solution. Usual intervals are annually or every two years.

Completely drain the cooling system by removing all drain plugs. A plug is located in the lower section of the radiator. There is also a drain plug in one side of the block, and some engines have a coolant drain plug in the engine oil cooler. The operator's manual should be used to identify the number of drain plugs and their location.

The system should then be filled with clean water, and the engine should be run until it reaches operating temperature and the thermostat opens. If the tractor is equipped with a cab heater, it should be turned on to allow the water to circulate through the entire system.

Stop the engine and remove the radiator cap and all the drain plugs. Drain the system completely. Care should be taken to protect your hands and arms from the hot water. If the cooling system contains rust or other deposits, a chemical radiator cleaner should be used to flush it. The manufacturer's recommendations should be followed. Normally the cleaner is added and the cooling system is filled with water. The engine should then be run for a predetermined period of time. Then the system is drained and a neutralizer added. When the process is completed, it is best to flush the system again with clean water before filling with coolant.

CAREFUL: **Never pour cold water into a hot engine!** The rapid cooling of engine parts could result in a crack in the head or engine block. The engine should always be allowed to cool before refilling the cooling system during the flushing and filling procedure.

After flushing has been completed, fill the cooling system with the recommended mixture of water and antifreeze. Check the overflow tube for obstructions. Clean dirt, insects, and other materials from the radiator core air passages, grille, and screens. Start the tractor engine and bring it to operating temperature so that the coolant will be circulated throughout the cooling system, and then recheck for leaks.

### Care of Hoses

Hoses should be visually inspected for signs of deterioration or damage. They should not be brittle and hard nor should they be soft or swollen. Hard, brittle hoses are prone to crack and leak. Soft and swollen hoses indicate structural failure and may have deteriorated on the inside. When this occurs, rubber particles may break off causing clogging in the radiator core, or the hose may burst under load.

If there is doubt as to the condition of a hose, replace it. This is also true for cab heater hoses on tractors and other equipment.

### Fan Belt Tension

The fan belt tension should be inspected periodically and proper adjustment should be maintained. Belts should not be allowed to become loose enough to slip in the sheaves. This will greatly affect the efficiency of the cooling system and can cause the engine to run hot due to insufficient air flow and coolant circulation. Excessive tension, however, will reduce the life of alternator, and water pump bearings.

Correct belt adjustment is determined by the amount of belt deflection between two specified sheaves. As a general rule, belts should have about ¼ inch [6.5 mm] of deflection per foot [30.48 cm] when 20 lb [9 kg] of force is applied to the belt. Check the operator's manual for the correct belt deflection and procedures for adjustment.

## DIESEL FUEL SYSTEMS

The agricultural tractor has recently become one of the primary users of diesel engines. Most new tractors sold today are diesel powered. For this reason it is important that people employed in agriculture know how to handle diesel fuel and perform preventive maintenance operations on diesel fuel systems.

### Diesel Fuel

The correct selection, storage, and handling of diesel fuel will help ensure that the agricultural tractor that you own, operate, or service will operate at maximum power for many trouble free hours.

**Grades of Diesel Fuel** There are two major grades of diesel fuel. The grades are defined by the American Society for Testing and Materials (ASTM). Both grades are used in diesel engines on farms and ranches.

Grade No. 1-D is a volatile diesel fuel used in engines where frequent changes in load and speed are required. It is also recommended for extremely cold weather operation.

Grade No. 2-D is a low volatility diesel fuel used in engines with relatively high loads and more uniform speeds.

Number 2 high speed diesel fuels are generally able to ignite and burn at the proper temperature and the proper rate in agricultural engines. However, No. 2 diesel fuel does not have a low enough pour point or cloud point in extremely low temperatures. Pour point is the temperature at which diesel fuel ceases to flow. Cloud point is the temperature at which paraffin and other solids begin to crystallize or separate and restrict the flow of diesel fuel.

No. 1 diesel fuel, which is refined to a higher degree, is recommended for cold weather operation. Some manufacturers recommend No. 1 diesel fuel at temperatures below 40°F [5°C] and others at temperatures below 0°F [−18°C]. Refer to the operator's manual to determine the grade recommended. It is important to remember that engine design and operating conditions affect the grade of fuel best suited for an engine.

**Cetane Number** The ignition quality of diesel fuel, the ease with which the fuel will ignite, and the manner in which it burns are measured by the cetane number. A high-cetane number diesel fuel permits an engine to be started at lower air temperatures, provides faster engine warm-up without misfiring, reduces carbon and varnish deposits, and helps eliminate engine knock. However, too high-cetane number diesel fuels may cause excessive smoking and incomplete combustion of the fuel.

The cetane numbers for diesel fuels marketed in the United States range from 33 to 64. The lowest cetane number diesel fuel recommended for use in the high-speed diesel engines used in agricultural tractors is 40. Some manufacturers recommend using Nos. 46 to 60. Always check the operator's manual to determine the correct cetane number for the engine, weather, temperature, and altitude.

**Sulfur Content** The presence of excessive sulfur in diesel fuel can increase piston ring and cylinder wear. Sulfur causes varnish formation on piston skirts and sludge formation in the crankcase. Diesel fuel containing less than 0.05 percent sulfur is recommended. Some manufacturers recommend that fuel with less than 1.0 percent sulfur be used. The operator's manual will specify sulfur levels for a particular tractor or engine.

## Storing and Handling Diesel Fuel

A major cause of trouble in diesel engines is contaminated fuel. A few small particles of foreign matter or a few drops of water can cause costly damage to the closely fitted precision parts of an injection pump or injection nozzle.

No fuel is satisfactory for use if it is dirty or contaminated. Water, scale, rust, dirt, and other foreign matter are contaminants that must be kept to a minimum in diesel fuel. Proper storage and handling is critical in maintaining an uncontaminated fuel supply.

Overhead storage tanks should be positioned to slope away from the outlet hose. Tanks should have a covered vent, a sediment and water drain plug in the lower end, and a fuel filter (Fig. 10-21). Diesel fuel should not be used immediately after a storage tank has been filled. It should be held in the tank at least 24 h to let the dirt particles and other contaminants settle to the bottom so that they will not be drawn out. All storage tanks should be cleaned at regular intervals. Extreme care should be followed to ensure that all fuel handling equipment be kept clean and covered when not in use. Check the operator's manual for additional recommendations which the manufacturer may have regarding fuel storage and handling.

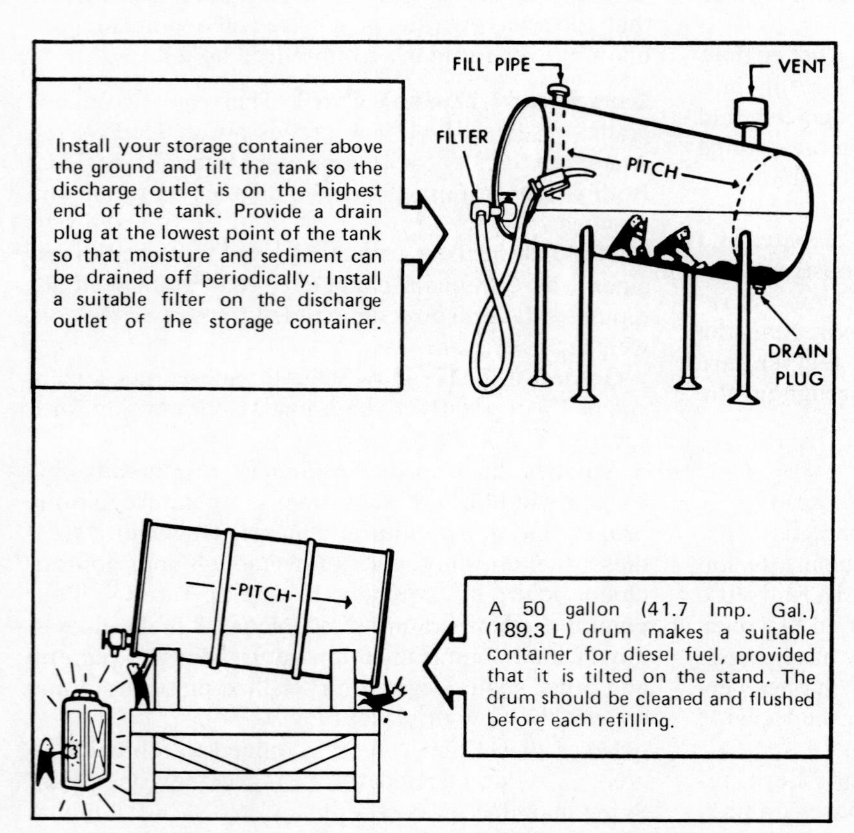


**Fig. 10-21.** Proper storage of diesel fuel is very important. (*Ford Tractor Operations.*)

## Refueling the Tractor

Most operator's manuals recommend that the diesel tank on a tractor be filled at the end, rather than at the start of the work day. This is recommended because as fuel is used air enters the tank. All air contains some moisture and the warmer the air the more moisture it can hold. If the fuel tank is left empty, the moisture in the air will condense into water as the tank cools and settle in the bottom of the tank. If the tank is filled at the end of the work day, the fuel will force most of the air out and greatly reduce the chances of water accumulating in the tank.

When using portable tanks, it is recommended to park the tank and bring the tractor to the tank for refueling. This practice will permit dirt particles and other contaminants to settle to the bottom of the portable tank and extend the life of the fuel filters.

## Water and Sediment

Check the operator's manual for recommendations for daily servicing of the fuel system. Since the presence of water is such a critical factor in diesel fuel, most tractors are equipped with clear sediment bowls that should be checked daily and drained whenever any water or sediment is visible.

If the fuel filter is equipped with a drain plug in the bottom, the water and sediment can be drained by removing the drain plug and loosening the vent plug (bleed screw) at the top of the filter. When the sediment and water have been removed, replace the drain plug and bleed the air out of the filter. If the tractor is equipped with a hand primer pump, use it to bleed the filter (Fig. 10-22). Close the vent plug (bleed screw) when a clear (air-free) stream of fuel is attained.

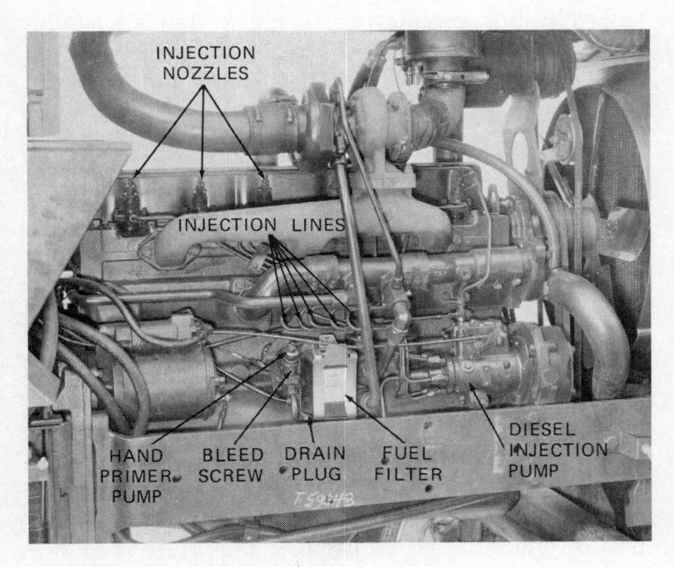


**Fig. 10-22.** Components of a diesel fuel system. (*Allis-Chalmers Manufacturing Co.*)

When there is water or excess sediment in the fuel transfer pump sediment bowl, it should be removed and cleaned. The fuel tank shut-off valve must be closed to prevent excessive fuel loss. Remove the bowl and strainer. Wash the bowl and strainer in clean diesel fuel and wipe them dry. Reinstall the bowl and strainer making sure that the gasket is in good condition and correctly seated. Open the tank shutoff valve and bleed the system at the filters (Fig. 10-22). Not all tractors have fuel transfer pumps, and some with pumps do not have a sediment bowl.

Many tractors are equipped with a water sump or trap in the bottom of the tank. This is especially true for tractors with large-volume fuel tanks. If the tractor is equipped with a sump, it should be drained daily. Whenever excess water is found in the fuel pump sediment bowl or the filters, all the water should be drained from the tank.

## Fuel Filter Service

The purpose of the fuel filter(s) is to remove water, sediment, and small abrasive particles before the fuel enters the fuel injection pump. Proper handling and storage of the fuel will increase the life of the fuel filter.

Refer to the operator's manual for the recommended filter change interval. Usually if one filter is used, it should be changed every 500 h of normal operation. When first-stage and second-stage filters are used, the first-stage filter is replaced every 500 h and the second-stage filter is replaced every 1000 h or each season.

Manufacturers use different types of filters. The operator's manual should be used to determine the procedure for removing and installing the filter (or filters). The basic procedure for changing filters is to close the fuel tank shutoff valve. Then clean any dirt from the outside of the filter and the surrounding area and drain the filter if it is equipped with a drain plug. Remove the filter in accordance with the instructions in the operator's manual. If the engine is equipped with a cartridge filter, clean the bowl with a brush or lint-free rag. It is important that the bowl be absolutely clean!

Replace the filter, using new gaskets if separate gaskets are used. Loosen the vent plug (bleed screw) and open the fuel tank shutoff valve. Bleed the air out of the filter using the hand primer pump if the tractor is equipped with one.

If, after the filter service has been completed, the tractor fails to start or runs poorly (misses) it probably has an *air lock*. Any air present in a diesel fuel system can prevent it from starting or running properly. When this happens, it is necessary to bleed the injection pump and the injection lines. This procedure is discussed in Chapter 11.

The diesel fuel system is highly reliable and almost maintenance-free if basic preventive maintenance is performed and clean, fresh diesel fuel is used. The service outlined here is general and it is important that the recommendations found in the operator's manual be followed.

## GASOLINE FUEL SYSTEMS

Even though most new agricultural tractors are powered by high speed diesel engines, gasoline engines continue to be an important part of agricultural production. Many tractors manufactured in the 1950s, 1960s, and even earlier years are still in use on farms and ranches today. Farm trucks, and small engines used on sprayers, elevators, and pumps, also play a vital role in agriculture and most of these are powered by gasoline engines. Some agricultural machines are still manufactured with gasoline engines.

### Gasoline

The common grades of gasoline sold today are regular, low-lead, unleaded, and premium unleaded. Gasoline is graded according to its *octane rating,* which is a measure of its ability to resist detonation. Detonation is uneven burning of the fuel in the combustion chamber after spark occurs. This uneven burning, or violent explosion of a part of the fuel, is a result of the extreme heat and pressure created by the rapidly burning fuel and air mixture. Common terms used for detonation are *combustion knock* and *ping*. It is usually noticed during rapid acceleration or when the engine is heavily loaded.

Detonation can cause severe damage to the engine. The violent explosion normally occurs on the side of the cylinder opposite the spark plug. It can erode the top of the piston, damage the top ring land, damage the top compression ring, or even be strong enough to crack the piston. Severe pressure is also exerted on engine bearings, valves, spark plugs, and gaskets. Other engine parts can also be damaged by detonation's occurrence.

Detonation may be caused by lean fuel mixtures, advanced engine timing, lugging of the engine, or overheating of the engine. The most common cause of detonation in engines that are properly maintained is an incorrect octane rating of the gasoline being used.

Most agricultural tractor engines are designed to operate on *regular*-grade gasoline. This means that the octane rating of regular gasoline is high enough to prevent detonation. If there is any question of whether the octane rating of the gasoline being used is correct, check the operator's manual. It will give the recommended grade and octane rating.

Light-duty trucks and other newer engines used in agricultural production use *unleaded* or *no-lead* gasoline. These fuels are required in order to meet the standards for exhaust emissions set by the Environmental Protection Agency (EPA). The fuel tank on these vehicles is equipped with an insert in the filler spout that is too small to accept a regular hose nozzle. A warning decal specifying "unleaded gasoline only" is displayed near the filler spout and on the fuel gauge.

Additives are essential in modern gasolines. Tetraethyl lead is used to raise the octane level in regular grade gasoline. Hydrocarbon fractions of high octane number are used to raise the octane level in unleaded gasoline. Other additives are used to resist spark plug fouling, gum formation, rust, carburetor icing, intake valve sticking, and other deposits in the fuel and intake system. In order to ensure that the gasoline you are using meets the manufacturer's specifications, purchase a reputable brand from a reliable supplier.

### Storing and Handling Gasoline

The three major problems in the on-farm storage of gasoline are losses due to evaporation, gum deposits, and contamination by dirt, water, and other foreign materials.

Evaporation is not a problem with underground storage tanks but may be sizable if aboveground tanks are not properly equipped. The color of the tank, the shade provided, and the type of vent installed all have a major effect on evaporation losses (Fig. 10-23).

*Note:* Pressure-vacuum relief valves are not legal in some states. To be sure if a valve can be used and what type is recommended check with your supplier or your state fire marshal.

Gum deposits, deposits that have a sticky or adhesive quality, are a greater problem when gasoline is stored in aboveground storage tanks. Refiners add inhibitors to protect the gasoline from gum deposits. Gum deposits can be avoided by not storing more gasoline than will be used in a 30-day period. Normally the inhibitors will provide adequate protection for up to 6 months when gasoline is properly stored.

Moisture in aboveground tanks can be controlled by the same methods used to reduce evaporation. A filter can be added to the outlet of the tank to prevent moisture and dirt from entering the fuel tank on the tractor or other agricultural engine. It is advisable to drain the moisture from aboveground storage tanks at least once each year. If water accumulates in an underground storage tank, it should be removed with a pump that has its inlet located at the lowest point in the tank.

Check with your fuel supplier for proper installation of underground tanks. It is important that they be correctly installed and that safety precautions be fulfilled.

The National Fire Protection Association (NFPA)

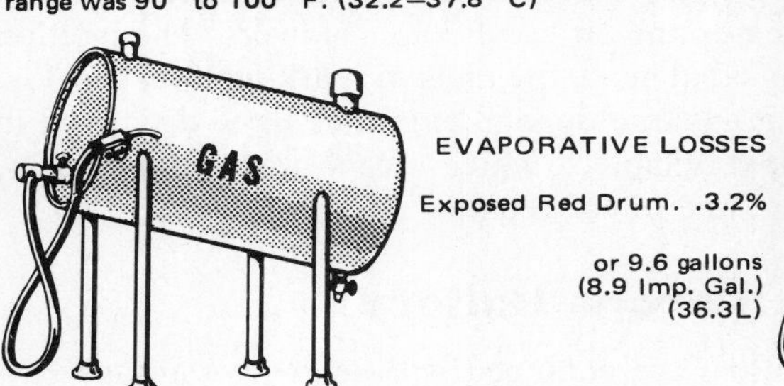

**Fig. 10-23.** Proper storage can reduce gasoline evaporation losses. (*Ford Tractor Operations.*)

standard for fuel storage requires that aboveground tanks which feed by gravity (overhead type) (Fig. 10-23) be equipped with self-closing valves. These are required in some states and are designed to cut off the flow of fuel from the tank in case of a fire while fuel is being drawn from the tank.

All gasoline tanks and pumps should be labeled GASOLINE with red letters to make certain that anyone can identify the type of fuel. This is very important where both gasoline and diesel are stored in similar tanks or drawn from similar pumps.

## Refueling the Tractor

Shut off the engine and turn off all electric switches before refueling the tractor. It is also recommended to let the engine cool off. Be sure that there are no open flames in the area and that no one is smoking.

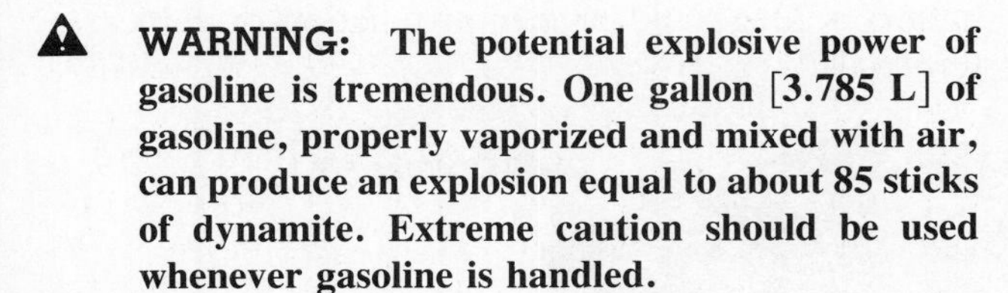

**WARNING: The potential explosive power of gasoline is tremendous. One gallon [3.785 L] of gasoline, properly vaporized and mixed with air, can produce an explosion equal to about 85 sticks of dynamite. Extreme caution should be used whenever gasoline is handled.**

## Fuel System Maintenance

Most agricultural tractors are equipped with gravity-flow fuel systems. The fuel tank outlet is located at a higher elevation than the carburetor fuel inlet and atmospheric pressure feeds the fuel to the carburetor. Trucks and some agricultural machines use a fuel transfer pump to move the fuel from the tank to the carburetor.

The primary service required on a gasoline tractor's fuel system is to periodically check the sediment bowl and fuel filter which are normally located in the bottom of the fuel tank. If water or other sediment has collected in the bowl, it should be removed and cleaned in a solvent other than gasoline. Clean the filter, check the gasket, and reinstall the filter and bowl. Some tractors are also equipped with in-line filters and/or carburetor inlet filters. These should be checked and serviced as specified in the operator's manual or when a power loss is detected.

Check the operator's manual for other services that may be necessary, such as carburetor adjustment or cleaning and governor service. A detailed procedure for servicing the sediment bowl and filters should be given in the operator's manual.

# LP GAS FUEL SYSTEMS

Liquified petroleum (LP) gas was a popular fuel for agricultural engines and tractors during the 1950s and 1960s. Early LP gas engines were gasoline engines that were converted to use LP gas as a fuel. This occurred in certain sections of the country where LP gas was priced much lower than gasoline.

Manufacturers offered tractors with LP gas fuel systems, but it was not until the early 1960s that they started to manufacture LP gas engines with higher compression ratios and improved exhaust valves. LP gas is not a major agricultural engine fuel and it will be discussed only briefly.

## LP Gas

LP gas is commonly referred to as *propane, butane,* or *bottled gas.* Actually, the LP gas used for tractor fuel today is 90 to 95 percent propane and only 5 to 10 percent butane. Propane and butane are closely related to gasoline but there is considerable difference in the way they are stored, handled, and used.

Most LP gas produced today meets either the octane specifications of either the Natural Gas Producers Association HD-5 or the octane specifications adopted by the ASTM referred to as *Special-Duty Propane.* The minimum research octane for each specification is about 110. Remember that the octane rating refers to a fuel's ability to resist the heat and pressures of combustion and burn evenly. The octane rating for LP gas is higher than the octane rating for regular grade gasoline.

## Storing and Handling LP Gas

LP gas must be stored in pressure type tanks. Rigid standards are maintained by the NFPA for the construction of LP gas tanks. Most states have laws regulating the location and installation of LP gas tanks.

Since LP gas must be kept under pressure to keep it in liquid form, the method of refueling a tractor is different than when gasoline or diesel fuels are used.

Two high-pressure lines with special fittings to fit the valves on the tractor tank are required. One line delivers liquid fuel from the storage tank to the tractor tank. The second line allows the vapor to flow from the tractor tank back to the storage tank. When the storage tank is lower than the tractor tank, a hand- or motor-operated pump must be used to pump the liquid from the supply tank. Check the operator's manual for complete procedures for refueling the tractor.

Remember that LP gas is always under pressure. Never loosen a line or fitting without first relieving the pressure. LP gas is heavier than air and all service and refueling should be done outdoors. Never smoke or use any open flame when refueling or servicing the system. Stop the tractor engine and turn off all electric switches when refueling.

▲ **WARNING: LP gas is stored at very high pressures. Use gloves and skin protection or frostbite might occur.**

## Fuel System Maintenance

The operator's manual should be read and understood before any maintenance is performed on an LP gas fuel system. Recommended maintenance is usually limited to cleaning the LP gas filter and replacing faulty valves and gauges. The manual should be followed exactly when any service is performed.

# THE ELECTRICAL SYSTEM

The electrical system on a modern diesel agricultural tractor provides energy to start the engine and operate electric accessories such as lights, gauges, and heater and air conditioner blowers. The ignition systems that ignite the fuels in spark-ignition gasoline and LP gas tractors and engines will be discussed in the next chapter, which deals with adjustments, tuneup, and minor repairs.

## The Storage Battery

Tractors are equipped with lead-acid storage batteries (Fig. 10-24). The lead-acid battery is an electrochemical device for storing energy in chemical form so that it can be released as electricity when needed to start the engine or operate electric components on the tractor or other agricultural equipment.

Both 6- and 12-volt (V) batteries are used on agricultural equipment. Some tractors use two 6-V batteries connected in series to produce 12-V current to start the engine and operate other electric accessories (see Fig. 10-25). Series connection produces voltage equal to the number of cells times 2 V per battery cell. Other tractors use one 12-V battery, and some use two or more 12-V batteries connected in parallel. When two or more 12-V batteries are connected in parallel (Fig. 10-25), the voltage remains at 12 V, but the capacity of the battery is increased to meet the needs of the tractor.

Both parallel and series connections are used in order to keep individual battery size and weight within reasonable limits, to permit safe battery handling, and to facilitate battery location and storage on the tractor. The operator's manual specifies correct battery voltage and proper installation of cables to terminals.

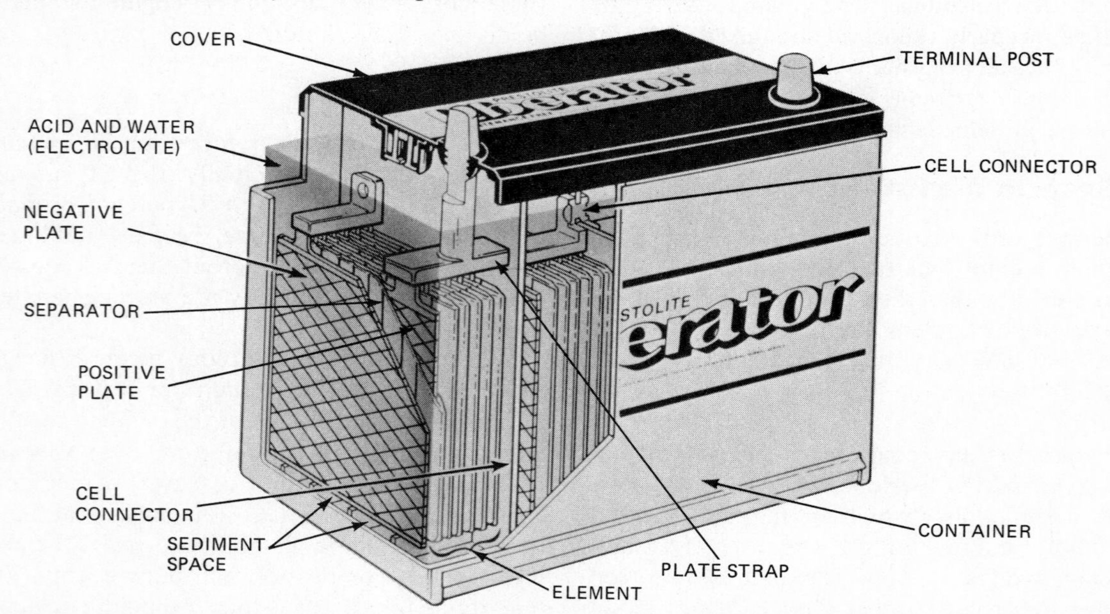


**Fig. 10-24.** Cutaway view of a maintenance-free lead-acid storage battery. (*Prestolite.*)

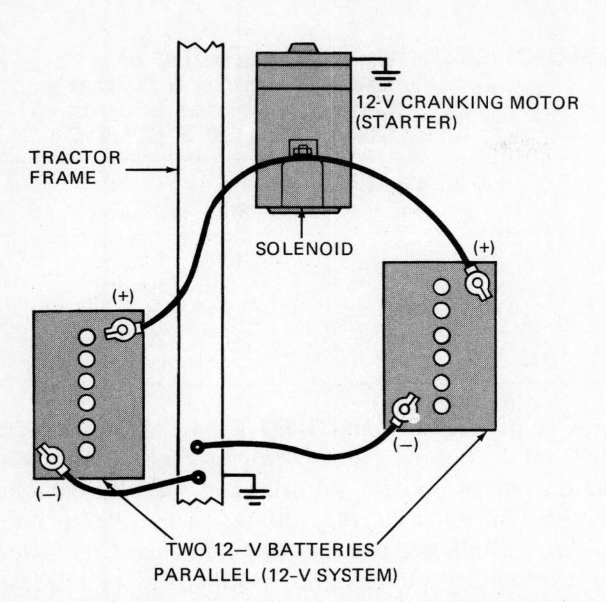


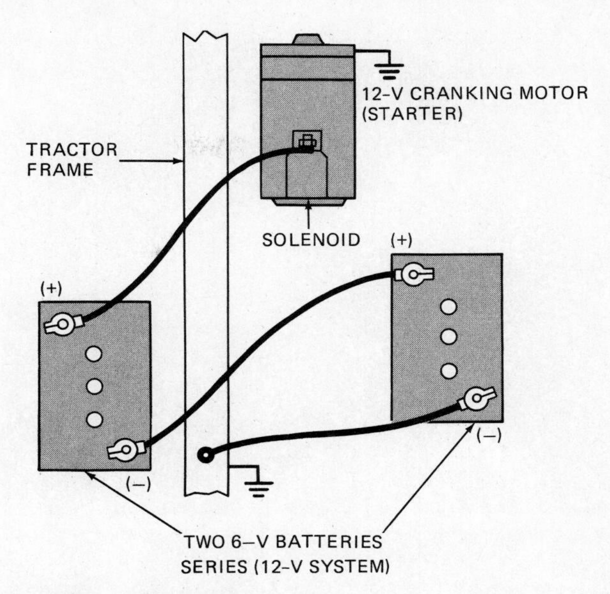


**Fig. 10-25.** Both series and parallel battery connections are found on agricultural tractors.

## Maintaining the Battery

The service recommendations given in the operator's manual should be followed in order to receive maximum battery life and to ensure that the tractor will start and operate properly. The service recommendations will give the location of the battery box and include instructions for gaining access to the battery or batteries. This is important because batteries in some agricultural tractors and machines are located under cab floors or in other remote areas. Some manufacturers are equipping new tractors and machines with sealed or maintenance-free batteries, and less frequent service is required.

**Electrolyte Level** Preventive maintenance includes checking the electrolyte level at recommended intervals that usually range from 50 to 200 h of operation. If the electrolyte level is low, the cells should be filled to the bottom of the vent plug's split ring with clean, soft water. Most manufacturers recommend that only distilled water be added to batteries. This is a good practice, because minerals and other impurities will build up on the plates in the battery and reduce battery life and performance. If water is added to a battery during cold weather, the engine should be running and allowed to run for at least 30 min so that the water added will be completely mixed with the electrolyte to prevent freezing.

**▲ WARNING Keep all open flames and sparks away from batteries, because an explosive gas, hydrogen, is released when the battery is being charged. Use caution to prevent sparks when connecting or disconnecting battery cables or booster cables. Always disconnect the grounding cable first and reconnect it last. On negative ground systems this is the negative cable. Some older tractors are positive-ground. When this is the case, the positive cable is the grounding cable.**

**Cleaning the Battery** The batteries and battery terminal connections should be kept clean. Some manufacturers recommend wiping the top of the battery after every 200 h of operation or whenever dirt appears excessive. Dirt and other foreign material on the top of a battery can absorb moisture and cause excessive battery drain or voltage loss between the terminals by conducting small electric charges across the battery top.

If corrosion deposits are present around the terminal connections, they can be removed and neutralized by using a solution containing ¼ pound (lb) [0.1 kilogram (kg)] of baking soda added to 1 quart (qt) [0.95 (L)] of water. This will also remove any acid that has accumulated on the battery, the battery compartment, and any surrounding areas. After cleaning, the solution should be thoroughly flushed with water. Care should be taken to prevent any of the cleaning solution from entering the battery cells.

Battery terminals should always be kept clean and tight. Corrosion which has built up between the battery post and the terminals that can not be washed away using the above procedure should be removed with a post and terminal cleaning tool (Fig. 10-26) or with sandpaper. After the battery posts and terminals are cleaned, they should be neutralized with a baking soda solution and flushed with water. A light coat of grease, petroleum jelly, or commercial coating material on the posts will help keep moisture out of the terminal connections.

**State of Charge** The state of charge in a battery can be determined by measuring the *specific gravity* or weight of the electrolyte in each battery cell. The specific gravity of pure water is 1.000. The specific

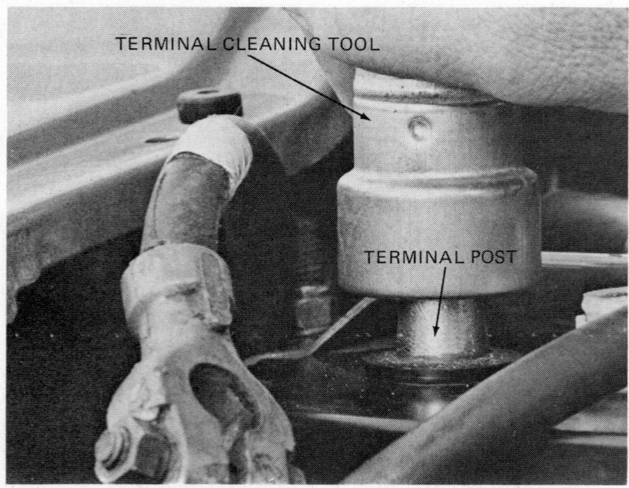


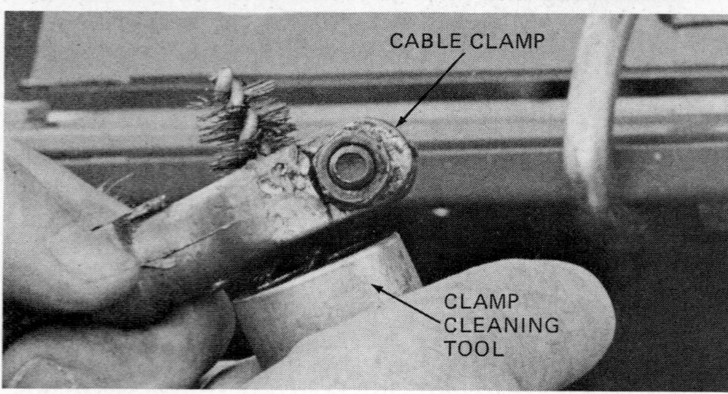


**Fig. 10-26.** Clean battery terminal posts and cable clamps with a cleaning tool to improve service life. (*Photo by Westen McCoy.*)

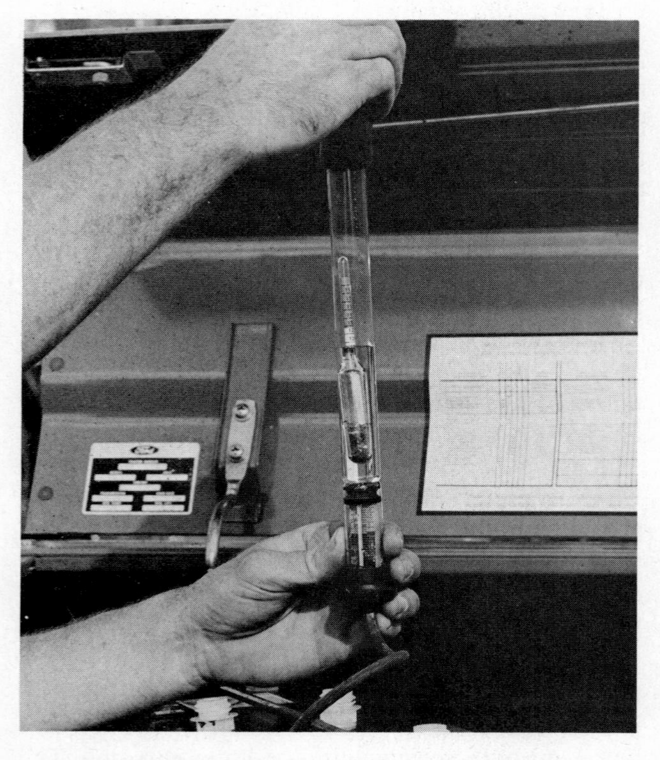

**Fig. 10-27.** Test the state of charge of a battery with a battery hydrometer. (*Photo by Westen McCoy.*)

**TABLE 10-2. Specific Gravity of Electrolyte in a Battery Cell at Various Stages of Charge at 80°F [26.7°C]**

| Specific Gravity Reading | State of Charge |
|---|---|
| 1.265 | 100% Charged |
| 1.225 | 75% Charged |
| 1.190 | 50% Charged |
| 1.155 | 25% Charged |
| 1.120 | Discharged |

gravity of sulfuric acid ($H_2SO_4$) is 1.835. Electrolyte in a fully charged battery cell is about 64 percent water and 36 percent sulfuric acid. A fully charged battery will have all the sulfates in the electrolyte, but in a discharged battery, the sulfates are deposited on the battery plates. The specific gravity of fresh electrolyte is 1.270. However, a battery is considered to be at full charge when the electrolyte has a specific gravity of 1.265, at half charge when the electrolyte has a specific gravity of 1.190 and discharged when the specific gravity is 1.120 (Table 10-2).

**Determining State of Charge** A battery hydrometer is used to determine the specific gravity of the electrolyte in a battery (Fig. 10-27). A hydrometer reading should not be taken immediately after water has been added to a battery. Adding water lowers the specific gravity of the electrolyte by diluting the solution. Draw some electrolyte from the battery cell being tested. Hold the hydrometer vertically and read the specific gravity on the float. Return the electrolyte to the same cell and repeat the procedure for each cell of the battery.

Battery manufacturers use 80°F [26.7°C] as a temperature reference to determine the specific gravity of the electrolyte. In order to get accurate readings at temperatures above or below 80°F [26.7°C] a temperature correction must be made. The correction for temperature is 0.004 specific gravity per 10°F [5.6°C] variation above or below 80°F [26.7°C]. The correction is added when the electrolyte temperature is above 80°F and subtracted when the temperature is lower (Fig. 10-28).

The operator's manual will specify the recommended interval for checking the specific gravity of the electrolyte in battery cells. Some recommend that it be checked weekly and others simply state that it should be checked and the battery recharged if the adjusted reading is below 1.215. If a battery fails to maintain a charge, be sure to inspect the V belt for proper tension.

**Storing batteries** It is advisable to remove the batteries from tractors and equipment that are not going to be used or stored for long periods of time. Batteries should be stored at full charge. A fully discharged battery will freeze at 18°F [−8°C], whereas a

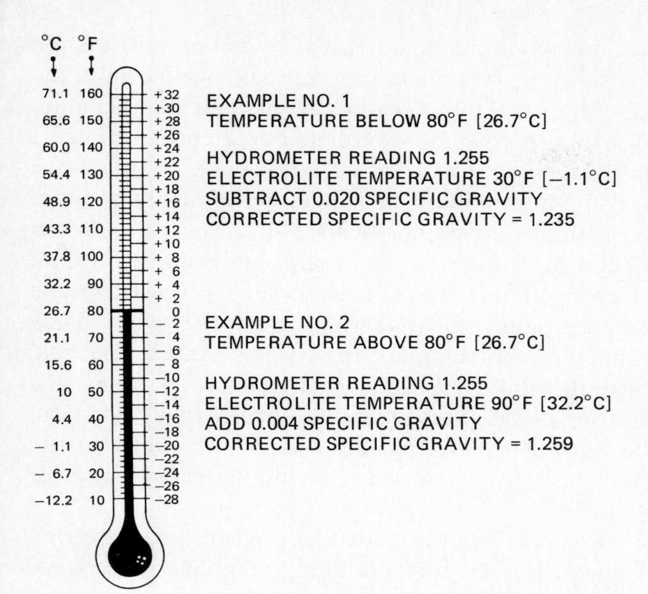


**Fig. 10-28.** A temperature correction chart is used to determine the specific gravity of an electrolyte when using a battery hydrometer.

fully charged battery is safe down to −70°F [−57°C]. Batteries should not be stored with the case sitting directly on a concrete floor because the self-discharge rate will be greatly increased. Check the operator's manual for specific details on battery storage, handling, and service.

### Maintaining the Alternator

Most late model agricultural tractors are equipped with alternators. Alternators require no lubrication and less maintenance than the generators used on older model tractors. They do require the correct belt tension to maintain the battery.

The battery ground cable should be disconnected when working near the alternator or regulator to prevent the possibility of a short across any terminals. Never operate a tractor equipped with an alternator with the battery (BAT) wire or any other wire disconnected. Disconnect cables from the positive battery terminals if the battery is to be recharged using a battery charger. For more specific details concerning alternator maintenance consult the operator's manual.

### Maintaining the Generator

Early model tractors are equipped with generators. Generators require frequent maintenance and some require external lubrication. The operator's manual should be consulted for recommended service and procedures.

### Maintaining the Starter

The *starter* or *starting motor* is similar to the generator in construction. It is designed to handle very heavy loads for up to 30 seconds. Starters require little maintenance if they are operated correctly. Most operator's manuals do not give service recommendations.

### Other Electric Components

Consult the operator's manual for service and maintenance of other electrical components such as circuit breakers, wiring, and other accessories.

## POWER TRAIN AND HYDRAULIC SYSTEM

The component parts that make up the power train on an agricultural tractor are the engine and power takeoff (PTO) clutches, the transmission, the differential, the final drive and the PTO. Refer to the Chap. 7, "Clutches, Transmissions, and Final Drives" for operation information.

The power train must absorb tremendous shocks and loads in a modern agricultural tractor. The periodic checks and maintenance recommendations given in the operator's manual should be followed.

### Lubricants

Many tractors use the same lubricant throughout the power train and hydraulic system. However, some tractors require different types of lubricants in each component. The operator's manual should be used to determine the recommended type.

**Gear Lubricants** Gear lubricants are assigned an SAE viscosity number. The classifications are different than for crankcase oil. SAE 80 is used to designate lighter oils and increases to SAE 140 for higher viscosity oils.

The API classifies gear lubricants for the types of service for which they are suited. API GL-1 is the lowest service grade and is recommended for slow gear speeds and low pressures. API GL-6 is for the high speeds and extreme pressure service. It is important that the correct viscosity *and* service grade be used. Check the operator's manual for the manufacturer's recommendations.

**Transmission-Hydraulic Fluids** Agricultural tractors that have a common reservoir for the transmission and hydraulic system use the same fluid to serve the gear train and the hydraulic system. These tractors require the use of special transmission-hydraulic fluids. Most manufacturers have developed fluids to meet these combined requirements, and they are sold by their dealers. It is best to use the oil specified by the manufacturer.

Many types of oil are used in tractors that have separate hydraulic reservoirs. The type recommended in the operator's manual should be used in these systems.

### Servicing Power Trains and Hydraulic Systems

Lubricants should be checked daily and maintained at their proper levels. Filters should be cleaned or replaced at recommended intervals. The lubricant should be changed as outlined in the specific operator's manual. Vents and breathers should also be checked and cleaned as suggested.

All lubricants should be kept clean and free from contaminants such as water and dirt. Funnels, openers, and other tools used to change lubricants should be kept clean. All dust, dirt, and other foreign material should be cleaned from around fill plugs, dipsticks, and drain plugs before they are removed.

Hydraulic hoses and connections should be checked for leaks and damage.

⚠ **WARNING** ***Never check for hydraulic leaks with your hand!*** **Use a piece of wood or cardboard, because the extreme pressure could cause a serious injury (Fig. 10-29).**

The operator's manual will give complete service data, including recommended intervals for service and service procedures. These should be understood and followed.

## OTHER MAINTENANCE

Many more service and maintenance recommendations are included in specific operator's manuals. Some of these refer to chassis lubrication, clutches, brakes, wheels and ballast, wheel bearings, and miscellaneous or optional equipment such as air conditioners, heaters, gauges, and lights. These systems should receive the service, maintenance, and adjustments detailed in the operator's manual.

**Fig. 10-29.** Using cardboard to check for hydraulic leaks. *Warning:* Never check for hydraulic leaks with your hand because the extreme pressure can cause a serious injury. (*Photo by Westen McCoy.*)

The operator's manual also includes information on storage, removal from storage, troubleshooting, safety, operation, and adjustments that should be understood before the tractor is operated or service is performed.

Time and money spent keeping a tractor or other agricultural machine clean will keep it in better operating condition and help maintain its value. Grease, oil, dirt, and chemicals such as fertilizers can damage paint, seals, bearings, and other parts. They reduce the efficiency of the cooling system and hold heat in vital areas such as transmission cases and bearing boxes. Some manuals specify that equipment be thoroughly cleaned after each operating season. Many operators clean their equipment on a regular schedule.

This text is not intended to replace the operator's manual. Rather, it is intended to emphasize the need for a complete preventive maintenance program and stress the importance of following the recommendations made by the manufacturer in the operator's manual. The tractor has been thoroughly tested, and these services and adjustments are essential for safe and trouble free operation.

## SUMMARY

Preventive maintenance is the performance of the scheduled routine services and checks specified in the operator's manual. Preventive maintenance will not totally eliminate equipment failure, but it can keep failure at a minimum.

The operator's manual is the key to a successful total preventive maintenance program. It covers operation, lubrication, safety, controls, adjustments, and other operational and maintenance procedures that should be read and understood before the tractor is operated.

Good service records are the key to a successful maintenance program. The operator's manual will give the recommended service intervals for the various systems and components on the tractor.

Air is the most abundant material in the production of power in the combustion chamber. Gasoline tractors have an air-fuel ratio of about 15 parts air : 1 part fuel. The ratio for diesels can go as high as 40 : 1.

The two most common types of air intake on agricultural tractors are the naturally aspirated and turbocharged systems. Turbochargers are used on larger tractors to increase the power output and efficiency.

Since air is used in great quantities to burn fuel, it must be cleaned. The dry-element air cleaner is used extensively on newer tractors. The oil-bath air cleaner is found on older tractors and some new tractors. Air cleaners must be serviced at recommended intervals for maximum engine life.

It is important that the proper type of lubricants be used in all systems in the tractor. Lubricants should be changed at the intervals recommended in the operator's manual. The lubricant levels should be checked on a regular schedule and filters should be serviced as recommended.

It is important that the type and grade lubricant specified in the operator's manual be used.

The engine cooling system regulates engine temperature. Water is the basic coolant but it has several limitations. These may be overcome by using ethylene-glycol-base-antifreeze and other inhibitors. The recommendations for service and maintenance of the cooling system given in the operator's manual should be followed.

Many agricultural tractors use diesel fuel. Proper storage and handling of diesel fuel is essential. Fuel filters are a vital part of the diesel fuel system. They should be serviced and changed at the intervals recommended in the operator's manual. Because of precision fitting parts, it is essential that water and dirt be kept out of the diesel fuel system.

Gasoline and LP gas are also used as fuel for agricultural tractors. Special equipment is needed to store LP gas and care must be taken when refueling LP gas tractors.

The battery produces the power to start the tractor engine as well as to operate electrical equipment. Batteries require special service and must be connected with the correct polarity. It is important that batteries be maintained at a high state of charge. The state of charge can be checked with a battery hydrometer.

Starters, alternators, and generators require little maintenance. The recommendations given in the operator's manual should be followed.

A complete preventive maintenance program includes keeping the tractor clean. Time and money spent keeping the tractor clean will keep it in better operating condition and help maintain its value.

## THINKING IT THROUGH

1. What is preventive maintenance?
2. Why is it recommended that a maintenance schedule be set up for each tractor owned?
3. Describe how air gets into the combustion chamber on a naturally aspirated gasoline engine.
4. Why are turbochargers and intercoolers added to larger tractor engines?
5. What is the purpose of the secondary filter used in many dry-element air cleaner systems?
6. How can a dry-element air cleaner be checked after it has been properly cleaned and dried?
7. How does an oil-bath air cleaner work?
8. Explain how an oil-bath air cleaner is serviced.
9. List five functions of engine oil.
10. What grade of oil is recommended for agricultural diesel tractors?
11. What is the primary function of the cooling system?
12. Explain why water alone is not a good coolant.
13. Why is ethylene-glycol-base-antifreeze referred to as permanent antifreeze?
14. What is the major cause of problems in diesel fuel systems?
15. Why is it recommended to refuel a diesel tractor at the end of the work period?
16. Explain how a hydrometer is used to check the state of charge of a lead-acid storage battery.
17. What is the specific gravity of the electrolyte in a fully charged battery?
18. How is it recommended to check for leaks in a hydraulic hose or connection?

# CHAPTER 11 ADJUSTMENT, TUNEUP, AND MINOR REPAIR

## CHAPTER GOALS

In this chapter your goals are to:

- Service diesel injection nozzles
- Service other diesel fuel system components and bleed the fuel system
- Describe how the spark-ignition system works and list the components
- Service distributors, spark plugs, and other spark-ignition system components
- Time a spark-ignition engine
- Adjust intake and exhaust valves
- Service ac alternators and describe how they work
- Service dc generators and describe how they work
- Discuss the purpose of voltage regulators
- Service starters or cranking motors
- Check and adjust toe-in
- Clean, pack, and adjust wheel bearings
- Discuss the operation and purpose of brakes on tractors

The process of making minor adjustments and repairs to maintain or restore engine performance is commonly referred to as a *tuneup*. A tuneup is a part of a complete preventive maintenance program. When the maintenance schedule provided in the operator's manual is followed, the tuneup is usually performed before there is a noticeable change in tractor performance or power.

Many operators wait until they notice a particular problem that affects operation or power before any consideration is given to tuneup adjustments. This is not a recommended practice and will normally result in more downtime during peak use and higher repair costs. Fuel costs will be higher, because an engine that is out of adjustment will use more fuel to do a job. Excess fuel that is not properly burned can wash the lubricant from the cylinder walls and contaminate the engine oil. This causes severe engine wear.

There are many employment opportunties in agricultural equipment sales and service for young men and women with skills and interests in equipment maintenance and service. Those who are preparing for a career in agricultural machinery service need to develop the ability to make repairs that exceed the maintenance level. Among other things, they need to be able to service and repair the diesel fuel system, the charging system, and the starting system.

## SERVICE SPECIFICATIONS

The operator's manual will give some of the specifications and procedures for tuneup and adjustments. However, it is recommended that a service manual be used when major tuneup and repairs are to be performed. Service manuals can be obtained from the dealer, from the manufacturer, or from publishing companies.

The service manual will give the manufacturer's recommended specifications and procedures. It will be easier to perform the job if you read the complete service recommendations in the service manual before beginning your work. Follow the procedures specific measurements, and settings. Specifications should be referred to each time service is performed. Trying to make adjustments from memory may result in poor performance and/or costly damage to engines and components.

The service specifications used in this book are general and should not be used in performing specific tuneups and adjustments.

## DIESEL FUEL SYSTEM

The service and maintenance of the diesel fuel system is a part of the daily preventive maintenance program. This was discussed in Chap. 10. The purpose of this chapter is to discuss other services and repairs of diesel fuel systems that are more technical and may require special tools and equipment.

### Diesel Injection Nozzles

Injection nozzles are used to atomize (break up) the diesel fuel and spray it into the combustion chamber so that it will ignite readily and burn completely. Nozzles are actually valves that use spring tension to oppose fuel flow. When the injection pump builds up enough pressure to overcome the spring tension, fuel is forced into the combustion chamber (Fig. 11-1).

Nozzles can be expected to last for many hours of operation if the diesel fuel is kept fresh and clean. One tractor manufacturer recommends that the nozzles be removed and tested every 600 h of operation. Some problems caused by dirty or faulty nozzles are hard starting or failure to start, irregular running (missing) or frequent stalling, lack of power, and black or gray exhaust smoke. A faulty spray pattern can be corrected by cleaning or replacing the nozzle body and nozzle valve.

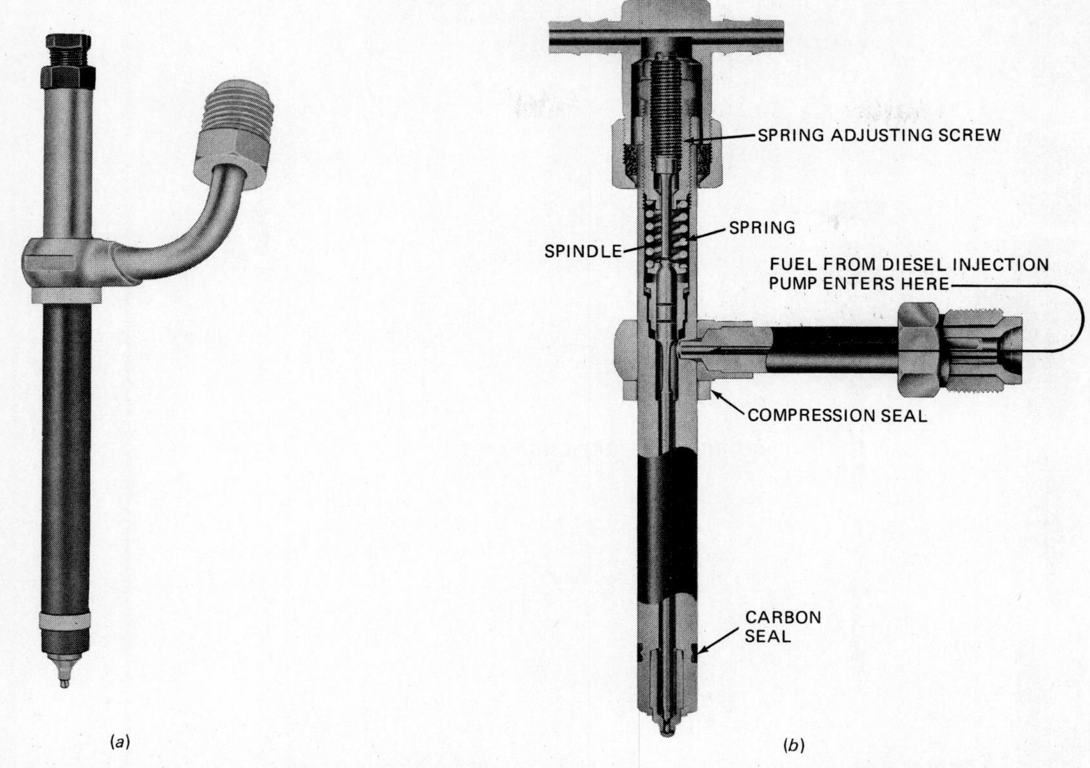


**Fig. 11-1.** A pencil-type diesel injection nozzle showing a cutaway view. (*Diesel Systems Group-Stanadyne, Inc., Hartford Division.*)

**Testing Injection Nozzles** A nozzle tester is required to service nozzles. Nozzles are attached using special adapters provided with the tester (Fig. 11-2). The operating instructions for the tester should be followed when performing nozzle tests. Several tests may be completed; the most common are tests to determine opening pressure, spray pattern, and valve leakage.

**Testing opening pressure** The tester is used to check *opening* or *cracking pressure*. Opening pressure is the pressure required to overcome the spring tension in the nozzle and allow fuel to enter into the combustion chamber. Incorrect opening pressure will allow the fuel to enter the combustion chamber at the wrong time. This has a similar effect on a diesel engine that incorrect ignition timing has on a spark-ignition engine. Incorrect spring tension can also cause the amount of fuel that is released into the combustion chamber to vary. The nozzle opening pressure is adjusted by loosening or tightening the spring-adjusting screw (Fig. 11-3).

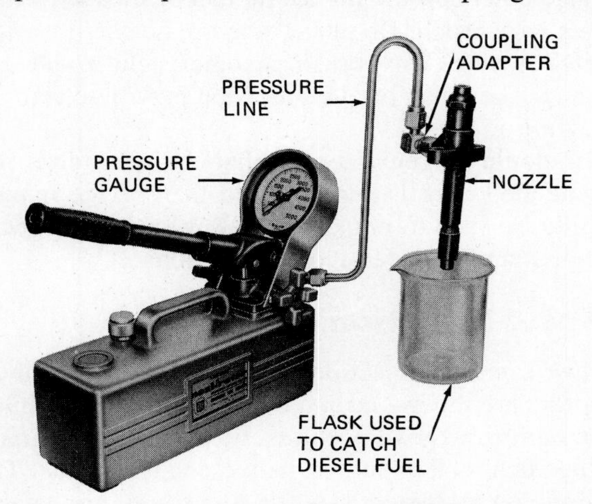


**Fig. 11-2.** A diesel tune up includes testing an injection nozzle on a nozzle tester. (*OTC Tools & Equipment Division of Owattona Tool Co.*)

**WARNING Keep hands away from the nozzle when checking opening pressure and spray pattern. The finely atomized fuel from the nozzle tip is ejected with such force that it can penetrate the skin and cause blood poisoning.**

**Testing spray pattern** The nozzle tester is also used to check the spray pattern (see Fig. 11-4). Check the service manual for the correct pattern. When the fuel is not correctly atomized (broken up), it will not completely burn. This can result in a loss of power, and the fuel that is not burned can wash the lubricant from the cylinder walls and dilute the crankcase oil. A faulty spray pattern can be corrected by cleaning or replacing the nozzle tip (Fig. 11-3).

**Testing for valve leakage** Valve leakage is tested while the nozzle is mounted on the tester. The pressure is raised slowly until it approaches the opening pressure. It is held for the time specified in the ser-

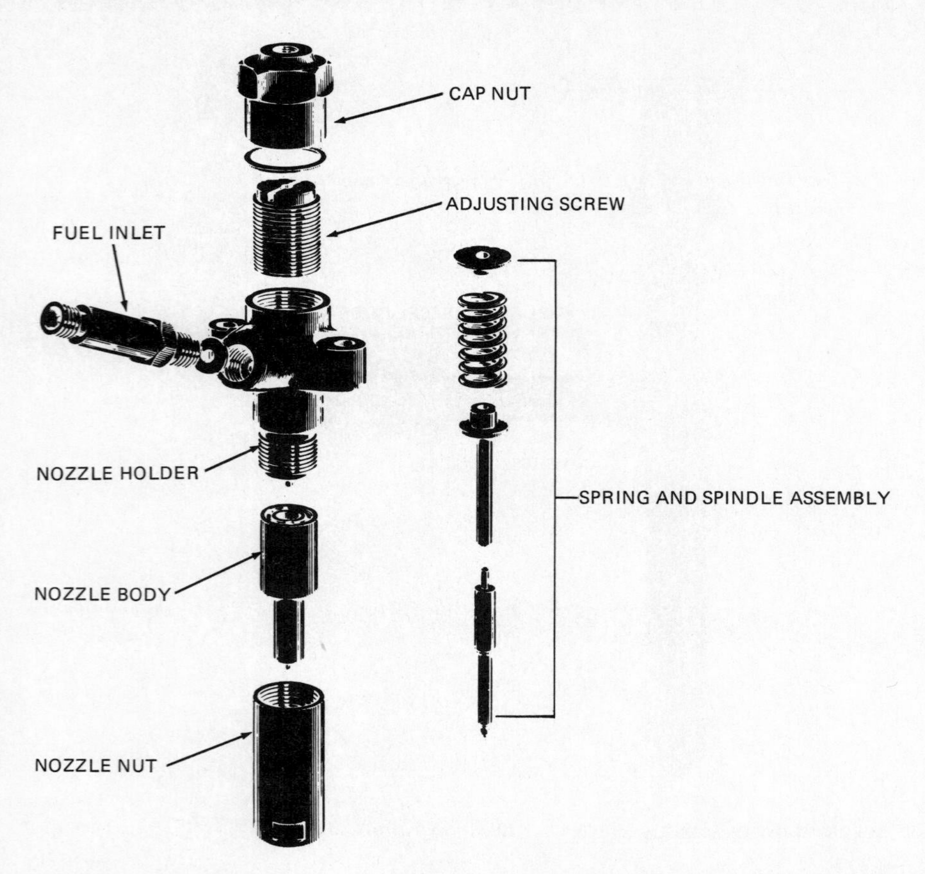


**Fig. 11-3.** Exploded view of a diesel injection nozzle. (*Ford Tractor Operations.*)

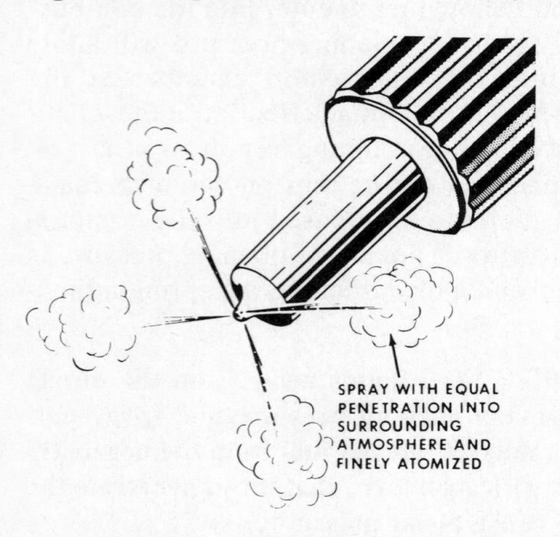


**Fig. 11-4.** Checking nozzle spray pattern. (*Schulz, Diesel Mechanics. Copyright © 1977 by McGraw-Hill, Inc. Reproduced with permission.*)

vice manual. If drops of fuel collect on the nozzle tip, the tip must be serviced or replaced.

Some manufacturers specify additional testing. The service manual contains specific testing procedures, specifications, and recommendations.

**Cleaning and Servicing Nozzles** When it is necessary to disassemble a nozzle for cleaning and service, it is of utmost importance that it be done on a clean workbench. Specks of dirt that are almost impossible to see can damage the nozzle parts and disrupt fuel delivery. For this reason, some service manuals recommend that the bench be covered with clean paper.

The whole nozzle assembly should be soaked in diesel fuel to remove all surface dirt. As the nozzle is disassembled, the parts should be placed in a container of clean diesel fuel to prevent contamination. If more than one nozzle is being cleaned at one time, be sure to keep the parts of each nozzle separate. The nozzle body and the nozzle tip can be cleaned with a brass wire brush. Do not scrape the nozzle tip with a knife. A steel wire brush or emery cloth will also damage the finish on the precision parts and ruin the nozzle tip.

It should be remembered that the procedure recommended here is general, and the service manual should be used to obtain special procedures, specifications, and service recommendations.

## Diesel Injection Pumps

Two common injection pumps used on agricultural tractors are the in-line and the rotary type. The injection pump and governor must be calibrated to supply a specific fuel flow at designated engine speeds. This is done on an injection pump test stand, and special tools are required for each type of pump. Some dealers have the equipment necessary to service pumps, while others send them to a service center that specializes in injection pump service.

When a faulty injection pump is suspected, it is advisable to take the tractor to a dealer or to remove the pump and take it in for service. Specific procedures for pump removal will be given in the service manual. If the pump is removed, it must be timed to the combustion cycle when it is reinstalled. Injection pump timing is discussed in Chap. 16.

## Fuel Transfer Pumps

Some agricultural tractors and engines are equipped with fuel transfer pumps (Fig. 11-5) designed to supply fuel to the injection pump. Some are equipped with hand levers that are used to bleed air from the diesel system when filters are changed or other service is performed.

A simple way to test the transfer pump is to disconnect the pump to filter line at the filter, set the engine stop control so that the engine will not start, and engage the starter. If fuel spurts from the line as the engine is cranked (turns over), the pump is operating properly. No fuel flow indicates that the pump is defective. Usually a defective transfer pump is replaced as a unit.

## Bleeding Diesel Fuel Systems

When air is allowed to enter a diesel fuel system, it must be removed to prevent air locks. Air locks prevent the normal supply of fuel from reaching the injection pump and the injection nozzles and may prevent the engine from starting. Air enters the fuel system when filters, transfer pumps, injection pumps, nozzles, or fuel lines are removed for service. Air also enters the fuel system when an engine runs out of fuel. The process of removing the air from the fuel system is referred to as *bleeding*. The procedure for bleeding will depend on the component or components removed for service.

**Fig. 11-5.** Some manufacturers use a fuel transfer pump that is equipped with a hand primer pump. (*Photo by Westen McCoy*).

**Bleeding Filters** If only the filters are removed, the fuel should be shut off by closing the fuel tank shutoff valve. When the filter is replaced, the filter vent plug (bleed screw) should be loosened. Then the air should be bled at the bleed screw. The hand primer pump is used if available. The bleed screw is closed when an air-free stream of fuel is attained. This should remove all the air in the system.

**Bleeding Fuel Transfer Pumps** When the fuel transfer pump is disconnected or removed, the fuel tank shutoff valve should be closed. After the transfer pump is reinstalled, the fuel line leading to the injection pump is loosened and the hand primer pump, if available, is used to pump the air out of the line. This should remove the air from the system.

*Note:* On tractors not equipped with hand-operated primer pumps, the fuel system is bled by the force of gravity on the fuel in the fuel tank. Most manufacturers recommend that the tank be full of fuel when the system is bled.

**Bleeding Injection Pumps** If the injection pump is removed and serviced, the air must be removed from the pump. Check the service manual for the location of the bleed screws. Loosen the bleed screws as indicated and use the hand primer pump, if equipped, to attain an air-free flow of fuel from the bleed screws. The filters, the transfer pump, and the fuel lines to the injection pump are bled without turning the engine over with the starter.

**Bleeding Injection Lines** When the injection nozzles or injection lines have been removed, the injection lines are bled at the nozzles. This is accomplished by loosening the lines one turn at the nozzles. Check the operator's manual or service manual for procedure, because some lines require the use of special wrenches to prevent damage to the lines.

After the line nuts are loosened, open the throttle control lever and crank the engine with the starter until fuel without foam flows from around the connectors. The engine stop knob or control must be in the run position. The line connections should then be tightened and free from leaks. It is a good practice to wipe all fuel from the filters, lines, and engine after the system has been bled. The engine should then start and run smoothly.

**WARNING The pressure in injection lines may exceed 2500 pounds per square inch (psi) [17,237 kPa] and can penetrate the skin. When bleeding injection lines, only open nuts one turn. Before any lines are disconnected, be sure that the pressure is relieved. Be sure that all lines are tightened after bleeding.**

The service manual is very important. The recommendations and procedures should be understood and followed. Skipping a recommended step can leave an air lock in the fuel system and prevent the engine from starting.

## ELECTRIC-IGNITION SYSTEM

Agricultural tractors and engines that burn highly volatile fuels such as gasoline and liquefied petroleum (LP) gas draw in both air and fuel on the intake stroke. The air-fuel mixture is compressed on the compression stroke, and a spark is used to ignite it. The spark is created by the electric ignition system.

### How the Ignition System Works

The major parts of spark-ignition system are the distributor, the breaker or contact points, the condenser, the coil, the battery, the primary and secondary wires, and the spark plugs (Fig. 11-6). The function of the system is to convert low voltage [6 or 12 volts (V)] to an ignition voltage of up to 20,000 V and deliver it at the correct time to ignite the air-fuel mixture in the combustion chamber.

The distributor (Fig. 11-7) houses the ignition advance mechanism, the breaker points, the condenser, and the rotor. The distributor cap completes the distributor assembly. Its purpose is to deliver the spark to the correct cylinder at the proper time in the combustion cycle.

The condenser is a safety device that helps prolong breaker point life and sustains the spark for better fuel ignition. It has no moving parts and takes on a charge when the breaker points open and releases it into the coil.

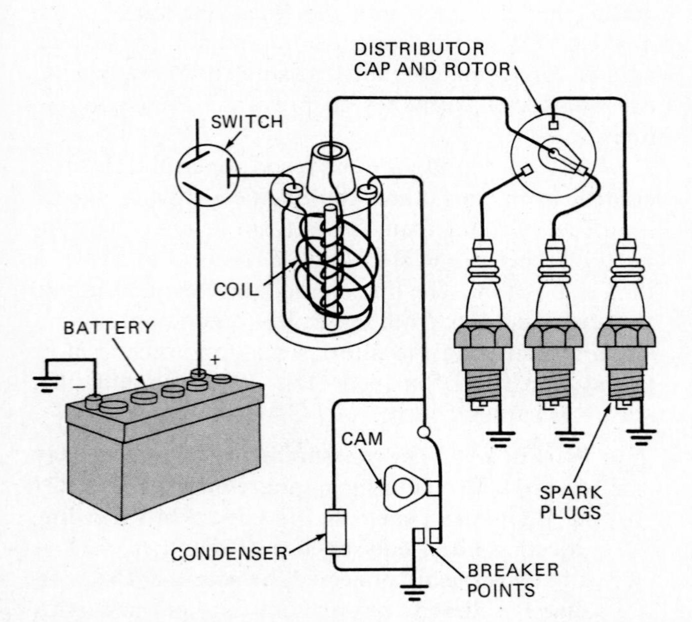


**Fig. 11-6.** The major parts of a spark-ignition system. (*Ford Tractor Operations.*)

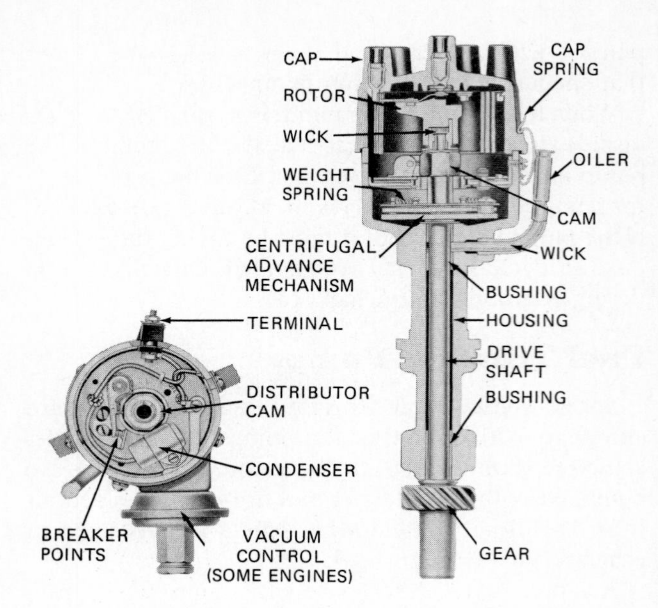


**Fig. 11-7.** A cutaway view shows the parts of a distributor. (*Delco Remy.*)

The coil is a voltage transformer that steps battery voltage up to ignition voltage.

**Primary and Secondary Ignition Circuits** The ignition system has two separate circuits. These are the primary, or low-voltage, circuit that functions when the breaker points are closed, and the secondary, or high-voltage, circuit that functions when the breaker points are open (Fig. 11-8).

The breaker points are opened and closed by the rotating distributor cam. While the breaker points are closed, voltage builds up in the primary windings of the coil and the current flows to ground through the breaker points.

When the breaker points open, part of the voltage that has been going to ground is stored in the condenser. The voltage that has built up in the coil primary windings is induced into the secondary windings. The secondary windings step the voltage up to an ignition voltage as high as 20,000 V. The voltage is carried from the coil to the center tower in the distributor cap by the coil secondary wire. It flows through the tower to a carbon contact button which contacts a spring clip on the rotor. The voltage is carried to the terminal posts in the distributor cap which connect to the spark plug towers through the conductor plate of the rotor. Rotor rotation is timed to the combustion cycle of the engine. The spark plug wires carry the current to the spark plug in the cylinder that is on the compression stroke. The current flows to ground by jumping across the electrode gap on the spark plug.

As the current jumps the electrode gap, a spark occurs. The voltage that was stored in the condenser is released to enhance the spark. This spark ignites the air-fuel mixture in the combustion chamber.

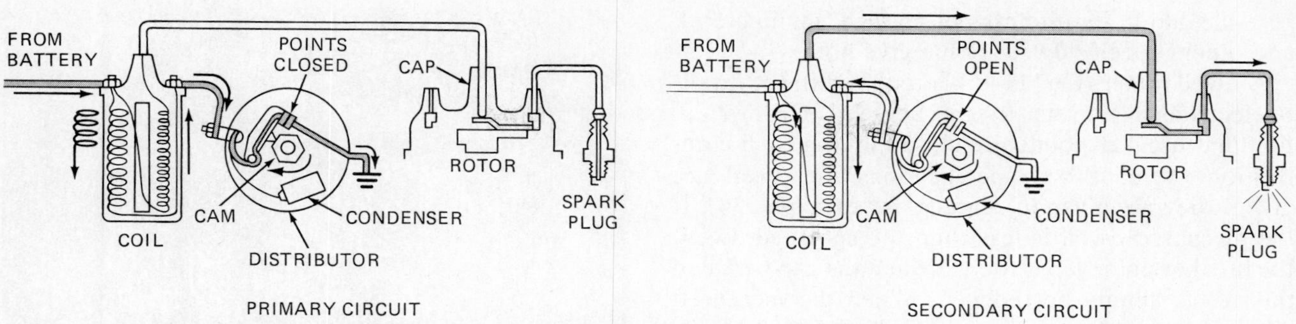

**Fig. 11-8.** The primary and secondary ignition circuits play an important role in tractor performance. Note that shaded line indicates current flow in primary and secondary circuits.

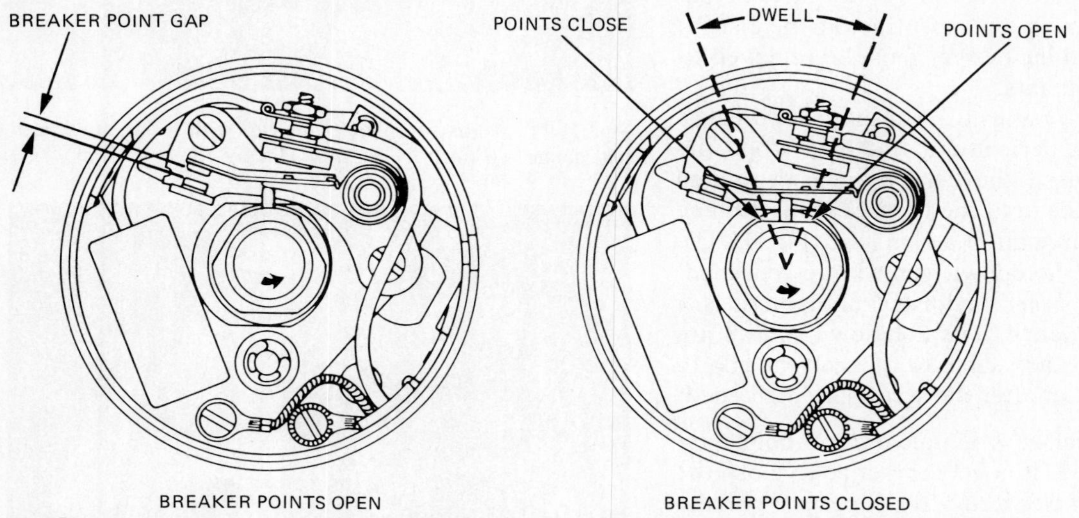


**Fig. 11-9.** Breaker point gap refers to the distance the points open. Dwell refers to the degrees of distributor cam rotation that the points are closed. (*Ford Tractor Operations.*)

**Electronic-Ignition Systems** Most spark-ignition engines produced now are equipped with electronic-ignition systems. Distributors in these systems have no breaker points. Some have an integral coil and module control, and others have an externally mounted module with the coil mounted in a conventional manner on the engine. Since many agricultural tractors are diesels and have no electric-ignition system, use of the electric-ignition system in production agriculture is primarily limited to trucks, small engines, and other agricultural equipment that is powered with gasoline engines.

## Distributor Service

It is impossible to obtain economical and satisfactory engine performance if the distributor is not properly maintained. The operator's manual will give the recommended service interval.

**Breaker Points** The breaker points are a critical part of the ignition system. Breaker points should be kept in adjustment and replaced when they show signs of wear (pitting or burning). Breaker points should be inspected each time adjustment is made.

Points should be aligned so that the contact surfaces come together flat and full surface contact is made. When properly adjusted, they will have the specified breaker point gap when they are held open by the distributor cam (Fig. 11-9).

When the breaker point gap is too wide, the breaker points open too soon and stay open too long. This reduces the time the breaker points are closed and will reduce the voltage buildup in the primary windings in the coil and may advance the timing. This may cause the engine to miss at high speeds because of insufficient secondary voltage to jump the spark plug gap and produce a spark.

When the breaker point gap is too narrow, the breaker points stay closed too long and open late. This may cause the points to burn and may also affect the secondary voltage and retard the timing.

**Adjusting breaker point gap** The two most common ways of setting breaker point gap are with a flat thickness leaf gauge or with a dwell meter. Both methods measure the degrees of rotation of the distributor cam while the breaker points are closed during each combustion cycle.

Check the operator's manual or service manual for recommended gap adjustment. Some manuals will give the gap in thousandths of an inch [millimeters], and others specify dwell. Some give both.

A dwell meter (Fig. 11-10) is required to set dwell. It electronically measures the degrees of cam rotation that breaker points are closed during each combustion cycle. When specifications for dwell are given, they are given in degrees, for example, 30°. If the measured dwell is less than the specified dwell, the breaker point gap is too wide and the gap between the points should be reduced. When the measured dwell is more than the specified dwell, the breaker point gap is too narrow. Some distributors are equipped with a window that allows dwell to be set with the engine running. On most tractors the distributor cap, rotor, and dust cover must be removed to adjust the breaker points.

The flat thickness leaf gauge is used to set breaker point gap on most agricultural tractors. When the thickness gauge is used, the distributor cam is rotated by turning the engine until the breaker point rubbing block is on the high point of a cam lobe (Fig. 11-11). The breaker point holddown screws are loosened, and the moveable arm is adjusted until there is a slight drag on the gauge. The holddown screws are then tightened. It is advisable to recheck the gap to see if it is still correct after the screws are tightened.

**Cleaning breaker points** Any time breaker points are installed or adjusted, it is good practice to close the points on a piece of clean, lint-free paper and pull the paper through them. This will remove any oil or grease that may be on the contact surfaces. It is also recommended to clean the cam and carefully apply a thin coat of special cam lubricant.

**Condenser Service** If the points are badly burned or pitted or if some metal is transferred from one contact surface to the other, it is recommended that the condenser be changed. If points do not have metal transfer, the condenser is usually suitable for continued use. When a new condenser is installed and the engine fails to start, reinstall the old condenser. If the engine starts, the new condenser is faulty. If it fails to start, reclean the breaker points and check to see that they are properly installed. A condenser tester may be used to check for capacity, shorting, or grounding (Fig. 11-12).

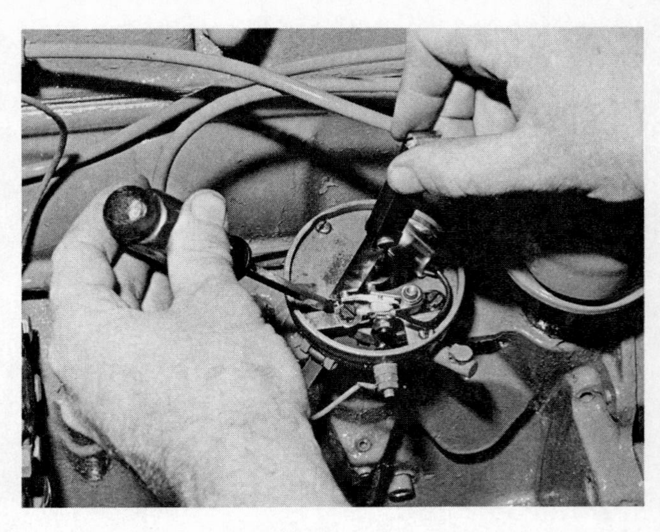

**Fig. 11-11.** Adjust breaker point gap with a flat, thickness leaf gauge. (*Photo by Westen McCoy.*)

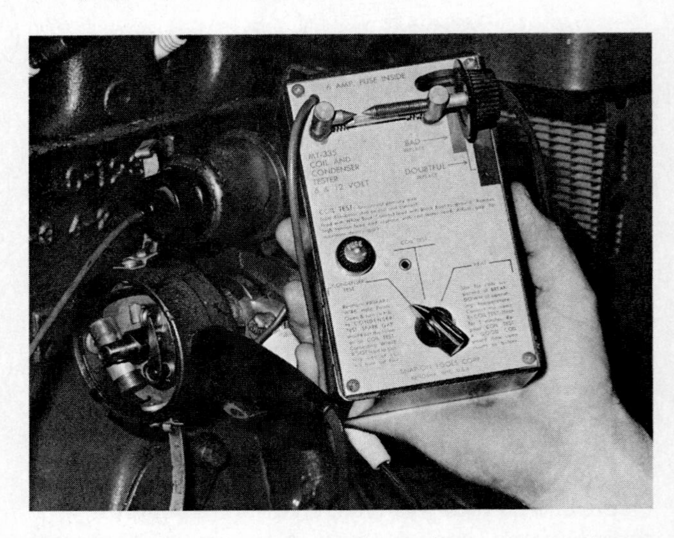

**Fig. 11-12.** A condenser tester is used to test a condenser on the engine. (*Photo by Westen McCoy.*)

**Fig. 11-10.** A dwell-tachometer set is used to measure engine rpm. The meter is also used to check and set dwell. (*Photo by Westen McCoy.*)

**Rotor and Distributor Cap** Rotors and distributor caps require little service. Check the contact strap on the rotor and the terminal posts in the cap for corrosion and pitting. Also, check the cap for hairline cracks and other physical damage.

## Servicing the Coil

The ignition coil normally requires no service. If an engine fails to start and spark is not produced when

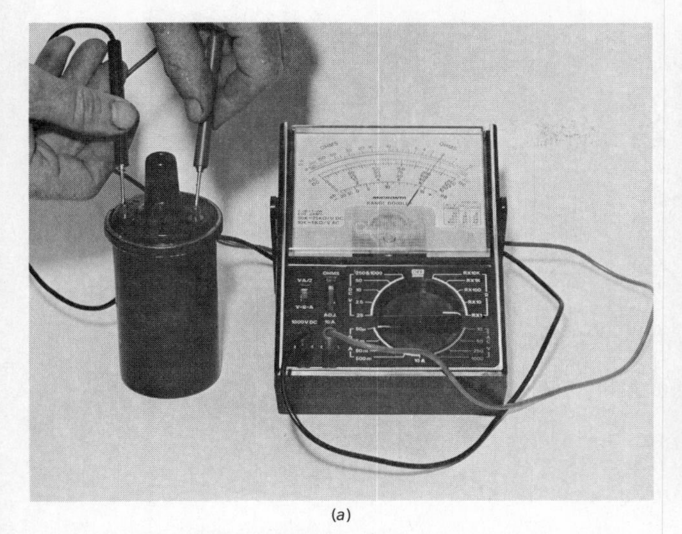

(a)

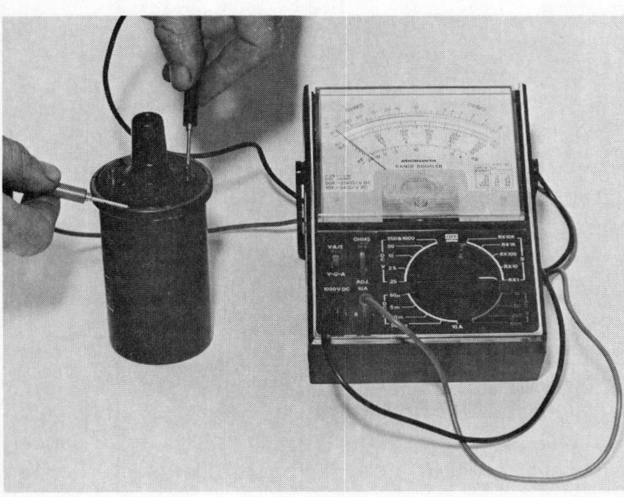

(b)

(c)

**Fig. 11-13.** Test a coil for an open circuit or ground with an ohmmeter. (*a*) Testing for open primary windings. Note the low resistance on the meter. If a test lamp is used, the lamp should light. (*b*) Testing primary windings for ground. Note the high resistance on the meter. If a test lamp is used, the lamp should not light. (*c*) Testing secondary windings for ground. Note the high resistance on the meter. If a test lamp is used, the lamp should not light. (*Westen McCoy.*)

**Fig. 11-14.** Test an ignition coil with a spark-gap-type coil tester. (*Photo by Westen McCoy.*)

the spark plug wires are removed and held close to ground as the engine is turned with the starter, the coil may be defective. A quick check may be made by removing the wire connecting the primary terminal on the coil to the distributor (Fig. 11-8). When the points are open, a spark should be produced when the wire is touched to ground. The ignition switch must be in the run position. No spark indicates that the coil may be defective and that it should be tested.

A coil may fail when it gets hot. When an engine stops after it reaches operating temperature and will not start again until it cools down, the problem may be the coil.

A coil can be checked for an open circuit or ground using a test lamp or an ohmmeter (Fig. 11-13). Special test equipment such as a spark gap tester or meter-type tester is required to test coil output accurately (Fig. 11-14)). A simple test can be made by removing a spark plug wire and allowing the spark to jump to ground with the engine running. The spark should jump a $^1/_4$-inch (in) [6-millimeters (mm)] gap. Be careful not to handle the uninsulated end of the wire, because the voltage produced varies and may be as high as 20,000 V.

**Coil Polarity** When a coil is installed, it is important that the polarity be correct. This means that the spark plugs fire from the center electrode to the ground electrode. When the coil polarity is incorrect, more voltage is required to jump the spark plug gap, and hard starting, missing, and poor performance may result.

Coil polarity may be checked by holding a lead pencil point between a spark plug wire end and a spark plug with the engine running. The flare should occur between the pencil lead and the spark plug. Reversed coil polarity will also result in an inverted wave pattern when an oscilloscope is used to check the ignition system.

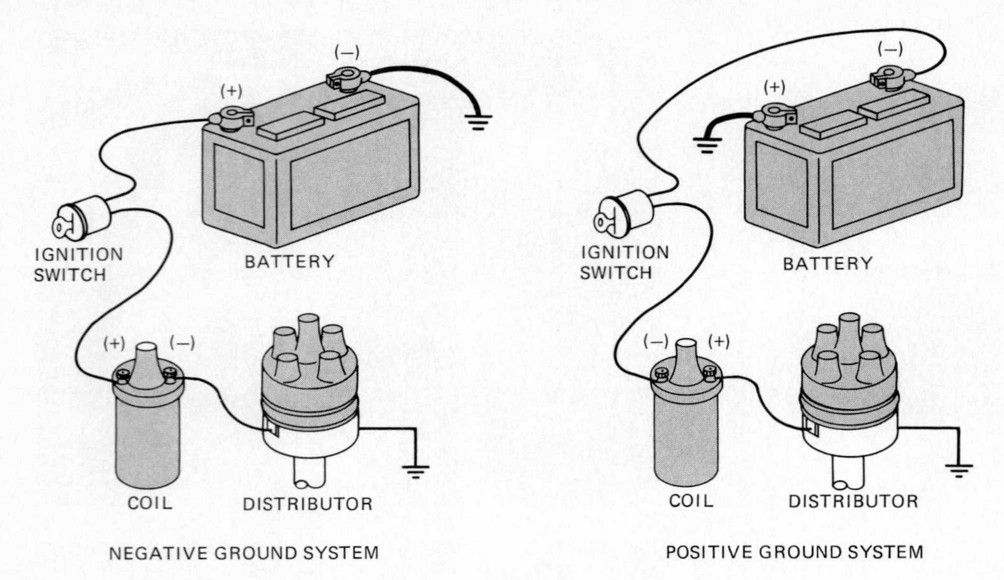


**Fig. 11-15.** Correct coil polarity can be attained by connecting the coil according to battery ground.

A simple way to attain correct coil polarity is to connect the coil according to the battery ground. If the battery is negative ground, attach the wire that connects the coil to the distributor to the negative coil terminal. On positive ground systems the distributor wire is attached to the positive terminal of the coil (Fig. 11-15).

## Servicing Spark Plugs

The recommendations in the operator's manual should be followed when servicing spark plugs. The manual may state that spark plugs should be cleaned after every 300 hours (h) of engine operation and that new spark plugs are to be installed every 600 h. It is a common practice to install new spark plugs each time service is recommended. The value of the labor required to clean spark plugs properly may equal or surpass the cost of new ones. The recommended procedures for cleaning spark plugs are discussed in Chap. 4.

The operator's manual will specify the type of spark plug to use. It is important that the correct type of spark plug with the correct heat range and reach be used. *Heat range* refers to the plug's ability to transfer heat from the center electrode to the cooling system. See Chap. 4 for a more detailed explanation of heat range.

Operating conditions and engine design determine the heat range of the plug to use. Engines that operate under heavy loads and at high rpm normally require a colder plug so that the heat will transfer faster. Engines that operate under light loads and at low rpm normally require a hot plug because the extra heat will help burn off deposits that occur under these operating conditions. Engines that operate under normal conditions usually require a normal heat range plug.

**Spark Plug Removal** Spark plug removal is usually a simple operation. However, there are some procedures that will result in better service. Spark plug wires should be removed by holding the terminal boot, never the wire. Care should be taken not to crack the insulator. Cracking can be prevented by using a special spark plug socket that contains a protective rubber shield. The area around the spark plugs should be cleaned to prevent dirt and other loose particles from falling into the cylinders when the plugs are removed. After the plugs are loosened one turn and before they are removed, clean the area with compressed air, or crank the engine for 30 seconds to blow out any dirt.

**Installing Spark Plugs** Spark plug gap should be adjusted to the specifications given in the operator's manual. Gap is the distance between the ground electrode and the center electrode. When the gap is set to the engine manufacturer's specifications, the spark will jump to ground under all operating conditions. If the gap is less than the specifications, the spark will jump too soon and may not be intense enough to ignite the air-fuel mixture. When the gap is wider than the specifications, the spark may not jump the gap under certain operating conditions.

The gap is adjusted by bending the ground electrode. It should be checked with a wire gauge. The recommended gap will be given in thousandths of an inch [millimeters].

When installing spark plugs, the threads should be clean and no oil should be used on them. However, there are several antiseize compounds available that are recommended. If old plugs are being reinstalled, it is recommended that a new sealing gasket be used. A properly tightened gasket conducts almost 50 percent of the heat flow from the plug to the cylinder head and engine coolant. Old gaskets may be flattened and may not fit tight. Spark plugs with tapered seats do not use sealing gaskets.

Spark plugs should be tightened to the torque specified in the operator's manual. If the plug is not tightened against the gasket, it will overheat, and if it is overtightened, the porcelain insulator may crack. Overtightening may also damage the threads in the cylinder head and can cause difficulty in removing the spark plugs at the next service interval.

**Spark Plug Wires** The spark plug wires should be inspected. If the insulation is cracked or extremely soft, the wires should be replaced. Damaged insulation can cause misfiring. The terminal ends and the spark plug wire connections in the distributor cap should be checked and cleaned if corroded. Continuity of wires can be checked with a continuity tester, test light, or ohmmeter. An ohmmeter is recommended for checking resistance-type wires. Other electronic test equipment such as an oscilloscope is very helpful in checking and servicing the ignition system.

## Timing the Ignition System

In order for the spark plug to ignite the air-fuel mixture at the correct time, the distributor rotation must be timed to the engine combustion cycle. When breaker points are reset or the distributor has been removed from the engine, the ignition timing should be checked and reset if necessary.

Check the operator's manual for timing specifications. Some manuals give low-speed (rpm) timing, and some give high-speed (rpm) timing.

Timing is checked and set with a stroboscopic timing light (Fig. 11-16). The three common timing lights available are the two-wire, three-wire, and 120-V ac power light. It is important to follow the instructions

**Fig. 11-16.** Set ignition timing on a spark-ignition engine with a stroboscopic timing light. Check the operator's manual for specifications. (*Photo by Westen McCoy.*)

that are provided by the manufacturer of the timing light for correct attachment.

For ignition timing to be accurate, it must be the last adjustment made on the electrical system. If the breaker point gap is changed, it will change ignition timing. Engine rpm also affects timing; therefore, a tachometer is needed. A tachometer measures engine rpm.

### Low-Speed (rpm) Timing Adjustment

Connect the timing light to the No. 1 spark plug. Connect the tachometer following the instructions provided by the manufacturer. Start the engine and allow it to reach operating temperature. Adjust engine speed to the rpm recommended in the operator's manual. Point the timing light at the indicator and timing marks located on the crankshaft pulley or flywheel.

The marks will appear stationary because the light produces a strobe effect. If the proper mark does not align with the indicator or pointer, loosen the holddown clamps on the distributor. Then slowly rotate the distributor housing clockwise or counterclockwise until the timing mark is in the correct position. Recheck engine rpm, because the change in ignition timing may cause it to vary. If this has happened, readjust the engine rpm and reset the timing. Tighten the distributor holddown clamps and recheck the timing marks.

**Checking Ignition Advance** When advance timing specifications are given, check ignition advance at this time. Set the engine rpm to the fast-idle no-load rpm recommended in the operator's manual. Point the timing light at the pointer. The full advance timing mark should now be in line with it. If the full advance timing mark is not in alignment with the pointer, the centrifugal advance system in the distributor is not operating properly.

A simple way to check the advance system is to remove the distributor cap and turn the rotor by hand. When the rotor is turned in the direction of rotation of the distributor cam, it should snap back to the original position when you release it. A slow return to the original position indicates that the advance mechanism needs cleaning. It should not turn in the opposite direction of rotation of the distributor cam. If it does, this indicates that the springs are too weak to pull the counterweights back or that excessive dirt or corrosion is not letting it operate freely. If there is looseness or noise as you turn the rotor, the springs should be checked.

The centrifugal advance system is usually located under the breaker point plate. To service this system, remove the breaker point plate and check for loose springs, wear in the assembly, or dirt and corrosion. Any of these can keep it from operating properly.

Some engines are equipped with a vacuum advance in addition to the centrifugal advance. These operate on intake manifold vacuum and advance the spark when there is an increase in intake manifold vacuum. These systems can be checked by removing the vacuum hose connected to the diaphragm and applying a vacuum. The vacuum should move the breaker point plate. The vacuum hose should also be checked for leaks.

The advance system is necessary on engines that operate at varying rpm. As engine rpm increases, the breaker points must open earlier in relation to piston location. This is because the crankshaft will rotate farther in the time it takes for spark and combustion to occur.

## Distributor Installation

When the distributor is removed from the engine for service or repair, the position of the rotor should be marked on the distributor. The position of the distributor housing in the engine block should also be marked or remembered.

After the service is completed, align the rotor with the mark on the distributor and the mark on the distributor housing with the mark on the engine block. As you push the distributor into place, the rotor may turn slightly as the gears mesh. When this happens, pull the distributor up enough to disengage the gears and move the rotor back far enough to compensate for the turning. Push the distributor back into place. Align the marks carefully and install the holddown clamps and bolts. Ignition timing should then be checked and adjusted.

If the engine is rotated while the distributor is removed, the procedure for reinstallation is more complicated. This procedure is discussed in Chap. 16.

# THE VALVE SYSTEM

Most agricultural tractors are powered by four-stroke-cycle engines. You may refer to Chap. 3 to review the principles of two-stroke- and four-stroke-cycle engine operation.

In four-stroke-cycle engines, the intake valve allows a fresh charge to enter the cylinder. Near the top of the compression stroke the charge is ignited. The piston is driven down by the expansion of the burning fuel. The exhaust valve allows burned gases to be removed from the cylinder during the exhaust stroke. The valves open and close the exhaust and intake ports to the cylinder. Both are classified as poppet-type valves.

## Valve Clearance

Valve clearance or lash is the gap between the end of the valve stem and the rocker arm when the valve is closed (Fig. 11-17). In order to maintain correct valve timing, valve clearance should be adjusted at

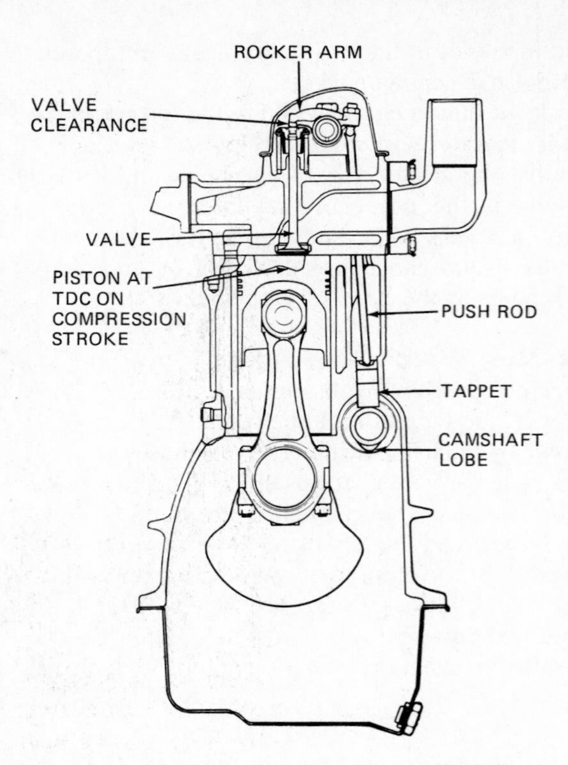


**Fig. 11-17.** Valve clearance must be adjusted when the engine is at the proper position in the combustion cycle. (*Ford Tractor Operations.*)

the service interval recommended in the operator's manual. Valve timing refers to the opening and closing of the valves at the proper time in the combustion cycle. The recommended service interval is usually every 500 or 600 h of engine operation.

**Checking and Adjusting Valve Clearance** Most manufacturers recommend that the torque or tightness of the cylinder head bolts be checked before valve clearance is checked and adjusted. The torque specification and exact torquing sequence or pattern should be obtained from the service manual for the particular make and model tractor (Fig. 11-18). Check the operator's or the service manual for recommended clearances and procedures. Often engine shrouds, fuel tanks, and other accessories must be removed. When fasteners are removed, they should be kept separately in labeled cups or metal containers for convenience and correct reassembly. Keep notes as necessary to identify where parts fit and label any lines or wires that must be disconnected.

Valve clearance should be set to the specified gap. The range is usually from 0.006 to 0.030 in [0.15 to 0.76 mm]. The manufacturer will specify the clearance for the intake and exhaust valves for each engine model. The specifications may be given as either a cold or a hot setting. If cold valve clearance specifications are given, the engine should be completely cooled down. When hot valve clearance specifica-

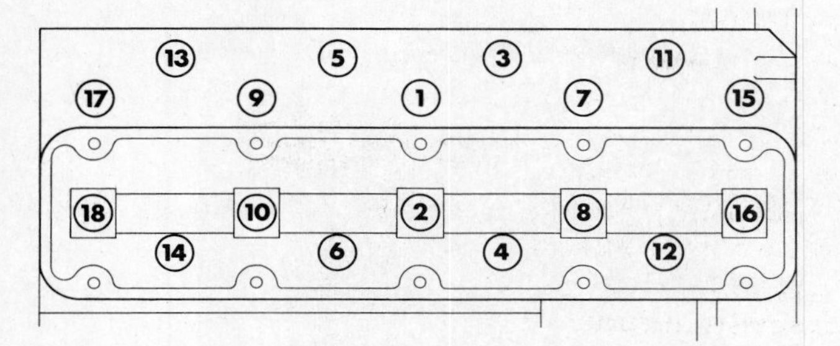


**Fig. 11-18.** A typical torquing sequence. The recommended sequence or pattern should be followed when tightening cylinder head bolts. (*Ford Tractor Operations.*)

tions are given, the engine should be at operating temperature.

In order to check and adjust valve clearance, the valve must be fully closed and the lobe on the camshaft that opens the valve must be in the proper position (Fig. 11-17). Both the intake and exhaust valves for a cylinder will be fully closed when the piston is at top dead center (TDC) on the compression stroke. This position can be determined by watching the rocker arms on the other cylinder in the pair and determining when that piston is coming up on the exhaust stroke. In a four-cylinder engine, No. 1 and No. 4 cylinders and No. 2 and No. 3 cylinders are paired. For six-cylinder engines, No. 1 and No. 6, No. 2 and No. 5, and No. 3 and No. 4 cylinders are paired.

When the exhaust valve begins to close, the piston is nearing TDC. Just before the piston reaches TDC, the intake valve begins to open. At this point both valves are moving. For several degrees of crankshaft rotation, both valves are slightly open between the intake and exhaust stroke. This may be referred to as valve overlap, or "point-of-rock," because if the engine were turned forward and backward at this point, the rocker arms would rock as the valves open and close.

On four-cylinder engines, No. 4 piston is completing the exhaust stroke when No. 1 piston is at TDC on the compression stroke. Valve clearance for the intake and exhaust valve for No. 1 cylinder can be adjusted when the engine is in this position. When the crankshaft is rotated 180°, or one-half turn, either No. 2 or No. 3 piston is completing the exhaust stroke and the other piston in that pair is at TDC on the compression stroke. The valves for that cylinder can be adjusted. When the crankshaft is rotated another half turn, No. 4 piston is at TDC on the compression stroke. The valves for No. 4 cylinder can be adjusted. Either No. 2 or No. 3 piston will be at TDC on the compression stroke when the crankshaft is rotated another half turn. The valves for this cylinder can then be adjusted. Six-cylinder engines require a one-third turn of the crankshaft to locate the next pair of pistons at TDC. Eight-cylinder engines are rotated a one-quarter turn to bring the next pair to TDC.

Pistons in three-cylinder engines do not come to TDC in pairs. If the point-of-rock method is used to adjust valves on a three-cylinder engine, the crankshaft must be rotated 360°, or one full turn, after point-of-rock has been determined for a cylinder before that particular piston is at TDC on the compression stroke.

Some operator's manuals and most service manuals give specific procedures for positioning the engine for valve adjustment. When this information is provided, it should always be followed.

**WARNING When working on a spark-ignition engine, the coil secondary wire should be removed from the center terminal on the distributor and grounded to prevent accidental starting of the engine.**

Valve clearance is checked by placing a flat thickness leaf gauge or a go no-go thickness gauge between the end of the valve and the rocker arm (Fig. 11-19). Valve clearance is adjusted by raising or lowering the adjusting screws. Some screws are self-locking, while others have locking nuts which must be loosened before adjustment and tightened after the adjustment has been made. Be sure that you identify the exhaust valves and the intake valves and set each to the specified clearance. Each valve is aligned with the port of the manifold. The exhaust valve will be in line with the exhaust manifold connection and the intake valve aligns with the intake manifold.

As you slide the correct thickness gauge in and out of the gap, you should feel a slight drag if the clearance is correct. If the next size thickness gauge is inserted in the gap, it should be hard to insert and tight. The next smaller size thickness gauge should be loose.

**Fig. 11-19.** Checking and adjusting valve clearance with a flat thickness gauge (2) and a box-end wrench (1) On some engines, a screwdriver is also needed. (*Ford Tractor Operations*)

After adjustment has been completed, start the engine so that you can check for adequate lubrication of the rocker arms. You may have to reinstall some of the parts that were removed in order to run the engine. If hot clearance specifications are given, the valve clearances can be rechecked when the engine reaches operating temperature.

Thoroughly clean the valve cover and the gasket surfaces on the head and cover. Use a new gasket and install the valve cover. Be sure the gasket is properly positioned to prevent oil leaks. Install all remaining parts.

## THE CHARGING AND CRANKING SYSTEM

Some agricultural tractors are equipped with direct current generators. However, in recent years tractors have been equipped with many electrical accessories that require more electric current when the engine is being operated at low rpm and at idle. Since the alternating current alternator can supply a higher continuous current at low engine rpm, it is used on most newer tractors. Alternators are more compact than generators of equal output, and less maintenance is required because of their relatively simple construction.

### How the Electrical System Works

Voltage produced by the alternator or generator is stored in the lead-acid storage battery. When an engine is started, the battery supplies all the voltage to the starter and to the ignition system on spark-ignition engines. The battery also supplies some of the current used during peak operation periods. During normal operation all the current needed to operate the electrical system is supplied by the alternator or generator, and the battery is recharged.

### Servicing Alternators

Alternators require very little service. Most operator's manuals limit service to keeping the wire terminal connections clean and tight, cleaning the alternator cooling fan occasionally, and adjusting the drive belt. Most manufacturers recommend that belt tension be checked and adjusted after every 150 to 200 h of operation. Specific instructions for belt adjustment are given in the operator's manual (also, see Chap. 8 for further discussion).

**Checking Alternator Output** A volt-ampere tester (VAT) can be used to check alternator output (Fig. 11-20). When a VAT is used, follow the procedures provided by the manufacturer. The VAT should permit you to isolate the faulty component. The service manual should be used to determine recommended tests and procedures.

**Fig. 11-20.** A voltampere tester (VAT) is used to check alternator output. (*Photo by Western McCoy*)

When all tests and checks have been made and it is determined that alternator service is needed, the alternator should be removed and disassembled for further testing. Be sure to disconnect the battery ground cable before alternator wires are removed.

### Alternator Operating Principles

Since most new engines on tractors and other equipment are equipped with alternators, anyone planning a career as a service person should be able to test and repair them. In order to do this, it is important to know the parts of an alternator and understand their function (Fig. 11-21).

**The Rotor Assembly** The alternator rotor assembly consists of a wire coil wrapped around an iron core. The coil is enclosed in two interlocking soft-iron sections and supported by a shaft. The ends of the coil are attached to two slip rings that are mounted on one end of the rotor shaft.

Two small carbon brushes ride on the slip rings. One brush is grounded, and the other is insulated and attached to the field terminal on the alternator slip ring end frame. The field terminal is connected through the regulator to the ignition switch.

When the ignition switch is turned on, a small amount of current flows from the battery through the regulator to the insulated brush. This current flows from the slip ring, into the coil, and out through the other slip ring to the grounded brush.

The current flowing through the coil creates a magnetic field and causes each finger on the rotor to become a magnetic pole. The rotor poles do not retain magnetism, and battery current must flow through

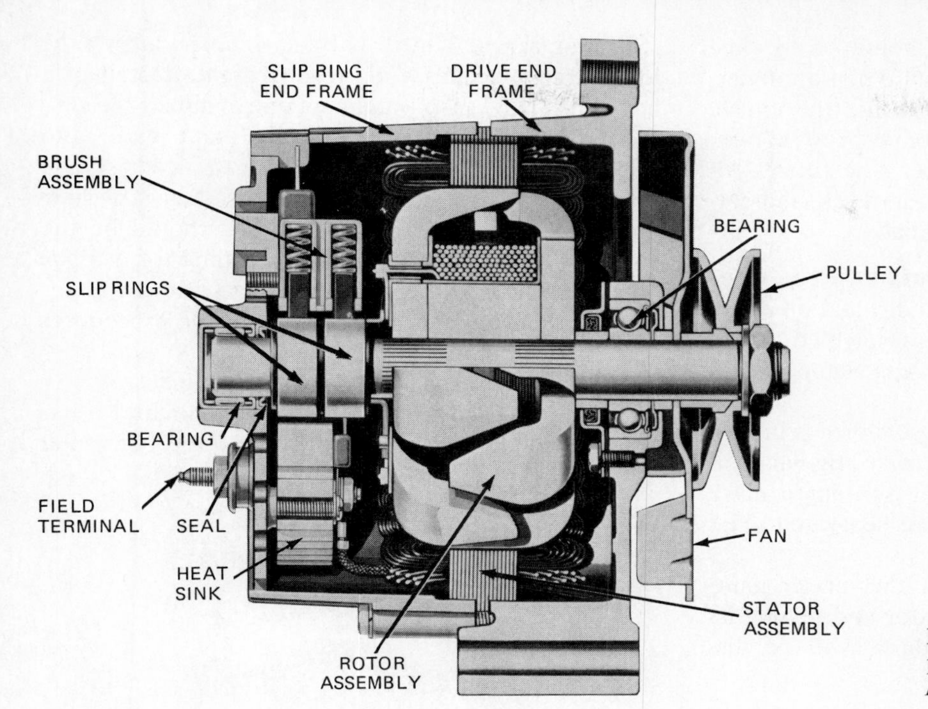


**Fig. 11-21.** A cutaway view shows the parts of an ac alternator. (*Delco Remy*)

the rotor coil before the alternator will charge. An alternator will not charge if the battery is completely discharged.

## The Stator Assembly

The stator assembly is a laminated soft-iron ring that has three groups of coils or windings held in slots (Fig. 11-21). One end of each stator winding is connected to a positive and negative diode. The other ends of the stator windings are attached together in a delta- or a Y-connection. The Y-connection produces a higher voltage with less amperage than the delta and is usually used on modern tractors and agricultural machines. The alternator produces three-phase alternating current.

**Diodes** Diodes are electrical devices that have the ability to pass electric current in one direction and resist current flow in the other direction. Three negative diodes are mounted in a heat sink or bracket that is attached to the slip ring end frame. Three positive diodes are mounted in an insulated heat sink that is also attached to the slip ring end frame. The diodes are connected and serve as a rectifier assembly that changes the alternating current produced in the stator windings to direct current.

Since diodes will only let current flow in one direction, they prevent the battery from discharging through the alternator when the tractor is not running. This eliminates the need for a cutout relay in the charging system.

**Service Precautions** Most operator's manuals list several service precautions that should be followed to prevent damage to the charging system when service is performed or repairs are made. It is important that these precautions be followed.

The battery ground cable *must* be removed before any part of the electrical system is serviced. All other electrical connections must be made before the battery ground cable is reconnected.

Battery polarity must be correct. Always be sure that the ground polarity of the battery and the alternator is the same. Most newer agricultural tractors equipped with alternators have negative ground systems, but check the operator's manual to be sure.

Do not short-circuit any terminals of the alternator or regulator. Any artificial circuit set up by purposely grounding or short-circuiting any terminals can cause serious electrical malfunctions.

Never attempt to start an engine or turn the ignition switch to the run position while a battery charger is hooked to the battery. It is good practice to disconnect the battery when charging or jump-starting another machine.

*Do not polarize an alternator!* It is unnecessary to polarize the alternator since the alternating current developed within the alternator is converted to direct current by the diodes.

Never operate the alternator on an open circuit. When no electric load is connected to the alternator, it can build up high voltages which can be extremely dangerous to anyone who might touch the alternator or the battery terminal.

*Check the condition of the battery before checking other electric components!* A short-circuited battery, or badly sulfated battery or a battery with a broken post will produce numerous problems. All terminal connections on wires and cables should be checked to be sure they are clean and tight. When an undercharged battery is found, try to locate the cause. Check for accessories that may have been left on. Check the alternator belt for proper adjustment. Alternators are harder to turn than generators, and correct drive belt adjustment is essential.

**Bench Testing and Service** Remember that it is important that all external dirt and oil be removed before any component is disassembled. Clean the alternator by wiping it with a rag damp with a suitable solvent.

The location of the end frame assemblies is important. They must be reassembled in their original position. This can be accomplished by scribing a mark across both frames before they are separated (Fig. 11-22).

Follow the procedure outlined in the service manual when disassembling an alternator and testing its components. General test procedures will be discussed here.

**Testing the brush assembly** Visually inspect the brush assembly. If the brushes show any visual defects or are worn to an exposed length of ¼ in [6.35 mm] or less, replace the brush assembly. Also check the insulated brush for grounding. An ohmmeter or test light can be used to make this test. When a test light is used, the lamp should not light. An ohmmeter should show a high reading. The grounded brush should have continuity to the bracket but not to the insulated brush.

**Testing the diodes** Diodes are responsible for many of the charging system problems in the alternator. Many times diodes are ruined by improper service to the charging system, as discussed previously. Faulty diodes may produce a humming or growling sound while the alternator is running. Usually the alternator will also have a low output.

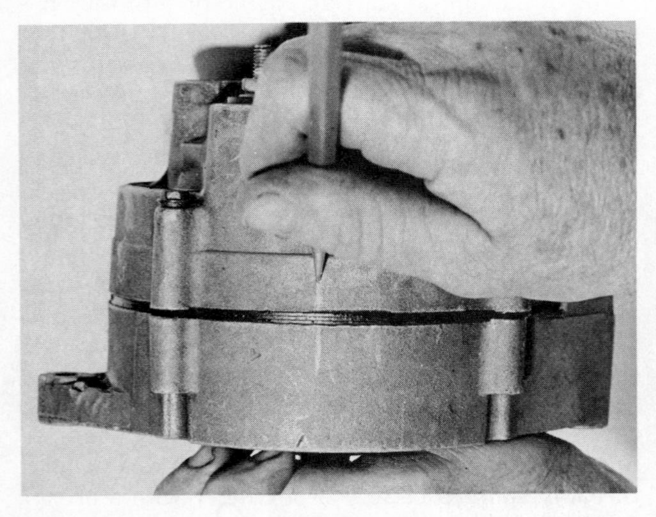

Fig. 11-22. Alternator end frames should be scribed before disassembly. (*Photo by Westen McCoy.*)

Diodes can be tested with an ohmmeter using the lowest-range scale. When testing diodes installed in heat sinks (Fig. 11-23), connect one ohmmeter lead to the diode case and the other to the diode lead. Note the ohmmeter reading and reverse the leads. Good diodes will give a high and a low reading. The high and low readings for all the diodes should be the same. A slight difference in the reading for a diode means that it is probably defective and should be replaced. Consult the service manual for procedures for replacing diodes.

Some alternators use a diode trio (Fig. 11-24). An ohmmeter is used to check the diode trio after it has been removed from the end frame. One ohmmeter

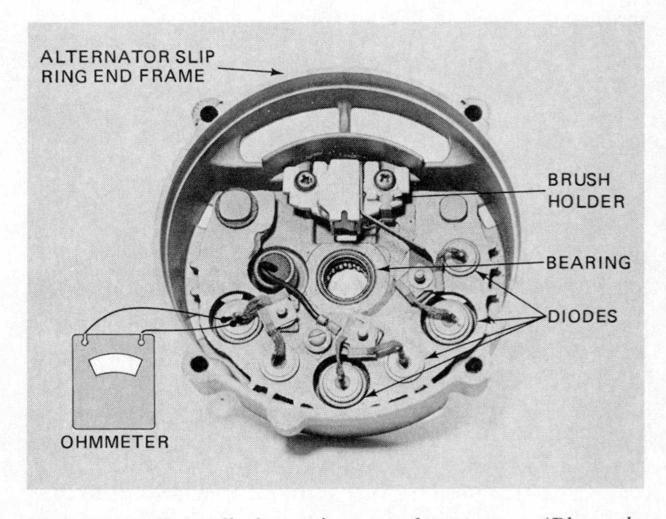


**Fig. 11-23.** Test diodes using an ohmmeter. (*Photo by Westen McCoy.*)

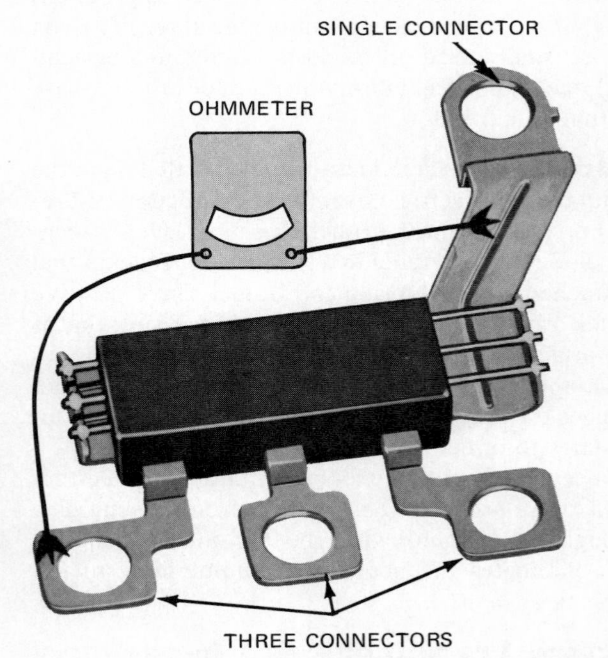


**Fig. 11-24.** Testing a diode trio requires six meter readings. (*Delco Remy.*)

lead is connected to the single connector, and the other is connected to one of the three connectors. They are then reversed. This is repeated for all three connectors. The high and low resistance readings should be the same for all three connectors.

Some alternators have a grounded heat sink and an insulated heat sink connected to the output terminal (Fig. 11-25). This is called a *rectifier bridge*. To check the rectifier bridge, connect the ohmmeter to the grounded heat sink and to one of the three tabs in the center of the insulated bar. Note the reading and connect to the other two tabs. Reverse the ohmmeter leads and repeat the check. If one or more of the readings for one diode are the same, the rectifier is defective. Now repeat the above procedure and check the insulated heat sink in the same manner. Twelve checks should be made in all.

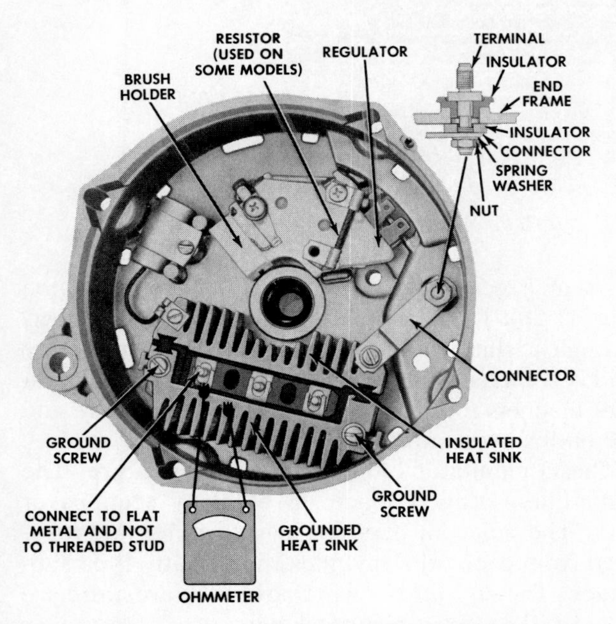


**Fig. 11-25.** Testing heat sinks or rectifier bridge requires twelve meter readings. (*Delco Remy.*)

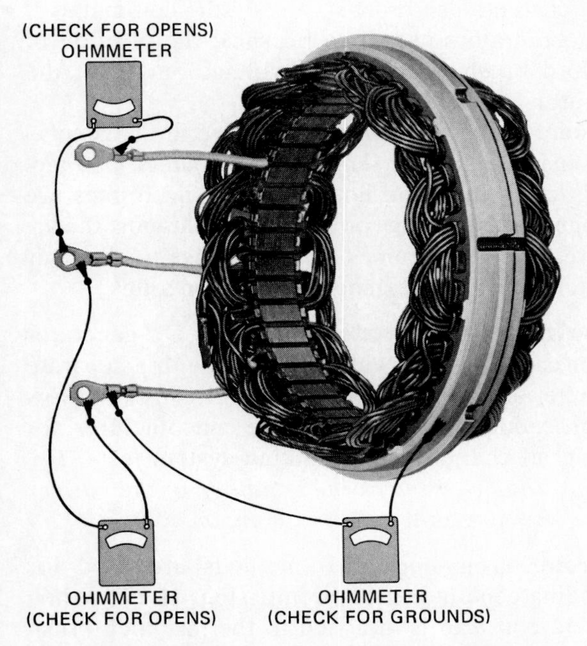


**Fig. 11-26.** Test an alternator stator for open and grounded circuits. (*Delco Remy.*)

**Testing the stator** A visual inspection of the stator assembly should be made (Fig. 11-26). Short circuits can usually be detected because the windings and insulation will be discolored. A 110-V test light or ohmmeter is recommended for testing the stator.

To check the stator windings for grounds, connect a test light or ohmmeter from a stator winding connector to the stator ring. The lamp should not light, and an ohmmeter should show a high reading. This one test will check all three legs of the windings. The test may vary on some stators, and so the service manual should be checked for recommended procedure.

The ohmmeter or test light must be connected from the center stator winding connector to each of the outside winding connectors to check for open windings (Fig. 11-26). If the light fails to light or the ohmmeter reading is high, the windings are open and the stator is defective. Remember that the test procedures may vary on some stators.

**Testing the rotor** To check the rotor for an open circuit, connect a test light or ohmmeter to each slip ring (Fig. 11-27). The test light should light, and an ohmmeter should show a low reading.

When checking a rotor for grounding, place one lead on one of the slip rings and the other on the rotor shaft. The test light should not light, and if an ohmmeter is used, it should have a high reading.

The slip rings must be smooth, round, and clean. Dirty slip rings can be cleaned using No. 00 sandpaper or 400-grain polishing cloth. *Never use emery cloth on electrical system components*. The metallic

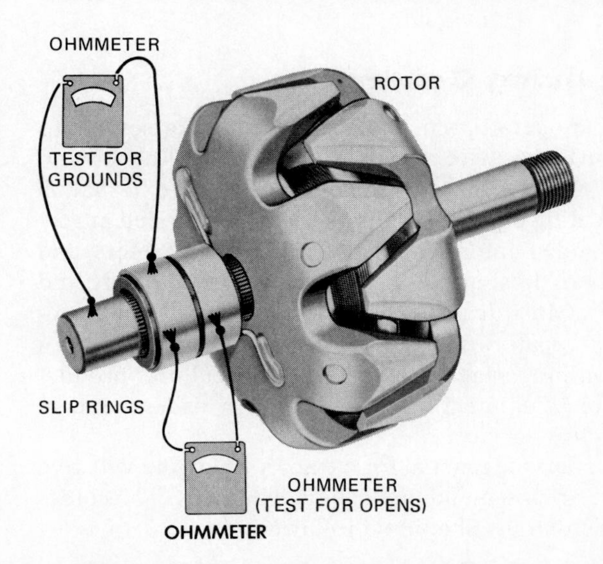


**Fig. 11-27.** Test an alternator rotor for open and grounded circuits. (*Delco Remy.*)

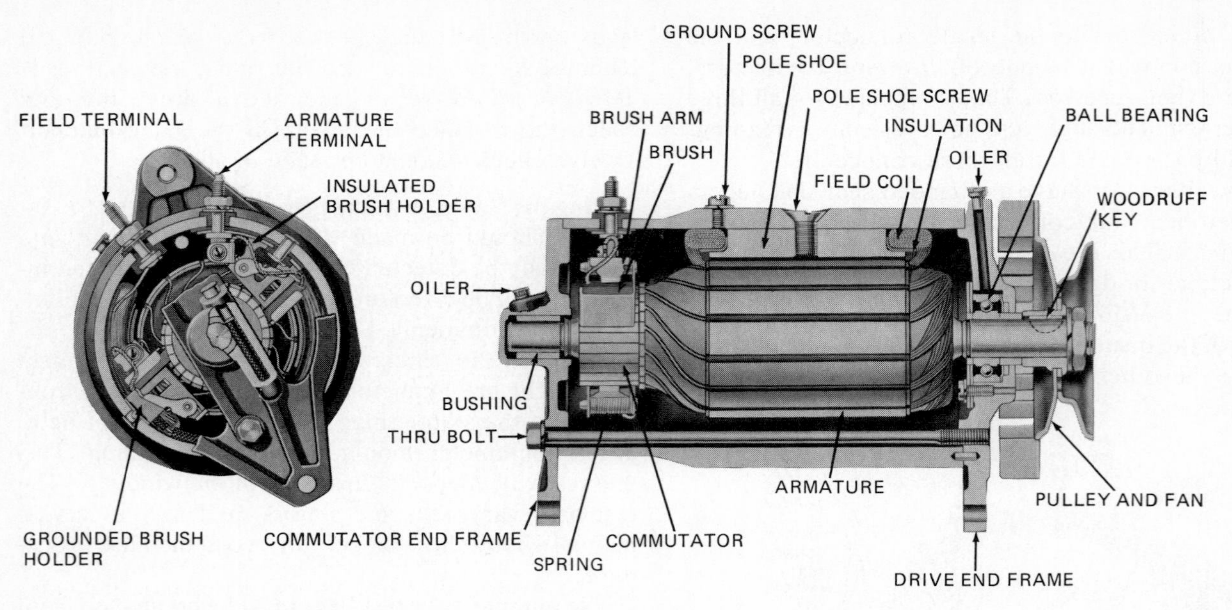


**Fig. 11-28.** A cutaway view shows the parts of a dc generator. (*Delco Remy.*)

particles from the emery cloth may cause a short circuit. If the slip rings are out-of-round or scored, they must be turned. This must be done on a lathe, and it is desirable to polish them after they have been turned.

**Alternator Bearings** Check the bearings in the end frames. The service manual will give recommendations for service and lubrication. It will also give the procedures for removing the bearings and for installing new ones.

**Alternator Reassembly** Reassemble the alternator in the reverse order of disassembly. Brushes are held up with a retaining pin to facilitate assembly. The pin must be removed after the alternator is assembled and the frame bolts are tightened. It is important that you follow the specific directions given in the service manual.

## Servicing Generators

Most operator's manuals limit generator service to keeping the wire terminal connections clean and tight. Some generators are equipped with external oil lubricating cups, and these should be serviced at recommended intervals. However, many farmers and ranchers do some of their own generator testing and repair. Most dealer shops and repair centers do generator repair work. Therefore, it is important for a person preparing for a career in agricultural production or equipment service to have a background in generator service.

The service manual for a tractor or engine will give service recommendations and techniques. These recommendations should be followed.

**Generator Operating Principles** In order to service a generator, it is important to know the parts of a generator and understand their function (Fig. 11-28). The armature is a wound rotor that turns through a stationary magnetic field. The stationary field is composed of the pole shoes, which are permanent magnets that are wrapped with copper wire and use battery voltage to increase the magnetic field.

The commutator is built onto the armature. The commutator ring is made up of many sections or bars. The adjacent bars are held together and insulated from each other by mica, hard, putty-type substance. The ends of the windings in the armature are connected to the commutator bars.

The brushes are held in brush holders and convert the alternating current generated by the generator to direct current. The brushes ride on the commutator. Some generators use three brushes; the position of the third brush regulates the voltage output of the generator.

Generator components are enclosed in the generator housing. The pole shoes, brush holders, and terminals connect to the housing. The end frames are equipped with bearings or bushings to support the armature. Most end frames have openings to allow air to enter and cool the generator as it operates.

**Checking Generator Output** DC generator output can be tested with a VAT or with a separate voltmeter and ammeter. When a VAT is used, follow the procedures provided by the manufacturer for making all charging and cranking system tests. *Remember to always check the condition of the battery before checking other electric components!*

**Generator grounding** Two methods are used for grounding generator field circuits (Fig. 11-29). When the field coil lead is attached to the insulated brush and the field coil does not ground in the generator, it is an externally grounded generator, or an *A-circuit.*

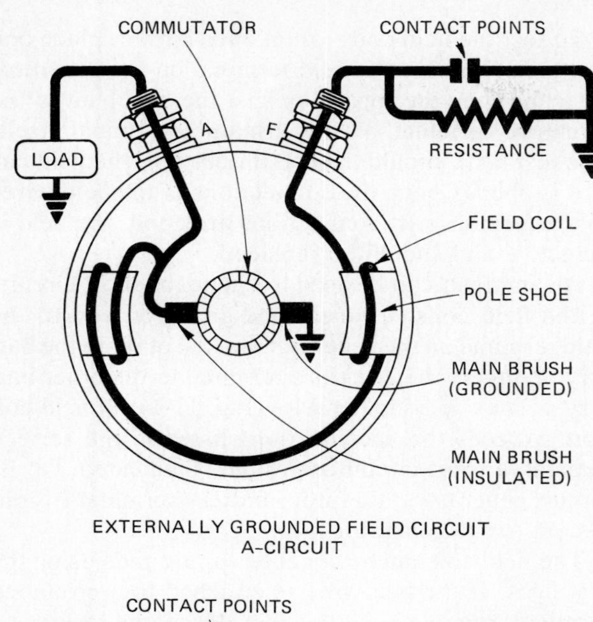


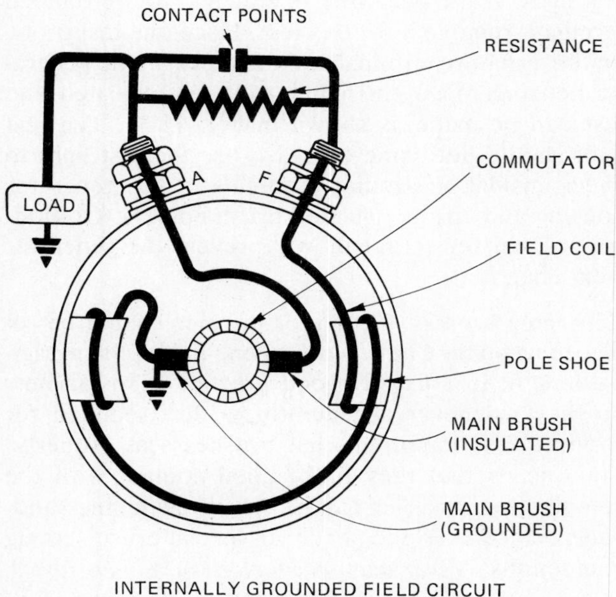


**Fig. 11-29.** Two methods are used for grounding generator field circuits. (*Delco Remy.*)

If the field coil lead attaches to the grounded brush, it is an internally grounded generator, or a *B-circuit*.

A voltmeter can be used to determine whether the generator field circuit is internally or externally grounded. To do this, disconnect the field wire from the generator. Leave it disconnected. Be careful not to let it ground to the tractor or generator. Connect the voltmeter from the generator field terminal to ground. Start the engine and run it at fast idle. If a volt reading registers on the voltmeter, it is an externally ground generator, or an A-circuit. No voltage reading indicates that it is an internally grounded generator, or B-circuit.

**Checking output on A-circuit generators** Output on A-circuit generators can be checked by removing the wire from the voltage regulator battery terminal (BAT) and connecting an ammeter in series between the terminal and the wire (Fig. 11-30). Disconnect the wire leading from the voltage regulator to the field terminal on the generator and use a jumper wire to ground the generator field terminal. Connect a voltmeter from the generator armature (ARM or GEN) terminal to ground.

Start the engine, turn on all electrical accessories, and increase engine rpm until a reading of about 35 amperes (A) (or the specified ampere reading) is obtained on the ammeter. A resistance may have to be placed across the battery terminals to attain this load on the battery. This can be done with a VAT.

If no amperage is indicated, check the voltage reading on the voltmeter. When the voltage reading is higher than normal battery voltage, a faulty cutout relay is indicated and the voltage regulator should be serviced or replaced. The cutout relay and voltage regulator are discussed later in this chapter. If there is no amperage reading and the voltmeter reading is lower than normal battery voltage, the generator is faulty and needs service.

**Checking output on B-circuit generators** Output on B-circuit generators can be checked (Fig. 11-31) by disconnecting the armature and field wires from the generator and placing a jumper wire between the terminals. Connect the ammeter positive lead to the armature terminal. The armature terminal is sometimes called the generator terminal.

Start the engine, and while it is running at idle, connect the negative ammeter lead to the positive battery post. This can be done by disconnecting the battery wire at the voltage regulator and attaching the ammeter lead to it. Increase engine speed to about 1500 revolutions per minute (r/min). The generator should be producing rated output at this speed. This

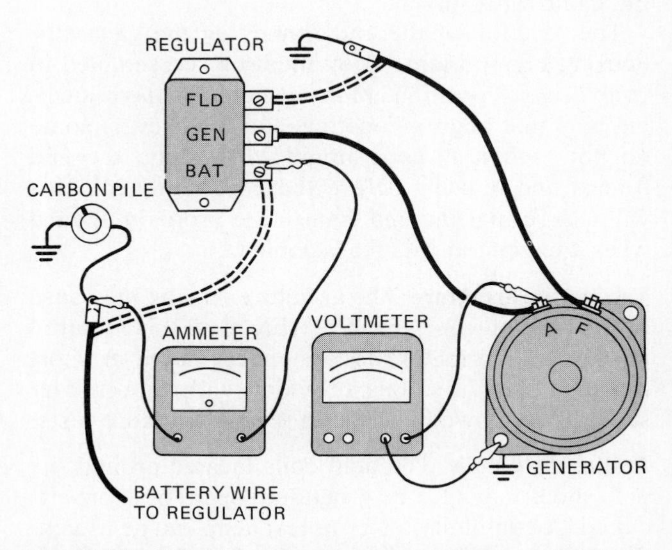


**Fig. 11-30.** Checking output on externally grounded A-circuit generators.

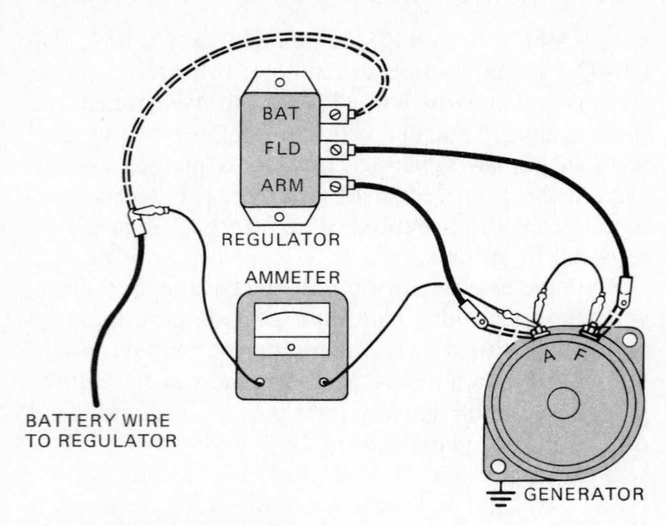


**Fig. 11-31.** Checking output on internally grounded B-circuit generators.

test should be conducted quickly, and the engine should be shut off immediately.

When no generator output is indicated, there are several conditions that should be checked. Check for worn or sticking brushes. There should be slight arcing or sparking at the brushes when the engine is running. Also check for corrosion or dirt on the commutator.

Check all the wires and terminal connections. Loose or corroded connections or grounded wires or terminals may be the reason the generator is not charging.

**Bench Testing and Service** When all tests and checks have been made and it is determined that the generator output is below specifications or that the generator is not charging, it should be removed and disassembled for further testing and service. All external dirt and oil should be removed before it is disassembled. Clean the generator by wiping it with a rag damp with solvent.

The location of the end frames on the generator housing is important. They must be reassembled in their original position. Many generators have locating pins that align in locating slots. However, some do not, and it is good practice to scribe the end frames and housing before they are removed. This will assure that the end frames are properly aligned when the generator is reassembled.

**Testing the armature** The armature may be the cause of the problem. A "growler" (Fig. 11-32) is required to check the armature for grounded, open, or short circuits. The directions provided with the growler should be followed when performing armature tests.

**Testing field coils** The field coils located around the pole shoes may also be grounded, open, or short-circuited. A continuity tester or test lamp can be used to check the condition of the field coils (Fig. 11-33).

To test the field coils for an open circuit, place one test probe against the field terminal on the generator case and the other probe against the field lead to the armature terminal or the armature terminal itself. The test lamp should light. If it does not, the field circuit is open. Check the connections of the field wires to the terminals. If connections are good, the field is defective and should be replaced.

An ammeter can be used to check for short circuits in the field coils. Connect one ammeter lead to the field terminal on the generator and the other to the battery. Connect the armature terminal to the other battery post using a jumper wire (Fig. 11-34). If field coil flow exceeds the specifications listed in the service manual, a short circuit or ground is indicated. On B-circuit generators, the other battery terminal is connected to the generator frame.

The field coil can be checked for grounds using the test light. If the field wire is attached to a grounded terminal, remove it for this test. Place one test probe on the armature terminal and the other on the generator housing. When the field terminal is insulated, the test can be made as shown in Fig. 11-33. The test light should not come on. Also use the test light to check insulation around terminals in the generator housing and check insulated brush holders. Grounding at insulated terminals will prevent the generator from charging.

**Generator brushes** Generator brushes should be replaced when they have worn to one-half of their original length. It is usually good practice to install new brushes whenever a generator is disassembled for service. It is important that brushes seat properly. This means that they make good contact with the commutator. Brushes can be seated using fine sandpaper, silicon carbide paper, or special brush seating compounds. *Never use emery cloth on generators.*

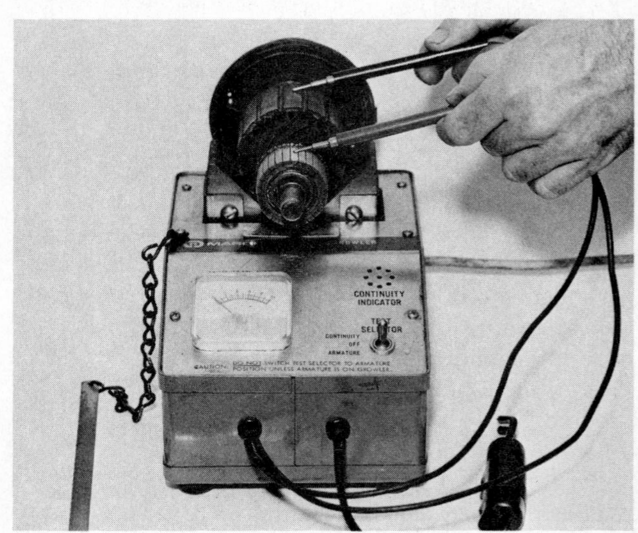


**Fig. 11-32.** Use a growler to test a generator armature for ground. (*Photo by Westen McCoy.*)

(a)

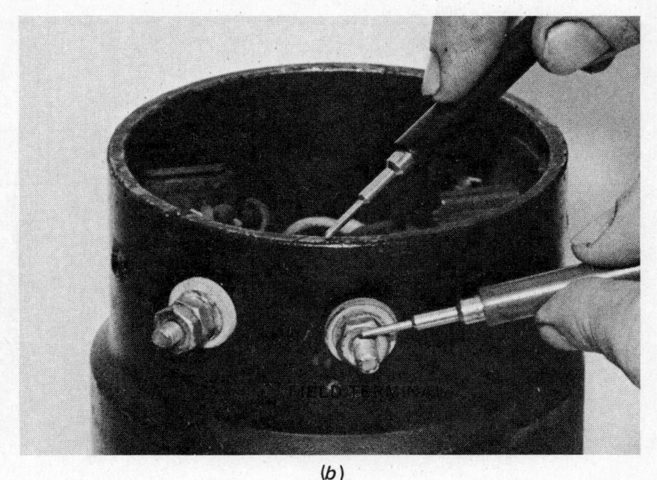


(b)

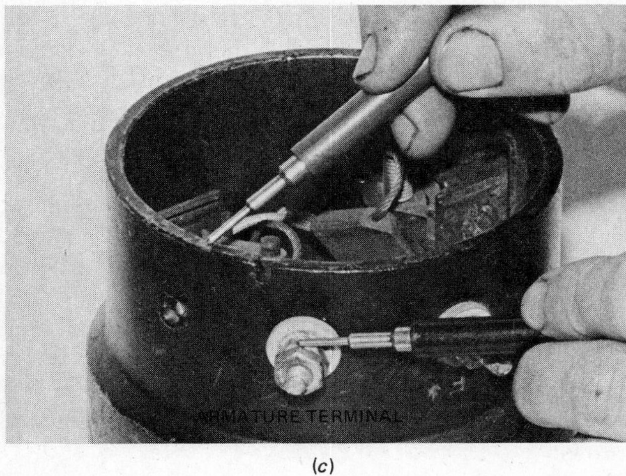


(c)

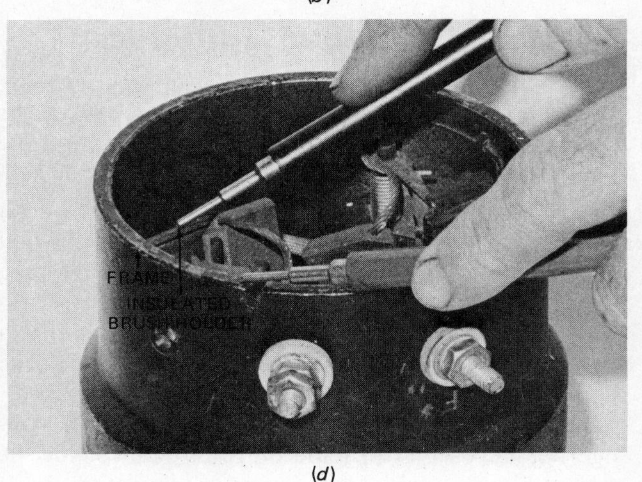


(d)

**Fig. 11-33.** A test lamp can be used to check the condition of the generator field circuits. (*a*) Testing field coils for an open circuit. (*b*) Testing field coils for ground. (*c*) Testing armature terminal for ground. Field wire should be removed. (*d*) Testing insulated brush holder for ground. (*Photo by Westen McCoy.*)

**Generator assembly** When the generator is reassembled, the bearings or bushings should be lubricated. Be careful not to get lubricant on any internal parts. Install the generator after reassembly and recheck the charging system.

The tests given here are general. Check the service manual for more complete testing information and procedures.

**Polarizing Generators** When any wires are disconnected from the starter, generator, or voltage regulator while service is being performed, the generator must be polarized before starting the engine. Polarizing is the process of sending a quick surge of current through the field windings of the generator to ensure that the pole shoes assume the correct polarity or that current flows in the same direction as in the battery.

On A-circuit generators (Fig. 11-35), a jumper wire is placed momentarily between the BAT terminal and the GEN or ARM terminal on the voltage regulator. This is done after all terminals and wires are reconnected.

On B-circuit generators disconnect the field (FLD) wire from the voltage regulator and momentarily touch it to the BAT terminal on the regulator. The wire is then reconnected. This is done after all terminals and wires have been reconnected.

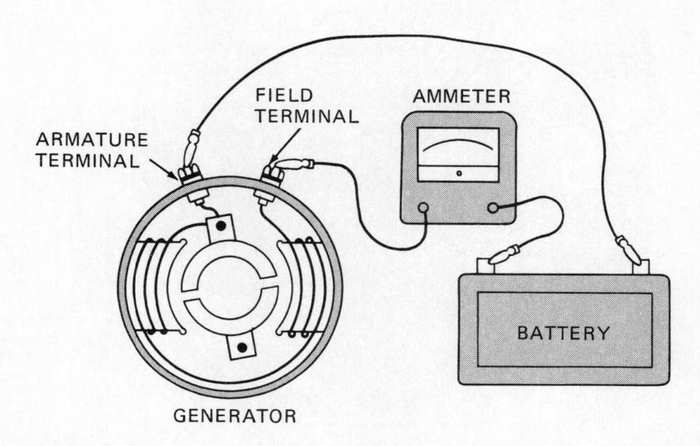


**Fig. 11-34.** An ammeter can be used to test for a short circuit in a field coil.

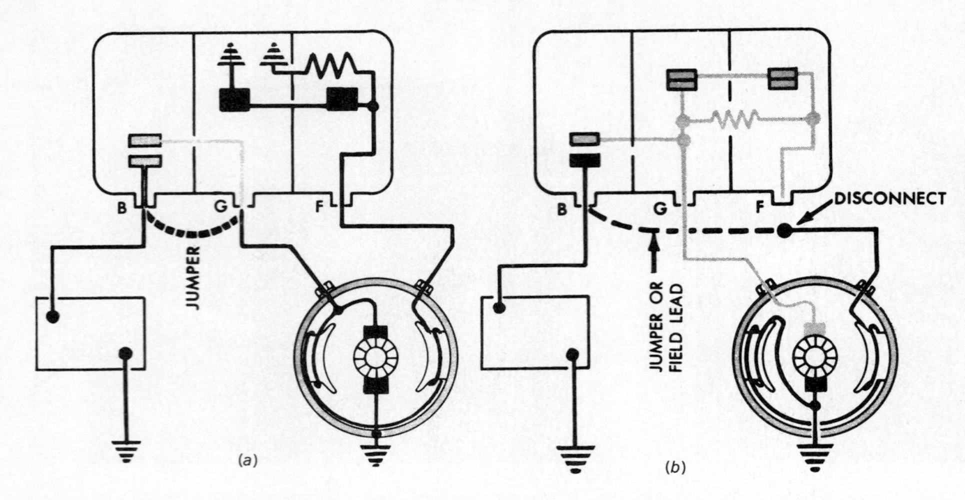


**Fig. 11-35.** Polarizing dc generators. (*a*) To polarize A-circuit generators, touch a jumper wire momentarily between the battery terminal and the generator or armature terminals on the voltage regulator. (*b*) To polarize B-circuit generators, disconnect the field wire from the voltage regulator and momentarily touch it to the battery terminal. (*Delco Remy.*)

## Servicing Voltage Regulators

DC generator and ac alternator charging systems are both equipped with voltage regulators. The voltage regulator controls the charging system and protects the other components (Fig. 11-36).

Some tractors with generators are equipped with cutout relays that open the battery circuit only when the engine is not running or when the battery voltage exceeds the generator output voltage. Tractors with this type of regulator are equipped with three-brush generators so that the generator voltage output can be adjusted.

Most tractors equipped with generators have voltage regulators that provide constant control for the generator. In addition to a cutout relay, they have both a voltage regulator that controls the maximum voltage output of the generator and a current regulator that limits maximum current output. Some regulators have a combined current-voltage control.

**Fig. 11-36.** A 6-V voltage regulator. (*Photo by Westen McCoy.*)

Voltage regulators require little service or maintenance. The tests discussed earlier for checking the generator will determine if the regulator is defective. Recommendations in the service manual for service and adjustments should be followed.

Most tractors equipped with alternators have transistorized regulators. They may be mounted internally in the alternator or mounted externally and attached to the engine or tractor chassis. These components are usually in a sealed case, and no service or maintenance is required. Some tractors are equipped with other types of alternator regulators, and the operator's or service manual should be consulted for maintenance and service recommendations.

## Servicing Starters or Cranking Motors

The starter is very similar to the generator in construction. Its purpose and operation, however, are entirely different. The generator is turned by mechanical power and generates electric energy which is stored in the battery. The starter converts electric energy supplied by the battery into mechanical energy to crank the engine (Fig. 11-37).

The starter is designed to operate for short intervals under great overload. On many agricultural tractors, the starter is activated by a solenoid, which is an electromagnet that closes the circuit between the battery and the starter. The electromagnet also engages the pinion gear on the starter driveshaft with the ring gear on the flywheel. When the engine starts and the ignition key is returned to the run position, the pinion disengages from the ring gear. Some tractors are equipped with inertia-type starter drives which engage when the starter starts to turn and disengage when the engine speed is faster than the starter motor speed.

Since the starter operates under an extreme overload when it cranks the engine, *it should not be run for more than 30 seconds* (*s*) *at a time*. If the engine

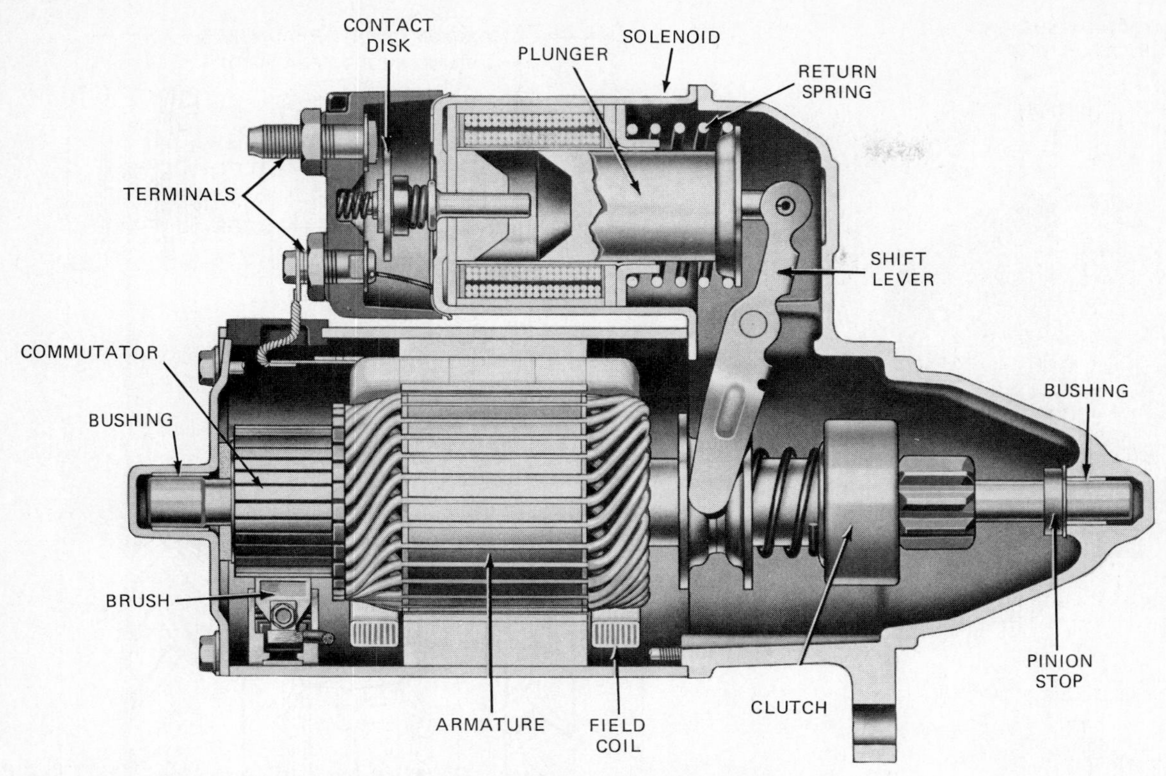


**Fig. 11-37.** A cutaway view shows the parts of a starter. (*Delco-Remy*)

fails to start in that period of time, the starter should be allowed to cool for 2 min before it is used again.

Before a starter is removed for service, check the condition of the battery. Be sure that it is fully charged and operating properly. Also check the battery posts and cable terminals to see that the connections are clean and tight. Inspect the wiring for damage and check all wire connections throughout the starter circuit. Check the battery ground cable connection to the frame of the tractor.

Inspect all switches and connect a jumper wire to bypass any switch suspected of being defective. A few minutes spent checking the system can prevent a lot of needless work and possibly reduce downtime.

A VAT can be used to check the starter and cranking system. Follow the instructions provided with the tester and compare the test results with the specifications given in the service manual.

**Bench Tests and Service** If the test indicates that the starter needs service, remove it from the engine. When it is removed, check the armature for freedom of rotation. Worn bearings, a bent armature shaft, or loose pole shoes can cause the armature to be hard to turn or to have drag. If the armature does not turn freely, the starter should be disassembled for service. Follow the disassembly procedures given in the service manual. Remember to scribe the starter case and end frames before disassembly.

When the starter is disassembled, the components can be checked with a growler and test light or continuity tester. Refer to the generator section of this chapter for similar testing procedures.

The starter drive and solenoid should be checked before the starter is reinstalled on the tractor. Starters equipped with an overrunning clutch-type starter drive should have the pinion clearance checked (Fig. 11-38). Check the service manual for specifications. Check inertia-type starter drives for visible wear such as a worn pinion gear or broken springs.

Never clean an overrunning clutch-type starter drive in solvent. It is packed with lubricant at the factory, and cleaning will damage it and cause it to fail prematurely. Inertia-type starter drives may be cleaned in solvent.

## THE ENGINE CLUTCH

A discussion of clutches is included in Chap. 7. Since clutch linkage adjustment is required on some tractors to maintain the correct clutch pedal free travel, it is mentioned here. The operator's manual should be used to determine recommended free travel, adjustment procedures, and service interval. Some manufacturers recommend that free travel be checked every 50 h of operation for maximum clutch life.

When adjustment will not correct clutch operation, the clutch disk, pressure plate assembly, and clutch release bearing will need to be removed and repaired or replaced. Read the service manual to obtain specific service procedures.

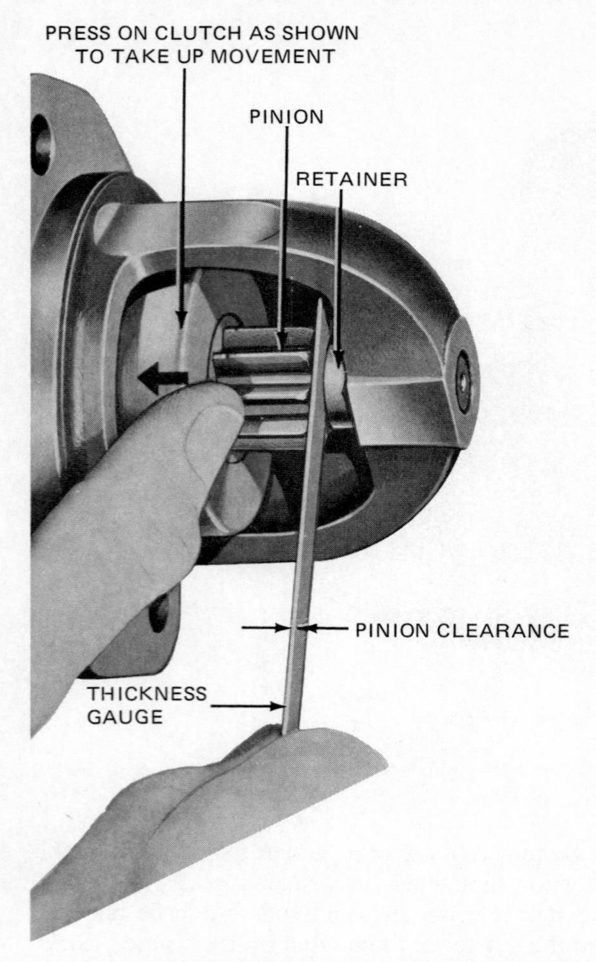


**Fig. 11-38.** Check pinion clearance before the starter is installed. (*Delco Remy.*)

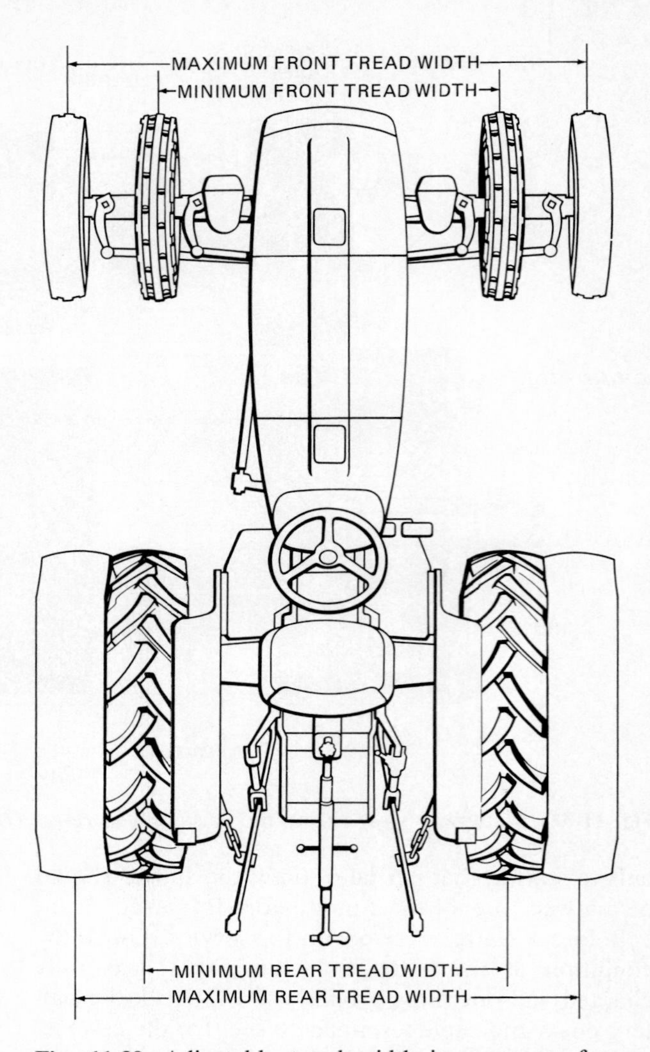


**Fig. 11-39.** Adjustable tread width is necessary for specific row crop spacing requirements. (*Allis-Chalmers Manufacturing Co.*)

## WHEEL AND BRAKE SERVICE

Servicing wheels and brakes is necessary for efficient and safe tractor operation. People who operate, service, or sell agricultural tractors should be able to adjust tread width and toe-in. They should also be able to service wheel bearings and brakes.

### Adjusting Tread Width

Most newer two-wheel-drive agricultural tractors are equipped with an adjustable front axle. Rear-wheel spacing is also adjustable, and many tractors can be equipped with dual wheels (Fig. 11-39). Front- and rear-wheel tread widths should be adjusted according to the work being done with the tractor. The operator's manual will give complete adjustment procedures and torque specifications.

⚠ WARNING **Agricultural tractors are heavy, and proper jacks, hoists, wood blocks, and jack stands are essential when adjusting wheel spacing! Check the operator's manual for safe procedures for lifting and supporting the tractor.**

### Toe-In

Toe-in refers to the adjustment of the front wheels on tractors equipped with adjustable front axles. *Toe-in* means that the distance between the wheels of the steering axle is slightly smaller at the front of the axle than at the rear. Proper toe-in is essential if the machine is to track or steer in a straight line. When toe-in is adjusted correctly, a tractor is easier to drive and there is less tire wear.

**Checking Toe-In** To check toe-in, turn the steering wheel so that the front wheels are in the straight-ahead position. Use a steel tape measure or a folding rule to measure the distance between the centerline of the tread at the front and then at the rear of the tires. The front measurement should be less than the rear measurement. The operator's manual will give the specified toe-in, which usually ranges from ⅛ to ⅜ in [3 to 10 mm].

**Adjusting Toe-In** Toe-in is adjusted by shortening or lengthening the tie rods. Tie rods are the adjustable rods that attach to the steering arms. The right- and left-hand tie rods should be adjusted the same amount so that the tractor will turn equally sharp in either direction. Follow the procedures outlined in the operator's manual when making adjustments.

## Wheel Bearing Service

Most agricultural tractors are equipped with tapered-roller bearings (Fig. 11-40). Some tractor hubs are equipped with grease fittings and are to be lubricated every 10 h of operation. Others do not have grease fittings. All front wheel bearings should be removed, cleaned, and repacked with wheel bearing grease at the intervals recommended in the operator's manual.

**Front Hub Assembly** An inner bearing and outer bearing are used in wheel bearing assemblies. The inner bearing is usually the larger of the two bearings. The bearings handle both radial and thrust loads.

A grease seal is installed in the spindle end of the hub. The seal retains the grease in the bearing and helps to keep the assembly free of moisture and dirt. Most operator's manuals recommend that the seals be replaced each time the wheels are repacked.

A thrust washer is used between the outer bearing and the spindle nut. Spindle nuts are usually held in place with a nut retainer and a cotter pin. Spindle nuts not equipped with nut retainers are usually "castle" nuts and are held in place with a cotter pin only.

The hubcap or dust cover seals the outer end of the hub. Tractors are equipped with hubcaps that bolt on, screw on, or press fit. The hubcap keeps dirt and moisture out of the assembly and should always be reinstalled after service.

**Removing Wheel Bearings** Use an appropriate jack to lift the tractor so that one front wheel is off the ground. Be sure that the tractor transmission is in park and that the brakes are set. It is good practice to block the rear wheels. Use wood blocks or safety stands to support the tractor. A jack should not be used as the only support.

Clean the dirt from around the hubcap and remove it. Remove the cotter pin and the spindle nut. Remove the thrust washer and the outer bearing. Place all parts in a clean container filled with solvent as they are removed.

The hub and wheel can now be removed. The inner bearing and seal will come off with the hub. To remove the bearing and seal, insert a soft punch or hardwood dowel pin from the front and gently tap them out of the back of the hub (Fig. 11-41). They should tap out easily. Be careful not to let them fall in the dirt.

**Cleaning Wheel Bearings** Clean the bearings in solvent or diesel fuel, *not gasoline*. Swirl the bearings in the solvent and use a stiff bristle brush to remove the grease deposits. Clean the spindle, the hub, the hubcap, and all other related parts.

It is important that all solvent be removed from the cleaned bearings and other parts of the hub assembly. Solvent left in the bearings will dilute the wheel bearing grease. Compressed air or a lint-free cloth can be used. If compressed air is used, be sure that it is dry, and do not spin the bearing. Spinning can damage the highly polished bearing surfaces causing early bearing failure. **Warning: spinning can cause a bearing to actually explode. This can cause extensive bodily injury.**

**Inspecting the Bearings** Inspect the cleaned bearings for damage, wear, rust, and corrosion. Examine the rollers and the bearing cups (races) carefully for pitting and signs of excessive heat. See Chap. 7 for a complete discussion on bearings.

When either a bearing or a cup shows damage, replace both parts. Also, replace the bearing and cup if the cage shows excessive looseness or is damaged.

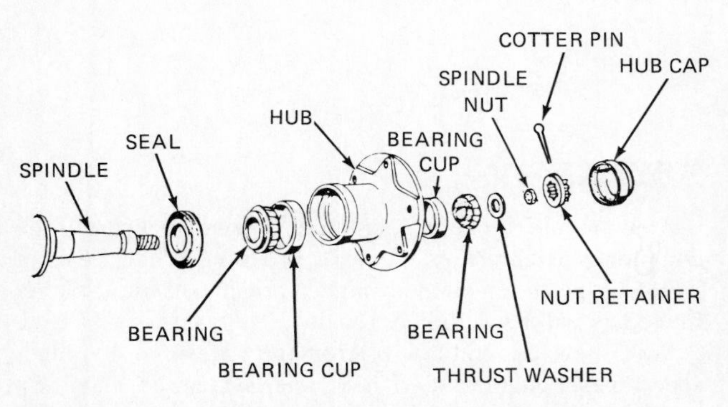


**Fig. 11-40.** The front hub assembly should be serviced at recommended intervals. (*Allis-Chalmers Manufacturing Co.*)

**Fig. 11-41.** Use a soft punch to remove the inner wheel bearing and seal from a hub. (*Photo by Westen McCoy.*)

**Packing Wheel Bearings** Use the grease recommended in the operator's manual to repack the bearings. Place a small amount of grease in the palm of your hand. Hold the bearing with the small end up and work the grease into the open end of the bearing. Continue to work the grease into the bearing until it forms a ring around the cage (Fig. 11-42). Rotate the bearing and continue to work grease into it until it is full. Larger bearings will hold several ounces of grease.

When the bearing is full of grease, coat the outer part of the bearing with a light coat of grease. Wrap the bearing in clean paper, cloth, or plastic and pack the other bearing. Commercial bearing packing equipment is used in many shops.

Coat the inside of the hub and the spindle with a light coat of grease. Replace the inside bearing in the hub and install a new grease seal. Place the hub on the spindle and install the outer bearing, thrust washer, and retainer nut.

**Adjusting Wheel Bearings** The procedure for adjusting wheel bearings is not the same in all operator's manuals. Some use a specific torquing sequence, and others recommend that adjustment be done by feel. Check the operator's manual for the recommended procedure.

One method is to tighten the spindle nut until there is pressure on the nut. This is to seat the bearings. Then loosen the nut until the wheel turns freely and retighten the nut until the wheel turns with a slight drag. Loosen the spindle castle nut until the nearest slot is aligned with a hole in the spindle and insert a new cotter pin. Be sure that the cotter pin is properly bent. Replace the hubcap.

Some tractor wheel assemblies may be different than the one discussed here, and service varies. When a different assembly is serviced, follow the recommendations and procedures given in the operator's manual.

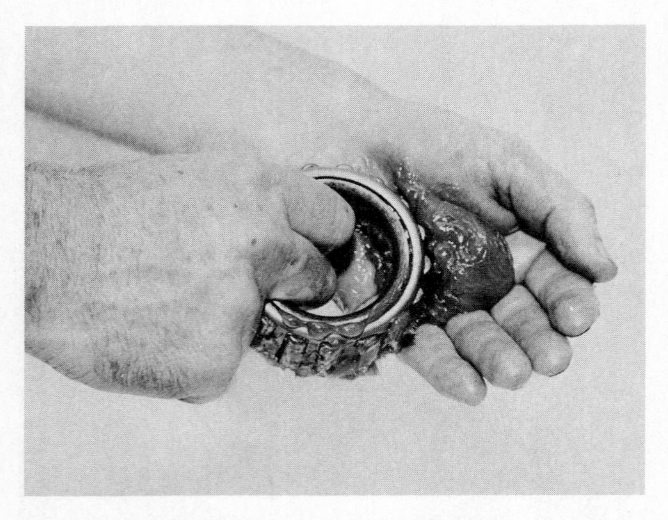

**Fig. 11-42.** When packing a wheel bearing, be sure the grease is forced into the cage. (*Photo by Westen McCoy.*)

## Brakes

Two-wheel-drive agricultural tractors are equipped with individual brakes for each drive wheel. Brakes are used to stop the tractor and to assist when turning. Both brakes are used when stopping and should be locked together when the tractor is operated at high speeds.

When the tractor is being operated at lower speeds, the brakes should be unlocked and used individually to assist in making turns. The brakes are actually a part of the steering system when a tractor is operated under field conditions.

**Types of Brakes** Agricultural tractors have been equipped with many types of brakes located in different parts of the drive train.

**Mechanical Brakes** Mechanical brakes may be disk brakes, shoe brakes, or band brakes. They are controlled by rods and cables. Mechanical brakes must be adjusted so that the pedals will exert equal pressure at a specified height.

**Hydraulic brakes** Hydraulic brakes may be shoe or disk brakes. They are controlled by hydraulic oil supplied by the brake master cylinder to the slave (wheel) cylinders, when the brake mechanism is actuated.

**Power brakes** Power brakes are disk-type brakes that are operated by hydraulic oil supplied by the tractor hydraulic pump. The brake pedal is attached to a control valve. When the pedal is depressed, hydraulic oil flows to the brake piston and forces the brake piston and pressure plate against the revolving disk.

**Brake Adjustment** Proper brake adjustment is important. Brakes are essential for safe operation of an agricultural tractor. It is important that they receive the service and maintenance recommended in the operator's manual.

## SUMMARY

Adjustments, tuneups, and minor repairs are a part of the complete preventive maintenance program. When a tractor is not properly maintained, higher repair costs and higher fuel costs will be the likely result.

Most new agricultural tractors are powered by high-speed diesel engines. The diesel fuel system consists of a fuel tank, transfer pump, fuel filters, injection pump, and nozzles or injectors. Fuel used in a diesel engine must be kept clean and free from moisture. Proper handling of fuel and fuel filter service at recommended intervals are essential for satisfactory performance.

Gasoline and LP gas tractors are equipped with electric-ignition systems to ignite the air-fuel mixture in the combustion chamber. Spark plugs and breaker points require frequent service and adjustment to keep the system in good operating condition.

The charging system supplies voltage that is stored in the battery. The two basic types of charging systems are the ac alternator system and the dc generator system. The generator system requires more service than the alternator system. Most new tractors are equipped with alternators.

Starters or cranking motors deliver the power to crank the engine. The engine is turned by the starter drive pinion gear which engages the ring gear on the flywheel. Starters and generators are similar units, but generators convert mechanical power to electric power while starters convert electric power to mechanical power.

Wheels are an important part of an agricultural tractor. Tread width is adjusted according to the work being done, and toe-in should be checked and adjusted at regular intervals. Front-wheel bearings should also be serviced as recommended to prevent failure.

Brakes are essential for safe operation. They are used to stop the tractor and to assist when turns are made at slow speeds.

It is recommended that a service manual be purchased for each machine if repairs are to be done on the farm or ranch. All dealerships and specialized repair shops use the service manual. The service manual will give the specific details, procedures, and adjustment specifications needed to do a good job.

## THINKING IT THROUGH

1. List the parts of a diesel fuel system.
2. How is opening pressure adjusted on diesel injection nozzles?
3. Describe how a diesel fuel system is bled.
4. List the parts in an electric-ignition system.
5. What is the purpose of the coil? the condenser? the breaker points?
6. What is dwell, and how is it adjusted?
7. When should a condenser be replaced?
8. How should the coil be wired to attain correct coil polarity?
9. Why is it important that the electrode gap be properly adjusted on spark plugs?
10. Explain how ignition timing is set on a gasoline tractor.
11. What is valve clearance? How is it adjusted?
12. List several advantages of alternators.
13. How is alternating current converted to direct current in an alternator?
14. Explain how diodes are tested.
15. How is alternating current converted to dc direct current in a generator?
16. Explain how A-circuit and B-circuit generators are polarized.
17. Why shouldn't an overrunning clutch-type starter drive be cleaned in solvent?
18. Why is tread width adjustable on tractors?
19. Why is it important that toe-in be adjusted properly?
20. Explain how wheel bearings are cleaned.
21. How are wheel bearings adjusted?
22. Why are two-wheel-drive tractors equipped with individual brakes on the rear wheels?

# CHAPTER 12 TESTING A TRACTOR ENGINE AND MAKING SERVICE RECOMMENDATIONS

## CHAPTER GOALS

In this chapter your goals are to:

- Use engine test equipment
- Evaluate the results of the tests performed
- Determine the service needed on a tractor

Almost anyone can be a parts changer. A parts changer replaces parts by a trial-and-error method until the problem disappears. The term *disappears* is used, because the cause of the problem may not have been found and it may *reappear* after only a few hours of operation. This process is often time-consuming and usually results in excessive parts, labor, and downtime costs for the tractor owner. There is little demand in the power and machiney industry for parts changers.

Test equipment is as important as a good set of tools. Test equipment is used to determine the condition of a tractor. This may be done to diagnose problems when service is needed or to establish the value of a tractor that is bought, sold, or traded.

The importance of throughly testing a tractor before service is recommended or performed cannot be overemphasized. A service worker should know how to operate test equipment, be able to evaluate the results of the tests as they are conducted, and be able to use the results as the basis for recommending and performing service. There are good employment opportunities for people who can accurately diagnose the condition of a tractor.

## OWNERS' AND OPERATORS' COMMENTS

It is good service practice to obtain as much information as possible about a tractor before any service is recommended or performed. If possible, talk to the owner or operator of the tractor.

Owners' and operators' comments can be of great value when diagnosing engine problems. They can provide information on maintenance, performance, and hours of operation. This information may indicate that service is needed because of normal wear that occurs after extended hours of operation.

The cause of the problem may be traced to improper maintenance or operation. When this is found, the owner should be informed and the recommended maintenance or operation procedures should be fully explained. It is very important that this be done. Failure to do so will probably result in premature failure after the service has been performed. This can create poor customer relations because the owner expects the tractor to provide good reliability and performance after it has been serviced.

## VISUAL INSPECTION

A close visual inspection may reveal problems that can be detected by sight, by unusual noises, or by unusual odors. It is good service procedure to perform a visual inspection *before* a tractor is cleaned.

### Oil Leaks

Look for oil leaks around the oil pan, drain plugs, valve covers, seals, and filters. External oil leaks can cause serious problems. A small leak can cause the oil level in the crankcase to drop below the safe operating level and may cause the engine to fail.

### Coolant Leaks

Look for damaged or loose hoses. Check the radiator and the radiator cap. Also, check the seal in the water pump and all gaskets where the cooling system components bolt to the cylinder head and block. Examine the cylinder head gasket and the expansion plugs in the cylinder block and head. If the coolant level is low, the engine may have overheated causing the problem.

### Fuel Leaks

Look for fuel leaks at the tank, the lines and connections, the filters, and the fuel pump. On diesel tractors, check the injection lines and the nozzles for leaks. Check for restrictions or water and other foreign matter in the fuel system.

### Air Intake System

Look for damaged or loose hoses and air tubes. Check the air cleaner for proper maintenance and operation. Check the precleaner. A failure in the intake system can cause extensive damage to an engine in a very few hours of operation.

### Clutch

Check the clutch pedal free travel and listen for unusual noises in the clutch. If the engine is removed for overhaul, the clutch should be repaired or replaced if it is defective.

## Electrical System

Check the wires and cables for damage and loose or corroded terminals. See if all wires are connected. Check the battery, starter, and alternator or generator.

## Ignition System

Look for damaged wires and loose or corroded connections. Check the distributor cap and the breaker points. Check the spark plugs for cracked insulators and external dirt. Spark plugs should be removed to check the condition of the electrodes. Check the coil for corrosion and dirt on the tower and inspect the high-tension wire and the boots.

## Engine Running

Listen for any unusual or loud engine noises. Look at the exhaust smoke. The color and amount of smoke may indicate major problems. Any unusual odor that is detected should be considered. Check all gauges to see if they are operating correctly.

## Other Checks

Check the condition of the coolant. Note whether rust or oil is present in the coolant. Check the steering system, the tires, and the fan belts. Check the oil in the crankcase. If the oil has water in it, it will have a milky color. Keep a list of any unusual conditions found.

# INSTRUMENT TESTS

The PTO dynamometer (Fig. 12-1) is used extensively to test, tuneup, and break in tractors. Dynamometers use a hydraulic pump and a hydraulic motor to load a tractor. They hook directly to the tractor power takeoff (PTO) shaft.

**Fig. 12-1.** PTO dynamometers are used to test, tuneup, and break in tractors. (*AW Dynamometer, Inc.*)

Dynamometers are equipped with load controls that enable the load to be varied. They are equipped with gauges that give PTO rpm and power output. The dynamometer can be used to check PTO power, which is the fundamental measure of engine efficiency. Fuel consumption can also be checked, and a complete dynamometer test will give other results that are useful in determining the condition of the engine.

Other instrument tests are also used. But the dynamometer test is used most. Some shops perform other tests only after an engine has failed a dynamometer test. Others rely primarily on the dynamometer test because it is fast and usually accurate.

## Dynamometer Testing

Follow the manufacturer's instructions and connect the dynamometer to the tractor PTO shaft. With the dynamometer connected and the load wheel or control in the no-load position, engage the PTO and bring the engine up to the no-load full-throttle rpm specified in the service manual. Compare the rpm attained with the specifications found in the service manual. If the no-load full-throttle rpm is above or below the manufacturer's recommendations, check the service manual for adjustment procedures and make the necessary adjustments before continuing with the test. Many tractors do not reach no-load full-throttle rpm because of restricted air intake systems.

As the engine is being brought up to operating temperature, check the PTO clutch. A PTO clutch in poor condition will slip when a load 10 to 15 percent above the maximum rated power is applied. If the clutch slips, check the adjustment before continuing the tests. Check the service manual for adjustment procedures.

When the engine reaches operating temperature, set the throttle at the recommended no-load full-throttle rpm. Load the dynamometer with the load wheel or load control until the recommended full-load PTO rpm is attained.

If the engine runs unevenly or stalls, a burned valve or valves could be the problem. Worn piston rings will allow excessive blowby. Blowby is a leakage of compression gases that escape past the piston and rings into the crankcase (Fig. 12-2). Blowby can be detected by watching for excessive vapor or smoke coming from the crankcase breather or oil filler spout. Excessive blowby indicates low compression.

Check the exhaust smoke. Excessive black or gray smoke may be caused by improper fuel injection or incorrect injection pump timing. Black smoke can

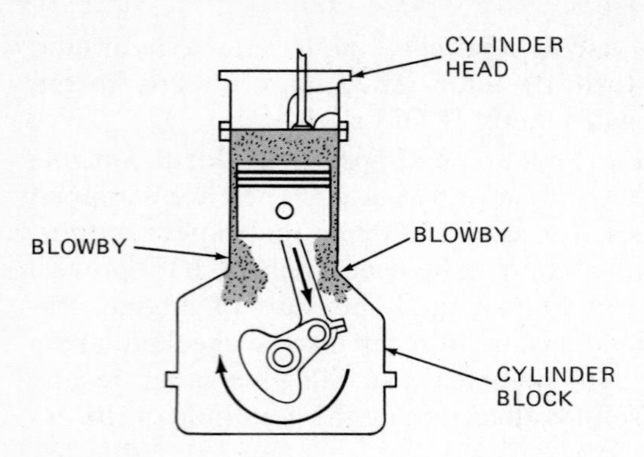


**Fig. 12-2.** Blowby is leakage of compression gases past the rings and piston into the crankcase.

also be caused by a restriction in the air intake system. White or bluish-white smoke can be caused by low operating temperature, incorrect injection pump timing, or low compression.

On spark-ignition engines, blue or gray smoke indicates excessive oil consumption. Black smoke indicates that the fuel mixture is too rich.

When a spark-ignition engine misfires under heavy load, the spark plugs may be the cause. The spark plugs should be serviced or replaced before any further dynamometer tests are conducted. If the misfiring continues, other ignition problems probably exist. Refer to Chap. 11 or the service manual for other possible causes.

Observe engine oil presure. Low oil pressure can be caused by excessive wear in crankshaft or camshaft bearings.

Generator or alternator output and engine operating temperature should be observed. Engine overheating may be caused by improper ignition timing or clogged coolant passages in the radiator.

A complete dynamometer test is the best way to troubleshoot an engine. Refer to the dynamometer manufacturer's instructions for complete test procedures and dynamometer operating procedures.

Record the results of the dynamometer test. Compare the results with the troubleshooting chart found in most service manuals. This will help determine the repairs needed to restore the engine.

## Compression Test

A compression gauge is used to perform a compression test. A compression test measures the amount of pressure that is developed in the combustion chamber on the compression stroke. Low compression readings indicate there is a problem in the combustion chamber and that the power output of the engine is low. To perform a compression test, run the engine until it reaches operating temperature.

**Checking Compression on Diesel Engines** On diesel tractors, check the service manual for recommended test procedures. Some manufacturers recommend that the test be performed with the engine running. This is done by checking one cylinder, stopping the tractor, reinstalling the injection nozzle, and checking the next cylinder (Fig. 12-3).

**Checking Compression on Gasoline Engines** To check compression on gasoline and liquefied petroleum (LP) gas tractors, all spark plugs should be removed. The carburetor throttle and choke should be in the wide-open position and the distributor wire to the coil should be removed and grounded.

Install the compression gauge in a spark plug port and turn the engine with the starter. The battery needs to be at a high state of charge because the cranking speed and the number of engine revolutions can cause the compression readings to vary. Turn the engine until the pressure gauge registers no further rise in pressure. As the engine turns, count the engine revolutions and use the same number of revolutions when checking the remaining cylinders. Revolutions can be counted by watching the movement of the gauge needle. The needle will fluctuate on each revolution.

Record the highest compression reading and check each of the remaining cylinders. Compare your readings with the specifications given in the service manual.

If the compression readings are very low, add 1 tablespoon of SAE 30W engine oil to the cylinder. Do not use too much oil. Adding the oil coats the moving parts and temporarily increases the sealing action between the pistons, the rings, and the cylinder walls. Check the compression again. If the compression goes up from the first reading a noticeable amount, ring wear is indicated. If no difference is found in the compression, bad valves, a blown head gasket, or piston damage may be the problem.

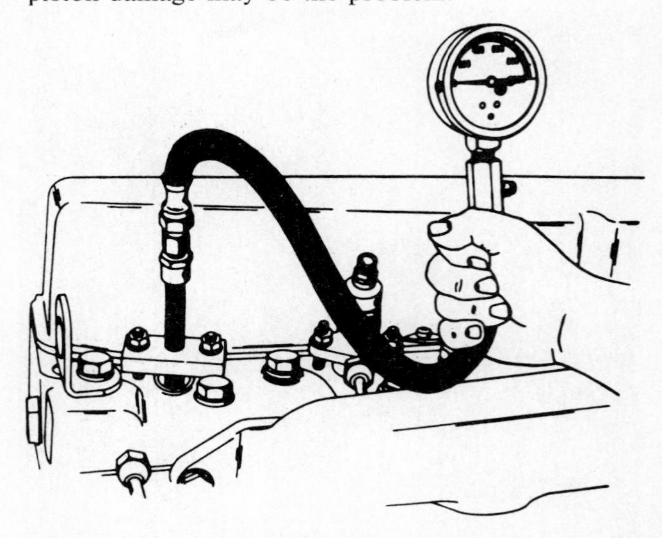

**Fig. 12-3.** A compression gauge is used to check the amount of pressure developed in the combustion chamber. The Test shown is on a diesel engine. (*Allis-Chalmers Manufacturing Co.*)

**Interpreting the compression test** The pressure readings for all cylinders should be near equal. The variation between the highest and lowest cylinder reading should not be over 10 percent. Variations between cylinders can have more effect upon engine performance than overall readings that are slightly below specifications. Low compression readings have a greater effect on diesel engine performance because diesels depend upon the heat of compression to ignite the fuel.

Low compression can be caused by bad valves, broken rings, worn rings, stuck rings, cylinder wear, damaged pistons, or blown head gaskets. If compression is too low or if there is excessive variation between cylinders, the cause should be determined and service should be performed.

The cause of low compression may be detected by watching the action of the compression gauge needle. When it rises only a small amount on the first compression stroke with small amounts of buildup on succeeding strokes, resulting in a very low reading, valves are usually faulty. They may be burned, warped, or sticking.

A low reading on the first stroke with a gradual buildup on succeeding strokes to a moderate pressure reading can mean that the rings are bad. They may be worn, stuck, or scored. Cylinder wear may also be a factor.

Low readings on adjacent cylinders can be caused by a blown head gasket or warped milled surfaces on the cylinder block or cylinder head.

Fig. 12-4. A cylinder leakage tester helps to determine the condition of the rings and valves. (*Photo by Westen McCoy.*)

## Cylinder Leakage Test

When a low compression reading is obtained on a cylinder, a cylinder leakage tester (Fig. 12-4) can be used to determine the cause. The leakage tester is attached to the cylinder port, and compressed air is used to make the test. The instructions provided with the tester should be followed.

To perform the test, the cylinder being tested is brought up to top dead center (TDC) on the compression stroke. The cylinder is then pressurized with air and the percentage of leakage is read on the gauge provided on the tester. Leakage exceeding 20 percent is normally considered excessive.

**Interpreting the Cylinder Leakage Test** While the cylinder is pressurized, listen for escaping air. An intake valve leak can be detected by listening at the air intake system inlet or the carburetor on a spark-ignition engine. If the exhaust valve is leaking, the escaping air can be heard at the open end of the exhaust system.

Leakage past the rings can be detected by listening at the crankcase breather or at the oil level dipstick tube. A blown head gasket will allow air to enter an adjacent cylinder. Listen at the cylinder ports of adjacent cylinders for escaping air.

Cracks in the engine block or cylinder head may allow the air to escape into the cooling system. Check the radiator for bubbles. This condition can also be caused by a leaking head gasket.

## Intake Vacuum Test

A vacuum gauge can be attached to the intake system to perform the intake vacuum test. On diesel engines, the vacuum gauge can be used to check for a restriction in the air intake system. The vacuum gauge can be used to detect many problems in spark-ignition engines.

**Testing Diesel Engines** To test a diesel engine, start the engine and bring it up to operating temperature. If the system is equipped with an intake restriction indicator, remove the indicator, put in a tee connector, and attach both the vacuum gauge and the restriction indicator. On intake systems that are not equipped with restriction indicators, connect the vacuum gauge to the intake manifold (Fig. 12-5).

Set the engine speed at fast idle. A high reading on the gauge indicates that there is a restriction in the air intake system. Check the service manual for recommended manifold vacuum or pressure.

**Testing Spark-Ignition Engines** To perform a vacuum test on spark-ignition engines, connect the vacuum gauge to the intake manifold between the carburetor and the engine and bring the engine up to operating temperature.

**Fig. 12-5.** The intake vacuum test is used to evaluate the condition of the engine. (*Photo by Westen McCoy.*)

**Interpreting the vacuum test** A low, constant reading indicates a loss of power in all cylinders. This can be due to low compression caused by bad valves or worn rings. It can also be caused by retarded ignition timing or an intake leak. Check the gasket between the carburetor and intake manifold.

A steady fluctuation or movement of the gauge needle indicates low compression in one or more cylinders. This can be caused by bad valves or rings in some cylinders, or it can be caused by ignition problems. A leaking head gasket can also cause low compression in one or more cylinders.

A slow fluctuation or back-and-forth movement of the needle when the engine is operating at idle usually indicates that the carburetor idle mixture is not properly adjusted. Adjusting the idle-mixture needle valve should stop the needle movement.

A random or uneven pressure drop may be found. When the needle remains steady but occasionally drops and returns to a steady reading, sticking valves or faulty ignition is usually the cause.

Back pressure in the exhaust system is indicated when the gauge needle slowly drops to 0 as engine rpm is increased. Excessive back pressure will cause a gradual drop in the pressure reading while the engine is idling. Back pressure can be caused by a clogged or damaged muffler or exhaust pipe.

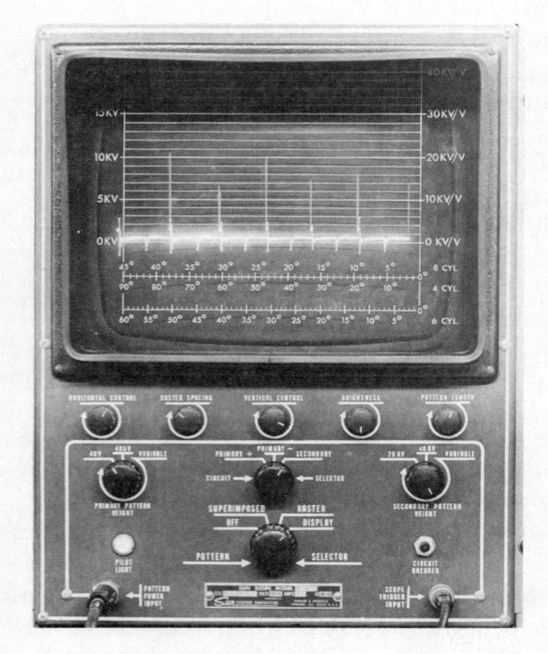


**Fig. 12-6.** An oscilloscope is used to show an ignition test pattern. (*Photo by Westen McCoy.*)

### Ignition System Test

An *oscilloscope* (Fig. 12-6) is an electronic instrument used to check the ignition system on spark-ignition engines. The instructions provided with the instrument should be followed when performing an oscilloscope test.

## MAKING SERVICE RECOMMENDATIONS

Service recommendations should be made when diagnosis and testing are completed. Many operator's manuals and most service manuals provide troubleshooting charts. These charts list common problems and probable causes. The manufacturers provide them as a service aid. They should be used as a reference when determining the service needed.

Service needs should be thoroughly discussed with the owner of the tractor. Parts and labor are expensive, and downtime can be even more expensive if service is required during a busy production season.

## SUMMARY

Test equipment should be used to determine the condition of an engine before service is recommended or performed. The equipment used most often to test tractors is the dynamometer. It provides a quick and accurate test of engine condition. Some shops perform other tests after a dynamometer test indicates service is needed. Test results should be compared with the troubleshooting charts found in most service manuals.

## THINKING IT THROUGH

1. When should a close visual inspection be made?
2. What test is the one most often used on tractors?
3. How can the PTO clutch be checked?
4. What does a compression test measure?
5. How is engine oil used when making a compression check on a gasoline engine?
6. What are some causes of low compression?
7. What is the purpose of a cylinder leakage test?
8. What test can be made with a vacuum gauge on a diesel engine?

# CHAPTER 13 PREPARATION FOR OVERHAUL

## CHAPTER GOALS

In this chapter your goals are to:

- Demonstrate how to prepare a tractor for overhaul
- Demonstrate how to prepare a tractor for engine removal
- Demonstrate how to remove the engine from a tractor

Under normal operating conditions and with proper maintenance, an agricultural engine will provide satisfactory performance for thousands of hours. However, moving parts and bearing surfaces in the engine do wear. Some parts wear even though they are properly lubricated and an established preventive maintenance and service schedule has been followed.

Engine overhaul is a common repair procedure that is performed to restore engine performance and dependability. Most tractor engines are designed so that all the major wear areas and bearing surfaces can be replaced with a minimum amount of machine work. This keeps the engine specifications standard and greatly increases the service life of a tractor. Many tractors in use today were manufactured in the 1940s and 1950s and they continue to be fully serviceable.

The parts and labor costs for an engine overhaul are high. However, the cost of overhauling an engine can be a good investment when it is compared to the cost of purchasing a new tractor. Overhaul can also reduce future costs due to downtime and lost production.

Since engine overhaul is a vital part of tractor service, there are good jobs available in dealerships and other shops that service and overhaul agricultural engines. Also, there is always a demand for good used tractors. Many dealers specialize in used tractors, and used tractor dealerships have been established. Many of these used tractors are rebuilt before they are sold.

A good understanding of basic engine components is necessary if a person is to become a competent employee in a dealership or service shop. The small gas engine discussed in Chaps. 3 and 4 has the same basic component parts found in large multicylinder engines. There are just more of some of the parts, and they are usually larger. A good understanding of small gas engine overhaul will be very helpful when working on large engines.

## PREPARING THE TRACTOR FOR OVERHAUL

When testing is completed and it is determined that engine overhaul is necessary, it is important that the proper steps be followed in order to ensure that a good job is done safely. The conscientious service worker will follow several basic procedures to ensure that the job is done correctly.

### Cleaning the Tractor

The importance of cleaning the tractor thoroughly cannot be overemphasized. Grease, oil, and dirt accumulate as the tractor is used. Corrosive chemicals such as fertilizers and insecticides may also accumulate on the tractor.

When a dirty engine is disassembled, much more time will be required to clean the internal parts properly because hands and tools will be harder to keep clean. It will be hard to keep dirt out of parts as service is performed and the engine is reassembled. Since dirt is a primary cause of engine failure, cleanliness is of utmost importance.

**Cleaning Equipment** Steam cleaners and high-pressure washers are recommended for cleaning agricultural equipment. Both systems are effective, and it is important that the instructions provided with the machine be followed. Improper use can cause serious bodily injury.

Commercial degreasers and solvents such as Varsol and diesel fuel can be used to loosen caked-on deposits. The directions for use should be followed when commercial degreasers are used. Some are highly flammable and are dangerous when used in a closed building or near an open flame. Caution should be taken to keep the chemicals off the skin and out of the eyes. Always make sure there is adequate ventilation and fire safety equipment available when cleaning machinery with solvents.

**Using Cleaning Equipment** When the tractor is not excessively dirty, a good cleaning with a steam cleaner or a high-pressure washer may be sufficient. Care should be taken not to spray hot steam on a diesel injection pump or to spray cold water on an injection pump that is hot. Disturbing the close metal-to-metal tolerances in the pump during cleaning could cause pump seizure.

Tractors that have an excessive buildup of caked dirt and oil should be soaked down with commercial degreaser or solvent. The solvent can be applied with a spray or with a paintbrush. The solvent should be allowed to set 15 to 20 minutes before being washed off. It may be necessary to use a putty knife, carbon scraper, or wire brush to remove some of the dirt.

The tractor should be allowed to dry before it is returned to the service area. Compressed air can be used to remove excess water.

It is important to use cleaning equipment correctly. When steam cleaners are used, they should be shut down correctly. Always follow the operating instructions provided with the equipment.

When cleaning is completed, make another visual inspection of the engine. Hidden cracks or other problems may be visible after the dirt and oil have been removed.

## Setting Up the Tractor in the Service Area

The tractor should be properly set up in the service area. The service area needs to be clean and well ventilated and have adequate lighting. Service should not be performed in an area with a dirt floor or in open areas where dirt or other foreign matter cannot be kept out of the disassembled engine.

**Tools and Equipment Needed** Adequate stands, jacks, hoists, and blocks are essential. An A-frame equipped with a chain hoist, an overhead crane, or a floor crane is needed to remove the tractor engine and lift other heavy components (Fig. 13-1). An engine stand (Fig. 13-2) is recommended for supporting the engine during overhaul. A workbench equipped with a bench vise and a floor jack are also needed.

A complete set of socket wrenches, end wrenches, and an assortment of other tools such as screwdrivers, punches, etc., are essential. Special tools and equipment will be discussed as they are needed.

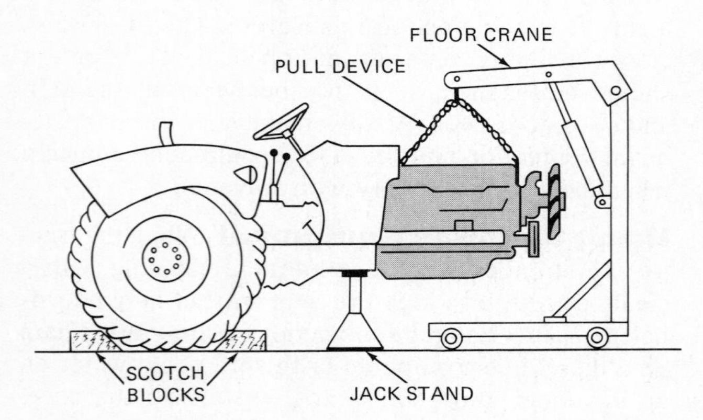


**Fig. 13-1.** A floor crane is used to remove a tractor engine. Always use a set of jack stands and blocks to prevent accidents.

**Fig. 13-2.** Engines mounted on a stand are much easier to service. (*OTC Tools & Equipment Co., Division of Owatonna Tool Co.*)

# PREPARATION FOR ENGINE REMOVAL

Before an engine is removed, proper precautions should be taken to ensure that all fluid systems are drained and that all connections to gauges and controls coupling the engine to the power train are disconnected. Refer to the technical service manual for engine removal instructions and procedures. A few minutes spent reading the manual before the job can prevent costly mistakes and save hours of down time.

Before any fasteners are removed, a good supply of small metal boxes, cans, or cups should be obtained to store them in. If all fasteners are kept in separately labeled containers, many problems will be eliminated when the engine is reassembled. It is advisable to label parts, draw diagrams, and make notes relating to the proper assembly of parts and components. Even the most experienced mechanics do not rely on memory alone during engine disassembly and overhaul. Time lost finding misplaced parts can be very expensive to the customer and must always be avoided.

## Drain Systems

A competent service worker will drain all systems that contain fluid when the engine is removed. This will permit inspection of the liquids for any unusual conditions. It will also minimize any spillage of oil and water on the service floor. Spills are hard to clean up and often cause accidents.

**Cooling System** Drain the entire cooling system. Only part of the coolant is removed when the radiator is drained. The water jacket in the engine block must also be drained.

**Crankcase** Remove the plug in the bottom of the oil pan and drain the crankcase oil. The oil filter should also be removed at this time.

**Hydraulic System** Study the service manual to determine the location of hydraulic lines and the proper method for draining and removal. On some tractors the entire hydraulic system is contained in the drive train. Most tractors manufactured in recent years have hydraulic pumps that are driven by the engine.

**Other Systems** Check with the owner or operator and determine if other systems such as the transmission, differential, and the complete hydraulic system should be drained. On older tractors, it is usually good procedure to replace all external seals in the drive train. These include the rear-axle, brake shaft, PTO, and transmission input shaft seals. This is also a good time to flush the fluid systems and replace all the filters.

### Remove External Parts and Accessories

The procedure for removing the hood and other sheet metal components such as heat shields and shrouds depends on the tractor make and model. The service manual should be used to determine the parts that should be removed and the recommended removal procedures.

**Front-End Assembly and Radiator** The engine block is a part of the frame on many tractors. Some manufacturers use side castings or structural iron such as channel iron or angle iron to extend from the power train to the front-end assembly. When side frames are used, it may not be necessary to remove the side frames or front-end assembly to remove the engine from the clutch or transmission housing.

The radiator is attached to the front-end assembly and usually must be removed to pull the engine. The radiator and radiator core are made of copper or aluminum and are easily damaged. Special care should be taken when removing and storing the radiator.

On most tractors, the front-end assembly must be removed to pull the engine. Special stands can be purchased or constructed to support the front-end assembly on larger tractors. When stands are used, it is not necessary to remove the radiator from the front-end assembly.

The steering linkage or power steering system must be disconnected on most tractors. Check the service manual for removal procedures and use tape and waterproof markers to identify the placement of any hoses removed.

Special jack stands or supports are needed to support the drive train, and the back wheels should be blocked (Fig. 13-1). Use a floor jack or hoist to lift the tractor and place a stand under the clutch housing or transmission case. *Use extreme care to locate the stand in an area that will support the load.* This is done before the front-end assembly is removed.

**Exhaust and Intake System** The exhaust and intake system should be removed. This includes the muffler and the hoses and tubes that connect to the air cleaner. Check these parts for wear and damage. Start a list of replacement parts, and include on your list any gaskets, hoses, or new parts needed.

**Control Linkages and Wires** All electric wires leading to the starter, alternator or generator, lights, sending units, or other accessories attached to the engine should be disconnected and labeled with tape and a waterproof marker. Also, disconnect all control linkages, lines to sending units, hydraulic hoses, and any other systems that connect the engine to the drive train. Place in marked containers any fasteners you remove and make notes to help when reinstalling all the linkages that were disconnected.

**Starter and Alternator or Generator** On many tractors, the starter bolts that hold the starter on the engine go through the block flange into the bell or clutch housing, which is a part of the drive train. When servicing these tractors, the starter must be removed before the engine is pulled. The alternator or generator and other external parts may be removed at this time.

**Fuel System** The fuel tank on some tractors is located over the engine and must be removed to pull the engine. A flare nut wrench should be used to remove the necessary fuel lines. Special precautions should be taken for safe storage of the fuel tank in an outdoor area.

## ENGINE REMOVAL

Special precautions should be taken when removing or pulling the engine. A nonslip-type pull chain, cable, or bar should be used. The pull device can be attached under cylinder head bolts or in threaded holes in the cylinder head. The bolts should be pulled up tight against the head.

On some tractors, it is better to remove the rocker arm assembly before the engine is pulled, and in some cases the cylinder head must be removed. Check the service manual for the correct procedures.

## Pulling the Engine

Attach a suitable hoist to the pull device and take up the slack. Be careful not to pick up the entire tractor and loosen the support under the clutch housing or transmission case.

Loosen and remove all the bolts holding the engine to the transmission housing. Pull the engine forward slowly and check for any remaining connections to the drive train. Do not apply excessive pressure. The engine should separate freely. If resistance is felt, check for remaining bolts and recheck the service manual for the correct procedure.

## Clutch Removal

The clutch and engine flywheel should be removed at this time. Since the flywheel is very heavy, it is recommended that the engine be lowered to the floor before its removal. This is a safety precaution.

It is necessary to put small punch marks on the flywheel and pressure plate. These punch marks will be used to reinstall the pressure plate in its original position on the flywheel (Fig. 13-3). The mark on the flywheel can be ignored if a new or rebuilt pressure plate is installed.

**Removing the Pressure Plate** All the bolts holding the pressure plate to the flywheel should be loosened a little at a time until the pressure plate loses its tension on the clutch disk. When the tension is released, finish removing the bolts and remove the pressure plate assembly.

Special tools and procedures are required for removing pressure plate assemblies on tractors equipped with two-stage clutches. Check the service manual for specific removal procedures.

**Fig. 13-3.** Small punch marks should be used to mark the flywheel and pressure plate to ensure that the pressure plate is reinstalled in its original position. (*Western McCoy.*)

Examine the pressure plate for cracks, ridges which indicate slippage, and overheating. Check the release fingers for wear, breakage, or bending. If the pressure plate assembly is suitable for continued use, store it properly. If it should be replaced, add it to the parts list.

**Clutch Disk** Examine the clutch disk. Look for worn or loose linings, glazed or oil-impregnated linings, loose damper springs or center hub, and signs of overheating. If any of these conditions exists, a new clutch disk should be added to the parts list.

Remember that some tractors are equipped with a two-stage clutch. The PTO clutch is inside the pressure plate assembly and should be replaced if it slipped during the dynamometer test.

**Flywheel Removal** The flywheel bolts directly to the end of the crankshaft. Many tractors are equipped with locking plates to prevent the flywheel from working loose during operation. When these are found, the tabs should be bent down before attempting to loosen the bolts or nuts. Other locking devices such as wire clips are also used.

The flywheel is exceedingly heavy. Use extreme caution when removing it. Place all fasteners and locking devices in a properly labeled container.

Check the clutch contact surface on the flywheel for excessive heat cracking and scoring. It is possible to have a flywheel refaced if the cracking is not excessive. Most manufacturers limit refacing to 0.040 in [1.02 mm] or the flywheel face becomes too soft.

Inspect the pilot bearing or bushing which is installed in the center of the flywheel or in the end of the crankshaft. Inspect the flywheel ring gear for wear. Add parts needed to the parts list.

On some engines, the rear crankshaft seal, hydraulic pump drive gear, or other components should be removed at this time. Be sure to check the service manual.

## Attaching Engine to Engine Stand

It is recommended that an engine stand (Fig. 13-2) be used to support the engine. This will make the engine easier to handle and permit it to be turned to different angles for easier access to internal components. A sturdy work table may be used in place of a stand.

## SUMMARY

Engine overhaul is a common repair procedure that is used to restore engine performance and dependability. Overhaul is necessary because moving parts and bearing surfaces wear after prolonged use.

Engines are designed so that the bearings and moving parts can be replaced at a nominal cost and with minimum machine work. Overhaul is expensive, but it is usually a good investment.

Many people are employed in dealerships and service shops to service and overhaul agricultural engines. A good understanding of basic engine components is necessary for a person desiring to gain employment in engine service.

## THINKING IT THROUGH

1. Explain why overhaul is a common repair procedure for agricultural engines.
2. Why is it important that a tractor be properly cleaned before it is serviced or repaired?
3. What processes can be used to loosen deposits that are caked on tractors?
4. What service procedure is recommended when electrical wires, control linkages, etc., are disconnected?
5. What should be done with fasteners as they are removed?
6. Why should special care be taken with the radiator?
7. Should an engine be hard to separate from a tractor after all the bolts are removed?
8. Why is it recommended that an engine be lowered to the floor before removing the clutch assembly and flywheel?
9. Why are the flywheel and pressure plate marked before the pressure plate is removed?
10. Where is the pilot bearing or bushing located?

# CHAPTER 14 ENGINE DISASSEMBLY

## CHAPTER GOALS

In this chapter your goals are to:

- Remove and service the cylinder head and valve assembly
- Remove and service piston and rod assemblies
- Remove the crankshaft, clean it, and check it for alignment
- Remove the camshaft, and evaluate its condition
- Clean and check cylinder blocks

Two types of engine overhaul are normally performed on agricultural engines: complete overhaul and partial overhaul. In complete overhaul, the engine is removed from the tractor, and the crankshaft, camshaft, and other internal parts are serviced. All bearing surfaces and wear areas are checked and replaced or serviced if excessive wear is found. The complete overhaul will be discussed in this chapter.

Many times a partial engine overhaul is performed. A partial overhaul consists of servicing the cylinders, pistons, and cylinder heads. New rings and connecting rod bearings are installed. Pistons and piston pins may be replaced if needed, and cylinder liners can be replaced on some engines. The connecting rods can be checked, and new piston pin bushings can be installed when needed. The cylinder head can be checked for warpage, and the valve train, which includes the valves, valve guides, valve springs, and the rocker arm assembly, can be reconditioned. Other components such as the engine oil pump can be checked and serviced.

This type of service is often called a *valve and ring job*. It is frequently done when the engine tests indicate low compression due to faulty valves, blown head gaskets, or worn rings in one or more cylinders.

Hours of operation on the engine or hours of service since the last complete overhaul should be used to determine if a complete overhaul is necessary. Service records and the condition of the air intake system and the fluids in the engine should also be considered.

## CYLINDER HEAD

The cylinder head (Fig. 14-1) on most agricultural engines is a one-piece alloy iron casting. Some air-cooled engines are equipped with aluminum alloy cylinder heads. The cylinder head is secured to the milled surface on the top of the engine block with heat-treated cap screws or stud bolts and nuts.

Intake ports are provided in the cylinder head for the intake of air on diesel engines and air and fuel on spark-ignition engines. Exhaust ports provide for the expulsion or removal of exhaust gases.

Cored passages are provided for the circulation of engine coolant to remove heat from around the valves and upper combustion chamber. A head gasket is used to seal the combustion chamber. It also seals all coolant and oil passages between the cylinder head and the engine block.

Most agricultural engines are valve-in-head design. In valve-in-head engines, the exhaust and intake valves are located in the cylinder head above each cylinder. This applies to four-stroke-cycle engines only. Larger two-stroke-cycle engines may have exhaust valves in the cylinder head, but they are usually equipped with intake and exhaust ports near the bottom of the cylinder.

Valve guides hold the valves in place and allow them to open and close. The valve seat or seat insert is located in line with the valve guide at the top of the combustion chamber. Valve springs, spring retainers, and locks or keepers hold the valves in place and allow them to open and close (Fig. 14-1).

Push rods extend through the cylinder head and into the cylinder block, where they ride on the cam followers or valve lifters which are operated by the lobes on the camshaft. The upper end of the push rods are concave, and the valve clearance adjusting screws on the rocker arms hold them in place.

The ports for spark plugs or injection nozzles are located in the cylinder head. The rocker arm assembly that opens and closes the valves is attached to the top of the head. A gasket and rocker arm cover seals the top of the head.

### Removing the Rocker Arm Assembly

The rocker arm shaft can be bent if it is improperly removed. To prevent damage to the shaft, start at one end of the shaft and loosen each bolt or nut in the rocker arm bracket two turns. Repeat this loosening procedure until the rocker arm assembly is free. Some rocker arm assemblies will expand and come apart when removed. Check the service manual for correct removal procedure.

Remember that all fasteners and parts removed should be placed in properly labeled containers. Note the position of oil lines and other components to be

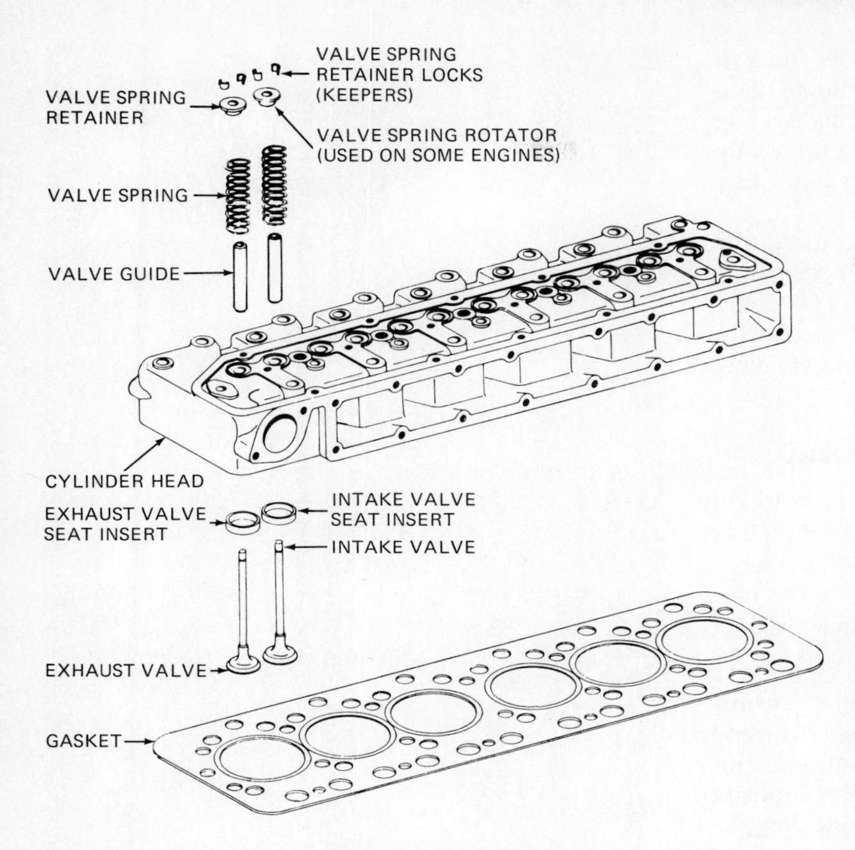


**Fig. 14-1.** The cylinder head assembly: gasket, exhaust valve, intake valve, exhaust valve seat insert, intake valve seat insert, cylinder head, valve guide, valve spring, valve spring retainer, valve spring retainer locks (keepers), valve spring rotator (used on some engines). *(Allis-Chalmers Manufacturing Co.)*

removed. They must be reinstalled in their original positions.

The push rods can be removed at this time. They should be stored so they can be kept in order. A shop-made storage rack (Fig. 14-2) is recommended for storing push rods, valves, valve springs, and other valve components.

## Removing Other Parts and Accessories

If the intake and exhaust manifold was not removed earlier, it should be removed at this time. Remove the diesel injection nozzles or spark plugs and any fuel lines, linkages, or oil lines that connect the cylinder head to other engine components.

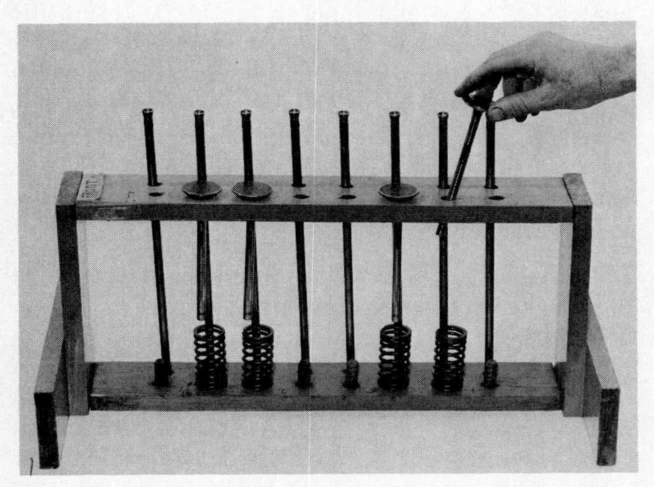

**Fig. 14-2.** A shop-made storage rack for valve components. *(Photograph by Westen McCoy)*

## Removing the Cylinder Head

Be sure that the engine has completely cooled down before loosening the cylinder head bolts. A cylinder head that is removed while the engine is still hot is likely to warp as it cools to room temperature.

Check the service manual for the recommended tightening sequence for the cylinder head bolts. The sequence should be reversed when loosening cylinder head bolts. If no sequence is given, start at the ends of the head and loosen the bolts in a circular pattern, moving toward the center of the head.

As bolts are removed, pay close attention to their length. If bolts of different length are used, it is recommended that a sketch be made showing their proper location. Note the location of any brackets or other accessories that mount under the bolts. Remember to place all bolts, nuts, and washers in properly labeled containers.

**Pulling the Head** The cylinder head on smaller engines can be removed by hand. However, cylinder heads on larger engines are very heavy, and it is recommended that a lift chain and hoist be used. The cylinder head should be lifted straight up off the cylinder block. This is very important because some cylinder heads are equipped with dowel pins to assure correct alignment.

It is not unusual for the head gasket to be stuck to the cylinder block and the cylinder head. Some manufacturers recommend the use of various sealing compounds. If a cylinder head does not lift easily from the cylinder block, check for head bolts that may not have been removed.

A brass hammer or mallet may be used to tap the head to help break the seal. Never force a pry bar or screwdriver between the milled surfaces. This could seriously damage the sealing surfaces. The use of a hoist to remove the cylinder head will eliminate most problems.

## Valve and Valve Guide Service

Use a valve spring compressor (Fig. 14-3) to remove the valves, springs, and keepers. As the parts are removed, place them in a rack (Fig. 14-2) to keep them in correct order. Remember that it is recommended that all parts be reinstalled in their original location due to their individual wear patterns.

**Checking the Valves** Check all the valves for burning, pitting, and heavy carbon deposits. Burned or pitted valves can be caused by distorted seats, insufficient valve clearance or lash, or carbon deposits on the face of the valve (Fig. 14-4). Burning and pitting can also be caused by weak valve springs, valves sticking in the valve guides, warped valve stems, detonation, preignition, or improper timing.

Detonation was discussed in Chap. 3. It is uneven burning of the fuel. Preignition is ignition of the fuel by hot spots, carbon deposits, spark plugs of the wrong heat range, loose spark plugs, or other conditions that cause the fuel to ignite before the spark occurs.

Heavy carbon deposits may be caused by worn valve guides or damaged seals. Excessive valve guide wear can cause carbon lines to form on the valve stem.

All valves should be checked for cracks or breaks. To check a valve for cracks, hold the valve by its stem and strike the end of the stem with a hammer. A sharp blow will cause the head to break off if the valve is cracked.

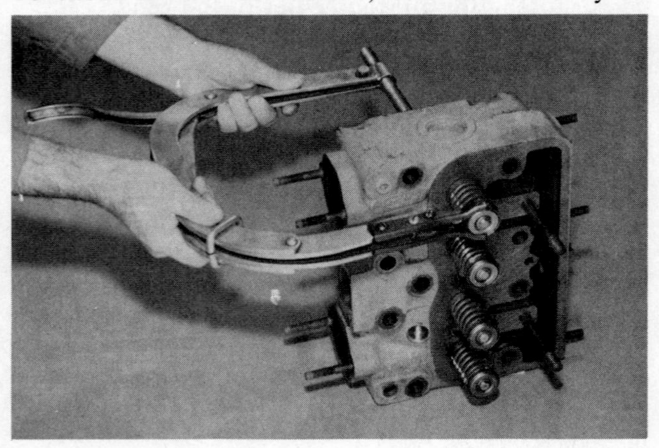

Fig. 14-3. A valve spring compressor is used when removing the valves. The spring compressor is also used to install valves. (*J.I. Case Co.*)

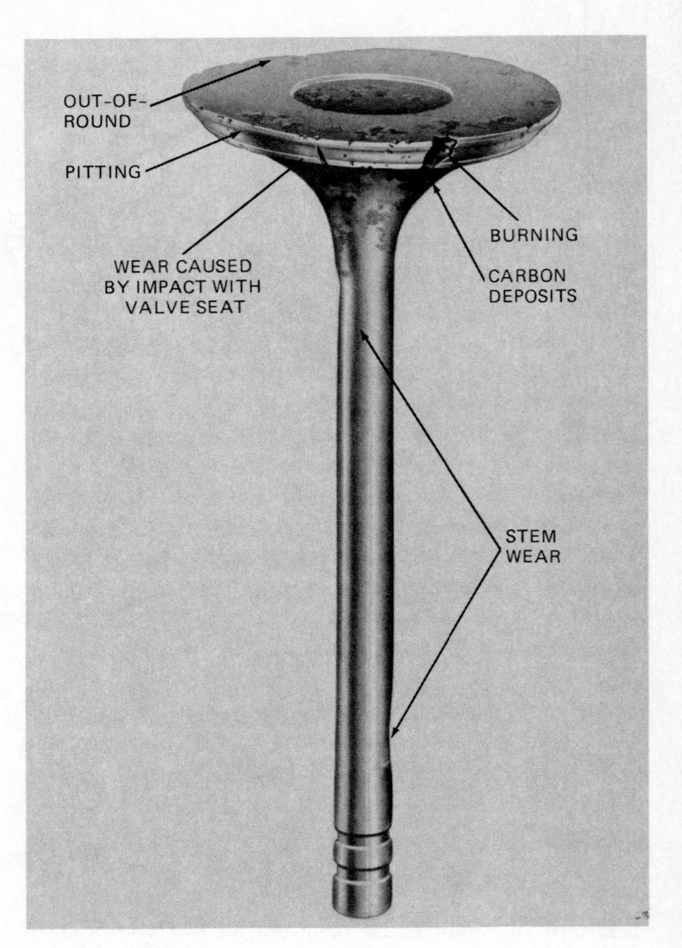


**Fig. 14-4.** Common causes of valve failure. (*Sioux Tools, Inc.*)

**Measuring Valve Guides for Wear** A small-hole gauge and an outside micrometer are needed to measure the inside diameter of the valve guides for wear and taper (Fig. 14-5). Insert the small-hole gauge into the guide and find the smallest diameter. Use the micrometer to determine the diameter. Then use the gauge to find the largest diameter, which will be near the ends of the guide. Determine the diameter with the micrometer and subtract the smallest diameter from the largest. This difference is valve guide taper.

**Measuring the Valve Stem** Use an outside micrometer to measure the valve stem for wear (Fig. 14-6). Measure the stem where it has been moving in the valve guide. This is the stem wear area. Subtract the valve stem diameter from the maximum valve guide diameter. This difference will give the stem-to-guide clearance.

Check the service manual for the recommended valve-stem-to-guide clearance and the allowable taper in valve guides. Also check for the specifica-

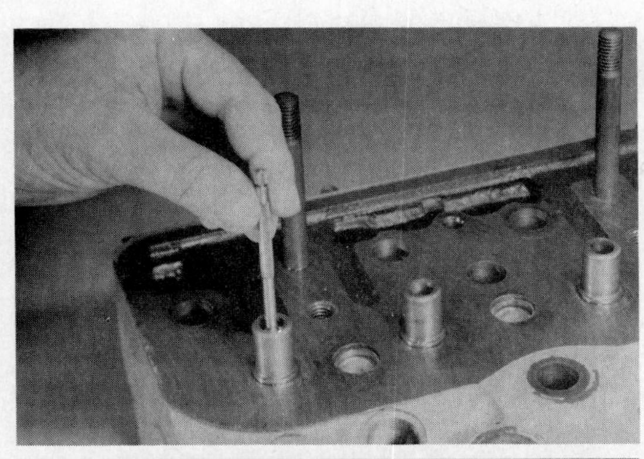

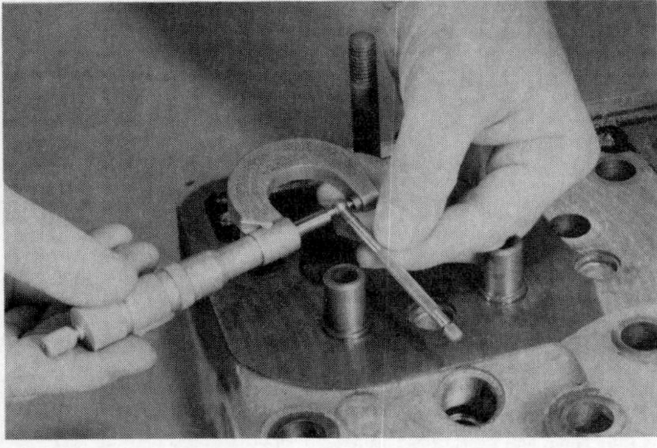

**Fig. 14-5.** A small-hole gauge and outside micrometer are used to check the valve guides for wear and specified guide-to-valve-stem clearance. (*J.I. Case Co.*)

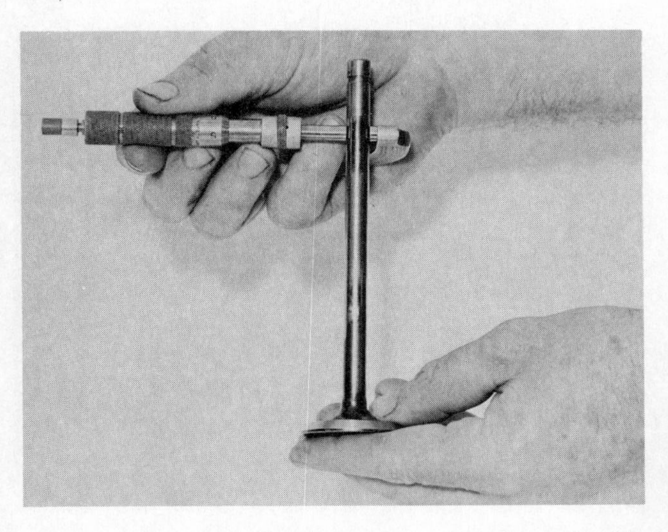

**Fig. 14-6.** Measure the valve stem for wear with an outside micrometer. (*Photograph by Westen McCoy*)

tions on valve stem diameter. When the valve stem wear exceeds specifications, the valve should be replaced.

**Valve Guide Reconditioning** If valve guide wear exceeds the limits given in the service manual, valve guide service is required. Many agricultural engines are equipped with renewable valve guides. When renewable valve guides are used, they can be removed by pressing them out the top of the cylinder head (Fig. 14-7). Never try to press a guide out the bottom of the head. The service manual should be consulted for the correct removal procedures.

**Installing new valve guides** Inspect the bores in the cylinder head to see that they are clean. Read the installation procedure in the service manual. Lubricate the guides as recommended. Engine oil, silicon spray, or white-lithium-base lubricant may be used. Valve guides in some engines have excessive interference fit and should be chilled in a freezer or container of dry ice before installation. Interference fit is when the diameter of the guide is larger than the bore in the cylinder head in which the valve guide fits.

New valve guides are pressed in from the top. Check the service manual to be sure that the guides are installed with the proper end up. Also, check for installation height above the cylinder head and be sure that the correct height is maintained (Fig. 14-8). A guide installing tool or brass block should be used to protect the guide as it is pressed in.

The valve guide bore should be checked after the new guides are installed. Guide bore tends to become smaller when guides are installed. This is due to the interference fit. A valve guide reamer (Fig. 14-9) can be used to ream or size the valve guides to manufacturer's specifications. The reamer should be turned clockwise at all times.

**Reaming valve guides** Worn valve guides can be reconditioned by reaming. When guides are reamed, new valves with oversized valve stems are installed.

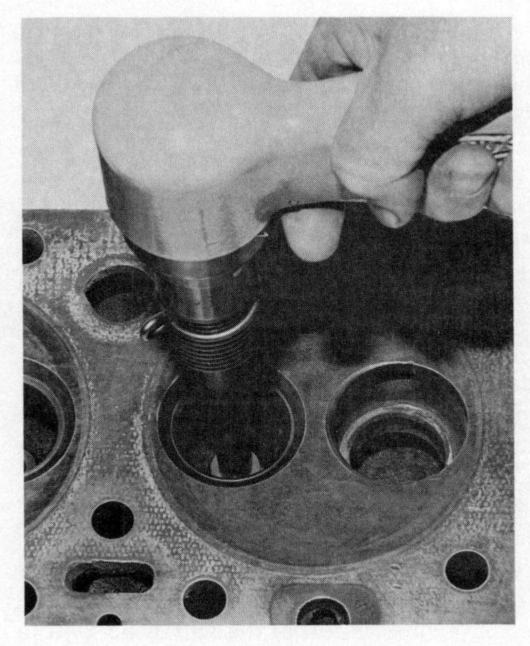

**Fig. 14-7.** Renewable valve guides may be removed with an air hammer equipped with a special valve guide tool. The tool may also be used to install guides. (*Photograph by Westen McCoy*)

**Fig. 14-8.** Checking the valve guides for proper installation height. (*Photograph by Westen McCoy*)

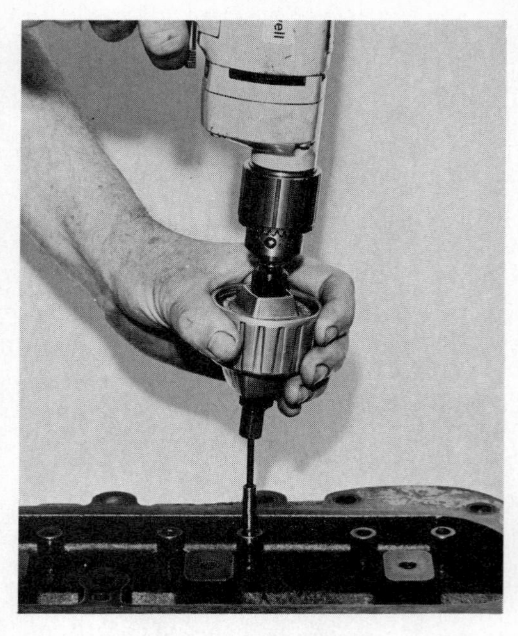

**Fig. 14-10.** Using a knurling tool to reduce the inside diameter of worn valve guides. The service manual should be checked to determine if knurling is recommended. (*Photograph by Westen McCoy*)

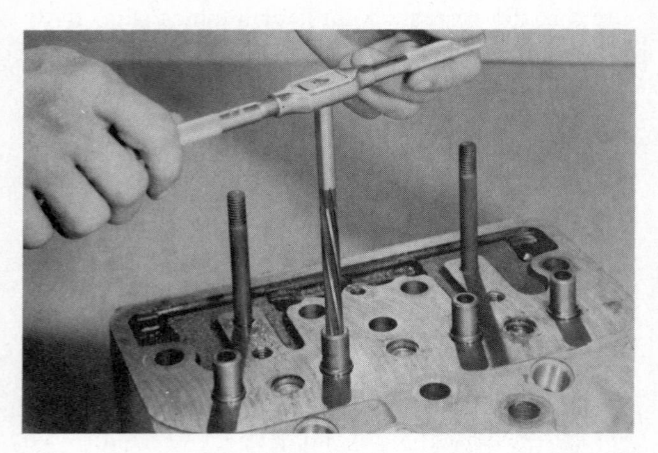

**Fig. 14-9.** A valve guide reamer is used to obtain specified valve stem-to-guide clearance when new guides are installed, guides are knurled, or when valves with oversized stem diameters are used. (*J.I. Case Co.*)

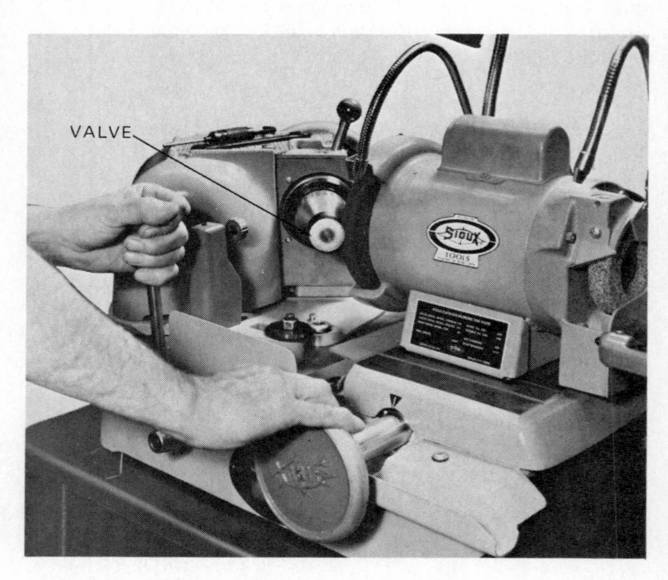


**Fig. 14-11.** Grinding the valve face. (*Sioux Tools, Inc.*)

**Knurling valve guides** Knurling is a process that reduces the inside diameter of the valve guide by displacing the metal. This is done with special knurling tools, and valves with original stem diameter can be used (Fig. 14-10). The instructions provided with the knurling kit should be followed.

**Cleaning Valves** Valves should be thoroughly cleaned. The carbon will be easier to remove if the valves are not soaked in solvent before cleaning. A glass bead machine or a power wire wheel can be used to remove the carbon. All traces of carbon should be removed from the valve head and stem. After the carbon has been removed, rinse the valves in solvent and blow them dry with compressed air. Remember, it is important that the valves be kept in their original order.

**Grinding Valves** A valve grinder (Fig. 14-11) is used to grind the face on the valves. It is important that the valve face be ground to the angle specified in the service manual.

Some manufacturers specify that valves be ground to an interference angle. An interference angle is attained when the valve face and valve seat are ground at slightly different angles (Fig. 14-12).

Set the valve grinding machine to the correct angle. Follow the manufacturer's instructions when operating the machine and always wear adequate eye protection. Be sure that the stone is properly dressed.

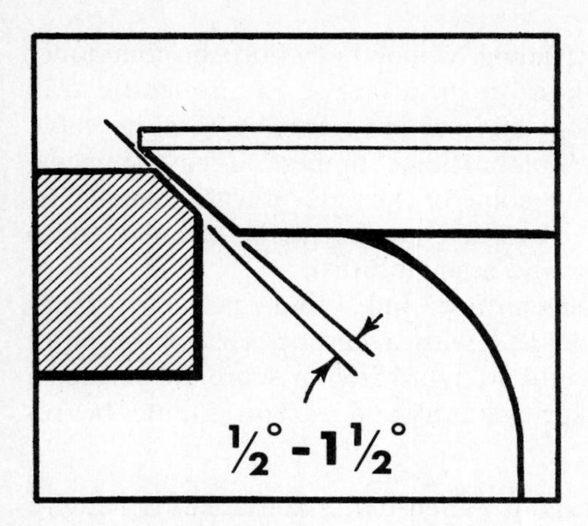


**Fig. 14-12.** An interference angle is attained when the valve face and seat are ground at slightly different angles. (*Sioux Tools, Inc.*)

Turn on the chuck and watch the valve head as it rotates. If a noticeable amount of runout, or wobble, is present, loosen the chuck, check it for cleanliness, and reposition the valve. If excessive runout is still noticed, the valve stem is warped and the valve must be replaced.

Remove only the metal necessary to clean the face of the valve. Stop the chuck and rotate the valve by hand to check for any places missed.

The margin should be checked. The margin is the area between the top of the valve head and the face (Fig. 14-13). It should be the same width all the way around the head of the valve. If the margin width is less than the width specified in the service manual, the valve should be replaced. A margin that is narrower than specifications may cause the valve to fail due to the intense heat and pressures it is exposed to during engine operation.

It is important to remember that the intake and exhaust valves are different. The heads are usually not the same diameter, and the recommended face angle may be different. Reset the machine when changing from intake to exhaust valves. It is also a good practice to resurface the stone.

**Grinding the end of the valve stem** The end of the valve stem may be pitted or worn. The valve grinder has attachments to reface the end of the stem. Check the chamfer on the end. It can be restored with the valve grinder (Fig. 14-14).

Check the keeper grooves on the valve stems. If the keeper grooves are damaged, the valve should be replaced.

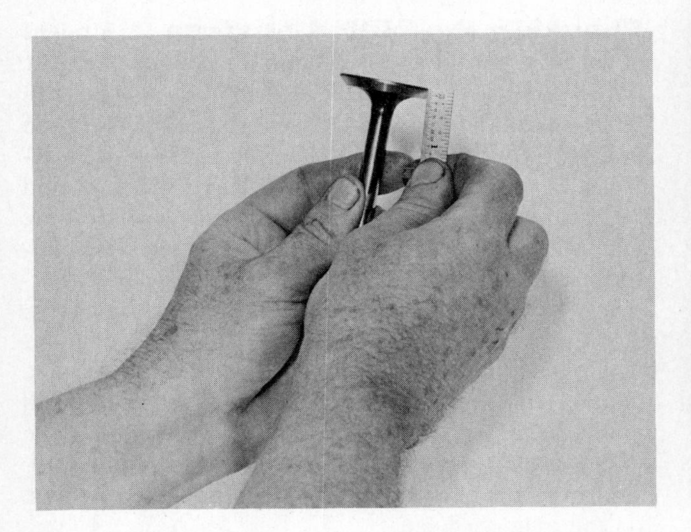

**Fig. 14-13.** Measure the valve margin after the valve face has been ground. (*Photograph by Westen McCoy*)

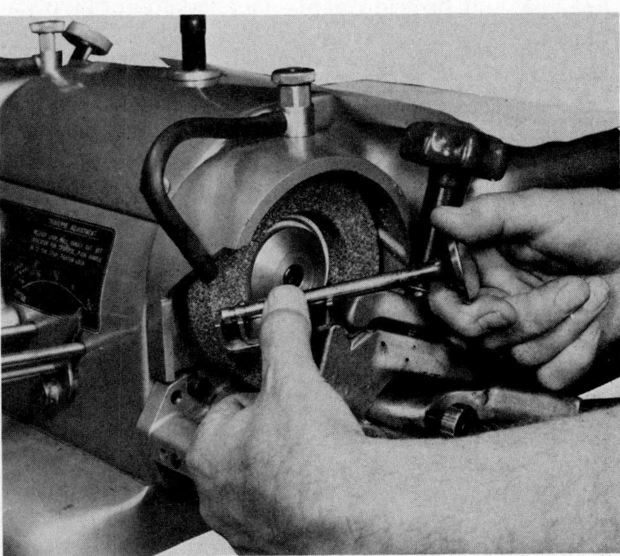

**Fig. 14-14.** Grinding the end of a valve stem. (*a*) Surfacing the end. (*b*) Chamfering the end. (*Sioux Tools, Inc.*)

When grinding is finished, the valves should be washed in solvent, blown dry, lightly oiled, and properly stored. Remember that it is important to keep all valves in their original order.

## Checking the Rocker Arms

Check the rocker arm ends that contact the valve stem. If they are worn, they should be replaced or ground to a smooth even curve. This can be done with a special attachment that is installed on the valve grinder (Fig. 14-15). Do not remove any more metal than is necessary. If the rocker arm assembly is disassembled, all parts must be reassembled in their original position after service is completed.

## Push Rods

Push rods should be kept in their original order. They should be washed in solvent and blown dry with compressed air.

Push rods that have tip wear or are bent should be replaced. A bent rod can be detected by rolling the push rods across a flat surface or with V-blocks and a dial indicator. Never attempt to straighten push rods. They will probably bend again if they are used. If push rods are needed, they should be added to the parts list.

## Cylinder Head Service

The cylinder head should be cleaned and checked for flatness of the milled surface. It should also be checked for cracks after all valves, injection nozzles, precombustion chambers, spark plugs, and other parts are removed.

**Cleaning the Cylinder Head** The cylinder head should be thoroughly cleaned to remove carbon, gasket material, and excessive deposits in the water jacket.

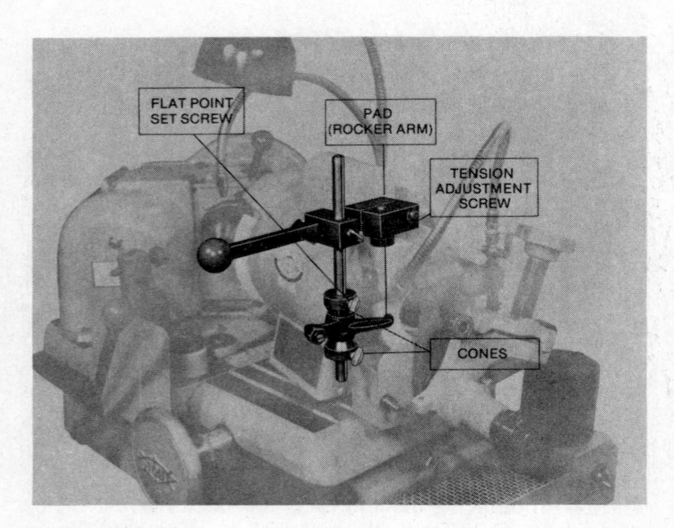


**Fig. 14-15.** Grinding the rocker arm contact surface. (*Sioux Tools, Inc.*)

**Mechanical cleaning** A small steel brush (commonly called a carbon brush) attached to an electric drill motor can be used to remove carbon from the valve ports and the combustion chambers. It may be necessary to remove some of the carbon with a small carbon scraper. The valve guides can be cleaned with a guide cleaner and a small brush.

The milled surface and other gasket surfaces should be cleaned with a carbon scraper or putty knife. Be careful not to scratch or score the surfaces. All old gasket material and carbon should be removed.

**Chemical cleaning** When the water jacket is heavily scaled, the cylinder head should be cleaned in a "hot tank." A *hot tank* is a heated vat containing an acid-type commercial cleaning agent.

The head should be soaked in the solution, then rinsed in an alkaline solution to neutralize the acid, and washed with steam or a high-pressure washer. Compressed air should be used to dry the head. The directions provided with the chemical cleaner should be followed.

**Glass bead cleaning** A glass bead machine may be used to clean the carbon from a cylinder head. The head should be rinsed in solvent and blown dry with compressed air to remove the oil film before it is placed in the glass bead machine.

Special precautions must be taken to clean the head properly after it is removed from the bead machine. All beads, loosened carbon, and other materials must be removed.

**Cleaning aluminum heads** Special precautions must be taken when cleaning aluminum cylinder heads. These heads should *never* be dipped in a hot tank. Check the service manual for recommended cleaning procedures.

**Checking the Milled Surface** A straight edge and a flat thickness gauge are used to check the milled surface for flatness (Fig. 14-16). The chamfered edge of the straight edge is placed against the milled surface, and the thickness gauge is used to detect any variation in the surface. The surface should be checked through the center and diagonally. Some manufacturers may specify a different testing pattern. Refer to the service manual for allowable warpage and exact instructions for performing the check.

If warpage exceeds specifications, the cylinder head must be resurfaced or replaced. The service manual will indicate if resurfacing is permitted and give specific instructions. On many engines, the valves must be set in to maintain the correct dimension from the head of the valve to the milled surface on the cylinder head.

**Checking for Cracks** Cylinder heads may be cracked due to freeze damage or overheating. Heads

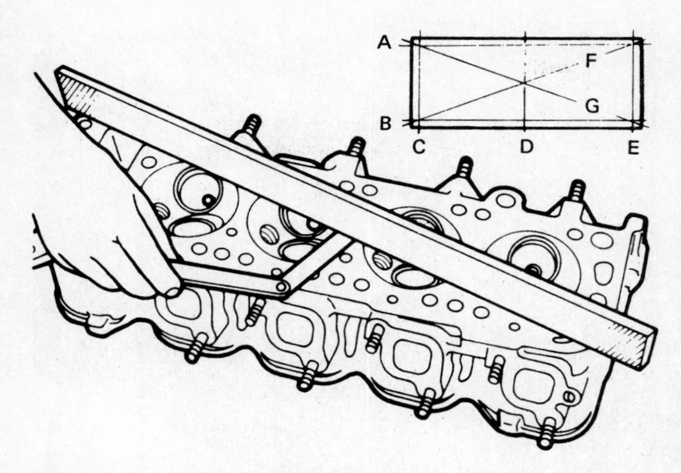


**Fig. 14-16.** A straight edge and a flat thickness gauge are used to check the flatness of all milled gasket surfaces on cylinder heads and engine blocks. (*Crouse/Anglin, Automotive Engines, 6th ed. Copyright © 1981 by McGraw-Hill, Inc. Reproduced with permission.*)

can be checked with a magnetic crack detector or can be pressure-tested. The magnetic crack detector is the method most widely used in service shops.

**Magnetic crack detector** The pole pieces of the magnetic tester are placed over the suspected area. They set up a magnetic field. A special magnetic powder is sprinkled over the area between the poles. The excess powder is then blown away, and if a crack is present, the remaining powder will follow the line of the crack (Fig. 14-17). This method will not work on aluminum heads as they will not conduct the magnetic fields.

## Valve Seat Service

Valve seats should be checked for excessive burning, pitting, or cracks. Seat inserts are used in many heads. They can be removed and replaced with the proper equipment. Heads with integral-type seats can be counterbored, and seats inserts can be installed. This is normally done at a machine shop.

Seats are ground with a stone that is driven by a special drive motor. The stone is mounted on a sleeve which is held in alignment with the valve guide by the stone sleeve pilot (Fig. 14-18).

**Selecting the Proper Stone** It is important to select the correct stone. Valve seats that are an integral part of the cylinder head may be ground with a fine finishing stone. Coarse-textured roughing stones can be used to make the initial or roughing cut on steel seat inserts. A fine finishing stone should be used to polish the seat.

Special stones are available for grinding seats made of Stellite or other hard metals. The information provided with the seat grinding kit should list the stones recommended for various makes and models of agricultural engines.

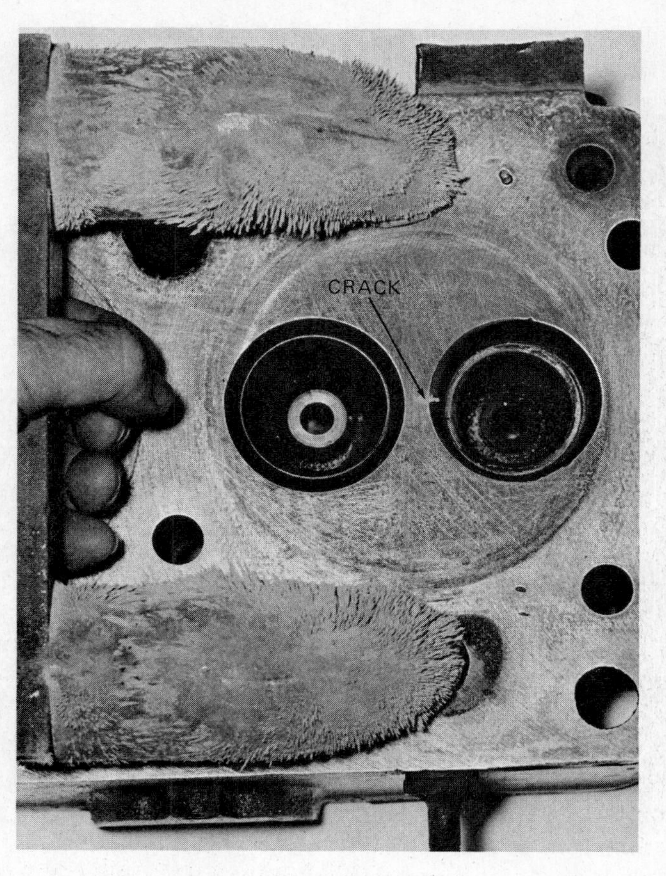


**Fig. 14-17.** A magnetic crack detector is used to check the cylinder head for cracks. Note the magnetic flux lines lined up with the crack. (*Photograph by Western McCoy*)

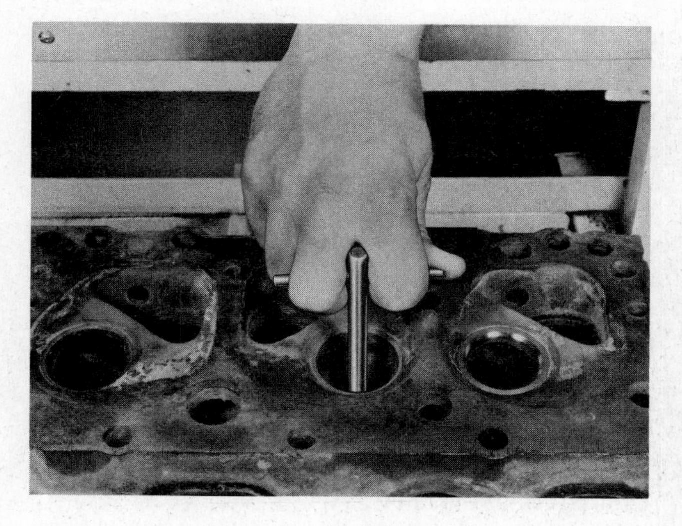

**Fig. 14-18.** A stone sleeve pilot installed in the valve guide. (*Sioux Tools, Inc.*)

**Dressing the stone** After the proper stone or stones are selected, they should be dressed. Check the service manual for the recommended seat angle. A stone dresser (Fig. 14-19) is used to dress stones. Redress the stone as needed to maintain the correct stone angle. On hard seat inserts, frequent dressing will be required.

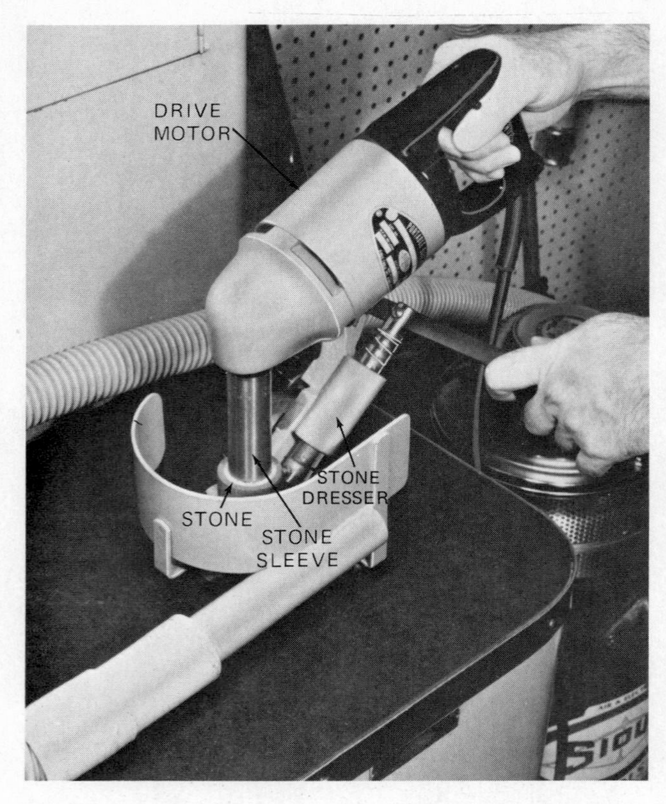


**Fig. 14-19.** Dressing a seat grinding stone. (*Sioux Tools, Inc.*)

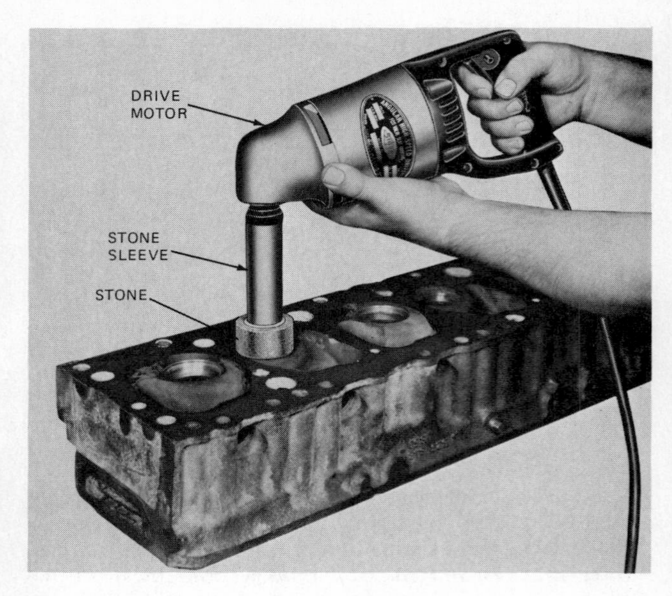


**Fig. 14-20.** Grinding valve seats. (*Sioux Tools, Inc.*)

**Fig. 14-21.** Seat width should be measured after it has been ground. (*Photograph by Westen McCoy*)

**Grinding the valve seat** Check to see that the valve guide is clean and that the seat is free from oil, carbon, and dirt. Select the correct sleeve pilot and insert it into the valve guide. The sleeve pilot must be tight in the guide.

Lubricate the sleeve pilot and install the sleeve with the correct stone attached (Fig. 14-20). Hold the drive motor so that no side pressure is placed on the sleeve. Run the motor for a few seconds and allow the stone to stop. Remove the motor and raise the stone. Examine the seat for remaining pits or burn marks. Repeat the grinding procedure until the seat is clear. Do not remove excess metal. Seats can easily be ruined by excessive grinding.

**Recommended seat width and position** When seats are ground, they are made wider. Check the service manual for recommended seat width and procedures for narrowing the seat. Measure the seat with a machinist's rule (Fig. 14-21).

If the seat is wider than recommended, it should be narrowed. Before any grinding is done, the valve face should be painted with a light coat of Prussian blue or marked with a soft-lead pencil to determine where the seat contacts the valve face. To determine the contact point, insert the valve and turn it one-quarter turn in the seat. The seat should be located at the center of the valve face (Fig. 14-22).

The position of the seat and the seat width can be adjusted by properly using the 15°, 45°, and 60° stones. A light coat of Prussian blue or soft-lead pencil marks on the seat will help identify the location and width of the seat (Fig. 14-23).

**Checking the Valve Seat** After the seats have been ground, they should be checked for concentricity or true roundness. This can be checked with a special valve seat dial indicator (Fig. 14-24). Runout or out-of-roundness should not exceed 0.002 inches (in) [0.05 millimeter (mm)].

Another way to check the seat and the valve face for concentricity is to place a thin coat of Prussian blue on the face of the valve. This test is accomplished by placing the valve in position, applying thumb pressure to the center of the valve head, and rotating the valve one-quarter turn and back to the original position.

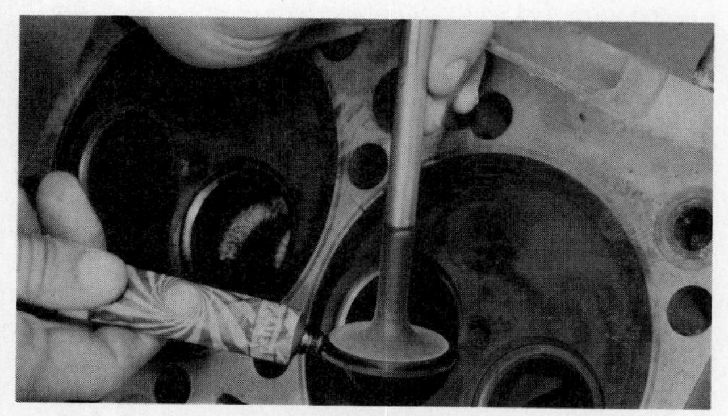

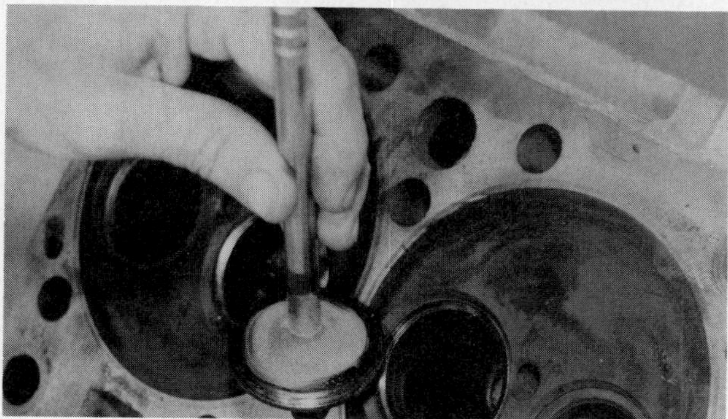

**Fig. 14-22.** Determine the valve seat location and width on a valve face with Prussian blue. (*J.I. Case Co.*)

Remove the valve to determine the condition of the seat and the valve. The seat should be marked with blue around its entire circumference. The seat should mark the valve near the center of the valve face.

### Checking Valve Springs

Valve springs should be cleaned in a solvent and blown dry. Never clean valve springs in a solution that will remove the paint. The paint keeps the springs from rusting, and discoloration of the paint indicates that the spring has overheated. Overheating can cause spring weakness.

Valve springs should be checked for squareness using a combination square (Fig. 14-25). The combination square can also be used to check the valve spring free length. Check the service manual for specifications on squareness and free length.

**Testing Spring Tension** A valve spring tester is used to check spring tension (Fig. 14-26). Since spring tension is important for proper valve action, each spring should be tested to see if it is within specifications.

To test spring tension, place the spring on the tester and compress it to the height specified in the service manual. If the tension, read in pound-feet [newton-meters], shown on the dial is less than the tension specified in the service manual, the spring should be replaced. Some manufacturers provide shims or spacers that can be used to adjust spring tension. Weak valve springs will not close the valves properly and keep them seated.

### Expansion Plugs

Expansion plugs or freeze plugs are often used in cylinder heads. Expansion plugs rust and corrode from the inside. Since they are inexpensive, and it is impossible to check their condition, they should be replaced when a cylinder head is serviced.

### Storing the Cylinder Head

The head, valves, and rocker arm assembly should be properly stored when service is completed. All exposed milled, polished, or machined surfaces should be lightly lubricated to prevent rusting. The head should be wrapped in paper or placed in a covered box to keep it clean while other service is being performed.

## PISTONS AND CONNECTING RODS

The piston receives the force of combustion, and the connecting rod transmits the power to the crankshaft. Combustion in a piston engine produces reciprocating motion or power. The connecting rod and the throw, or offset, in the crankshaft convert the reciprocating motion to rotary motion which propels the machine and delivers the power to do work.

### Piston and Connecting Rod Assembly

A complete piston and connecting rod assembly (Fig. 14-27) includes the piston, the piston pin, the rings, the connecting rod, the rod bearing, and the piston pin bushing.

**The Piston** Pistons are usually made of cast iron, iron alloys, or aluminum alloys. They are precision-made and must withstand the heat and pressure created during combustion. Pistons are made to fit snugly in the cylinder, but they must have proper clearance so that they will move up and down freely.

**Parts of a piston** The main parts of a piston (Fig. 14-28) are the head, the lands, the ring grooves, the skirt, and the piston pin bosses. The *head* is the top surface of the piston. Piston heads may be designed to create turbulence, which aids in the mixing of the air and fuel for proper combustion. Some diesel engines are equipped with pistons that have a precombustion chamber recessed into the head.

Pistons may be designed to go in the engine in a specific position. The front is usually identified by a small arrow, a small depression, or the word *front*. Always look for these marks when removing and installing pistons.

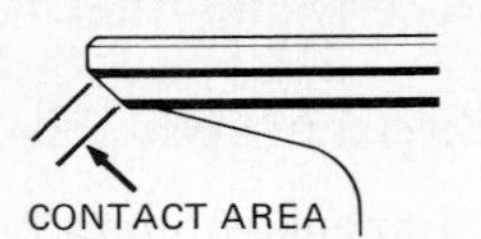


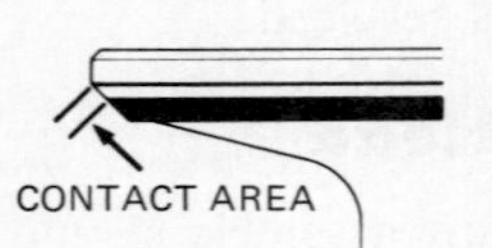


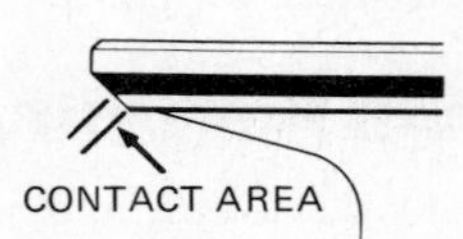


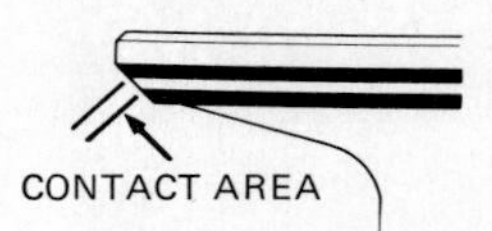


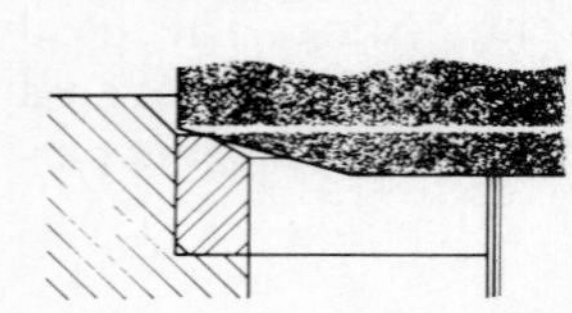


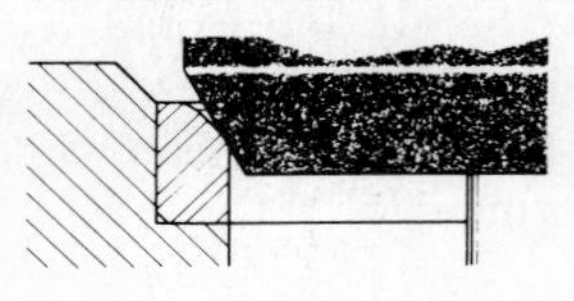


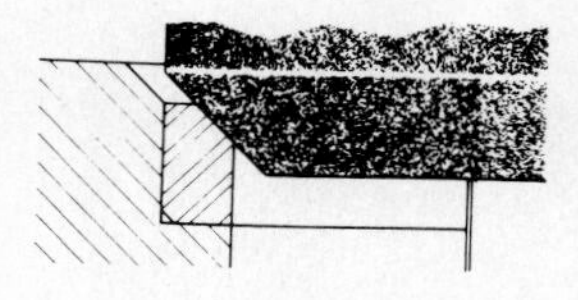


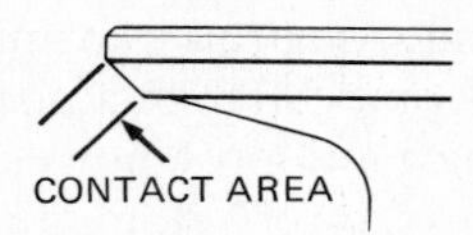


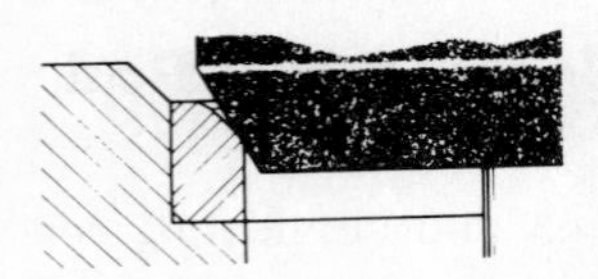


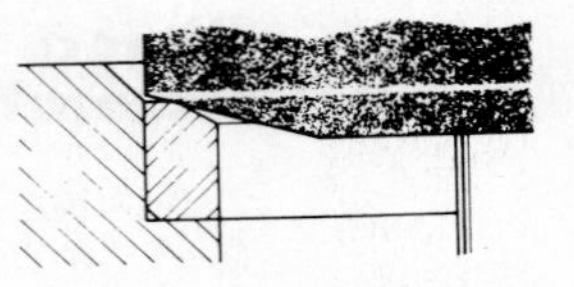


**Fig. 14-23.** Adjusting the position and width of valve seats. (*J.I. Case Co.*)

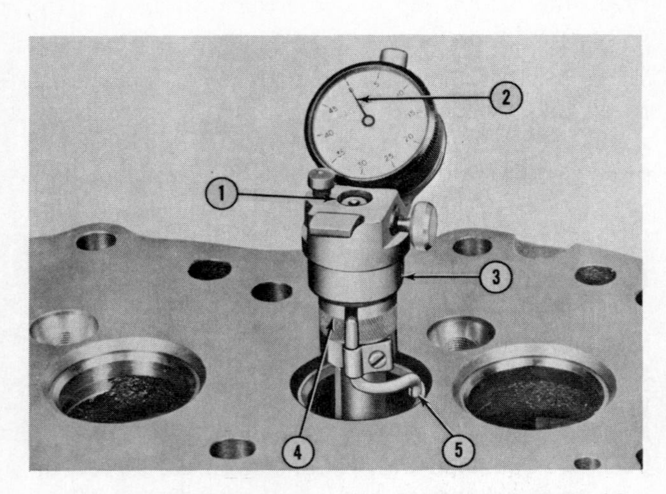


**Fig. 14-24.** Valve seat concentricity can be checked with a valve seat dial indicator. The rotating mount (3) is centered in the guide by the pilot (4) which is secured by the adjusting screw (1). The contact point (5) is adjusted to contact the seat. The face on the dial indicator (2) is turned so that the indicator pointer is set at 0. As the indicator is rotated, out-of-roundness is determined by watching the indicator pointer. (*Ford Tractor Operations*)

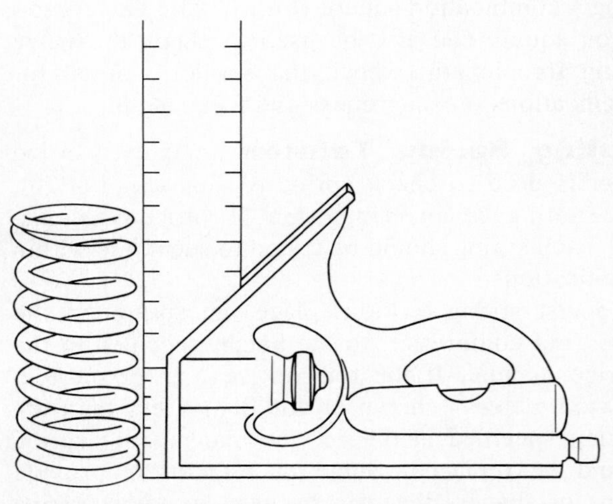

**Fig. 14-25.** Check the valve spring for squareness. (*Ford Tractor Operations*)

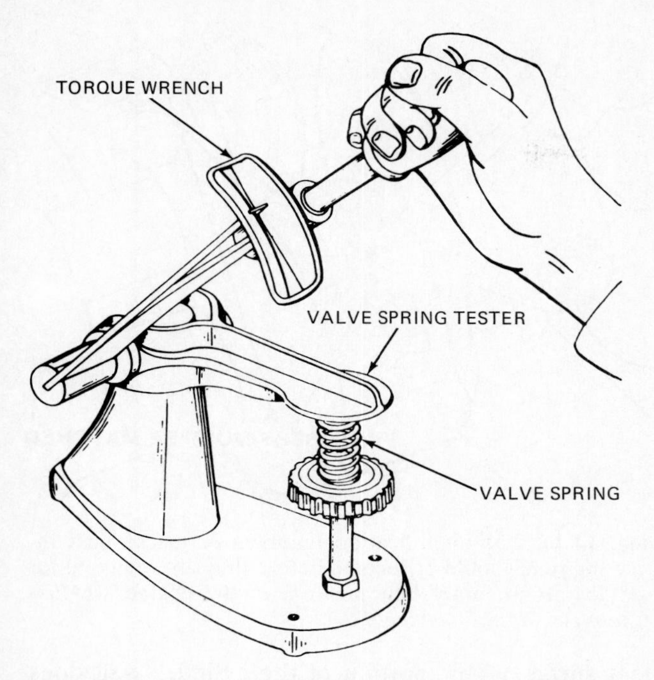


**Fig. 14-26.** Use a valve spring tester to check valve spring tension. (*Allis-Chalmers Manufacturing Co.*)

*Lands* are the flat surfaces between the ring grooves. *Ring grooves* are the machined areas in the pistons where the rings are installed. There are grooves for compression rings and oil control rings.

The *skirt* is the lower part of the piston that forms a bearing area with the cylinder wall and helps hold the piston in place. *Piston pin bosses* are the heavier part of the skirt that has been machined to hold the piston pin. The pin allows the piston to move as the position of the connecting rod changes due to crankshaft rotation.

**The Rings** Piston rings provide a moving seal that prevents combustion gases from passing into the crankcase and oil from passing into the combustion chamber in excessive quantities. Rings also help to cool the piston by transferring heat to the cylinder wall.

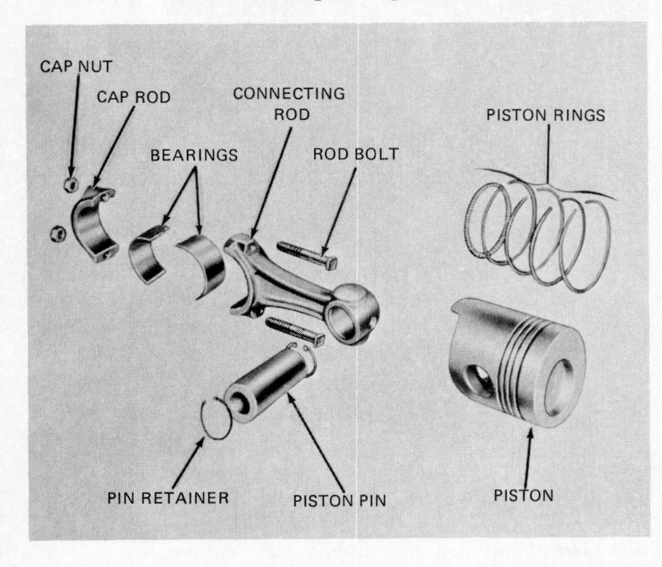


**Fig. 14-27.** Piston and connecting rod assembly. (*Ford Tractor Operations*)

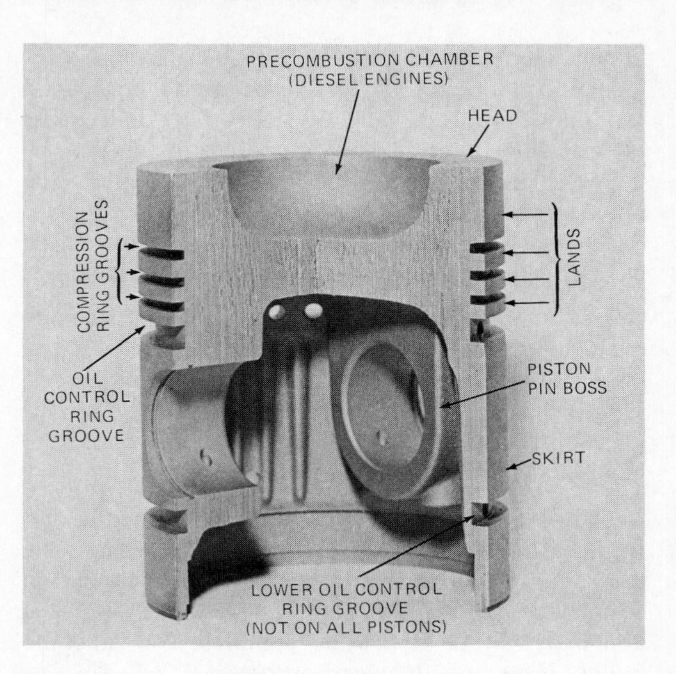


**Fig. 14-28.** A cutaway view of a piston. (*Photograph by Westen McCoy*)

Piston rings are cast iron or steel. Some are filled with special metals or are chrome-plated. They must be self-tensioned so that they will change diameter rapidly to conform to any variations in the cylinder wall.

**Compression rings** Compression rings seal the combustion chamber. The number of compression rings used is determined by the design of the piston and the rings.

The top ring acts as both a compression ring and the final oil control ring. The top ring must seal between the sides of the ring and the ring groove. It must also seal between the face of the ring and the cylinder wall.

The top ring groove wears more rapidly than the other grooves because it is exposed to the highest temperature and pressures. Any dirt or other airborne abrasives that enter the combustion chamber also tend to wear the top ring and groove.

Worn grooves and rings will pump oil. As the piston is forced down, the ring moves to the top of the groove and oil enters under it. When the piston moves back up, some of the oil is forced to the top of the ring. As the piston moves back down, the oil is forced into the combustion chamber.

**Oil control rings** Oil control rings are commonly referred to as *oil rings*. There are several types of oil control rings used.

Cylinder walls are lubricated by throw-off oil from the connecting rod bearings as the piston goes up on the compression and exhaust strokes. Much more oil is normally thrown on the cylinder walls than is required for lubrication. The excess oil must be wiped off by the oil ring, which has two wiping surfaces and is located in a slotted ring groove on the piston.

The excess oil that is wiped from the cylinder walls helps to keep the cylinders, rings, and pistons clean. Particles of carbon, dirt, and metal due to engine wear are trapped in the oil and flow to the crankcase.

**The Connecting Rod** The connecting rods must be strong enough to transmit the thrust of the piston to the crankshaft. They are made as light as feasible to reduce reciprocating weight. The primary parts of a connecting rod are the eye, the shank, the head, and the cap (Fig. 14-29).

**The head and cap** Connecting rods are formed in one piece, and special machines are used to prepare the rods for use in an engine. Since the parts are made as a unit, it is extremely important that connecting rods and caps not be mixed. A complete rod must be bought if a cap is damaged or lost.

Friction-type rod bearings provide a renewable wear surface in the head and cap. They are normally referred to as *rod bearings* or *inserts*.

**The rod eye** The rod eye is the small bore at the top of the connecting rod that attaches to the piston pin. A piston pin bushing is installed in the eye to provide a renewable wear surface. The bushing is honed to provide the correct pin clearance.

## Connecting Rod and Piston Service

When the oil pan is removed, the crankshaft is readily accessible. Check the connecting rod heads and caps for identifying numbers or marks (Fig. 14-30). This is important because all rods and caps must be reinstalled with the marks or numbers matched and facing in the original position. It is good procedure to make a note indicating the position of the connecting rods to be used as a reference when the engine is reassembled.

**Removing the Ring Ridge** There is an area at the top of the cylinder that is above the ring contact surface. This portion of the cylinder wall does not wear, and carbon tends to build up on it. This build-up is referred to as the *ring ridge*.

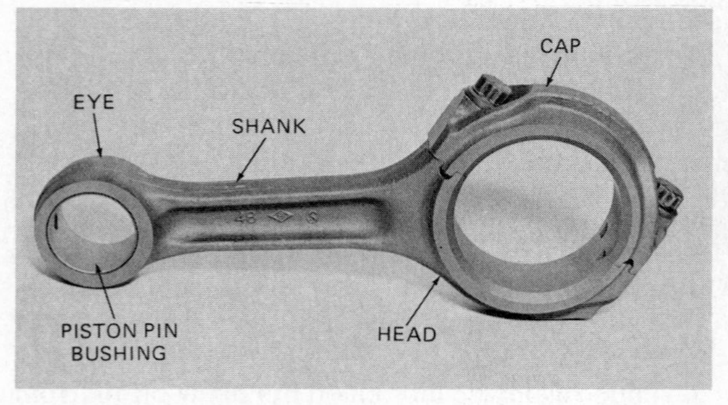


**Fig. 14-29.** A connecting rod assembly. (*Photograph by Westen McCoy*)

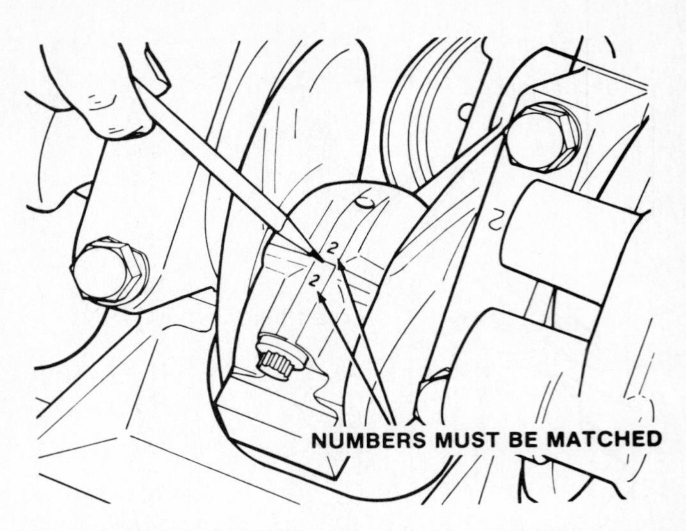


**Fig. 14-30.** The identification numbers or marks on connecting rods should be located before they are removed for service. Also, note their position in the engine. (*Allis-Chalmers Manufacturing Co.*)

The ring ridge can be felt by feeling the inside top of the cylinder wall. The ridge should be removed with a cylinder ridge reamer (Fig. 14-31) before the pistons are removed. Failure to remove the ring ridge can result in broken rings or damage to the pistons as they are removed.

Check the cylinder carefully as the ridge is being removed. Only the ridge should be removed. It is possible to cut too deep and damage the cylinder

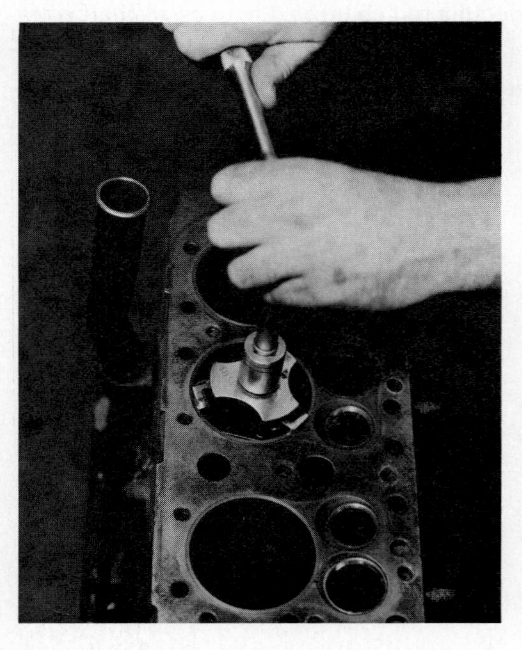

**Fig. 14-31.** The ring ridge should be removed from the block with a ridge cutting tool before the piston and rod assemblies are removed from the cylinder block. (*Photograph by Westen McCoy*)

wall. When a ridge is properly removed, the area will blend in smoothly with the rest of the cylinder.

**Removing the Pistons and Rods** To remove a connecting rod, rotate the engine until the rod journal is near bottom dead center (BDC). Remove the rod cap. If shims are used under the cap, they should be removed and marked so that they can be placed in their original location when the engine is reassembled.

Some rod caps are attached to the rod with cap screws. Other engines used rod bolts and nuts. When bolts are used, it is a good practice to cover the bolts to prevent them from scratching the crankshaft journal as they are removed. Commercial journal protectors or pieces of plastic or rubber hose may be used.

The piston rod assembly is removed by pushing or tapping it out through the top of the cylinder. A hardwood stick or hammer handle may be used. When the rings clear the cylinder, the rod will be free. Care should be taken not to let it drop on the floor.

After the rod is removed, the rod cap should be replaced to protect the bearing bore. This will also help keep the rods and caps together.

After the pistons are removed, they should be checked to determine if abnormal conditions exist that could have led to excessive oil consumption or low compression.

Some conditions that can be detected by inspection are scuffing and scoring, corrosion, preignition damage, detonation damage, and broken or stuck rings (Fig. 14-32).

**Fig. 14-32.** The causes of piston damage include: (*a*) preignition damage; (*b*) detonation damage; (*c*) scuffing and scoring damage; (*d*) corrosion damage. (*Dana Corp.*)

**Cleaning Pistons** The rings should be removed before the pistons are cleaned. Pistons may be cleaned by soaking them in a carbon-removing chemical or by a glass bead machine. The service manual should be checked to determine the recommended cleaning procedure.

**Chemical cleaning** When pistons are cleaned in a chemical cleaner, they should be washed in solvent to remove the oil film. The pistons should then be soaked in the cleaner for the time recommended in the instructions provided with the chemical. Read the instructions before using the chemical. Some chemicals will seriously damage aluminum alloy pistons.

If carbon is still present, the pistons may be soaked a second time. After the pistons are removed from the chemical, they should be rinsed with water and blown dry with compressed air. Some light scraping may be necessary to remove any carbon that is left.

**Glass bead cleaning** If a glass bead machine is available, it will do a good job of cleaning pistons. Pistons should be washed in solvent and blown dry with compressed air before cleaning.

It is possible to damage pistons with a glass bead machine. The bead blast should not be held in one place too long because the metal can be eroded. This can be prevented by keeping the blast moving.

If pistons are cleaned with the connecting rods attached, they should be removed after cleaning. Care should be taken to remove all the glass and carbon particles from all parts.

**Ring groove cleaning** Ring grooves can be cleaned with a ring groove cleaner (Fig. 14-33) or the square end of one of the old rings that has been broken. This second method is not recommended by some manufacturers because hard scratching can damage the finely machined surfaces on the piston.

A wire brush *should not* be used to clean pistons. The brush can scratch the fine surfaces on the piston.

**Checking the Connecting Rods** Connecting rods should be checked for bending and twisting with a rod alignment tool (Fig. 14-34). Any rods that are out of alignment should be replaced or straightened. Heavy connecting rods used on most agricultural engines should be replaced if any bend or twist is detected.

Proper rod alignment is essential, because the piston must work squarely in the cylinder in order for the rings to seal properly. Misaligned connecting rods can cause blowby and excessive oil consumption. They also increase the pressures on rod bearings, pistons, piston pins, and cylinder walls.

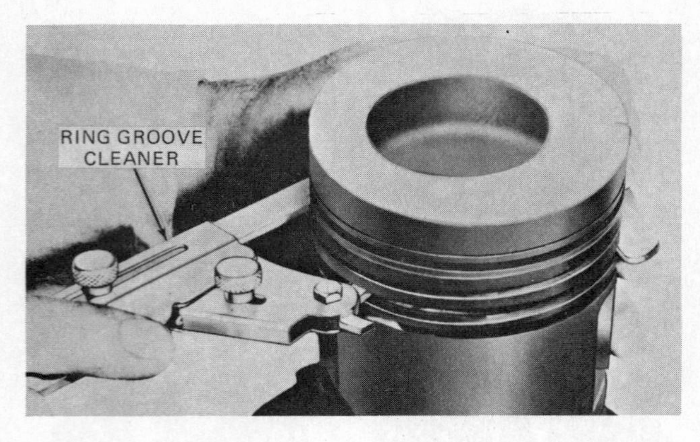


**Fig. 14-33.** Cleaning ring grooves with a ring groove cleaner. *(Ford Tractor Operations)*

**Fig. 14-34.** Use a rod alignment tool to check the connecting rod. *(Photograph by Westen McCoy)*

**Checking the Rod Bearings** A close visual inspection of the rod bearings may identify specific problems that caused engine failure or that should be corrected during overhaul. The bearings should be cleaned in solvent and blown dry. It is good practice to keep bearings in original order, because some problems may exist in only part of the bearings. It is also recommended to keep the top halves separate from the bottom halves.

Some conditions that can be detected by visual inspection are oil starvation, acid corrosion, dirt and other foreign matter, and overheating. Bearing wear can also be used to detect journal taper and bent connecting rods (Fig. 14-35).

**Checking the Piston Pin Bushing** The bushing can be checked for excessive clearance with the piston mounted on the rod. If any side movement can be detected between the piston pin and the bushing, the clearance is excessive.

A telescoping gauge and an outside micrometer are used to check pin clearance. This procedure will be discussed in chapter 15.

## CAMSHAFT

The camshaft has journals that support it in the cylinder block and lobes that open the valves and allow them to close. The camshaft is driven by the camshaft gear, which is driven by the crankshaft gear or idler gear. The camshaft gear has exactly twice as many teeth as the crankshaft gear. This allows the camshaft to turn at one-half the rpm of the crankshaft in four-stroke-cycle engines.

### Checking the Camshaft Gears

Backlash should be checked. *Backlash* is the distance the camshaft gear will move without moving the crankshaft gear. It is actually the play or slack between the gears as they mesh. Excessive backlash will cause valve timing to be off and can cause excessive camshaft vibration.

When the crankshaft pulley and timing gear cover are removed, the camshaft gear, crankshaft gear, and other drive gears on the front of the engine are exposed. The service manual should be consulted to determine the recommended procedure for removing the crankshaft pulley. The pulley and the crankshaft can be damaged if improper procedures are used.

**Checking Backlash with a Thickness Gauge** A thickness gauge can be used to check the backlash in the gears. To check backlash with a thickness gauge, select a gauge that exceeds the maximum allowable backlash by 0.001 in [0.025 mm]. Rotate the camshaft gear until two gear teeth are firmly against each other. With the gears held in this position, try to insert the thickness gauge on the slack side of the teeth (Fig. 14-36). It is a good practice to check the backlash in several places.

**Checking Backlash with a Dial Indicator** A dial indicator can be used to determine the backlash in gears. The indicator should be set up so that the contact stem is as parallel to tooth travel as possible (Fig. 14-37).

To determine the backlash, rotate the camshaft gear until two teeth are firmly against each other. Set the indicator dial face so that the 0 mark is in line with the indicator needle. Then carefully rotate the camshaft gear in the opposite direction. The distance the gear moves can be read on the dial. Be careful not to move the crankshaft gear. It is a good practice to check the backlash on several gear teeth.

Consult the service manual to determine backlash specifications. If backlash exceeds specificiations or

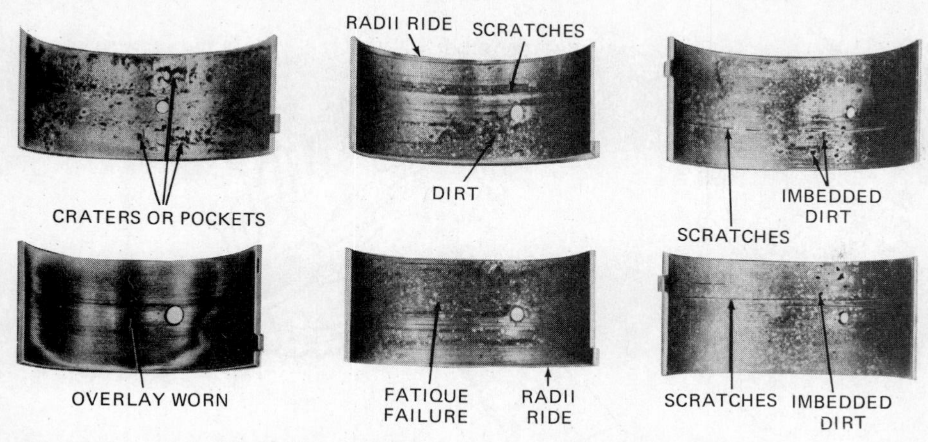


**Fig. 14-35.** Typical defective bearings. (*Ford Tractor Operations*)

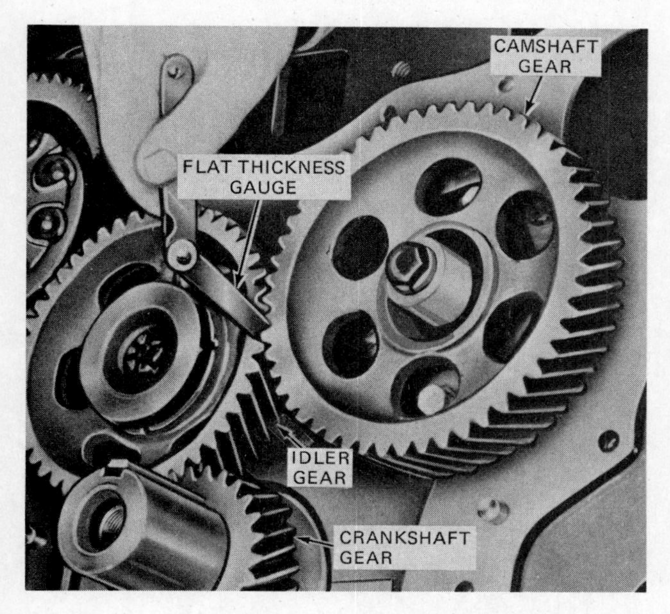


**Fig. 14-36.** Check the timing gear backlash with a thickness gauge. (*Ford Tractor Operations*)

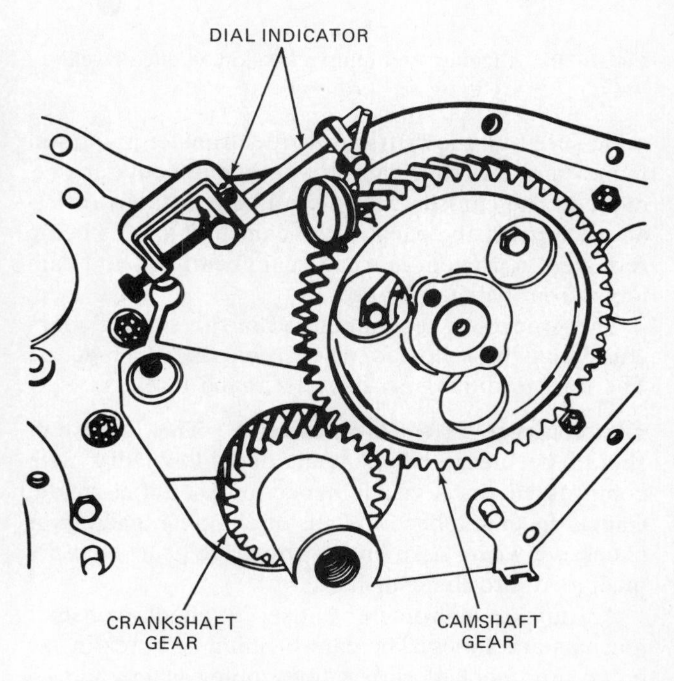


**Fig. 14-37.** Checking the timing gear backlash with a dial indicator. (*Allis-Chalmers Manufacturing Co.*)

if the gears show wear or defects, they should be replaced. It is advisable to replace both gears. Failure to replace one of the gears can result in gear chatter and rapid wear. This is good procedure to follow in any gear drive.

## Camshaft Service

It is a good service procedure to locate all timing marks before removing any gears in the engine gear train (Fig. 14-38). The service manual will give the location of timing marks.

Some agricultural engines do not use timing marks, and occasionally timing chains are used instead of gears. The service manual should be read and the timing procedure should be understood before any components are removed.

**Camshaft Thrust Plates** Determine how the camshaft is held in position. Many camshafts on agricultural engines are held in place by thrust plates that are installed behind the camshaft gear. When a thrust plate is used, it is attached to the front of the cylinder block with cap screws.

The cap screws must be removed through the openings in the camshaft gear. Rotate the engine to align the cap screws with the openings so that they can be removed. Note the shape of the thrust plate and its position. Some thrust plates can be reinstalled in the wrong position. This can result in a loss of engine oil pressure or other problems. Remember to put all fasteners in properly labeled containers.

**Removing the Camshaft** The engine should be positioned so that the oil pan side is up. This will prevent the cam followers or valve lifters from falling out as the camshaft is removed. If lifters come out from the top, they should be removed before the engine is rotated.

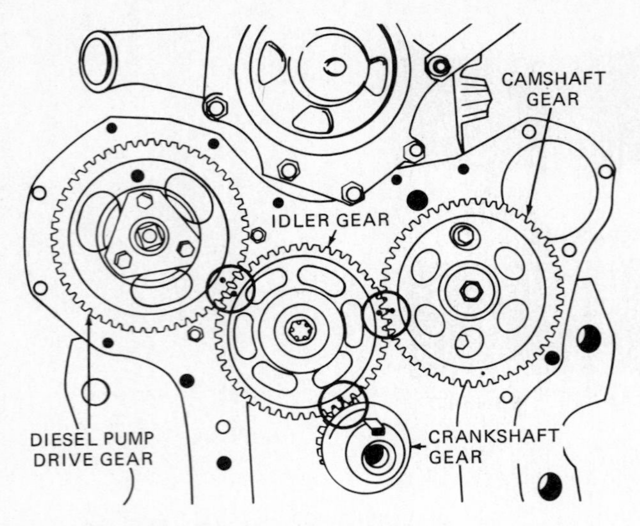


**Fig. 14-38.** Aligning the timing marks on an engine gear train. (*Ford Tractor Operations*)

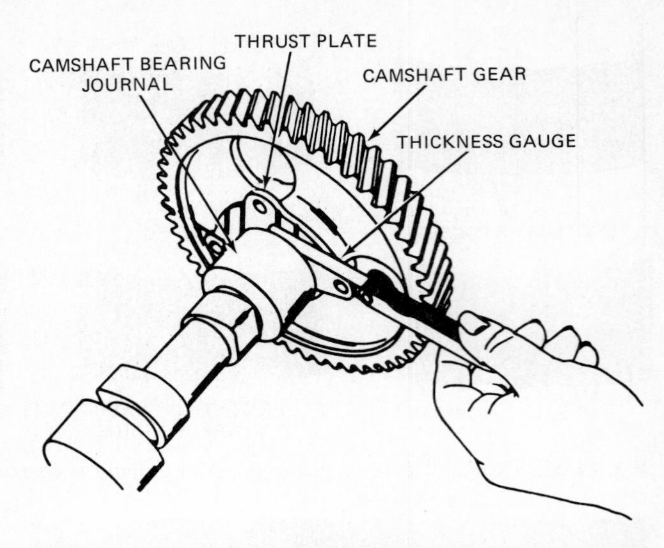


**Fig. 14-39.** Checking camshaft end play with a flat thickness gauge. (*Allis-Chalmers Manufacturing Co.*)

Be sure that the distributor, fuel transfer pump, oil pump, and all other cam-driven components are removed. The camshaft should pull out without resistance. Support the back of the camshaft as it is being removed to prevent journals, cam bearings, and cam lobes from being damaged.

The procedure recommended in the service manual should be followed when removing idler gears. The retainer bolts may have left-hand threads.

**Checking the Camshaft** The camshaft should be cleaned in solvent and blown dry with compressed air. A visual inspection should be made. Check the cam lobes for signs of chipping, galling, or excessive wear. Also check any drive gears that are machined into the camshaft.

An outside micrometer is used to check camshaft journals and lobes. The cam bushing or bores in the block are checked with a telescoping gauge and an outside micrometer. These checks will be discussed in Chap. 15.

**Checking camshaft end play** A flat thickness gauge can be used to check camshaft end play. The end play is controlled by the thrust plate. To check the end play, attempt to insert a thickness gauge 0.001 in [0.025 mm] thicker than the maximum end play allowed by the manufacturer between the thrust plate and the camshaft bearing journal (Fig. 14-39). If the thickness gauge cannot be inserted, camshaft end play is within specifications.

End play can be checked with a dial indicator before the camshaft is removed. This is done by mounting a dial indicator on the front of the cylinder block, as shown in Fig. 14-40. To check the end play, force the camshaft to the rear of the engine and set the dial face to 0. Then force the camshaft forward and note the movement. This will give the amount of camshaft end play.

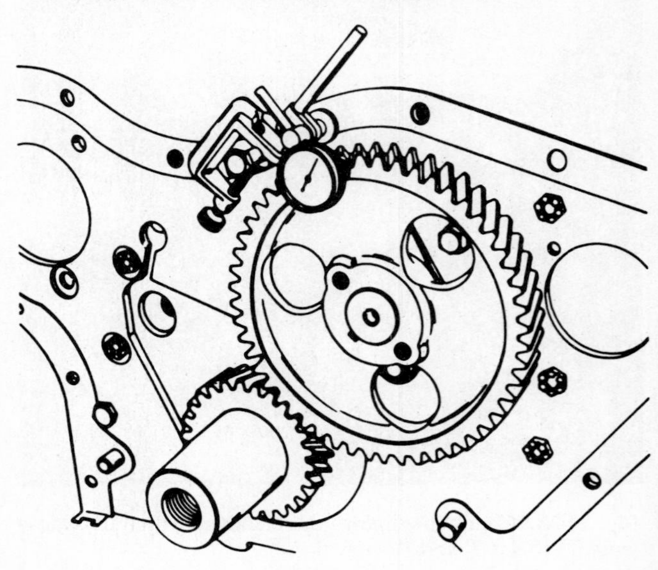

**Fig. 14-40.** Checking camshaft end play with a dial indicator. This setup can also be used to check camshaft gear runout. (*Allis-Chalmers Manufacturing Co.*)

When camshaft end play exceeds specifications, the thrust plate should be replaced. The end play should then be rechecked. If end play still exceeds specifications, the cam gear should be replaced (Fig. 14-41).

## Cam Followers

The cam followers or valve lifters ride the intake and exhaust lobes on the camshaft, changing the rotary motion of the camshaft to a reciprocating motion to open the valves. This reciprocating motion reaches the valves through the push rods and rocker arms.

**Solid Cam Followers** The mushroom-type solid cam followers are commonly used in agricul-

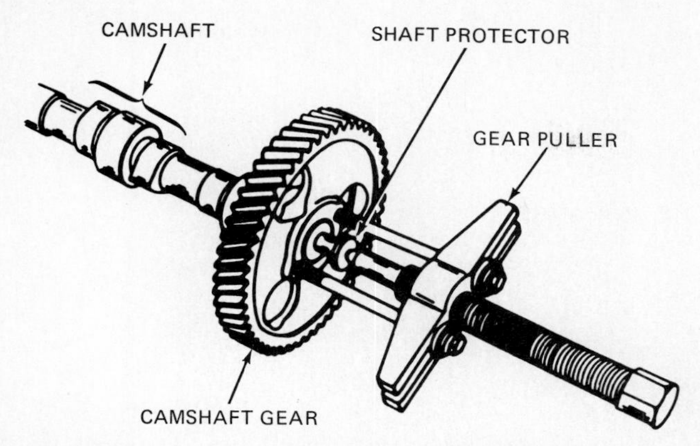


**Fig. 14-41.** Removing the camshaft gear with a gear puller. The camshaft gear may also be removed with a floor arbor press. (*Allis-Chalmers Manufacturing Co.*)

tural engines. They are positioned in bores in the cylinder block that are located above the cam lobes. They can be easily removed after the camshaft has been removed.

**Checking Cam Followers** Cam followers should be kept in original order. They should be washed in solvent and blown dry with compressed air.

Very little wear takes place on cam follower stems or bores in the block. The face is commonly hardened and ground to provide a long-wearing surface against the cam lobes. When operating properly, the cam follower will rotate in the bore.

Examine the face carefully for roughness, scuffing, or excessive wear. If the cam follower has not been rotating properly, there will be a definite wear area on the face. Replace any cam followers that show wear.

**Checking Cam Follower Bores** Inspect the bores in the cylinder block. They should be clean, smooth, and free from score marks. Any score marks or corrosion found in the bores should be removed. It is a good practice to see that all followers work freely in the bores.

## CRANKSHAFT

The crankshaft is supported in the engine block by the main bearings. It converts the power produced by the pistons to rotary power. This power is transmitted to the drive train by the flywheel.

Crankshafts used in agricultural engines are drop-forged steel or steel alloys. They are heat-treated to assure strength and durability. Crankshafts are balanced to reduce engine vibration.

The primary parts of a crankshaft assembly are the main bearing journals, rod bearing or crankpin journals, crank arms, and oil channels to carry oil to the bearings. Friction-type bearings are used to reduce friction and provide a renewable wear surface in the main bearing bores and connecting rod heads (Fig. 14-42).

### Crankshaft End Play

End play is the distance the crankshaft can move forward and backward in the cylinder block. A small amount of end play is necessary for the crankshaft to turn freely. Check the service manual for recommended end play.

It is good procedure to check end play before removing the crankshaft. End play can be checked with a thickness gauge or a dial indicator.

**Using the Thickness Gauge** To check crankshaft end play with a thickness gauge, pry the shaft to the front or rear of the engine. Select a thickness gauge that is 0.001 in [0.025 mm] thicker than the maximum end play specified in the service manual. Try to insert the thickness gauge between the thrust surface on the crankshaft and the thrust flange on the main bearing (Fig. 14-43).

If the thickness gauge can be inserted, the end play is excessive. The end play can be determined by finding a thickness gauge that has a slight drag as it is inserted.

**Using the Dial Indicator** To check end play with a dial indicator, mount the dial indicator on the cylinder block (Fig. 14-44). Force the crankshaft forward and set the contact stem on the end of the crankshaft. Set the dial face to 0. Then force the crankshaft to the rear of the engine and note the amount of indicator movement. This will give the amount of end play in the crankshaft.

### Removing the Crankshaft

Position the engine so that the main bearing caps are up. Check the main bearing caps for identifying marks. It is essential that all main bearing caps be reinstalled in their original position. Mixing or reversing the caps even though they look identical and have the same casting number can cause serious misalignment.

When no marks are found on the main caps, they should be marked with a center punch or a numbering stamp. Do not attempt to mark caps near the center. This could distort the cap. Caps should be marked on a heavy section near the bottom. Be sure to make adequate notes so that they can be used during reassembly.

**Remove the Main Bearing Caps** Locking devices are often used to keep the cap screws from working loose. These should be removed or bent down. Loosen the cap screws and remove the main bearing caps and the lower half of the main bearing.

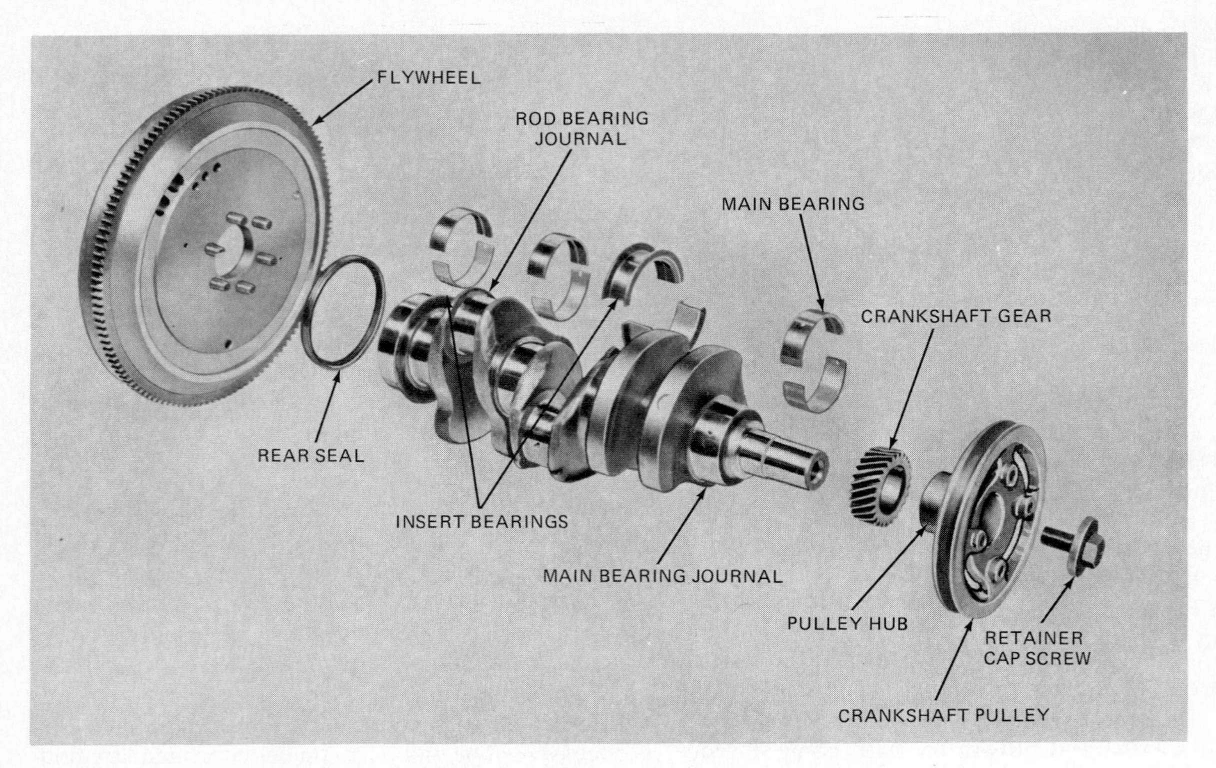


**Fig. 14-42.** The crankshaft assembly. (*Ford Tractor Operations*)

**Fig. 14-43.** Checking crankshaft end play with a flat thickness gauge. (*Photograph by Westen McCoy*)

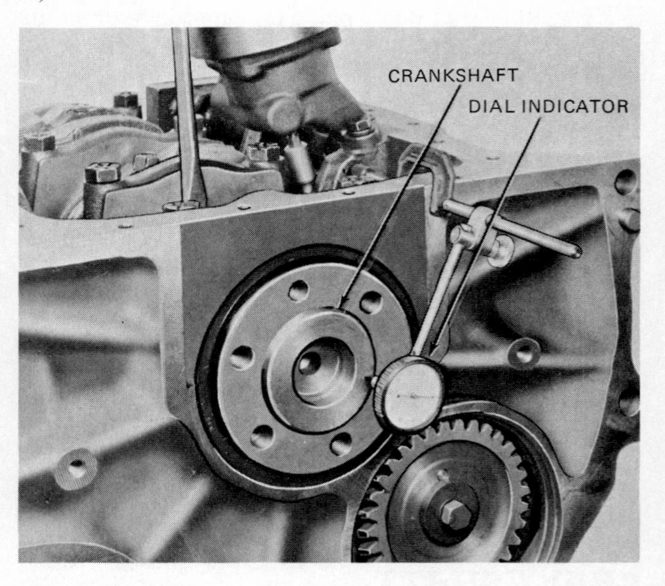


**Fig. 14-44.** Checking crankshaft end play with a dial indicator. (*Ford Tractor Operations*)

Check the cap screws as they are removed to see that they are all the same length and diameter.

Some main caps lift off easily. Others are hard to remove. A plastic hammer can be used to loosen caps. Dowel pins are used in some main caps to ensure proper alignment. All caps should be lifted straight up as they are removed.

**Engine Balancers** Some agricultural engines are equipped with engine balancers that are driven by the crankshaft. Balancers use two rotating gears with counterweights to balance the engine mechanically.

Before the balancer unit is removed, rotate the engine and look for timing marks on the crankshaft gear and the balancer drive gear. The balancer is timed to the crankshaft. Check the service manual to determine any service recommendations.

**Lifting the Crankshaft** The crankshaft is removed from the engine block by lifting it straight up. Be careful not to damage journals or thrust surfaces.

Crankshafts in smaller engines can be removed by hand. A lift swing can be used to lift heavy crankshafts.

## Checking the Crankshaft

Clean the crankshaft in solvent. A valve guide or rifle cleaning brush can be used to clean the oil channels. Use compressed air to blow out the oil channels and dry the shaft. All journals should be lightly oiled to protect the highly polished surfaces from rust.

Check the journals for scoring, chipping, or signs of overheating. Overheating will cause discoloration or bluing of the crankshaft journals. If overheating has occurred, the journals should be checked for cracks.

An outside micrometer is used to check the journals for wear. The procedure for checking journals is discussed in the next chapter.

**Checking Crankshaft Alignment** Crankshaft alignment or runout can be checked on a stand equipped with V-blocks. If V-blocks are not available, alignment may be checked in the cylinder block (Fig. 14-45).

To check alignment in the cylinder block, remove the center main bearings. Mount a dial indicator on the block and set the contact stem on the center of a journal. Turn the dial face and align the 0 mark with the dial indicator needle. When the crankshaft is turned, the runout or wobble detected should be less than the maximum running clearance given in the service manual.

Check all journals. The end journals can be checked by removing the end bearing and installing a center bearing. Wear in bearings and journals that are out-of-round can affect the alignment test.

If misalignment is indicated, the crankshaft should be rechecked with new bearings. This should be done after the crankshaft journals have been checked for wear.

**Storing the Crankshaft** It is important that the crankshaft be properly stored while it is out of the engine. The crankshaft may be stored on its end. It is good practice to secure the shaft to a table leg so that it will not fall if this method is used. Soft rope should be used to secure the shaft. V-blocks can also be used to support a crankshaft while it is out of the engine. Never lay a crankshaft on a flat surface. This can cause misalignment.

**Fig. 14-45.** Checking crankshaft alignment with a dial indicator. (*Photograph by Westen McCoy*)

## Checking Main Bearing Bores

The main bearing bores are align-bored and should be in alignment. Bores should be checked for alignment and distortion. Remove the bearings and clean the bore and main caps before any tests are made.

**Checking Alignment** To check bore alignment, place a straight edge across the center of the bores (Fig. 14-46). If a 0.0015-in [0.038-mm] thickness gauge can be inserted between the straight edge and any bore, align-boring or honing will be required. Check the service manual for alignment specifications.

**Checking Distortion** To check bores for distortion, remove the bearings, install the main caps, and torque them to the specifications given in the service manual. Use a telescoping gauge or an inside micrometer (Fig. 14-47) to check the bores for out-of-roundness. If out-of-roundness exceeds the limits given in the service manual, align-boring or honing will be required.

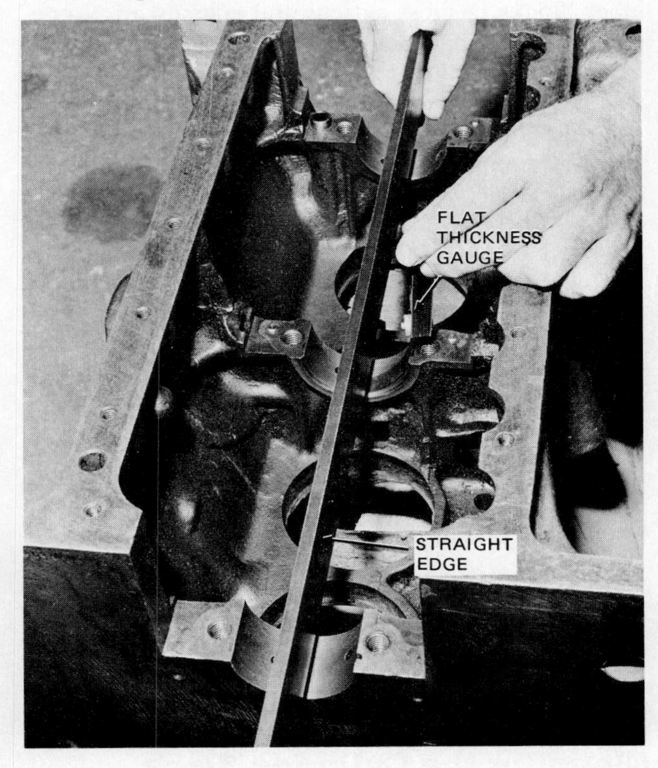


**Fig. 14-46.** Checking the main bearing bore alignment with a straightedge and a flat thickness gauge. (*Photograph by Westen McCoy*)

**Fig. 14-47.** Check the main bearing bores for distortion using a telescoping gauge and outside micrometer or an inside micrometer. Measure bore at three positions as indicated by the arrows. (*Photograph by Westen McCoy*)

Align-boring and honing are done to realign the main bearing bores. Special tools are required for both processes, and the work is usually done in a machine shop. Align-boring is not commonly needed on agricultural engines.

## CYLINDER BLOCK

The cylinder or engine block is the foundation for all the other basic engine parts and components. Cylinder blocks in most agricultural engines are a one-piece casting made of gray cast iron or alloy cast iron. Die-cast aluminum alloy blocks are used in some air-cooled engines.

Cylinder blocks are cast with end walls and center webs to support the crankshaft and camshaft. The main oil gallery in the block supplies engine oil to all moving parts. Coolant is circulated in the water jacket to cool internal parts.

Cylinders may be cast into the block with the block machined or bored and used as the cylinder wall. Many agricultural engines are equipped with cylinder liners to provide a renewable wear surface in cylinders. This allows the cylinder to be reconditioned without reboring when cylinder wear exceeds specifications.

### Cylinder Block Service

All components such as the diesel injection pump, idler gears, expansion plugs, and oil gallery plugs should be removed. Old gasket material should be scraped off the milled surfaces and other gasket flanges.

### Cleaning the Cylinder Block

The cylinder block can be cleaned with a steam cleaner. Solvent should be used in the water to dissolve sludge and other foreign material that has collected. After cleaning, the block should be flushed with water or treated with solvent-free steam to remove the solvent.

**Hot Tank Cleaning** If the water jacket is heavily scaled, the cylinder block should be cleaned in a heated vat containing a commercial cleaning agent. After cleaning, the block should be rinsed to remove the acid solution and dipped in an alkaline tank to neutralize the cleaning agent.

When blocks are cleaned in a hot tank, all bearings and plugs should be replaced. This type of cleaning will dissolve cam bearings and expansion plugs. O-rings on wet cylinder liners will also be dissolved.

The cylinder block should be dried with compresser air after either in steam cleaning or after cleaning in a hot tank. A light coat of oil should be applied to the cylinder walls, gasket surfaces, and bearing bores to prevent rusting. Remember that it is good procedure to install new expansion plugs.

**Cleaning Aluminum Blocks** Special precautions must be taken when cleaning aluminum cylinder blocks. Check the service manual for recommended cleaning procedures.

### Check Cylinder Block for Cracks

Both overheating and freezing can cause the cylinder block to crack. Check the milled surface for cracks between cylinders and between the water jacket and oil passages.

The cylinder walls or cylinder liners and the lower bosses between the cylinders should be checked. Also check for external cracks in the water jacket.

**Magnetic Crack Detector** A magnetic crack detector can be used to check suspected areas on the milled surface.

**Pressure Testing** The cylinder block can be pressure-tested for cracks. When pressure testing, all water openings in the block are sealed. The block is immersed in a tank of hot water and allowed to soak for 15 to 20 minutes (min). The block is then pressurized with compressed air. Any leaks will produce bubbles in the water.

### Checking for Warpage

The milled gasket surface on the top of the cylinder block should be checked for warpage. A straight edge and a flat thickness gauge should be used to check for any uneveness.

The surface should be checked through the center and diagonally. If any warpage is found, refer to the service manual to determine the manufacturer's specifications.

When warpage exceeds specifications, the cylinder block must be resurfaced. Special milling equipment is needed to resurface blocks, and that job is usually done in a machine shop.

## SUMMARY

A complete engine overhaul is performed when the engine is removed from the tractor and all internal parts are checked for wear and are reconditioned or replaced.

The service manual should be used as a guide during engine disassembly. It will give the specific recommendations and procedures for disassembling, checking, and servicing the various components.

All parts should be cleaned and inspected for wear or unusual conditions. A parts list should be started, and any parts found to be defective should be included on the list.

All fasteners should be stored in properly labeled containers to make reassembly easier. Internal engine parts should be kept in original order. This will enable them to be reinstalled in their original position in the engine.

## THINKING IT THROUGH

1. Why should an engine be allowed to cool down before the cylinder head is removed?
2. How is a valve guide checked for wear?
3. Discuss three ways worn valve guides can be repaired.
4. What is the valve margin?
5. Explain how the milled surface on a cylinder head is checked for warpage.
6. Explain how valve seat width and valve contact area can be corrected.
7. How is valve spring tension tested?
8. What is the purpose of compression rings? Oil rings?
9. How do worn rings and ring grooves cause an engine to pump oil?
10. Explain why it is essential that connecting rod caps be reinstalled in their original position.
11. What is the ring ridge?
12. Explain why a wire brush should not be used to clean pistons.
13. Explain what backlash is and tell how it is checked.
14. How is the camshaft timed to the crankshaft?
15. Explain the purpose of cam followers.
16. What is crankshaft end play?
17. Why is it essential that main bearing caps be marked?
18. How is crankshaft alignment checked?
19. Explain the recommended procedure for storing a crankshaft while it is out of the engine.
20. How is main bearing bore alignment checked?
21. Why are cylinder liners used in many agricultural engines?

# CHAPTER 15 DETERMINING PART NEEDS AND ORDERING PARTS

## CHAPTER GOALS

In this chapter your goals are to:

- Use precision measuring tools and measure engine components for wear
- Use manufacturer's specifications and make service recommendations
- Make out a parts list using measurements taken and specifications given in the service manual

One way to recondition an engine is to replace all bearings and replaceable parts included in an overhaul kit. This method is expensive and may result in poor performance or early engine failure.

Skilled agricultural machinery service workers use precision measuring tools to accurately determine the condition of all engine components. They use the specifications given in the service manual to determine if the parts should be reused, reconditioned, or replaced.

Successful service people take the time to read service manuals and become familiar with a machine before service is performed. They also have a good understanding of basic operating principles.

There is a good demand for skilled service workers in the agricultural industry because their work is dependable. Dealers like to build a reputation for dependable service because agricultural producers expect good service. Many producers consider service when purchasing tractors and other agricultural equipment.

## PRECISION MEASURING TOOLS

Precision measuring tools are as vital as a good set of wrenches. They should be used to check internal engine parts for proper size, clearance, and alignment.

The precision measuring tools commonly used in engine service are the outside micrometer, inside micrometer, small-hole gauge, thickness gauge, telescoping gauge, dial indicator, and straight edge. Precision measuring tools that read in thousandths of an inch (U.S. Customary System) or millimeters (metric system) may be purchased. Some precision tools provide both scales.

Precision tools require special handling. They should be wiped clean with a lightly oiled, lint-free cloth after each use and stored in a protective case. It is good procedure to check them for accuracy each time they are used.

Refer to Chap. 5 for a review on how to use precision measuring tools.

## ENGINE COMPONENT APPRAISAL

It is essential that all bores, cylinders, and moving engine components be checked for size, clearance, and alignment. Some checks such as timing gear backlash, crankshaft end play, and crankshaft runout were checked as the engine was being disassembled.

The cylinder block and cylinder head were checked for physical damage, trueness of milled surfaces, and main bearing bore alignment. Valve guides were checked to determine taper and inside diameter. Guides that exceeded wear limits given in the service manual were reconditioned by knurling, reaming, or replacement.

Valves were checked for visual damage and stem wear. The valves that were within specifications for stem wear and had no visual damage were reground. Rocker arms were visually inspected, and rocker arms with wear on the contact pad were reconditioned.

Connecting rods were checked for proper alignment. The piston pin and piston pin bushing were checked for wear. Pistons were checked for abnormal wear and physical damage. The camshaft was checked for excessive wear and galling or pitting of the cam lobes.

Other internal parts such as rod and main bearings were checked to see if wear was normal or abnormal. All parts were cleaned, properly lubricated, and stored.

In this chapter, the appraisal of engine components using common precision measuring tools will be discussed. Manufacturers often recommend the use of special tools to make specific checks. Since these tools are special tools and usually available only in dealer shops, they will not be discussed. If it is necessary to use one of these tools, the instructions provided in the service manual should be followed. It should be noted that most checks done with special tools can also be done with the common measuring tools.

## Manufacturer's Specifications

The service manual should be used to determine the tolerance ranges or wear limits and specific clearances recommended by the manufacturer for internal engine components. The use of rule-of-thumb specifications can lead to component failure in modern high-speed diesel engines. A good service worker looks up all specifications each time service is performed.

## Rocker Arm Assembly

The rocker arm shaft and the rocker arm bushings should be checked for excessive wear (Fig. 15-1). Use an outside micrometer to check the shaft surfaces where the rocker arms work. A small telescoping gauge and outside micrometer are used to check the bushings in the rocker arms.

When the bushings in the rocker arms are worn beyond specifications, it is usually recommended that the rocker arm assembly be replaced. Check the service manual for the manufacturer's recommendations.

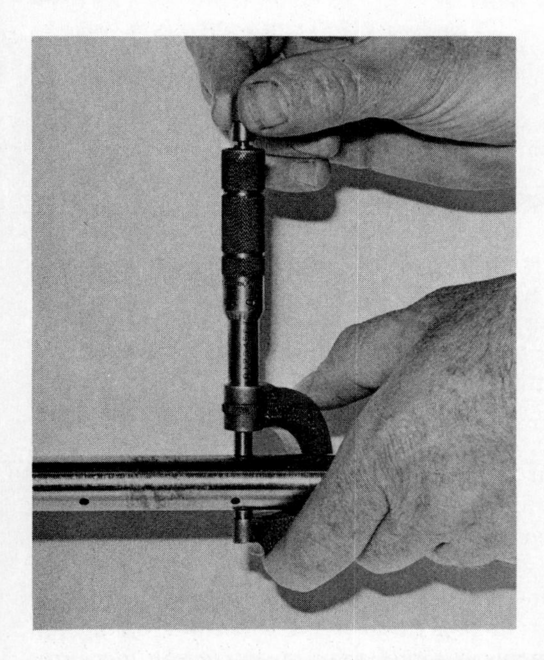


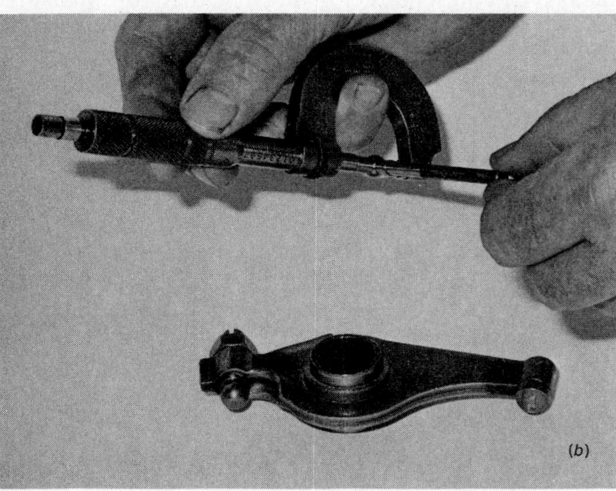


**Fig. 15-1.** (*a*) Measuring a rocker arm shaft. (b) Rocker arm bushing.' (*Photograph by Westen McCoy*)

## Cylinders

Many tractor engines are equipped with cylinder liners. Cylinder liners may be the dry-type or the wet-type (Fig. 15-2). Cylinder liners provide a renewable wear surface in the cylinder block. They can also be made of special alloys which provide a longer-lasting cylinder.

**Dry Cylinder Liners** Dry cylinder liners are sleeves which are installed in a cylinder that was machined for the liner. They usually have a sealing lip at the top that helps to position the liner and hold it in place.

Some agricultural engines that have integral cylinders can be reconditioned by boring the cylinders and installing dry liners. Manufacturers normally stock liners for these engines.

**Wet Cylinder Liners** Wet liners are installed in cylinder blocks that are cored so that the outside surfaces of the liner form a part of the water jacket. Wet liners usually have a sealing lip at the top and are sealed at the bottom by O-rings. The O-rings may be installed in grooves in the liner or in the cylinder block.

Some manufacturers use an O-ring to seal the top of the liners. The cylinder head gasket and the sealing lip are used to seal the top of most wet liners.

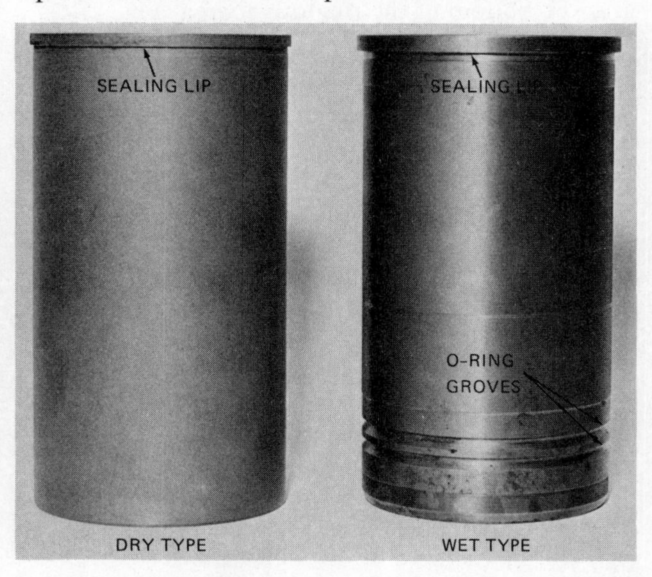


**Fig. 15-2.** Dry- and wet-type cylinder liners. (*Photograph by Westen McCoy*)

## Checking Cylinders

Some wear occurs in the cylinders as the rings move up and down the cylinder walls. The amount of wear depends on the type of materials the rings and cylinders are made of, hours of service, and maintenance.

The cylinders should be inspected in the ring travel area. When the appearance is good and there are no scuff or score marks deep enough to require that the liners be replaced or the block rebored or honed, the cylinders should be checked for taper and out-of-roundness.

**Cylinder Taper** Cylinders tend to wear more at the top of the ring travel area than they do at the bottom (Fig. 15-3). This tends to produce taper in the cylinder walls. When cylinder walls become tapered, the rings must move in and out as the piston moves up and down.

**Out-of-Roundness** Cylinders tend to wear more on the thrust side. This is caused by the thrust forces exerted on the piston as it goes down on the power stroke. The thrust side on most agricultural engines is the left side of the cylinder block as it is viewed from the flywheel end.

**Measuring Cylinder Wear** Cylinder inside diameter can be measured with an inside micrometer or a telesoping gauge and an outside micrometer. It is easier to get accurate readings if a telescoping gauge and outside micrometer are used (Fig. 15-4).

Each cylinder should be measured parallel to the crankshaft near the top of the ring travel and measured again at a right angle to the crankshaft. The cylinders should then be measured in the same positions near the bottom of the ring travel area (Fig. 15-4).

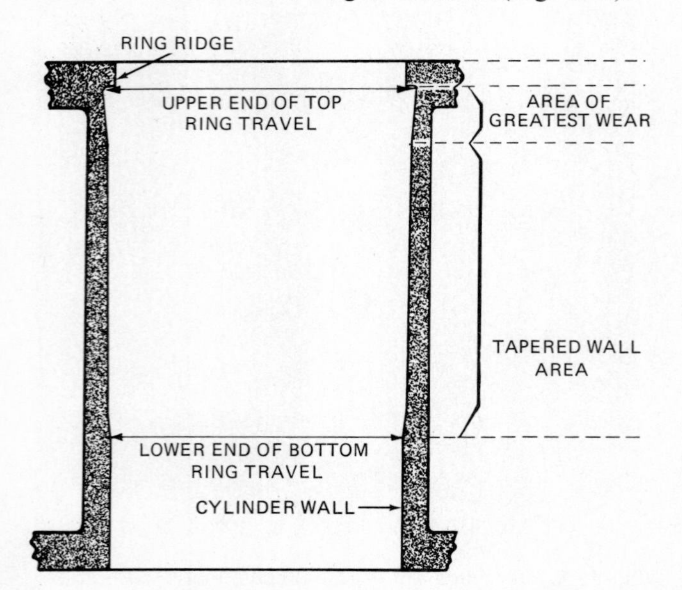


**Fig. 15-3.** Cylinders tend to wear more near the top of the ring travel area. (*Dana Corp.*)

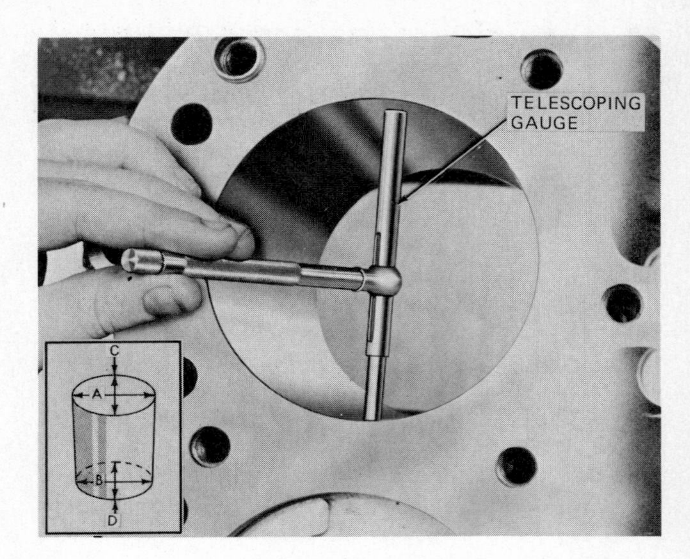


**Fig. 15-4.** At least four cylinder wear measurements are taken at right angles to each other at both the top (A and C) and bottom (B and D) of the ring travel area. These measurements are taken with a telescoping gauge and an outside micrometer. (*Ford Tractor Operations*).

**Calculating Cylinder Wear** As the measurements are made, they should be recorded. A table is helpful when recording measurements and calculating wear (see Table 15-1).

Cylinder taper is present when the lengthwise or crosswise measurement is greater at the top than it is at the bottom. To determine taper, subtract the bottom reading from the top reading (Table 15-1). The crosswise measurements for cylinder No. 1 are 4.236 inches (in) [107.59 millimeters (mm)] and 4.227 in [107.37 mm]. Taper is 0.009 in [0.23 mm]. Since this is the largest difference calculated in measurements obtained at clockwise or lengthwise positions, maximum cylinder taper is 0.009 in.

Out-of-roundness is present when one of the top measurements is larger than the other top measurement. Out-of-roundness may also be detected at the bottom of the ring travel area. To determine out-of-roundness, subtract the smaller measurements from the larger measurements. (Table 15-1). The top measurements for cylinder No. 1 are 4.229 in [107.42 mm] and 4.236 in [107.59 mm]. The difference in the two measurements is 0.007 in [0.18 mm]. Since this is the largest difference calculated in measurements obtained at the top or bottom of the ring travel area, maximum cylinder out-of-roundness is 0.007 in [0.18 mm].

When taper or out-of-roundness exceeds specifications on one cylinder, it is a common practice to replace all cylinder liners or recondition all cylinders. Therefore, the maximum taper and out-of-roundness found for all cylinders is used to determine wear. Check the service manual and compare measurements taken and wear calculated to manufacturer's specifications. If wear exceeds the manufacturer's

**TABLE 15-1. Determining Cylinder Wear***

| | U.S. CUSTOMARY MEASUREMENTS | | | |
|---|---|---|---|---|
| | Lengthwise Measurement | Crosswise Measurement | Amount of Out-of-round Wear | Amount of Taper Wear |
| No. 1 cylinder | | | | |
| Top of ring travel | 4.229 | 4.236 | 0.007† | 0.003 |
| Bottom of ring travel | 4.226 | 4.227 | 0.001 | 0.009† |
| No. 2 cylinder | | | | |
| Top of ring travel | 4.230 | 4.234 | 0.004 | 0.004 |
| Bottom of ring travel | 4.226 | 4.227 | 0.001 | 0.007† |
| No. 3 cylinder | | | | |
| Top of ring travel | 4.230 | 4.233 | 0.003 | 0.004 |
| Bottom of ring travel | 4.226 | 4.227 | 0.001 | 0.006† |
| No. 4 cylinder | | | | |
| Top of ring travel | 4.230 | 4.235 | 0.005 | 0.004 |
| Bottom of ring travel | 4.226 | 4.227 | 0.001 | 0.008† |
| No. 5 cylinder | | | | |
| Top of ring travel | 4.231 | 4.235 | 0.004 | 0.005 |
| Bottom of ring travel | 4.226 | 4.227 | 0.001 | 0.008† |
| No. 6 cylinder | | | | |
| Top of ring travel | 4.230 | 4.232 | 0.002 | 0.004 |
| Bottom of ring travel | 4.226 | 4.227 | 0.001 | 0.005 |

* Standard cylinder bore is 4.225, allowable out-of-roundness is 0.005, and allowable taper is 0.005.
† Exceeds maximum wear.

specified limits, the cylinders should be reconditioned.

**Using the cylinder gauge** A cylinder gauge (Fig. 15-5) can be used to make a quick check of cylinder condition. It can be used to detect taper and out-of-roundness but will not give cylinder diameter.

A piston ring and a flat thickness gauge can also be used to detect taper. A difference in the gap between the ends of a ring at the top and bottom of the cylinder indicates taper.

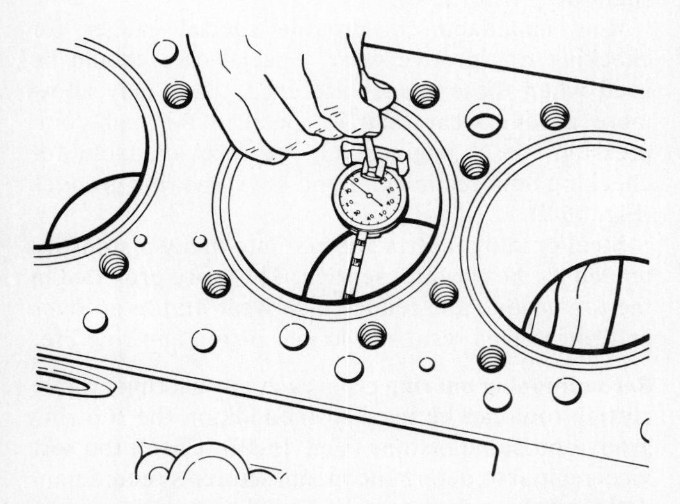

**Fig. 15-5.** A cylinder gauge equipped with a dial indicator can be used to determine if cylinders are worn. (*Allis-Chalmers Manufacturing Co.*)

## Pistons and Piston Pins

New pistons and pins are normally installed when cylinder liners are replaced or the cylinders are rebored. Pistons and pins may be worn beyond manufacturer's specifications or have physical damage even though the cylinders are within specifications.

**Measure Piston Skirt Wear** To determine piston skirt wear, measure the diameter of the piston skirts across the thrust faces which are at a right angle to the piston pin bosses. The measurement is made with an outside micrometer (Fig. 15-6). Measure the pistons at the top and bottom of the skirt. The pistons should be replaced if wear exceeds the specifications given in the service manual.

**Piston Clearance** Piston clearance is the clearance between the cylinder and the piston. When new cylinder liners and pistons are installed, piston clearance should be checked for each piston in the cylinder it is to be installed in. Insufficient clearance can result in premature failure of the pistons, cylinder liners, or both.

Clearance should also be checked if original cylinders are to be used. Pistons tend to wear with service, and the clearance may be excessive. Excessive

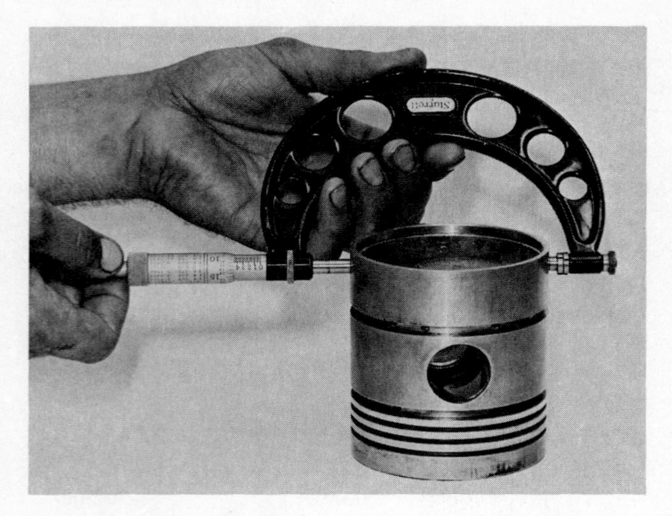

**Fig. 15-6.** Measuring piston skirt diameter with an outside micrometer. (*Photograph by Westen McCoy*)

clearance allows the piston to tilt in the cylinder and may result in oil consumption and blowby.

**Measuring piston clearance** Use the telescoping gauge and outside micrometer to measure the cylinder diameters. If original cylinders are used, the measurements should be taken at the lower end of the ring travel area and at right angles to the crankshaft.

Measure the pistons across the thrust faces with an outside micrometer. Subtract the piston diameters from the cylinder diameters. The difference is piston clearance.

Piston clearance can be checked with a feeler ribbon. This check is done by inserting the pistons and a 1/2-in- [12.7-mm-] wide feeler ribbon of specified thickness into the cylinders. The withdrawal resistance of the feeler ribbon is determined with a spring scale (Fig. 15-7). Check the service manual for feeler ribbon thickness, withdrawal resistance, and test procedures.

## Ring Grooves

The ring grooves hold the rings parallel to the cylinder walls. The sides of a ring groove should be flat, parallel, and smooth, and ring side clearance should be within specifications.

A new ring installed in a worn ring groove will sag. This causes the upper edge of the ring face to come into contact with the cylinder wall and results in oil being wiped into the combustion chamber. Excessive ring side clearance will cause the edges of the ring face to wear.

**Checking ring grooves** Ring grooves can be checked with a new piston ring and a flat thickness gauge. Check the service manual for recommended side clearance. Note that different side clearances are often specified for the top compression rings, lower compression rings, and oil rings.

Insert the new ring in a properly cleaned groove and select a thickness gauge 0.001 in [0.025 mm]

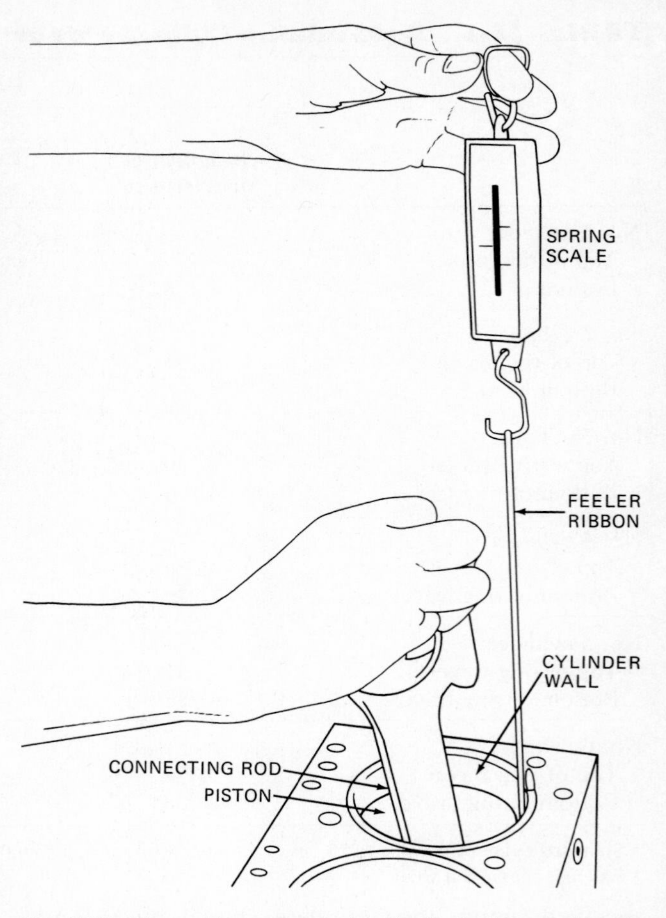


**Fig. 15-7.** Piston clearance can be checked with a feeler ribbon and a spring scale. (*International Harvester Co.*)

thicker than the recommended side clearance. If the gauge can be inserted in the groove, clearance is excessive. Since ring grooves tend to wear unevenly, it is good procedure to check the groove at several points. Groove clearance can be determined by selecting the thickness gauge that can be inserted with a slight drag (Fig. 15-8).

Ring manufacturers provide special gauges for checking ring groove wear. Special care should be used when these gauges are used. They may allow more groove wear than is specified for high-compression diesel engines. Gauges are available for checking both rectangular and keystone ring grooves (Fig. 15-9).

Steel or alloy inserts are cast into many pistons intended for heavy-duty service. These are provided in the top groove and reduce side wear on top grooves and rings. This results in longer piston and ring life.

**Reconditioning top ring grooves** A special ring groove cutting tool can be used to recondition the top ring groove on some pistons (Fig. 15-10). Check the service manual to determine manufacturer's recommendations before grooves are reconditioned.

When the top grooves are reconditioned, a flat steel spacer is used to compensate for the metal re-

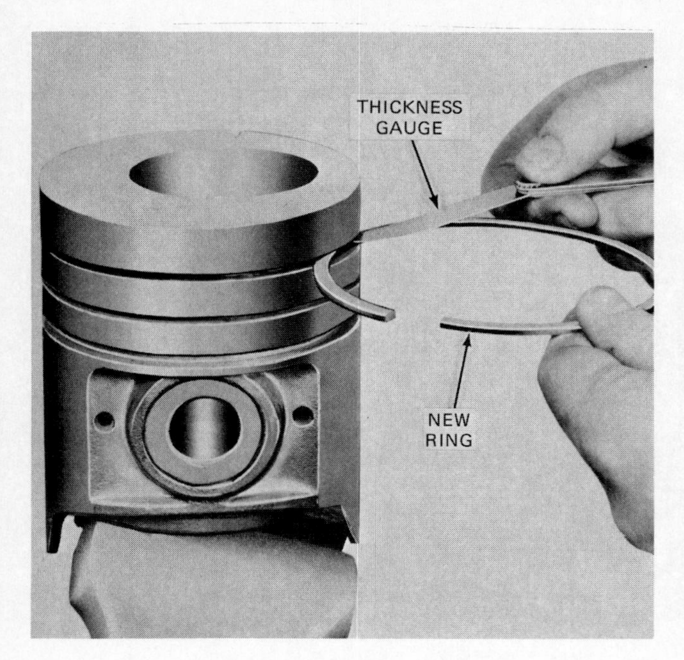


**Fig. 15-8.** A thickness gauge and a new ring are used to determine ring groove clearance. (*Ford Tractor Operations*)

moved. It is installed above the ring and provides a smooth wear surface.

**Piston Pins and Bushings** The piston and piston pin must be disassembled from the connecting rod to accurately check the piston pin, pin bosses, and the piston pin bushing. Before attempting to remove the piston, check the service manual to determine recommended removal procedures.

The full-floating piston pin that is held in place by internal snap rings is the most commonly used pin in agricultural engines. Fixed pins and semifloating pins are also used.

Some engines are equipped with piston pins that are interference-fitted in the piston pin bushing. Special procedures should be followed when removing and installing interference-fit pins. Check the service manual for specific service recommendations. Remember that an interference fit is where the pin is slightly larger than the bore, or in this case the bushing, it fits in.

**Removing piston pins** Before removing piston pins, be certain that the front of the pistons are marked so that they can be reassembled in the same position if they are reused. Also, locate the number on the connecting rod and make notes to be used during reassembly.

Follow the removal procedures provided in the service manual. Soaking piston assemblies equipped with full floating pins in 180 degrees Fahrenheit (°F) [82 degrees Celsius (°C)] water for 5 minutes (min) will make pins easier to remove. Pins should be removed while the piston is still hot.

Keep the pistons, pins, connecting rods, and snap rings together. It is important that all parts be reassembled in their original positions.

**Measuring piston pin clearance** Piston pin clearances are usually extremely small. Check the service manual for recommended clearance. Typical clearances are 0.0001 to 0.0009 in [0.0025 to 0.0229 mm].

Since clearances are so small, accurate measurement is essential. A micrometer equipped with a vernier scale is recommended.

Micrometers equipped with a vernier scale can be used to measure a component accurately to the nearest 0.0001 in or 0.001 mm. The vernier scale is read by determining which line on the thimble aligns with one of the scale lines on the sleeve (Fig. 15-11). (Also see Chap. 5 for further discussion on reading vernier scales.)

Measure the piston pin with a micrometer to determine exact diameter. Use a telescoping gauge and a micrometer to measure the inside diameter of piston

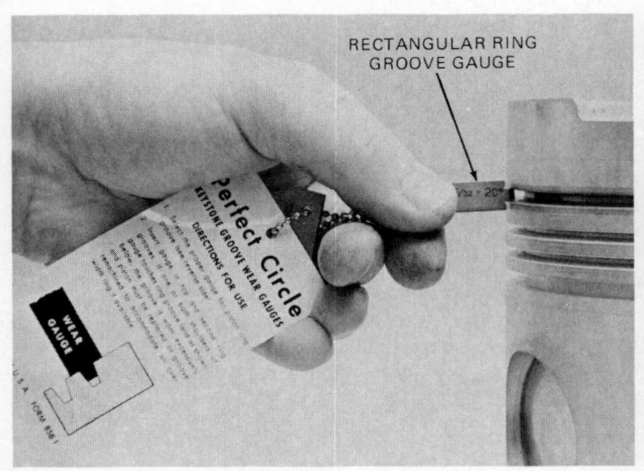


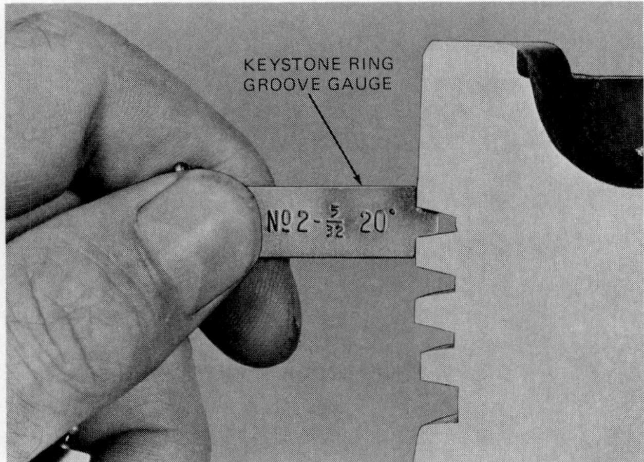


**Fig. 15-9.** Wear gauges can be used to check keystone and rectangular ring grooves for wear. (*Dana Corp.*)

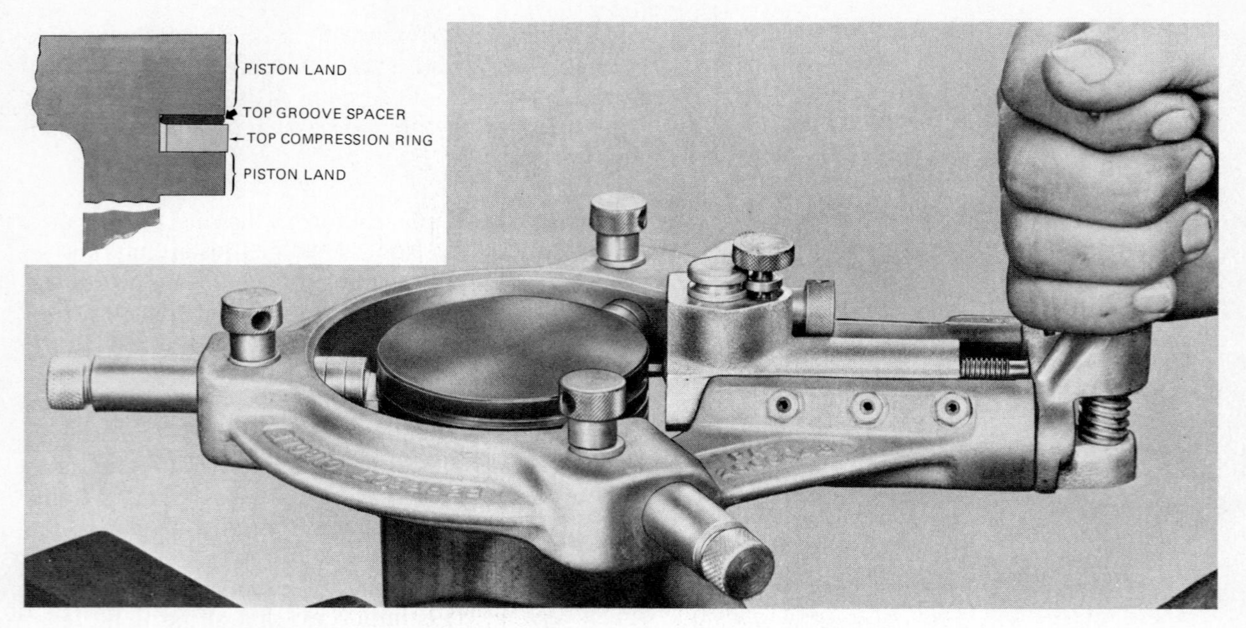


**Fig. 15-10.** Reconditioning a worn top ring groove and using a flat steel spacer above the new ring. (*Dana Corp.*)

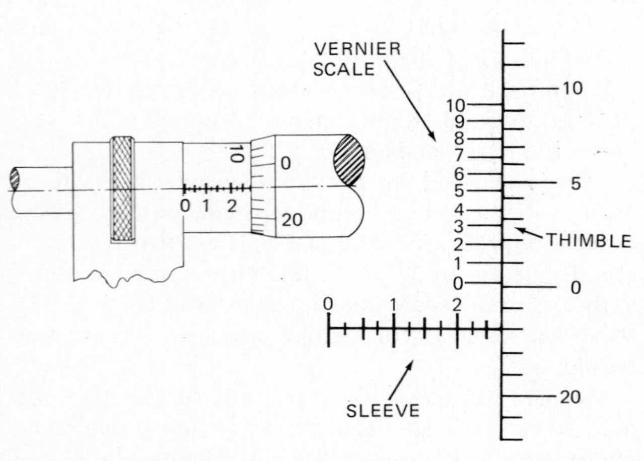


**Fig. 15-11.** Reading a vernier scale on a micrometer. This measurement is 1.2731 if this is a 1- to 2-in micrometer.

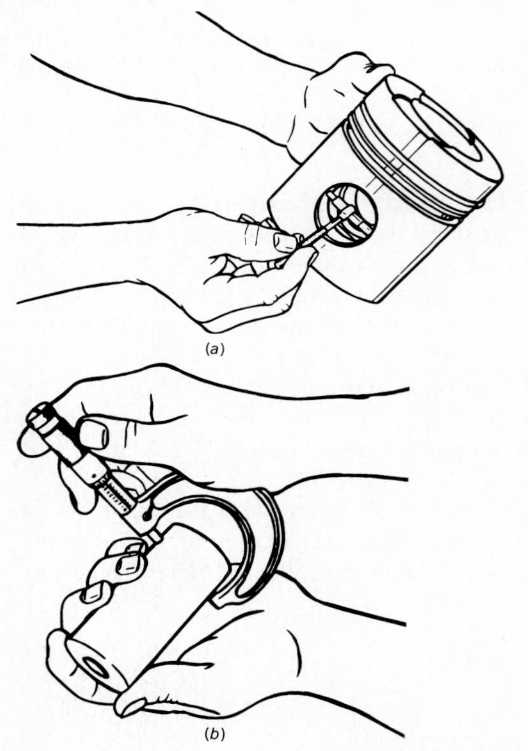


**Fig. 15-12.** (*a*) The measured piston pin and piston pin bore are used to determine piston pin clearance. (*b*) A micrometer equipped with a vernier scale is recommended. (*Allis-Chalmers Manufacturing Co.*)

pin bosses or bores in the piston. Check the bosses for taper and out-of-roundness (Fig. 15-12).

Subtract the pin diameter from the piston boss diameter to determine pin clearance in the piston. When clearance exceeds the manufacturer's specifications, the piston and pin should be replaced.

**Piston pin bushing** The pin bushing in the eye of the connecting rod is checked using the same procedures used to check pin clearance. If the bushing is worn or if clearance is excessive, the bushing should be replaced.

Piston pin bushings can be replaced by pressing the old bushings out and pressing in the new ones. The bushings are then reamed or honed to attain proper pin clearance. Piston pins bushings are often replaced and fitted in a machine shop. Honing machines are used to get the close fit specified by the manufacturer (Fig. 15-13).

## Crankshaft

The journals on a crankshaft must be smooth and round. The crankshaft should be reground if the jour-

**Fig. 15-13.** Honing a piston pin bushing. (*Photograph by Westen McCoy*)

**Fig. 15-14.** Measuring crankshaft journals. (*Photograph by Westen McCoy*)

nals are rough, ridged, or scored. Regrinding will also be required if one or more of the journals exceed manufacturer's limits for wear.

**Measuring Crankshaft Journals** Crankshaft journals are measured with an outside micrometer. The journals must be clean and free from oil if accurate measurements are to be obtained. Since crankshaft journals tend to wear unevenly, it is recommended that measurements be taken in several locations. Measurements should be taken at the front and rear of all journals at each location (Fig. 15-14).

The crankshaft journals can be measured in a systematic manner if the cross section of a journal is viewed as the face of a clock and measurements are made at the 12, 1:30, 3, and 4:30 o'clock positions (Fig. 15-15). This will result in a total of eight measurements for each journal, four for the front of the journal and four for the rear.

Tables can be used to record the measurements as they are taken. (Table 15-2). When all the measurements or diameters are recorded, crankshaft journal condition can be determined.

**Journal taper** A crankshaft journal is tapered when it is a larger diameter on one end than it is on the other end (Fig. 15-15). When journals are tapered, running clearance for the bearing will not be equal across the full width of the journal. This can result in areas of excessive clearance and can cause the bearing to wear more on one edge. Excessive journal taper can result in premature bearing failure.

To determine taper, subtract the smallest measurement obtained at each position from the largest (Table 15-2). The 12 o'clock measurements for the No. 1 main journal are 3.246 in [82.448 mm] for the

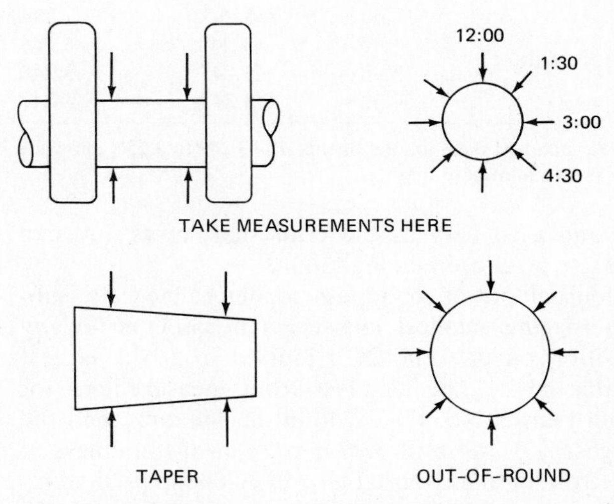


**Fig. 15-15.** Measuring crankshaft journals to check for taper and out-of-roundness.

front of the journal and 3.248 in [82.499 mm] for the rear of the journal. Taper in this journal at the 12 o'clock position is 0.002 in [0.05 mm]. Since this is the largest difference calculated in measurements obtained at one position, maximum main journal taper is 0.002 in [0.05 mm].

The same procedure is used to determine rod or crankpin journal taper (Table 15-3). If the taper for one or more main or crankpin journals exceeds the wear limits specified in the service manual, the crankshaft must be reground or replaced.

**Journal out-of-roundness** A crankshaft journal is out-of-round when there is a difference in the measurements taken at different positions at one end of a journal (Fig. 15-15). This causes bearing running

**TABLE 15-2. Determining Main Journal Wear***

| Journal | | U.S. CUSTOMARY MEASUREMENTS | | Out-of-Roundness | Taper | Maximum Diameter |
|---|---|---|---|---|---|---|
| | | Front of Journal | Back of Journal | | | |
| No. 1 Main | 12:00 | 3.246 | 3.248 | 0.002† | 0.002† | 3.248 |
| | 1:30 | 3.247 | 3.248 | | | |
| | 3:00 | 3.248 | 3.248 | | | |
| | 4:30 | 3.247 | 3.248 | | | |
| No. 2 Main | 12:00 | 3.2465 | 3.248 | 0.0015† | 0.0015† | 3.248 |
| | 1:30 | 3.247 | 3.247 | | | |
| | 3:00 | 3.248 | 3.248 | | | |
| | 4:30 | 3.247 | 3.248 | | | |
| No. 3 Main | 12:00 | 3.248 | 3.248 | 0.0005 | 0.0005 | 3.2485 |
| | 1:30 | 3.248 | 3.248 | | | |
| | 3:00 | 3.248 | 3.248 | | | |
| | 4:30 | 3.248 | 3.2485 | | | |
| No. 4 Main | 12:00 | 3.247 | 3.247 | 0.001 | 0.001 | 3.248 |
| | 1:30 | 3.247 | 3.248 | | | |
| | 3:00 | 3.247 | 3.247 | | | |
| | 4:30 | 3.248 | 3.248 | | | |
| No. 5 Main | 12:00 | 3.247 | 3.248 | 0.001 | 0.001 | 3.248 |
| | 1:30 | 3.248 | 3.248 | | | |
| | 3:00 | 3.247 | 3.248 | | | |
| | 4:30 | 3.247 | 3.247 | | | |

* The standard main journal diameter is 3.249 to 3.250, allowable out-of-roundness is 0.001, and allowable taper is 0.001.
† Exceeds allowable wear.

clearance to vary as the crankshaft turns and can cause premature bearing failure.

Journal out-of-roundness is determined by subtracting the smallest measurement obtained in any position on one end of a journal from the largest (Table 15-2). The smallest front measurement for main journal No. 1 is 3.246 in [82.448 mm], and the largest is 3.248 in [82.499 mm]. Out-of-roundness at the front of main journal No. 1 is 0.002 in [0.05 mm]. Since this is the largest difference calculated in dimensions obtained at one end of a main journal, maximum out-of-roundness for the main journals is 0.002 in [0.05 mm].

The same procedure is used to determine rod or crankpin journal out-of-roundness (Table 15-3). If the out-of-roundness for one or more main or crankpin journals exceeds the wear limits specified in the service manual, the crankshaft must be reground or replaced.

**Maximum diameter** If the crankshaft is within manufacturer's specifications for taper and out-of-roundness, maximum diameter is used to determine journal size. Maximum diameter is the largest measurement taken for the main journals or rod journals. It is sometimes referred to as minimum wear. Engine and bearing manufacturers supply undersize bearings to compensate for crankshaft wear if diameter is less than the standard journal size.

**Undersize bearings** Undersize bearings have the outside diameter of standard bearings and are used in connecting rods with standard bore diameter. The inside diameter of the bearings is smaller.

Common undersizes available for diesel agricultural engines are 0.002 in [0.05 mm], 0.010 in [0.25 mm], and 0.020 in [0.51 mm]. For lower-compression gasoline engines, 0.030-in [0.76-mm] and 0.040-in [1.02-mm] undersize bearings also are available. The 0.002-in undersize bearings are intended for use on journals where the maximum diameter is 0.0015 in [0.038 mm] to 0.0025 in [0.064 mm] less than the standard journal diameter. This is to compensate for wear and to keep the running clearance of the bearing within specifications. To determine journal wear, subtract the maximum diameter measured from the standard journal diameter given in the service manual (see Tables 15-2 and 15-3).

The 0.010-in and 0.020-in undersize bearings are to be used when the crankshaft is reground due to excessive wear. It is not recommended to regrind a crankshaft that will require undersize bearings smaller than the maximum undersize listed in the service manual.

**TABLE 15-3. Determining Rod Journal Wear***

| Journal | | U.S. CUSTOMARY MEASUREMENTS | | Out-of-Roundness | Taper | Maximum Diameter |
|---|---|---|---|---|---|---|
| | | Front of Journal | Back of Journal | | | |
| No. 1 Rod | 12:00 | 2.745 | 2.7455 | 0.001 | 0.0005 | 2.746 |
| | 1:30 | 2.746 | 2.746 | | | |
| | 3:00 | 2.746 | 2.7455 | | | |
| | 4:30 | 2.746 | 2.746 | | | |
| No. 2 Rod | 12:00 | 2.7455 | 2.745 | 0.0015 | 0.0005 | 2.7465 |
| | 1:30 | 2.746 | 2.7465 | | | |
| | 3:00 | 2.7455 | 2.745 | | | |
| | 4:30 | 2.746 | 2.7455 | | | |
| No. 3 Rod | 12:00 | 2.746 | 2.746 | 0.0005 | 0.000 | 2.746 |
| | 1:30 | 2.7455 | 2.7455 | | | |
| | 3:00 | 2.746 | 2.746 | | | |
| | 4:30 | 2.746 | 2.746 | | | |
| No. 4 Rod | 12:00 | 2.745 | 2.745 | 0.0015 | 0.001 | 2.7465 |
| | 1:30 | 2.746 | 2.7465 | | | |
| | 3:00 | 2.746 | 2.746 | | | |
| | 4:30 | 2.745 | 2.746 | | | |
| No. 5 Rod | 12:00 | 2.745 | 2.745 | 0.001 | 0.0005 | 2.746 |
| | 1:30 | 2.7455 | 2.7455 | | | |
| | 3:00 | 2.746 | 2.746 | | | |
| | 4:30 | 2.7455 | 2.745 | | | |
| No. 6 Rod | 12:00 | 2.745 | 2.745 | 0.001 | 0.0005 | 2.746 |
| | 1:30 | 2.746 | 2.746 | | | |
| | 3:00 | 2.746 | 2.746 | | | |
| | 4:30 | 2.7455 | 2.746 | | | |

* Standard rod or crankpin journal diameter is 2.745 to 2.7465, allowable out-of-roundness is 0.002, and allowable taper is 0.002.

When the crankshaft journals are worn past the limits found in the service manual, the crankshaft must be replaced or rebuilt. Rebuilding is a specialized job where special metals are added to the journals. A crankshaft should not be rebuilt unless the procedure is recommended by the manufacturer.

## Camshaft

The clearance between the camshaft journals and bearings should be checked. To measure the clearance, measure the camshaft journals with an outside micrometer (Fig. 15-16). The diameters of the camshaft bores can be measured with a telescoping gauge and an outside micrometer (Fig. 15-17).

Check the service manual for recommended camshaft clearance, camshaft journal diameter, and camshaft journal bore diameter. If camshaft clearance exceeds specifications, the camshaft journals, camshaft journal bore, or both are worn.

Many agricultural engines are equipped with camshaft bushings. When bushings are worn beyond manufacturer's specifications, they should be replaced. Some manufacturer's supply undersize camshaft bushings that can be installed to compensate for camshaft journal wear. If camshaft journals are tapered or out-of-round, the camshaft should be replaced.

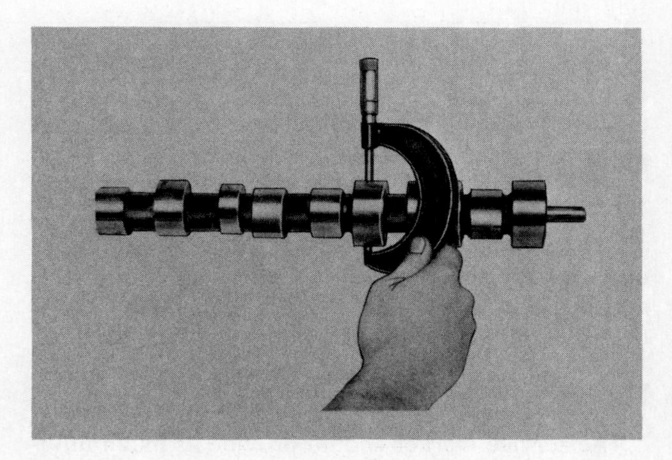

**Fig. 15-16.** Measuring camshaft journals. (*Deere & Co.*)

**TABLE 15-4.**

PARTS LIST
Make Ag Trac
Model-Year A-100 1981
Serial No. US 7543C

| Part | Part No. | Size | No. Needed |
|---|---|---|---|
| Overhaul gasket set | | | 1 |
| Front crankshaft seal | 13438 FS | | 1 |
| Rear crankshaft seal | 26438 RS | | 1 |
| Exhaust valves | | Std. | 4 |
| Intake valves | | Std. | 1 |
| Valve guides (intake) | | Std. | 4 |
| Valve guides (exhaust) | | Std. | 4 |
| Intake valve keepers | | | 4 sets |
| Exhaust valve keepers | | | 4 sets |
| Piston rings* | | Std. | complete set |
| Pistons* | | Std. | 4 |
| Cylinder liners* | | Std. | 4 |
| Piston pins* | | Std. | 4 |
| Piston pin bushings | | Std. | 4 |
| Main bearings | | 0.010 undersize | complete set |
| Rod bearings | | 0.010 undersize | complete set |
| Pilot bearing | | | 1 |
| Clutch release bearing | 78654 BF | | 1 |
| Clutch disk | | 11 in. | 1 |
| Water pump kit | | | 1 |
| Oil pump kit | | | 1 |
| Upper radiator hose | | | 1 |
| Lower radiator hose | | | 1 |
| Fan belt | | | 1 |
| Oil filter | | | 1 |
| Fuel filters | | | 2 |
| Engine oil | | | 7 qt |
| Hydraulic oil | | | 8 gal |

* Usually ordered as sleeve assemblies.

## ORDERING PARTS

When all engine components have been inspected and measured to determine exact size, parts may be ordered. The specifications given in the service manual should be used to determine if parts are reusable or should be replaced.

### Parts List

The service worker should provide as much information as possible with the parts list. A complete parts list will include the correct name for all parts needed. The exploded views in the service manual can be used to identify specific parts of an assembly. It should be indicated if the parts ordered are to be standard, oversize, or undersize. If part numbers are visible, they should be included.

The tractor make, model, and serial number should be given. The service manual can be used to determine the location of the serial number. This information will help the parts worker to identify the engine and order the correct parts (Table 15-4). Failure to

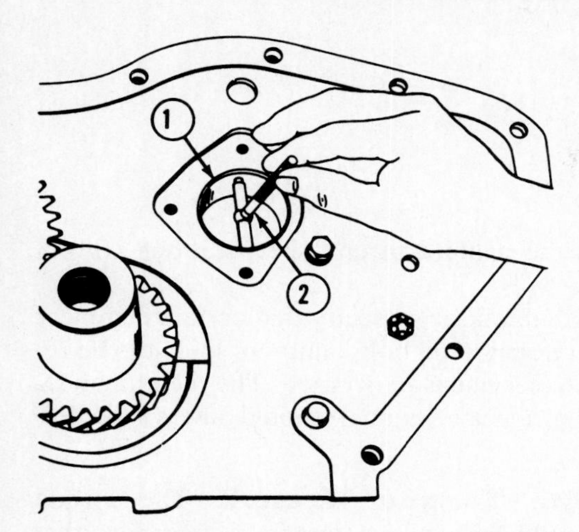


**Fig. 15-17.** (1) Measuring the bore diameter of a camshaft journal with a telescoping gauge. (2) The gauge is then measured with an outside micrometer. (*Allis-Chalmers Manufacturing Co.*)

provide adequate information can lead to incorrect parts and costly delays while service is being performed.

## SUMMARY

Precision measuring tools are as important as a good set of wrenches when engine service is performed. A successful agricultural machinery service worker will use precision tools to measure all internal engine parts for wear.

When parts are worn beyond the tolerance ranges given in the service manual, they are reconditioned or replaced. However, parts that are within specifications are reused. This results in good engine service and reduces the repair costs.

A complete parts list will include adequate information to identify the tractor and the parts needed. This will help the parts worker to order the correct parts needed and will reduce service downtime.

## THINKING IT THROUGH

1. Select the precision measuring tool or tools listed on the left that would be used to measure the parts listed on the right:

| | |
|---|---|
| (a) Small-hole gauge | ___ Crankshaft journals |
| (b) Outside micrometer | ___ Piston pin bushing |
| (c) Thickness gauge | ___ Cylinder bore |
| (d) Straight edge | ___ Cam lobes |
| (e) Telescoping gauge | ___ Camshaft journal bore |
| | ___ Camshaft journals |
| | ___ Valve guide bore |
| | ___ Cylinder head milled surface |

2. What is the difference between outside and inside micrometers?
3. When is a micrometer with a vernier scale needed?
4. Explain the difference between wet cylinder liners and dry cylinder liners.
5. Where is the greatest amount of cylinder wear usually found?
6. What is taper? Out-of-roundness?
7. What happens when piston clearance is excessive?
8. How is piston clearance measured?
9. Explain how maximum journal diameter is use when new main and rod bearings are ordered.
10. What is an undersize bearing?
11. What information should be included on a parts list?

# CHAPTER 16 ENGINE REASSEMBLY

## CHAPTER GOALS

In this chapter your goals are to:

- Remove and install wet and dry cylinder liners
- Install engine components and make appropriate checks
- Reassemble an agricultural tractor engine after overhaul
- Set ignition timing on spark-ignition engines
- Time a diesel injection pump

As an engine is reassembled, it is important that all parts be clean and properly lubricated. The service manual will give the manufacturer's specifications for all clearances. It will also give the recommended procedures for reassembly and bolt torque specifications for critical engine components. A competent service worker should be able to reassemble an engine that was disassembled by someone else.

## FASTENERS

Some fasteners commonly used in engine assembly are machine bolts and screws, stud bolts, cap screws, and nuts. They are all vital parts of an engine and each may be exposed to heavy loads, severe stress, and vibration. The failure of one fastener, such as a connecting rod cap screw, can result in extensive engine damage.

### Cleaning Fasteners

Deposits in threads can cause false torque readings as bolts and other fasteners are tightened. All fasteners should be cleaned in solvent to remove grease and other deposits. Rust, carbon, and scale can be removed with a wire brush. Taps and dies can be used to recondition threads that are slightly damaged (Fig. 16-1). Remember to keep all fasteners in properly labeled containers.

### Tightening Fasteners

Fasteners holding vital engine parts together are usually special bolts that should be tightened to the torque specified by the manufacturer. The recommended torque for these bolts is given in the service manual. If the torque is not given, general Society of Automotive Engineers (SAE) and American Society for Testing and Materials (ASTM) torque specifications may be used to determine the approximate torque (Table 16-1). These torques should be used only when the manufacturer's specifications are not available.

Some manufacturer's specify that certain fasteners such as connecting rod bolts, nuts, or locknuts be replaced when service is performed. The recommendations in the service manual should always be followed.

**Using the Torque Wrench** The torque wrench should be used to tighten all fasteners that have a specified torque. Torque wrenches measure the torque (twisting force) that is applied to the fastener. They are most accurate around the middle of their range. A 0 to 150 pound-feet (lb·ft) [0 to 200 newton-meters (N·m)] wrench will give accurate readings from about 40 to 110 lb·ft [55 to 150 N·m]. For this reason, it is a good idea to have several torque wrenches of differing capacities.

It is generally recommended to torque or tighten fasteners in five steps. This is especially important where several fasteners are used to secure a component such as a cylinder head. (1) Snug the fasteners in a random pattern a little at a time. This will allow the component to seat properly and prevent binding.

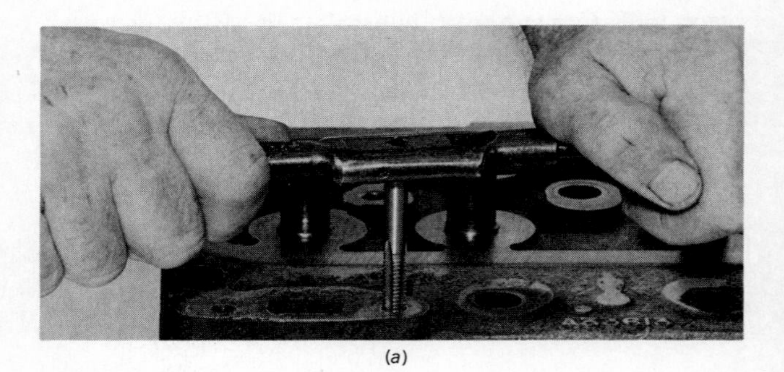

(*a*)

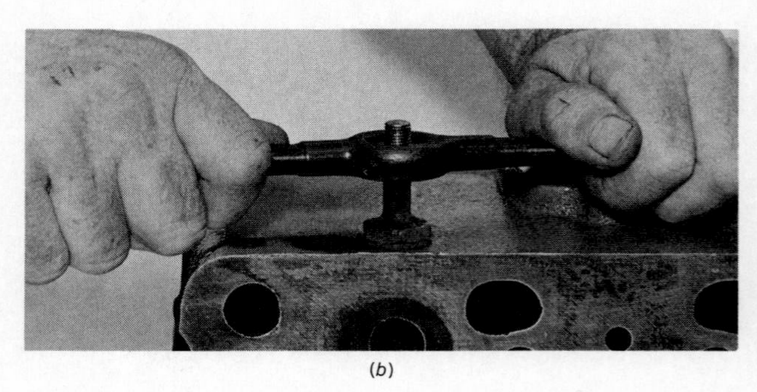

(*b*)

**Fig. 16-1.** (*a*) Using a tap to recondition internal threads and (*b*) a die to recondition external threads. This cleans the connections and ensures that proper torque can be attained. Clean threads lessen the possibility of stripping or twisting off bolts. (*Photograph by Westen McCoy*)

**TABLE 16-1. Maximum Torque Valves for Four Grades of Steel Bolts (U.S. Customary Measure)**

| GRADE MARKINGS | | | | |
|---|---|---|---|---|
| SAE GRADE | 1 OR 2 | 5 | 6 | 8 |
| Size Threads per in | Maximum Tightening Torque in Pound-feet | | | |
| 1/4 in | | | | |
| 20 | 4 | 6 | 8 | 9 |
| 28 | 5 | 7 | 9 | 10 |
| 5/16 in | | | | |
| 18 | 8 | 13 | 16 | 18 |
| 24 | 9 | 14 | 18 | 20 |
| 3/8 in | | | | |
| 16 | 15 | 23 | 30 | 35 |
| 24 | 17 | 25 | 35 | 40 |
| 7/16 in | | | | |
| 14 | 24 | 35 | 45 | 55 |
| 20 | 25 | 40 | 50 | 60 |
| 1/2 in | | | | |
| 13 | 35 | 55 | 70 | 80 |
| 20 | 40 | 65 | 80 | 90 |
| 9/16 in | | | | |
| 12 | 55 | 80 | 100 | 110 |
| 18 | 60 | 90 | 110 | 130 |
| 5/8 in | | | | |
| 11 | 75 | 110 | 140 | 170 |
| 18 | 85 | 130 | 160 | 180 |
| 3/4 in | | | | |
| 10 | 130 | 200 | 240 | 280 |
| 16 | 145 | 220 | 280 | 320 |
| 7/8 in | | | | |
| 9 | 125 | 320 | 400 | 460 |
| 14 | 140 | 350 | 440 | 500 |
| 1 in | | | | |
| 8 | 190 | 480 | 600 | 680 |
| 12 | 200 | 530 | 660 | 740 |

*Note:* The torque specifications are all given in pound-feet. To convert to pound-inches, multiply the rating given by 12.

(2) Set the torque wrench to one-third of the specified torque and tighten each fastener in the sequence recommended in the service manual. (3) Repeat this process at two-thirds the recommended torque and (4) at full recommended torque. (5) Then go over the fasteners at least one more time at full torque.

## Removing Broken Bolts

When bolts are properly cleaned and tightened according to torque recommendations, few should be twisted off during reassembly. However, it is not uncommon to twist off bolts during disassembly. When a bolt cannot turn itself out, extreme caution should be taken when trying to remove it.

Penetrating oil and silicon sprays can be used to loosen the rust and carbon that may seal the threads. Heat can be used if it will not damage the component. When heat is used, be certain that the component is not near a fuel line or the fuel tank.

A cap screw that breaks off with a portion of it exposed above the component may be removed with locking pliers or a small pipe wrench. Another method is to arc-weld a nut of the proper size to the end of the bolt and remove it with a wrench (Fig. 16-2). Special electrodes are available for this purpose.

If the cap screw is broken off flush with the component, it may be removed with a small punch or chisel. It can also be drilled and a screw extractor used to remove it (Fig. 16-2). A center punch is used to mark the center of the screw. Be careful not to apply enough force to the screw extractor to break it. It is very difficult to remove a screw extractor or twist drill that has been broken off in a bolt or stud.

When these methods fail, use a tap drill size chart (Table 16-2) to determine the drill size recommended for the bolt. Start drilling through the screw. Remove the drill and check the threads. If the screw hole was centered, the threads should not be damaged. Complete the drilling and use the proper tap to clean the threads (Fig. 16-1).

All broken-off bolts should be removed before parts are reassembled. When the threads are damaged during removal, the component may have to be taken to a machine shop for repair. If a larger bolt can

(*a*)

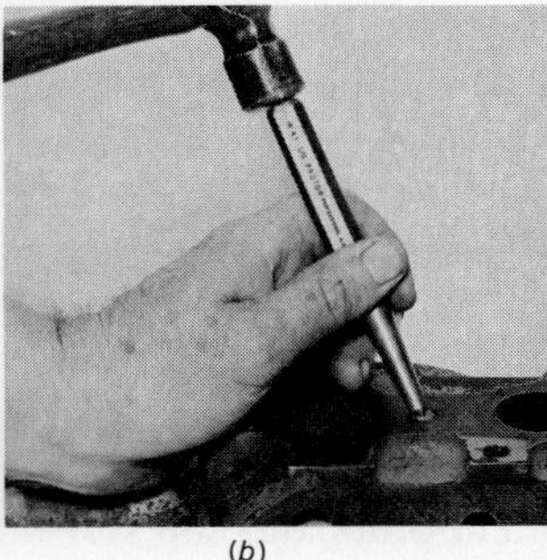

(*b*)

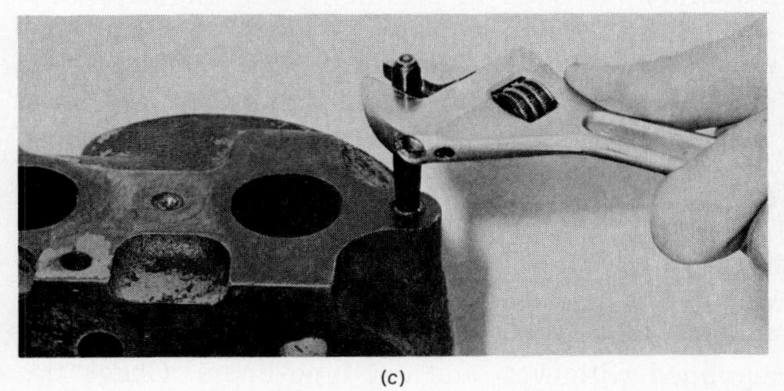

(*c*)

**Fig. 16-2.** Three methods for extracting broken bolts and studs include: (*a*) arc-welding a nut to the exposed end; (*b*) tapping with a center punch or chisel; and (*c*) drilling and installing a reverse-threaded screw extractor. (*Photograph by Westen McCoy*)

**TABLE 16-2. Recommended Tap and Drill Size (U.S. Customary Measure)**

| | NATIONAL COURSE | | NATIONAL FINE | | SCREW EXTRACTOR | |
|---|---|---|---|---|---|---|
| Bolt Diameter, in | Threads Per Inch | Drill Size | Threads Per Inch | Drill Size | Screw No. | Drill Size |
| 1/4 | 20 | 13/64 | 28 | 7/32 | 1 | 5/64 |
| 5/16 | 18 | 1/4 | 24 | 17/64 | 2 | 7/64 |
| 3/8 | 16 | 5/16 | 24 | 21/64 | 3 | 5/32 |
| 7/16 | 14 | 23/64 | 20 | 25/64 | 3 | 5/32 |
| 1/2 | 13 | 27/64 | 20 | 29/64 | 4 | 1/4 |
| 9/16 | 12 | 31/64 | 18 | 33/64 | 4 | 1/4 |
| 5/8 | 11 | 17/32 | 18 | 37/64 | 5 | 17/64 |
| 3/4 | 10 | 21/32 | 16 | 11/16 | 5 | 17/64 |
| 7/8 | 9 | 49/64 | 14 | 13/16 | 6 | 13/32 |
| 1 | 8 | 7/8 | 12 | 15/16 | 6 | 13/32 |

be used, a larger tap can be used to make new threads. It is important to remember that metric fasteners are used on some agricultural engines. This must be considered when selecting twist drills and taps.

## Selecting Replacement Bolts and Cap Screws

Bolts and caps screws with national coarse (NC), national fine (NF), international standard organization coarse (ISO coarse), and international standard organization fine (ISO fine) threads are used in the assembly of agricultural engines. Since there are different types of threads, it is important that bolts, cap screws, and nuts with the correct thread type be used when replacements are required.

Bolts and cap screws are available in a variety of strengths. Table 16-3 gives the ASTM and ASE grade markings, specifications, composition, and tensile strength for steel bolts. It is essential that replacement bolts and cap screws be the same grade that the manufacturer used in producing the engine. Some special bolts, such as cylinder head bolts and connecting rod bolts, must be obtained from the manufacturer.

# CYLINDER LINERS

When cylinder wear exceeds the specifications or wear limits given in the service manual, the liners should be replaced. Agricultural engines are equipped with wet- and dry-type liners. Check the service manual to determine the type used.

## Dry Cylinder Liners

A sleeve puller is recommended for removing dry cylinder liners (Fig. 16-3). Use the information supplied with the puller to select the correct puller plate. Adjust the support bars on the puller to fit the block. The support bars should set on solid sections of the block. Pullers may be the hydraulic or screw-threaded type.

Dry liners tend to "freeze" or stick after long service. If liners do not move when pressure is applied, it may be advisable to clean the block in a hot tank to remove rust and carbon. Steam can also be used to heat the block so that the liners can be removed.

**Installing Dry Liners** When the liners are removed, the bore in the block should be cleaned. A deglazer (Fig. 16-4) can be used to clean the bore. The carbon in the counterbore at the top of the bore must also be removed. A small pocketknife can be used. If any carbon is left, it can cause the new liner to distort or not seat properly (Fig. 16-5). The bores should be washed with steam or detergent and water after they have been cleaned.

Check the service manual for recommended installation procedures. New liners should be washed in solvent to remove any protective coating that may be on them. Some manufacturers recommend chilling liners in dry ice or a freezer to shrink them. An installer bar can be used to install the liners (Fig. 16-6).

After liners are installed, they should be checked with a straight edge and flat thickness gauge to determine standout. *Standout* is the distance the liners extend past the top of the block (Fig. 16-7). Check the service manual for recommended standout.

## Wet Cylinder Liners

Wet cylinder liners are usually easier to remove than dry liners. They can be removed with a sleeve puller or with a block of hardwood and a hammer.

Wet liners seal at the bottom with O-rings. The sealing surfaces in the block should be cleaned. If the

**TABLE 16-3. ASTM and SAE Grade Markings for Steel Bolts**

| | | | Physical Properties | | |
|---|---|---|---|---|---|
| Grade Marking | Specification | Material | Bolt Size (in) | Proof Load min. (psi) | Tensile Strength min. (psi) |
| | SAE—grade 0 | Steel | All sizes | .... | .... |
| | SAE—grade 1<br>ASTM—A 307 | Low carbon steel | All sizes | .... | 55,000 |
| | SAE—grade 2 | Low carbon steel | Up to ½ | 55,000 | 69,000 |
| | | | Over ½ to ¾ | 52,000 | 64,000 |
| | | | Over ¾ to 1½ | 28,000 | 55,000 |
| | SAE—grade 3 | Medium carbon steel, cold-worked | Up to ½ | 85,000 | 110,000 |
| | | | Over ½ to ⅝ | 80,000 | 110,000 |
| | SAE—grade 5 | Medium carbon steel, quenched and tempered | Up to ¾ | 85,000 | 120,000 |
| | | | Over ¾ to 1 | 78,000 | 115,000 |
| | | | Over 1 to 1½ | 74,000 | 105,000 |
| | ASTM—A 325 | | Up to ¾ | 85,000 | 120,000 |
| | | | Over ¾ to 1 | 78,000 | 115,000 |
| | | | Over 1 to 1½ | 74,000 | 105,000 |
| | | | Over 1½ to 3 | 55,000 | 90,000 |
| BB | ASTM—A 354 grade BB | Low alloy steel, quenched and tempered (medium carbon steel, quenched and tempered may be substituted where possible) | Up to 2½ | 80,000 | 105,000 |
| | | | Over 2½ to 4 | 75,000 | 100,000 |
| BC | ASTM—A 354 grade BC | Low alloy steel, quenched and tempered (medium carbon steel, quenched and tempered may be substituted where possible) | Up to 2½ | 105,000 | 125,000 |
| | | | Over 2½ to 4 | 95,000 | 115,000 |
| | SAE—grade 6 | Medium carbon steel, quenched and tempered | Up to ⅝ | 110,000 | 140,000 |
| | | | Over ⅝ to ¾ | 105,000 | 133,000 |
| | SAE—grade 7 | Medium carbon alloy steel quenched and tempered, roll threaded after heat treatment | Up to 1½ | 105,000 | 133,000 |
| | SAE—grade 8<br>ASTM—A 354 grade BD | Medium carbon alloy steel, quenched and tempered | Up to 1½ | 120,000 | 150,000 |

**ASTM Specifications:**
A 307—Low carbon steel externally and internally threaded standard fasteners.
A 325—Quenched and tempered steel bolts and studs with suitable nuts and plain, hardened washers.
A 354—Quenched and tempered alloy steel bolts and studs with suitable nuts.
**SAE Specification:**
Physical requirements for threaded fasteners.

O-ring is installed in the block, it must be removed and the groove or grooves cleaned. The counterbore at the top of the bore must also be cleaned (Fig. 16-8).

**Installing Wet Liners** Wash the liners in solvent to remove any protective coating. Before installing O-rings, insert the liner into the bore of the cylinder block to make sure it can be pushed down into place and turned in the bore by hand pressure. If the liner cannot be installed by hand, more cleaning is necessary.

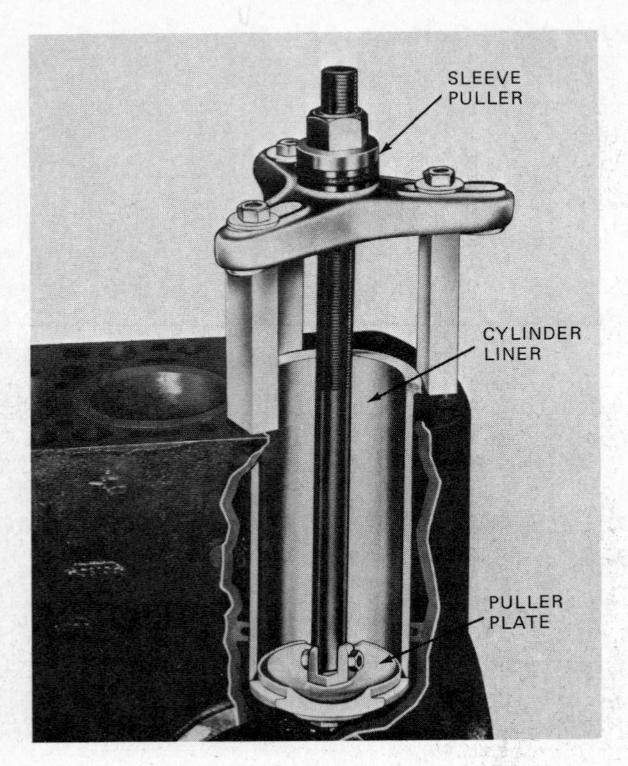


Fig. 16-3. Pulling dry-type cylinder liners with a manually operated sleeve puller. (*OTC Tools & Equipment Co., Division of Owatonna Tool Co.*)

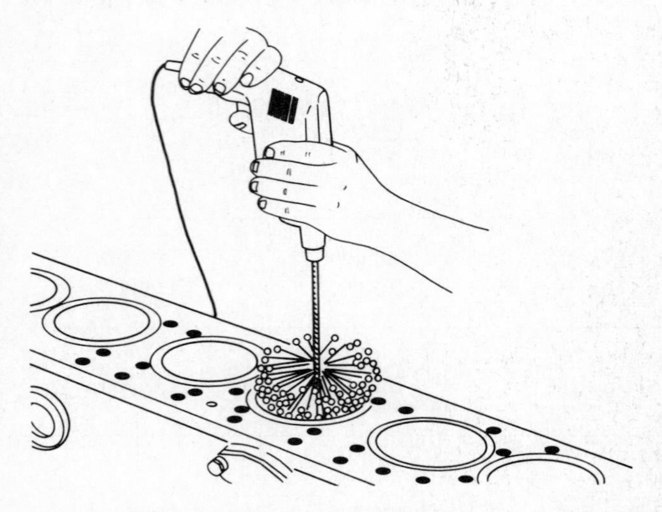

Fig. 16-4. Deglazing cylinder bores before installing dry cylinder lines.

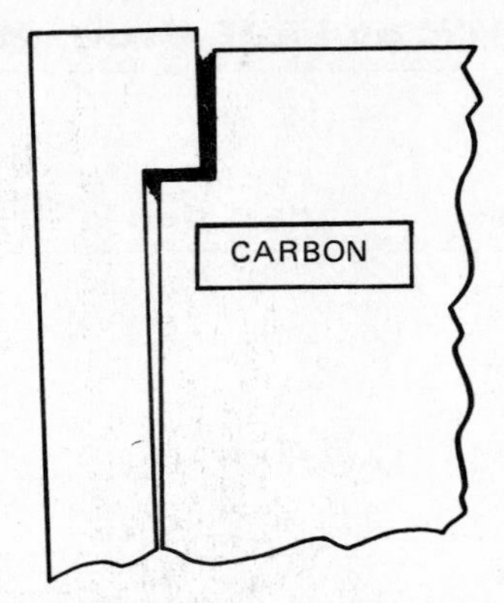


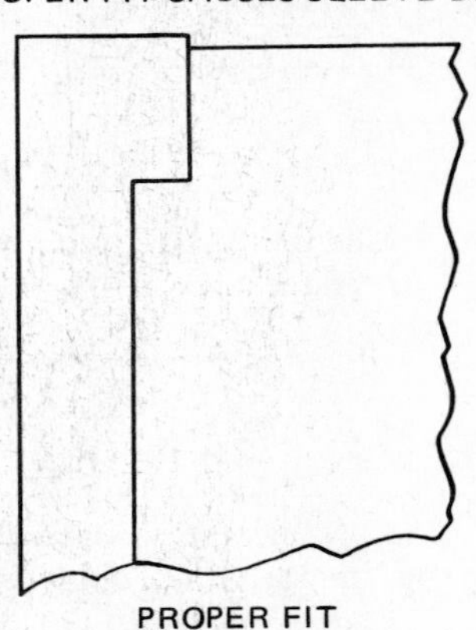


Fig. 16-5. Carbon must be removed from the counterbore in the block to prevent distortion of new cylinder liners. (*Dana Corp.*)

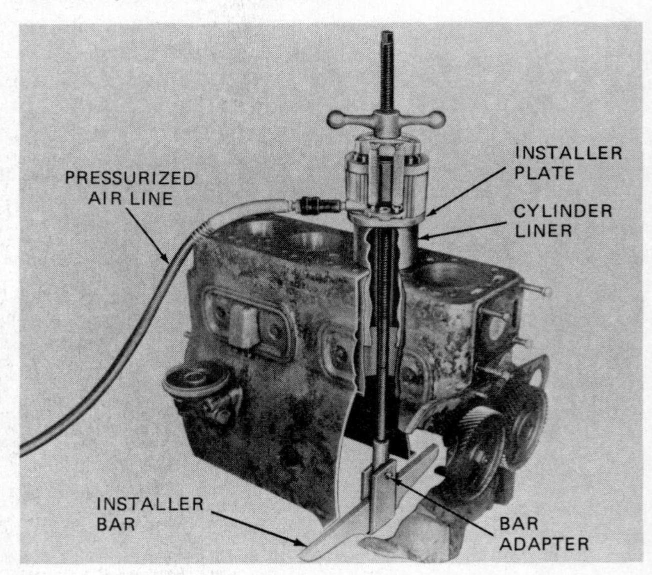


Fig. 16-6. Installing a dry-type cylinder liner. (*OTC Tools & Equipment Co., Division of Owatonna Tool Co.*)

Use a straight edge and flat thickness gauge to determine liner standout. If standout is not within specifications, check the service manual for recommended procedures for correcting standout. Many manufacturers provide cylinder sleeve shims for this purpose.

When standout is within specifications, the liners may be installed. Install the O-rings as specified in the service manual. Some manufacturers use different O-rings on each liner, and it is important that they be installed in the correct groove. Use the lubricant specified in the service manual to lubricate the O-rings, liner, and block. Some commonly used lubricants are petroleum jelly, liquid soap, and vegetable oil. Lubricant should be applied as the liners are installed.

Some manufacturers supply special-purpose lubricants for liner installation. Check the service manual for specific recommendations.

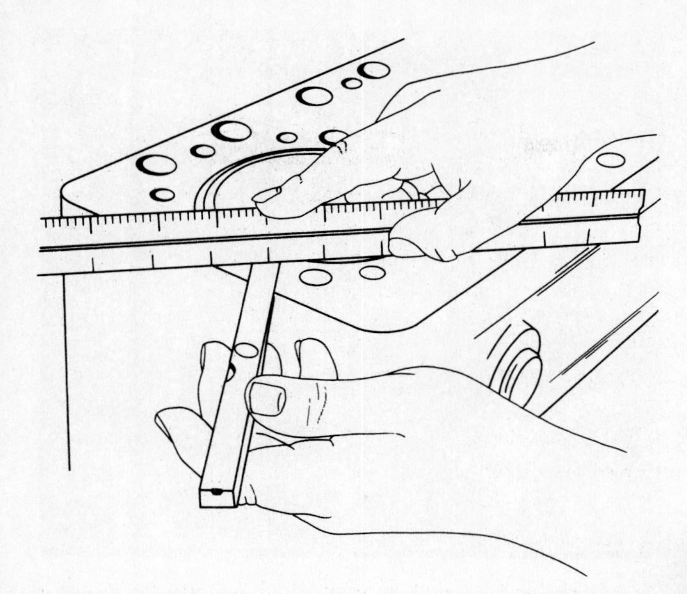

Fig. 16-7. Checking cylinder liner standout. (*Allis-Chalmers Manufacturing Co.*)

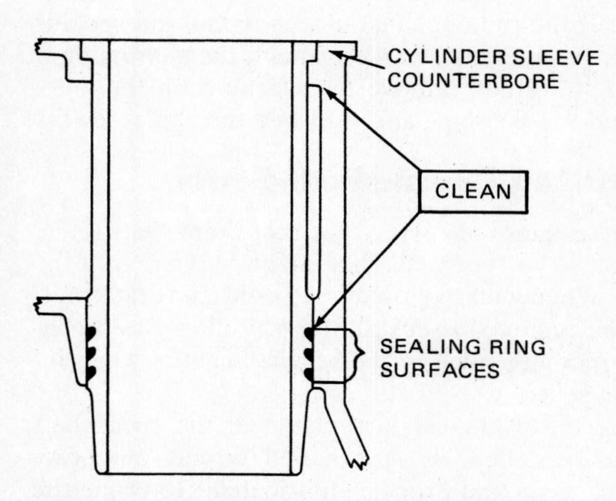


**Fig. 16-8.** The counterbore and lower sealing ring surfaces in the block must be cleaned to prevent liner distortion and coolant leakage. (*Dana Corp.*)

Carefully place the liner in the bore and push it into place. A block of hardwood and a hammer may be used to seat the liners. Recheck the liner standout.

Since wet liners are not press-fit, the manufacturer may recommend that bolts and washers be used to hold the liners in place during engine assembly (Fig. 16-9). Check the service manual for recommendations.

## Deglazing Cylinders

When liners are not replaced, it is a recommended practice to use a hone or a deglazer to provide a crosshatch pattern on the liner surface (Fig. 16-10). Deglazing gives a fine finish and does not enlarge the cylinder diameter. Cylinders are deglazed to provide cavities for holding oil during piston ring break-in.

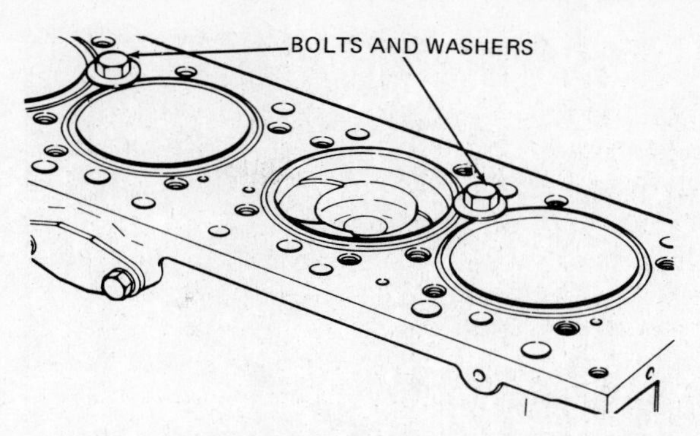


**Fig. 16-9.** Bolts and washers are used to hold wet cylinder liners in place during engine assembly. (*Allis-Chalmers Manufacturing Co.*)

**Fig.16-10.** Cylinder or liner walls should have a crosshatch pattern. (*Dana Corp.*)

Some manufacturers recommend honing or deglazing new liners. Check the service manual for recommended procedures.

Cylinders should be properly cleaned after deglazing or honing. Lightweight oil on a cheap shop towel can be used to swab out the abrasive materials. Washing with detergent and water is the preferred practice. After the cylinders are washed, they should be blown dry and lightly oiled to prevent rust.

# INSTALLING THE CRANKSHAFT

The crankshaft journals and main bearing bores should be clean. Be sure that the bearing bores and caps are free from oil. Insert the upper halves of the main bearing shells. Be sure that the locking lips are in the lip slots (Fig. 16-11). Also check to see that oil holes are properly aligned. Some main bearings have a top half and a bottom half. Failure to install bearings in the proper locations can result in oil stoppage to vital engine parts. Keep bearing halves together

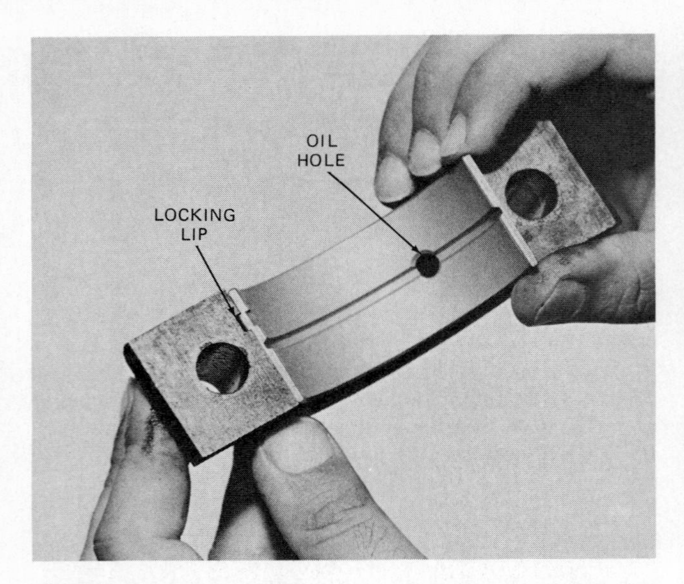


**Fig. 16-11.** Installing bearing shells. Be certain locking lips and oil holes are properly aligned. (*Photograph by Westen McCoy*)

**Fig. 16-12.** Placing Plastigage across a crankshaft journal in preparation for determining bearing running clearance. (*Photograph by Westen McCoy*)

and be sure they are installed in the proper bore. Bearing surfaces should not be handled. Fingerprints can cause corrosion.

Use engine oil or white-lithium-base lubricant to lubricate the bearings and the main journals during reassembly. Set the crankshaft in place. Install the bearing shells in the bearing caps, lubricate them, and install the caps. Be sure that the caps are installed in their original position. The identification marks or numbers should be used to position the caps.

## Torquing Main Caps

Tighten the main bearing bolts to the torque specified in the service manual. It is recommended that bolts or nuts be torqued in stages. Each time a bearing is torqued a stage, turn the crankshaft to check for any binding. If binding occurs, the cause must be determined. When all main caps have been torqued down, spin the crankshaft several revolutions. It should turn freely.

## Checking End Play

Use a dial indicator or flat thickness gauge to check crankshaft end play. Refer to Chap. 14 for a review on checking end play.

## Checking Running Clearance

Running clearance should be checked with Plastigage. To check running clearance, remove the main caps. Wipe the lubricant from the top of the journals and the bearing inserts. Place a strip of Plastigage across the journals (Fig. 16-12), install the caps, and retorque them. *Do not turn the crankshaft.*

Remove the caps and check the width of the Plastigage with the scale provided on the package (Fig. 16-13). If the running clearance is within the specifications given in the service manual, the bearing fit is correct. Carefully remove the plastic from the journals and the bearings and retorque the cap screws.

## Installing Crankshaft Seals

Some engines use wick-type rear crankshaft seals. Others use lip-type seals that attach to the rear of the block. When wick-type seals are used, the crankshaft must be removed to install them. Follow the procedures recommended in the service manual when installing seals.

After the wick seals are installed, the crankshaft can be reinstalled. Be sure that all journals and bearings are clean and properly lubricated. Retorque the bolts and secure them with the locking devices recommended by the manufacturer.

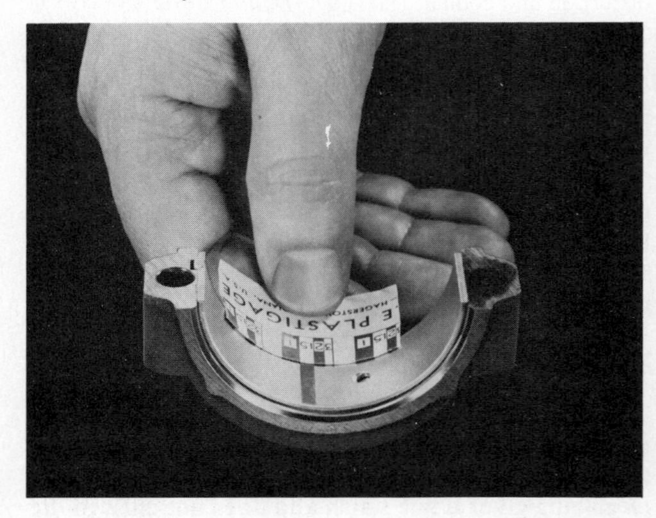


**Fig. 16-13.** Checking the width of the crushed Plastigage to determine bearing running clearance. (*Dana Corp.*)

## INSTALLING THE CAMSHAFT

If the camshaft gear is to be replaced, it should be done at this time. An arbor press can be used to remove the old gear and install the new one. When the new gear is installed, be sure that the timing mark faces away from the cam lobes. Check the service manual for recommended procedures. Remember that it is good procedure to replace the crankshaft gear when the cam gear is replaced.

If the engine is equipped with mushroom-type cam followers, they should be installed at this time. Lubricate the cam followers and the bores in the block with oil or white-lithium-base lubricant and install them in their original position. All the cam followers should move freely.

Lubricate the camshaft journals and lobes. Also, lubricate the cam bore or bushings in the block. Carefully slide the camshaft into the block. Support the back of the camshaft to prevent damage to lobes, journals, and bushings. If the core hole plug or welch plug in the rear camshaft bore was removed, a new one should be installed at this time.

### Timing the Valves

Turn the crankshaft so that the timing mark points toward the camshaft. As the camshaft is pushed into place, align the marks on both gears. Check the service manual for timing procedures. It is a common practice on diesel engines to use an idler gear between the crankshaft gear and the injection pump drive gear. The idler gear must also be timed to the crankshaft gear (see Fig. 14-38).

### Checking Backlash

A thickness gauge or dial indicator should be used to check the backlash between the cam gear and the crankshaft gear. Refer to Chap. 14 for a complete discussion on checking backlash.

### Checking End Play

The procedure for checking end play was discussed in Chap. 14. It is good procedure to check end play when the camshaft is reinstalled.

### Checking Cam Gear Runout

Some manufacturers give specifications for runout (wobble) in the cam gear as it is rotated. When these specifications are given, runout is checked by mounting the dial indicator in the same position that was used to check camshaft end play. To check runout, force the cam gear against the thrust plate and rotate the camshaft. The amount of runout is determined by watching the pointer on the dial indicator face. Check the service manual for service recommendations.

## INSTALLING PISTON ASSEMBLIES

Pistons are installed on the connecting rods when they have been removed. Remember that the connecting rods must be installed with the identifying marks or numbers in the proper position. Many pistons also have a front. The pistons should be installed on the connecting rods that correspond to the cylinder liners the pistons were matched with. For example, No. 1 piston should be installed on No. 1 connecting rod.

Lubricate the piston pin, the pin bosses in the piston, and the pin bushing. Insert the pin and secure it in place with the snap rings (Fig. 16-14). The connecting rod should move freely on the piston pin, but there should not be any slack in the bushing.

### Checking Ring End Gap

Before rings are installed on the pistons, the ring end gap should be checked. To check ring end gap, push a ring down into the cylinder (Fig. 16-15). Use a flat thickness gauge to determine the gap between the ends of the rings. Check the service manual for the manufacturer's recommended gap.

When rings are to be installed in cylinders that have original liners or have not been rebored, the ring end gap should be checked in the lower portion of the cylinder below the ring travel area.

Ring end gap must be within the specification given

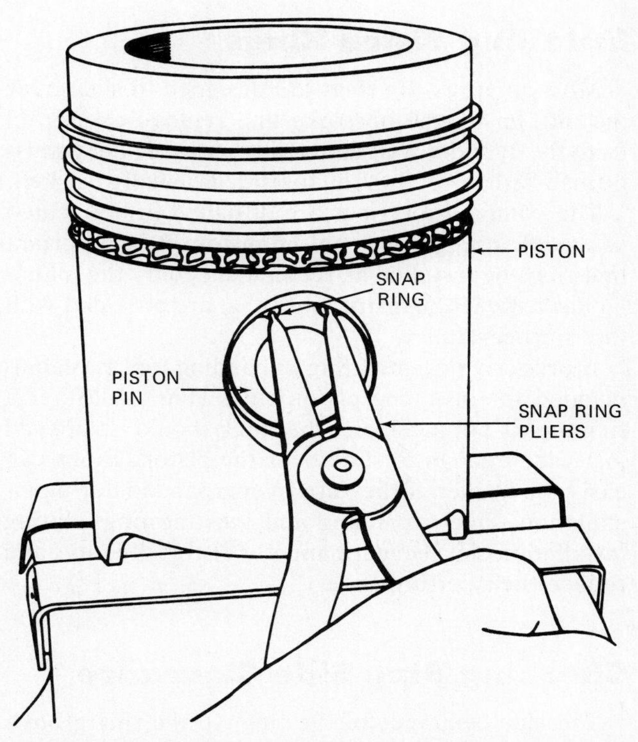


**Fig. 16-14.** Installing a snap ring to secure a piston pin. (*Allis-Chalmers Manufacturing Co.*)

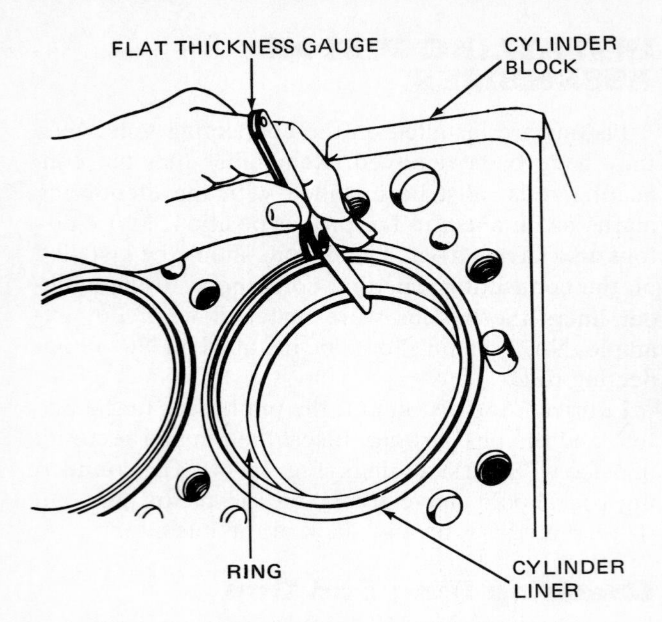


**Fig. 16-15.** Checking ring end gap in the cylinder. (*Allis-Chalmers Manufacturing Co.*)

in the service manual. As the rings heat up, they expand. Insufficient end gap can result in continued ring expansion after the ends touch. This creates excessive pressure on the cylinder walls and can cause ring, cylinder, or piston damage. Excessive ring end gap results in a large opening between the ends of the rings after the engine reaches operating temperature. This will result in excessive blowby or loss of compression.

## Installing Piston Rings

Most compression rings are designed to slide over the oil film on the upstroke and remove excess oil from the cylinder wall on the downstroke. They must be installed correctly. The top side is usually marked.

The compression rings are designed to be located in specific ring grooves on the piston. It is important that rings be installed as recommended by the manufacturer. Read the instructions that are provided with the rings carefully.

A properly designed ring-expanding tool is recommended for installing piston rings (Fig. 16-16). If a ring tool is not available, the rings should be opened just wide enough to slide over the piston. Rings can easily be broken if they are overexpanded during installation. Overexpanding and twisting rings during installation can also permanently distort the rings and reduce their performance.

## Checking Ring Side Clearance

The side clearance of the rings in the ring groove should be checked. Check the service manual for the manufacturer's recommendations.

**Fig. 16-16.** Using a ring-expanding tool to install piston rings. (*Photograph by Westen McCoy*)

## Installing Pistons

The pistons, rings, and piston pins should be thoroughly lubricated with SAE 30W engine oil or a white-lithium-base lubricant. The ring end gaps should be staggered around the piston. Some manufacturers give specific locations for end gaps. Check the service manual for recommendations.

Ring compressors are used to compress the rings on the pistons so that they can be installed in the cylinders (Fig. 16-17). Some manufacturers recommend a specific type of compressor. Check the service manual for recommendations.

To install the ring compressor on the piston, secure the connecting rod in a vise equipped with soft jaws. Lubricate the inside of the compressor and fit it over the rings. Be sure that the compressor is right side up and tighten it snugly on the piston.

**Fig. 16-17.** Using a ring compressor (1) to install the piston assembly in the cylinder. Note push stick (2) used to push or tap the piston into the cylinder. Arrow (3) indicates the front of the engine. Ratchet lock (4) secures the compressor on the rings and is used to release the compressor after installation is completed. (*Ford Tractor Operations*)

When the ring compressor is properly installed, the piston assembly can be installed. Insert the bearing shell in a clean dry rod bore. Be sure that the locking lip is in the lip slot and that oil holes are properly aligned. Lubricate the bearing, crankshaft journal, and cylinder wall. Position the piston so that the front of the rod and piston assembly is facing the front of the engine.

Place journal protectors on the rod bolts and push or tap the piston into the cylinder. If the piston sticks as it is being installed, remove it and determine the problem. The crankshaft journal should be in the bottom dead center (BDC) or top dead center (TDC) position.

Place the lower rod bearing in the rod cap and lubricate it. Install the rod cap and torque the capscrew to specification. Check the service manual to see if new bolts, nuts, or locknuts are recommended.

It is a good practice to turn the crankshaft each time a cap is tightened. It will take more force to turn the engine as pistons are installed, but if a bind is felt, the bearing should be checked.

It is important that all rods be tightened to the torque recommended in the service manual. When rod caps are not tightened as recommended, they will be out-of-round. This is due to the crush-fit of the bearings which is designed to lock the bearing in the rod bore (Fig. 16-18).

### Checking Running Clearance

After all piston assemblies have been installed, the crankshaft should be turned several revolutions. The rod caps should then be removed, wiped free of oil, and the bearing running clearance should be checked with Plastigage. Check the service manual for recommended running clearance.

If running clearance is within specifications, the journals and bearings should be lubricated and reinstalled. The locking devices recommended by the manufacturer should be installed. This extra check of the running clearances will result in a longer service life of the engine.

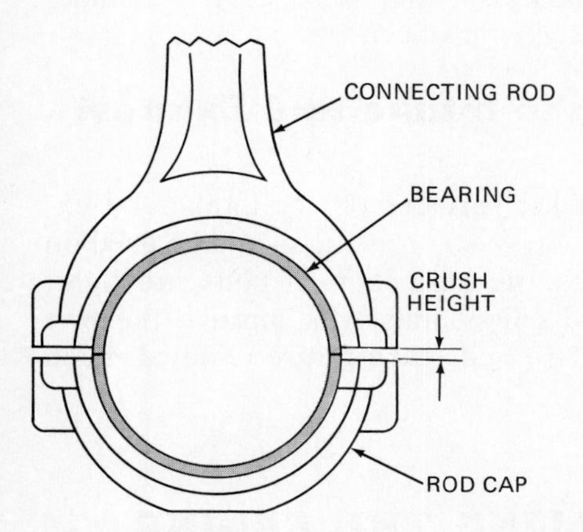


**Fig. 16-18.** Bearings have added crush height and become round when properly torqued on the crankshaft.

## INSTALLING THE OIL PAN

The oil pump and engine balancer should be installed. Some oil pumps must be timed. Engine balancers are timed to the crankshaft. This is done with timing marks. Check the service manual for details.

It is good procedure to check the engine oil pump before it is installed. Follow the procedures given in the service manual. If the pump gears, wear plate, or bushings are worn beyond specifications, the pump should be rebuilt or replaced.

Recheck to see that all rod and main bearing bolts are properly locked. Also, check the rear crankshaft seal. Some engines that are equipped with lip-type seals have bolts that must be installed from inside the block.

Before installing the oil pan, be sure that all gasket surfaces are properly cleaned. Check the gasket for proper fit. Pan gaskets are usually cork or paper. If the gasket is too small, it can be stretched by soaking it in water a few minutes. Gaskets that are too large can be shrunk by placing them in an oven set on low heat a few minutes.

New gaskets, properly installed between two clean engine parts, should provide a good seal. Gasket sealant may be used to supplement the gasket. If sealant is used, the nonhardening types are preferred, and they should be used sparingly.

When installing the oil pan, be careful to pull all the bolts up evenly. The bolts are usually small, and they should not be overtightened. Special care must be taken when the bolts screw into aluminum alloy components such as timing gear covers.

## INSTALLING THE TIMING GEAR COVER

On many engines, the diesel injection pump and drive gear should be installed before the timing gear cover is installed. Be sure that the timing marks on the drive gear are properly aligned. It is good procedure to recheck all timing marks. On some engines, the governor is located inside the timing gear cover. Check the service manual to be sure that all components are properly installed.

Replace the front crankshaft seal. Lip seals are commonly used to seal the front of the crankshaft (Fig. 16-19). Be sure that all gasket surfaces are properly cleaned and that the gasket fits properly. If gasket sealant is used, use it sparingly. Install the cover and pull all bolts up evenly. It is common for different-length bolts to be used. Check carefully to see that all bolts are in their proper location.

The hydraulic pump, governor, crankshaft pulley, water pump, and components may be installed at this

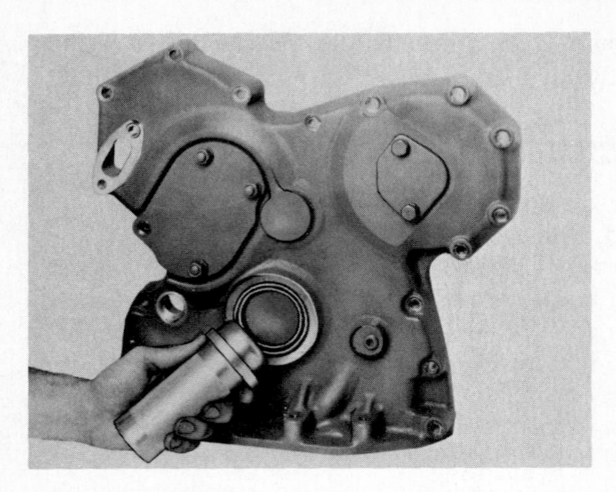

**Fig. 16-19.** Installing a new lip seal in the timing gear cover with a bearing seal drift. (*Deere & Co.*)

time. Check the service manual for correct installation procedures.

## INSTALLING THE CYLINDER HEAD

Check to see that the cylinder head and valve components are clean. If they have been sitting in the shop for a period of time, they should be rinsed in solvent and blown dry.

Be sure that valves, springs, and retainers are installed in their original position and that the springs have the correct end up. Check the service manual to see if new keepers should be installed. If original keepers are used, they should be checked for wear.

Lubricate the valve stems and the valve guides as they are assembled. Special valve guide lubricants, engine oil or a white-lithium-base lubricant can be used. Use the valve spring compressor to compress the retainers and springs. A drop of grease on the inside of the keepers will help hold them in place while the spring is being released.

Be sure new valve stem seals are properly installed on the intake valves. If valve stem seals are left off, engine oil will be drawn into the combustion chamber on the intake stroke.

Check the milled surfaces on the cylinder head and block to be sure that they are properly cleaned. Examine the gasket to determine which side should face up. Some gaskets are marked *top*. Be sure that no oil holes or water passages are covered by the gasket. Unless it is specified in the service manual, no gasket sealant should be used on head gaskets.

Carefully set the cylinder head on the block. Heavy cylinder heads should be lowered into place with a hoist or floor crane. Be sure that the head sits down flat on the gasket. If dowel pins are used, make sure that they align properly. Some heads are secured with stud bolts. They should be in place before the head is installed. Antisieze thread compounds may be used on cylinder head bolts.

Place all bolts in the head. Remember to place long bolts in their proper location and to replace any brackets that are held in place by the head bolts. Use the recommended sequence for torquing head bolts and tighten them in the recommended stages. The stages are snug, one-third torque, two-thirds torque, full torque, and repeat at full torque.

### Installing Push Rods

The push rods can be installed at this time. Be sure that the push rods are seated properly in the cam followers and that the cup end is up. The lower ends of the push rods should be lubricated before they are installed. Push rods should be installed in their original position.

### Installing the Rocker Arm Assembly

If rotators or caps are used on any of the valves, they should be installed at this time. Position the rocker arm assembly on the cylinder head and align the rocker arm adjustment screws in the cup ends of the push rods. Install the retainer bolts and tighten them a little at a time, starting at the center and working alternately toward each end. Tighten the bolts to the torque specified in the service manual. Be sure to reconnect all oil lines.

Some rocker arm assemblies are held in place by head bolts. When this system is used, the rocker arms must be in place when the cylinder head is torqued.

### Adjusting Valve Clearance

Valve clearance should be adjusted at this time. Refer to Chap. 11 for a discussion on valve adjustment. The service manual should be checked to determine the recommended valve clearance and adjustment procedures. Do not tighten the rocker arm cover securely at this time. It is good procedure to check valve adjustment and retorque the cylinder head after the engine has been run.

### Installing the Intake and Exhaust System

Check the gasket surfaces on the intake and exhaust manifold. No gasket sealant should be used on manifold gaskets. Be sure that all bolts are tight. Antisieze thread compounds will improve the bolt torquing procedure and make future removal much easier.

## REINSTALLING THE ENGINE

Attach the pull device to the engine and support it with a hoist or floor crane. Remove the engine stand.

### Installing the Flywheel

Some engines have a mounting plate that is installed between the flywheel and the engine block. Hydraulic pump drive gears and other components that attach to the rear of the engine block should also be installed. Check the bearing core hole plug in the rear camshaft bore.

If the flywheel ring gear is worn, it should be replaced at this time. Most ring gears are shrink-fit. Shrink-fit gears should be heated to expand them so that they will slip over the flywheel. They are usually heated to around 400 degrees Fahrenheit (°F) [204 degrees Celsius (°C)]. Check the service manual for the manufacturer's recommendations.

Install the flywheel and torque the bolts or nuts. Check the service manual for recommended torque. Remember to torque the bolts in stages. Be sure to secure the bolts with the locking devices recommended by the manufacturer.

### Installing the Clutch and Pressure Plate

If the pilot bearing or bushing is worn, it should be replaced at this time. An aligning tool is required to line up the clutch disk with the crankshaft (Fig. 16-20). Be sure that the clutch disk is installed with the correct side facing the flywheel.

Start the bolts in the pressure plate and pull them up evenly until they are tightened to the torque recommended in the service manual. If special bolts have been installed in two-stage pressure plates to hold the power takeoff (PTO) clutch disk in place, they may be removed at this time. Check the adjustment of the pressure plate release fingers.

The clutch release bearing and the transmission input shaft seal should be serviced at this time. It is good procedure to replace the transmission seal on older tractors.

**Fig. 16-20.** Using a pilot tool to install the clutch and pressure plate. (*Photograph by Westen McCoy*)

### Attaching the Engine to the Drive Train

The engine can now be reattached to the drive train. Adjust the hoist or floor crane so that the engine is in line with the clutch housing. Long bolts or rods can be used as guide pins to align the engine.

Push the engine carefully into place. It may be necessary to turn the crankshaft or PTO shaft to align the splines on the transmission shaft with the splines in the clutch disk or disks. The engine should pull into place with little resistance. When the engine is in place, all bolts should be installed and tightened.

### Front-End Assembly and Radiator

When all the components such as the water pump and governor are attached to the front of the engine, the front-end assembly and radiator can be reinstalled. The steering linkage or power steering system can be reconnected. The jack stand under the transmission housing can be removed.

### Exhaust and Intake System

The air cleaner and all hoses and air tubes that connect it to the intake manifold can be reconnected. All hoses, tubes, or clamps that are damaged or show excessive wear should be replaced.

The exhaust system can be connected to the exhaust manifold. Be sure to install a new gasket at the manifold connection and replace all burned-out components.

### External Components and Accessories

The starter, alternator or generator, hydraulic pump, and other components can be reinstalled. Connect all electric wires, circuits to gauges, hydraulic lines, and fuel lines. If the tractor is equipped with a generator, the generator should be polarized. Refer to Chap. 11 or the service manual for procedures.

## TIMING THE ENGINE

Valve timing was set when the timing marks on the camshaft and crankshaft gears were aligned. Ignition timing on spark-ignition engines and diesel injection pump timing on diesel engines must also be set.

### Timing Spark-Ignition Engines

The distributor must be timed to the crankshaft in order for spark to occur at the proper time. When installing the distributor, the crankshaft should be turned until No. 1 piston is at TDC on the compression stroke. Number 1 piston is at TDC when the timing mark on the flywheel or crankshaft pulley is in line with the pointer. However, No. 1 piston is only at TDC on compression every other time the timing marks align.

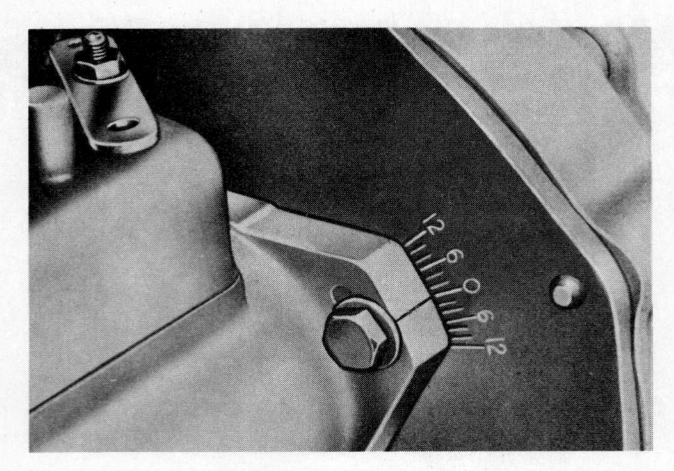


**Fig. 16-21.** Diesel injection pump timing is indicated by the reference marks on the pump casting and the timing gear cover. Adjustments are made by loosening the adjuster nuts and moving the pump body. *(Ford Tractor Operations)*

There are several ways to determine if No. 1 piston is up on the compression stroke. One method is to remove the spark plug from No. 1 cylinder. Hold your finger or thumb over the spark plug port and turn the engine over slowly. When pressure is felt in the cylinder, the piston is coming up on compression. Continue turning the engine until the timing marks align. This method will work on all four-stroke-cycle engines.

The valve overlap or "point-of-rock" method was discussed in Chap. 11. With this method, No. 1 piston is at TDC on the compression stroke when the other piston that is at TDC is at "point-of-rock." On four-cylinder engines No. 4 piston would be at point-of-rock, and on six-cylinder engines No. 6 piston would be at point-of-rock. This method will not work on three-cylinder engines.

With the engine in this position, install the distributor so that the rotor points to the No. 1 spark plug tower on the distributor cap. The rotor may turn as the distributor drive gear meshes with the gear on the camshaft. If it does, pull the distributor up and move the rotor enough to compensate for the movement and then reinstall the distributor. The distributor housing can be turned several degrees to complete the alignment of the rotor with the No. 1 spark plug tower.

When the distributor is installed, the holddown clamps should be installed and the retainer bolts tightened. Install the No. 1 spark plug wire in the No. 1 tower on the distributor cap. Determine the direction of rotation for the rotor and install the remaining spark plug wires in the proper firing order. Firing order is the numerical sequence in which the spark plugs fire. This can be found in the service manual.

This should time the engine close enough for it to start. Accurate adjustment of timing is done with the engine running. This procedure was discussed in Chap. 11.

## Timing Diesel Engines

Injection pumps are timed with marks. When the mark on the injection pump drive gear was aligned on the front of the engine, the pump should have been timed. Most pumps have several degrees of adjustment that can be made by turning the pump (Fig. 16-21). The pump should be reinstalled in the same position it was in when it was removed.

If the engine will not start or excessive black or gray smoke is observed, injection pump timing may be off. The procedure given in the service manual should be followed to check the timing accurately. It is not recommended to adjust diesel injection pump timing with the engine running because the excessive torque required to operate the pump could cause damage or personal injury.

## SUMMARY

When all machine work is completed and new parts are acquired, the engine can be reassembled. All parts must be clean, and properly lubricated as they are reassembled.

As parts are assembled, they should be checked for proper fit and running clearance. The service manual should be used to determine manufacturer's specifications and recommendations.

It is important that valves, diesel systems, and ignition systems be timed to the crankshaft. Most timing is done with marks on the gears at the front of the engine.

When the engine is completely assembled, it can be reattached to the tractor drive train.

After the engine has been reattached to the drive train, the front-end assembly and all other components removed during engine removal can be reinstalled. All electrical systems, hydraulic systems, and other controls should be reconnected. Notes made during engine removal will help make the job easier.

## THINKING IT THROUGH

1. Why is it important that all bolts and other fasteners be cleaned before engine reassembly?
2. Explain the procedure recommended for torquing engine bolts.
3. How can broken bolts be removed?
4. How is cylinder liner standout checked?
5. Discuss the installation procedures recommended for wet cylinder liners.
6. Why is it recommended to deglaze some cylinders?
7. Explain the methods recommended for cleaning cylinders after deglazing or honing.
8. Discuss the use of Plastigage to check bearing running clearance.
9. How is correct valve timing achieved as an engine is reassembled?
10. Why is it important that ring end gap be within specifications? How is it checked?
11. How is the clutch disk aligned on the flywheel?
12. Describe the installation of the distributor on spark-ignition engines.

# CHAPTER 17 STARTING, BREAK-IN PROCEDURES, AND PAINTING

## CHAPTER GOALS

In this chapter your goals are to:

- Prepare a tractor properly for initial starting after overhaul
- Start the engine and make the necessary adjustments that will provide performance and economy
- Use a dynamometer to break in an engine
- Prepare a tractor for painting and operate spray-painting equipment

An engine overhaul should not be considered complete until the engine has been properly adjusted and break-in has been completed. The first few hours of engine operation are very important. Lubrication, engine rpm, operating temperature, and load should be carefully monitored and regulated.

When proper break-in procedures are followed, the piston rings, bearings, and other engine parts should wear in properly. The rings should seat to the cylinder walls and form a good seal to prevent blowby and seal in the forces of combustion. The bearings and other internal parts should be polished, and friction should be at the normal level. If a tractor is delivered to the owner or operator without proper break-in, extensive engine damage may result.

It is recommended that a power takeoff- (PTO-) driven dynamometer be used to break in a tractor engine. The dynamometer allows the service worker to regulate the engine rpm and load accurately. While the engine is being run, a close watch can be kept on all systems and minor problems can be located that could lead to a service call after the tractor is delivered to the owner.

## PRESTARTING CHECKS

Before the engine is started, all systems should be properly serviced and checked. Failure to check all systems could cause extensive damage to the tractor. It would be hard to justify unnecessary repair costs to the owner.

### Fill the Cooling System

The radiator and cooling system should be filled to the cold coolant level with the water and the antifreeze mixture that will provide the freeze protection needed. The National Weather Service can be contacted to determine the freeze protection needed in different sections of the country.

Most manufacturers recommend that antifreeze be used year-round. It is never recommended to run water alone in a cooling system.

### Service the Fuel System

On diesel engines, it is good procedure to drain the fuel tank to remove water and sediment that may have collected in the tank. A steam cleaner can be used to clean dirty tanks. The tank should be thoroughly dried after cleaning.

New fuel filters should be installed. The tank should be filled with clean, fresh fuel. On diesel engines with gravity flow to the diesel injection pump, it is important that the fuel tank be full. The pressure of the fuel in the tank makes bleeding the system easier. Most manufacturers recommend that the tank be over one-half full when the system is bled.

**Bleeding the Diesel Fuel System** All the air must be removed from the diesel fuel system before the engine will start and run. The procedure for bleeding diesel fuel systems is discussed in Chap. 11. The service manual will describe the recommended bleeding procedures. Remember that the pressure in a diesel fuel system is very high and that proper safety precautions should be taken to prevent injury.

**Carburetor Adjustment** Carburetor adjustment should be checked on gasoline engines. The idle and load adjusting needles should be set as recommended in the service manual. To set the adjusting needles, screw them in slowly until they seat lightly and then back the needles out the number of turns recommended. Do not overtighten the adjusting needles. Overtightening will damage the needle tips and make proper adjustment of the carburetor impossible. If adjustment recommendations are not available, back the needles out 1 to 1½ turns for initial starting.

### Service the Crankcase

The crankcase should be filled with the grade and viscosity of oil recommended by the manufacturer, and a new oil filter should be installed. The service manual should be checked to determine the quantity of oil needed. It is good procedure to pour at least 1 quart [0.9 liter (L)] of the oil on the rocker arm assembly. This will help lubricate the rocker arms, valves, and cam lobes.

When an engine is started the first time after overhaul, it will operate for several seconds before the oil pump can prime itself and supply oil to the moving parts in the engine. This is the primary reason for lubricating all parts as they are assembled. Turbocharged diesel engines require special attention when starting to assure that the turbocharger is properly lubricated. When an engine is started without proper lubrication, a dry start will occur. This will result in severe engine wear in a short period of time.

### Charge the Lubrication System

One way to protect against dry starts is to use a lubrication system pressurizer to charge the lubrication system. The pressurizer can be installed at any opening to the engine lubrication system. It can be installed in the oil-sending unit port or in any plug hole that can be located on the oil filter or in the oil gallery.

When a pressurizer is used, the crankcase oil is poured into an oil reservoir tank. The tank is then sealed, and compressed air is used to pressurize it. The tank pressure should be equal to the engine oil pressure recommended in the service manual for the engine.

To add the oil to the engine, open the line valve on the tank. The oil will be forced through the engine lubrication system, and the oil pump will be primed. This should be done just before cranking the engine. It is still good procedure to pour 1 quart [0.9 L] of the oil over the rocker arm assembly. Be sure not to overfill the crankcase.

### Check the Battery

Use a battery hydrometer or a voltmeter to check the state of charge in the battery. If the electrolyte level is low, water should be added. Use a battery charger to bring the battery up to a good state of charge.

Be sure that the battery terminal posts are clean and that the cable terminals are tight. Also, check the terminal connections at the solenoid and at the starter.

### Check the Hydraulic System

If the hydraulic pump was removed, the hydraulic oil level will be low. The hydraulic filter should be changed if the system was drained. Remember that it is important to use the hydraulic oil recommended by the manufacturer.

Some hydraulic pumps must be bled or primed when they are reinstalled. If a pump is not self-priming and it is not primed, *cavitation* may occur. Cavitation is caused by air that is trapped in the hydraulic system. Cavitation causes abnormal pump noise and vibration. If cavitation is allowed to continue, it can cause serious damage to the pump. Check the service manual for the manufacturer's recommendations.

**Power Steering** Some tractors are equipped with separate power steering pumps. If the tractor has a separate pump, the reservoir should be filled to the recommended level. Check the service manual to determine service recommendations.

### Transmission, Differential, and Final Drive

Check the lubricant levels in the transmission, differential, and final drive. If they were drained during disassembly or if any oil is being added during reassembly, it is important that the oil specified by the manufacturer be used.

Some tractors have separate fluid reservoirs for the PTO and final drive. Check the service manual and be sure that all systems are filled to their recommended level.

## STARTING THE ENGINE

It is good procedure to double-check all systems and make sure all the adjustments are made before starting the engine. As a safety precaution, a pressurized dry chemical fire extinguisher should be available.

When starting the tractor, the service worker should be seated on the operator's platform or in the cab. The transmission should be in neutral or park, and the brakes should be set.

### The Starter

An engine that is properly adjusted and timed should start easily. If the engine is hard to start, the starter should not be operated longer than 30 second(s) at a time. After it has been operated for 30 s, allow it to cool for 2 minutes (min) before attempting to start the engine again. If the engine fails to start after several tries, recheck the fuel and ignition systems.

### Check Engine Oil Pressure

When the engine starts, monitor the oil pressure. If the engine is equipped with a gauge, watch for pressure rise. On engines equipped with indicator lights, the oil pressure warning lights should go off. On engines not equipped with a gauge, it is good procedure to install a temporary gauge so that engine oil pressure can be determined (Fig. 17-1). If oil pressure is not indicated after approximately 10 to 15 s, stop the engine immediately and determine the cause.

**Turbochargers** On engines equipped with turbochargers, engine oil pressure must be attained before the engine is started. Check the service manual for starting recommendations.

**Fig. 17-1.** An oil pressure gauge may be temporarily installed in the sending unit port to determine operating oil pressure. (*Photograph by Westen McCoy*)

## Check Other Systems

Check the ammeter or charging indicator light. Keep a close watch on the temperature gauge. Check the hydraulic system. If the hydraulic pump is not self-priming, it should be primed. On self-priming pumps, the hydraulic lift should be cycled several times to remove any air that may be trapped in the system. If a system is not operating properly, it should be serviced before engine operation is continued.

## Check the Engine

The engine should be allowed to run at about one-third throttle for several minutes. This will provide operating oil pressure, and bearing throw-off will be sufficient to provide adequate lubrication to the cylinder walls (Fig. 17-2).

When a gasoline engine reaches operating temperature, the carburetor should be adjusted and the ignition timing should be set. It may be necessary to make some adjustments before operating temperature is reached. Keep a close watch on the operating temperature.

Check closely for any oil, coolant, or fuel leaks. All leaks should be stopped. Remember that when trying to stop leaks in the hydraulic system the engine should not be running.

# ENGINE BREAK-IN

It is a good practice to break in an engine on a PTO dynamometer. When using a dynamometer, follow the manufacturer's operating instructions. Be sure that the dynamometer is properly attached to the tractor and that the PTO shaft is as straight as possible. It is essential that the tractor be in a well-ventilated area. Remember that exhaust fumes are deadly.

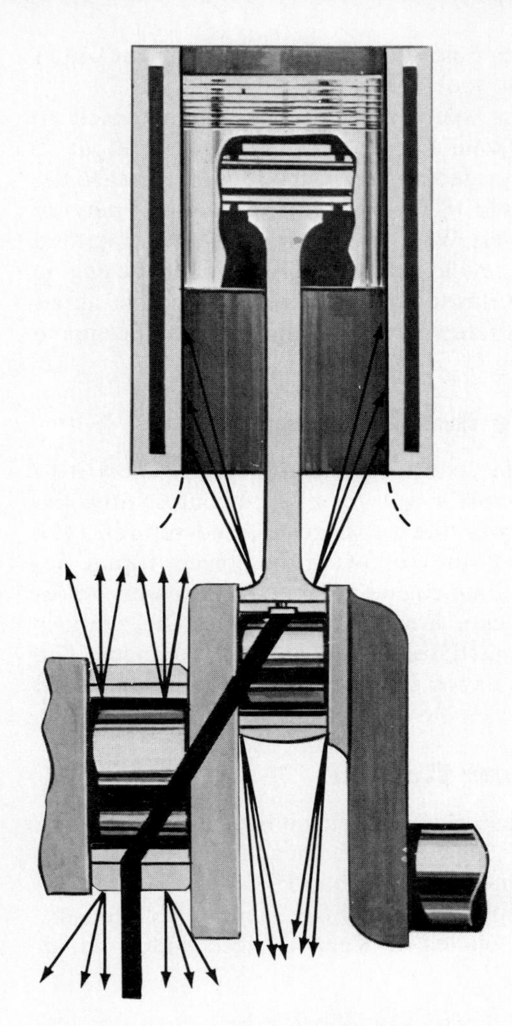

**Fig. 17-2.** Oil throw-off from the rod and main bearings lubricates the cylinder walls and piston rings. (*Dana Corp.*)

## Dynamometer Run-In

Bring the engine to operating temperature. This should be done with the engine operating at 600 to 800 revolutions per min (r/min) and with no load applied.

As the engine is being brought to operating temperature, the fast-idle rpm should be checked and adjusted. Fast-idle rpm is the full-throttle no-load rpm. Since diesel injection pumps are calibrated and adjusted on a test stand, no diesel adjustment should be necessary. However, it is not uncommon for governor adjustments to be needed on gasoline engines. Check the service manual for adjustment procedures.

When the engine reaches operating temperature, increase the engine speed to about 1200 r/min and load the engine to about 25 percent of its rated power. Run the engine at this rpm and power load for

30 min. Keep a close watch on all gauges and watch for any oil or coolant leaks.

The engine speed should then be increased to about 1600 r/min and the load increased to about 50 percent of its rated power. After 30 min, increase the engine speed to 1800 r/min and the load to 75 percent of rated power. When the engine has been operated at this speed and load for 30 min, increase the engine speed to full throttle and maintain the load at about 75 percent of rated power. Continue to run the engine at this setting for 60 min.

### Retorque the Cylinder Head

Remove the load from the engine and allow it to run at 600 to 800 r/min for a few minutes. Stop the engine. Remove the valve cover, and retorque the cylinder head bolts following the correct torque sequence. On some engines, it may be necessary to remove the rocker arm assembly. After the cylinder head is retorqued, readjust the valve clearances. The valve cover can then be reinstalled and tightened down.

### Full Power Run-In

Restart the engine and bring it back up to operating temperature. Set the engine speed at full throttle and load the engine to the full load PTO rpm. Run the engine at this speed and load for 5 min. Check for any coolant and oil leaks. Keep a close watch on all gauges.

## DYNAMOMETER TESTS

The dynamometer can be used to set the ignition timing and adjust the carburetor on gasoline engines. To make these adjustments the engine should be at operating temperature.

### Ignition Timing

Set the engine speed at full throttle and load the engine until the manufacturer's recommended full-load PTO rpm is reached. Loosen the distributor hold-down clamp and move the distributor in both directions until the greatest power reading is attained on the power gauge.

After timing has been set with the dynamometer, recheck the timing with a timing light. This procedure is discussed in Chap. 11. If low-rpm timing is off several degrees, the advance system in the distributor is not working properly. The distributor should be repaired or replaced and the ignition timing reset.

### Carburetor Adjustment

To check the carburetor adjustment, install a flowmeter in the fuel line leading from the fuel tank to the carburetor inlet (Fig. 17-3). Set the engine speed at full throttle and use the dynamometer to load the engine until the manufacturer's recommended full-load PTO rpm is reached. The load adjusting needle on the carburetor should be open about two turns.

Observe the PTO power that the tractor is producing and check the flowmeter to determine the fuel consumption rate in gallons [liters] per hour. When the fuel consumption rate is within the manufacturer's recommendations, the carburetor is probably operating properly. However, if the fuel consumption is below the specifications, the carburetor needs to be repaired or replaced before proceeding with the adjustment.

When the carburetor is operating properly, the load adjusting needle should be turned in as far as possible without reducing the maximum power output more than 1 horsepower (hp). This can be done by closely observing the power rating needle on the dynamometer power gauge. This single adjustment can result in tremendous fuel savings for the owner.

### Check PTO Power

The service manual will indicate the manufacturer's rated PTO power. Set the engine speed at full throttle and load the engine to the full-load PTO rpm. Compare the power attained with the manufacturer's rated power. If the engine has been properly over-

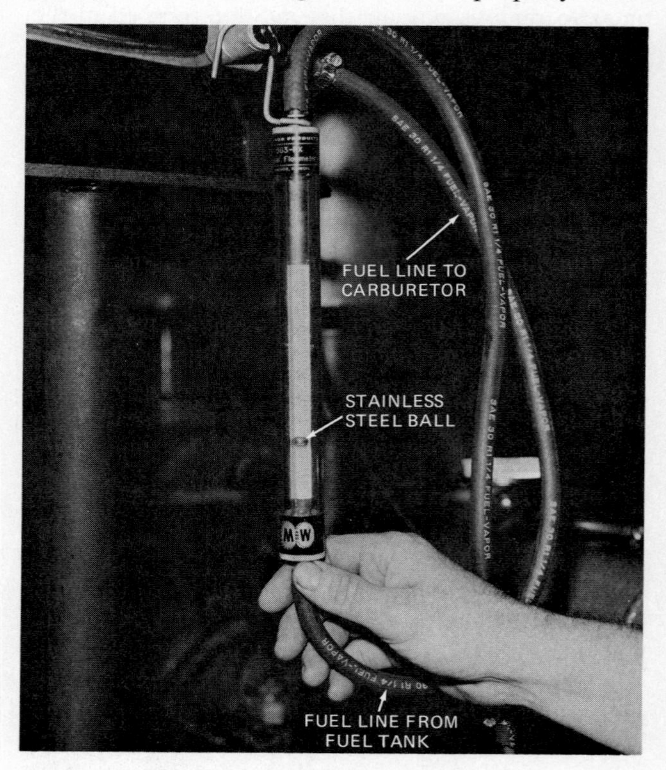


**Fig. 17-3.** A flowmeter being used to measure fuel consumption in gallons (liters) per hour. Flow rate is indicated by the position of the stainless steel ball. (*Photograph by Westen McCoy*)

hauled and adjusted, full recommended power should be attained. Follow the instructions provided with the dynamometer to determine power output.

## PAINTING THE TRACTOR

Since agricultural tractors maintain their productive value over a long period of years, it is good procedure to repaint them when the paint deteriorates and body rust begins to appear. The sheet metal parts on a tractor will be damaged or ruined if they are not protected from rust. Drive train and engine components are not normally affected by rust. The tractor's value will be more easily maintained if it is kept painted.

### Cleaning the Tractor

The tractor should be thoroughly cleaned before it is painted. A steam cleaner or high-pressure washer can be used to clean the tractor. A suitable solvent can be applied to remove caked-on dirt and oil. It may be necessary to use a scraper or wire brush to remove some of the dirt. Water and detergent should then be used to remove the solvent. It is good procedure to sand any rusty spots on the drive train.

**Using a Sandblaster** A sandblaster can be used to remove old paint and rust from sheet metal parts such as hoods, fenders, and cowlings. Proper protective equipment should be worn when using a sandblaster. The fine sand comes out under extreme pressure and can cause painful injury to exposed skin. It is essential that the operator's head be protected with an approved hood.

It is not a good procedure to sandblast assembled parts. The sand can damage seals and get inside the components. If engine blocks and drive train components are to be sandblasted, they should be completely disassembled and all bearing and gasket surfaces should be protected. This can be done by using masking tape and heavy paper to cover them. Always be sure that all sand is removed after sandblasting.

### Preparing the Tractor for Painting

Dents in the body sheet metal can be straightened using two hammers or a hammer and a body "dolly." Breaks and cracks can be repaired by welding. Be careful not to weld near fuel tanks.

**Using Body Putty** Body putty can be used to repair rusted areas and smooth out dents. The instructions on the putty container should be followed when mixing, applying, and smoothing the body putty. All paint and rust must be removed from the metal before any body putty is applied.

**Sanding** When sheet metal parts are not excessively rusted and the paint is relatively smooth, sandpaper can be used to prepare it for painting. Wet-or-dry sandpaper used with water will produce suitable results. The water will lubricate the surface and will prevent the paint that is removed from loading or filling the pores in the sandpaper. Fine-grit sandpaper should be used to prevent scratching and to produce a smooth surface.

**Preparing for Paintings** In order to get to all parts of the tractor, it is usually best to remove the wheels. Appropriate stands should be used to support the tractor. Sheet metal parts can be secured on spray stands or jigs (Fig. 17-4).

**Removing Film** It is good procedure to clean all painted parts with a commercial cleaning solvent designed for this purpose. This will remove wax, grease, or polish that could cause the primer and paint to peel.

**Metal Conditioners** Commercial metal conditioners can be used to remove any rust that may remain on bare metal that has been sanded or sandblasted. Metal conditioners are acids which change iron oxide (rust) into other compounds. These compounds are stable, and if the surface is sealed with primer as soon as the conditioner has dried, rust formation should be stopped. The conditioner also etches the metal and helps the primer to adhere to the metal.

Since metal conditioners are acid, it is recommended that neoprene-coated gloves be worn when a conditioner is being applied. Conditioners are normally applied with a brush or a clean white rag.

Commercial metal conditioners are concentrated and are mixed with water before use. It is extremely important that the proper ratio of water be used. The directions on the container should be followed. Always use conditioners in well-ventilated areas where there are no open flames.

**Fig. 17-4.** A shop-made stand may be used to support sheet metal, fenders, and small parts while they are being painted. *(Photograph by Westen McCoy)*

**Masking Tape** Masking tape and paper can be used to cover gauges, steering wheels, tires, and other components that are not to be painted. Masking tape should be applied when the tractor is ready to paint and should be removed as soon as possible after painting is finished.

## Selecting Paints and Primers

The two basic finishes used to protect metal are enamel and lacquer. Enamel is used on agricultural equipment because it provides a good finish and dries to a luster.

It is a common practice to repaint tractors and other agricultural equipment their original color. However, if the manufacturer has changed colors, some prefer to use the new colors.

## Primer Undercoats

Primer, or primer-surfacer, is an undercoat used on bare metal to obtain a surface for adhesion of the topcoat paint to the metal. Primer also helps prevent rust or corrosion.

**Enamel Primer** Enamel primer is preferred for tractors and other agricultural equipment. It is tougher, more flexible, and provides a better adhesion surface than lacquer primer. The drying time and sanding requirements vary with the type used. The directions on the container should be read and followed.

**Applying Primer** When tractor components such as fenders and hoods are sandblasted, they should be primed immediately. Rust will form if the bare metal is exposed overnight. Also, as mentioned earlier, it is good procedure to use a commercial metal conditioner.

Primer should be applied to all bare component surfaces where the paint was not completely removed. This can be done by spot priming or by priming the entire tractor.

## Enamel Topcoats

Primer has a dull finish, and since its primary function is to adhere to the base metal, it does not weather well. A topcoat of enamel is used to produce a durable and attractive finish.

**Synthetic Enamels** Synthetic enamels are extensively used to repaint tractors and other agricultural equipment. Alkyd enamel is most used, although acrylic enamel may be used. Alkyd and acrylic are both vehicles used in enamels. Vehicles include both the solvents and binders. Solvents thin the paint so that it can be applied evenly on the surface of the metal. As the paint cures or dries, the solvents evaporate and the binders go through a chemical reaction and form a durable high-gloss finish. The pigment in the enamel provides the color.

## Reducers and Thinners

Primers and enamels are manufactured at a specific viscosity. Viscosity is a measure of resistance to flow.

Both primers and enamels are manufactured at higher viscosities than they are applied. This is done to reduce settling of the ingredients during storage. The part that settles is the pigment. It is essential that it be thoroughly mixed with the solvents and binders before the paint is used.

Viscosity is affected by temperature. When the temperature is low, paint viscosity is high, and at high temperature the viscosity is low. Reduction or thinning should be the same at all temperatures, because the rate of evaporation of the reducer or thinner varies.

**Thinning Paint and Primer** The only accurate way to thin paint and primer is to follow the directions on the container carefully. The amount of thinner or reducer recommended should be carefully measured, and the amount of reduction should be the same regardless of the temperature. A kitchen measuring cup can be used to measure paint, primer, and reducers or thinners accurately.

It is good procedure to thin only the paint or primer needed to do a job. When thinned paint or primer is left over, it should not be mixed with paint or primer that was not thinned. Mixing the thinned with the unthinned will cause excessive settling of the pigment and prevent accurate thinning of the remaining paint or primer. If more than one can of paint is needed, the color will be more uniform if the paints are mixed together before painting is begun.

The paint or primer should be mixed thoroughly. When the pigment has settled out and has formed a hard layer in the bottom of the can, the liquid part should be poured off and the settled part well broken up. The liquid can then be poured back in slowly and mixed with the pigment. If any thinner or reducer is used to help break up the pigment, it should be measured and deducted from the total amount added.

Paint or primer should not be overthinned. Excessive thinning reduces the thickness of the paint and can cause sags or runs. Overthinning is expensive because as the paint cures or dries, the thinner or reducer evaporates. Overspray or paint drift is also increased.

## Spray Painting

Enamel can be applied with a brush or a spray gun. When a spray gun is available, it should be used. Paint can be applied faster and more evenly with a spray gun. The results will be a more professional looking job and the value of the tractor or equipment will be increased.

**Spray Booth** Spray painting should only be done in a spray booth. A spray booth will provide ample lighting and ventilation. The air entering a properly designed spray booth will be free from dust and other foreign particles which can settle in the wet paint and cause a rough finish. It is not recommended to spray paint in areas where welders or other equipment are being operated. Some paints are highly explosive, and a serious accident could result.

Most states have strict worker, fire, safety, and building protection codes regarding the installation and operation of commercial spray booths using volatile liquids. These codes require the use of shielded electrical systems to eliminate the possibility of sparks, adequate filtering of interior and exterior particle dispersion, safe disposal of solvent residues, extensive sprinkler and fire control systems, and the use of protective clothing and masks. These laws must be actively followed by the competent service worker.

**Types of Spray Guns** Several types of spray guns are available. Bleeder-type guns are used with small air compressors and with continuously running compressors. They can be used to paint tractors and other agricultural equipment, but they are slow.

The nonbleeder-type spray gun is preferred (Fig. 17-5*a*). Most nonbleeder-type spray guns can be used for either siphon- or pressure-feed. Since this spray gun is used most, it will be further discussed.

**Parts of a Spray Gun** In order to use and care for a spray gun properly, a service worker should have a thorough understanding of the component parts and their function.

**Air cap** The air cap screws onto the gun body (Fig. 17-5*b*). It directs compressed air through small holes in the horns. These small jets of air atomize the paint and form the spray pattern. The streams of paint and air mix outside the gun.

**Fluid nozzle or tip** The fluid nozzle meters (regulates the flow of) the paint and directs it into the airstream. It also forms a seat for the fluid needle to shut off the flow of paint when the trigger is released (Fig. 17-5*b*).

**Fluid needle** The fluid needle shuts off the paint flow through the fluid nozzle. When the spray gun is operated siphon-feed, it regulates the paint flowing through the fluid nozzle. The fluid control valve regulates the distance the fluid needle moves away from the seat in the fluid tip. It is normally opened until one thread is showing (Fig. 17-5*a*). Fluid tips and needles are selected to meet the spraying needs and must be the same size.

**Pattern control valve** The size and shape of the spray pattern are controlled by the pattern control valve (Fig. 17-5*a*). As the pattern control valve is closed, the spray pattern becomes circular. When the pattern control valve is opened, the pattern becomes oblong. The pattern control valve should be open when painting large surfaces.

**Air valve** The flow of air through the spray gun is controlled by the air valve (Fig. 17-5*a*). When the spray gun is operated, the fluid needle closes before the air valve. Bleeder-type spray guns do not have an air valve.

**Trigger** The trigger operates the air valve and the fluid needle (Fig. 17-5*a*). When the trigger is pulled part way, the air valve is opened. As the trigger is pulled further back, the fluid valve is opened.

**Using the Spray Gun** The spray gun should always be clean and properly assembled. Air pressure at the spray gun should be set at 45 to 55 pounds per square inch (psi) [310 to 380 kilopascals (kPa)] when alkyd enamels are used. There will be a pressure drop between the air transformer and the spray gun. Table 17-1 can be used to determine the approximate setting of the air transformer to gain the desired pressure at the gun. The pressure on a pressure feedgun paint cup should not exceed 50 psi [345 kPa].

The cup or pressure tank should be filled with properly thinned primer or paint. Remember that all bare metal should be primed. All paint and primer should be strained as they are poured into the spray containers. Commercial paint strainers are available at most paint dealers.

**Test the spray pattern** A piece of heavy paper or slick cardboard is ideal for testing the spray pattern. To test the pattern, hold the air nozzle 6 to 10 inches (in) [15 to 25 centimeters (cm)] from the paper. With the spray gun level, pull the trigger all the way back and release it at once.

If the pattern appears coarse or too wet, increase the air pressure at the air transformer about 5 psi [34 kPa]. Decrease the air pressure about 5 psi or open the fluid valve slightly if the pattern is too fine or dry looking.

The shape of the pattern can be adjusted with the pattern control. When painting tractors and other agricultural equipment, the pattern should be about 9 to 10 in [23 to 25 cm] wide.

A crescent-shaped pattern will occur when one or more holes in the air horn are plugged. Holes should be cleaned with a broom straw or a toothpick. Do not use metal wire to clean the air horns. This may damage the air passages.

A pattern that is heavy in certain areas or has a spatter effect is caused by insufficient atomizing

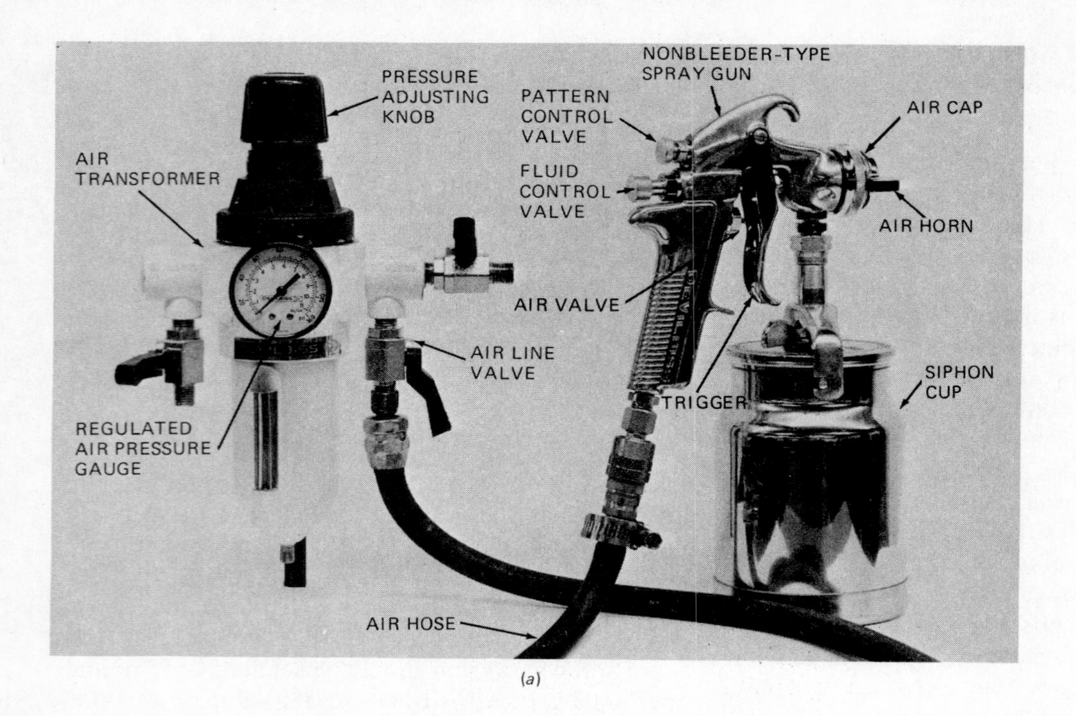


(*a*)

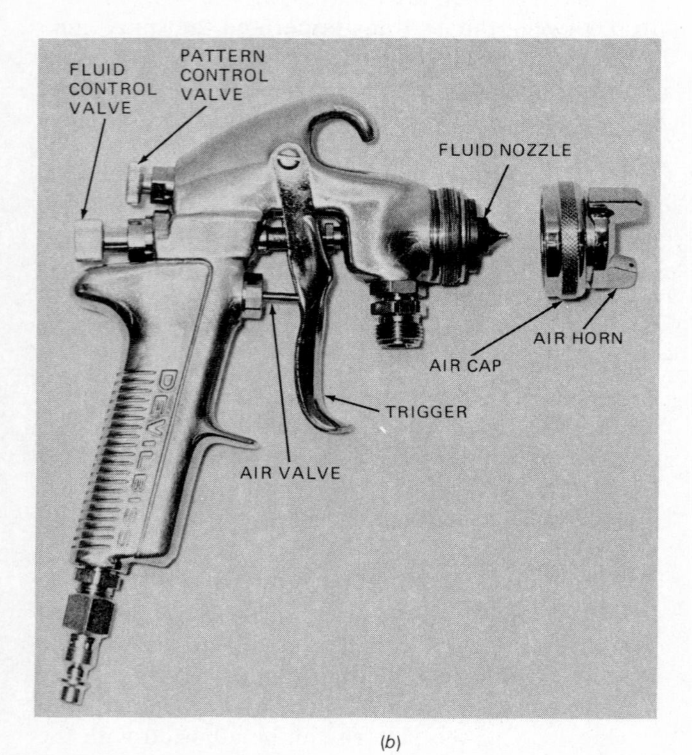


(*b*)

**Fig. 17-5.** (*a*) A nonbleeder-type spray gun complete with siphon cup and air transformer. (*b*) Spray gun fluid nozzle, air cap, and air horn. (*Photograph by Westen McCoy*)

pressure. This can be corrected by increasing the pressure at the air transformer.

A loose air cap or paint that has dried around the outside of the fluid tip will cause the pattern to be wide or heavy at the top or bottom of the spray pattern. When this occurs, the air cap and fluid tip should be removed and cleaned with a suitable thinner. A small amount of thinner in a cup and a small brush can be used to clean the fluid tip and air cap.

The spray gun must be operated properly if a good paint job is to be attained. Paper or cardboard should be used to practice on.

**TABLE 17-1. Estimated Air Pressure At Spray Gun**

| Transformer Pressure Setting | PRESSURE AT SPRAY GUN FOR LENGTH OF HOSE | | | | |
|---|---|---|---|---|---|
| | 10 ft | 15 ft | 20 ft | 25 ft | 50 ft |
| 1/4-in hose | | | | | |
| 40 psi | 32 | 31 | 29 | 27 | 16 |
| 50 psi | 40 | 38 | 36 | 34 | 22 |
| 60 psi | 48 | 46 | 43 | 41 | 29 |
| 70 psi | 56 | 53 | 51 | 48 | 36 |
| 80 psi | 64 | 61 | 58 | 55 | 43 |
| 90 psi | 71 | 68 | 65 | 61 | 51 |
| 5/16-in hose | | | | | |
| 40 psi | 37 | 37 | 37 | 36 | 32 |
| 50 psi | 48 | 46 | 46 | 45 | 40 |
| 60 psi | 56 | 55 | 55 | 54 | 49 |
| 70 psi | 65 | 64 | 63 | 63 | 57 |

**Distance from the work** The spray gun should be held 6 to 10 in [15 to 25 cm] from the object being painted. It is also important that the spray gun be held at right angles to the work so the spray pattern hits the surface squarely (Fig. 17-6).

**Trigger operation** The gun should be moving into the paint area when the trigger is pulled, and the trigger should be released as the gun is leaving the paint area. Long flat surfaces should be painted in sections, and the trigger should be pulled as the gun reaches an unpainted section and feathered or released slowly as it reaches a painted section. The trigger should be pulled and released on each stroke.

**Overlap** Each stroke should overlap the preceding stroke about one-half (Fig. 17-7). This results in single coat coverage.

**First coat** The first coat should not be expected to cover the primer or the old paint completely. This coat should be allowed to dry until it gets tacky. It should be ready for the second coat in 10 to 20 min. The time required depends on the air temperature and humidity.

**Second coat** The second coat should be applied in the same manner. It should look wet as it goes on, and there should be complete coverage. Each coat should be put on lightly and evenly to prevent sags and runs. Primer will dull as it dries, but enamel should retain its shine.

**Additional coats** Additional coats can be applied in the same manner. Be sure to allow the paint to flash between coats. *Flash* is the first stage of drying when part of the reducer works its way to the surface and evaporates. The paint should have a normal gloss when flash has occurred.

When the paint film is excessive, drying time is much longer. A thick film of enamel is not more dura-

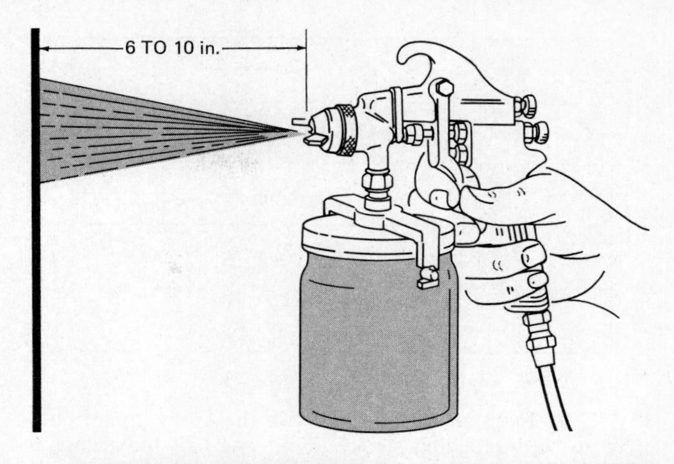


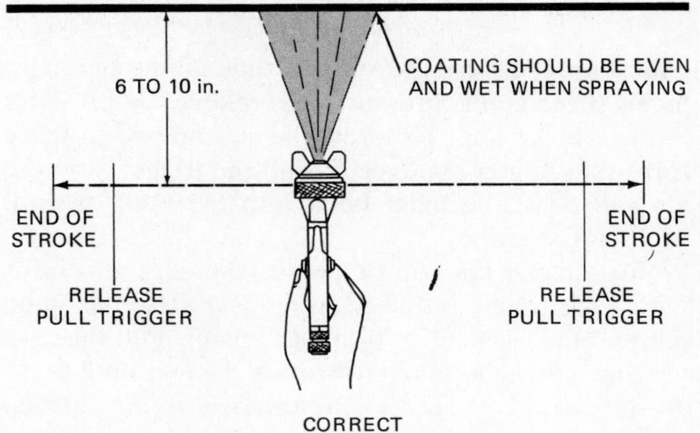


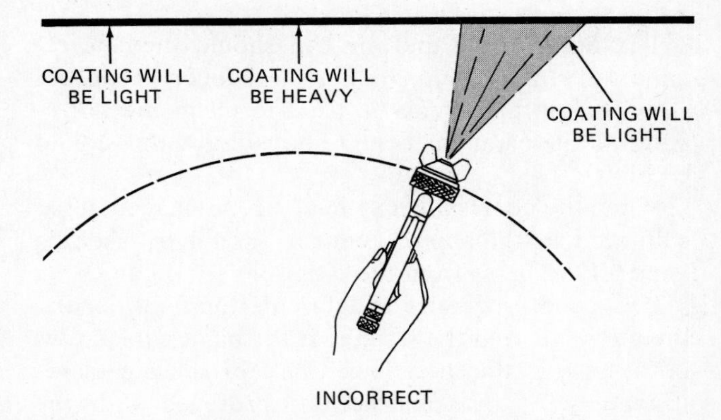


**Fig. 17-6.** Good spray-painting techniques will ensure a good paint job.

ble than a normal coat, and excessive enamel will not cover dents in the surface. Also, a heavy coat tends to sag and run.

## Cleaning the Spray Gun

A spray gun should be cleaned as soon as the paint job is finished. Most problems with spray guns result from improper cleaning and care.

When a spray gun is not properly cleaned, paint can partially or completely clog the nozzle and the fluid tubes and valves. This can cause the gun to spit or form an irregularly shaped spray pattern the next time it is used.

To clean the gun, loosen the spray gun from the

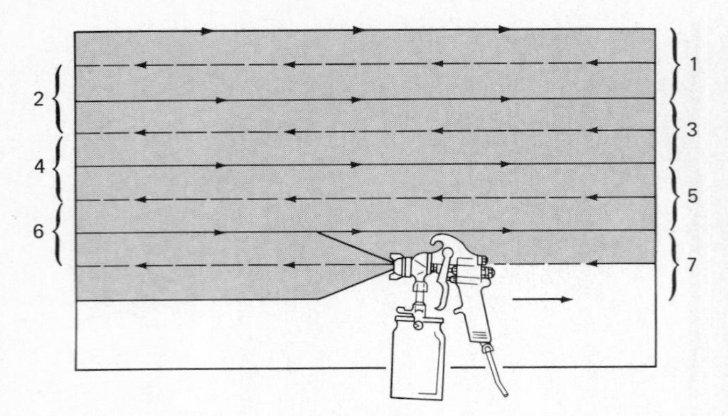


Fig. 17-7. Each stroke applied with the spray gun should overlap the preceding stroke about one-half for a single-coat application.

paint cup and raise the suction tube above the paint level. If the gun is pressure-fed, relieve the pressure in the tank or cup. Unscrew the air cap two to three turns and hold a rag over it. Pull the trigger, and the air will push the paint back into the tank or cup. Tighten the air cap.

Next, clean the cup or pressure tank and refill it with a small amount of clean reducer. Use the same reducer that was used to thin the paint. Pull the trigger and spray the thinner through the gun until there are no traces of paint in the thinner. Again, loosen the air cap and hold a rag over it. Pull the trigger for a few seconds.

The fluid nozzle and air cap should then be removed and soaked in thinner or reducer. A broom straw or toothpick can be used to clean the holes. After all the paint has been removed, the unit should be blown dry.

When the nozzle and cap have been replaced, a rag saturated in thinner or reducer should be used to wipe all the paint from the exterior of the gun.

It is good procedure to lubricate the fluid needle, the air valve, and the trigger. If the paint cup screws onto the gun, the threads can be lubricated with petroleum jelly. The instructions provided with the spray gun should be followed.

**Reassembling the Tractor** When the paint has dried, the tractor can be reassembled (Fig. 17-8). Touch-up paint can be used to paint fasteners that were missed.

**Applying decals** Original equipment decals can be purchased for most agricultural tractors and equipment. Decals help return the tractor to its original appearance and give proper safety and operating rules (Fig. 17-8). The paint should be allowed to dry overnight before applying decals.

### Painting Safety

Paints, primers, thinners, and reducers are flammable in varying degrees, and their flammability increases as they are sprayed into the atmosphere. All solvent containers should be kept closed, except when pouring, and all painting and cleaning should be done in a spray booth.

Fig. 17-8. This tractor is over 15 years old. After a complete overhaul and a new paint job, it is ready for many more years of service. (*Photograph by Westen McCoy*)

Keep all open electric motors and open flames out of the painting area. Under the right conditions, the spark from the motor of an air compressor, electric drill, or cigarette can cause an explosion.

Personal safety protection such as safety glasses and respirators should be worn. Rubber or neoprene-coated gloves should be worn when handling metal conditioners and paint removers. Safety shoes are recommended.

A pressurized dry chemical fire extinguisher should be placed in an easy-to-reach location. It should be checked periodically to be sure it is charged.

## TRACTOR DELIVERY

Even though the tractor engine has been run on a dynamometer for 3 or 4 hours, it is good procedure to follow the engine break-in procedures given in the operator's manual. The service worker should go over these recommendations with the owner when the tractor is delivered and make sure the owner fully understands them.

## SUMMARY

An engine overhaul is not complete until the engine has been properly adjusted and a dynamometer break-in has been completed. The dynamometer can be used to wear in engine parts under controlled conditions. Proper break-in can reduce the chances of engine failure or service calls after delivery.

The value of tractors and other agricultural equipment can be maintained by painting. When painting tractors, it is essential that the tractor be properly cleaned. Enamel primers and alkyd enamel are usually used on agricultural equipment.

Spray guns must be clean and properly adjusted. It is important that they be cleaned and stored after each use.

## THINKING IT THROUGH

1. What machine can be used to break in a tractor engine in the repair shop?
2. How are the idle and load adjustment needles set on a carburetor before the engine is started?
3. Why is it recommended to pour at least 1 quart [0.95 L] of oil in an overhauled engine over the rocker arms?
4. How is a pressurizer used to put oil in an overhauled engine?
5. What causes cavitation?
6. Why should an engine be run at about one-third throttle for several minutes when it is first started?
7. When should the cylinder head be retorqued during break-in?
8. How can a dynamometer be used to adjust ignition timing on gasoline engines?
9. Why is it not a good practice to sandblast assembled parts?
10. When should a metal conditioner be used when repainting a tractor?
11. What type of primer is recommended for tractors and other agricultural equipment?
12. What type of paint is most used?
13. Why is it important not to thin paints and primers excessively?
14. List the major parts of a spray gun.
15. What spray gun pressure should be used to spray enamels?
16. How far should a spray gun be held from the object being painted?
17. How long should enamel dry before another coat is applied?
18. Describe the procedure for cleaning a spray gun.
19. List the safety precautions that should be observed when spray painting.

# UNIT V SOIL PREPARATION AND CROP PRODUCTION EQUIPMENT

| Competencies | Agricultural Machinery Service Manager | Agricultural Machinery Parts Person/ Manager | Agricultural Tractor and Machinery Mechanic | Agricultural Machinery Setup and Delivery Mechanic | Agricultural Tractor and Machinery Mechanic's Helper | Small-Engine Mechanic | Farmer, Grower, and Rancher |
|---|---|---|---|---|---|---|---|
| Select the proper types of primary and secondary tillage machinery | Essential | Desirable | Desirable | Essential | Desirable | Not necessary | Essential |
| Select the proper planting and seeding equipment | Desirable | Desirable | Desirable | Essential | Desirable | Not necessary | Essential |
| Select the proper chemical application equipment | Desirable | Desirable | Desirable | Essential | Desirable | Not necessary | Essential |
| Assemble and adjust soil preparation and crop production equipment | Essential | Essential | Essential | Essential | Essential | Desirable | Essential |
| Determine proper hitching to obtain correct line of draft | Essential | Essential | Essential | Essential | Desirable | Not necessary | Essential |
| Calibrate planting, seeding, and chemical application equipment | Essential | Desirable | Essential | Essential | Desirable | Desirable | Essential |
| Make field service adjustments | Essential | Desirable | Essential | Essential | Not necessary | Desirable | Essential |

**Essential**

**Desirable**

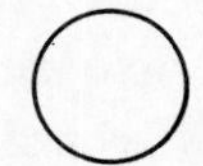
**Not necessary**

For centuries humans have tilled the soil and planted seeds to feed themselves. In fact the industry of agriculture is based upon the production, processing, and distribution of food to feed the people of the world. When farming depended on human and animal power and simple hand tools, production per worker was very low. Through the application of the knowledge and skills of many people in agribusiness, including those in agricultural plant, animal, and engineering sciences, and those involved in education and research, the production capacity has increased to a level where an agricultural producer in the United States is feeding him or herself, as well as 64 other people in the world.

Without mechanization this tremendous productive capacity would not have been possible. Agricultural production is machine-oriented. It is predicted that in the future a crop must be capable of being planted, nurtured, and harvested by machine, or it will disappear from the food market.

As a worker in agricultural power and machinery, you have a tremendous responsibility for assuring the proper operation of production machinery. Your knowledge and skills are needed to service, adjust, and make repairs if the machinery is to work efficiently and produce a high-quality product. For example, the productive value of fertile soil is lost if soil preparation machinery fails to prepare a satisfactory seedbed. Or a harvesting machine that fails to collect, separate, and clean the seed has destroyed the productive capacity of a high-yielding grain crop. As a service representative you will be called upon to troubleshoot machine operating problems, arrive at practical solutions, and make recommendations to owners/operators. As a salesperson you will need to know the purpose of a machine and how to interpret specifications and choose the proper size and type of attachments. The mechanic must be able to assemble, operate, adjust, and test new machines and make repairs in the field or service center.

Soil preparation and crop production machinery is one segment of agricultural power and machinery. The equipment includes specialized machines that are used to (1) stir the soil, (2) place the seed, and (3) nurture the growth of plants through the growing stage at which time weed and insect control becomes necessary.

This unit is designed to help you understand the purpose of soil preparation and crop production machinery and to develop the skills and abilities for selecting, adjusting, and maintaining the equipment.

# CHAPTER 18 AGRICULTURAL TILLAGE MACHINERY

## CHAPTER GOALS

In this chapter your goals are to:

- Identify the types and purposes of tillage machinery
- Describe the principles of draft
- Make shop and field adjustment of tillage machinery

The primary purpose of agricultural tillage machinery is to "stir" (till) the soil and provide a suitable condition for seed germination at planting time. Proper tillage conditions the soil and assists in breaking up compacted layers so that air, moisture, and nutrients can reach the roots. Tillage operations are especially important as a method of conserving soil moisture, controlling weeds, and preventing soil erosion. Excessive tillage of the soil not only is expensive and wasteful of power, but may reduce crop yield as a result of destroying root growth.

The specialist in power and machinery understands that tillage machinery requires a tremendous amount of power to move the equipment through the soil and achieve the stirring action. It is estimated that tillage operations require over 60 percent of the power used on American farms. More than 300 billion tons of soil are moved in the United States each year for the production of agricultural products. For example, cultivation of an acre of ground for controlling weeds will move approximately 300,000 pounds (lb) [136,000 kilograms (kg)] of soil. A tillage tool out of adjustment may require up to twice the power and still do an unsatisfactory job. The method of operation, accuracy of adjustment, and condition of repair affect the amount of power required and the quality of tillage produced.

## PURPOSE AND CLASSIFICATION OF TILLAGE

Crop production processes include a number of specific operations. These are (1) preparing a seedbed, (2) planting the seed, (3) controlling weeds, insects, and disease, and (4) maintaining a suitable soil surface condition. Of these operations the key purposes of tillage are to provide a desirable seedbed to achieve proper seed germination and to nurture the growing crop to maturity.

Tillage is classified into primary and secondary operations. *Primary tillage* is identified as those operations which cut and shatter the soil with relatively deep penetrating tools, leaving a rough surface texture. Depending upon the operation, penetration may be from 6 to 30 inches (in) [152 to 762 millimeters (mm)]. The purpose of primary tillage operations is to aerate the soil, improve the soil structure, incorporate organic matter and fertilizers, improve water penetration, encourage root development, and control weeds and insects. *Secondary tillage* processes use machinery that works the soil to a depth of 2 to 6 in [51 to 152 mm]. Where the objective of primary tillage is to work the soil in the root zone, secondary tillage is used to pulverize, level, and firm the soil in order to prepare a good seedbed and to control weeds and conserve moisture.

## PRIMARY TILLAGE MACHINERY

Machines which are used to perform primary tillage are classified into three major groups: (1) plows, (2) listers and bedders, and (3) rotary tillers.

### Plows

Plows are used to cut and aerate the soil and to incorporate organic matter into the soil. The three principal types of plows are (1) moldboard, (2) plow disk, and (3) chisel plow (Fig. 18-1).

**Moldboard Plows** Moldboard plows (Fig. 18-1*a*) are designed to cut, lift, and invert a 14- to 20-in [36- to 51 centimeters (cm)] section of soil. This action buries the trash and crop residue where it is exposed to soil microorganisms, resulting in a release of soil nutrients from the process of decay. This is accomplished by a specially shaped piece of steel called a *moldboard*. The moldboard lifts, turns, and pulverizes the soil. It is one of five specific parts of the plow *bottom* (Fig. 18-2). The *moldboard, shin, share,* and *landside* are bolted to the *frog*. These parts form a three-sided wedge. Upon entering the soil, the *share* serves as a flat cutting edge to cut the furrow slice from the unplowed land. The *landside* is a flat plate which runs against the furrow wall and keeps the bottom moving straight ahead by absorbing the side force of the share and moldboard. The *shin* is actually a replaceable part of the moldboard. It separates the furrow slice from the unplowed land. It is made replaceable because wear occurs more rapidly at this point of the moldboard.

The bottom is bolted to a vertical riser called a *standard* or *beam* (Fig. 18-2). The standard is attached to the plow frame. Safety release mechanisms are built into the standard to protect the bottom and

(a)

(b)

(c)

**Fig. 18-1.** Principal types of plows: (*a*) moldboard; (*b*) plow disk; (*c*) chisel plow. (*Ford Tractor Operations; International Harvester Co.*)

standard from damage should the bottom contact a hidden obstruction (Fig. 18-3).

The cutting and wedging action of the plow bottom moving through the soil may vary in depth from 3 to 16 in [76 to 406 mm], depending on the type of bottom and plow. As the moldboard lifts the soil upward, block segments of the layer are sheared at regular intervals, resulting in a broken surface (Fig. 18-4). These blocks must slip against one another, causing a crumbling action, which very effectively pulverizes the furrow slice. As the soil layer is inverted, it is moved toward the previous furrow slice. The crop residue is trapped between the two surfaces. This material forms a channel for entry of air and water (Fig. 18-5).

**Types of bottoms** The shape of the bottom and share is designed to produce the best results in certain soil or operating conditions (Fig. 18-6). A *general-purpose bottom* works well at speeds of 3 to 4 miles per hour (mi/h) [5 to 6 kilometers per hour (km/h)] and turns sod, stubble stalks, and covers trash. The general-purpose bottom penetrates the soil better than any other type bottom. *Stubble bottoms* produce the best trash coverage and pulverization of the soil. They also scour the best in sticky stubble ground at 3 to 5 mi/h [5 to 8 km/h]. The high waist turns furrows uphill at slow speeds, making it ideal for plowing on hillsides. A *high-speed bottom* will operate to $6^1/_2$ to 7 mi/h [10.5 to 11 km/h] because the moldboard has a longer wing and more curvature. *Slatted moldboards* have approximately 50 percent open area, thereby offering less friction to turn sticky and waxy clay soils. *Deep-tillage bottoms* have high moldboards to permit turning a furrow slice when plowing as deep as 16 in [406 mm]. They are used in some irrigated sections of the country. *Heavy-duty bottoms* are designed to crumble the soil and leave it smooth when operated at 5 to 6 mi/h [8 to 10 km/h]. They throw extra soil from the 18- to 20-in cut [457 to 508 mm] to leave more room for large tractor tires.

**Types of shares** Shares are designed to pull the bottom into the soil and to lift the furrow slice. The most

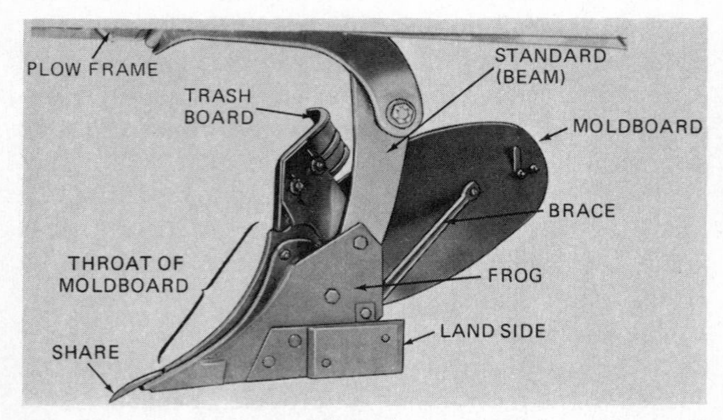


**Fig. 18-2.** Parts of a moldboard plow bottom. (*Deere & Co.*)

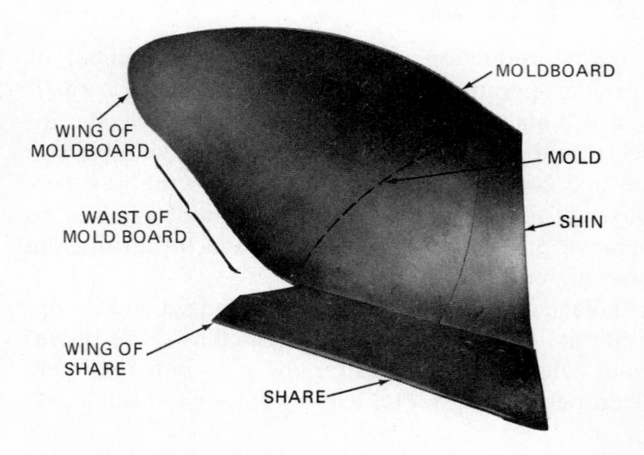

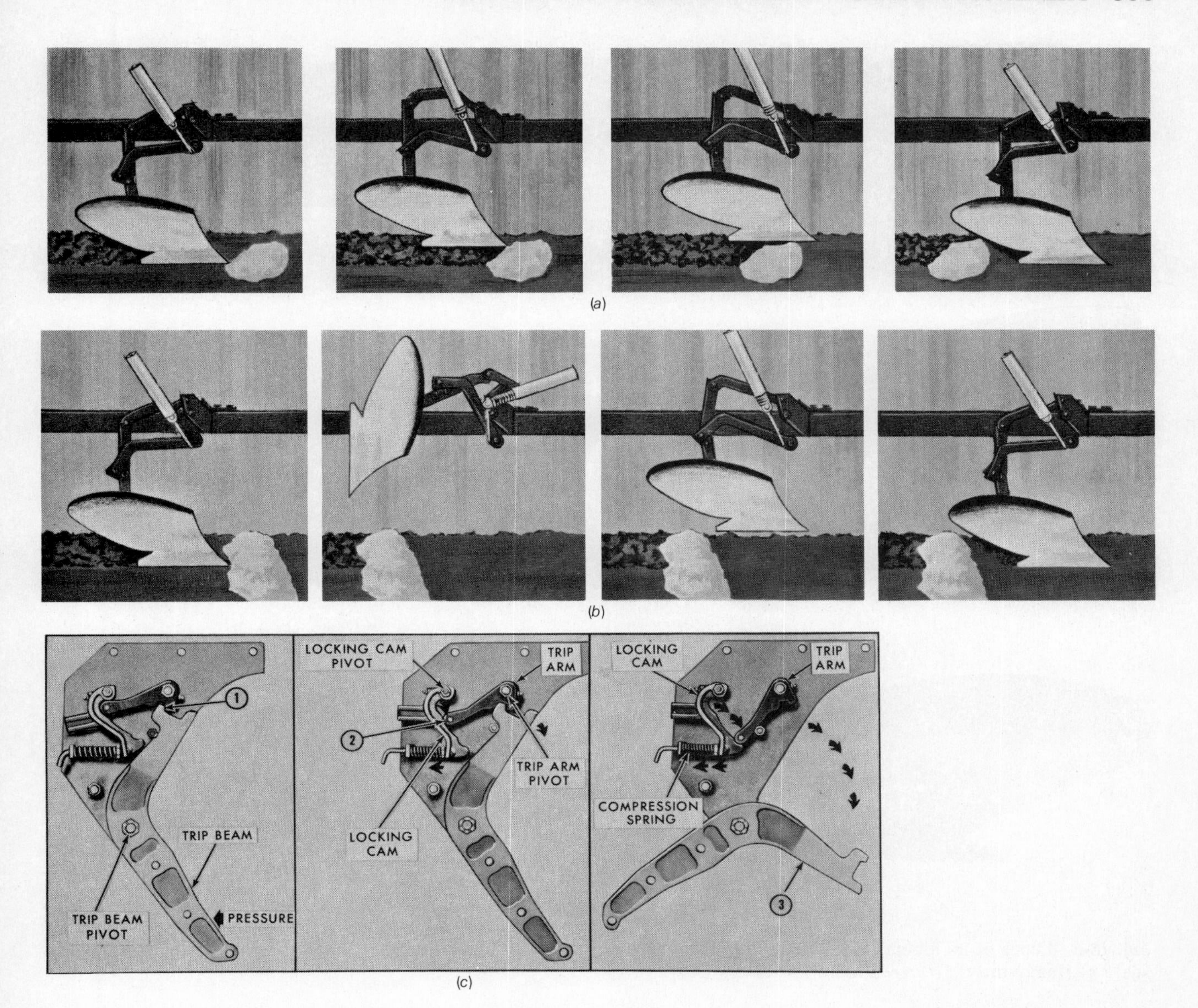


**Fig. 18-3.** The safety release protects the standard and bottom from damage when striking hidden obstructions. In (*a*) the share strikes stone and clears itself by moving up and over and then returning to the normal plowing position. The large obstruction in (*b*) causes the safety release to trip, allowing the bottom to clear the obstruction. Spring action returns the bottom to the normal position and depth of plowing without stopping forward movement. The safety trip mechanism in (*c*) is actuated when pressure on the shank transmits pressure to the lock (1); the trip arm is forced downward on the locking cam (2) until the shank is released from the trip arm lock (3). The plow must be stopped and raised or backed up to reset. (*Ford Tractor Operations; International Harvester Co.*)

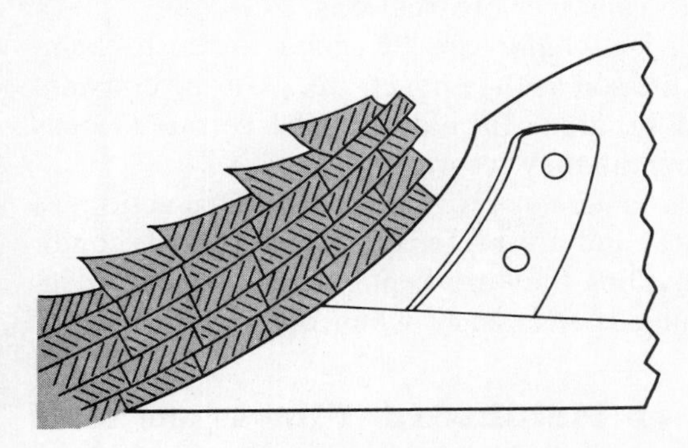

**Fig. 18-4.** The lifting action of the soil causes sections to be sheared. Soil passing over the moldboard breaks into blocks.

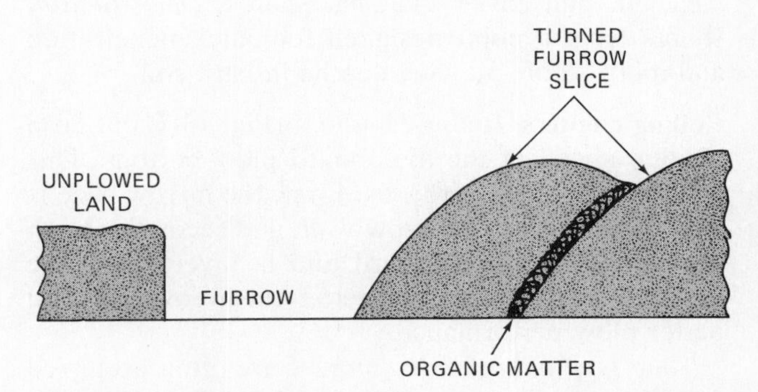


**Fig. 18-5.** A properly turned furrow traps organic matter between furrow slices where it can act as a wick taking water into the soil.

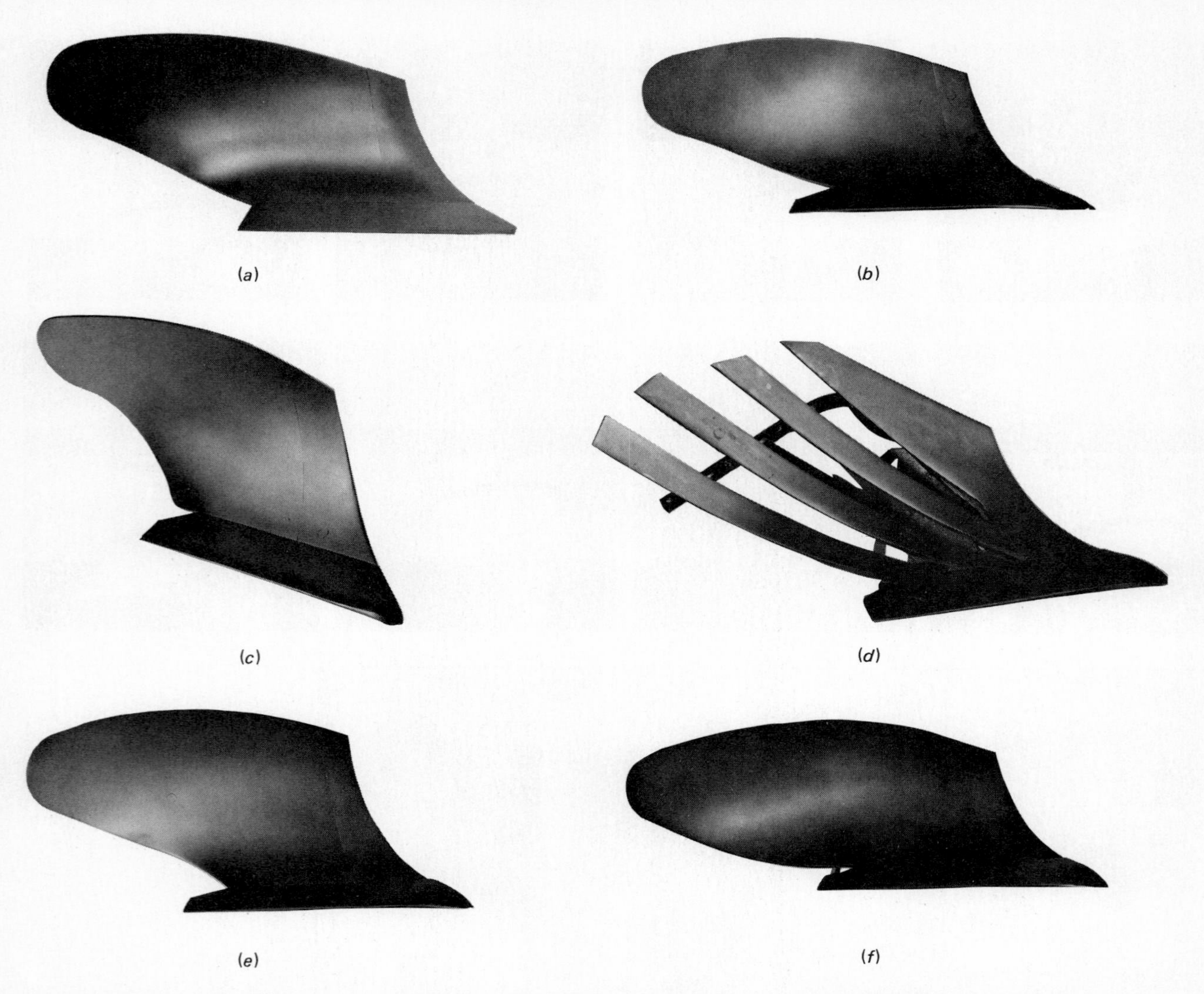


**Fig. 18-6.** Types of moldboards. (*a*) General purpose. (*b*) Stubble. (*c*) High speed. (*d*) Slatted. (*e*) Deep tillage. (*f*) Heavy duty. (*International Harvester Co.*)

common types are the flat, curved, and deep-suction. *Flat shares* are considered a standard type for most operations. They are low-cost, full-width, and sever the entire furrow slice in order to cut all roots. *Curved shares* are used in sticky, hard scouring soils. The curve matches the moldboard. They penetrate well and pull easier than flat shares. *Deep-suction* shares have a point designed for quick penetration and more stable plowing depths in hard soil.

**Rolling coulters** Rolling coulters (Fig. 18-7) cut crop residue ahead of the moldboard plow bottom. This action serves two purposes. First, the furrow slice is cleanly cut from the furrow wall, and secondly, trash clears the bottom standard and is inverted by the moldboard. The result is better trash coverage and better plow performance.

*Yoke-type plain blade coulters* are often preferred for deep plowing and are economical. The blades are available in 17-, 20-, and 22-in [431-, 508-, and 558-mm] diameters. They must be kept sharp to cut damp and tough residues.

*Disk coulters* are concave or cup-shaped and 20 in [508 mm] in diameter. The operating angle is adjustable to provide the ultimate in cutting and handling trash. The shank is spring-cushioned to allow the disk to ride over obstructions.

*Ripple-edge blades* are designed to mesh themselves to the soil as they penetrate, assuring constant full-speed rotation. In addition, the serrated edges self-sharpen as they wear.

*Serrated coulters* provide a chopping action as they rotate, and are preferred in heavy trash conditions, providing they are kept sharp. A spring-cushioned shank is necessary when operating in rocky soils.

**Types of Moldboard Plows** Moldboard plows are classified according to the method by which they are attached to the tractor and whether

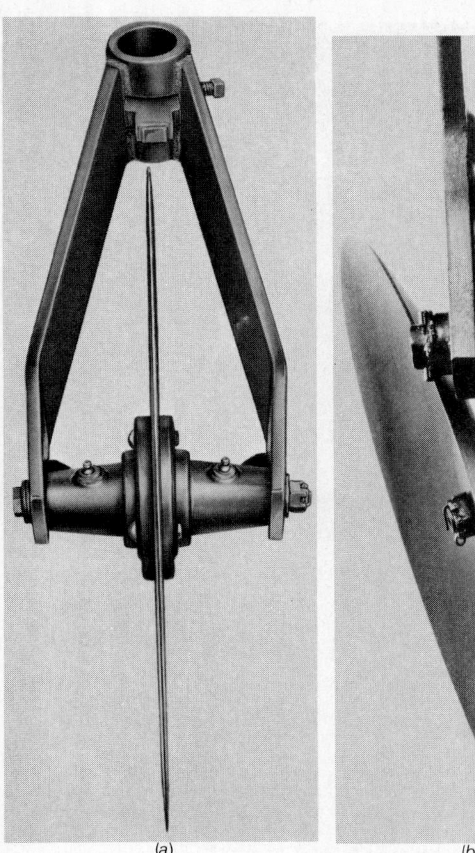

(a)

(b)

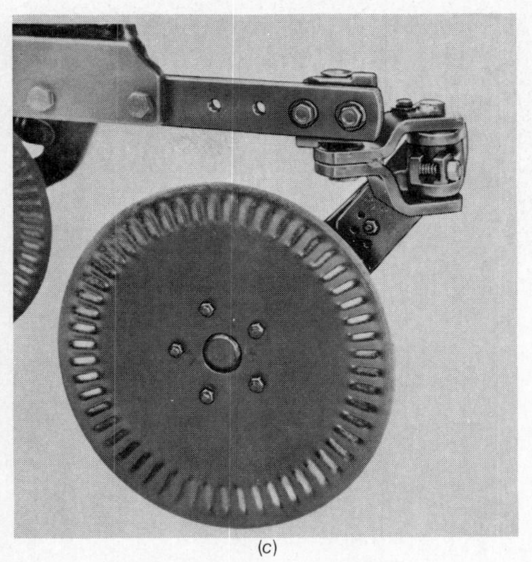

(c)

(d)

**Fig. 18-7** Types of coulters. (*a*) Yoke-type, plain blade coulter. (*International Harvester Co.*) (*b*) Concave disk coulter. (*International Harvester Co.*) (*c*) Ripple-edge coulter. (*Deere & Co.*) (*d*) Serrated coulter. (*International Harvester Co.*)

the bottoms are reversible. The five types are (1) drawn or pull, (2) integral attached, (3) semi-integral attached, (4) semi-integral, variable width, and (5) reversible or two-way.

**Drawn moldboard plows** While all moldboard plows are "drawn" through the soil, the typical drawn or pull-type plow is equipped with three wheels: a front and rear furrow wheel and a land wheel. The drawn plow is a complete unit which can be attached to the tractor drawbar (Fig. 18-8). Before the development of the three-point hitch, all plows were of the drawn type. Large four-wheel-drive tractors provide sufficient power to pull two or more drawn plows by use of a flex or tandem hitch. The modern drawn plow has an adjustable hitch which permits the tractor to be operated with the drive wheels on unplowed land or with the right tractor wheels operating in the furrow. These two arrangements are referred to as *on-land* and *in-furrow* hitches (Fig. 18-9).

Remote hydraulic cylinders are used to raise the plow out of and lower the plow into the ground and to regulate plowing depth by acting through the three support wheels. Depth control of a drawn plow is

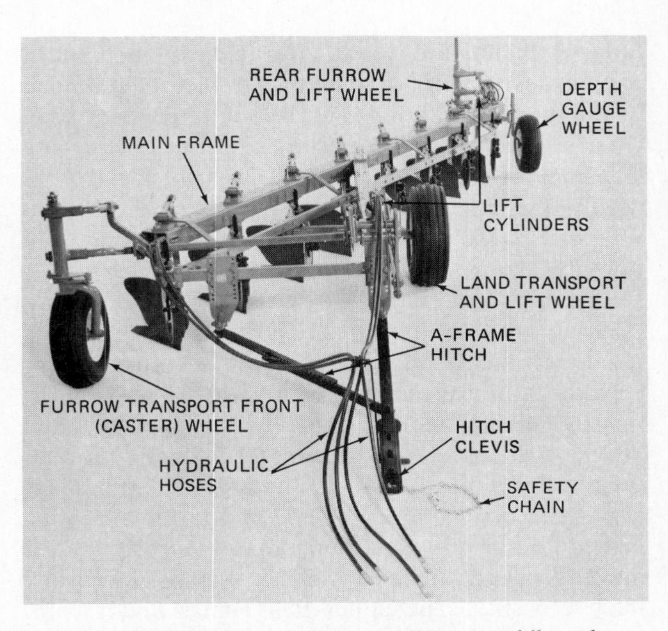


**Fig. 18-8.** Eight-bottom drawn or pull-type moldboard plow. (*Allis-Chalmers Manufacturing Co.*)

**Fig. 18-9.** Position of tractor operating (*a*) in-furrow. (*International Harvester Co.*) (*b*) On-land. (*Ford Tractor Operations*)

precise because the wheels also serve as gauge wheels.

**Integral moldboard plows** An integral moldboard plow is designed for attachment to tractors equipped with a three-point hitch (Fig. 18-10). It does not have wheels, except that it may have an adjustable rolling landside (Fig. 18-10). This type of landside can be used to increase or decrease the pressure on the furrow wall to make the plow trail straight behind the tractor.

Raising or lowering of the plow is controlled by the hydraulic sensing system of the tractor. The maximum-size integral plow is usually five-bottom. More bottoms add excessive weight when the plow is raised, making steering difficult. This type of hitch provides the most weight transfer from the plow to the rear traction wheels. As illustrated in Fig. 18-11, the upper connection, or *link,* of the tractor is attached to the plow A-frame mast. As the tractor moves forward, the lower draft links pull on the plow hitch crossbar. The upper link comes under compression because the plow wants to roll end over end, but is stabilized by the top link. This action places a downward force on the front of the tractor. At the same time the lower drag link arms are attempting to raise the plow to prevent excessive penetration and maintain a preset depth of plowing. The lifting action applies downward pressure to the rear of the tractor. Thus, part of the weight of the plow and its downward force is transferred to the drive wheels of the tractor. Since the plow is carried by the tractor and there are no wheels to develop rolling resistance, this type of plow pulls relatively easily.

**Semi-integral moldboard plows** Semi-integral moldboard plows are hitched to the lower hitch links of a three-point draft sensing-type tractor. The front (hitch) end of this type of plow is carried and controlled by the tractor linkage (Fig. 18-12). The rear of the plow rides on a furrow transport wheel. The rear wheel is steered by a rod connected to the front hitch point, enabling the plow to follow the tractor closely on turns. Most semi-integral plows use a gauge wheel to maintain a constant plowing depth in variable ground conditions (Fig. 18-13).

The front of the plow is raised and lowered by the tractor linkage, while the rear of the plow is con-

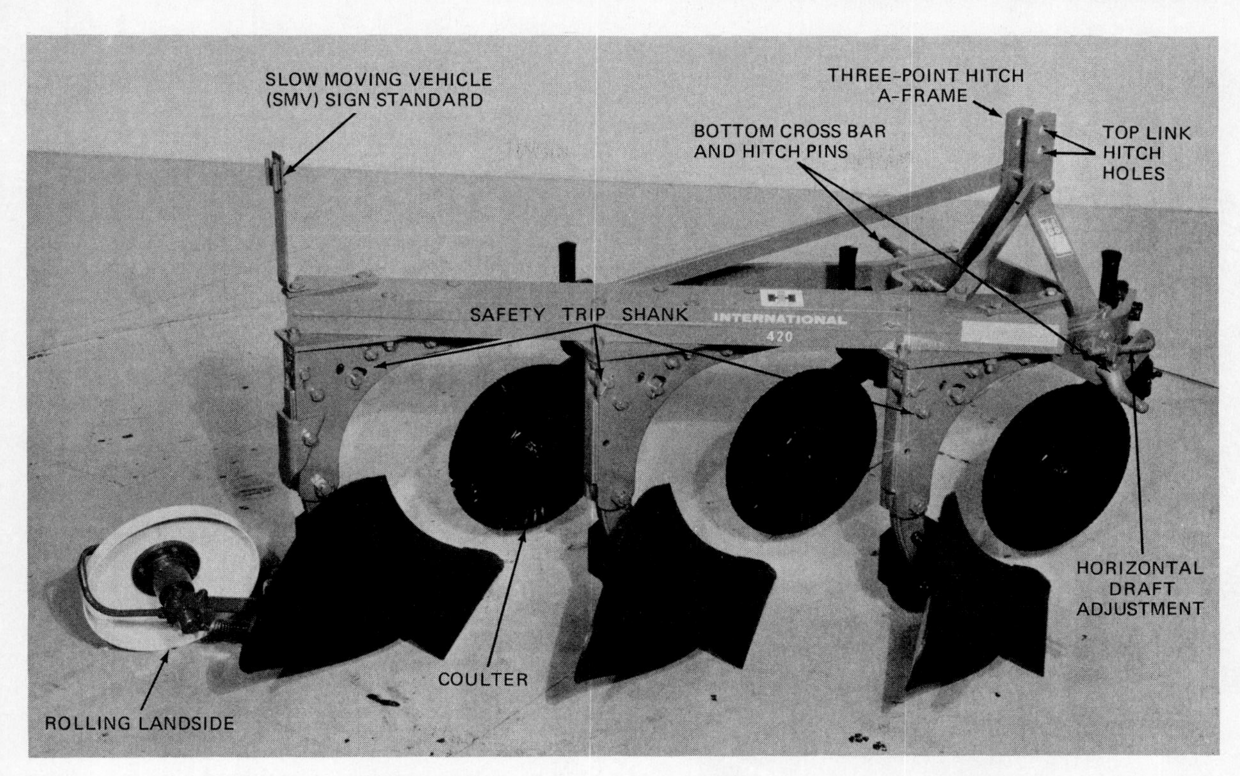


**Fig. 18-10.** Three-bottom integral-hitch moldboard plow with rolling landside. (*International Harvester Co.*)

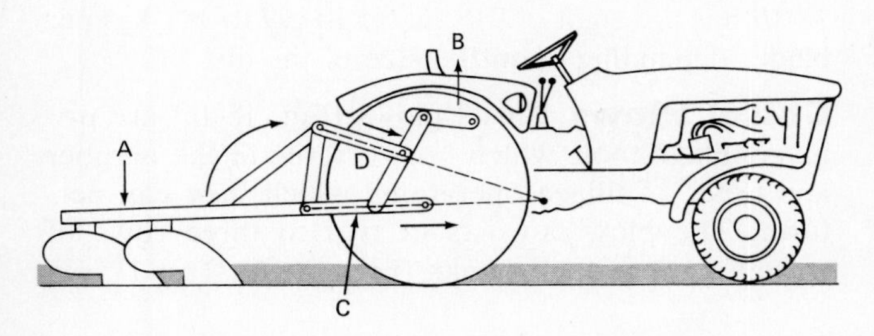


**Fig. 18-11.** The integrally mounted moldboard plow is hitched to the tractor at "three points." The connecting linkage provides weight transfer to the tractor drive wheels. The weight of the plow and the downward suction of the bottoms pull the plow into soil (A). The hydraulic lift arm of the tractor compensates by lifting the plow (B). Tractor pull on the lower hitch links (C) causes the plow to rotate against the top link (D). The lifting effect of (B) and downward force (D) transfer weight to the rear drive wheels.

**Fig. 18-12.** Six-bottom semi-integral moldboard plow with flexible hitch and front furrow wheel for on-land tractor operation. (*International Harvester Co.*)

trolled by a remote hydraulic cylinder. The independent action of raising and lowering provides precise control to the front and rear bottoms of the plow.

Semi-integral plows can be operated with the tractor in-furrow or on-land. On-land plow hitches are flexible rather than rigid—this is because a front furrow wheel carries the plow and controls the depth of the front bottoms.

**Variable width (cut) semi-integral plow** A moldboard plow that can be adjusted during operation to change its cutting width is referred to as a *variable-width* or *variable-cut* plow. These plows are semi-integrally mounted to the tractor using an in-furrow rigid hitch. By hydraulically shifting the frame of the plow and hitch point to the tractor, the operator is able to change the width of cut from 14 to 22 in [356 to 558 mm] for each bottom without leaving the tractor seat (Fig. 18-14). The plow beams pivot on the frame to maintain proper alignment. This type of plow is becoming very popular, because it can be set to achieve the kind of plowing that is needed. Complete trash coverage and level plowing or slabby, trash-laced ridges for erosion control can be produced. In addition it provides the following advantages: (1) It adjusts to match tractor power to soil conditions, (2) it simplifies closing of dead furrows and finishing head lands, (3) the width can be reduced when pulled for transport on the road, and (4) it makes plowing of terraced, contoured, or point-row fields much easier.

**Reversible or two-way plows** Two-way plows have sets of both right- and left-hand bottoms which are used alternately as the plow is pulled in the field.

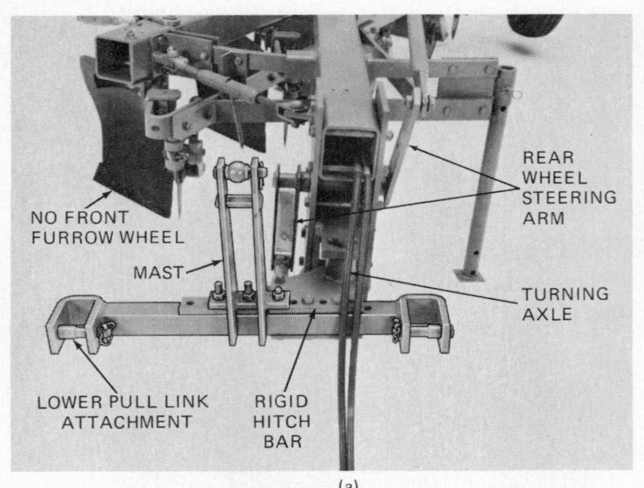


(a)

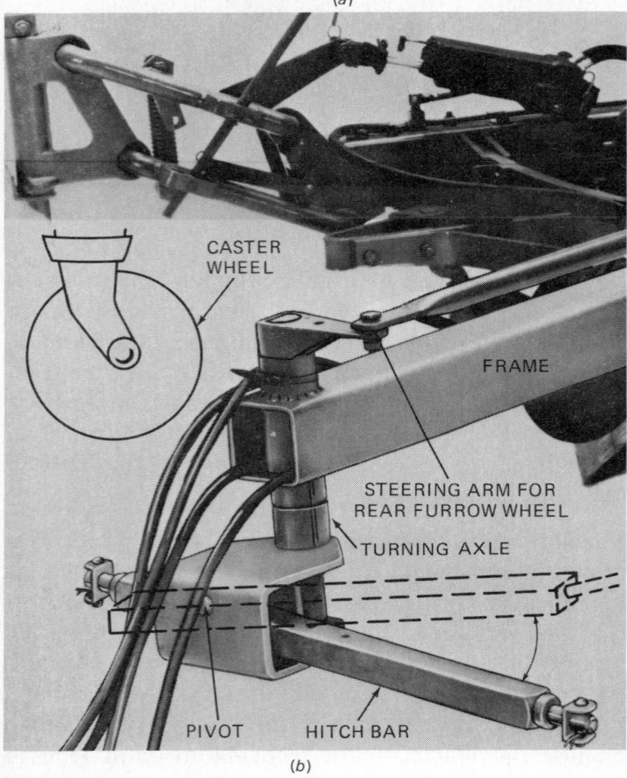


(b)

**Fig. 18-13.** (*a*) Semi-integral moldboard plow with rigid hitch for in-furrow tractor operation. (*b*) Flexible hitch for on-land operation. (*Deere & Co.*)

This plow turns all furrows in the same direction (Fig. 18-15). Two-way plows are often used to till irrigation land, because they create no dead, or back furrows to impede water flow. They are also effective in reducing travel time when plowing point rows in contour farming, or when it is desirable to throw all the furrows uphill or downhill.

Most two-way plows are integral-hitched to the tractor. At the end of each furrow the plow is lifted out of the ground by the three-point hydraulic linkage. As the turn is made, the bottoms are rotated, using a hydraulic cylinder (Fig. 18-15) and the plow is ready for the return trip across the field.

**Power Requirements for Moldboard Plows** The number of bottoms which are attached to a moldboard plow frame identifies the capacity and power required to move it through the soil. Table 18-1 illustrates the amount of power takeoff (PTO), horsepower required by the tractor to pull one bottom according to the size of the bottom, the depth of plowing, and the type and firmness of the soil. For example, assume that you want to know the PTO power needed to pull a five-bottom, 16-in [406 mm] moldboard plowing 9 in [229 mm] deep at 4.5 mi/h [7.2 km/h] in firm loam soil. The data in Table 18-1 indicate that each bottom requires 26 PTO horsepower (hp) [19.4 kilowatts (kW)]. Thus, the five bottoms require a tractor having 130 PTO hp [97 kW].

**Plow Disk** Plow disks (Fig. 18-1*b*) are heavy-duty offset plows that have 24 to 32 in [610 to 813 mm] notched, cone-shaped disks. These plows are specifically designed for rough ground and heavy trash conditions where complete trash coverage is not desired. The cone-shaped blades are spaced 9 to 12 in [229 to 305 mm] on the gang spool. Their aggressive action chops and mixes trash into 6 to 10 in [152 to 254 mm] of soil. The surface mulch of residue helps to reduce wind and water erosion.

The frame of the plow disk is exceptionally heavy and strong. For example, the plow illustrated in Fig. 18-1*b* has a weight of 218 to 245 lb [99 to 111 kg] per blade, depending upon the size of the disk.

**Chisel Plows** Chisel plows (Fig. 18-1*c*) are primary tillage tools which are versatile in the number and types of tillage operations which they can perform. The plow consists of two to three rows of curved spring steel shanks (Fig. 18-16), 1 to $1^1/_2$ in

**TABLE 18-1. PTO Horsepower Requirements per Bottom (4.5 mi/h [7.2 km/h]—Firm Soil*)**

| | Depth, in | Gumbo | Clay | Loam | Sandy Loam | Sand |
|---|---|---|---|---|---|---|
| 14 in GP bottoms | 5 | 17 | 15 | 13 | 9 | 4 |
| | 6 | 21 | 18 | 15 | 11 | 5 |
| | 7 | 24 | 21 | 18 | 12 | 6 |
| | 8 | 27 | 24 | 20 | 14 | 7 |
| 16 in GP bottoms | 6 | 23 | 21 | 17 | 12 | 6 |
| | 7 | 27 | 24 | 20 | 14 | 7 |
| | 8 | 31 | 27 | 23 | 16 | 8 |
| | 9 | 35 | 31 | 26 | 18 | 9 |
| 18 in GP bottoms | 7 | 31 | 27 | 23 | 16 | 8 |
| | 8 | 35 | 31 | 26 | 18 | 9 |
| | 9 | 39 | 35 | 29 | 20 | 10 |
| | 10 | 44 | 39 | 32 | 23 | 11 |

* For loose or tilled soil multiply PTO horsepower by 1.2.

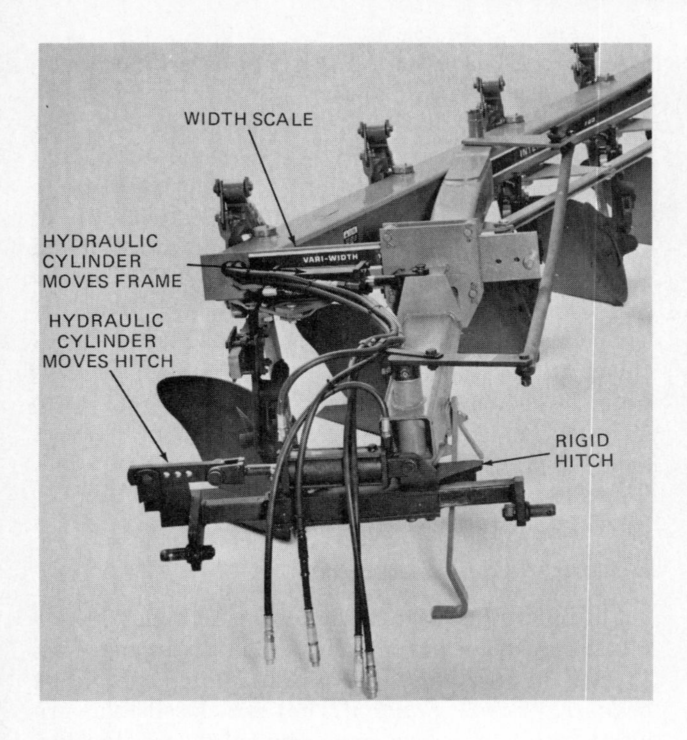


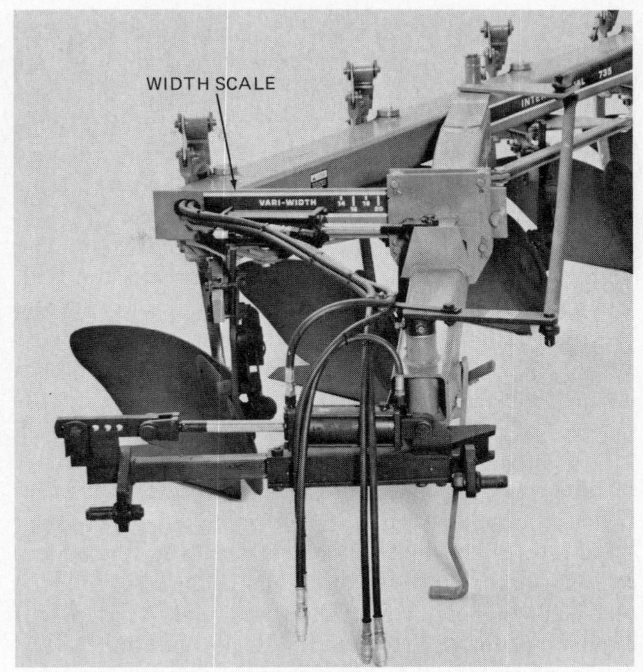


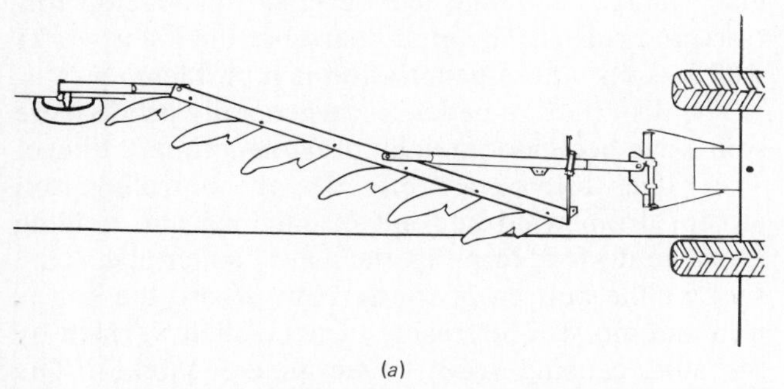

(a)

(b)

**Fig. 18-14.** A variable-width (cut) semi-integral moldboard plow. (*a*) Narrow-cut position—the plow frame is in the narrow position with the hitch point centered. (*b*) Wide-cut position—hydraulic cylinders are extended to move the frame and hitch point for proper line of draft. (*International Harvester Co.*)

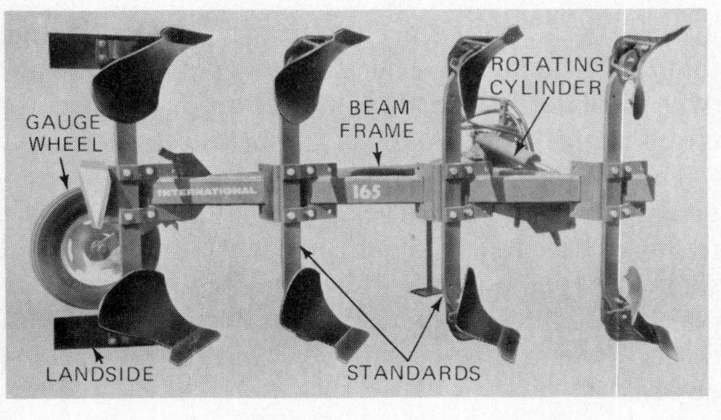


**Fig. 18-15.** The two-way moldboard plow. Allows the operator to turn furrows in one direction. The hydraulic cylinder rotates the indexing shaft 180° to change the positions of the bottoms. (*International Harvester Co.*)

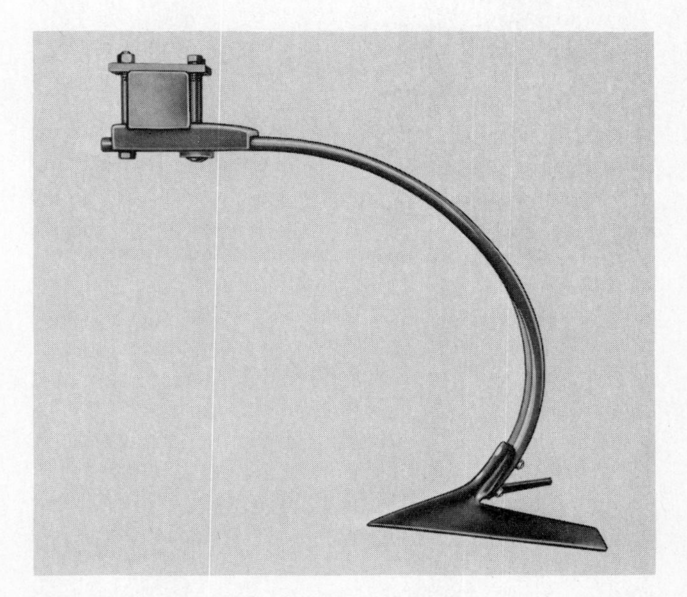

**Fig. 18-16.** A rigid-type chisel plow shank equipped with V-sweep. The size of the shank may vary from 1 × 2 × 26 in to 1½ × 2 × 32 in [25 × 51 × 660 mm to 38 × 51 × 813 mm] in size. Choice of size depends upon soil conditions and tillage requirements. The 1½ × 2 × 32 in [38 × 51 × 813 mm] is usually recommended.

[25 to 38 mm] thick and 2 in [51 mm] wide, attached to a very strong box steel frame. The shanks are arranged in staggered rows to allow for good trash flow. The shanks may be equipped with various types of points or shovels (Fig. 18-17). Narrow shovels are used to penetrate to a depth of 15 in [380 mm]. Normal operation is 8 to 10 in [200 to 254 mm], or just deep enough to break up a moldboard plow pan. The spring action of the shank lifts and breaks the soil, leaving approximately 75 percent of the crop residue on the surface. By fitting the shank with wide sweeps, the plow can be operated just deep enough to cut off weeds with a minimum of surface disturbance to provide excellent stubble-mulch or mulch tillage land management associated with dryland farming. These plows work best when the soil is firm and dry, by shattering the compacted layers and breaking up the large clods. If too wet, the shank opens a narrow groove and rolls up large chunks of wet soil which dry hard. An attachment to the shovel makes it possible to inject anhydrous ammonia ($NH_3$) gas fertilizer into the soil at root zone depth in preparation for planting or seeding (Fig. 18-16*b*).

Chisel plows are drawn or integral-hitched to the tractor. Drawn plows wider than 20 feet (ft) [6 meters (m)] have flexible outrigger sections which can be folded for transport. Integral plows are 5 to 20 ft [1.5 to 6 m] in width. Most plows have shanks which provide 28 to 32 in [711 to 813 mm] of clearance between frame and shank.

**Fig. 18-17.** Types of points and shovels for chisel plows: (*a*) twisted shovel, (*b*) anhydrous ammonia fertilizer boot, (*c*) 2-in [51 mm] wide fertilizer boot, (*d*) V-sweep shovel. (*Deere & Co.; Sun Flower Mfg. Co.*)

The draft of a chisel plow is approximately half that of a moldboard plow of the same width and working at the same depth. The travel speed of 6 mi/h [9.7 km/h] and a plow width of 10 to 45 ft [3 to 13.7 m] provide exceptionally high field capacity of 6 to 27 acres/h [2.4 to 10.9 hectares (ha/h)].

Subsoilers are a form of chisel plow which are designed to penetrate to a greater depth (Fig. 18-18). A chisel plow can work effectively at 10 in [254 mm] deep. Subsoilers, if necessary, will penetrate 20 to 30 in [508 to 762 mm] to break up compacted layers of soil to permit better root and moisture penetration. Chisel plows do not stir and mix subsoil with top soil; subsoilers do not bring subsoil to the surface.

## Listers and Bedders

Listers are furrow-making tools which may have a moldboard-type bottom or a small disk gang (Fig. 18-19). They are sometimes called *middlebusters, disk bedders,* or *lister planters*. They are classified as primary tillage tools and are designed to develop furrows and ridges. The moldboard bottom lister works well in soils where penetration is a problem.

The disk bottom bedders are generally used where soils have been loosened by previous primary tillage. The ridges formed are effective in controlling soil erosion from wind and aid in catching and holding snow. In dryland farming, the lister planter places the seed at the bottom of the furrow, where the soil is firm and moist. The freshly turned soil is warmed by the sun, causing seeds to germinate quickly. The ridge protects the young seedling from damage by wind and blowing soil. In areas of high rainfall, planting is done on the ridge where the bed dries and warms faster than in the furrow.

Listers and bedders are extensively used to "form" the furrows for furrow irrigation. In this type of agricultural crop production, seed placement is usually on the side of the ridge to control salt concentration problems in the root zone.

The angle and spacing of disk bottoms can be adjusted to produce the desired shape of bedding ridge for the soil condition. The advantages of disk bedders over the lister are that they work at a travel speed of 6 to 7 mi/h [10 to 11 km/h] and have lighter draft.

The *tool bar* implement provides the means of attaching the lister or disk bottoms integrally with the tractor. The tool bar also permits special "bed shapers" (Fig. 18-20) to form the bed for precision control planting. Tools can be clamped on the bar, thus obtaining a number of row and furrow spacing options.

## Rotary Tillers

Rotary tillers used for primary tillage consist of a set of rotating blades within a housing. The rotating

**Fig. 18-18.** A subsoiler plow can penetrate the soil 20 to 30 in [508 to 762 mm] deep to break up hard layers of soil in the root zone. (*Orthman Manufacturing Co.*)

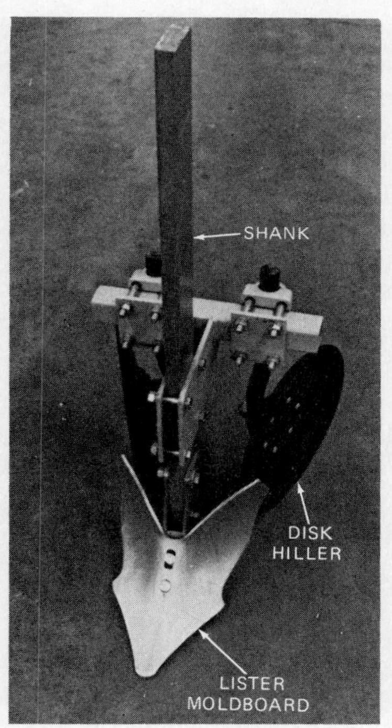


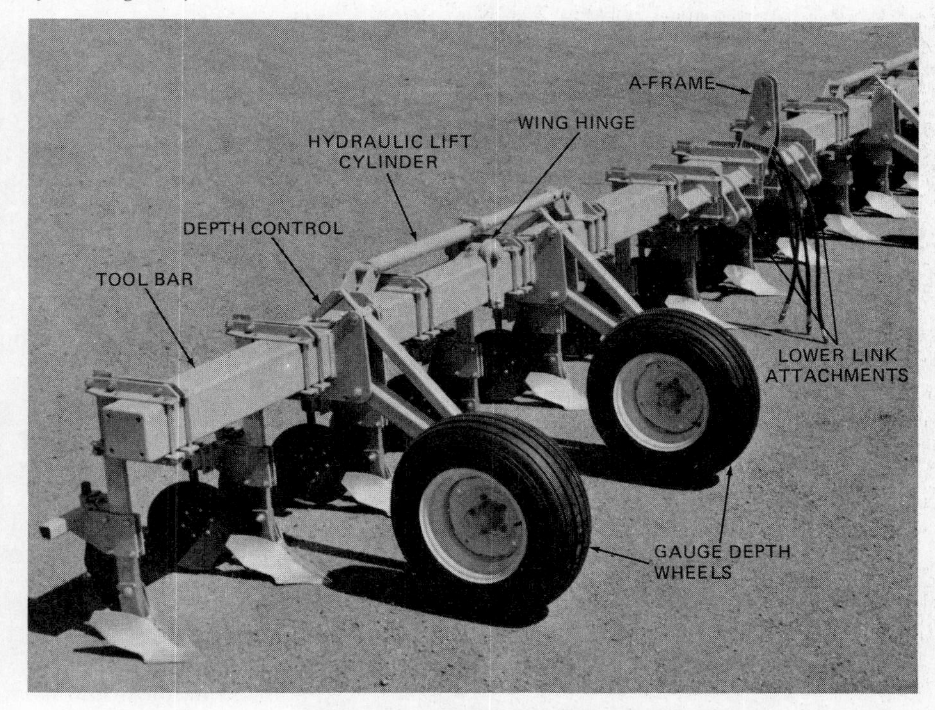


**Fig. 18-19.** A row lister bedder with lister moldboard and disk tiller attachment. (*Orthman Manufacturing Co.*)

action chops, pulverizes, levels, and mixes the soil and crop residue in a "once-over" operation.

All rotary tillers produce *negative draft*. That is, the rotating blades tend to pull the tiller along the ground surface and become a rotating power wheel.

The most common type of primary tillage machine is the *integral-mounted PTO drive tiller* with a maximum size of 12 to 15 ft [3.7 to 4.6 m] (Fig. 18-21).

Rotary tillers are also popular for nursery and vegetable production as well as for small gardening and landscaping operations.

**Fig. 18-20.** Special bed shapers are used to prepare the seedbed for precision planting.

(a) (b)

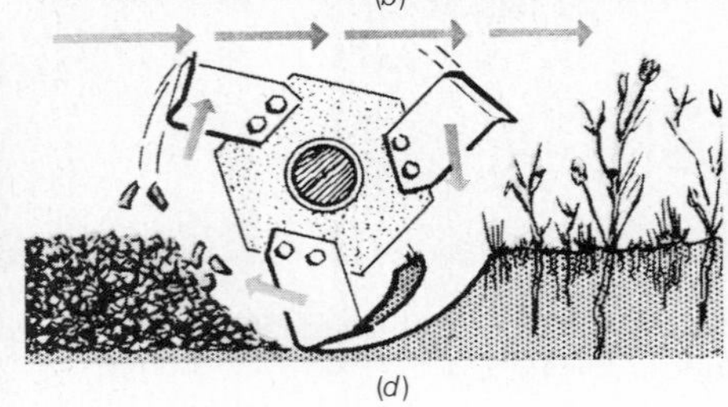

(c) (d)

**Fig. 18-21.** Rotary tillers powered by the PTO drive from the tractor perform both (*a*) primary and (*b*) secondary tillage. (*c*) Primary rotary tillage blades (*d*) chop and incorporate crop residue. (*Howard Rotavator Co.*)

Small hand-controlled garden tillers have a capacity of 12 to 28 in [305 to 711 mm]. Garden tillers are available in three models: a rear-tine self-propelled tiller, a front-tine tiller, and a small tractor-powered unit (Fig. 18-22). The tractor-powered and the small self-propelled garden tillers must actually regulate the forward travel of the tool instead of pulling it. Likewise, the front-tined garden tiller must be allowed to proceed forward slowly for the blades to chop the soil. This is accomplished by a small subsoiler which is pulled behind the blades. The operator controls the forward travel and the depth of penetration of the rotating blades by the amount of resistance of the subsoiler. Pushing down on the control handles retards forward movement and allows deep penetration. Raising the handles increases forward travel and reduces penetration. Maximum depth of tillage is up to 9 in [229 mm] for large tillers to 7 in [178 mm] penetration for the small garden type.

(a) (b)

(c)

**Fig. 18-22.** Rotary garden-type tillers. (*a*) Rear tine. (*b*) Front tine. (*c*) Riding garden tractor types. (*Ariens Co.*)

## TYPES AND PURPOSES OF SECONDARY TILLAGE MACHINERY

Secondary tillage tools are grouped into the following types: disk harrows, cultivators, and rotary tillers.

### Disk Harrows

The plow and harrow are some of the oldest tillage implements known. While the plow was used to stir the soil, the harrow provided a method for covering seeds for protection and for obtaining better germination.

Lightweight to medium-weight disk harrows are some of the most commonly used secondary tillage tools. Disk harrows are adaptable to nearly every soil and cultivation condition. They are especially effective in penetrating hard soil and in working rocky and root- or stump-infested land. Disk harrows are classified as *single-action, tandem,* and *offset* types. Each type may be either integral-hitch or wheel type. Each blade is 18 in [457 mm] in diameter and spaced $7^1/_2$ in [190 mm] on light disk harrows, to 20-in [508 mm] diameter × 9-in [229 mm] spacing on

**Fig. 18-23.** The tandem disk harrow with offset and overlapping disk gangs. (*Krause Plow Corp.*)

medium-duty disks. Modern drawn disks are generally wheel disks; that is, transport wheels are an integral part of the machine (Fig. 18-23). The wheels raise the disks out of the ground for transport and are also used to control the depth of penetration.

*Single-action disks* (Fig. 18-24), when mounted on a tool bar, are used to build borders for flood irrigation and to perform other, similar tasks.

Most *tandem disk harrows* are arranged with offset and overlapping disk gangs (Fig. 18-23). The front two gangs throw the soil out; the rear gangs throw it in. The offset of each of the front and rear gangs eliminates the furrow ridge in the center of the front gangs, while the rear gangs level the work of the front gangs.

*Offset* disk harrows have a front gang moving soil in one direction and a rear gang turning soil in the opposite direction. The term *offset* identifies that the hitch is offset to one side of the centerline of the disk and the tilled strip. This operation makes offset disk harrows very effective for working under low-hanging branches of orchards. For this reason they may sometimes be called *orchard disks* (Fig. 18-25). Most offset disks are the drawn-wheel type.

**Fig. 18-24.** Forming borders of furrows for flood or furrow irrigation. The disks can be reversed to tear down old borders.

**Fig. 18-25.** The offset-type disk harrow. (*Sunflower Manufacturing Co.*)

## Cultivators

The control of weeds in crops is the main purpose of cultivation. Other purposes, including stirring of the soil to reduce crusting, to increase aeration, and to improve water intake, may be equally important. Cultivation encompasses a number of cultural practices including stubble mulch, summer fallow, control of wind and water erosion, and tending of row-crop plants. Tests have demonstrated that row crops respond by increased yields when cultivated at least once.

Types of cultivation equipment include (1) row-crop cultivators, (2) field cultivators, (3) rotary hoes, and (4) rotary tillers.

**Row-Crop Cultivator** Row-crop cultivators operate *between* the rows of crops from the time of emergence until they can no longer safely fit beneath the tractor axle and cultivator frame. Special sweep-type units, depending upon the type of crop, are con-

(a)

(b)

**Fig. 18-26.** Row-crop cultivators: (*a*) rear-mounted and (*b*) front-mounted. (*International Harvester Co.*)

nected to a tool bar. The tool bar may be rear- or front-mounted (Fig. 18-26). Control of the cultivator on hillsides is easier with the front-mounted type.

**Field Cultivator** Field cultivators are similar in use and appearance to chisel plows, except that the frame is much lighter and they are intended to operate at a depth of approximately 4 in [102 mm] (Fig. 18-27). In contrast with a row-crop cultivator, they are used for preparing a seedbed in soil that has been previously worked with a plow disk or chisel plow. In some instances planting equipment is mounted on a field cultivator for till-plant operation.

The shank of a field cultivator is spring-cushioned. The spring provides aggressive, vibrating action for excellent trash clearance (Fig. 18-27*a*).

Special coil spring shanks equipped with anhydrous ammonia soil points or knives are used to apply fertilizer and other chemicals 4 to 5 in [102 to 127 mm] below the surface, where it will be accessible to plant roots (Fig. 18-28).

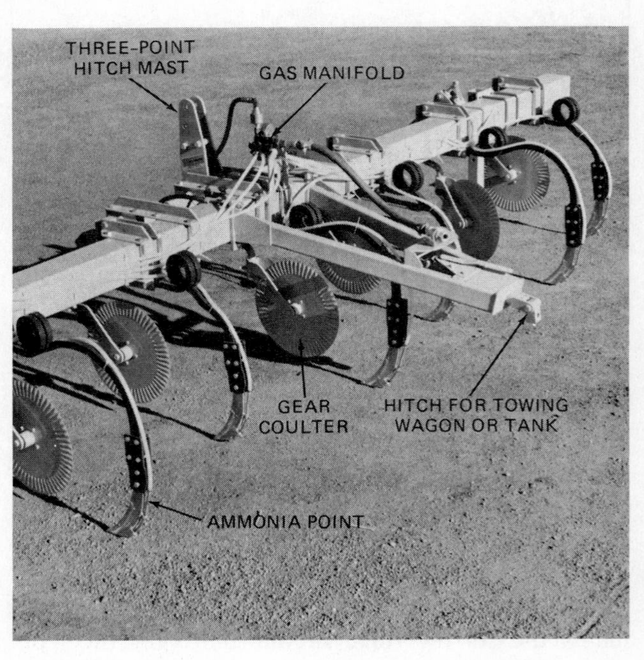


**Fig. 18-28.** A special type of mounted field cultivator equipped for anhydrous ammonia fertilizer application. Note the coil shank with ammonia points or knives. (*Orthman Manufacturing Co.*)

(a) (b)

**Fig. 18-27.** Pull-type field cultivators have become very popular secondary tillage tools for field finishing or seedbed preparation working 2 to 5 in [50 to 125 mm] deep. Outside wings fold for transport and flex for even penetration across the full width of the unit. (*a*) Shanks are spring-cushioned (insert) for vibrating action and trash removal. Shank points are $1^3/_4 \times {}^1/_4 \times 11$ in [$44.45 \times 6.35 \times 279.4$ mm] and are used for shattering crust ahead of planting. (*b*) V-sweeps are 4 to 12 in [101.6 to 304.8 mm] wide and are used for weed control, leaving crop residue on or near the surface. (*Deere & Co.*)

**Fig. 18-29.** The rotary hoe used for the removal of small weed seedlings in row crops. (*Deere & Co.*)

**Rotary Hoe** A rotary hoe consists of 10 to 16 teeth that form spoked wheels approximately 20 in [508 mm] in diameter that are spaced 6 to 8 in [152 to 203 mm] apart on each axle. The teeth are curved so that as the hoe wheel turns, the teeth enter the soil vertically. Upon leaving the soil, the tooth bursts out of the ground, throwing a small quantity of soil and small weed seedling with it (Fig. 18-29). Early cultivation of row crops (before emergence and until the plants are approximately 2 in [51 mm] tall) is possible with the rotary hoe.

**Rotary Tiller** The rotary tiller for secondary tillage consists of a rotor having a spiral arrangement of spikes instead of knives (Fig. 18-30). The spikes penetrate previously plowed or chiseled soil to a depth of 2 to 5 in [51 to 127 mm] breaking clods and leveling the surface. A special ''crumble'' roller following the rotary tiller firms the soil to ready it for till-plant field operations.

Rotary tillers are effective for seedbed preparation and cultivation of standing crops. Care must be observed to prevent excessive rotary tillage, which may destroy soil structure, aerate the soil too much, or cause the soil to crust badly after rains.

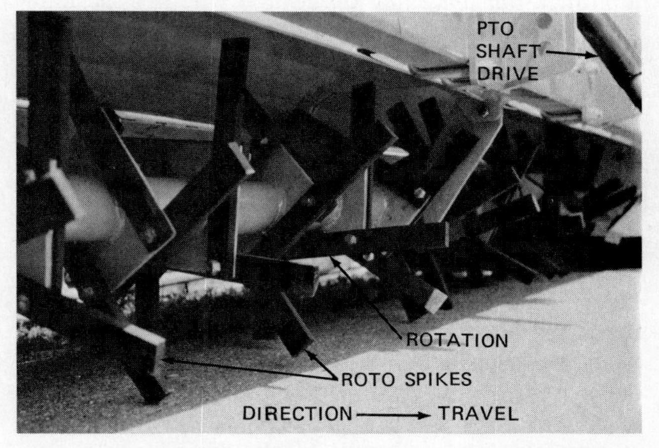


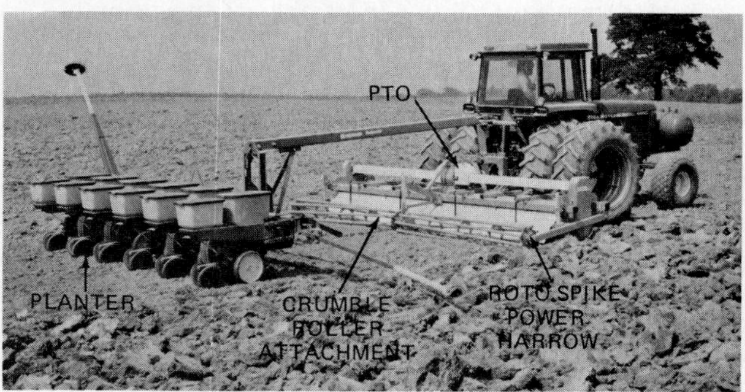


**Fig. 18-30.** The secondary tillage power harrow prepares a seedbed 180 in [4.57 m] wide at 6 mi/h [9.7 km/h]. (*Howard Rotovator Co.*)

## ADJUSTING TILLAGE MACHINERY

The quality of work which a tillage machine is capable of producing is dependent to a large degree, upon how well it has been assembled and adjusted. A machine out of adjustment does poor-quality work, requires excessive power to pull, and creates undesirable wear on certain working parts.

Manufacturers of tillage equipment have spent countless hours testing each machine to assure proper operation. The implement dealer is especially concerned that a new machine performs well, because the business depends upon sales and service. The farmer, landscape contractor, or groundskeeper expects the best from their investment in the machine. As a specialist in power and machinery, you will be called upon to use your knowledge and skill to properly assemble and make the necessary adjustments to tillage equipment. You are the person who will make a satisfied customer, a business profitable for the dealer, and a trouble-free machine.

## Setting Up the New Machine

Some fully assembled tillage machinery will weigh more than 10,000 lb [4536 kg]. For convenience in handling such heavy units, some machines are shipped from the factory in bundles broken down into components or subassemblies. It is the responsibility of the machinery dealer to assemble the machine and make the necessary final shop adjustments.

An assembly and predelivery instruction manual and an operator's manual are provided with each machine. These manuals contain specifications which the setup worker must have for assembly and adjustment. Some manufacturers include all the data in one manual.

Before starting the assembly, locate the machine identification plate and determine whether the manuals are for the correct model of machine. The setup worker should make it a standard procedure to write the serial number found on the identification plate in the place provided in the operator's manual for the new owner's use.

## Assembling Components

Start the setup by locating the main frame assembly of the machine on a level shop floor. Position the main frame to the correct working height as identified in the assembly instructions. Level the frame with a carpenter's level. Note that the tool bar of a row-crop cultivator (Fig. 18-31) is placed on sturdy machinery jack stands about 30 in [762 mm] off the floor.

Expanded assembly drawings and parts lists are a part of most manuals. This information assists the setup worker in identifying parts by number and common name. Fasteners such as cap screws, U-bolts, pins, washers, and additional hardware are also more accurately identified and installed.

**Fasteners** Cap screws and bolts are a common method of connecting components. The setup worker must select the correct diameter, length, and number of threads per inch specified for each fastener. Most assembly instructions specify the particular grade of bolt to use. *Grade* identifies the breaking strength of steel fasteners. Most machinery is assembled using three grades of bolts, namely, Society of Automotive Engineers (SAE) grades 1, 5, and 8.

Grade 5 has approximately twice the strength of grade 1. Grade 8 is $2^1/_3$ times stronger than grade 1. Grades 5 and 8 are also more resistant to shearing (cutting) than grade 1 because the fasteners are heat-treated. Grade markings are found on the head of the bolt (Fig. 18-32). A grade 1 cap screw has no markings; grade 5 is marked with three radial lines and grade 8 with six radial lines. (*Note:* The metric equivalents for these three SAE grades are, respectively 4.6, 8.8, and 10.9.)

It is important that the proper grade of bolt be installed as specified in the assembly instruction. For example, the moldboard plow hitch assembly (Fig. 18-33) indicates that flanges A and B are attached to the adjusting plate H with eight $^3/_4$ in [19 mm] × $2^1/_2$ in [64 mm] cap screws. The four upper cap screws are to be grade 5; grade 8 cap screws are to be used in the lower holes.

Bolts or cap screws should not be fully tightened until all parts of the frame have been completely assembled. This will permit minor alignment adjustments to be made and prevent unnecessary distortion of the parts from uneven tightening. When all parts are in place, it is very important that bolts and cap screws be tightened to the proper specification with a torque wrench. Properly torqued bolts will prevent broken parts and poor machine performance. For example, SAE grades 5 and 8 cap screws (Fig. 18-33) have different torque requirements. The $^3/_4$-in [19 mm] grade 5 plated cap screws are tightened to a torque of 272 pound-feet (lb·ft) [369 newton-meters (N·m)]; the grade 8 plated cap screws are torqued to 383 lb·ft [519 N·m].

An adapter may be required to increase the range of a torque wrench (Fig. 18-34). For example, assume a torque wrench has a maximum capacity of 300 lb·ft [407 N·m] and it is necessary to torque the SAE grade 8 cap screws to 383 lb·ft [519 N·m]. An adapter ($A$) can extend the length of the wrench. However, a correction in the scale reading for torque must be made. If the wrench has an effective length of 1.5 ft [45.7 cm] and the adapter is 1 ft [30.5 cm] long, the desired wrench reading ($Tw$) is determined as follows:

$$Tw = \frac{Ta \times L}{L + A}$$

$$= \frac{383 \times 1.5}{(1.5 + 1)}$$

$$= \frac{574.5}{2.5}$$

$$= 230 \text{ lb·ft } [312 \text{ N·m}]$$

where $Tw$ = reading of torque wrench (desired reading)
$L$ = length of torque wrench
$A$ = length of adapter
$Ta$ = desired torque value on fastener

Special wrenches must sometimes be shop-constructed. For example, to tighten the disk harrow gang bolt nut to the specified torque of 850 lb·ft [1152

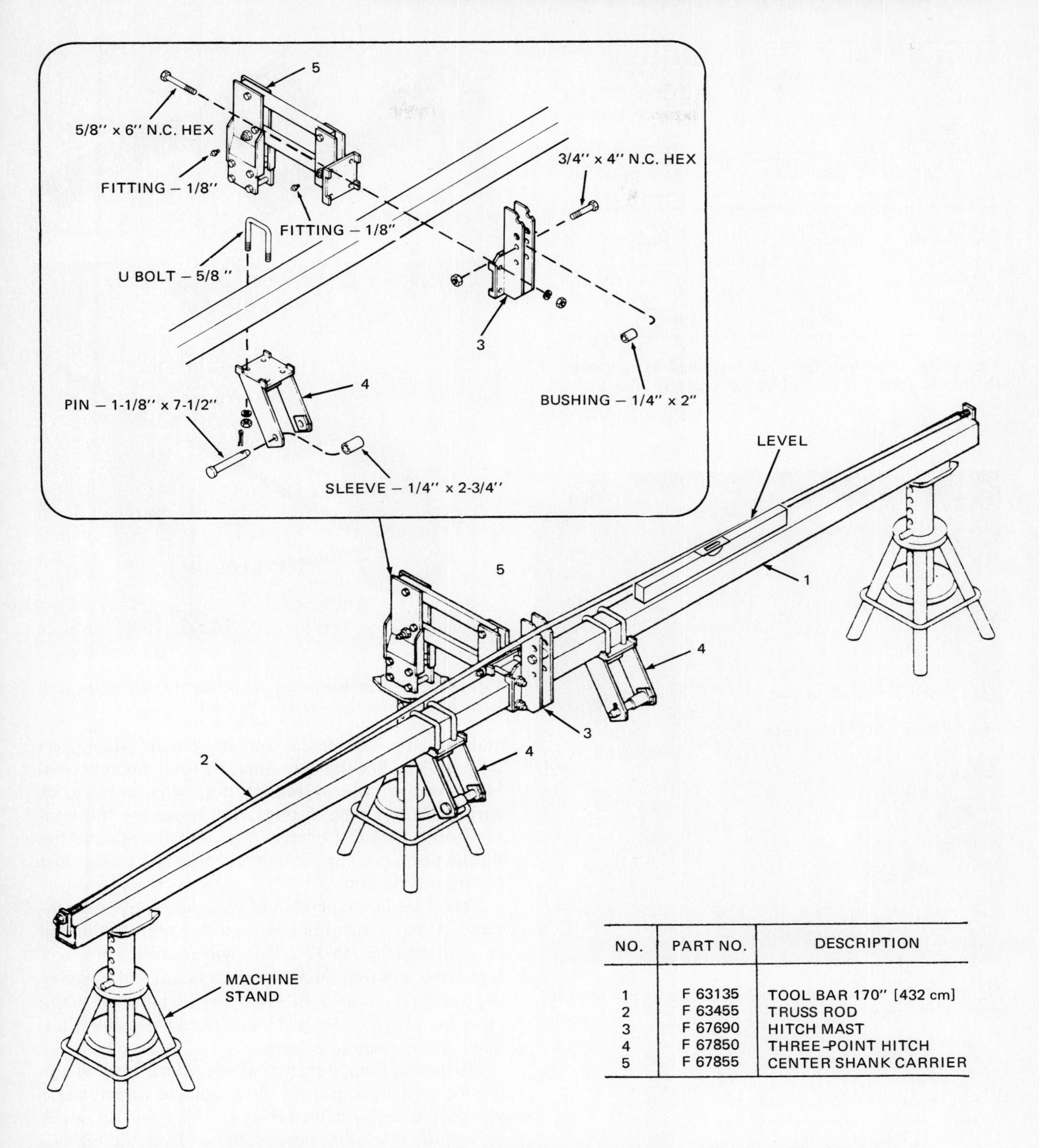


| NO. | PART NO. | DESCRIPTION |
|---|---|---|
| 1 | F 63135 | TOOL BAR 170" [432 cm] |
| 2 | F 63455 | TRUSS ROD |
| 3 | F 67690 | HITCH MAST |
| 4 | F 67850 | THREE-POINT HITCH |
| 5 | F 67855 | CENTER SHANK CARRIER |

**Fig. 18-31.** Setup procedure and details for assembling the tool bar of a row-crop cultivator. (*Ford Tractor Operations*)

N·m], special wrenches are available from the manufacturer (Fig. 18-35). In order to use the two wrenches, they must be fitted with pipe handles (Fig. 18-35). Then, the setup worker applies a downward force of 106 lb [48 kg] to the end of the 8-ft [2.4-m] wrench to develop 848 lb·ft [1150 N·m] of required torque (8 ft × 106 lb = 848 lb·ft).

**Hydraulic System** Modern tillage equipment utilizes the hydraulic system of the tractor to lower

| SAE GRADE 1 (NO MARKS) LOW CARBON STEEL | SAE GRADE 5 (THREE MARKS) MEDIUM CARBON STEEL HEAT TREATED | SAE GRADE 8 (SIX MARKS) MEDIUM CARBON-ALLOY STEEL; QUENCH TEMPERED |
|---|---|---|
| | | |
| METRIC EQUIVALENT | METRIC EQUIVALENT | METRIC EQUIVALENT |
| 4.6 mm | 8.8 mm | 10.9 mm |

Fig. 18-32. Grades of bolts are identified by markings on the heads. (See Torque Table, Appendix C-5B.)

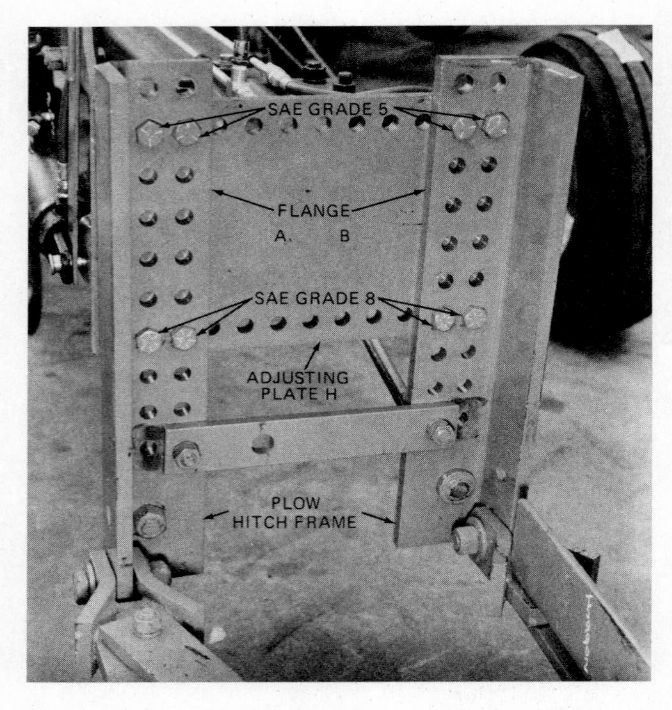


Fig. 18-33. Instructions for location of cap screws by bolt grade in the plow hitch.

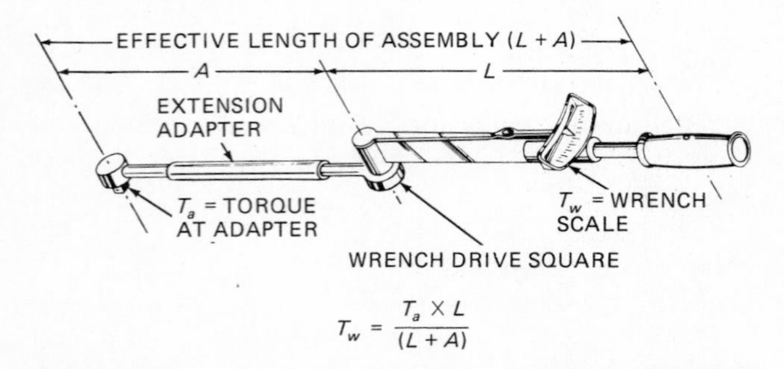


Fig. 18-34. A torque wrench extension is used to increase the capacity of the wrench.

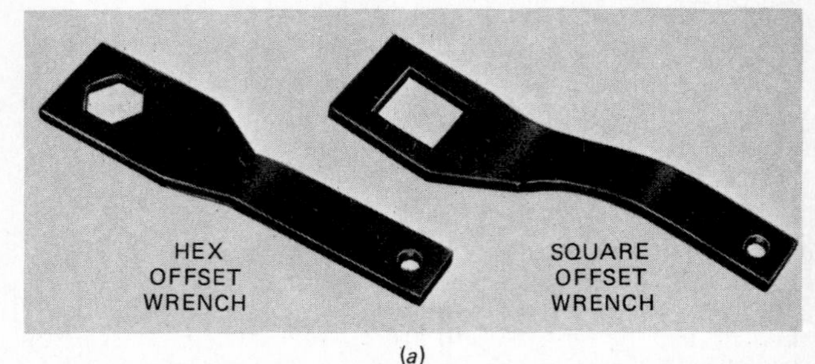


(a)

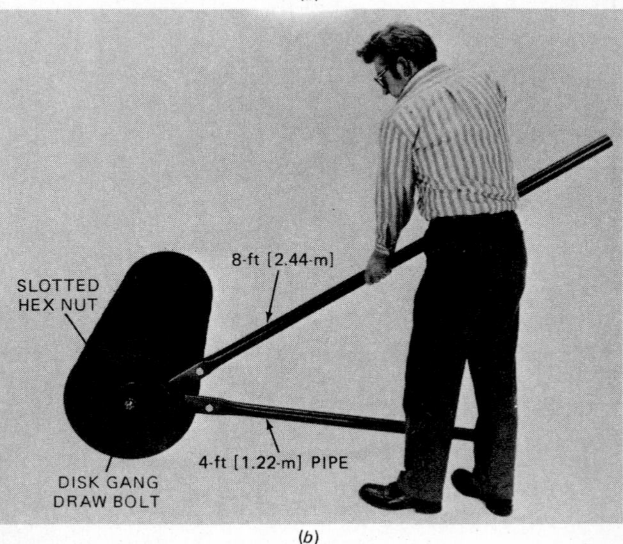


(b)

Fig. 18-35. Special wrenches are needed to tighten the disk harrow gang bolt nut. (*Deere & Co.*)

and raise the equipment and to actuate other parts (Fig. 18-36). Special care must be used when assembling the hydraulic system of tillage equipment to be sure that the piping, fittings, and hoses are not contaminated with dirt or moisture. Always remove the plastic protector caps from the piping and fittings just before installation.

Teflon sealing tape should be used on all male fittings. Always start the tape one thread from the end of a fitting (Fig. 18-37). This will prevent shreds of tape from entering the hydraulic system and plugging the oil filter. Use 3 in [76 mm] of tape for 1/2-in [13-mm] fittings. Wrap it in a clockwise direction, making one complete wrap.

Most tillage implements which utilize remote hydraulic cylinders include flow control and resistor valves in certain oil lines (Fig. 18-36). Resistor valves reduce the rate of oil flowing to and from the lift cylinders to allow slow raising and lowering of equipment for safety. Special care must be taken to see that the hydraulic system is bled free of air before operating the cylinders. Air is removed by attaching the supply lines to the tractor and operating the cylinder through several cycles. The cylinder can then be safely attached to the anchor and lift supports. Air in the cylinder or lines might allow the frame to fall dangerously fast.

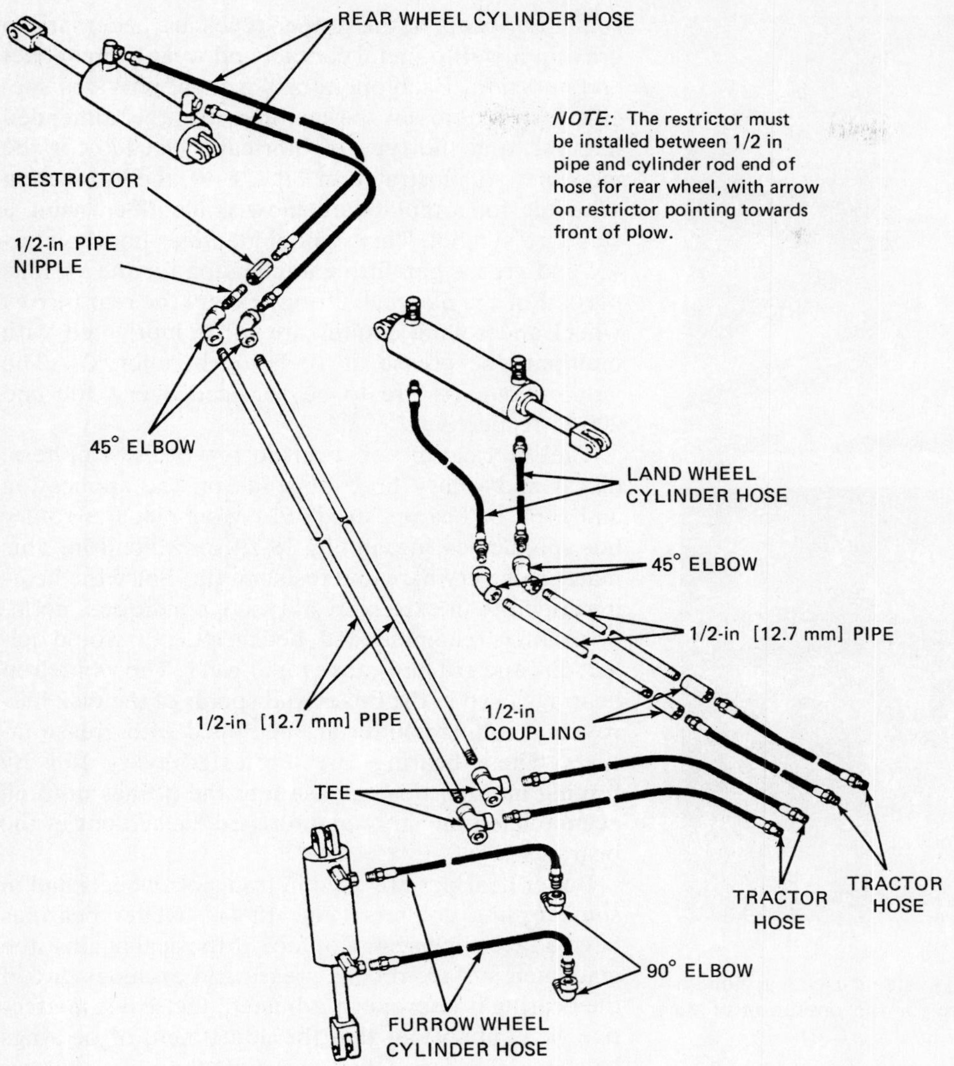

**Fig. 18-36.** Hydraulic system schematic for the lifting wing of a moldboard plow. (*Ford Tractor Operations*)

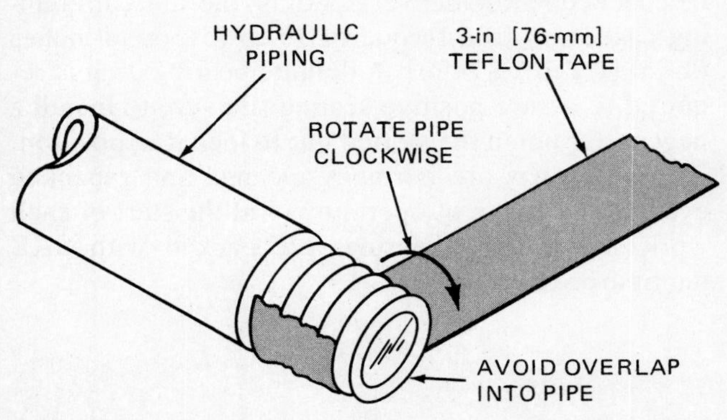


**Fig. 18-37.** Teflon sealing tape must be properly applied to hydraulic fittings to avoid contamination and leakage.

Moldboard plows utilize various types of mechanical safety trip devices to protect plow bottoms, standards, and the plow frame from damage when hitting an obstruction. Automatic reset standards (Fig. 18-38) trip and reset themselves by using springs or hydraulic cylinders. Some hydraulic reset systems require an *accumulator* to operate the system independent of the tractor. A bladder-type accumulator (Fig. 18-39) is charged with 1350 to 1700 psi [9300 to 12,000 kPa] of nitrogen. The amount of pressure used depends upon soil conditions. An oil line from the tractor hydraulic system is temporarily connected to the accumulator to charge the accumulator system with oil. The amount of hydraulic pressure is about 200 psi [1379 kPa] more than the gas pressure.

When a bottom strikes a solid object, oil is forced into the accumulator by the retracting cylinder. The oil compresses the bladder against the gas. As soon as the bottom clears itself, oil is forced back into the cylinder by the gas pressure in the accumulator. A restrictor in the accumulator controls the rate of cylinder travel upon reset.

**CAREFUL** Always follow printed instructions carefully when servicing accumulators.

**Lubrication** The various moving parts of tillage equipment are subjected to a tremendous amount of stress and vibration when cutting, lifting, and moving

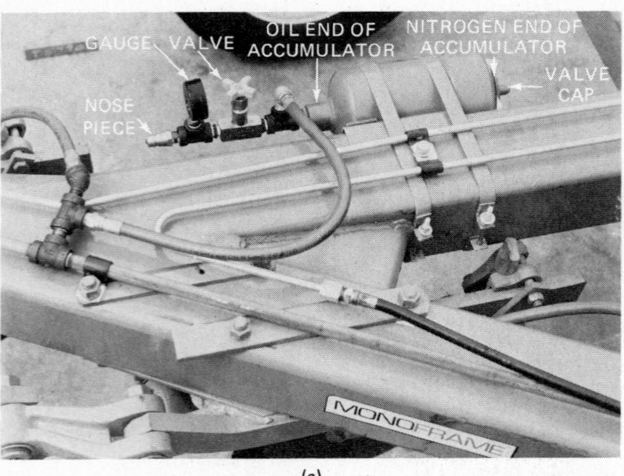


(*a*)

(*b*)

**Fig. 18-38.** Many hydraulic reset systems use an accumulator to supply hydraulic pressure for the operation of the system. (*Allis-Chalmers Manufacturing Co.*)

ACCUMULATOR
CHARGING LINE VALVE
CHARGING LINE GAUGE
ORIFICE
CHARGING LINE
HIGH PRESSURE OIL
HIGH PRESSURE NITROGEN
HYDRAULIC CYLINDERS

**Fig. 18-39.** Bladder-type accumulator. The nitrogen pressure in the bladder produces pressure on the oil in the hydraulic cylinders. (*Deere & Co.*)

soil. Lubricants of the proper types are necessary to prevent metal-to-metal contact and wear from impact and abrasion. Each operator's manual provides specific instructions on the location, the recommended interval, and the type of lubricants needed for the machine. As illustrated in Fig. 18-40, the lubrication schedule for a moldboard plow is identified using a standard symbol. This symbol identifies hourly interval and grease gun fitting application for the various parts. For the plow, all fittings except the rear furrow wheel and rolling coulter are to be lubricated with multipurpose grease at 10 hour (h) intervals. The latter two parts are to be serviced every 100 and 200 h, respectively.

The lubrication of transport wheels, coulters, disks, and rotary hoes depends on the application and type of bearing used. For example, the rotary hoe spindle bearings in Fig. 18-29 are white iron. This material is very hard and resists wear. Since the bearings operate in extremely abrasive conditions, no lubrication is recommended, because the oil would collect dirt and grit and cause rapid wear. The white iron bearings used in the boxes and spools of the disk harrow and rolling coulter are equipped with grease fittings. These bearings are lubricated every 10 h by forcing multipurpose grease into the fittings until all accumulated moisture and dirt are flushed out of the bearings.

Roller bearings are used in transport wheels and in some rolling coulters (Fig. 18-41). Roller bearings have a face-type seal. Contact of the seal against the machined surface retains grease and excludes dirt; if the bearing is improperly adjusted, the seal is ineffective. It is important that the adjustment of bearings be checked before delivery. Adjust the slotted retaining castle nut to a torque of 10 to 25 pound-inches (lb·in) [1.1 to 2.8 N·m]. A definite bearing drag is required to assure positive sealing. Be sure to install a new cotter pin in the slotted nut to lock it in position. These bearings are normally cleaned and repacked every 100 to 200 h of operation or at the start of each work season. The bearings are packed with SAE multipurpose grease.

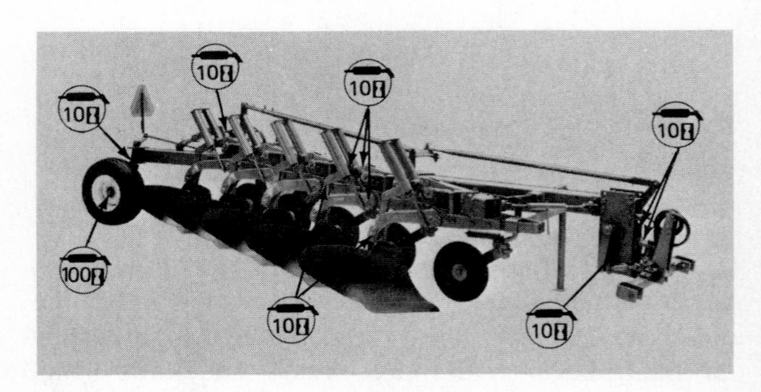


**Fig. 18-40.** Moldboard plow lubrication points and scheduled intervals. (*Deere & Co.*)

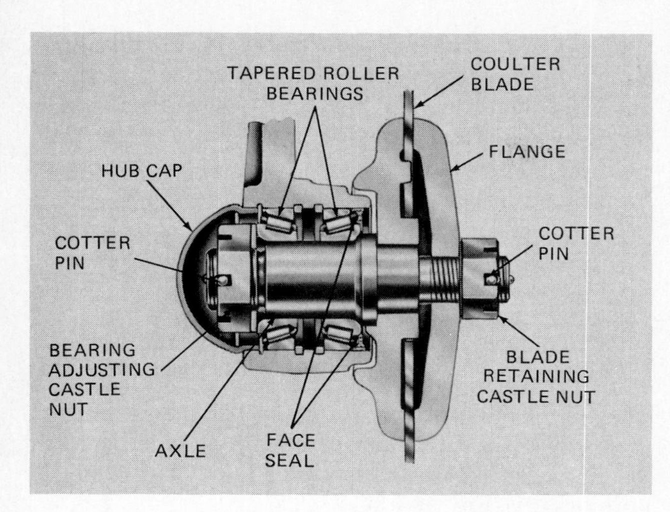


**Fig. 18-41.** Tapered roller bearings are often used in rolling coulters. (*Deere & Co.*)

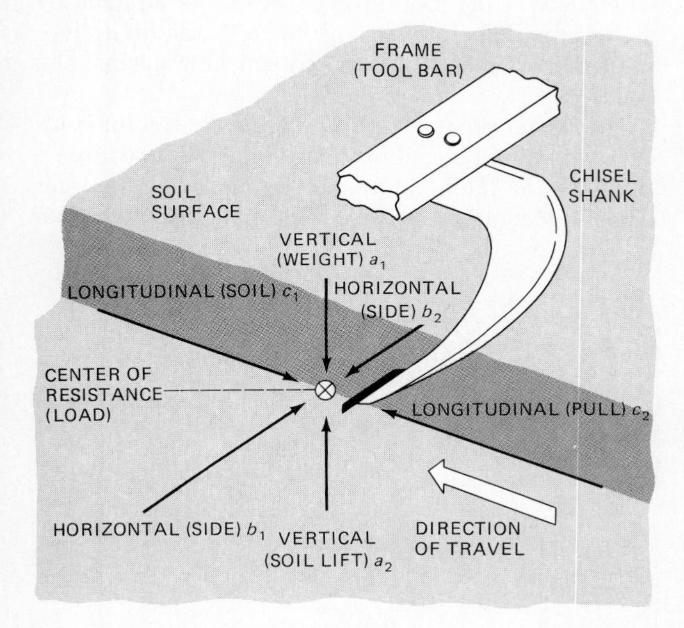


**Fig. 18-42.** Forces act upon tillage tools in three planes: (*a*) vertical, (*b*) horizontal, and (*c*) longitudinal.

**Transport Wheel Tire Pressure** As soon as the transport wheels have been attached, the tires should be checked for the correct pressure. Tires not properly inflated affect the rolling resistance, weight-carrying ability, and draft alignment of the machine. Use the pressures indicated in the operator's manual.

## Hitching Tillage Machinery

Have you observed that a tractor pulling a tillage tool through a field seems to labor under the heavy load? Perhaps you have noticed that the implement either followed the tractor smoothly or strained as it penetrated, cut, and lifted the soil. Generally, the way in which a tillage tool is assembled and hitched to a tractor determines how the machine performs its intended job. Each machine was designed by agricultural engineers to operate with the greatest possible efficiency when properly hitched. As a specialist in power and machinery, you must be able to hitch implements properly in accordance with the instructions which have been provided, and analyze hitch problems which may develop during field operations.

**Principles of Hitching** When a tillage machine enters the soil, levels off at an established depth, and runs in a stable way, *forces* which affect its operation are said to be in balance. *Force* is the action of one body upon another attempting to change its position. The tillage tool is one body, and the soil is another. While the tractor exerts a force to pull the tool, the soil offers resistance against the forward travel. The soil also offers resistance to the plow entering the soil. Thus the forces which act upon the tillage machine come from three different sources: (1) the weight or gravity of the machine, (2) the pull applied by the tractor, and (3) the resistance of the soil against the tool. These forces act through three planes as follows (see Fig. 18-42):

1. Vertical—the *downward* (weight or gravity of the tool) and the *upward* (soil resistance) forces
2. Horizontal—*side forces* (lifting and turning of soil, direction of pull)
3. Longitudinal—the pull of the tractor and the resistance of the soil as the tool is moved

**Center of resistance** When all these forces are equal, they are in balance. They concentrate on a point called the *center of resistance,* or *load.* It is at this point that all the forces are in equilibrium. However, to make the chisel plow move through the soil, the longitudinal force of pull ($c_2$) must exceed the longitudinal force of soil resistance ($c_1$). For example, the many forces which act upon a moldboard plow bottom can be summarized as follows (Fig. 18-42):

1. Vertical forces
   ($a_1$) Weight of the plow (downward force)
   Lifting of the soil (downward force)
   ($a_2$) Soil's resistance to penetration (upward force)
   Upward pull of the tractor (upward force)
2. Horizontal (side) forces
   ($b_1$) Side pull from the tractor
   Side push of the landside or furrow wheel
   ($b_2$) Movement of the soil to open furrow
   Friction of the soil on the sloping bottom
   Cutting and wedging action of the share
3. Longitudinal forces
   ($c_1$) Resistance of the soil to cutting
   Friction of the share and landside
   Friction of the soil on the moldboard
   ($c_2$) Pull in the direction of travel

For the moldboard plow, these forces have been determined to be in equilibrium approximately at a point which is (Fig. 18-43):

1. One-half the depth of plowing (vertical forces)
2. One-fourth the width of cut of the bottom measured from the shin of the moldboard (horizontal forces)
3. Twelve to fifteen inches [305 to 381 mm] back from the point of the share for a 16-in bottom

A rule of thumb for locating the center of resistance for a single plow bottom is to locate a point which is one-half the depth of cut and one-fourth the width of cut from the shin of the moldboard. In the example in Fig. 18-43, the center of resistance (load) is 4 in [102 mm] to the right of the shin and 12 to 15 in [305 to 381 mm] from the point of the share for a 16 in [406-mm] bottom.

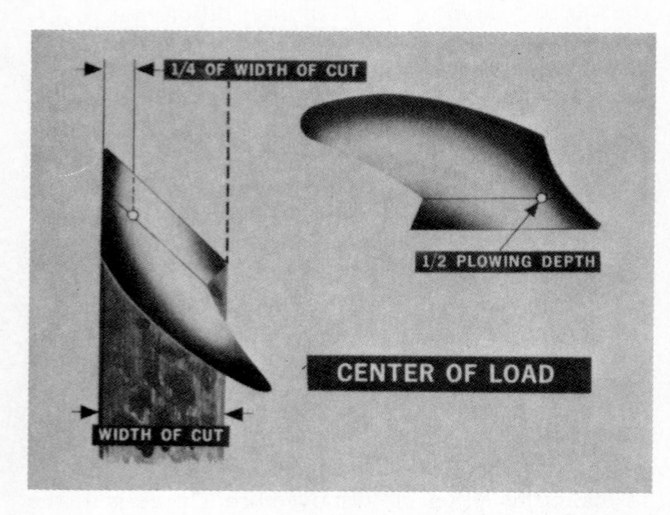


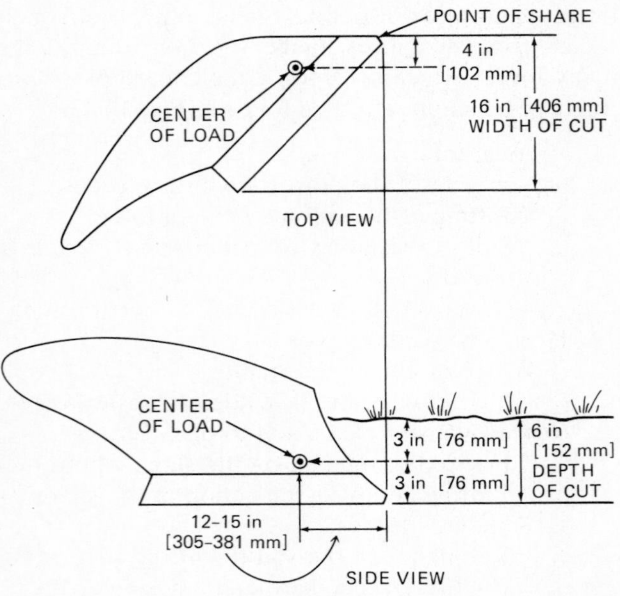


Fig. 18-43. Determining the center of load on a 16-in [406-mm] plow bottom.

The center of resistance (load) for a multibottom plow is determined by adding one-fourth the width of one bottom to one-half the total width of cut measured to the left of the wing of the first bottom. For example, the center of resistance for a five-bottom 16-in [406 mm] plow is as follows (Fig. 18-44):

1. $^1/_4 \times 16$ in = 4 in [102 mm]
2. $^1/_2 \times 5 \times 16$ in = 40 in [1016 mm]
3. Center of resistance (load) = 44 in [1118 mm] measured from wing of first bottom

The center of resistance for disk tillage tools such as the plow disk is located at the intercept of the full depth cut vertically and one-half the width of cut horizontally (Fig. 18-45).

**Center of pull** The longitudinal force ($c_2$) (Fig. 18-42) which moves the tool through the soil is supplied by the tractor. This force is developed by the rotation of the drive wheels of the tractor pushing against the soil.

The center point of pull (force) on the tractor is located at a point ahead of the rear axle, depending on the type of hitch, and midway between the rear wheels. Most tractors are equipped to pull tillage ma-

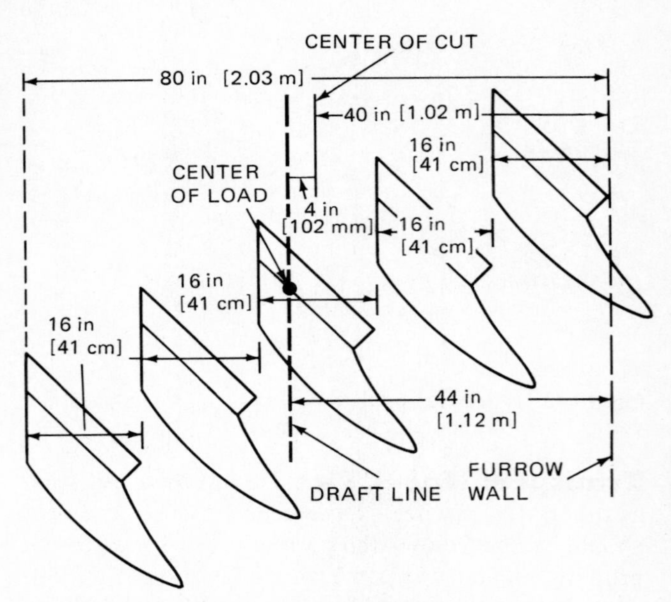


Fig. 18-44. Location of the center of load for a five-bottom 16-in [406-mm] moldboard plow.

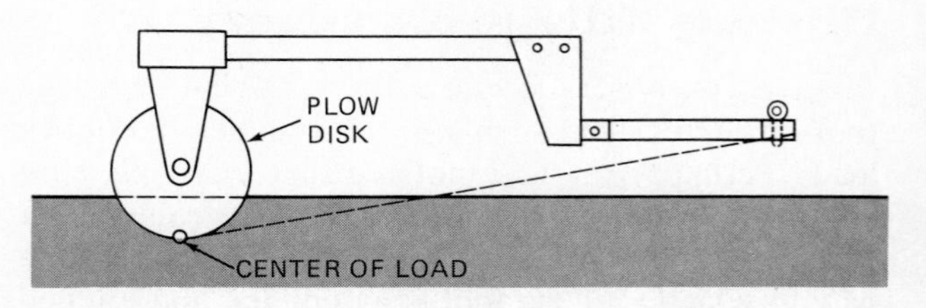


Fig. 18-45. Center of load for a plow disk.

chinery by attaching to either the drawbar or the three-point hitch linkage.

The drawbar is a heavy steel bar located under the rear axle housing of the tractor. It is attached at a single point approximately 6 in [152 mm] ahead of the rear axle on the centerline of the chassis. The remaining portion of the drawbar extends rearward under the tractor. The hitch point for attaching implements is supported at a height of 15 to 21 in [381 to 533 mm] from the ground (Fig. 18-46). The support is designed to allow the drawbar to swing freely from side to side or to be pinned in place in one of several positions. A tractor can be turned more easily if the drawbar is allowed to swing when pulling a load.

A three-point hitch permits the implement to be attached directly to the tractor at three points. The two lower links are fixed in length and are supported by adjustable rods. The links are raised and lowered by

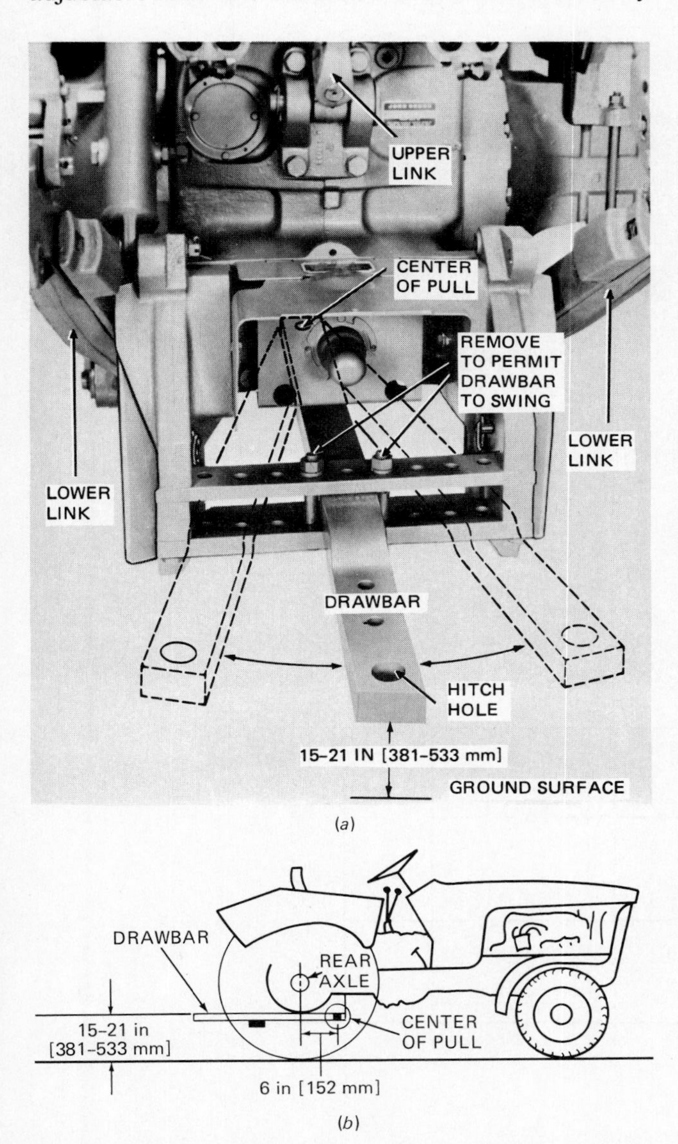


**Fig. 18-46.** Center of pull on a tractor when the load is attached to the drawbar. (*Deere & Co.*)

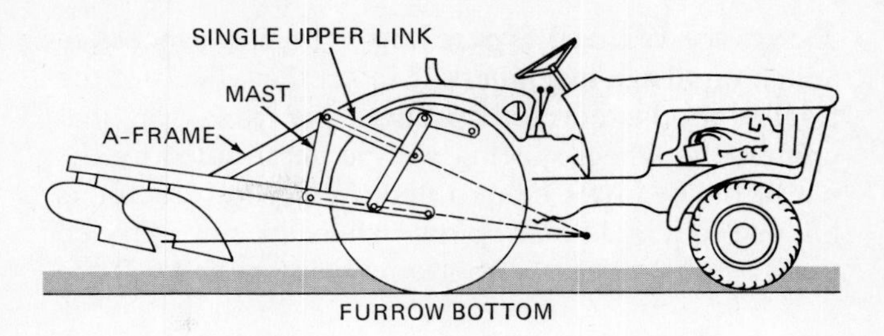


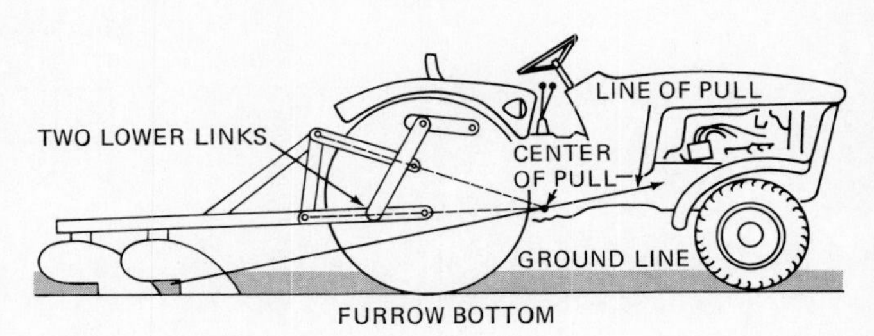


**Fig. 18-47.** The center of pull for an integral hitch on a tractor is located at a point where lines parallel to the upper and lower links converge. The point varies with the position of the implement.

a rocker shaft arm connected to a hydraulic cylinder. The single upper link connects to the A-frame mast of the implement and to a fixed point on the tractor. When the implement is lifted by the lower links, the upper link causes the implement to raise evenly and be carried by the tractor. The center of pull for the three-point hitch is at the point where lines extended through the center axis of the upper and lower links converge (Fig. 18-47).

Three-point link hitch systems are sized according to the power range of the tractor into five categories: 0, I, II, III, and IV. As identified in Fig. 18-48, the dimensions of the frame and hitch pins increase to accommodate the power output of the tractor. The professional in power and machinery will therefore be able to apply the information supplied by assembly instruction manuals when reference is made to one of the five categories of a three-point hitch.

## Shop Adjustment

Before a new tillage machine is delivered to the purchaser, the necessary shop adjustments to obtain the approximate alignment and adjusments of the various parts will be made. For example, the hitch of a moldboard plow may be checked for the correct line of draft, or a row-crop cultivator may be adjusted for row spacing.

Most shop adjustments require that approximate operating positions for the machine be simulated. This is accomplished by placing blocks of proper thickness under the wheels of the tractor to obtain working depth. Working depth for plowing might be 6 to 8 in [152 to 203 mm], while a row-crop cultivator might be set for a depth of 2 in [51 mm].

**Tractor Wheel Spacing** The first process in adjusting tillage machinery is to set the wheels of the tractor to the correct spacing. For row-crop machines, the wheel spacing will be determined by the width of the rows being tilled. When the tractor is hitched to a moldboard plow, wheel spacing is based upon (1) the number of bottoms being pulled, (2) the type of plow (integral, semi-integral, or drawn), and (3) whether the drive wheels are operated on-land or in-furrow.

To set the row spacing of a row-crop planter or cultivator accurately, it is suggested that a diagram of the desired spacing be marked on the shop floor with chalk (Fig. 18-49). The diagram shows lines for sev-

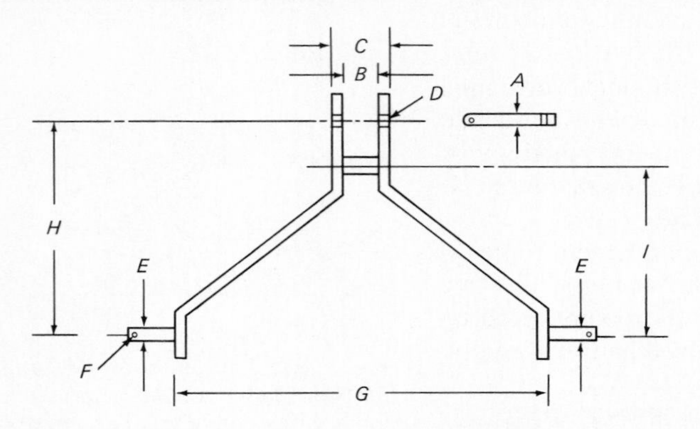


STANDARD MEASUREMENTS
THREE-POINT IMPLEMENT HITCHES BY CATEGORIES, in [mm]

| LETTER | ITEM | CATEGORY 0 | I | II | III | IV |
|---|---|---|---|---|---|---|
| | MAXIMUM DRAWBAR HORSEPOWER | 20 or LESS | 20-45 | 40-100 | 80-225 | 180-400 |
| | UPPER HITCH POINT | | | | | |
| A | PIN DIAMETER | 5/8 [15.9] | 3/4 [19] | 1 [25.4] | 1-1/4 [31.8] | 1-3/4 [44.5] |
| B | WIDTH, INSIDE MAST | 1-1/8 [28.6] | 1-3/4 [44.5] | 2-1/16 [52.4] | 2-1/16 [52.4] | 2-9/16 [65.1] |
| C | WIDTH, OUTSIDE MAST | 1-15/16 [49.2] | 3-11/16 [93.7] | 3-3/4 [95.3] | 3-3/4 [95.3] | 5-13/16 [147.6] |
| D | HOLE DIAMETER | 21/32 [16.7] | 49/64 [19.4] | 1-1/64 [25.8] | 1/7/16 [36.5] | 1-51/64 [45.6] |
| | LOWER HITCH POINT | | | | | |
| E | STUD DIAMETER | 5/8 [15.9] | 7/8 [22.2] | 1-1/8 [28.6] | 1-7/16 [36.5] | 2 [50.8] |
| F | LINCHPIN HOLE DIAMETER | 19/64 [7.5] | 31/64 [12.3] | 31/64 [12.3] | 31/64 [12.3] | 23/32 [18.3] |
| G | LOWER HITCH POINT SPREAD | 20 [508] | 27 [685.8] | 32-1/2 [825.5] | 38 [965.2] | 45-7/8 [1165.2] |
| H | MAST HEIGHT | 12 [305] | 18 [457.2] | 18 [457.2] * | 22 [558.8] | 27 [685.8] |
| I | MAST HEIGHT – FAST COUPLER | | 15 [381] | 15 [381] | 19 [482.6] | 27 [685.8] |

*22 in OPTIONAL

**Fig. 18-48.** Standard measurement dimensions for five categories of three-point hitches.

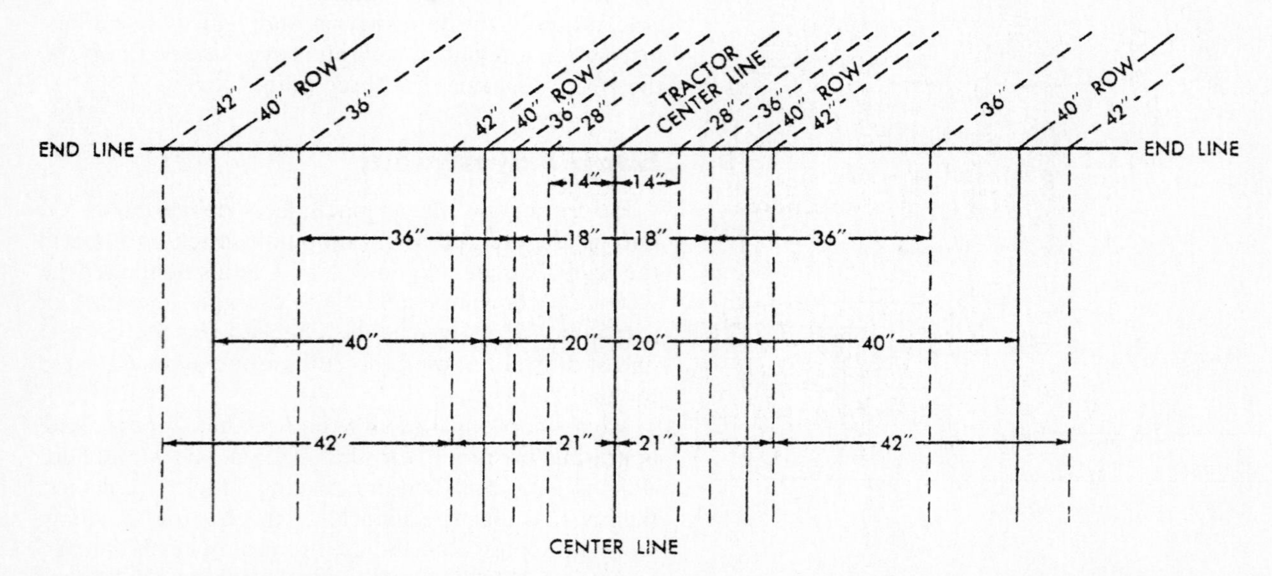


**Fig. 18-49.** Lines are located on the shop floor to represent rows of crops to be planted or cultivated. The end line is for locating the front wheels of the tractor. The tractor centerline is used to position the tractor for adjusting the width of the wheels to row spacing.

eral row spacings. However, only the end line, the tractor centerline, and lines representing the rows required are necessary. The tractor is driven directly over the diagram centerline with the front wheels positioned on the end line. The front and rear wheels of the tractor are positioned to operate directly between the rows.

A chalk line can be used to represent the furrow wall when positioning tractor wheels before adjusting the moldboard plow in the shop. For in-furrow operations, position the tractor furrow wheel 2 in [51 mm] from the chalk line. Wheel spacing is then determined by measuring from the furrow wall to the center of pull on the tractor (Fig. 18-50). When making on-land wheel adjustments, space the right outside tire of a dual-wheel tractor 4 in [102 mm] from the chalk line. Wheel spacing is then measured to the outside edge of the wheel.

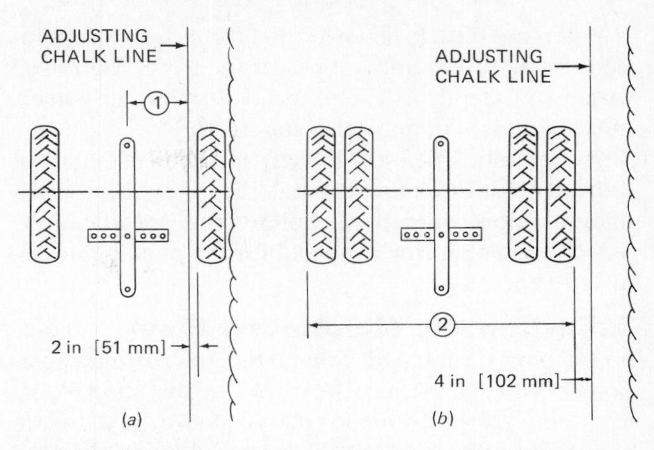


**Fig. 18-50.** Tractor may be operated in-furrow (*a*) with wheel spacing measured from the furrow to the center of pull (1), or on-land (*b*) with the right outside wheel 4 in [102 mm] from the furrow wall and spacing of wheels measured to the outside edge of the outside wheels (2).

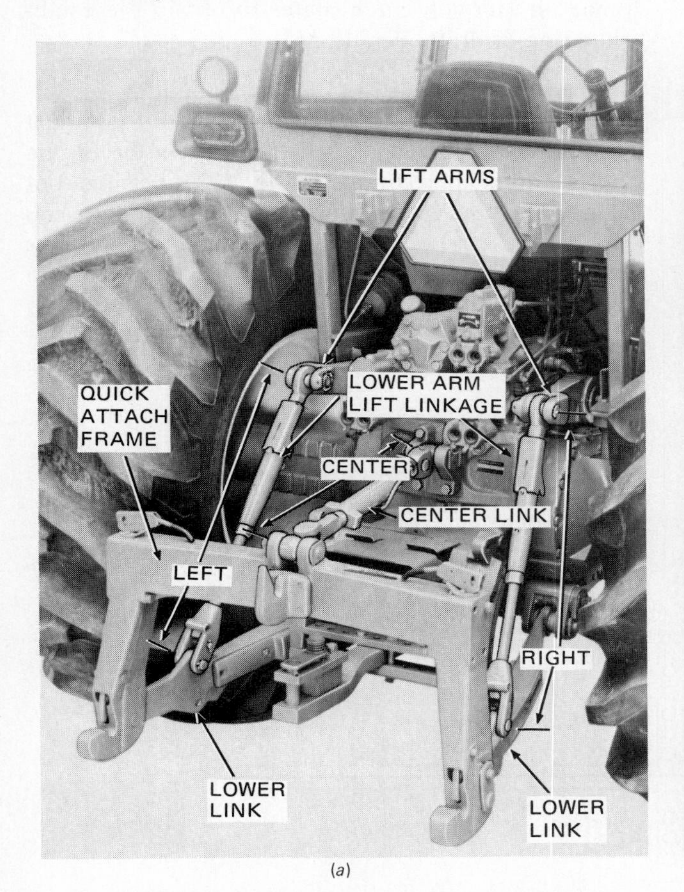

| TRACTOR | IN-FURROW LEFT LIFT LINK (DIMENSION "1") | IN-FURROW RIGHT LIFT LINK (DIMENSION "2") | IN-FURROW CENTER LINK (DIMENSION "3") | ON-LAND LEFT LIFT LINK (DIMENSION "1")- | ON-LAND RIGHT LIFT LINK (DIMENSION "2") | ON-LAND CENTER LINK (DIMENSION "3") |
|---|---|---|---|---|---|---|
| 4030,4230<br>4040,4240 | 30 in<br>[762 mm] | 29 in<br>[737 mm] | 26-1/2 in<br>[673 mm] | 30 in<br>[762 mm] | 30 in<br>[762 mm] | 26-1/2 in<br>[673 mm] |
| 4430<br>4440 | 34 in<br>[864 mm] | 33 in<br>[838 mm] | 26-1/2 in<br>[673 mm] | 34 in<br>[864 mm] | 34 in<br>[864 mm] | 26-1/2 in<br>[673 mm] |

(*c*)

**Fig. 18-51.** (*a*) Lift links for the semi-integral mounted plow are adjusted for the specific tractor and for in-furrow or on-land operations. (*b*) Link adjustments are made by turning the screw linkage. (*c*) Typical lift link dimensions. (*Deere & Co.*)

When the tractor is positioned for in-furrow operation, the left front and left rear wheels must be raised to plowing depth. For on-land operation, all wheels must be blocked up to plowing depth.

Proper adjustments of the left and center linkage of a three-point hitch are essential for integral or semi-integral plow operation. Follow the specifications listed in the operator's manual for the plow being adjusted (Fig. 18-51).

**Adjusting the Moldboard Plow** A moldboard plow is hitched behind the tractor and positioned with the bottoms lowered to the shop floor. The wing of the front bottom is placed next to the chalk line. The land wheel of the plow is elevated 6 in, using a block. A carpenter's level is placed on the main frame and leveled horizontally across the bottom (cross-leveling) and fore and aft (front to rear).

A drawn plow is leveled by adjusting the land wheel and rear furrow wheel. For integral or semi-integral hitch plows, the frame is cross-leveled by use of the right leveling crank of the tractor's lower lift link. Also, the lift links are set to the length listed in the operator's manual (Fig. 18-51*a*), and in accordance with whether the tractor is operated in-furrow or on-land. Fore and aft leveling is accomplished by rotating the top link adjustment turnbuckle (Fig. 18-51*b*) for mounted plows.

**Checking line of draft** The approximate adjustment of the hitch for drawn and semi-integrally hitched plows can be established by shop adjustments.

The hitch is adjusted both horizontally (as viewed from the top, Figs. 18-52 and 18-53) and vertically (as viewed from the side, Fig. 18-54). With the plow properly positioned behind the tractor, attach a string tautly between the center of load on the plow and the center of pull on the tractor. Adjust the hitch of the plow so that the hitch pin of the drawn plow or the pivot point of the semi-integral- and integral-mounted plows falls on the line both horizontally and vertically.

Horizontal hitch adjustment of semi-integral plows is accomplished by positioning the hitch tube and the hitch crossbar coupler for the number of bottoms, tractor wheel spacing, and hitch category (Fig. 18-55). Vertical hitch adjustment of this type of plow is made by moving hitch plates to obtain the established line of draft (Fig. 18-56).

**Adjusting Coulters** Rolling coulters are adjusted to satisfy average operating conditions. Adjust the coulter fore and aft so that the center of the coulter hub is approximately 4 in [102 mm] ahead of the point of share. The horizontal adjustment (top view) of the coulter is made by rotating the coulter shank until the blade is 1/2 to 5/8 in [13 to 16 mm] to the left of the moldboard's shin (Fig. 18-57). When

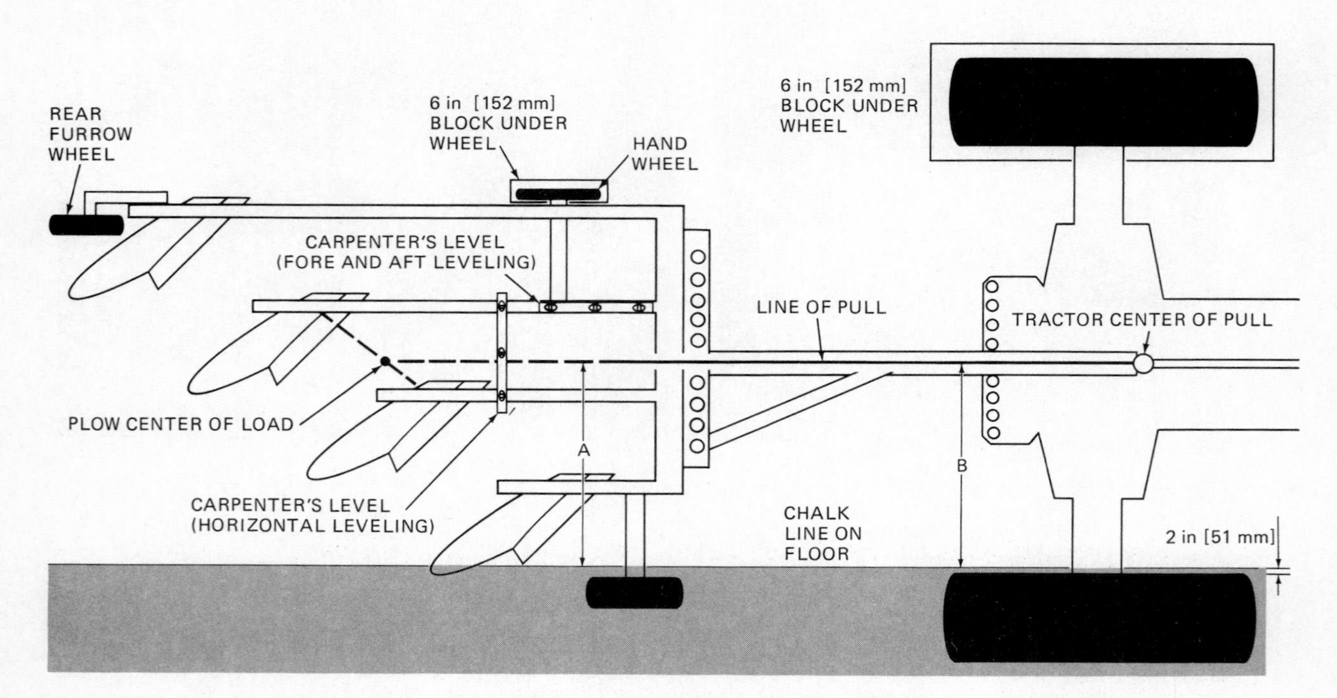


**Fig. 18-52.** Horizontal line of draft for a pull-type plow. When a pull plow is hitched to the drawbar of a tractor, the line of pull will be parallel to the furrow wall. The plow load center (A) and tractor wheel spacing (B) should be the same dimension. Level the plow frame horizontally and fore and aft with a carpenter's level.

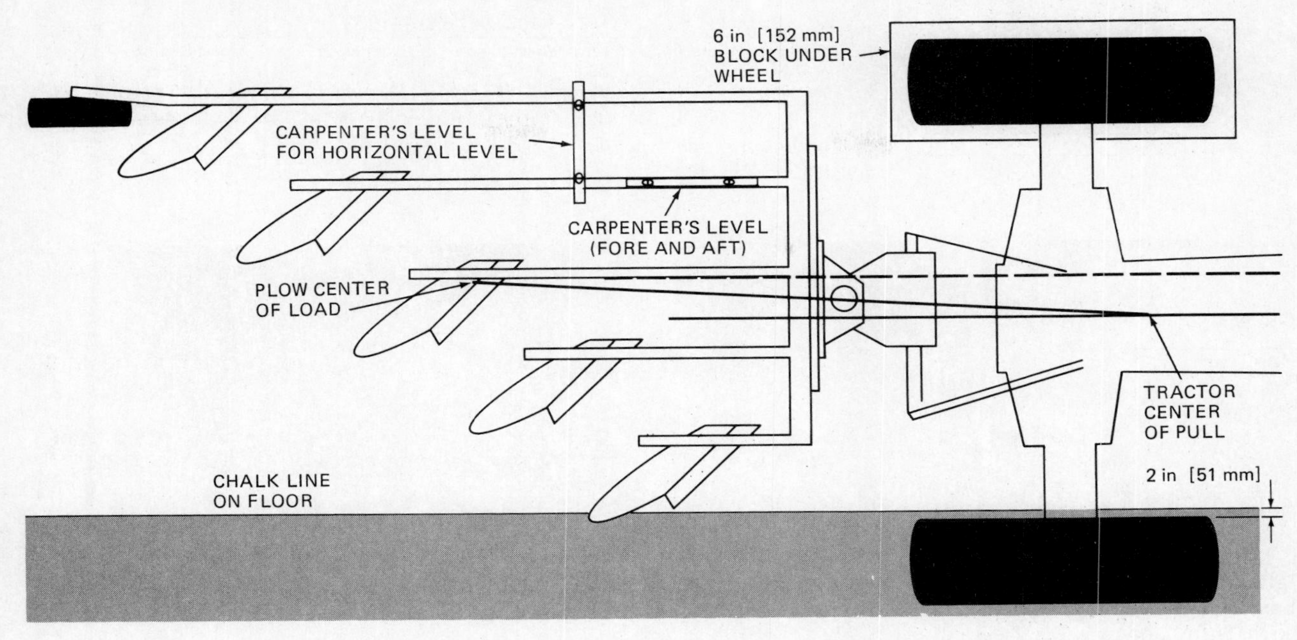


**Fig. 18-53.** Shop adjustment of a semi-integral plow. The line of draft will be a straight line from the plow center of load to the tractor center of pull. The line of draft will usually not be parallel to the furrow wall.

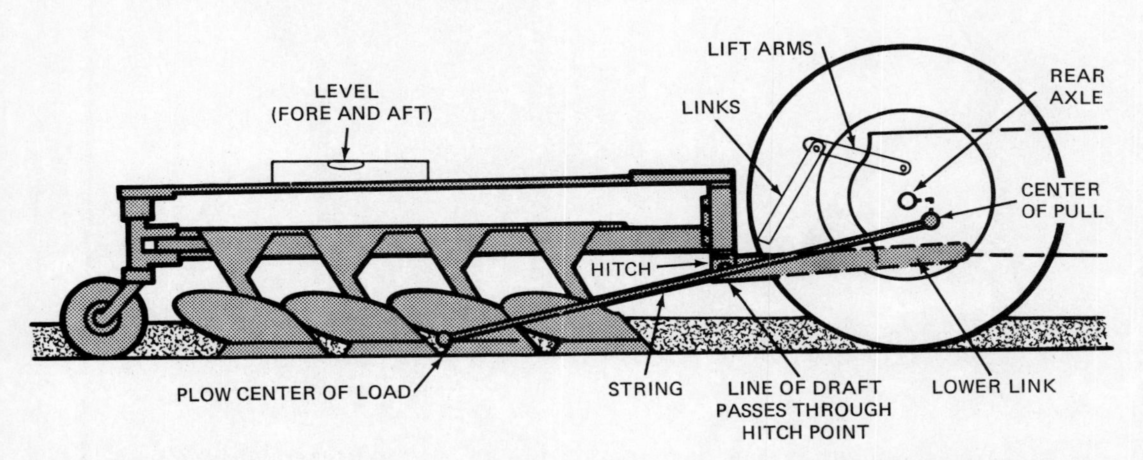


**Fig. 18-54.** Locating the vertical line of draft. The vertical line of draft is a straight line running from the plow center of load to the center of pull on the tractor. Center of pull is 4 in [102 mm] ahead and below the tractor rear axle. The line of draft should pass through the hitch point for pull, semi-integral, and integrally mounted plows for both in-furrow or on-land operation.

the coulter has been set to cut to a depth of 3 in [76 mm], the adjusting bolts are tightened to the recommended torque values.

## Field Adjustment

After proper shop setup, final adjustment of tillage tools must be performed under actual working conditions. The process is called *field adjustment*. There is a definite order in which field adjustments are made. These steps are identified in the operator's manual. The service worker who delivers the new tillage tool and assists in making the final adjustments must realize that several of the operations are interrelated. Therefore it is usually necessary to repeat several of the steps a number of times, just as when fine tuning an engine.

Making field adjustments requires the experience of working with someone who is skilled in the various operations. However, an understanding of shop adjustments and how to apply operator's manual information will speed your development into the role of a specialist in power and machinery.

**Checking Horizontal Hitch of Moldboard Plow** Proper horizontal hitch setting will cause the plow to run straight ahead, make the front bottom take a full cut, and help reduce plow draft.

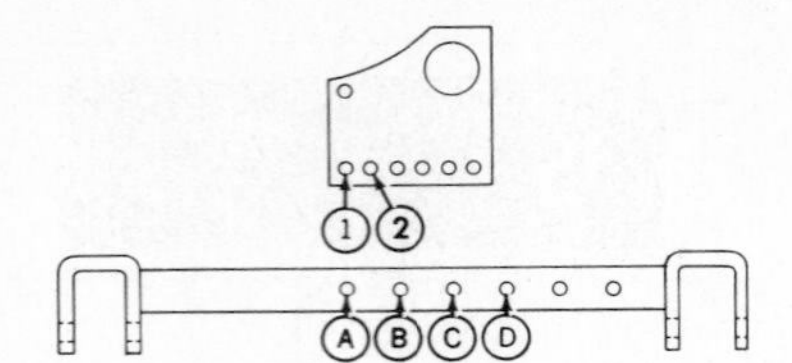


HITCH CROSSBAR SETTINGS

| Number of Bottoms | Category 2, Tractor Wheel Settings 28 in [711 mm] | 30 in [762 mm] | 32 in [813 mm] | Category 3, Tractor Wheel Settings 28 in [711 mm] | 30 in [762 mm] | 32 in [813 mm] |
|---|---|---|---|---|---|---|
| 4 | 1–B | 2–B | 2–B | NR | NR | NR |
| 5 | 1–C | 1–C | 2–C | 1–A | 2–A | 2–A |
| 6 | 1–D | 2–D | 2–D | 1–B | 2–B | 2–B |

(*a*)

QUIK COUPLER

(*b*)

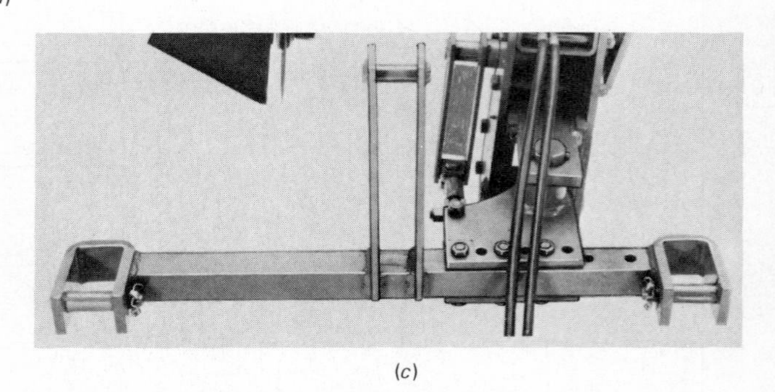

(*c*)

**Fig. 18-55.** (*a*) Adjustments for semimounted plow hitch crossbar when operated in-furrow according to the number of bottoms. (*b*) The crossbar shown is for category 2 with or without Quik-Coupler. (*c*) The crossbar shown is for category 3. (*Deere & Co.*)

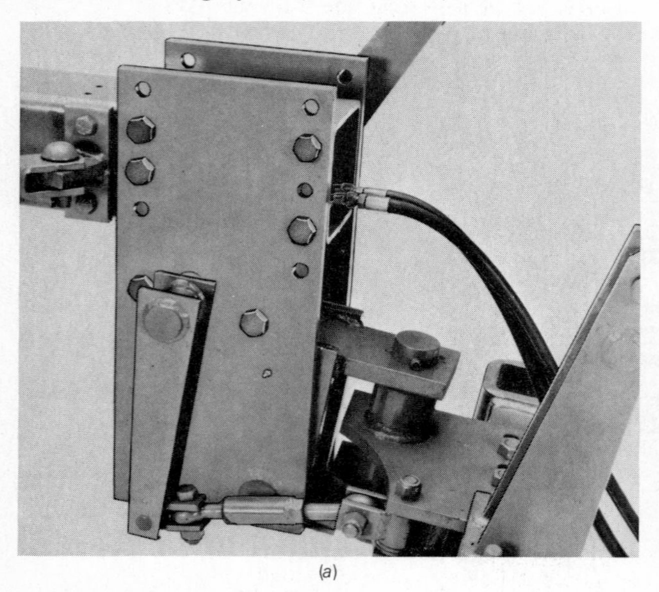

(*a*)

(*b*)

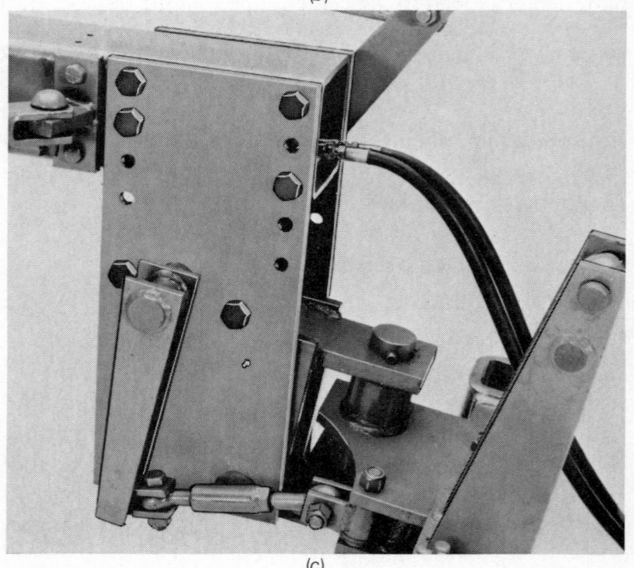

(*c*)

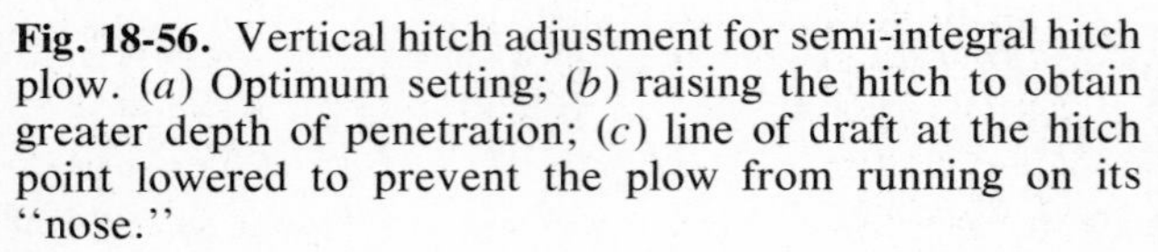

**Fig. 18-56.** Vertical hitch adjustment for semi-integral hitch plow. (*a*) Optimum setting; (*b*) raising the hitch to obtain greater depth of penetration; (*c*) line of draft at the hitch point lowered to prevent the plow from running on its "nose."

The best procedure for determining proper horizontal hitching is to lower the plow to the approximate plowing depth and plow to open a furrow. On the next round, plow a short distance with the tractor running in the desired relation to the furrow wall. Stop and measure the width of cut made by the front bottom. Measure from the front coulter to the furrow wall and determine if the distance is the same as the size of the plow bottom.

Figure 18-58*a* illustrates a plow that is running straight ahead and taking the proper width of cut. If, however, the amount of cut is too little, the hitch on the plow is too far to the left of the line of draft (Fig.

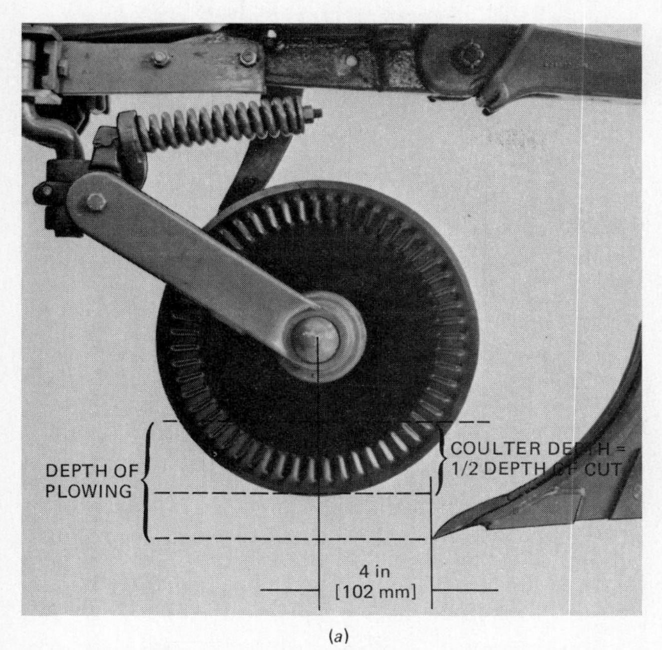

(a)

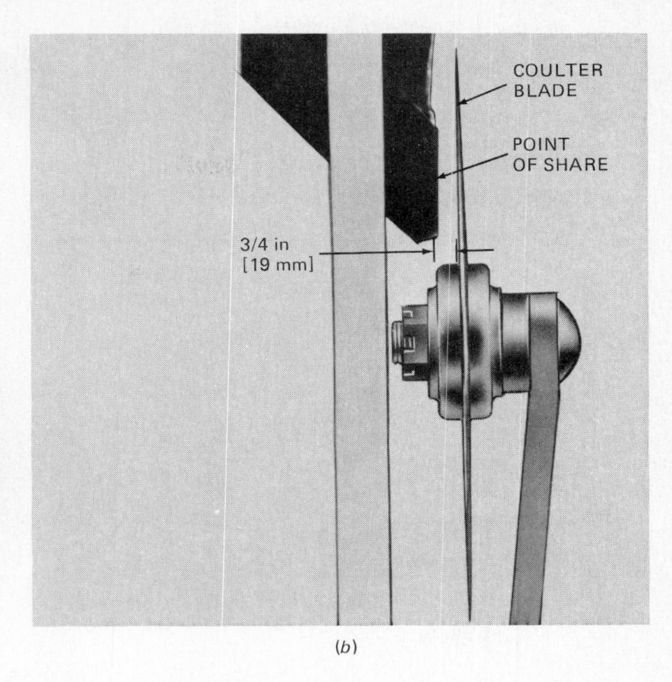

(b)

**Fig. 18-57.** Adjust the rolling coulter. (*Deere & Co.*)

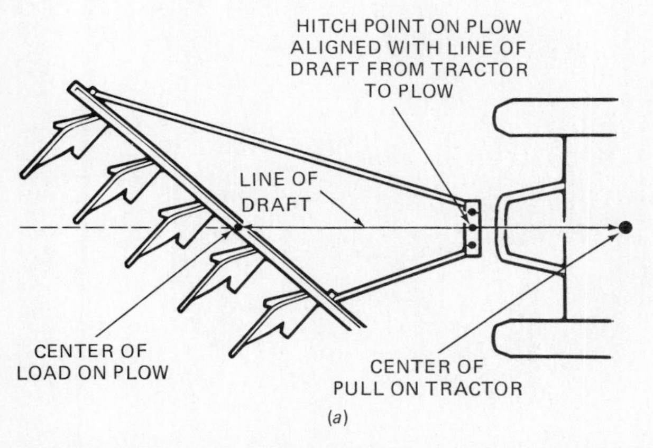

(a)

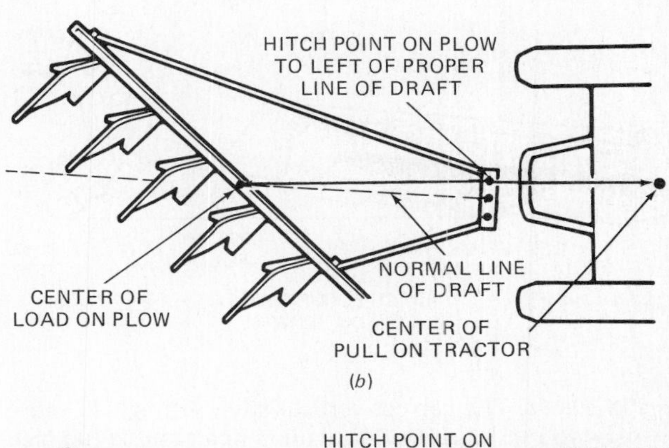

(b)

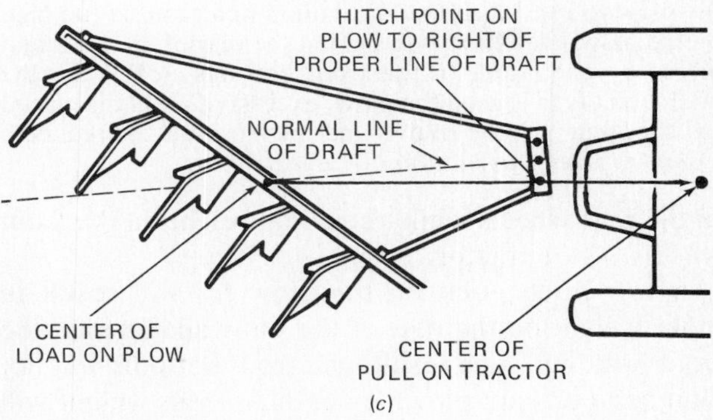

(c)

**Fig. 18-58.** (*a*) The correct horizontal hitch setting on a plow will make it run straight ahead and take the proper width cut. (*b*) When the hitch is too far left, the plow undercuts. (*c*) With the hitch at the right, overcutting results. (*Adapted from International Harvester Co.*)

18-58). Continued use of this hitching will result in poor efficiency, narrow furrow configuration (shape), and excessive landside wear. Overcutting (Fig. 18-58*c*) results when the hitch on the plow is too far to the right of the line of draft. The result will be increased draft, uneven depth, and poor furrow appearance.

Correcting the horizontal hitch adjustments on the plow may not solve all problems. When tractor steering or front furrow width is adversely affected, it may also be necessary to adjust the landing of the tractor drawbar and the tractor wheel tread width.

**Checking Vertical Hitch of Moldboard Plow** Proper vertical hitch setting will make the plow run level, maintain uniform plowing depth, scour better, and reduce the amount of draft to a minimum. The two factors in determining proper vertical hitch are (1) plowing depth and (2) height of the tractor pull point.

In Fig. 18-59*a*, the plow runs level and maintains a uniform depth. If the hitch point on the plow is too high (*b*), the effect is to lighten the rear of the plow and increase penetration of the front bottoms. This may result in failure of the rear bottoms to maintain constant depth, overcutting, excessive share wear, and damage or excessive wear to the front wheel of pull-type plows. Also more weight will be transferred

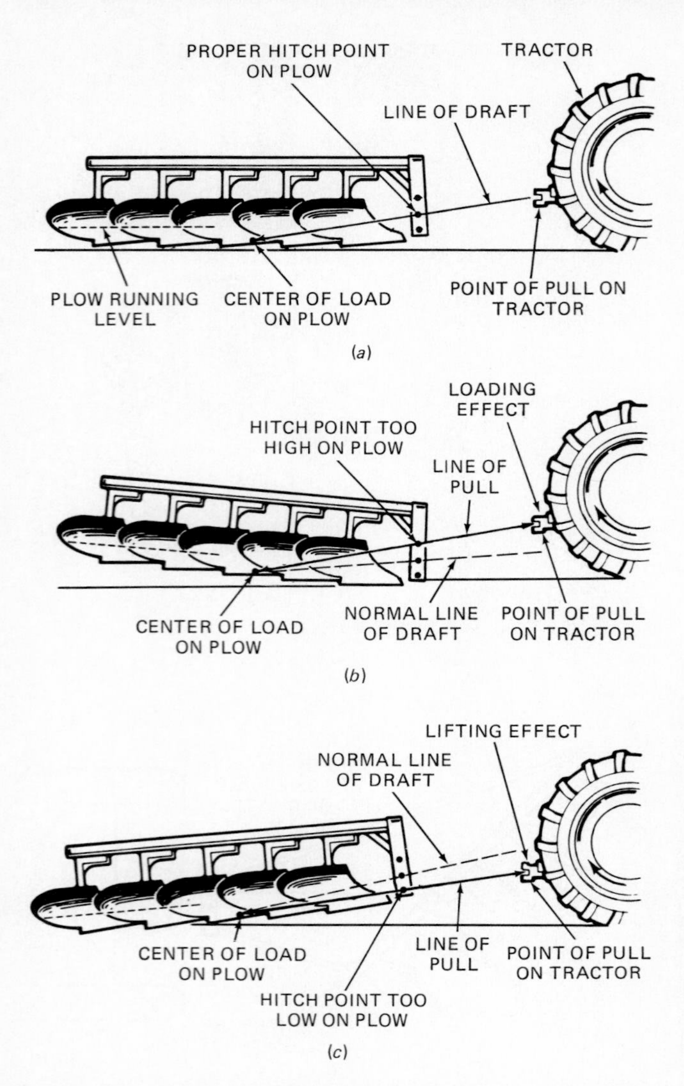


**Fig. 18-59.** (*a*) The correct vertical hitch setting will cause the plow to run level. (*b*) When the hitch point is too high on the plow, the effect is to lighten the rear of the plow and increase penetration of the front bottoms. (*c*) When the hitch point is to low on the plow, excessive weight is placed on the rear of the plow and penetration is reduced. (*Adapted from International Harvester Co.*)

to the rear wheels while reducing weight on the front wheels of the tractor.

A low hitch point on the plow (*c*) will result in more weight on the rear of the plow and less on the front bottoms. As a result, and front bottoms will not maintain constant plowing depth and less weight will be transferred to the rear wheels and more will go to the front wheels of the tractor.

Field setting of a mounted plow requires adjustment of the center link for attaining proper penetration and of the right linkage for levelness of the bottoms (Fig. 18-60*a*). The horizontal line of draft is made through the adjustment of the cross-shaft rotation (Fig. 18-60*b*).

**Checking Rear Furrow Wheel or Rolling Landside Setting** Final setting of the rear furrow wheel should be made in the field. Normally, the inside edge of the tire should clear the furrow wall by 3/4 in [19 mm] (Fig. 18-61). If adjustment is required, check the operator's manual for the correct procedure.

Final setting of the rolling landslide can be made during field operation. Check the distance from the bottom of the furrow to the bottom of the landside. There should be approximately 1/4 in [6 mm] of space. When the vertical position of the rolling landside is set properly, the impression left by the heel in the bottom of the furrow should be barely visible (Fig. 18-62). If adjustment is necessary, consult the operator's manual for details.

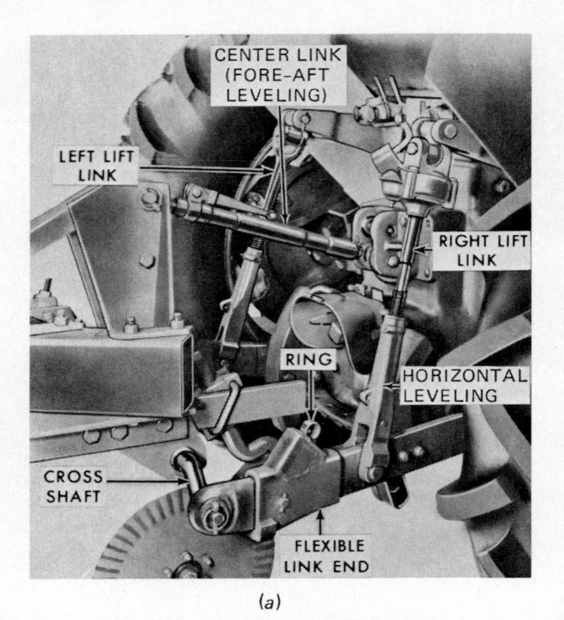


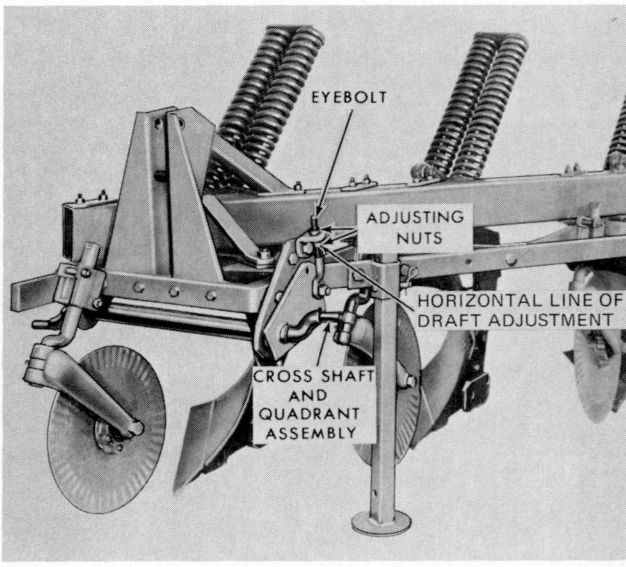


**Fig. 18-60.** (*a*) Adjustments for leveling a mounted moldboard plow for fore, aft, and horizontal leveling. (*b*) Adjustment for correcting horizontal line of draft under field conditions. (*Deere & Co.*)

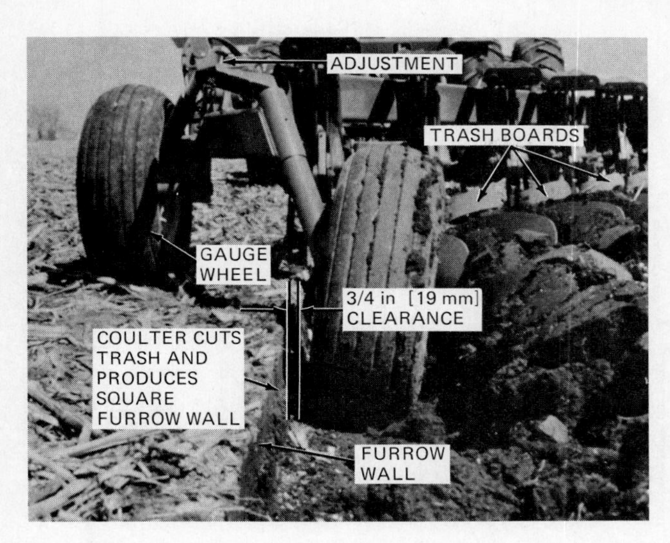


**Fig. 18-61.** The proper plow hitching and adjustment of coulters and rear furrow wheel, or rolling landside, result in clean furrows and good trash coverage.

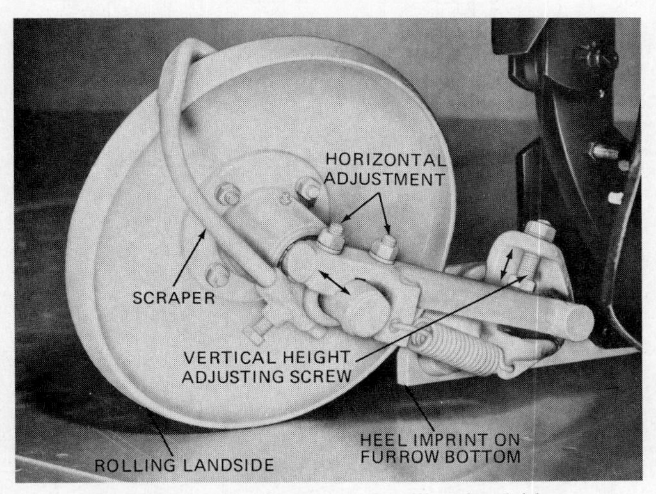


**Fig. 18-62.** The correct setting of rolling landside or rear furrow wheel. (*Deere & Co.*)

**Setting the Offset Plow Disk and Disk Harrow** The offset plow disk and harrow have considerable side draft. This action is created by the offset of the two disk gangs. Thus, it is necessary to offset the hitch from the centerline of the frame.

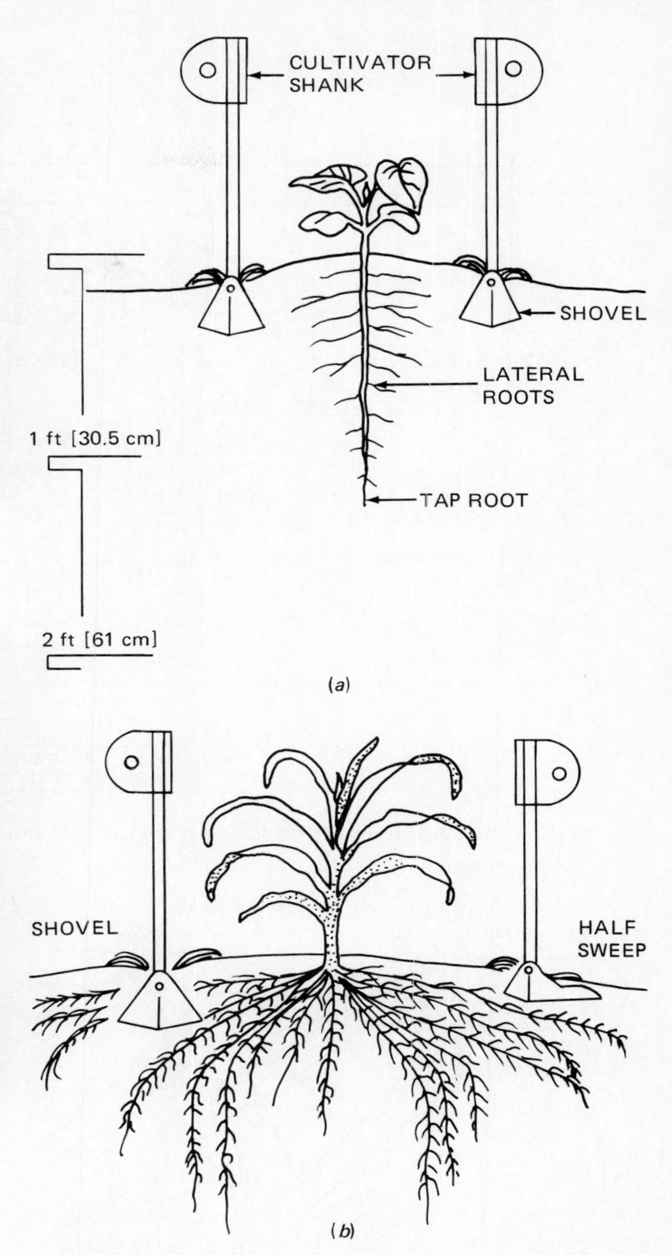


**Fig. 18-64.** Proper setup and adjustment of row-crop tillage equipment are determined by the type and growth characteristics of the root system. (*a*) Cotton plant. (*b*) Sorghum or corn plant.

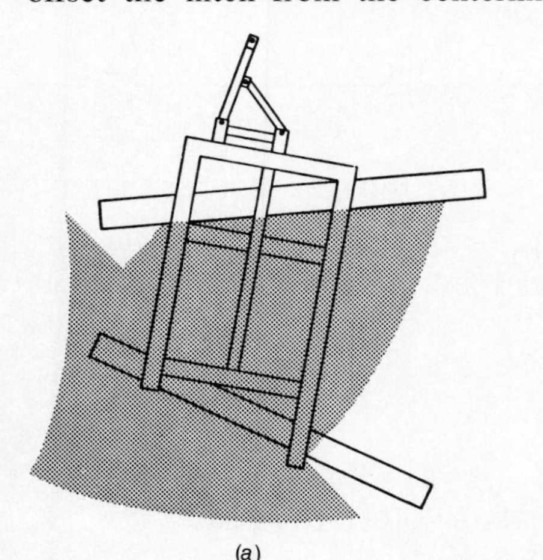


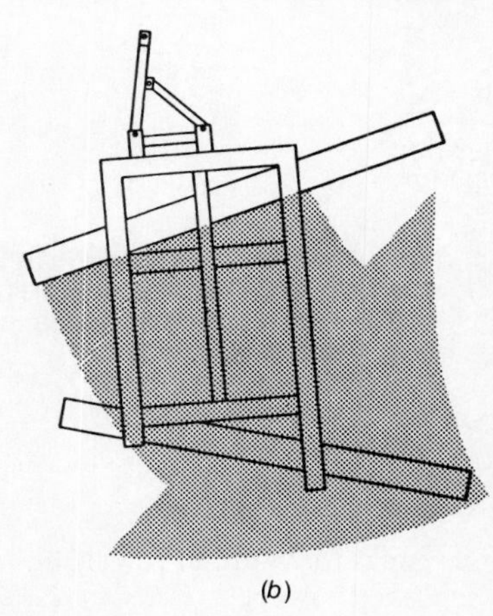


**Fig. 18-63.** Side draft in offset plow disk or disk harrow. In (*a*) the offset disk runs to the left and in (*b*) to the right. (See the text for proper adjustment and refer to the operator's manual for details.)

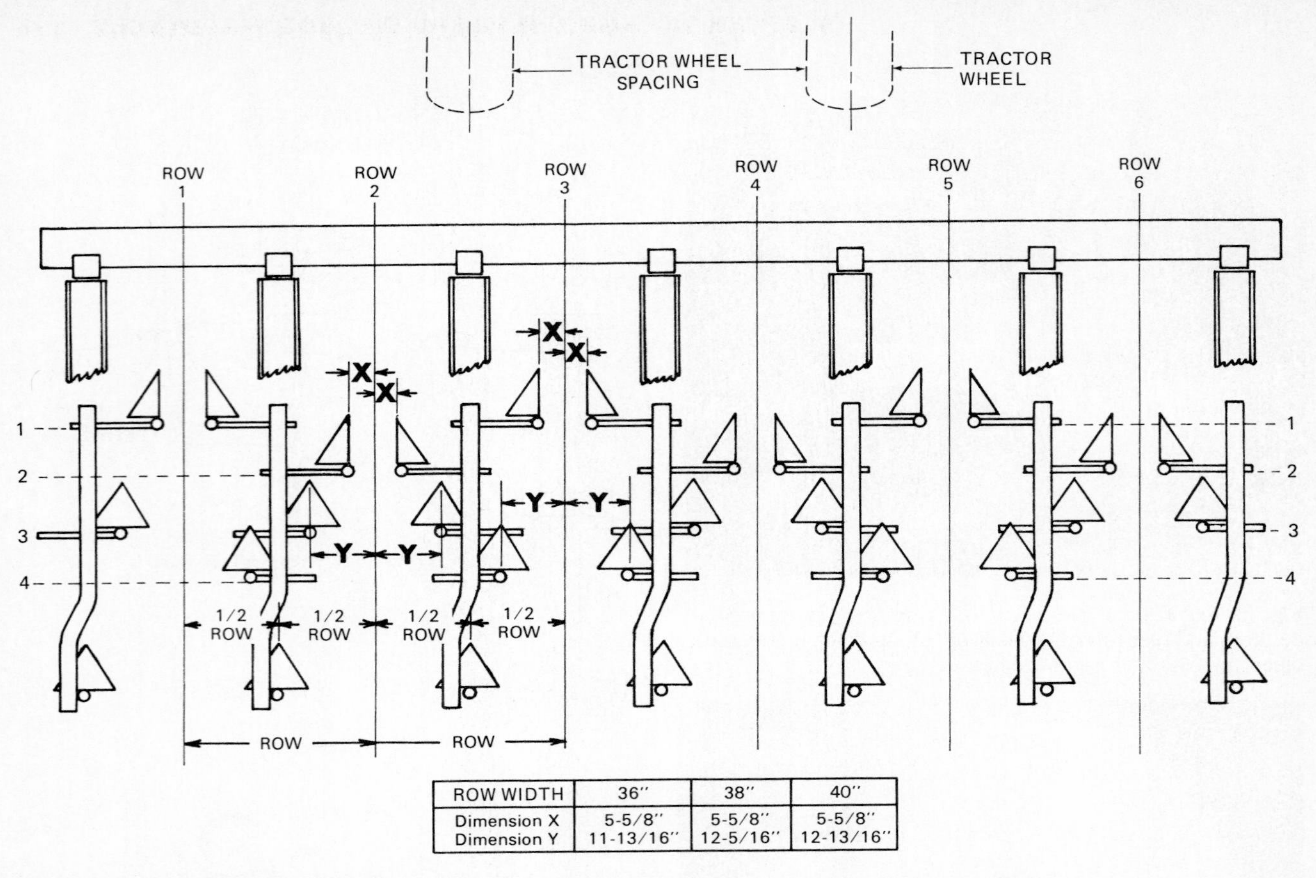


| ROW WIDTH | 36″ | 38″ | 40″ |
|---|---|---|---|
| Dimension X | 5-5/8″ | 5-5/8″ | 5-5/8″ |
| Dimension Y | 11-13/16″ | 12-5/16″ | 12-13/16″ |

**Fig. 18-65.** Spacing of tractor wheels and row-crop cultivator shanks. (*Massey-Ferguson, Inc.*)

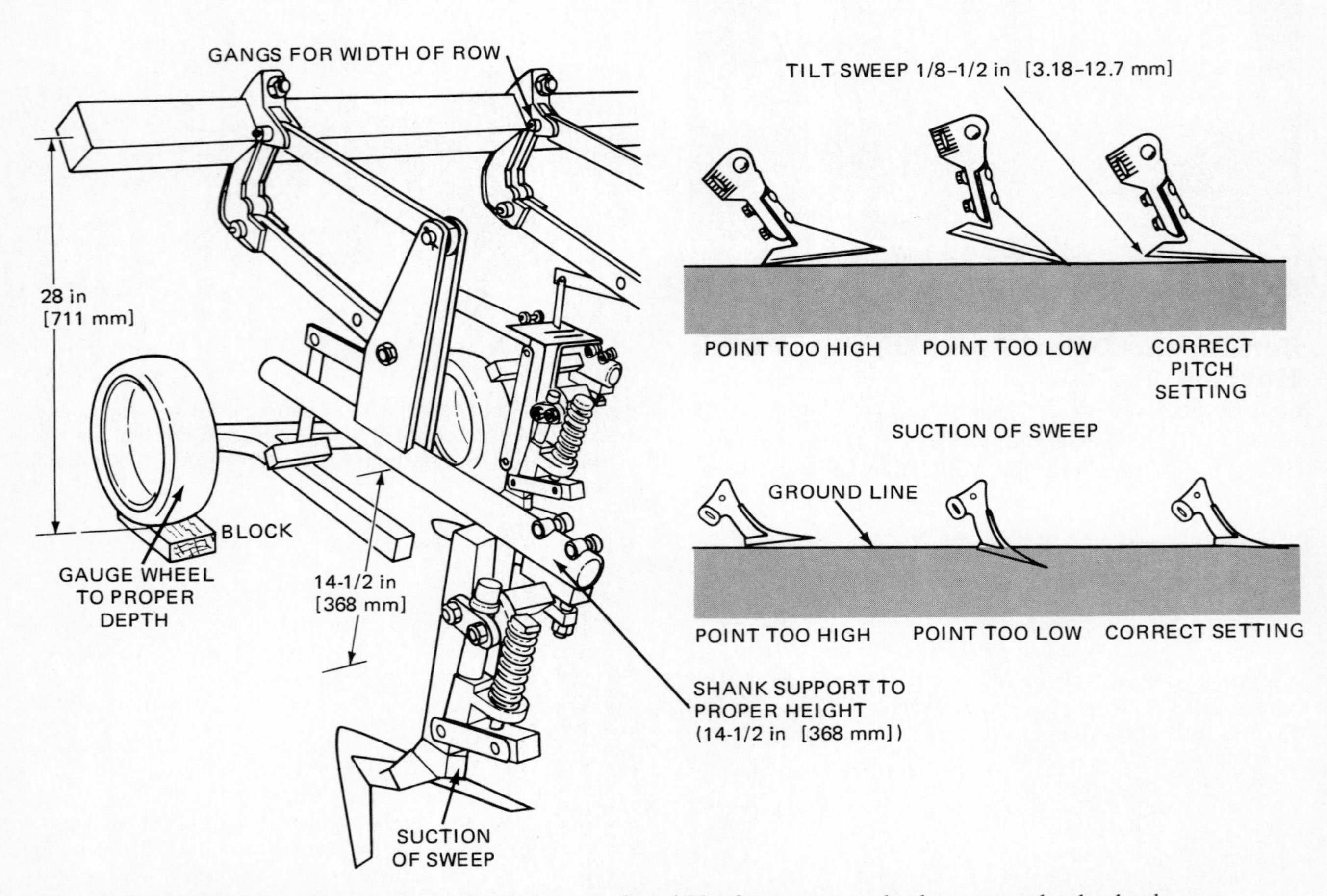


**Fig. 18-66.** Adjustments of a row-crop cultivator: gangs for width of row; gauge wheel to proper depth; shank support to proper height ($14^1/_2$ in [368 mm]); suction of sweep. (*Massey-Ferguson, Inc.*)

Field adjustment of the offset disk must be made in order to compensate for the side draft and have the disk pull in line with the tractor. Adjustments should be made one at a time so their effects can be evaluated.

As illustrated in Fig. 18-63, if the disk pulls to the left, it is necessary to do one or all of the following: (1) raise the front gang and lower the rear gang with the leveling screws, (2) decrease the angle of the front gang, (3) increase the angle of the rear gang, and (4) move the hitch on the disk plow to the right.

When the offset disk has a tendency to pull to the right, it is necessary to (1) increase the working depth of the front gang and raise the rear gang with the leveling screw, (2) increase the angle of the front gang, (3) decrease the rear angle, or (4) move the hitch on the disk to the left.

**Setting the Row-Crop Cultivator** Setup and adjustment of a row-crop cultivator are determined by the maturity and growth characteristics of the plant's root system. As illustrated in Fig. 18-64*a*, plants having a taproot system are less likely to be damaged by the type of shovel or depth of cultivation. The shallow and widespread fibrous root system of crops such as sorghums and corn (Fig. 18-64*b*) are easily pruned (cut) by shovels operated too deeply or too near the plant. When inside cultivator shanks are properly adjusted, half sweeps should penetrate 1 to 2 in [25 to 51 mm] to reduce root damage.

Before final field adjustment, a new row-crop cultivator is assembled and adjusted by the machinery setup worker. The cultivator is attached to the tool bar and tractor in the front- or rear-mounted position with the tractor wheels adjusted to twice the row spacing (Fig. 18-65). For example, if the row spacings are 36 in [914 mm], the wheels are set to 72 in [1.82 m]. The tractor wheels are then raised 2 inches [51 mm] by driving up on blocks. The cultivator gangs and tool bar are supported by blocks to the specified level identified in the assembly manual (Fig. 18-66). The cultivator sweeps are then adjusted to obtain the correct suction (downward pull).

## SUMMARY

Nearly 60 percent of the power used in agricultural production is used to till the soil. As a specialist in agricultural power and machinery, you have a responsibility to help the agricultural industry in utilizing power for tillage operations efficiently. Your role as a service technician to assemble and place equipment into proper adjustment will provide the basic component for making efficient use of tractor power.

Tillage machinery provides the basis for establishing a system of producing food and fiber. Primary tillage tools are designed to work the soil in the root zone of the plant providing an avenue for air, moisture, and nutrients to be obtained by the roots and microorganisms of the soil. Secondary tillage equipment works the soil to form the desirable environment for seed germination. These machines firm the soil, destroy weeds, and nurture the growing plant through cultivation.

Manufacturers rely upon your knowledge of the principles of hitching and the skill required to make adjustments in the field where the machines are being used. Your ability to apply the principles of operation and analyze performance problems is an invaluable service to the agricultural industry and one of which you can be proud.

## THINKING IT THROUGH

1. What is the primary purpose of tilling the soil?
2. What are six functions of primary tillage tools?
3. How does secondary tillage differ from primary tillage?
4. Name the five parts of the moldboard plow bottom and the function of each.
5. What is the difference between a chisel plow and a plow disk?
6. What is the purpose of a two-way plow?
7. What is the purpose of a one-way disk plow?
8. Why are chisel plows used for mulch tillage?
9. When is a lister or bedder used for primary tillage?
10. Identify three secondary tillage machines that can remove small weeds in row crops.
11. List four practices that are necessary in preparing to assemble a new machine.
12. Why is it important to be able to identify the grade of bolt when assembling machinery?
13. How can the capacity of a torque wrench be increased?
14. Why is an accumulator type of hydraulic system well adapted to automatic-reset safety trip plow bottoms?
15. Why is correct tire inflation pressure important for the transport wheels of tillage tools?
16. What is the difference between the center of load and the center of pull in hitching tillage tools?
17. How should the line of draft pass through the hitch point?
18. What is the difference in operating a tractor in-furrow and on-land?
19. Why are the wheels of a tractor elevated by blocks when making shop adjustments of tillage machines?
20. Why is a carpenter's level a useful tool in making shop adjustments of tillage machines?
21. Where are the adjustments made to correct horizontal line of draft for a semi-integral plow?
22. How should a rolling coulter be adjusted?

# CHAPTER 19 PLANTING AND SEEDING EQUIPMENT

## CHAPTER GOALS

In this chapter your goals are to:

- Identify the type and characteristics of planting and seeding equipment
- Perform the adjustment and calibration of planting and seeding equipment

Have you ever driven along a road and noticed the even height and spacing of fields of growing wheat and thought, "That's a good stand"? Perhaps you have also observed the opposite condition—skips in the row or uneven height of growth. In either case, the success of plant development often depends on the seeding practices that were used. To be a specialist in power and machinery, you need to understand the causes of loss in crop production due to mechanical failures and how they may be controlled through proper setup, adjustment, and maintenance practices.

## TYPES AND CHARACTERISTICS OF PLANTERS AND DRILLS

The purpose of seeding machinery is to place seeds uniformly and at the correct rate in rows (excluding broadcast planting equipment). These machines are designed to perform five important functions: (1) to develop a seedbed by opening a furrow, (2) to meter seed at a controlled rate to obtain the correct population of plants for maximum yield, (3) to place the seed at the proper depth and spacing, (4) to cover the seed with soil to the proper depth, and (5) to firm the soil around the planted seed for excellent soil and moisture contact. The two classifications of machines which perform these functions are *row-crop planters* and *grain drills*.

### Row-Crop Planters

The word *planter* identifies a machine which can be adjusted to place the seeds of a great variety of crops in the soil with a high degree of accuracy for rate, depth, and spacing. Row-crop planters are used for those crops that can be cultivated to control weeds, and can be harvested most effectively by machines when in rows. Spacing of the rows and plants in the row must be precise to allow clean cultivation and the desired plant population.

**Planter Parts** Row-crop planters are sized according to the number of rows and the distance between rows. The most common sizes are those that have four to eight planter units spaced at intervals equal to the row widths desired. Common row spacing is 30 to 38 in [760 to 965 mm], although a spacing of 18 to 42 in [457 to 1067 mm] is also used. The planter unit is a self-contained planting and metering device and involves the following basic parts: (1) furrow openers, (2) seed metering and placement system, (3) seed depth control, (4) seed covering and packing device, and (5) row marker.

A modern row-crop planting unit is illustrated in Fig. 19-1*a* to *e*. The depth of seed placement is controlled by double-gauge wheels which run on each side of double-disk furrow openers to level and firm the soil (*a*). A firming point produces a well-defined V-groove (*b*) for catching the seed from the metering and placement system (*c*). Twin disks positioned directly behind the openers move moist soil over the seed (*d*). Then a firming wheel removes air and voids from around the seed (*e*) while the center V-rib forms a fracture slit in the soil profile to aid emergence of the seedlings (*e*).

**Furrow openers** Furrow openers place a definite groove in the soil, into which the seed is placed at the correct depth (Fig. 19-2). The most common types of openers are the runner, double disk, lister, and V-wing.

Runner openers are used when planting in soil that has been previously tilled. The runner is slightly V-shaped and cuts a groove (Fig. 19-3). The V shape compresses the soil below and on each side of the groove. Seed falls to the bottom of the groove, thus making contact with firm soil.

V-disk (double disk) openers consist of two relatively flat disks that are angled to run close together at the bottom and approximately 2 in [51 mm] apart at the top (Fig. 19-4). As they are moved forward, a notch is left in the soil, similar to that made by the runner. V-disk openers cut through trash and give more uniform planting depth in minimum tillage plantings (Fig. 19-5).

A double-disk planting unit with one disk removed is presented in Fig. 19-6. Note the relationship of the furrow firming point and the position of the seed drop tube to the disk opener.

The planting unit illustrated in Fig. 19-7 is designed with gauge wheels to control planting depth. Furrow openers may be attached to the unit. These openers are used for several purposes, the most common of which are (1) to move away the trash, clods, and dried topsoil from the area where the planter

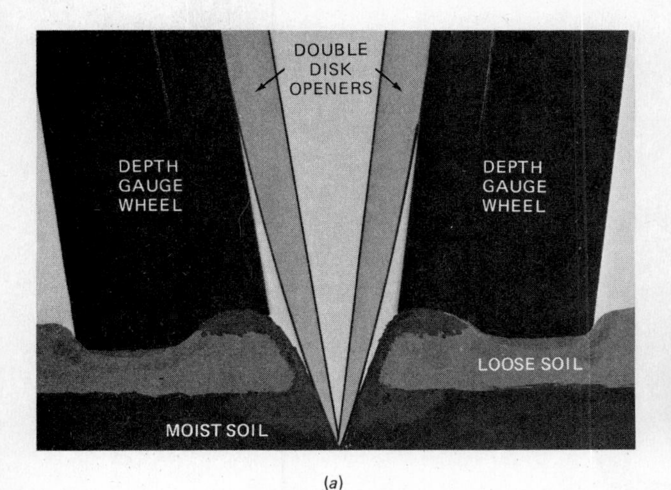

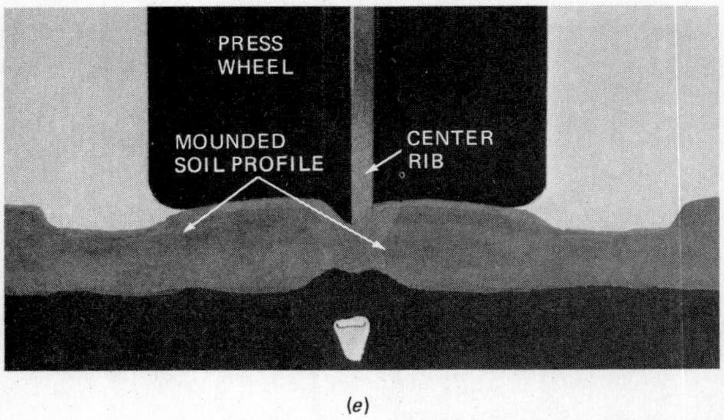

(a)

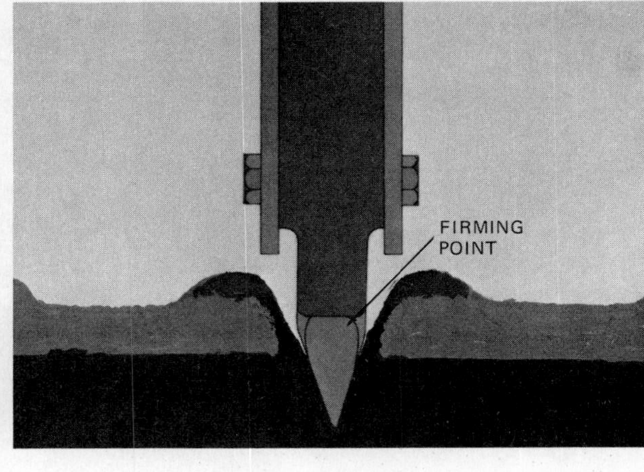


(b)

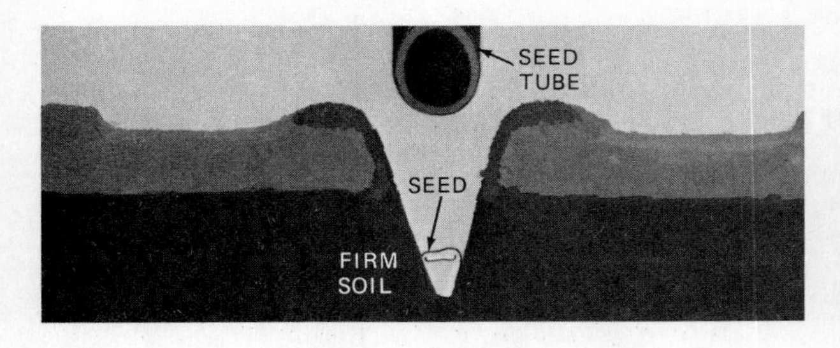


(c)

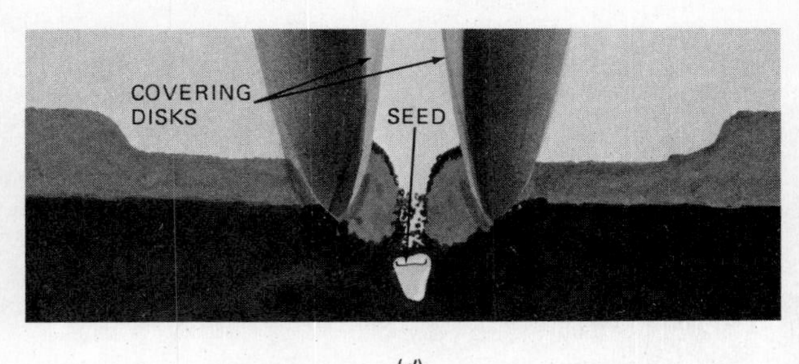


(d)

(e)

**Fig. 19-1.** Sequence of operation with a row-crop unit. (*International Harvester Co.*)

opener will operate, and (2) to place the planter opener deep enough to reach soil that is moist enough to ensure seed germination. These devices include double disk and V-wing (Fig. 19-7). Double-disk furrow openers are used with both runner and V-disk seed openers.

**Seed metering and placement** The purpose of the seed metering and placement system of a planter is to deliver the seed to the seed tube at the selected rate to obtain the desired plant population at the proper interval in the seedbed. The metering devices are classified into two types: seed plate and plateless.

**Seed plate planters** The seed plate type of metering consists of plastic or iron seed plate which rotates at the bottom of the seed hopper (Fig. 19-8). The plate has openings called *seed cells* at or near the outer edge. The size and shape of the cells in the plate are selected for the seed to be planted. As the plate rotates, a seed (or seeds) fills each cell. At the proper moment the seed is ejected from the cell into the dis-

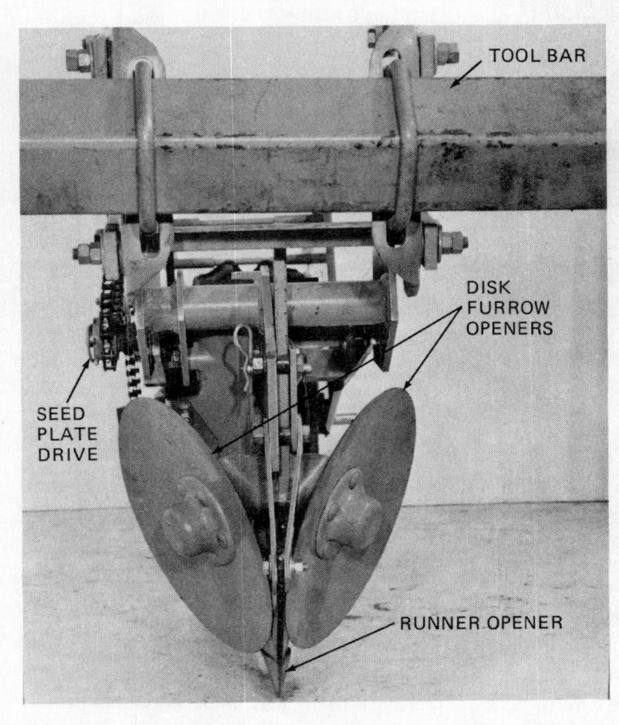


**Fig. 19-2.** The double-disk furrow opener is a common planter type used for row crops. (*Allis Chalmers Co.*)

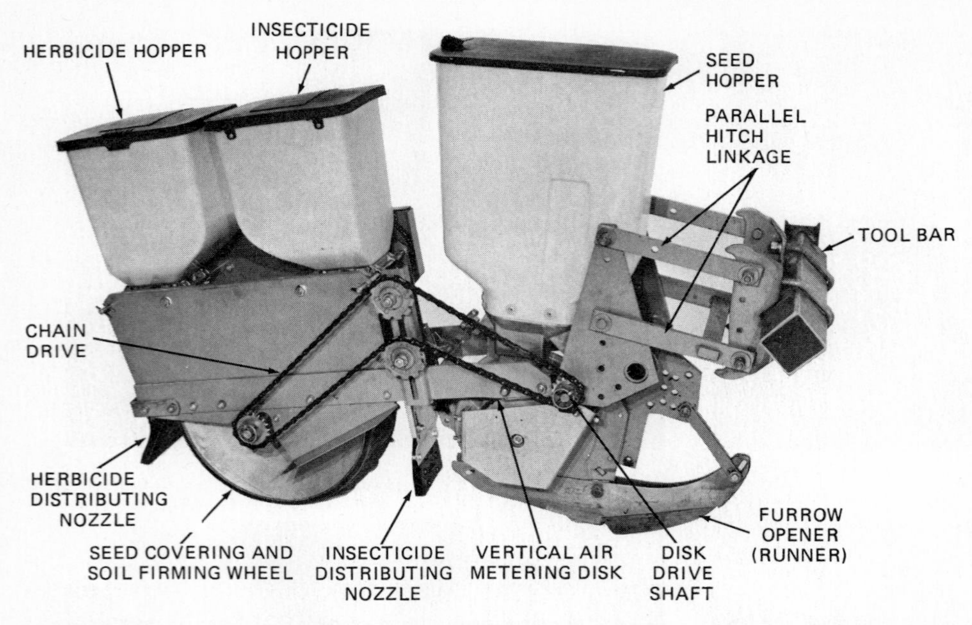


**Fig. 19-3.** The single-unit vertical disk air planter may use a runner furrow opener. The seed-covering and firming wheel drives the metering disk with a chain. (*Allis Chalmers Co.*)

**Fig. 19-4.** This V-disk (double-disk) furrow opener planter has a seed plate metering system. (*Amoco Oil Co.*)

charge opening by the knockout mechanism (Fig. 19-9). A cutoff pawl riding on the top of the seed plate prevents other seeds from entering the cell (Fig. 19-10).

Seed plates vary widely in the number of cells; the range is from as few as 8 cells to as many as 82. For corn the 16- and 24-cell plates are most common. In addition, the shape determines the number and orientation of seeds held in the cell. The basic types of corn plates are (1) edge drop, (2) flat drop, and (3) hill drop (Fig. 19-11).

Edge-drop plates are used to plant corn seeds which have been screen-selected for size and shape. With this plate, one seed stands on edge in each cell. Flat-drop corn plates are best suited to kernels which are uniform in thickness. The plate requires a filler ring to compensate for thickness. Filler rings must be used with all seed plates less than 1/4-in [6.4-mm] in thickness. Filler rings are not interchangeable between seed plates since there is a specific ring for each seed plate.

Seed plates are matched to the seed to be planted. Seed corn, for example, will be found in many different widths, thicknesses, and lengths and may vary

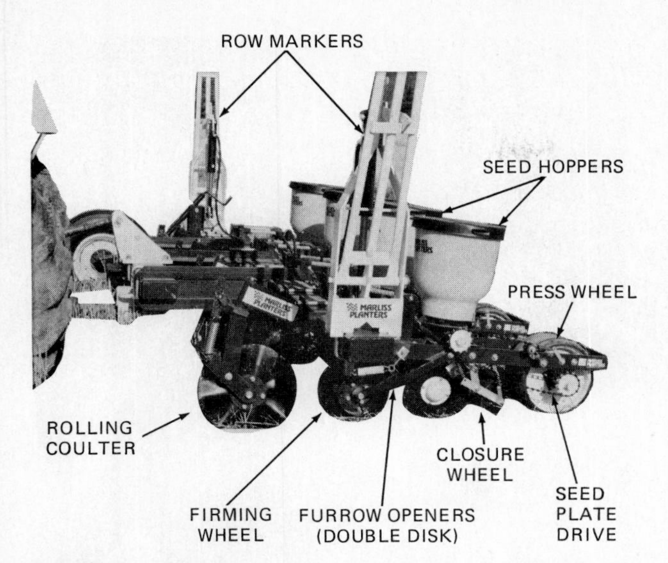


**Fig. 19-5.** This row-crop planter is designed for minimum tillage planting. Specially designed rolling coulters cut through trash or crop residue in untilled land. (*Marliss Planters.*)

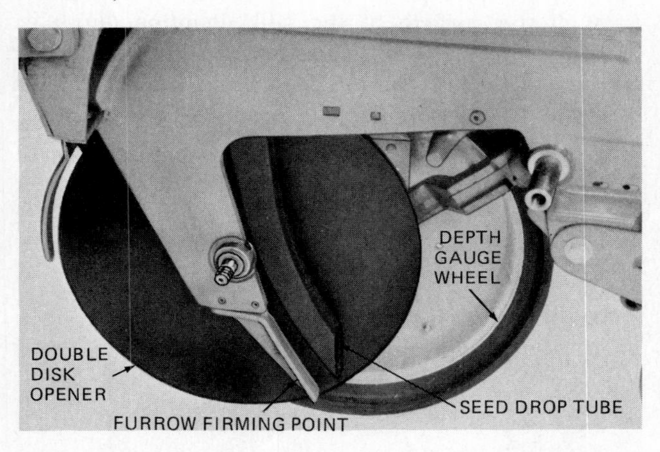


**Fig. 19-6.** A double-disk opener with one disk removed shows the location of the furrow firming point and seed drop tube. (*Deere & Co.*)

from year to year, depending on weather conditions and other factors. The seed plates are specifically designed by each manufacturer of planters. In some cases a plate cross-reference by part number is listed for other planters. Seed producers usually recommend the proper plate by type and number (Table 19-1). However, variation can occur in the drop rate of seed and cell fill as the number of cells varies. For example a 24-cell plate will fill better than a 16-cell plate because it turns 50 percent slower to achieve the same planting rate. As a result, the 24-cell plate will usually produce a greater drop rate.

Vegetable planters require specially designed seed plates and ejection systems. The planter in Fig. 19-10 uses a toothed roller-type knockout pawl for positive discharge of seeds from the plate.

**Plateless planters** Plateless planters operate by finger pickup metering, by feed-cup metering, or by one of several air metering methods. The finger pickup system is specifically designed for corn. It will pick up individual kernels of various sizes and shapes, reducing the problem of having the incorrect-size plate for the kernel size.

The finger pickup assembly has 12 spring-loaded fingers that are opened and closed by a cam as they rotate. The assembly rotates in a vertical position. A finger closes upon a kernel of corn and carries it clockwise until it reaches a discharge hole where it is ejected into the placement mechanism. Here the seed is confined to one of the openings created by the paddles on the seed belt (Fig. 19-12). The rotation of the belt spaces the kernels in an exact pattern as they are released into the seed tube.

The feed-cup mechanism (Fig. 19-13) is designed with four interchangeable precision feed cups and corresponding seed guide for planting a variety of types and sizes of seeds. Selection of the proper cup is based upon the type of seed to be planted. As the cup rotates, seeds are metered through the seed guide into the seed tube at the furrow opener.

The principle of air metering utilizes air under pressure or the presence of vacuum to pick up and meter seeds. One manufacturer utilizes a perforated drum which is pressurized with air from a fan driven by the tractor. The seed drum has rows of holes around the surface for each row being planted (Fig. 19-14). The drum is driven by ground wheels which maintain the accuracy of seed placement according to speed of travel. The seed is fed to the rotating drum by gravity, and the drum is charged with air by a blower. Air escapes through the seed pockets and manifold. The seed attempts to plug the holes and stop escaping air (like water through a floor drain or a minnow bucket). Drum rotation carries the seed past the brush, and the excess is removed. The seed passes under the cutoff wheels, which stops air flow, causing seeds to drop into the manifold. A constant flow of air escaping through the manifold carries the seed to the furrow. The constant air flow also eliminates trash problems.

Another manufacturer uses a vertical rotating disk mounted on each planter unit which is pressurized by an electric motor-driven fan. The air is blown into the seed supply cup where it escapes through the openings in the seed disk. As the disk rotates, the seeds shut off the escaping air. Air pressure holds the seed in the hole unit the disk rotates to the air cutoff point. At this point the seed drops freely into the seed distribution tube and into the furrow opener (Figs. 19-15 and 19-16).

**Seed monitors** Electronic seed monitors (Fig. 19-17) are placed in the seed tube of the planter unit to count the seed drop. As the seed interrupts a light beam of a photoelectric cell (electric eye), an electric signal counts the seed drop. Some monitors only inform the

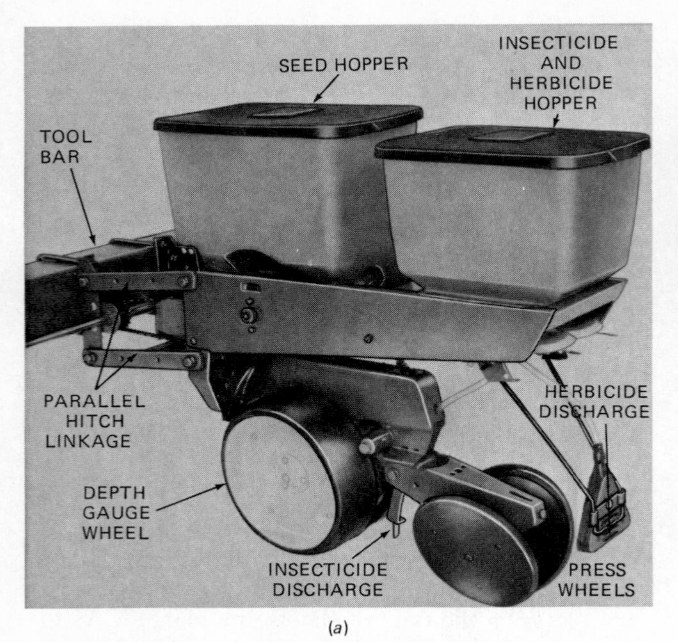


(*a*)

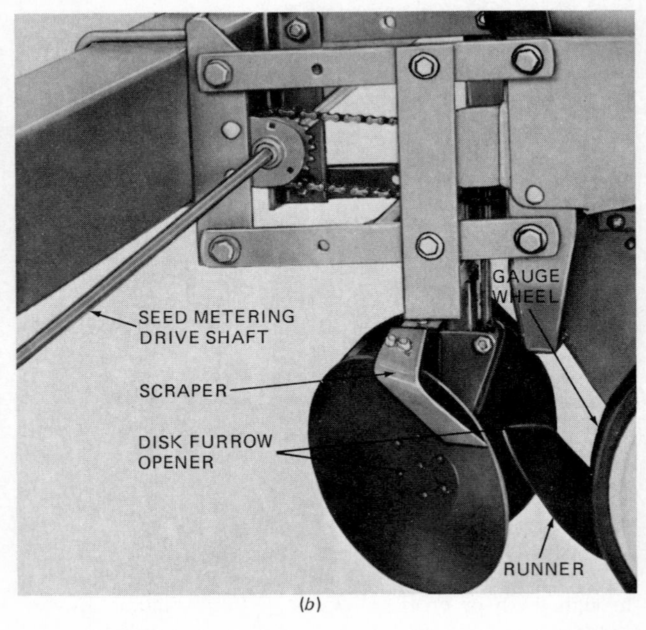


(*b*)

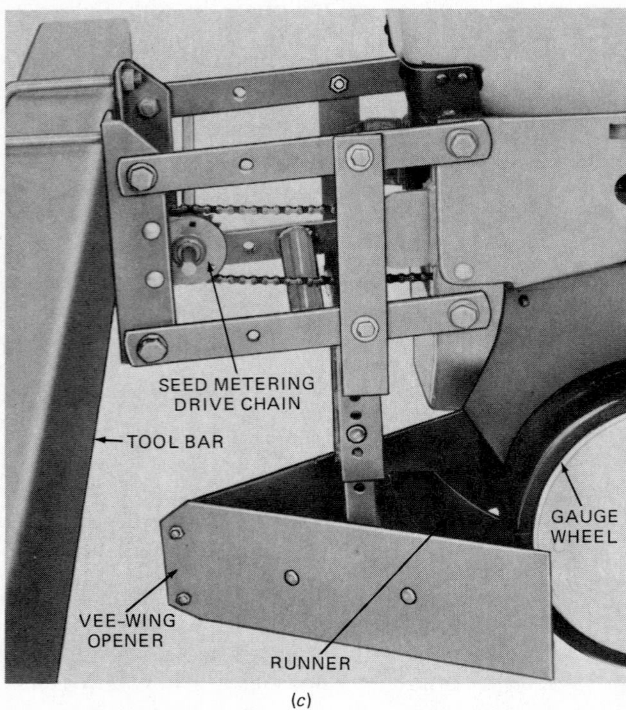


(*c*)

**Fig. 19-7.** Most planter units are (*a*) equipped with gauge wheels for controlling planting depth. Disk furrow openers (*b*) are used to remove trash and soil from in front of the runner. A V-wing opener (*c*) smooths soil ahead of the runner for more uniform depth control. (*Deere & Co.*)

operator that each row is planting. In the event of failure in a unit, an alarm will sound. Other monitors are capable of computing the planting population from the distance covered and the number of seeds dropped.

**Planting Depth** The deeper a seed is placed in the soil, the longer it takes the seedling to break through the surface of the soil. Planting depth is closely associated with the size of seed and the amount of food stored in the endosperm. Table 19-2 presents the relationship between the type of seed and the seeding depth.

**Fig. 19-8.** The plate-type planter uses a seed plate as the metering device. The type of seed plate that is inserted into the bottom of the seed hopper depends on the size and shape of seed to be planted. (*International Harvester Co.*)

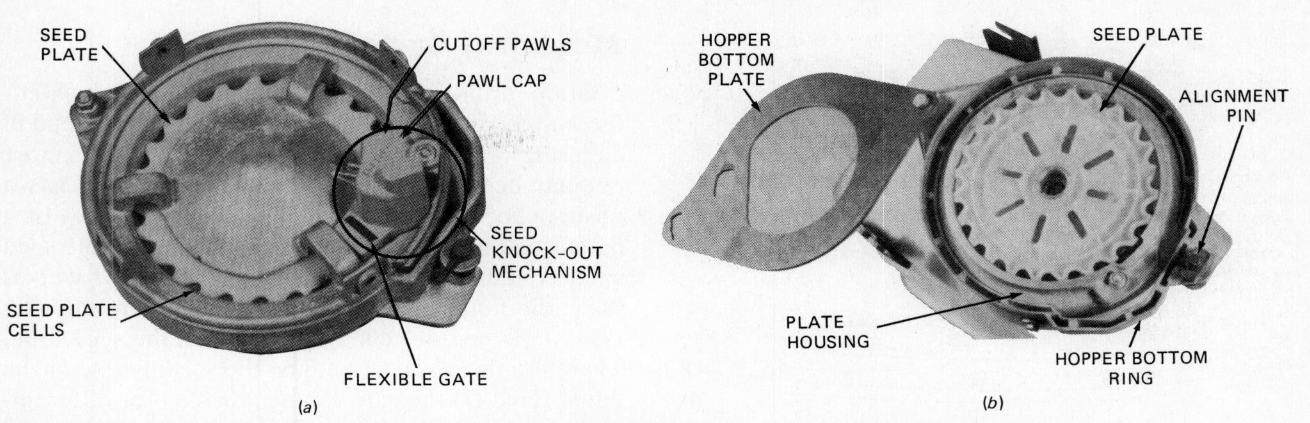


**Fig. 19-9.** A close-up view shows the parts of a seed plate metering system. (*a*) Top view. (*b*) Bottom view. (*International Harvester Co.*)

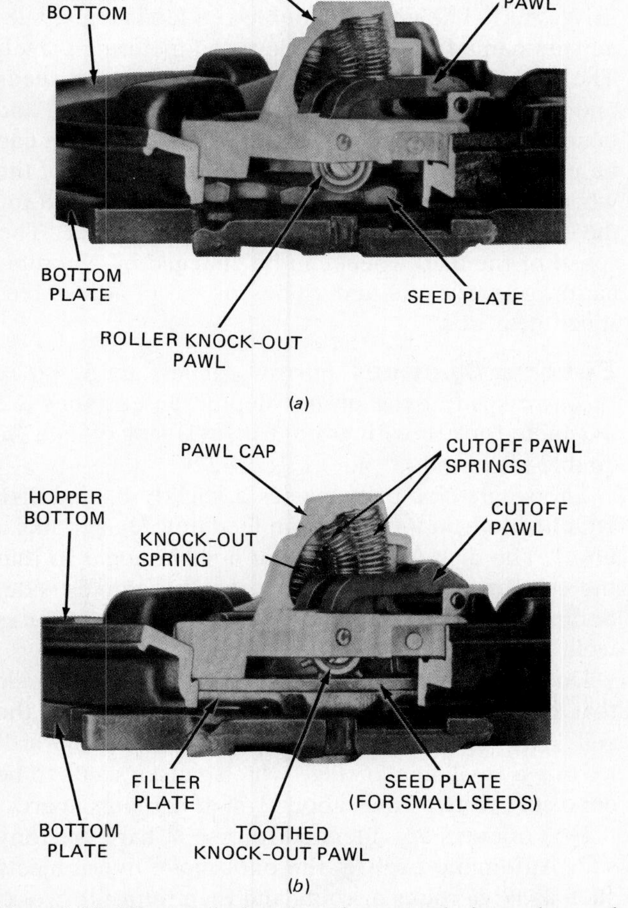


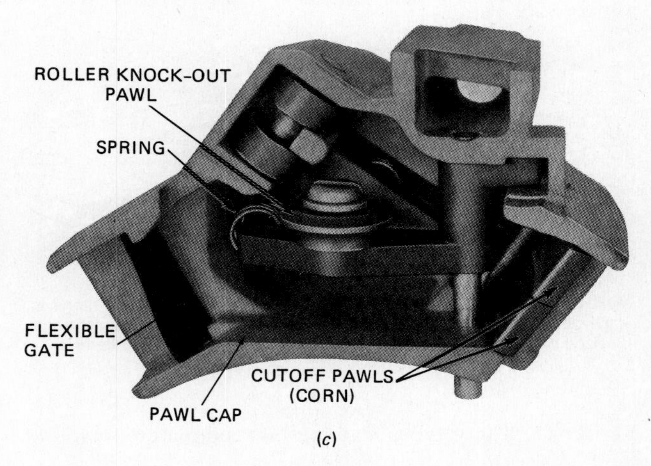


**Fig. 19-10.** The seed plate knockout pawl ejects the seed into the discharge tube. (*a*) Cross section of roller knock-out pawl for corn seed. (*b*) Cross section of toothed roller knock-out pawl for small seeds such as beets and sorghum. (*c*) Bottom view of knock-out pawl and operating parts. (*International Harvester Co.*)

The three primary methods of controlling the depth of the planter's furrow opener are (1) position of the rear press wheel, (2) position of the gauge wheels, and (3) depth bands or shoes. The most common system for controlling depth is position of the press wheel. This may be accomplished by adjusting a threaded screw which raises or lowers the furrow opener or a notched lever controlling a cam.

## GRAIN DRILLS

Drill and broadcast are two seeding methods that provide solid coverage of plants over the field. These planting methods are classified as *solid planting* as compared with the row-crop planting method described previously. High-population, high-yielding grain and hay crops such as wheat, oats, barley, rice, alfalfa, clover, and many other grass and legume crops are solid-planted.

Planting by drill describes a mechanical process that results in an accurate distribution of seeds at a uniform depth in rows that are too narrow for cultivation.

### Classification and Sizing

Grain drills are classified by type, according to the method of transport, the number of grain boxes, and the method of metering and placing the seed. Grain drills are constructed to be either carried by the three-point hitch of the tractor or pulled from the

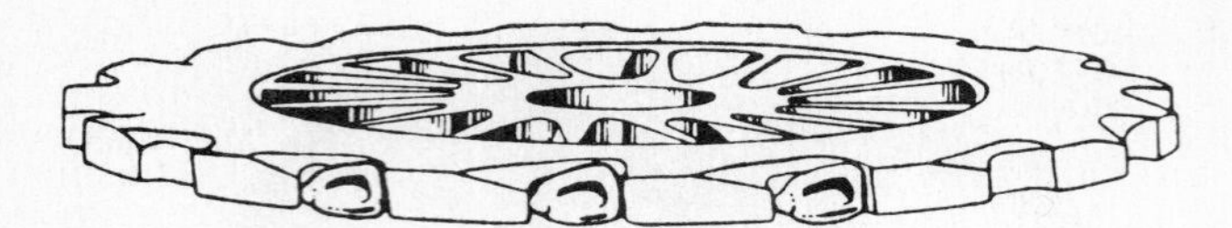

An **EDGE-DROP** plate is used to plant properly sized seed, both hybrid and open-pollenated. With this type plate, one seed stands on edge in each cell. Edge-drop plates take maximum advantage of accurate seed selections

(*a*)

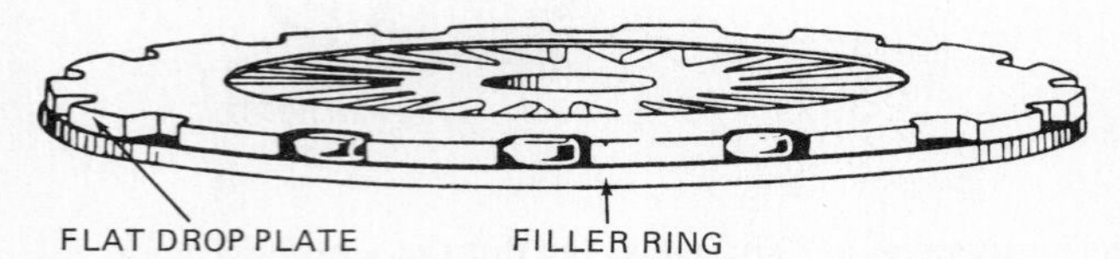


A **FLAT-DROP** plate is best suited for flat kernels. A filler ring is required with this plate.

(*b*)

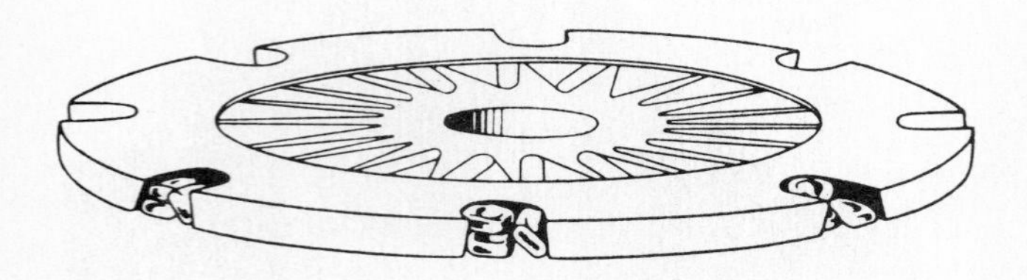

A **HILL-DROP** plate has larger cells, so that several seeds are gathered in each cell to plant an entire hill. It is well adapted for hill-drop planting of unsized corn, irregularly shaped seed, and hybrid butt and tip kernels.

(*c*)

**Fig. 19-11.** Three types and uses of seed plates. (*International Harvester Co.*)

drawbar. The trail-behind (trailing) type is the most common because it is more adapted to multiple hitching and accurate metering. Trailing-type drills may be supported by *end wheels* at each end of the main frame. The wheels are used for transport, for powering the feed and metering system, and for controlling the depth of planting. Trailing *press-wheel* drills are supported by press wheels at the rear of each placement unit and by the hitch bar when attached to the tractor, or by casters in the front (Fig. 19-18).

Grain drills are also designated as plain or fertilizer drills, depending on whether the supply box is constructed to distribute fertilizer as well as seed. Occasionally grass-seeding units are added to grain drills as attachments (Fig. 19-19).

The size of a grain drill is identified by the number of furrow openers, by the distance between openers, or by the number of feet [meters] width. End-wheel drills may have 12 to 24 openers with a spacing of 6 to 10 in [152 to 254 mm] between each. Press-wheel drills have from 6 to 24 openers which may be spaced 6 to 16 in [152 to 406 mm] apart. A specification of 21 × 6 identifies 21 openers spaced 6 in [152 mm] apart. A machine of this size would seed a width of 10 feet (ft) 6 in [3.2 meters (m)].

## Metering System

Grain drills may be purchased with the *fluted-forced* feed or the *internal double-run* feed method of metering. The fluted-forced feed (Fig. 19-20) is most popular because it is simple in construction, has a positive feed, and is easy to clean. It consists of a fluted feed roll, a feed cutoff, and an adjustable seed gate. A square shaft powered by the ground wheels turns the fluted feed roll, forcing the grain out over the gate, where it falls by gravity into the seed tube. There are three ways to adjust the seeding rate of the fluted feed: (1) the gate can be adjusted for different-sized seeds, (2) the feed roll can be moved along the drive shaft to expose more or less of the fluted roll, and (3) the speed of the fluted feed roll can be varied by interchanging appropriate gears at each end of the drive shaft. The internal double-run feed (Fig. 19-28) gets its name from the double-faced metering wheel. The wheel has a small side for handling small seeds and a large side for planting wheat, oats, peas, and beans. A feed cover plate located in the grain box can be positioned to expose the small or large side of the wheel. A feed gate (Fig. 19-21) can be positioned for the size and quantity of grain to be planted. The speed of the feed wheel can be changed by a combination of sprockets and chains or by an adjustable-speed gear box.

**Furrow Openers** Furrow openers are designed to place seed at the proper depth. Three types are available for use with grain drills: (1) single disk, (2) double-disk, and (3) hoe (Fig. 19-22).

The single-disk opener uses a slightly dished disk attached to a boot (a cast-iron feed tube shaped like a boot). The disk is set to run at a slight angle so that the seed drops from the boot on the convex side. Scrapers keep the disk clean. This opener penetrates well, cuts trash, and does not easily clog.

Double-disk openers are two disks at a slight angle that form a cutting edge where they penetrate the soil. The disks rotate as they are moved forward, leaving a small center ridge which causes seeds to be deposited in two rows about 1 inch [25 mm] apart.

Hoe openers are intended for use in hard or stony soil. Automatic trips let the boot move over objects such as large rocks or roots and then return to working position.

One of the newer types of seeding equipment is designed for minimum tillage planting. The rotary coulters (Fig. 19-23) cut a groove into sod or untilled soil where seed is deposited. Packer wheels compress the soil around the seed. Agricultural research work is being conducted to study minimum and no-tillage planting methods.

Advantages of no-till farming include (1) less wind and water erosion, (2) less water runoff, and (3) reduced labor and fuel costs.

**TABLE 19-1. Seed Plate Sizes and Cross-Reference for Corn**

Flat Drop Plates

| 16-Cell Plate | 24-Cell Plate | CELL DIMENSIONS Length, in | Width, in | Thickness, in | Uses Other Than Corn |
|---|---|---|---|---|---|
| 3002-A** | | 40/64 | 14/64 | 9/64 | |
| 1853-AA | 480 191 R1 | 24/64 | 14/64 | 12/64 | |
| 1854-AA | 493 518 R1 (Plastic) | 24/64 | 18/64 | 12/64 | Beets and beans |
| 3365-A | | 26/64 | 20/64 | 12/64 | |
| 3214-A | | 28/64 | 18/64 | 12/64 | Beans |
| 3329-A | | 28/64 | 22/64 | 12/64 | |
| 3350-AA | | 28/64 | 28/64 | 12/64 | |
| 3266-A | | 30/64 | 24/64 | 12/64 | Peas and beans |
| 1855-A | | 32/64 | 24/64 | 12/64 | |
| 3330-AA | | 32/64 | 31/64 | 12/64 | |
| | 1930-A | 34/64 | 16/64 | 12/64 | Whole beets |
| 3321-A | | 34/64 | 18/64 | 12/64 | |
| | 1931-A | 38/64 | 20/64 | 12/64 | Beans |

Seed Plate Cross-Reference*

| Corn Size | MF 20 CELL EDGE DROP Plate No. | MF Cell Size† | JD 16 CELL EDGE DROP Plate No. | JD Cell Size† | IH CELL EDGE DROP Plate No. | IH Cell Size† |
|---|---|---|---|---|---|---|
| Large Flat | 440 792 M2 | 44-12-20 | H 696 B | 44-12-20 | 1978A | 40-13-20 |
| Large Round | 442 994 M2 | 44-23-20 | H 2820 B | 42-23-20 | 462356R1 46235R1 | 42-22-20 |

* Partial listing of commonly used seed plates.
† Edge-drop cell size in length, depth, and thickness in 64th inch.
*Note:* The above chart should be used as a *guide* only, with final selection based on actual planting checks made with the planter operated at the desired planting speed.
*Source:* International Harvester Co.

# ADJUSTING AND CALIBRATING PLANTERS AND DRILLS

As a professional in power and machinery you may be employed as a representative for an agricultural machinery dealership. The responsibilities will require firsthand knowledge of predelivery adjustments of planting and seeding machinery. You will be assisting the owner/operator to achieve the best possible performance from the equipment.

One of the important tools for making adjustments of machinery is the operator's manual. This "tool" contains specific information which the manufacturer considers essential to obtain best results. Only general concepts are presented in this chapter regarding adjustment and calibration of planting and seeding equipment. Specific details must be obtained from the operator's manual for the particular make and model.

## Characteristics of Planters and Drills

Planters and drills are two of the most highly refined machines used in agricultural crop production. As a result, small differences in adjustments or increases in clearance of moving parts due to wear can affect the discharge, placement, and coverage of seeds. For example, a high degree of accuracy in metering corn kernels is required to obtain a discharge of approximately 19,000 kernels per acre. At a planting speed of 4 miles per hour (mi/h) [6.4 kilometers per hour (km/h)], 9 kernels must drop every second; at 6 mi/h [9.7 km/h] 13 kernels must be metered.

According to agricultural research workers at the University of Illinois, worn plates and hopper bottoms can reduce seeding accuracy by 20 percent. The data in Table 19-3 identify the changes in metering accuracy by replacement of worn metering parts.

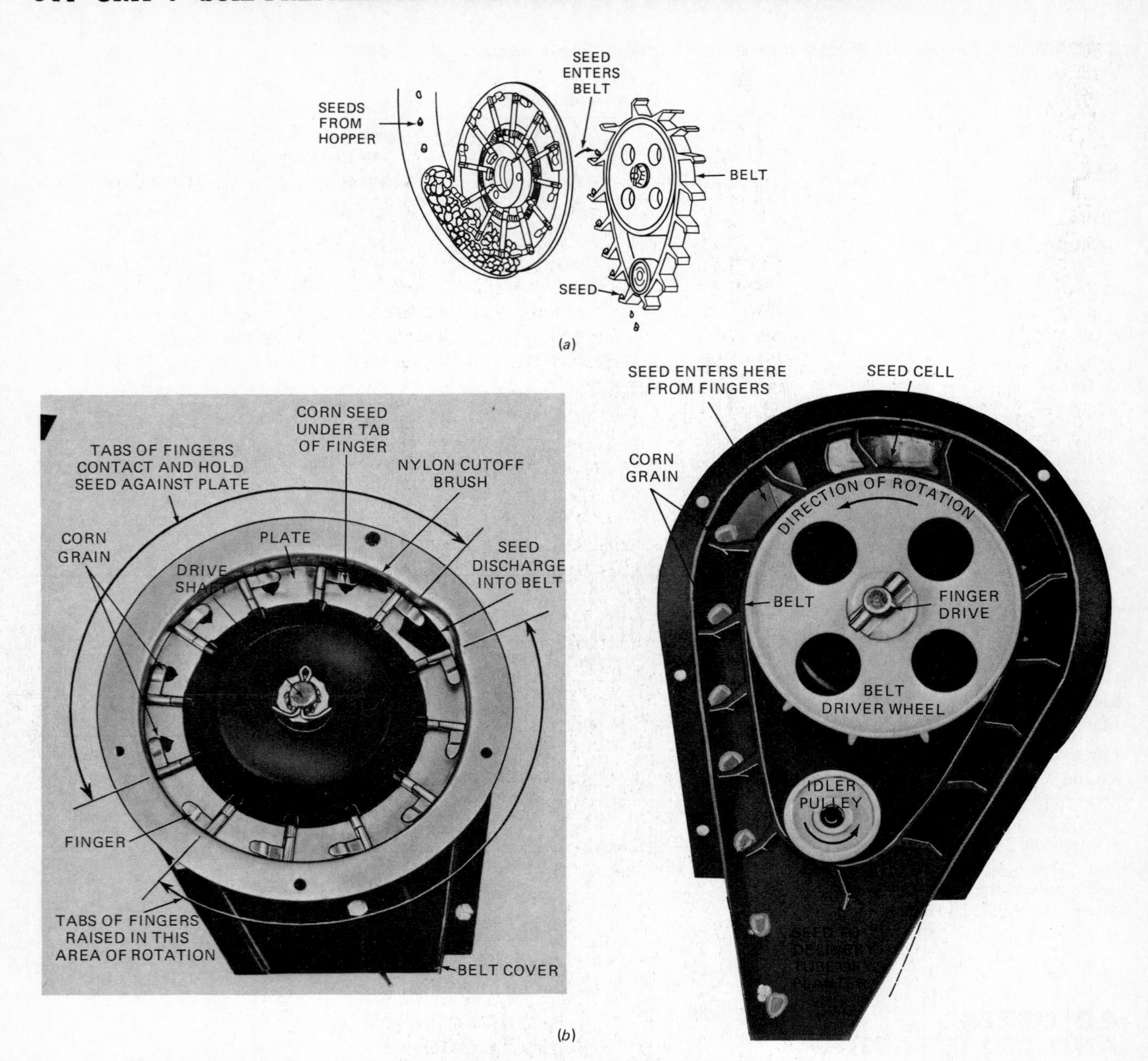


**Fig. 19-12.** One plateless planter uses a finger pickup design. Fingers pick up a single kernel of corn and eject it into the belt seed cell for placement into the seed discharge tube. (*Deere & Co.*)

## Seed Metering Rate

The rate of planting or drilling seeds depends on the particular crop and the climatic and cultural practices for the region of the country where it is grown. The rate may be expressed in pounds per acre [kilograms per hectare] or in desired plant population per acre [per hectare]. Table 19-4 gives the planting and seeding practices for seven common crops with typical rates for plant population and pounds per acre. It is common to identify the planting rate for corn by 18,000 to 30,000 plants per acre [44,000 to 74,000 per hectare (ha)], as shown in Table 19-4. This is also referred to as the emergence rate. Loss factors due to germination, insects, disease, and planter inaccuracies require higher seeding rates. For example, seeding rates are usually figured on the basis of 60 to 90 percent of the seeds planted actually developing into plants.

A second concept regarding the rate of seeding involves the relationship between the type of metering system and the standard practices of selling seed. For example, hybrid seed corn is sold in bags on a kernel-count basis. A single-cross hybrid has 80,000 kernels per bag, while double-cross is sold with a kernel count of 60,000 per bag. A bag weighs approximately 50 pounds (lb) [22.7 kilograms (kg)], depending on variety.

Wheat is sold by the bushel and has a standard weight of 60 lb [27.2 kg]. The size of wheat grains de-

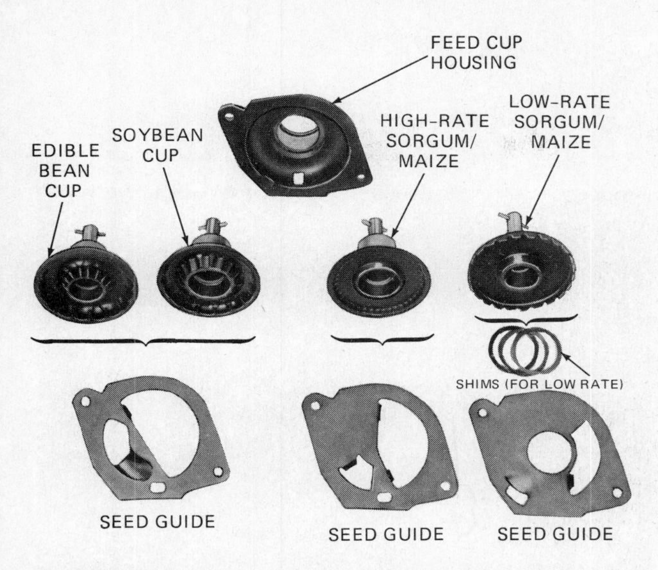


**Fig. 19-13.** The feed-cup method of metering uses a vertical cup rotating in a housing driven by a horizontal shaft. (*Deere & Co.*)

pends on its variety. A 60-lb [27.2-kg] bushel of large-grain wheat will require a larger bag than 60 lb of small-grain wheat. The object when planting grain and forage crops is to place the *correct* number of kernels at the *correct* interval in uniformly spaced rows to achieve the desired plant population. The planting or seeding machine must meter individual seeds with weight or size of kernels properly compensated for by the adjustment of the mechanism. Wheat, for example, is to be planted by a grain drill adjusted to meter at a rate of 80 to 100 lb/acre [90 to 112 kg/ha] (Table 19-4). However, the grain drill measures *volume* and not *weight* of seed. Thus, turning the drive shaft of the metering feed one revolution meters the same *volume* of large or small wheat seeds. If the feed wheel is set to meter a specified number of pounds of seed, the seed count (number) of large seeds that are metered will weigh less than the amount of small seeds. The result will be fewer pounds of large seeds planted. It is always desirable to prepare the equipment properly and to check the quantity of seeds metered before planting begins.

## Preparing Planting and Seeding Machinery

The processes required to prepare planters and drills for field operation are similar. The operations include assembling or "setting up" the machine according to the manuals provided with the planter or drill. The important operations include (1) leveling, (2) tire inflation, (3) lubrication, and (4) setting and checking of the seeding rate of the metering system.

**Leveling** The frame of the planter and drill must be leveled to ensure proper penetration of the furrow openers and uniform depth of seeding. For example, the tool bar to which a planter unit is integrally attached must be level and at the correct height (Fig. 19-24). A grain drill must be hitched to the tractor drawbar such that the grain box will be level (Fig. 19-25). The hitch can be positioned to accommodate drawbar heights of 15 and 17 in [381 and 432 mm] by reversing the plates of the clevis.

**Tire Inflation** Tire inflation pressure is important for pull-type planters and drills. The tires not only carry the weight of the machine, but also power the metering system. The amount of inflation affects the rolling radius of the wheel. If the radius is different than what the machine was designed for, the seeding rate will be different from that listed in the operator's manual.

The operator's manual provides information about inflation pressure according to size of tire used. Some manuals list specific pressure; other manufacturers identify the height the axle should raise above the floor level to obtain the correct rolling radius of the wheels (Fig. 19-25).

**Lubrication** Lubrication of new or repaired planters and drills includes grease fittings, gear cases, and chain drives in accordance with the recommendations found in the operator's manual. Grease fittings are lubricated using a pressure lubrication gun. Sufficient multipurpose lithium-base grease is applied to show around the bearing surfaces. The gear cases must be filled to the correct level with the type of lubricant specified in the operator's manual (Fig. 19-26).

New roller chains contain sufficient lubricant without further attention. Used chains should be checked for wear; if satisfactory, soak them in light machine oil for 24 hours (h) before installation.

## Setting and Checking of the Metering System

Setting the planter or drill to place the correct quantity of seed in the soil is not a difficult task, providing the desired seeding rate is known. For example, assume you wish to plant corn at a rate suggested in Table 19-4. The plant population that is suggested for corn is 18,000 to 30,000 plants per acre [44,000 to 74,000 per ha]. Assume a population of 21,000 plants per acre [52,000 per ha] is desired, in rows spaced 30 in [762 mm] apart. A 10 to 15 percent overplant of seeds is recommended to compensate for seed that does not germinate or seedlings that do not live. If a 15 percent overplant is desired, the number of seeds required per acre is as follows:

21,000 plants/acre
× 1.15 overplant factor
24,000 seeds/acre (approximate)

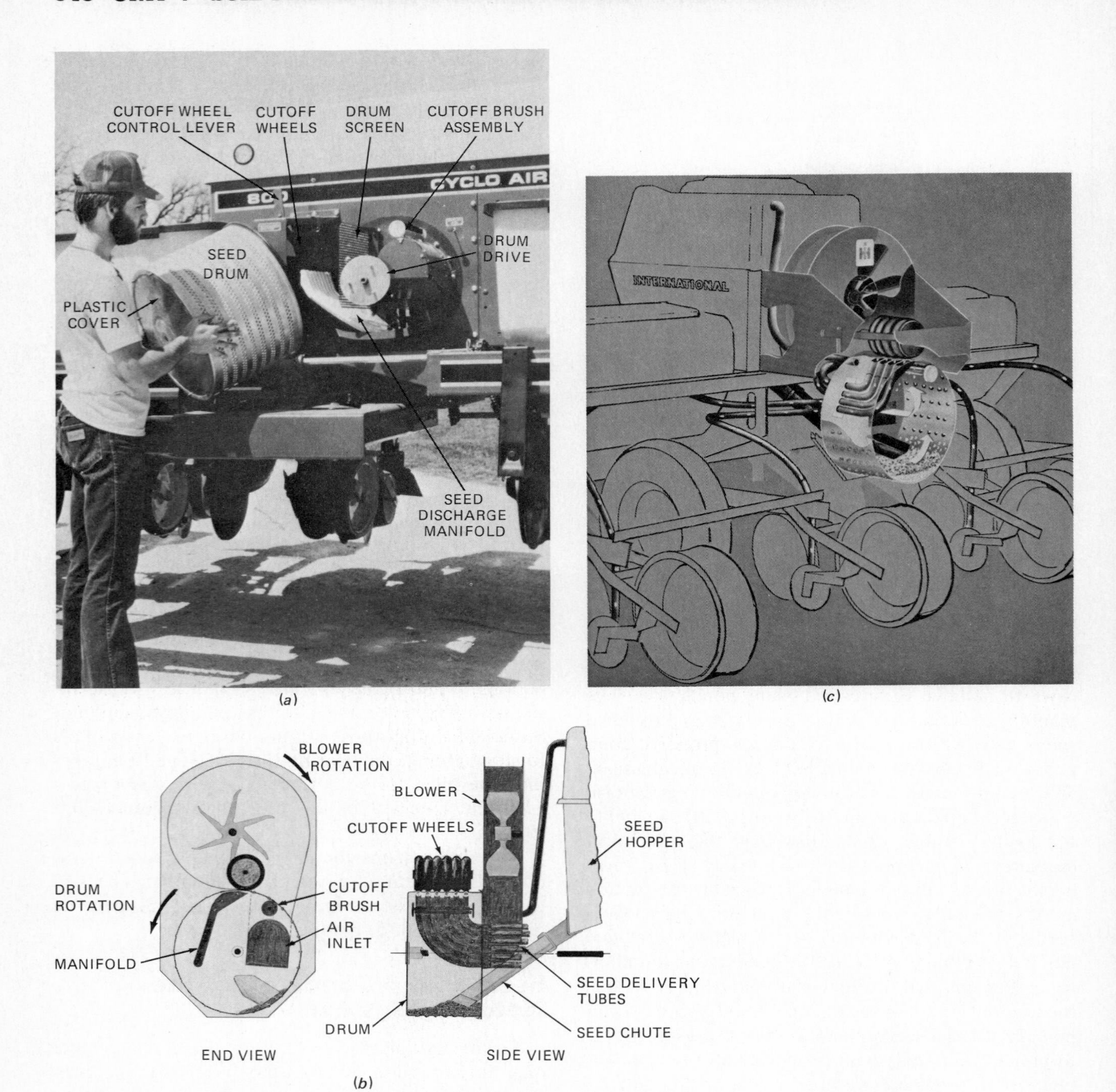


**Fig. 19-14.** Some air planters use a rotating-drum metering system. Seed leaves master seed hopper and enters the ground-driven revolving seed drum through a delivery chute. A PTO-powered blower supplies equalizing air pressure to seed hopper and drum. Seeds are held in pockets of drum by constant air pressure. As seeds ride to top of drum, a brush removes the excess. Rubber cutoff wheels momentarily close drum's holes, releasing seed into discharge manifold for delivery to the rows. (*International Harvester Co.*)

To achieve a seeding rate of 24,000 corn kernels per acre [59,280 per ha] in 30-in [762 mm] rows, the planter must be set to drop kernels at a definite spacing in inches in the row. This interval of spacing is determined by computing the number of inches [mm] of row required to make an acre at 30-in [762 mm] spacing. This can be calculated as follows:

$$\text{Kernel spacing, in} = \frac{43{,}560\ \text{ft}^2/\text{acre}}{\dfrac{30\text{-in rows}}{12\ \text{in/ft}}} \times 12\ \text{in/ft} \div 24{,}000\ \text{kernels/acre}$$

$$= \frac{43{,}560}{2.5} \times 12\ \text{in} \div 24{,}000\ \text{kernels}$$

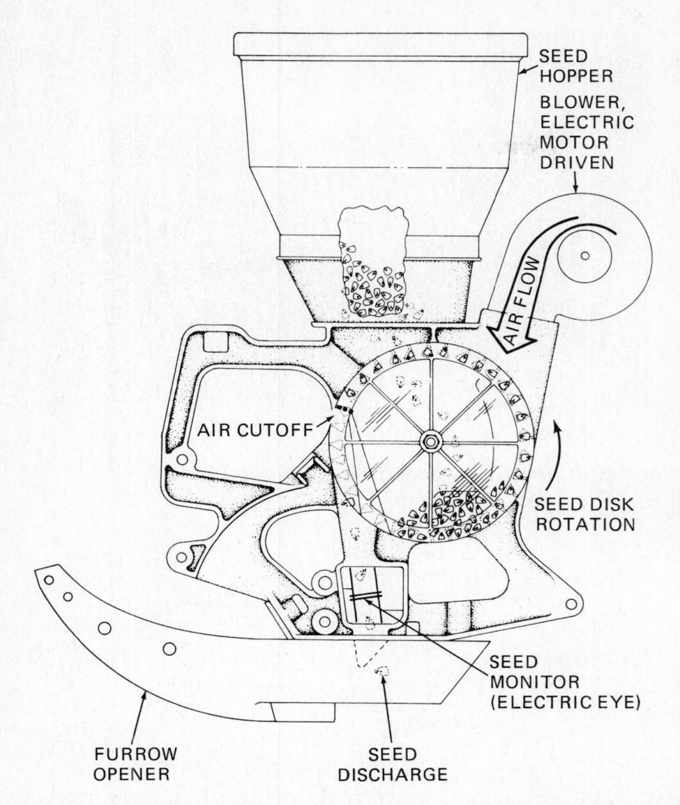


**Fig. 19-15.** Vertical disk-type plateless air planter with seed monitor. (*Ford Tractor Operations.*)

$$= \frac{209{,}088 \text{ in of row/acre}}{24{,}000 \text{ kernels/acre}}$$
$$= 8.71 \text{ in [221 mm] spacing/kernel}$$

The data presented in Table 19-5 confirm the spacing of 8.7 in [221 mm] required to obtain a seeding rate of 24,000 kernels per acre [59,280 per ha] in 30-in [762 mm] rows.

**Selecting Speed Ratio** The next step in setting the planter for planting corn at a spacing of 8.7 in [221 mm] is to select the correct size of drive chain sprockets. The seed plate must be rotated at the correct speed in accordance with the number of cells in the seed plate. For most plate planter units the correct sprockets on driver No. 1, driven No. 1, driver No. 2, and driven No. 2 (Fig. 19-27) can be selected from Appendix B, Table B-1. The table is designed for use with a 16-cell seed plate. You should follow the seed spacing column of 8.69 in [220.73 mm] (the nearest to 8.7), and find that the following sprocket sizes are required:

| POWER SHAFT SPROCKET | COUNTER SHAFT SPROCKETS | | FEED SHAFT SPROCKET |
|---|---|---|---|
| Driver No. 1 | Driven No. 1 | Driver No. 2 | Driven No. 2 |
| 16 | 10 | 14 | 17 |

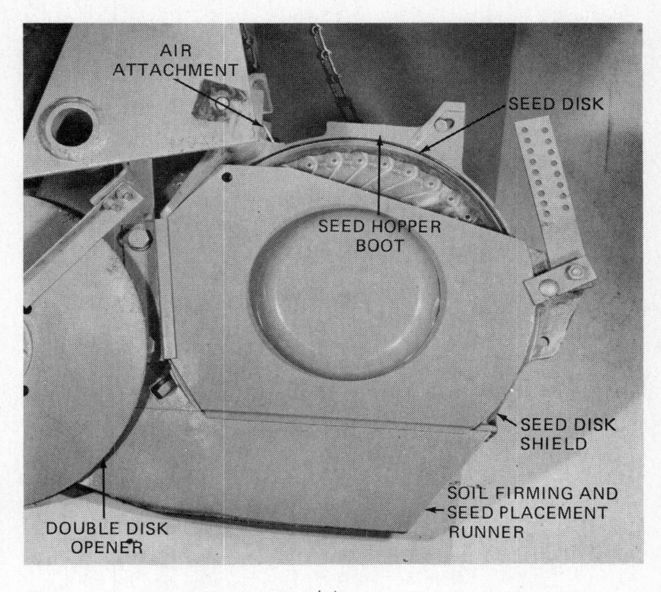


(*a*)

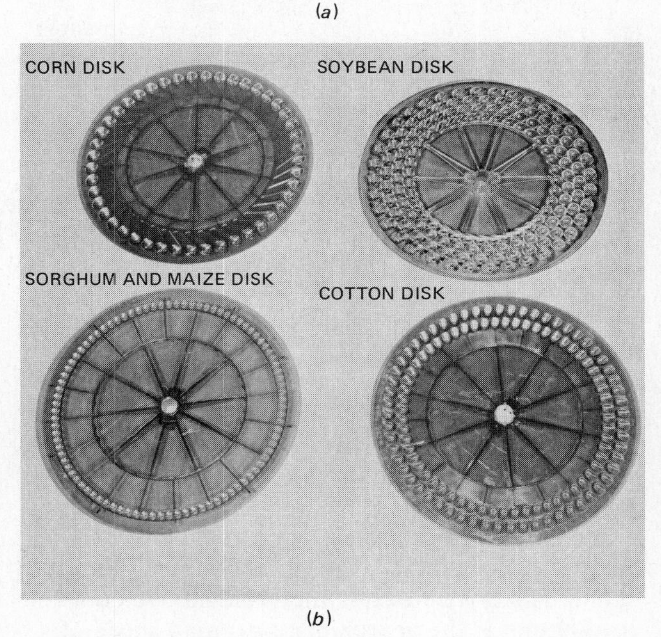


(*b*)

**Fig. 19-16.** Air planter and types of disks for metering various seeds. (*Allis Chalmers Co.*)

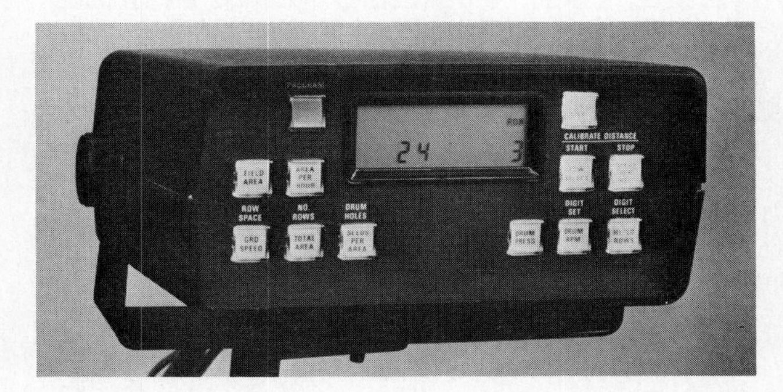


**Fig. 19-17.** An electronic seed monitor provides an operator with as many as 10 different types of information during planting. (*International Harvester Co.*)

**Fig. 19-18.** Two types of trail-behind grain drills. (*a*) Press-wheel drill. (*b*) End-wheel drill. (*International Harvester Co.*)

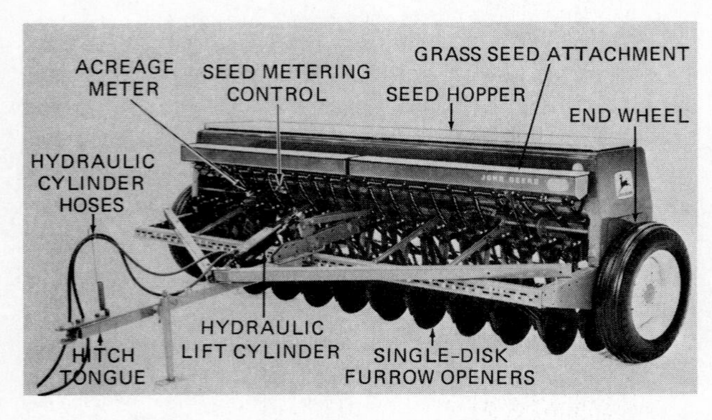


**Fig. 19-19.** An end-wheel grain drill with a grass seed attachment as an option. (*Deere & Co.*)

**Calibrating the Planter** After the proper seed plate for the size and shape of corn kernel has been installed the planter unit should be checked for the actual seed interval and population rate. Both operations can be performed by raising the planter unit so that a container can be placed under the seed spout. The drive wheel is then rotated five complete revolutions at the approximate planting speed of 4.4 mi/h [7.1 km/h] (approximately five turns in 6 seconds). The effective diameter of the 7.60 × 14 in drive wheel is 28.0 in [711 mm]. The effective diameter is less than the actual diameter, because the rubber tire compresses slightly under the downward pressure of the tire in the loose soil. Thus, the effective circumference of the wheel is 88 in (28.0 in × 3.1416 = 88.0 in, or 2235 mm). In five turns of the wheel the planter would have moved forward 440 in [5 turns × 88 in= 440 in, or 1118 centimeters (cm)].

**TABLE 19-2. Typical Depth of Planting Seeds of Selected Agricultural Crops**

| Crop | DEPTH | |
|---|---|---|
| | Inches | Millimeters |
| Alfalfa | ¼–½ | 6–13 |
| Barley | 1½–2 | 38–51 |
| Corn | 2–2½ | 51–64 |
| Cotton | 1–1½ | 25–38 |
| Oats | 1½–2 | 38–51 |
| Sorghum | 1–2 | 25–51 |
| Soybeans | 1–2 | 25–51 |
| Wheat | 1–2 | 25–51 |

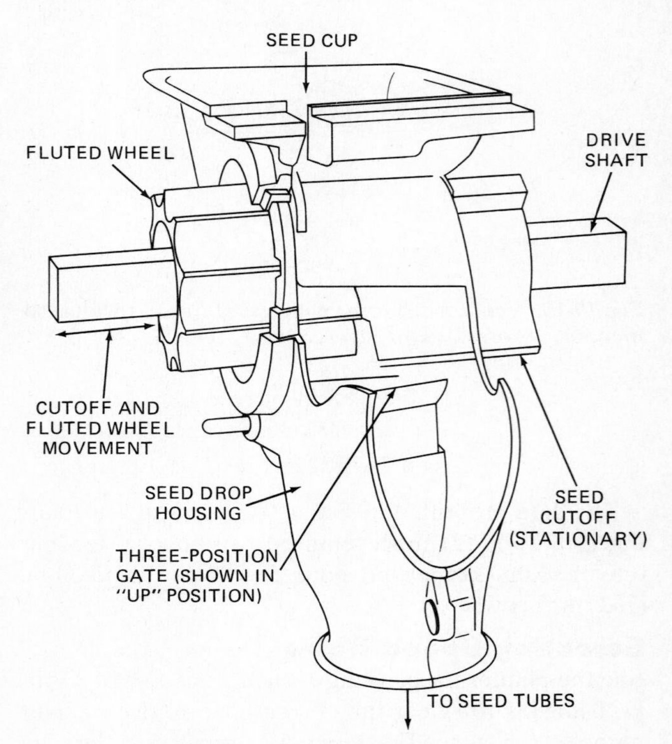


**Fig. 19-20.** The fluted-forced-feed drill metering system is very popular. A fluted wheel feeds grain into the seed drop housing. The metering rate is determined by the length of fluted wheel that is exposed to the seed. The position of the seed gate depends on the seeding rate and size of seeds. (*International Harvester Co.*)

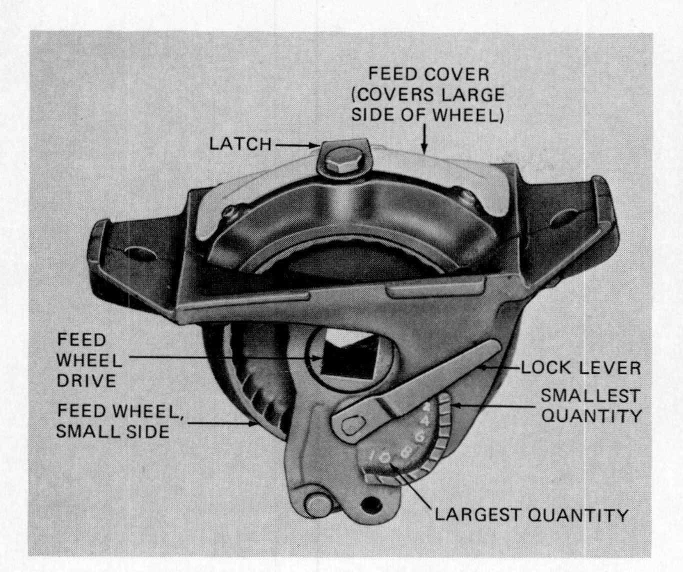


**Fig. 19-21.** The double-run metering system is available for grain drills. Seeding rate is controlled by the speed of the feed wheel, position of the lock lever, and side of the wheel in use. (*Deere & Co.*)

**Fig. 19-23.** A three-point minimum tillage drill is popular for pasture reseedings. (*Marliss Planters.*)

To determine the drilling distance, divide 440 in by the number of seeds caught. Assume the number was 50. This would give a drilling distance of 8.8 in [224 mm] between kernels ($^{440}/_{50} \times 8.8$ in [224 mm]). This is the spacing provided by selection of the proper sprocket combination and plate cells. The seed population will also be approximately 24,060 per acre [59,428 per ha].

Any differences which result after a check of the planter seeding rate must be corrected. Irregularities in kernal discharge may be due to wrong plate (number of cells or type for shape of kernel), improper-size sprockets, or too fast a speed of travel during planting. Assume the number of seeds caught was 55. This would provide a seed spacing of 8.0 in ($^{440}/_{55}$ in) [20 cm]. The seeding rate is approximately 10 percent too great. To reduce the rate, change the driven No. 1 sprockets to obtain a seed spacing rate of 9.73 in [24 cm], and repeat the trial operation.

**Calibrating the Grain Drill** A shop procedure similar to that discussed for the corn planter is used to calibrate the grain drill. The seeding rate for

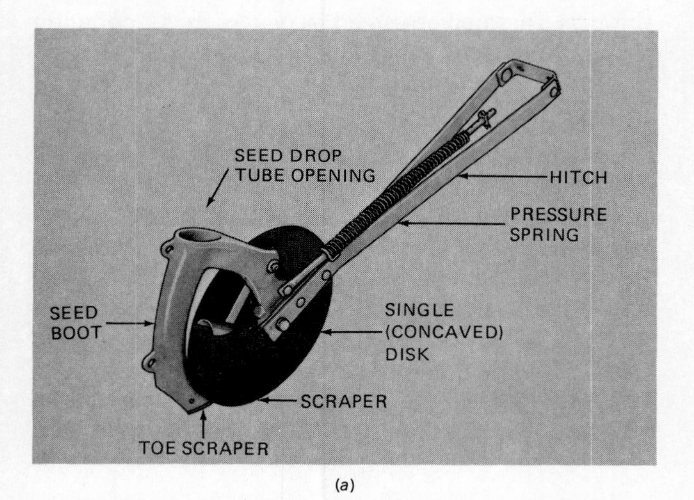


(*a*)

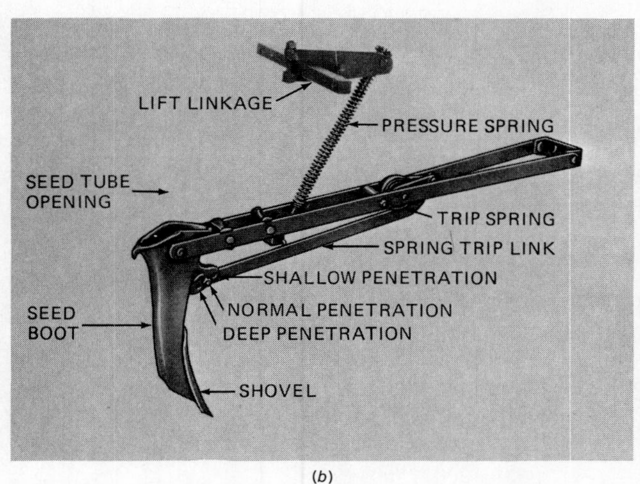


(*b*)

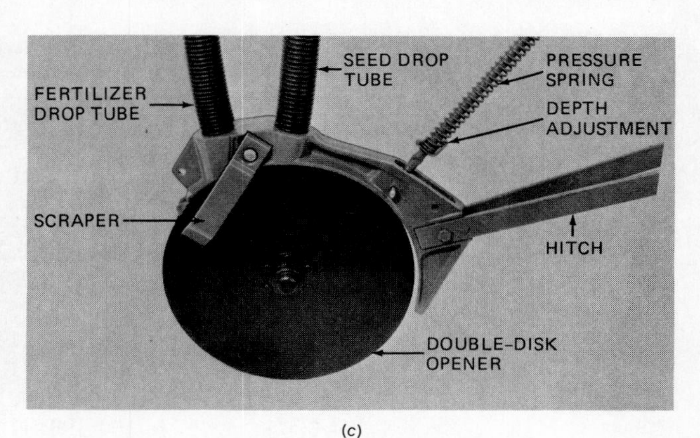


(*c*)

**Fig. 19-22.** Three types of furrow openers are used on grain drills. (*a*) Single-disk for drilling in hard trashy soil conditions. (*b*) Spring trip hoe opener used in hard stony conditions. This unit is not commonly seen. (*c*) Double-disk opener for planting in well-prepared seedbeds and at a high speed of travel. (*Deere & Co.*)

**TABLE 19-3. Effect of Worn Parts on Metering Accuracy (Percent Planter Plate Fill) of Flat and Round Kernels of Corn (100% Cell Fill = Optimum)**

| CONDITION, METERING PARTS | | | | |
|---|---|---|---|---|
| Plate Ring | Hopper Wall | Strike-off Pawl | Kernel Shape | Cell Fill, % |
| New | New | New | Flat | 113 |
| New | Worn | New | Flat | 120 |
| New | Worn | Worn | Flat | 124 |
| Worn | Worn | Worn | Flat | 131 |
| New | New | New | Round | 104 |
| New | Worn | New | Round | 108 |
| New | Worn | Worn | Round | 116 |
| Worn | Worn | Worn | Round | 124 |

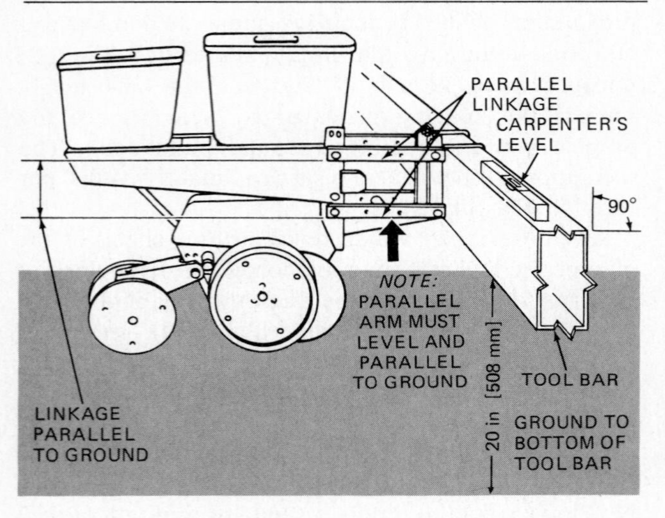


**Fig. 19-24.** The tool bar must be level and adjusted to the correct height. (*Deere & Co.*)

the crop to be planted by grain drill is based on the number of pounds of seed to be applied per acre. The data in Table 19-4 suggest that wheat should be planted at a rate of 80 to 100 lb [36 to 45 kg] per acre. Assume a rate of 80 lb [36 kg] per acre of wheat is selected and the method of metering is fluted-feed. To set the drill to seed at this rate, refer to the operator's manual for the seeding data chart. The chart (see Table 19-6 for a typical chart) provides information on the row spacing [6 in (152 mm)] and the notch setting on the seed index selector lever. The number of pounds per acre listed in Table 19-6 for each of the crops is based on a drill set in slow-speed drive and equipped with properly inflated 6.70 × 15 tires. Tires of other size or type must be compensated for as indicated in the chart. Set the lever to notch 16, which meters 88 lb [40 kg] of wheat per acre (Fig. 19-28).

The *feed gate* of the fluted-feed method of metering is set to the No. 1 position (Fig. 19-28*a*) before adding wheat seed to the hopper. The No. 1 setting of the feed gate is used to seed small grain such as wheat, oats, barley, rye, flax, and rice. The No. 2 position is for medium-sized seeds such as common beans, while No. 3 is for large seeds such as soybeans and lima beans.

If the drill is a double-run feed, the proper side of the feed wheel gate and lock lever setting (Fig. 19-21) must be used to set the metering rate. Also, the correct sprocket position for the feed shaft drive must be selected (Fig. 19-28*b*). To obtain a seeding ratio of approximately 80 lb [36 kg] of wheat per acre with a double-run metering system, the large side of the feed wheel is used with the gate lock lever set in No. 11 with the No. 6 sprocket drive position. This setting will meter an indicated 84 lb [38 kg] of wheat per acre when drilled in 6-in [152 mm] row spacings (see Appendix B, Table B-2).

**TABLE 19-4. Typical Plant Population and Seeding Rate for Common Crops**

| Crop | Plant Population* per acre [ha] × 1000 | SEEDING RATE‡ Pounds [kg] per acre [ha] | Seeds per acre [ha] × 1000 | Approx. No. Seeds per lb [kg] × 1000 |
|---|---|---|---|---|
| Barley | 80–104 [198–257] | 80–100 [90–112] | 1000–1300 [2700–3211] | 13 [5.9] |
| Corn† | 18–30 [44–74] | 15–25 [17–28] | 22–36 [54–89] | 1.5 [0.68] |
| Cotton | 48–72 [119–179] | 15–22 [17–25] | 60–90 [148–222] | 4 [1.8] |
| Oats | 80–104 [198–257] | 80–100 [90–112] | 1000–1300 [2700–3211] | 13 [5.9] |
| Sorghum | 60–100 [148–254] | 6–10 [6.7–11.2] | 72–120 [178–296] | 12 [5.5] |
| Soybeans | 110–190 [272–479] | 45–80 [50–90] | 135–240 [333–593] | 3 [1.4] |
| Wheat | 80–96 [198–237] | 80–100 [90–112] | 1000–1200 [2700–2964] | 12 [5.5] |

* Based upon 80% emergence.
† Sold 80,000 kernels per bag.
‡ Varies for climatic and cultural practices.

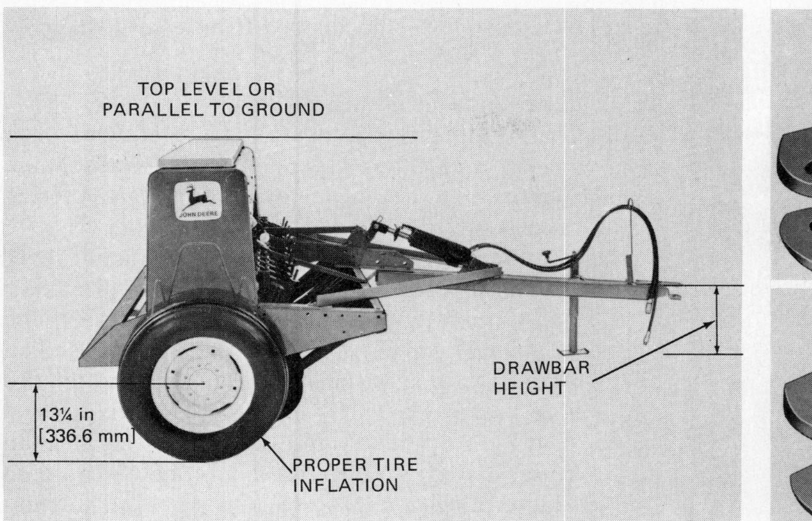


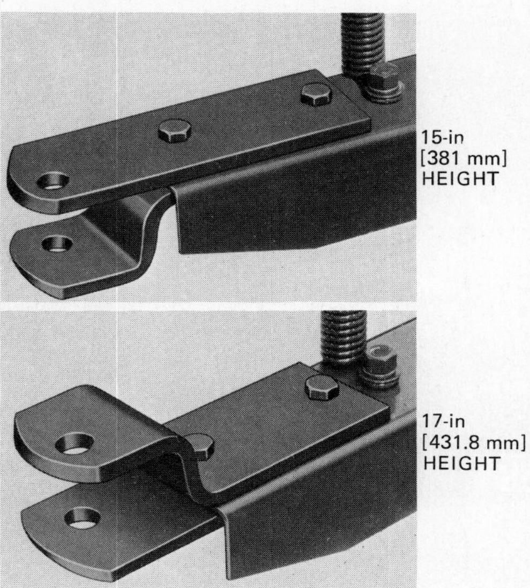


**Fig. 19-25.** Check the levelness of the grain drill and the inflation pressure of tires. The hitch clevis is positioned for tractor drawbar heights of 15 and 17 in. Tire size and inflation pressure are important to provide the correct rolling radius of drive wheels for the seed metering system. (*Deere & Co.*)

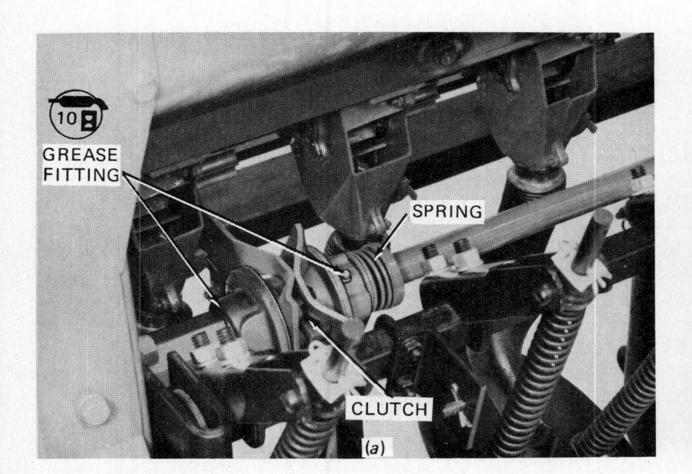

**Fig. 19-26.** Follow the recommendations listed in the operator's manual when lubricating (*a*) grease fittings and (*b*) gear cases. (*Deere & Co.*)

## TABLE 19-5. Distance Between Corn Kernels in Row for Different Planting Rates and Row Spacings*

| | SPACING OF KERNELS IN ROW FOR DRILLED PLANTING RATES, IN [MM] | | | | | |
|---|---|---|---|---|---|---|
| **Row Spacing, in [mm]** | **20,000/acre [49,400/ha]** | **22,000/acre [54,340/ha]** | **24,000/acre [59,280/ha]** | **26,000/acre [64,220/ha]** | **28,000/acre [69,160/ha]** | **30,000/acre [74,100/ha]** |
| 20 [508] | 15.7 [399] | 14.3 [363] | 13.7 [348] | 12.1 [307] | 11.2 [284] | 10.5 [267] |
| 30 [762] | 10.5 [267] | 9.5 [241] | 8.7 [221] | 8.0 [203] | 7.5 [190] | 7.0 [179] |
| 36 [914] | 8.7 [221] | 7.9 [200] | 7.3 [185] | 6.7 [170] | 6.2 [157] | 5.8 [147] |
| 38 [965] | 8.3 [211] | 7.5 [190] | 6.9 [175] | 6.3 [160] | 5.9 [150] | 5.5 [140] |
| 40 [1016] | 7.8 [198] | 7.1 [180] | 6.5 [165] | 6.0 [152] | 5.6 [142] | 5.2 [132] |

* Adapted from Dennis, Robert, *Irrigated Corn for Grain or Silage in Arizona,* Bulletin Q372, Cooperative Extension Service, College of Agriculture, University of Arizona.

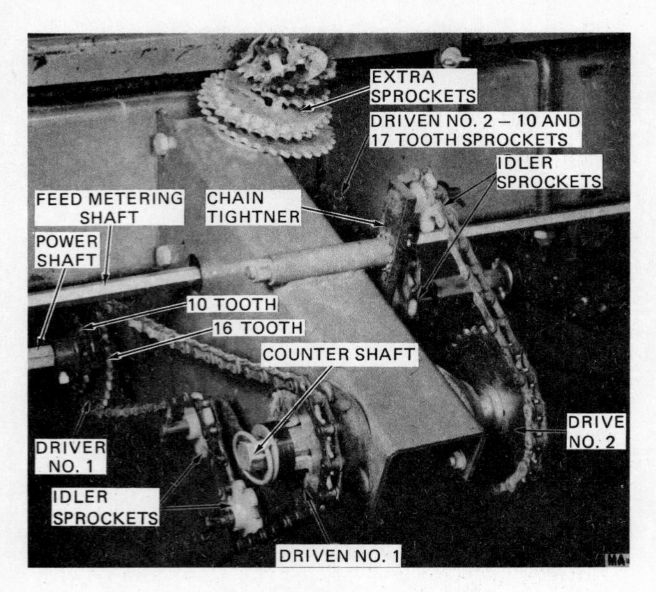


**Fig. 19-27.** A metering plate drive for a planter. The sprockets are selected according to published seeding charts. (*International Harvester Co.*)

To determine the accuracy of these settings, proceed as follows:

1. Jack up the drive wheel.
2. Remove the feed tubes from the boot of each opener and attach a paper bag to the discharge end to catch the grain that falls from the metering wheels.
3. Turn the drive wheel the correct number of revolutions to seed $^1/_{20}$ (0.05) of an acre [0.02 ha]. Most operators' manuals provide data on the number of wheel revolutions required to seed an acre according to the size of the drill and the tire size (Table 19-7). For example, the drive wheel of a 21 × 6 grain would be rotated 566 turns to cover an acre [1398 turns per hectare] when equipped with 6.70 × 15, 4-ply tires. The rotations for $^1/_{20}$ acre are as follows:

$$\frac{566 \text{ turns}}{20} = 28.3 \qquad \text{or } 28\tfrac{1}{3} \text{ turns}$$

**TABLE 19-6. Seeding Chart for Fluted-Wheel Metering* (6-in [152 mm] Spacing, lb/acre)**

| Notches on Seed Index | | | 4 | | 8 | | 12 | | 16 | | 20 | | 24 | 28 | 32 | 36 | 40 | 44 | 48 | 52 | 56 | 60 |
|---|---|---|---|---|---|---|---|---|---|---|---|---|---|---|---|---|---|---|---|---|---|---|
| Wheat | | | 22 | 32 | 42 | 53 | 64 | 76 | 88 | 101 | 114 | 128 | 142 | 171 | 200 | 230 | 261 | 292 | 322 | 353 | | |
| Barley | | | | 22 | 30 | 38 | 46 | 54 | 63 | 71 | 80 | 89 | 98 | 116 | 134 | 153 | 172 | 192 | 212 | 233 | 254 | |
| Oats or safflower | | | | | 24 | 30 | 37 | 43 | 50 | 56 | 63 | 69 | 76 | 90 | 104 | 118 | 133 | 148 | 163 | 178 | 193 | 208 |
| Rye | | | | 29 | 38 | 48 | 58 | 67 | 77 | 87 | 98 | 108 | 118 | 139 | 160 | 181 | 202 | 224 | 246 | 269 | | |
| Rice—short kernel | | | | | | | | | 53 | 63 | 72 | 80 | 87 | 100 | 114 | 131 | 150 | 172 | 196 | | | |
| Rice—long kernel | | | | | | | | | 46 | 54 | 61 | 67 | 74 | 87 | 100 | 114 | 131 | 149 | 170 | | | |
| Peas | | | | | | | 70 | 87 | 105 | 123 | 141 | 160 | 179 | 219 | 260 | 303 | 348 | 394 | 442 | 492 | 543 | |
| Soybeans or Navy beans | | | | | 33 | 46 | 61 | 76 | 92 | 109 | 127 | 145 | 163 | 201 | 241 | 281 | 321 | 360 | 399 | 436 | 471 | 504 |
| Buckwheat | | | | | 30 | 37 | 45 | 54 | 63 | 73 | 83 | 93 | 104 | 125 | 146 | 168 | 193 | | | | | |
| Sorghum or vetch | | | 19 | 28 | 37 | 47 | 57 | 68 | 79 | 90 | 102 | 114 | 126 | 152 | | | | | | | | |
| Crested wheat grass | | | | | 11 | 13 | 16 | 19 | 22 | 25 | 28 | 31 | 34 | 41 | | | | | | | | |
| Alfalfa or rape | | 10 | 17 | 24 | 32 | 40 | 48 | 56 | 65 | | | | | | | | | | | | | |
| Millet | | 11 | 19 | 27 | 34 | 42 | 50 | 59 | 69 | | | | | | | | | | | | | |
| Flax or Sudan grass | | | 15 | 23 | 31 | 39 | 47 | 55 | 63 | 71 | 80 | | | | | | | | | | | |

* The pounds per acre on the seed rate charts are based on drills with slow-speed drive and 6.70 × 15 and 7.60 × 15-in rib-implement tires and 7.50 × 20-in double-rib tires. Improper tire size or soil conditions will affect seed rates.

To convert the seed chart rates to proper rates for 7.60 × 15-in double-rib tires and 7.50 × 20-in rib-implement tires, use the following conversion factors:

7.60 × 15 - 6 PR double-rib—seed chart reading × 0.94 = adjusted rate
7.50 × 20 - 4 PR rib-implement—seed chart reading × 1.03 = adjusted rate

To convert seed rates to proper rates for drills with fast-speed drive, double the rates shown on the chart. After setting your drill to accomodate the following charts, check the quantity of seed drilled.

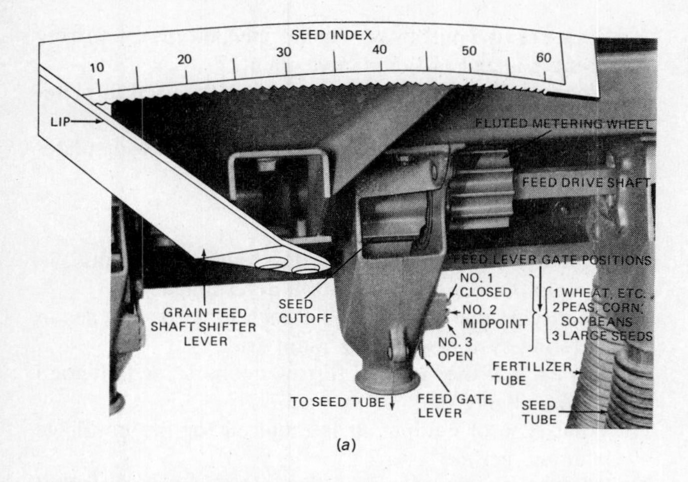


(a)

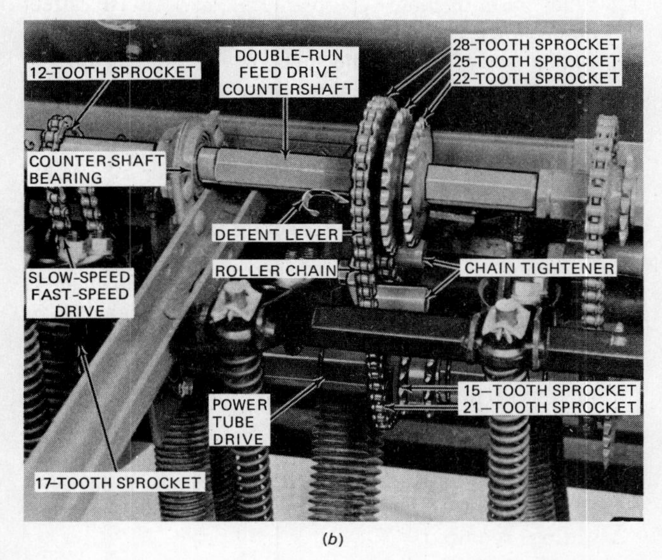


(b)

**Fig. 19-28.** Always refer to the operator's manual for the procedures used to set the grain drill for the seeding rate. (*a*) For a fluted-feed drill, the shift lever and feed gate are moved to the proper settings. (*b*) For a double-run drill select the proper sprocket size according to Appendix B, Table B-2. (*Deere & Co.*)

If the operator's manual does not provide the necessary data, use the following formula:

$$\text{Wheel rotations per } 1/20 \text{ acre} = \frac{1}{20} \times \frac{43{,}560 \text{ ft}^2/\text{A} \times 12 \text{ in/ft}}{W \times D \times 3.1416}$$

where $W$ = width of drill, ft

$D$ = diameter of wheel, in

**Example**

$$W = 21 \times 6 \text{ drill} = \frac{21 \times 6}{12 \text{ in/ft}} = 10.50 \text{ ft}$$

$$D = 6.70 \times 15 \text{ tire} = 2\,(6.70) + 15 = 28.4 \text{ in}$$

or 28.0 in effective diameter

$$\text{Wheel rotation per } 1/20 \text{ acre} = \frac{1}{20} \times \frac{43{,}560 \times 12 \text{ in}}{10.50 \times 28.0 \text{ in} \times 3.14}$$

$$= \frac{1}{20} \times \frac{522{,}720}{923.16}$$

$$= \frac{1}{20} \times 566.23$$

$$= 28.3 \quad \text{or} \quad 28\ 1/3 \text{ turns}$$

4. Combine and carefully weigh all the seed collected and compare to the setting on the seed chart. Remember to divide the weight on the grain chart (Appendix B, Table B-2) by 20, because the calibration was for 1/20 acre.

$$\text{Fluted-feed rate} = \frac{88 \text{ lb}}{20} = 4.4 \text{ lb } [1.9 \text{ kg}]$$

$$\text{Double-run rate} = \frac{84}{20} = 4.2 \text{ lb } [1.9 \text{ kg}]$$

For example, assume that a weight of 5.0 lb [2.26 kg] of wheat was collected from the fluted-feed drill. This is an additional 0.6 lb [0.27 kg] for the 1/20 acre, or 12 lb [5.4 kg] (0.6 × 20) more than the 88 lb [39.9 kg] the shifter lever was set to deliver. The total amount of seed delivered would be 100 lb (88 lb + 12 lbs) [45.36 kg]. To correct the metering rate of the drill, the feed

**TABLE 19-7. Drive Wheel Revolutions per Acre [Hectare] and Specifications for Size of Grain Drill***

| Tire Size | Wheel Revolutions per Acre [Hectare]† | | | | | | | | | |
|---|---|---|---|---|---|---|---|---|---|---|
| 6.70 × 15 4-ply rib impl. | 566 [1398] | 496 [1225] | 458 [1131] | 728 [1798] | 566 [1398] | 485 [1198] | 443 [1094] | 566 [1398] | 446 [1102] | 446 [1102] |
| 7.50 × 20 4-ply rib impl. | 435 [1075] | 379 [936] | 348 [860] | 563 [1391] | 435 [1075] | 379 [936] | 348 [860] | 435 [1075] | 348 [860] | 348 [860] |
| Number of feeds | 21 | 24 | 26 | 14 | 18 | 21 | 23 | 16 | 20 | 16 |
| Spacing of feeds, in | 6 | 6 | 6 | 7 | 7 | 7 | 7 | 8 | 8 | 10 |

* Use the data in the operator's manual for the specific make and model of drill.

† Properly inflated pneumatic tire.

shifter lever would be closed one notch (between 12 and 16). The chart indicates that 76 lb [34.5 kg] of seed would be delivered at this setting. If, however, the drill is seeding 12 lb [5.4 kg] too much, it is possible that the setting would meter 88 lb (76 lb + 12 lb) [39.9 kg].

Repeat the calibration operation after making the adjustment to verify the actual seeding rate.

The example illustrates the importance of calibrating the drill. In this case the result would have been overseeding, causing unnecessary seed costs and possible reduction in harvest yield.

## SUMMARY

The production of food and fiber as supplied by plants is fundamental to the industry of agriculture. Plant growth starts with the placement of the seed in the soil. Accuracy of planting requires the use of highly specialized agricultural machinery. Crops that require cultivation are placed in rows by planters. Drills are used to obtain solid seeding of high plant population crops.

Accuracy of planting and seeding includes placement of seed at the proper depth and rate. Depth of planting varies with the size of seed and the characteristics of seedling growth. Rate of seeding is usually based on the number of pounds [kilograms] of seeds planted per acre [hectare]. Corn, however, is seeded to obtain a specified number of plants per acre [ha]. In all cases, it is important that the equipment be calibrated to assure accuracy of the planting and seeding operation.

As a professional in power and machinery, you will be called upon to assist machine operators in properly adjusting and calibrating planters and drills.

## THINKING IT THROUGH

1. What are the primary and secondary functions of planters and drills?
2. What is the difference between seeding with a planter and seeding with a drill?
3. Identify the primary use of each of the three types of furrow openers used on planters.
4. What are two purposes of using special types of furrow openers ahead of the planter opener?
5. What are the three basic types of plates used in plate-type planters? When is each used.
6. Identify the operating principles of the following plateless planters:
   (a) Finger pickup
   (b) Air, pressure type
7. What are the three ways of classifying grain drills?
8. How is the size of a grain drill determined?
9. Identifying one advantage for each of the two types of metering systems used on grain drills.
10. List each of the types of furrow openers used on grain drills and give one use for each.
11. What type of equipment is required for the no-tillage planting method?
12. What parts of plate-type planters can seriously affect filling of the cell and the subsequent planting rate?
13. Explain how the seeding rate of a grain drill is controlled in each of the following systems:
    (a) Double-run wheel feed
    (b) Fluted wheel feed
14. Does a grain drill meter grain by volume or by weight? Explain your answer.
15. Why should a grain drill operate with the seed box level?
16. What is meant by the *effective diameter* of the wheel of a grain drill? How is effective diameter influenced by tire inflation pressure?
17. Identify the two adjustments that must be made to set a plate-type planter for the correct number of corn kernels per acre [ha].
18. Assume you wish to seed barley at a rate of 100 lb/acre [45.4 kg/0.405 ha], with a fluted-feed grain drill equipped with 6.70 × 15 tires. Identify the following settings required:
    (a) Grain shaft speed ___
    (b) Shifter lever setting — lb
    (c) Feed gate setting No. ___
19. If in the above question, the drill is a double-run, provide the following settings:
    (a) Sprocket drive, No. ___
    (b) Lock lever setting, No. ___
20. If the size of the grain drill is 26 × 6, how many revolutions of the wheel would be required to seed $^1/_{20}$ acre?

# CHAPTER 20 WEED, INSECT, AND SOIL FERTILITY CONTROL EQUIPMENT

## CHAPTER GOALS

In this chapter your goals are to:

- Identify the types of field machines used in production agriculture for applying chemicals for weed and insect control and for applying commercial fertilizers
- Calibrate equipment for the safe and effective distribution of chemicals used in crop production
- Know how to service, maintain, and repair machinery which distributes crop chemicals

Backpack hiking is great fun. Combined with a camping and fishing trip, it gives you an opportunity to become very close to plant, animal, and insect life and the wonders of natural surroundings. The well-equipped backpacker will have water, food, and first-aid essentials, including an insect repellent and a snakebite kit. In a similar way the producer of agricultural products must be equipped to administer chemicals for insect and weed control and supplemental nutrients for proper plant growth.

Machinery used for these operations does not have the structural strength, mechanical complexity, and high-energy requirements of tillage, planting, or harvest equipment. It is, however, designed to be "finely tuned" to properly place and distribute chemicals at a specified rate for maximum benefits to production. Your job as a specialist in power and machinery will include working with producers of agricultural crops and custom operators who use this equipment. You will be expected to help producers make proper selection of components and give assistance in calibrating, servicing, and repairing the equipment.

## PURPOSE OF CROP PRODUCTION CHEMICALS

The previous two chapters identified types of soil preparation and crop production equipment which are used to prepare a seedbed and to plant the seed. For the farmer, the payoff for using high-quality seed planted at a rate to obtain the desired plant population comes when each seedling develops into a mature, healthy plant, capable of producing a high yield. This period of time is one in which weeds, insects, and weather conditions can materially affect crop production. A rapidly growing plant needs a vigorous root system to obtain the necessary amount of moisture and soil nutrients. Thus the soil must be fertile and free of soilborne insects and disease.

Weeds compete with growing crops for the same moisture, soil nutrients, and sunlight. The natural growth vigor of weeds will often cause them to overtake crops grown in rows and solid plantings. Besides reducing crop yields, weeds increase the difficulty of harvest and contaminate grain crops with weed seeds. Weeds also decrease the quality of hay and forage crops. According to research conducted in 1979 by the University of Illinois and the U.S. Department of Agriculture, 100 nutsedge plants per square meter [1.2 square yards] reduced corn yields an average of 8 percent. In some test plots, the weeds reduced yields more than 50 percent.

Insects, being a part of the animal kingdom, compete with humans for food. The ravage of insects has been one of the most often recorded historical events of the human struggle in producing crops for food. While a number of insects are beneficial and complementary to crop production, those that are classified as harmful destroy billions of dollars' worth of crops per year. In addition, millions of dollars are spent annually to control insects.

Through the process of photosynthesis the growing crop has the ability to convert the energy of the sun and nutrients from the air, water, and soil into food for humans. When expressed in amounts of nutrients which plants take from the soil, the conversion process resembles the input to a manufacturing plant. For example, it requires approximately 3800 tons [3447 metric tons (t)] of water and 540 pounds (lb) [245 kilograms (kg)] of soil nutrients from an acre [0.405 hectare (ha)] of soil to produce 100 bushels (bu) of corn per acre [8.70 cubic meters per hectare ($m^3$/ha)]. Of the total soil nutrients required, approximately 420 lb [191 kg] comprise the primary elements of nitrogen (N), phosphorus (P), and potassium (K). Commercial fertilizers are applied near the time of planting with various types of machines to assure that these elements are always available for maximum utilization by the plants.

Agricultural chemicals are an exceptionally important tool for the modern farm. For the entire agricultural industry, approximately $^1/_2$ million tons of chemicals are used yearly for insect control. The annual use of chemicals for commercial fertilizer is 50 million tons and over 2 million tons for weed control. The many types of chemicals used for weed and insect control, commonly referred to as *pesticides,* are subject to many environmental and health regulations. The information in this chapter will be limited to the types and characteristics of machinery and components and the calibration, service, and maintenance of the equipment.

## TYPES OF MACHINERY

The proper agricultural chemicals must (1) be used to satisfy a specific need and (2) be applied at the correct time, rate, and concentration to achieve the maximum return for the money invested. Application machinery therefore is an important factor in achieving timeliness and the precise distribution rate.

The classification of machinery for application of agricultural chemicals is based on the two forms in which they are distributed, *liquid* and *dry*. Liquid chemicals are usually mixed with a diluent such as water or diesel fuel and applied as a spray or mist. Some chemical fertilizers, such as anhydrous ammonia, are stored as liquid under pressure and then released into the soil as gaseous vapor. The advantages of liquid application are that the chemical becomes immediately available and the diluent (when it is water) is low cost.

In the dry form, the chemicals are combined with an inert material that may allow a slower release of the active agent. This can extend the effectiveness of the chemical. Chemicals in the dry form may consist of either grains or a dry powder for dust application.

### Liquid Application Equipment

The sprayer is one of the most common machines used to apply liquid chemicals for weed and insect control. Liquid fertilizers can also be formulated and applied by spray. The type of agricultural spray equipment varies according to method of delivery, the chemical to be applied, and the spraying pressure used. The delivery system may be by (1) hand- or power-operated knapsnack, (2) tractor-, trailer-, or truck-mounted equipment, or (3) airplane or helicopter. Hand-operated types are generally limited to small areas which are difficult or impractical to reach with other equipment. Hand sprayers may be used to spot-treat small areas, for example, weeds in fence rows or around buildings and small patches of persistent weeds in large fields.

Ground equipment sprayers are classified into four major types: (1) low-pressure, (2) high-pressure, (3) mist, and (4) recirculating. Although a rope wick applicator is not a sprayer, it is also used to apply herbicide.

**Low-Pressure Sprayers** Low-pressure sprayers operate at pressures of 20 to 50 pounds per square inch (psi) [138 to 345 kilopascals (kPa)]. They are widely used and versatile agricultural machines. This is because the various components which make up the sprayer are relatively inexpensive and are adapted to the control of weeds, insects, and disease in crops. They may also be equipped to spray live-

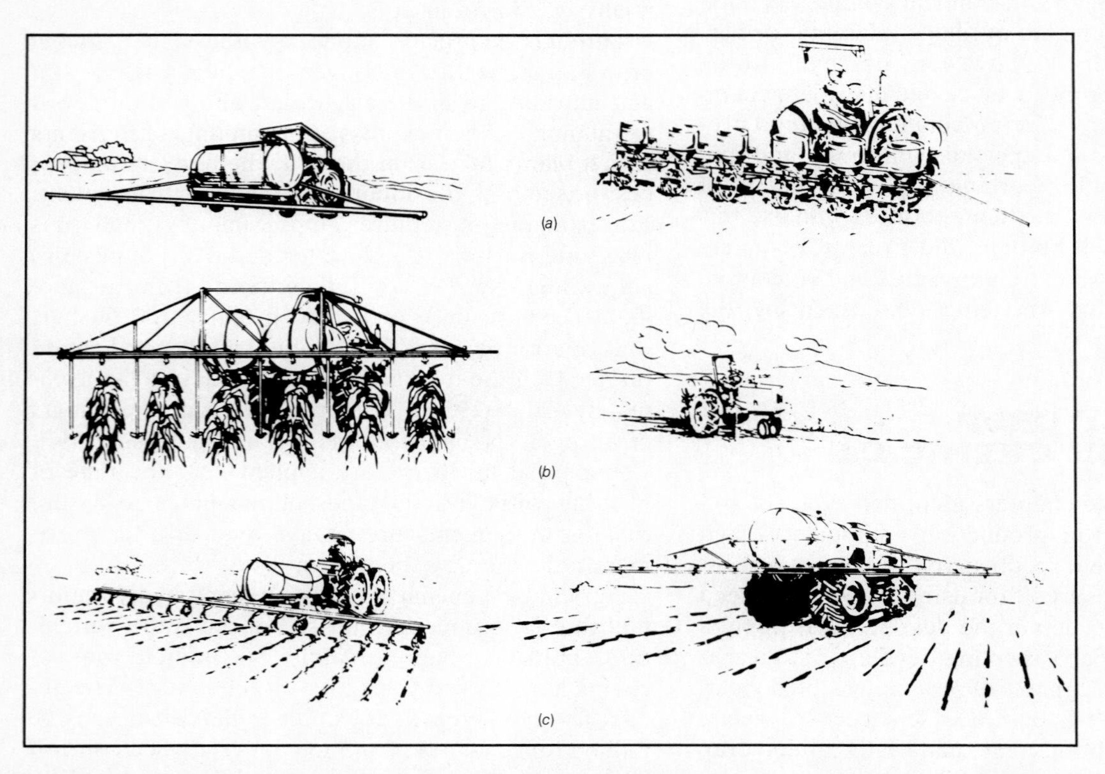


**Fig. 20-1.** Types of low-pressure spray equipment. (*a*) Pre-plant emergence sprayers. (*b*) Post-plant emergence sprayers. (*c*) Liquid fertilizer sprayers. (*Spray Systems Co.*)

stock for the control of insects and disease and to apply paint. Low-pressure spray is especially adapted to the application of preemergence and post-emergence weed and insect control and fertilizer chemicals in crop production (Fig. 20-1*a* to *c*). The equipment may be attached to secondary tillage tools and row-crop planters to apply insecticides and herbicides. This allows the chemical to be worked into the soil before or during the time of planting. Row-crop cultivators can be equipped with low-pressure spray equipment to increase the post-emergence effectiveness of weed and insect control. Low-pressure sprayers are used to apply liquid ammonia fertilizer immediately ahead of a disk harrow or field cultivator so that it will be incorporated into the soil to prevent loss through vaporization.

With the increased use of no-tillage or minimum tillage farming, using planting machinery similar to that illustrated in Fig. 19-15, it is necessary to use contact herbicides applied with a low-pressure sprayer. A conversion to ''no-till'' farming means a substitution of the sprayer and a contact herbicide for some of the primary and secondary tillage equipment generally used for weed control.

**High-Pressure Sprayers** High-pressure sprayers (Fig. 20-2) are similar in construction to low-pressure types but operate at pressures in excess of 100 psi [689 kPa]. The high pressure provides a driving force to penetrate more completely into the foliage of trees, the hair coats of animals, and the crevices of buildings when applying fungicides or insecticides. Fogging nozzles can be used with these sprayers. The sprayer is more costly than the low-pressure type. This is because of a special pump and equipment which will develop and withstand the high pressures.

**Mist Applicators** Mist applicators are a type of low-pressure sprayer. The machine uses a low-pressure pump and spray nozzles to direct insecticide into the discharge airstream of a high-volume squirrel cage fan. The air from the fan creates a mist which will carry for a considerable distance on a calm day. The direction of the mist can be controlled by the position of the air discharge. A mist sprayer (Figure 20-3) is a popular tool for the livestock producer because the components are low cost and the misting process is effective when properly used. The spray can be directed on livestock without disturbing them. The mist is used to treat buildings and feedlots for fly control and to spray roadsides and fence rows for brush and weed control when drift is not a problem. Rate of discharge is based on the number and size of nozzles and rate of travel.

Another important mist-type blower-sprayer is used in orchards (Fig. 20-4). This is a large trailer- or truck-mounted unit equipped with a high-capacity fan driven by a tractor power takeoff (PTO) or an internal combustion engine. The unit produces air-speed velocities of over 100 miles per hour (mi/h) [161 kilometer per hour (km/h)]. Spray nozzles inject the pesticide solution into the rapidly moving airstream. The mist that is formed travels to all parts of the trees when the sprayer is moved through the orchard. The conventional orchard sprayer is not adapted to field crop use because the control of drift is too difficult.

**Recirculating Sprayers** One of the recent developments of low-pressure sprayer components is the recirculating type, also known as an RCS. This sprayer is designed for the specific purpose of applying contact herbicides in such a manner that unused spray is collected and returned to the supply tank,

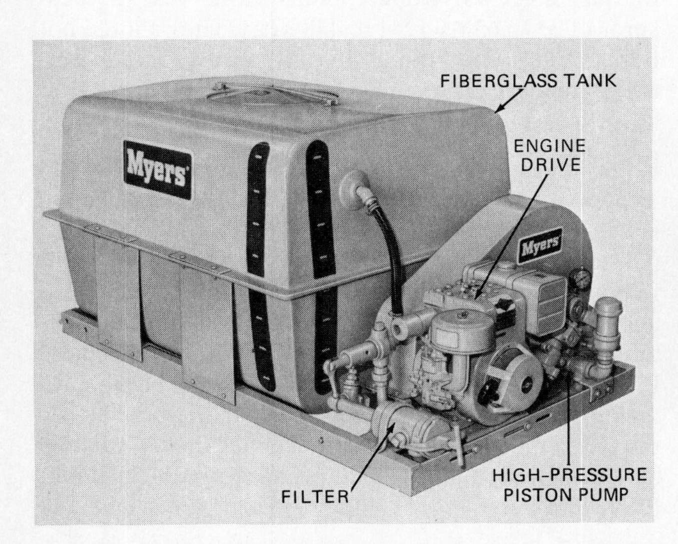


**Fig. 20-2.** High-pressure sprayer. (*F. E. Myers and Bro. Co.*)

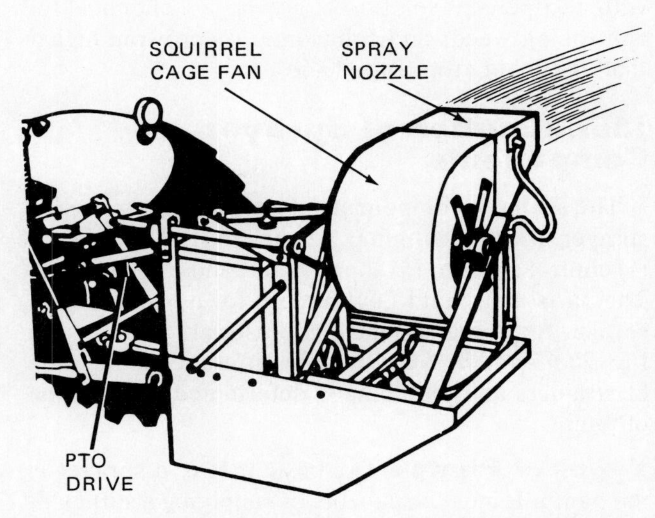


**Fig. 20-3.** Low-pressure, high-volume mist-type applicator (sprayer).

**Fig. 20-4.** Tractor PTO-powered mist-type orchard sprayer. (*FMC Corp., AG Mach. Div. and Spray Systems Co.*)

thus avoiding waste. This type of sprayer makes it economically possible to control the growth of volunteer corn, Johnson grass, and tall-growing weeds in soybeans, cotton, and other short-growing row crops. The RCS applies herbicide only to the tops of tall weeds and avoids herbicide contact with the shorter crop.

The *box-type* RCS (Fig. 20-5*a*) directs three or four streams of herbicide horizontally across the row above the crop canopy. Spray that does not come in contact with a weed is caught by a box-shaped container on the opposite side. RCS units that apply herbicides reduce the cost per acre for weed control.

**Rope Applicator** The *rope* or wick-type applicator is not a sprayer but is a popular type of herbicide dispenser. The rope is saturated by wick action with a 33 percent solution of glyphosate chemical for controlling weeds and Johnson grass growing higher than the main crop (Fig. 20-5*b*).

## Identification of Sprayer Components

The six basic components of a typical low-pressure sprayer are the (1) pump, (2) nozzles, (3) strainers, (4) control system, (5) supply tank, and (6) plumbing. The parts and liquid flow pattern for a low-pressure sprayer for boom and hand gun uses are identified in Fig. 20-6*a* and *b*. Note that the difference in location of strainers and plumbing is determined by the type of pump.

**Types of Pumps** The basic part of a sprayer is the pump. It must be capable of supplying a sufficient volume of liquid at the proper pressure. The three most common types used in agricultural sprayers are the roller, piston, and centrifugal pumps. Roller and piston pumps are classified as positive-displacement types because the volume of liquid pumped is nearly constant over a wide range of pressures. In comparison, the centrifugal pump produces maximum output at pressures up to 20 psi [138 kPa]. As the pressure gets higher, the volume discharged gets smaller. At a pressure of approximately 70 psi [483 kPa] there is no discharge.

**Roller pumps** The roller pump consists of a slotted rotor which revolves in an eccentric case. Rollers made of nylon or rubber move in and out of the slots as the rotor is turned. They seal the space between the rotor and the wall of the case. Liquid is pulled into the housing as the gap between the housing and the rotor widens, and it is forced out the discharge as the gap narrows. Roller pumps work best at pressures of 35 to 65 psi [241 to 448 kPa]. They should not be used to spray liquids that are abrasive, like wettable powders. Higher pressures up to 250 psi [1725 kPa] can be used occasionally for hand-gun spraying.

**Piston pumps** Piston pumps provide high pressures of approximately 600 psi [4137 kPa]. A piston pump is required to spray at pressures more than 100 psi [689 kPa]. The high-pressure discharge makes it effective when using the hand gun to clean animal pens and wash machinery. It is also a good pump for wettable powder suspension and other abrasive liquids. Piston pumps that deliver up to 10 gallons per minute (gal/min) [38 liters per minute (L/min)] can be attached to the PTO of a tractor. Special drives must be used with piston pumps that produce up to 25 gal/min [95 L/min].

**Centrifugal pumps** Centrifugal pumps are popular for low-pressure sprayers because they are long-lasting and can handle wettable powders and abrasive liq-

Fig. 20-5. Recirculating sprayer (*a*) and ropewick applicator (*b*) for control of tall-growing weeds. (*Southern Weed Science Laboratory, USDA, SEA.*)

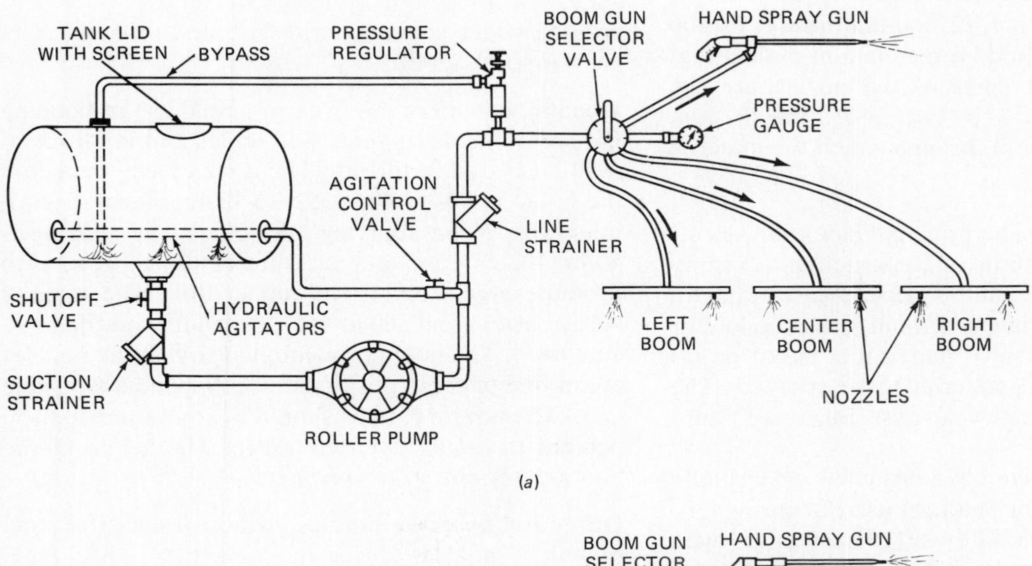


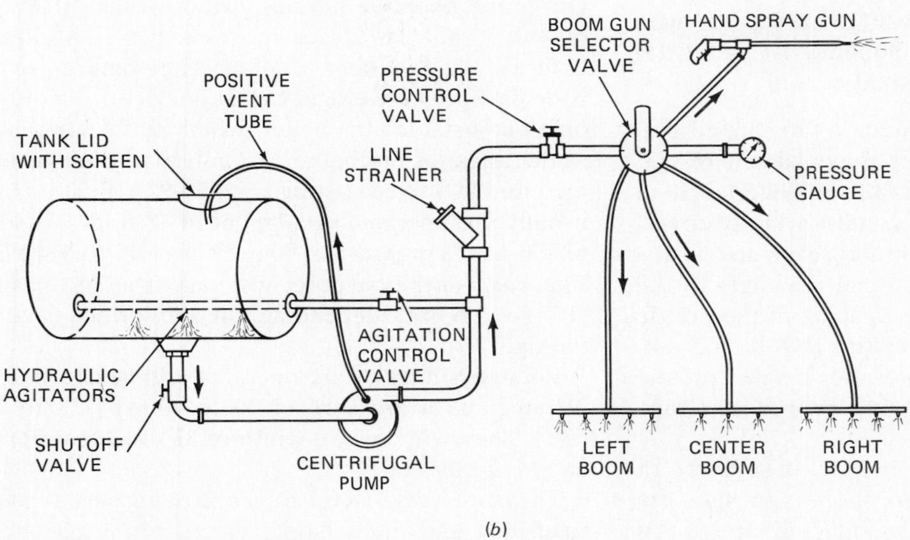


Fig. 20-6. (*a*) Basic sprayer system for a positive displacement pump. (*b*) Basic sprayer system for a centrifugal pump.

uids well. The high pumping discharge capacity has an added advantage because the unused liquid can be returned to the supply tank to provide hydraulic agitation for keeping the solids in suspension. Centrifugal pumps must operate at 3000 to 4000 revolutions per minute (r/min). A step-up gear or belt or a hydraulic motor drive is required when driven from the tractor's PTO at 540 or 1000 r/min. A hydraulic motor operated from the hydraulic system of the tractor frees the PTO for other uses and helps maintain uniform pump speed.

Centrifugal pumps should be mounted lower than the supply tank so that gravity will aid in priming the pump. A small plastic vent tube leading from the pump housing to the supply tank opening prevents air from being trapped and allows the pump to prime itself.

A summary of the advantages and disadvantages of each of the three pumps is presented in Table 20-1.

**Nozzles** Nozzles are the business end of a sprayer. They are an important factor in controlling (1) the rate of application, (2) the uniformity, (3) the surface area covered, and (4) the amount of drift that occurs. The principal parts of the nozzle are the body, strainer, tip, and cap (Fig. 20-7). The tip contains the orifice (opening) through which the material passes.

**Selection of Tip** The principal factors in selecting a tip are (1) the material of construction, (2) spray pattern, (3) spray angle, and (4) discharge rate. A tip may be constructed using aluminum, brass, nylon, or stainless steel. Selection of material is based on (1) cost, (2) wear, and (3) material to be sprayed. The wear, initial cost, and wear-cost ratio are summarized in Table 20-2.

Although the tips have been designed for virtually every purpose, the spray pattern used to spray agricultural chemicals is as follows: (1) flat fan, (2) even flat fan, (3) hollow cone, (4) flooding flat spray, (5) offset and boomless, and (6) solid stream.

**Flat-fan tip** *Flat-fan* tips produce a fan-shaped pattern with tapered edges. The tip is available in 65, 73, 80, and 110° spray angles. The fan must overlap 30 to 50 percent at the edges to give a uniform spray coverage. This is accomplished by rotating each nozzle at a 12 to 15° angle on the boom so that the edges of the adjacent fan spray do not hit. Spacing of the nozzle on the boom is 20 or 30 inches (in) [508 to 762 millimeters (mm)]. Accordingly, the height of the boom must be adjusted to give proper coverage (see Appendix C, Table C-1).

The flat-fan tip is used for broadcast spraying of herbicides and insecticides on boom-type sprayers and as an attachment to planters and cultivators. The tip should operate at pressures of 15 to 30 psi [103 to 207 kPa]—never over 40 psi [276 kPa]. It is used when penetration and complete coverage of the foliage are not necessary.

**Flat-even fan tip** The flat-even fan tip is designed to apply an even coverage across the spray pattern. While it is very similar to the flat-fan tip, it is used only for the band application of both herbicides and insecticides (see Appendix C, Table C-2).

**Hollow-cone tip** Hollow-cone tips (Fig. 20-8) are used to apply insecticides, defoliants, growth inhibitors, and fungicides at pressures of 40 to 120 psi [276 to 827 kPa]. This tip produces a finely atomized hollow-cone penetrating spray which gives overall coverage of plant leaf surfaces. Considerable drift is produced with these tips; therefore they are not recommended for herbicides. Coverage from the side below plant leaves is obtained by arranging the nozzles on boom *drops*. The tips are available in 45, 70, and 75° angles. The 75° tip is designed to operate at a height of 10 in [254 mm] in order to produce a band 7 in [178 mm] wide. The 70° nozzle is suspended on boom drops to provide wider coverage with two and three nozzles per row (Fig. 20-8).

**Flooding flat-spray tips** The principal use of flooding flat-spray tips is to apply herbicides and mixtures of herbicides and fertilizers. The nozzles are spaced on the boom at 20- to 40-in [52- to 102-centimeter (cm)] intervals. When applying herbicides, the spacing is 40 in [102 cm] and at a pressure of 10 psi [69 kPa] to produce large droplets and reduce drift. The angle of spray varies from 100 to 145°, depending on the pressure used. The boom is positioned so that the nozzles are in one of three positions (see Appendix C, Table C-3). The nozzle pattern should overlap from 10 to 50 percent to assure even coverage. The height of the boom determines the spacing.

**Offset and boomless nozzles** Although the offset and boomless nozzles differ in capacity of discharge, both are used without a boom. They emit an extra-wide flat spray coverage. Offset nozzles are mounted on an adjustable attachment which permits a change in the angle of discharge. Boomless tips are usually mounted in a fixed casting (Fig. 20-9). Offset nozzles usually are operated at a height of 18 or 24 in [46 or 62 cm] and a pressure of 30 to 60 psi [207 to 414 kPa]. The width of the coverage will vary from 68 to 143 in [1.7 to 3.6 m], depending on the orifice size of the tip.

Boomless nozzles are operated at a height of 36 in [91 cm] and a pressure of 20 to 40 psi [138 to 276 kPa]. They will cover a width of 33 to 63 feet (ft) [10 to 19.2 meters (m)].

They are very useful for controlling weeds along roadsides and ditch banks where fence posts and other obstacles make boom sprayers impractical.

**TABLE 20-1. Characteristics of Sprayer Pumps**

| Type Pump: | Roller | Centrifugal | Piston |
|---|---|---|---|
| | 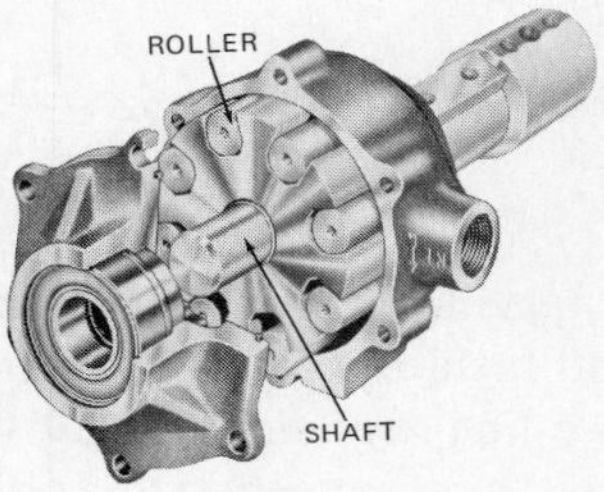 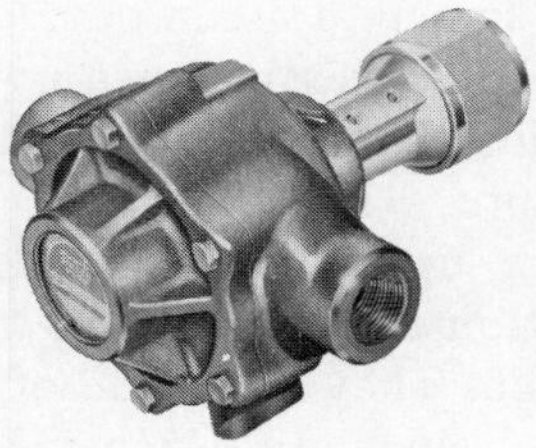 ROLLER<br>SHAFT | 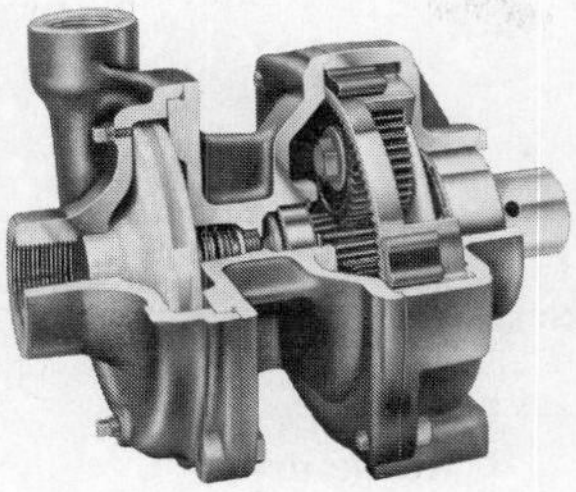 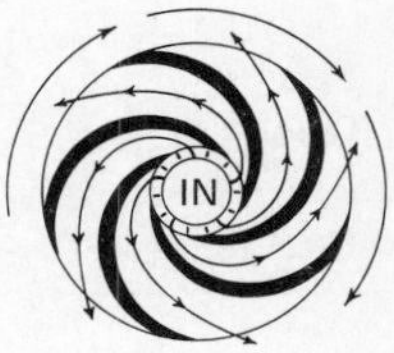  IN | 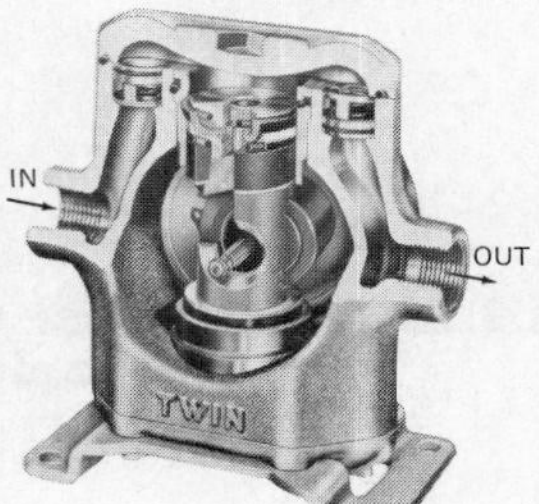 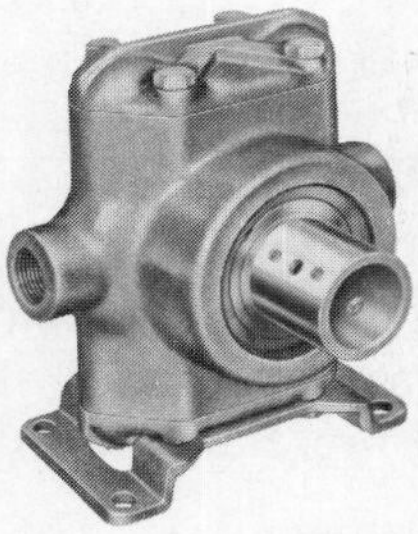 IN<br>OUT<br>TWIN |
| Operating Pressures, psi [kPa] | 0–300<br>[0–2068] | 0–50<br>[0–345] | 0–600<br>[0–4137] |
| Type Displacement | Positive (Variable) | Variable | Positive |
| Type Relief Valve Required | Bypass | None (bypass desirable) | Bypass (unloader valve for pressures over 200 psi [1400 kPa] |
| Output, gal/min [L/min] at 50 psi [345 kPa] | 10–25<br>[38–95] | 50–110<br>[190–416] | PTO mount limited to 10<br>[38] |
| Advantages | Low initial cost; low maintenance cost; effective at PTO speeds; corrosion resistant; good self-priming | Long life; pumps all formulations of pesticides; low maintenance; bypass provides tank agitation | Wear resistant; pumps all formulations of pesticides; parts easily replaced; good priming. |
| Disadvantages | Short life when:<br>wettable powders<br>sand<br>barrel scale | High speed required; not self-priming unless vented; limited to 50 psi [345 kPa] pressure applications. | Expensive; pulsations require surge tank or pulsation damper. |

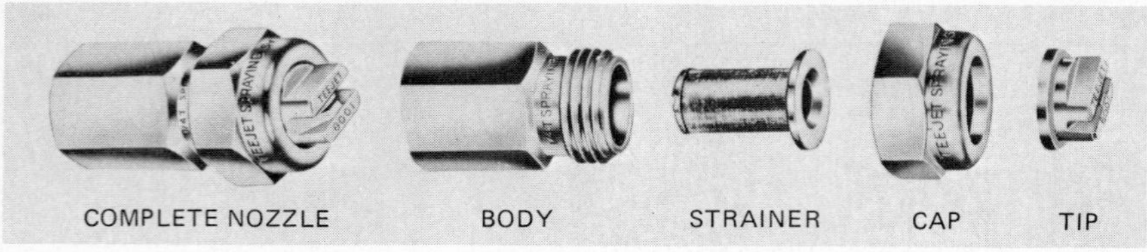

COMPLETE NOZZLE BODY STRAINER CAP TIP

**Fig. 20-7.** Spray nozzle parts. (*Spray Systems Co.*)

### TABLE 20-2. Wear-Cost Ratio of Nozzle Tips

| Material | Wear | Cost | Wear-Cost Ratio | Resistance to Wear/Corrosion; Wettable Powder/ Liquid Fertilizer |
|---|---|---|---|---|
| Aluminum | 1 | 1 | 1:1 | Very poor |
| Brass | 1½ | 1 | 1.5:1 | Poor |
| Nylon | 4 | 0.85 | 4.7:1 | Good |
| Stainless steel | 5 | 3 | 1.66:1 | Excellent |

These tips do not provide the uniform coverage of the flat-fan tip, and they should not be used when drift would present a hazard.

**Solid-stream tips** The principal use of the solid-stream tip (Fig. 20-10) is with the box-type recirculating sprayers. This tip directs a single or multiple jet to a distant target (Fig. 20-10). For this reason it may also be used in hand guns and to inject liquids into the soil.

**Strainers** Strainers are a required part of agricultural spray equipment. They prevent large particles from lodging in the small orifice of the tip. The location and types of strainers depend on the type of pump used. A suction-line strainer (Fig. 20-11) is required when using a roller pump to remove harmful abrasives before they enter the pump. Centrifugal pumps require a high-capacity suction-line strainer to prevent starving the pump, causing air to enter the pump, and resulting in loss of prime.

Pressure-line strainers (Fig. 20-12) should include a screen that has a mesh size equivalent to the size used in the nozzle strainer. For example, most-low discharge nozzles of less than 3 gal/min [11.4 L/min] require a 200-mesh-screen strainer; a nozzle having a discharge of up to 10 gal/min [37.9 L/min] uses a 100-mesh screen. Larger nozzles use 50-mesh-size strainers. High-pressure piston pumps do not use a pressure-line strainer since it might rupture under high pressure.

There are several types of nozzle strainers (Fig. 20-13). The strainers are available in 50-, 100-, and 200-mesh screen. They are made of Monel or stainless-steel wires, which are expensive but have a long life if handled properly. One type contains a check valve which opens at 5-psi [34.5-kPa] pressure. The valve closes when the pressure drops, to prevent nozzle drip during turning or moving across fields. It also prevents unnecessary loss of chemicals when the control valve is closed or the pump is stopped.

**Controls** The control system of a sprayer requires a pressure gauge and control valves. The *pressure gauge* (Fig. 20-14) is one of the most important parts of a sprayer. Its function is to measure the liquid pressure in the pressure supply lines. Pressure is important because it influences the discharge rate of the nozzles and affects the atomization or size of the spray droplets. Doubling the pressure will increase the flow rate 40 percent. For example, increasing the pressure from 20 to 40 psi [138 to 276 kPa] will change the discharge of a nozzle rated at 1 gal/min [3.8 L/min] to 1.4 gal/min [5.3 L/min]. The resulting decrease in size of the droplets may cause spray drift.

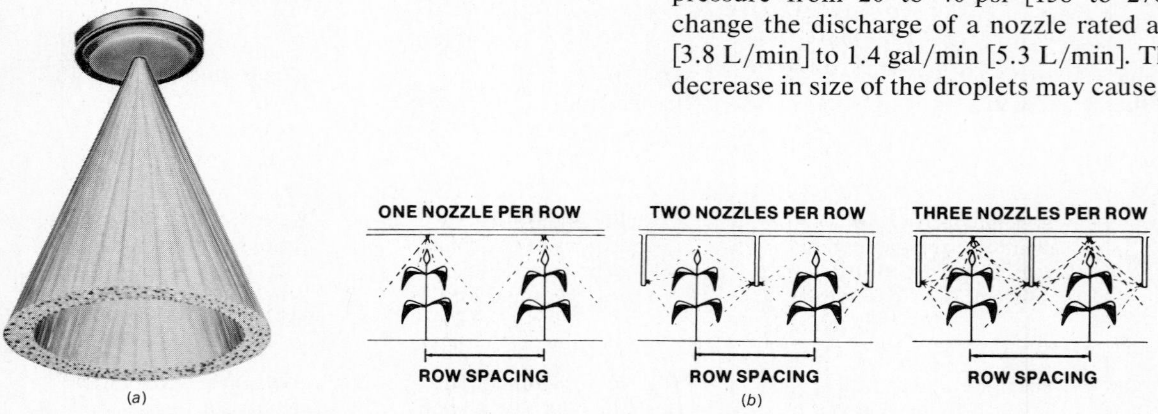


**Fig. 20-8.** Hollow-cone tip for row-crop spraying of insecticides at high pressures and low rates. Often used in airblast-type sprayers. Good when using abrasive materials such as wettable powers. (*Delevan Corp. and Spray Systems Co.*)

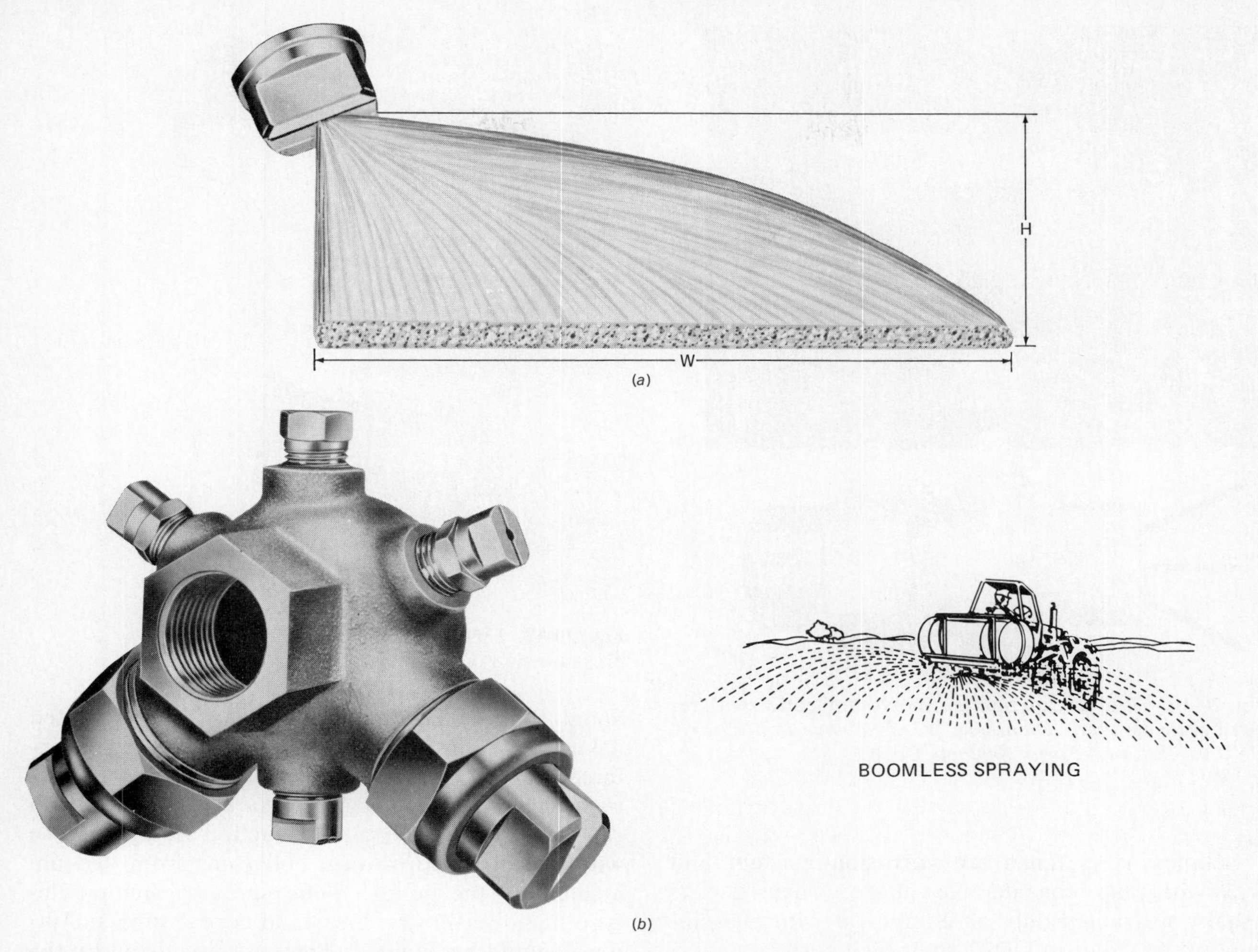


**Fig. 20-9.** Offset flooding fan nozzle (*a*) and boomless nozzle (*b*). (*Spray Systems Co.*)

Control valves include (1) main shutoff, (2) pressure regulating, and (3) distribution control. Main shutoff valves are used to control the flow from the bottom of the supply tank (Fig. 20-15). Some sprayers omit this valve by inserting the suction hose into the top of the supply tank.

Pressure regulating valves (Fig. 20-14) include the pressure relief and unloading-type valves. Pressure relief valves are used to regulate the pressure on low-pressure sprayers equipped with positive-displacement pumps. A bypass connection returns excess liquid to the supply tank.

The unloading valve (Fig. 20-16) is used with high-pressure piston pump sprayers. It allows fluid to bypass to the tank at low pressure and reduces the power requirement to the pump when the discharge is shut off. These valves are more sensitive to rapid pressure change and have less wear than relief valves.

Distribution control valves (Fig. 20-17) make it possible to select the discharge to sprayers equipped with more than one boom. They also function as a shutoff to the nozzle supply lines.

**Agitators** The liquid solution of certain wettable powder chemicals is difficult to keep in suspension within the supply tank without mechanical agitation. The most common method of stirring the tank contents is hydraulic. The extra fluid that is bypassed from the pump is directed through an agitator. The venturi principle creates a large amount of turbulence which stirs and helps to keep solids in suspension.

**Supply Tank** The supply tanks used for sprayers range in size and cost from used, relatively low-cost 55-gal steel barrels to expensive stainless-steel units. The desirable tank should be corrosion-resistant and easy to clean and fill. Polyethylene tanks are popular because they are inexpensive and are noncorrosive with liquid fertilizers (except ammonium phosphate). They are durable and resist puncture, but if cracked or broken, cannot be repaired satisfactorily. These tanks must be stored out of the sun because ultraviolet light causes polyethylene to break down.

Fiberglass tanks are resistant to most chemicals but may crack from impact. Kits are available to repair these tanks.

Fig. 20-10. Solid-stream nozzle. Shows application with recirculating sprayer. (*Southern Weed Science Laboratory, USDA, SEA, and Spray Systems Co.*)

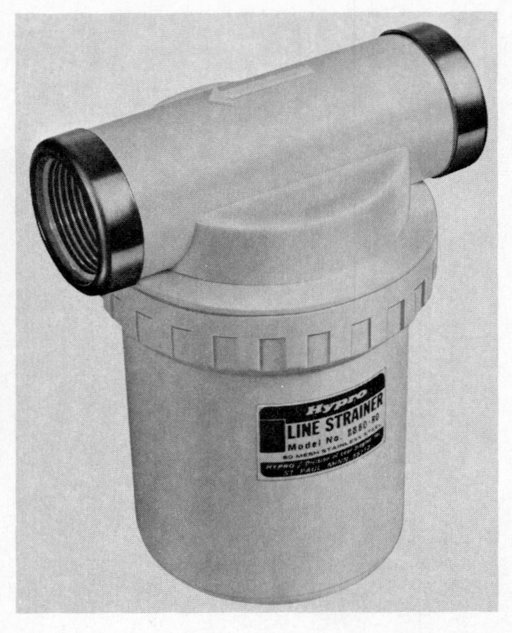


Fig. 20-12. Line strainers are located in the pressure side of the pump. (*Hypro*)

Stainless-steel tanks are corrosion-resistant and will withstand considerable abuse. Their use is usually warranted only on equipment with high annual use in order to justify their high cost.

**Plumbing** Pipes and hoses are used to transfer liquids from the tank to nozzles. Selection of the material to use depends on its resistance to corrosion, on its size—it must be big enough to prevent pressure drop—and on its resistance to bursting under pressure.

Galvanized pipe is often used as a *wet boom*. A wet boom is capable of supporting itself and distributing liquid. It is corrosion-resistant, but the zinc often flakes off the inside, stopping up the screens. Reinforced rubber hose of a quality which will resist corrosive materials and high pressure is often used in *dry-boom* sprayers. Rubber suction hose must be wire-reinforced to prevent collapsing from vacuum created by the pump. When spraying ammonium-type liquid fertilizers, brass and copper pipe and fittings should be avoided. They are weakened by the corrosive action and may create a safety hazard.

**Specialized Metering Pumps** Liquid fertilizer is applied by (1) spray ahead of a disk harrow or field cultivator, (2) gravity drop from a drip tube into a furrow opener, and (3) pumping into knife or

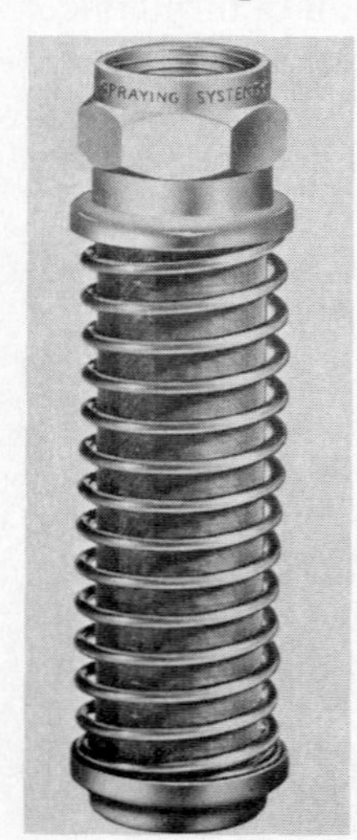


Fig. 20-11. Suction-line strainer with 50-mesh screen. (*Spray Systems Co.*)

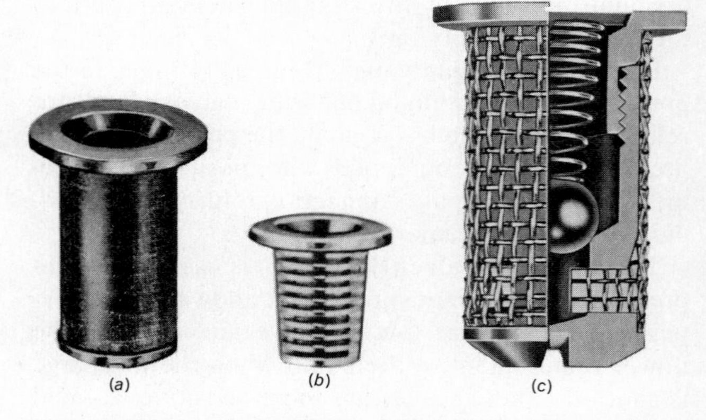


Fig. 20-13. Nozzle strainers. Metal (Monel or stainless steel) strainer (*a*), with 100- to 200-mesh screen; slotted strainer (*b*) for use with liquids containing suspended solids, 25- to 50-mesh equivalent; combination strainer and check value (*c*) to prevent nozzle dripping when pressure drops below 5 psi [34 kPa], 24- to 200-mesh screen. (*Spray Systems Co.*)

(a)

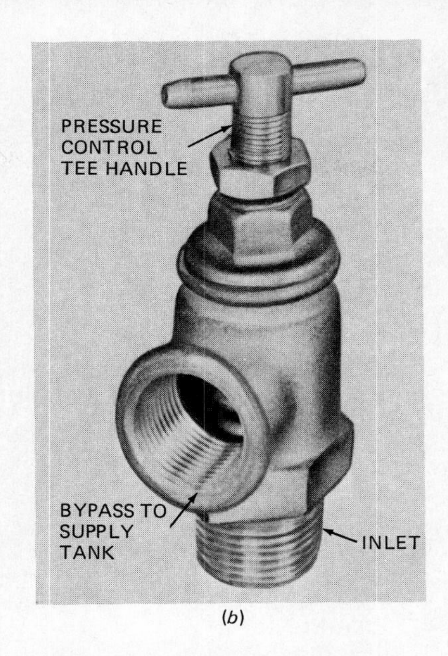


(b)

**Fig. 20-14.** Pressure control system for sprayers includes pressure gauge (*a*) and pressure relief value (*b*). (*Hypro*)

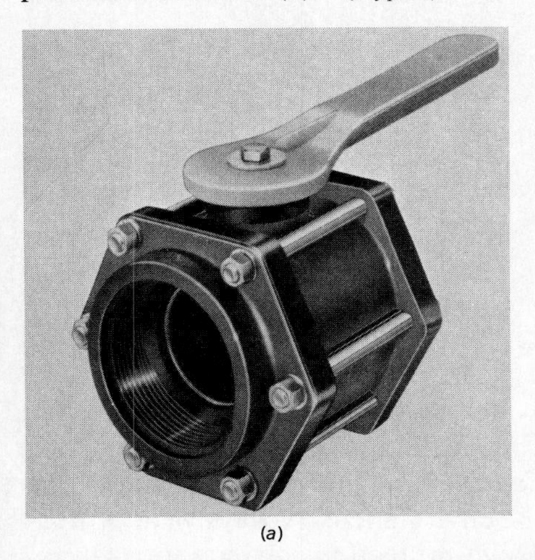

(a)

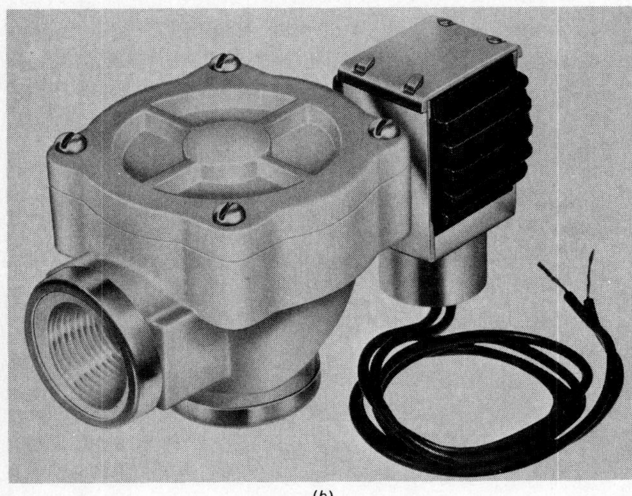

(b)

**Fig. 20-15.** Shutoff valves—manual type (*a*) located at the outlet end of the supply tank. Electric (solenoid) valve (*b*) in the supply line to the boom permits operation from the tractor seat or cockpit. (*Hypro*)

chisel openers. The spray unit uses conventional spray equipment.

A *hose pump* is used as the metering device for the gravity system. The hose pump (Fig. 20-18), consists of a rack of flexible plastic hoses which are squeezed by a roller assembly. The number of hoses equals the number of crop rows being planted. A drive from the ground wheel rotates the roller assembly at a speed which will give the required discharge. The corrosive fertilizer is contained within the plastic tubes of the pump.

A special positive-displacement piston pump (Fig. 20-19) is used to place liquid fertilizer into knife and chisel openers at a metered rate. The pump is driven from a ground wheel. Rate of discharge is controlled by the sprocket ratio. The pump parts are made of corrosive-resistant material.

## Dry Chemical Application Equipment

Agricultural chemicals which are applied dry are distributed as dust or granules. The use of equipment to apply insecticides in a dust form has declined except for home and garden use. As a specialist in agricultural power and machinery, much of your contact will be with equipment designed to distribute granular pesticides and fertilizers.

Modern planters and drills may be equipped with granular attachments for distributing granular herbicides and insecticides (Fig. 20-20).

These machines use positive-type metering systems which dispense granules by volume. The feed mechanism uses fluted- or traction-type feed wheels. The speed of the feed wheel and the opening of the metering gate regulate the rate of distribution.

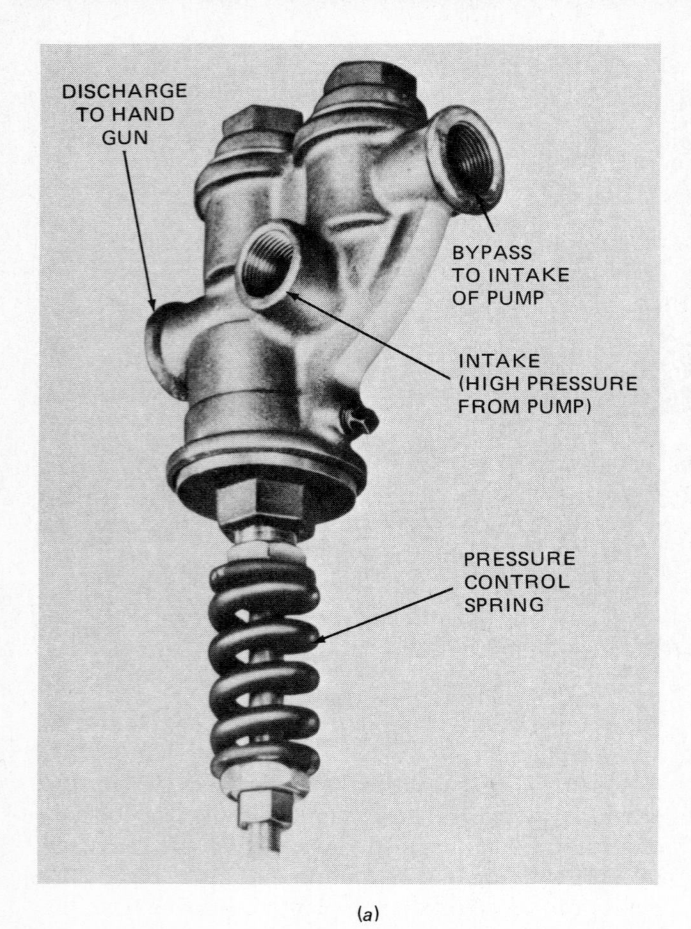


(a)

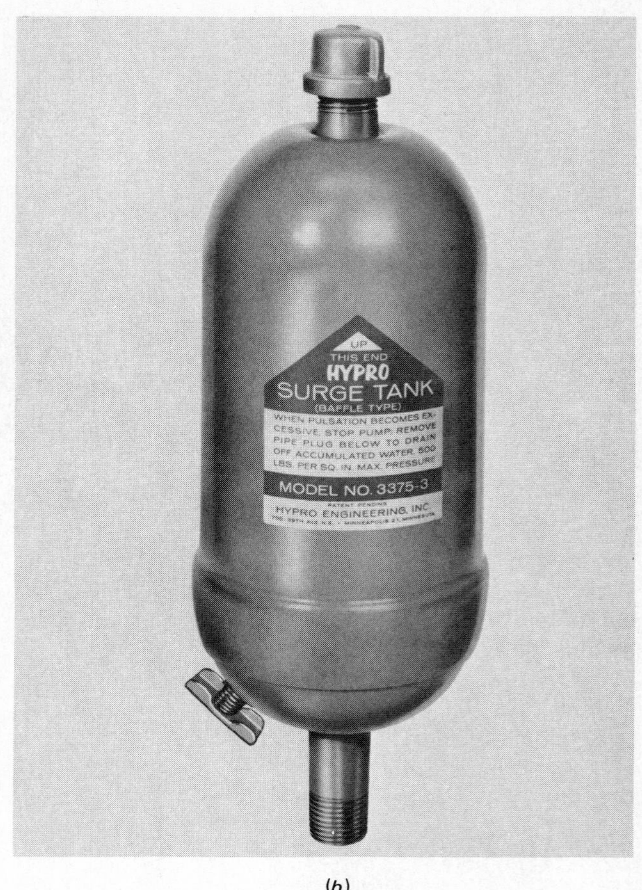


(b)

**Fig. 20-16.** Unloader valve (*a*) and surge tank (*b*) are necessary when using a piston pump at pressures over 200 psi [1380 KPa]. The unloader valve reduces pressure in the pump when there is no discharge. The surge tank reduces pump pulsations. (*Hypro*)

Insecticides and herbicides which are metered by the planter unit are spread to the width recommended by the chemical manufacturer. The granules fall from the metering gate into a feed tube and are spread evenly over the width of the band by a diffuser *fan* (Fig. 20-20). Insecticide is placed immediately following the furrow opener, and herbicide is placed following the press wheel covering the grain.

Fertilizer drop tubes on a grain drill fall to the side and rear of the furrow opener. This prevents the seed from coming in direct contact with the fertilizer.

## CALIBRATING THE EQUIPMENT

The specialist in agricultural power and machinery will be working directly with people who use machinery to apply agricultural chemicals. As a service person, you also have the responsibility to set up and test new equipment and make repairs. If you are a service manager, you will supervise mechanics and service personnel who will be working with used chemical application machinery. It is important that service workers understand the potential health hazards that exist when being exposed to both pesticides and fertilizer. Use caution and protect yourself and others from being severly injured or poisoned, and avoid harming the environment. Always wear protective equipment (eye protection, respirators, gloves, and protective clothing) when it is necessary for you to come in close contact with chemicals. For example, anhydrous ammonia fertilizer equipment regulates the gas vapor injected into the soil. When working with this equipment you must wear protective goggles, gloves, and skin protection. Protective paraphenalia is necessary to prevent accidental freezing of the skin and dehydration of the eye, which can cause blindness.

Enactment of Public Law 94-140 established federal regulation through the Environmental Protection Agency (EPA). This agency set minimum regulations to be met, before certifying private and commercial applicators to work with pesticides.

**⚠ WARNING It is not possible in this chapter to identify all the chemicals and the hazards that are present. Regard all chemicals, as well as machinery which has been exposed to pesticides and has not been decontaminated, as hazardous.**

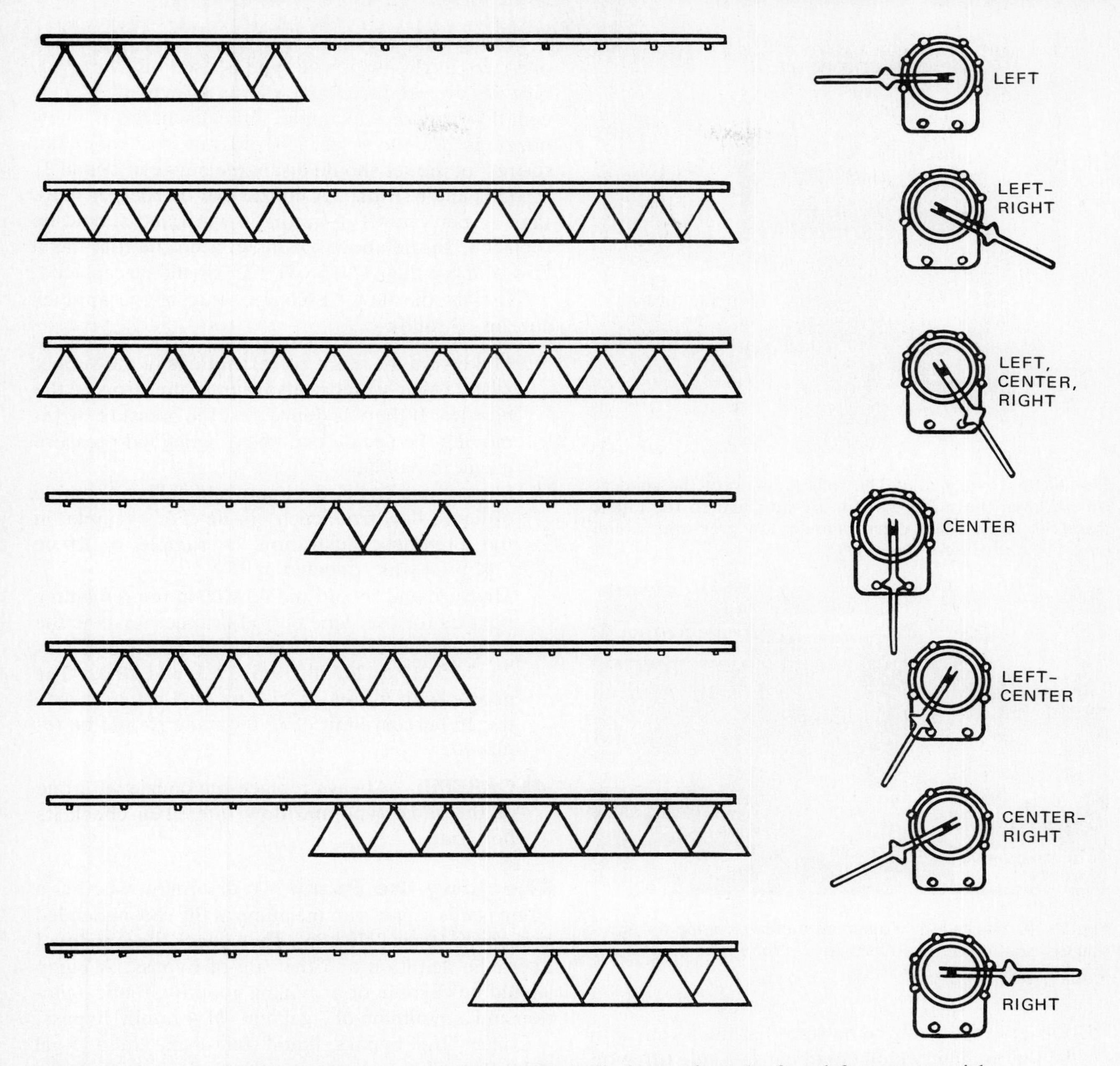


**Fig. 20-17.** The sprayer boom selector valve operator controls discharge of nozzles from left, center, or right boom. (*Spray Systems Co.*)

## Preparing the Equipment

Specific problems that relate to equipment malfunctions which appear to be minor can develop into serious application errors. The margin of safety in using pesticides is narrow. The difference between the amount of herbicide to control weeds and that which will damage a crop is very small; it can be neither too much nor too litte. Recommendations of the chemical manufacturer must be followed because each product is specific as to time and rate of application.

When working with sprayers, check out the equipment by cleaning and inspecting all the screens and strainers as follows:

**⚠ WARNING Never place screens or tips to your mouth to blow them clean of sediment.**

1. Use soapy water, a special brush, reverse flushing with water, and a wooden toothpick to clean a clogged tip orifice.
2. Remove sediment from the tank and inspect it for rust or foreign material. Rinse if necessary.
3. Determine if the pump rotates freely before starting the PTO drive. Make sure safety shields are in place and the pump torque arm and drive are properly secured.
4. Place water in the tank and slowly operate the pump. (Roller and piston pumps are lubricated by the liquid and should not be operated dry.)

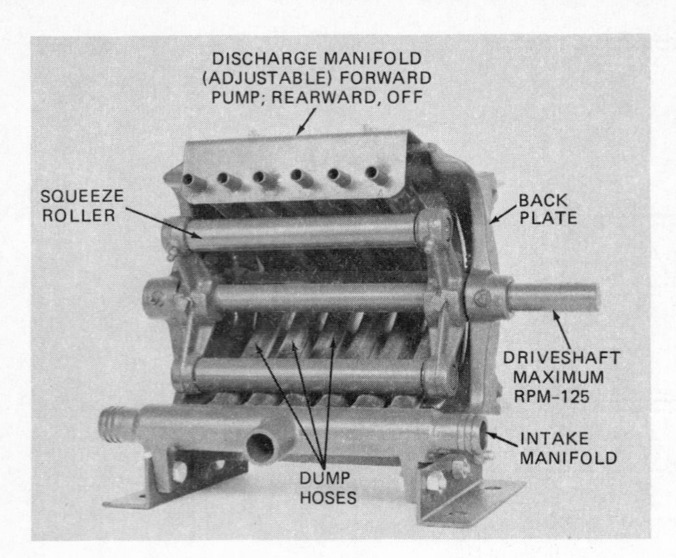


**Fig. 20-18.** Hose pump. The rolling action of the squeeze rollers upon the hoses (which are attached to the intake manifold) meter exact volumes under low pressure. (*John Blue Co. and George Stewart*)

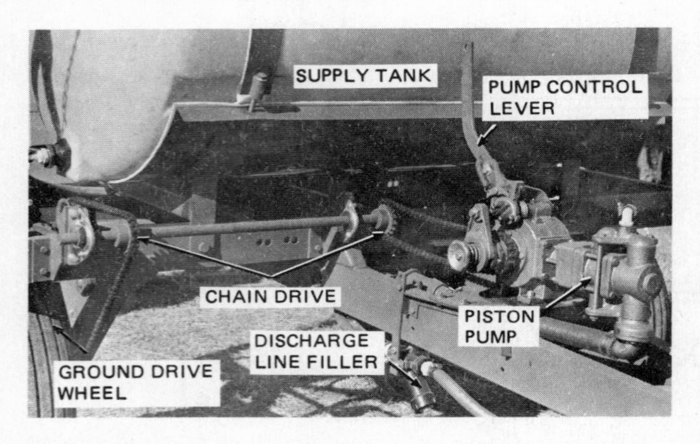


**Fig. 20-19.** Positive-displacement metering pump for accurate distribution of liquid fertilizer. (*John Blue Co.; David Johnson, photographer.*)

5. Observe the pump to be sure that the torque arm is not binding, which could cause a side force on the pump bearing.

**Checking the Nozzles** As soon as the pump is primed, operate at standard PTO speed. Set the pressure regulator for a gauge pressure reading of approximately 40 psi [276 kPa]. Check the system for leaks and replace leaking hose or tighten fittings. Inspect the discharge from each of the nozzles to evaluate the spray pattern. If the pattern differs, the tip may be worn. It could also be a different type or have a different distribution rate. Check the manufacturer's ID located on the end of the tip for angle of spray, type of spray and discharge rate in gallons (or liters) per minute (see Appendix C, Table C-4).

The amount of nozzle wear is impossible to check by visual inspection. Wear is determined by measuring and comparing the amount of discharge for an equal period of time and pressure setting.

Testing also reduces the chances that more than one size, angle, or nozzle type is used on a sprayer. Nozzles are produced to a tolerance level of ±5 percent flow rate. For example, if the discharge of a new nozzle is 20 ounces (oz) [591 mL] in 15 seconds (s), the rest of the set should discharge between 19 and 21 oz. [562 and 621 mL]. A nozzle that discharges more than 10 percent off its rating is worn and should be replaced. In the above example, a nozzle that has a flow of more than 22 oz [651 mL] should be replaced.

Evaluate the flow (discharge) rate of the sprayer nozzles as follows:

1. Make sure there are no restrictions in the supply system that would cause a pressure drop at the nozzles. If there is doubt, test the pressure by removing the nozzle and inserting a good pressure gauge in its place.
2. Start at one end of the boom and collect a sample of liquid flow from each nozzle. For example, in the previous illustration, a sample of 20 oz [591 mL] was collected in 15 s.
3. Measure and record the flow from the remaining nozzles for the same period of time. Assume the volume collected for the remaining nozzles was 20, 20.5, 19.5, 20, 20, 19.5, 22.5, and 20 oz. The nozzle with a flow of 22.5 oz [665 mL] exceeds the 10 percent limit of 22.0 oz and should be replaced.

**⚠ CAREFUL** Always replace the nozzle with one of the same type and flow rate; then check its flow rate.

**Checking the Pump** To determine whether a pump needs repair, run the pump at the recommended rpm and at 40 psi [276 kPa]. Then check the amount of hydraulic agitation and the rate of bypass. A pump should be capable of providing good hydraulic agitation and a minimum of 3 gal/min [11.4 L/min] bypass.

Collect the bypass liquid for 1 min in a 5-gal [18.9 L] bucket and evaluate the pump wear from the amount collected.

## Calibrating Sprayers

A sprayer is calibrated to evaluate whether the proper volume of liquid is being applied to a specific area of ground. The factors affecting the discharge rate of a sprayer are:

1. Nozzle tip flow (discharge) rate. The flow rate varies with (1) the size of the orifice in the tip, (2) the viscosity of the fluid, and (3) the pressure on the fluid. Appendix C, Table C-4 presents the flow rate of one nozzle in gallons per minute and in gallons per acre at various speeds in miles per hour.
2. The solution density (viscosity). The conversion factors for density of solution are based on its

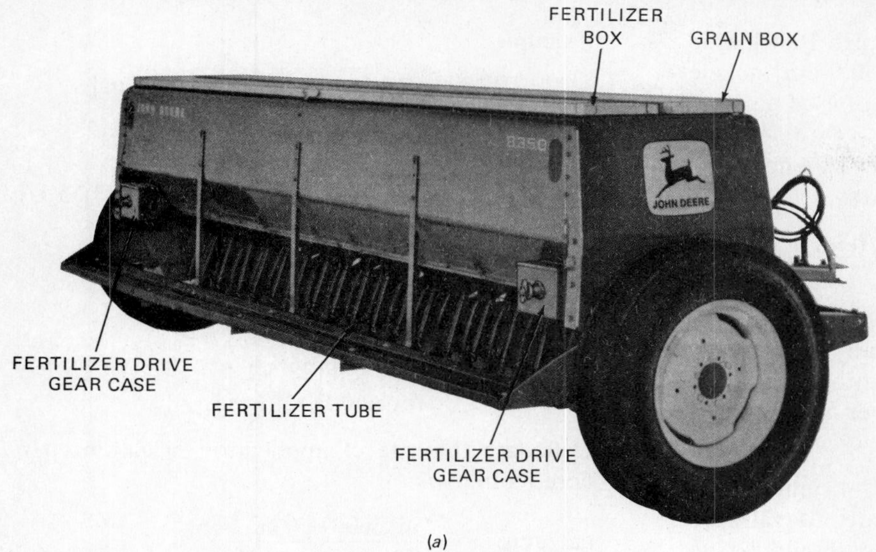


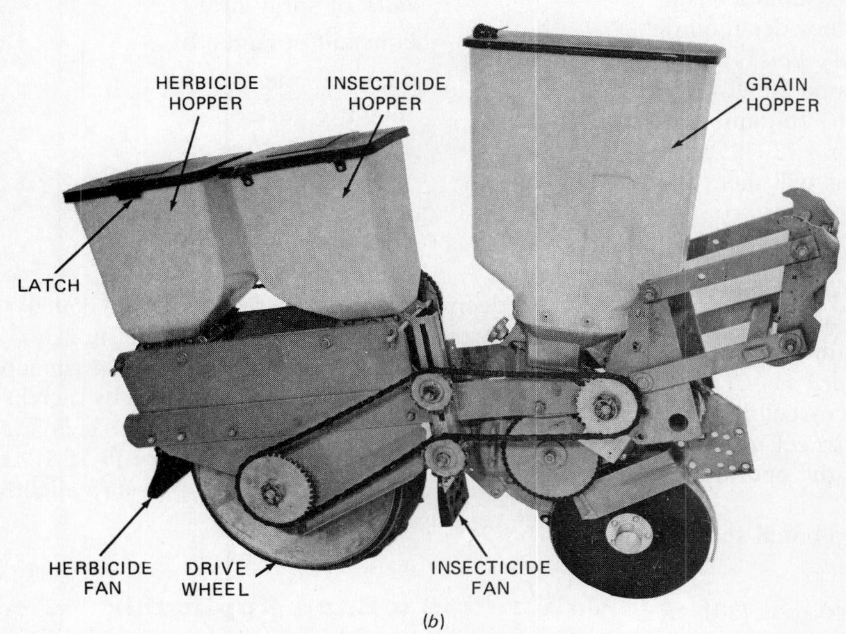


**Fig. 20-20.** Fertilizer drill (*a*) (*Deere & Co.*) and planter (*b*) with pesticide hoppers (*c*) are typical of granular application machinery. (*Allis-Chalmers Co.*)

weight per gallon and specific gravity; see Appendix C, Table C-4). Note that water is the standard base with a value of 1.00.)

3. The pressure of the fluid. The effect of pressure on the flow rate of the nozzle is listed in capacity measured in gallons per minute (GPM) (Appendix C, Table C-4).

**Choosing the Nozzle Flow Rate** You may have to assist a customer in selection of the proper size of nozzle for the distribution rate needed. The rating stamped on the nozzle is in gal/min (Appendix C, Table C-4) at a pressure of 40 psi [276 kPa].

Distribution rate on a label of a spray chemical is usually listed in the number of gallons of liquid per acre. If you wish to help the customer determine the amount of liquid in gallons per acre (gal/acre) a sprayer will distribute, proceed as follows:

$$\text{gal/acre} = \frac{5940 \times \text{gal/min per nozzle}}{\text{mi/h} \times W}$$

where $W$ = nozzle spacing on the boom, or width of the spray swath in boomless spraying (in inches)

5940 = constant determined by

$$\frac{43{,}560 \text{ ft}^2\text{/acre} \times 12 \text{ in/ft}}{88}$$

88 = 88 ft/min = 1 mi/h [1.6 km/h]

**Example** A TeeJet 730154 (Delevan LF-1.54) nozzle has a discharge of 0.154 gal/min [0.6 L/min] at 40 psi [276 kPa]. The customer wishes to spray at 4 mi/h [6.4 km/h], with 20-in [50.8-cm] spacing. What is the discharge rate of the nozzle in gallons per acre?

$$\text{gal/acre} = \frac{5940 \times \text{gal/min per nozzle}}{\text{mi/h} \times W}$$

$$= \frac{5940 \times 0.154}{4 \times 20}$$

$$= 11.4 \text{ gal/acre [106.6 L/ha]}$$

**Solution** Operate the sprayer at 4 mi/h [6.4 km/h] and 40 psi [275.8 kPa], with 20-in [50.8-cm] nozzle spacing, and apply liquid at the rate of 11.4 gal/acre [106.6 L/ha].

If it is necessary to select the proper-size nozzle in gallons per minute, proceed as follows:

$$\text{gal/min per nozzle} = \frac{\text{gal/acre} \times \text{mi/h} \times W}{5940}$$

**Example** Assume the customer wishes to apply 20 gal/acre [187 L/ha] at 5 mi/h [8.05 km/h], with a nozzle spacing of 20 in [50.8 cm]. Determine the nozzle flow rate required in gallons [liters] per minute using a 73° angle flat-spray pattern nozzle.

**Solution** Using Appendix C, Table C-4, enter the 5 mi/h [8.05 km/h] column and locate 20 gal/acre [187 L/ha]. Reading left, the nozzle should be an LF-3.08 or a 730308. However, these nozzles require a working pressure of 50 psi [344.7 kPa]. This is above the desired pressure. Locate the second 20 gal/acre [187 L/ha] in the 5-mi/h [8.05 km/h] column. The recommended nozzle is an LF-3.85 or a 730385 with 30 psi [206.8]. Either of these nozzles will meet the customer's exact needs.

**Checking Sprayer Discharge** The actual discharge of a spray rig must be checked and correction made if the rate in gallons per acre [liters per hectare] varies from the desired amount. The procedure is referred to as *calibrating the sprayer*. Calibration is required in order to obtain the correct distribution by evaluating the speed of travel and the pressure setting that will be used by the operator in the field.

The shop procedure for calibrating a boom sprayer is as follows:

1. Measure a distance of 660 ft or 40 rods [201 m] on a level surface. Erect stakes at each end.
2. Drive the spray rig over the 660-ft course at the gear and throttle setting that will be used during the time the chemicals will be applied. Accurately measure and record the time required to cover the course.
3. Move the rig to a level spot and set the brakes.
4. Operate the sprayer at the pressure that will be used in the field.
5. Catch the discharge from three or four nozzles (one nozzle from each boom of a three-boom rig) for the exact length of time that it took to travel the 660 ft in step 2.
6. Record the volume of liquid obtained from the test nozzles and calculate the volume that would have been delivered by all nozzles as follows:

$$\begin{array}{c}\text{gal applied over}\\ \text{660 ft [201 m]}\end{array} = \frac{\begin{array}{c}\text{gal}\\ \text{caught}\end{array} \times \begin{array}{c}\text{number of nozzles}\\ \text{on sprayer}\end{array}}{\begin{array}{c}\text{number of nozzles}\\ \text{used in test}\end{array}}$$

**Example**

Width of spray rig = 40 ft [12.2 m]
Total number of nozzles = 24
Number of nozzles tested = 3
Total gal [L] caught in test = 3.75 quarts [3.55 L] or 0.94 gal

**Solution**

$$\text{gal applied} = \frac{0.94 \times 24}{3}$$
$$= 7.52 \text{ gal [28.5 L] for 660 ft [201 m]}$$

7. Calculate the rate of application in gallons per acre as follows:

$$\text{gal/acre} = \frac{\text{gal/660 ft} \times 66}{\text{width of spray in ft}}$$

where 66 = constant obtained by

$$\frac{43{,}560 \text{ ft}^2\text{/acre}}{\text{660-ft course}}$$

**Solution**

$$\text{gal/acre} = \frac{7.52 \text{ gal} \times 66}{\text{40-ft spray width}}$$
$$= 12.4 \text{ gal/acre [116 L/ha]}$$

If the desired application rate was 11.4 gal/acre (as in the previous example, of the TeeJet nozzle), a reduction in the discharge from 12.4 to 11.4 gal/acre [116 to 106.6 L/ha] could be accomplished by increasing the speed or decreasing the pressure. It is suggested that the reduction of 1.0 gal/acre [9.35 L/ha] be accomplished by decreasing the pressure slightly and then repeating the test.

**Calibrating a Band Applicator** The procedure for calibrating the sprayer for band spraying is the same as broadcast spraying, except that the width of the band and the width of the row must be taken into consideration. The chemical manufacturer may list the application rate as *broadcast rate* in gallons per acre [liters per hectare] rather than *field rate* in gallons per acre [liters per hectare]. Field rate considers the width of the band and the row spacing. Most band spraying is done with even flat-fan spray tips. To determine the size nozzle to use, convert the broadcast rate to the field rate as follows:

$$\begin{array}{c}\text{gal/acre,}\\ \text{field rate}\end{array} = \frac{\text{band width (in)}}{\text{row spacing (in)}} \times \begin{array}{c}\text{gal/acre,}\\ \text{broadcast rate}\end{array}$$

**Example** The broadcast rate is 20 gal/acre [187 L/ha], the band width is 14 in [35.6 cm], and the row width is 40 in [101.6 cm]:

$$\begin{array}{c}\text{gal/acre,}\\ \text{field rate}\end{array} = \frac{14}{40} \times 20$$
$$= 7.0 \text{ gal/acre [65.5 L/ha]}$$

The proper tip for the even flat-fan nozzle is obtained from the manufacturer's tip selection data such as are provided in Appendix C, Table C-5. For other row or band widths, the conversion factors at the bottom of Table C-5 are used. The appropriate conversion number of gallons per acre [liters per hectare] in the treated area in gallons per acre [liters per hectare]. For example, for 30-in [76.2-cm] rows, the number of gallons per acre [liters per/hectare] in the table is multiplied by 1.33.

**Calibrating Granular Applicators** The process for calibrating granular applicators is not as complex as for sprayers. The process is well outlined in the operator's manual. Always follow the steps the manual provides, because the method of making adjustments varies with each machine.

**Grain drill fertilizer attachment** To calibrate the grain drill fertilizer attachment, proceed as follows:

1. Raise the drill off the ground so that the wheels can be rotated.
2. Place a container under each feed tube to catch the fertilizer.
3. Set the gear and chain drive settings on the fertilizer attachment (see Appendix C, Table C-6).

   For example, 100 lb [45.4 kg] of fertilizer per acre is to be distributed in 7-in [17.8-cm] rows. The density of the fertilizer affects the discharge rate. Therefore apply the conversion factor to density. If the fertilizer density rating is 60, use the settings to deliver 110 lb/acre [123.2 kg/ha] (100 × 1.10 conversion factor = 110 lb [50 kg]). This would require that the chain drive be on Drive 1 and the gear box set to E-3 to give 113 lb/acre [126.7 kg/ha].
4. Add fertilizer to the box, jack up the wheels, and rotate the ground wheels the correct number of turns to distribute 1/20 of an acre. (Refer to Table 19-7 in Chap. 19: a 7 × 14 grain drill equipped with 6.70 × 15 tires requires 728 turns of the wheel to cover an acre. For 1/20 acre, the wheel would be rotated 36.4 turns.)
5. Carefully weigh the fertilizer in all the containers and multiply by 20. For example, assume 5.1 lb of [2.31 kg] fertilizer was collected. This would equal 102 lb [46.3 kg] of fertilizer per acre (5.1 lb × 20 = 102 lb [46.3 kg]). The desired rate was 110 lb [50 kg].
6. To increase the flow, adjust the gear case in the fertilizer drive to the D5 gear setting and repeat the test.

**Differences in Calibrating Band Applicators** Calibrating of band applicators such as are used on planters (Fig. 20-20*b*) differs from drill applicators of fertilizers in the following ways:

First, the mechanical feed involves the adjustment of a metering gate by adjusting a screw or lever.

Second, the chemical manufacturer may recommend the rate of application of granular chemicals in the following ways:

1. Pounds per acre [kilograms per hectare] for a given band width and row spacing
2. Pounds per acre [kg/ha] for complete (broadcast) coverage

**Band application** When the chemical manufacturer recommends pounds per acre for band width and row spacing, it is necessary to follow the steps listed in the operator's manual to set the *metering gate*. (See Appendix C, Table C-7 and Fig. 20-21 for examples of equipment manufacturer's recommended settings of the metering gate.)

If the chemical manufacturer recommends pounds per acre for complete (broadcast) coverage only, the pounds per acre for band width and row spacing must be determined. The band width will provide the approximate pounds of the chemical concentration per square inch [square centimeter] as recommended for complete (broadcast) coverage. The following formula is used to find the pounds per acre for band width and row spacing:

$$\text{lb/acre (band width and row spacing)} = \frac{\text{lb/acre (broadcast rate)} \times \text{band width (in)}}{\text{row spacing (in)}}$$

**Example** The chemical manufacturer recommends 20 lb/acre [22.4 kg/ha] for complete (broadcast) coverage. The band width is 14 in [35.6 cm], and row spacing is 40 in [101.6 cm].

**Solution**

$$\text{lb/acre (band width and row spacing)} = \frac{20 \text{ lb} \times 14 \text{ in}}{40 \text{ in}} = 7 \text{ lb/acre [7.8 kg/ha]}$$

Delivery rate for 14-in [36 cm] bands and 40-in [102 cm] rows would be at a rate of 7 lb/acre [7.8 kg/ha] (broadcast) when 20 lb/acre [22.4 kg/ha] is recommended.

**Calibrating the applicator** The rate of delivery may be affected by the temperature, humidity, ground conditions, speed of travel, or obstructions in the metering gate. Large variations exist in the "flowability" of chemicals even within the same product.

The calibration must be performed in the field at planting speed. (*Note:* The press wheel driving the unit must not be turned by hand because field conditions cannot be simulated accurately.)

To check the application rate, proceed as follows:

1. Set the metering gate to the proper rate as determined from the operator's manual.

   **Example** Appendix C, Table C-7 lists gates and dial numbers to obtain the pounds per acre at 5-

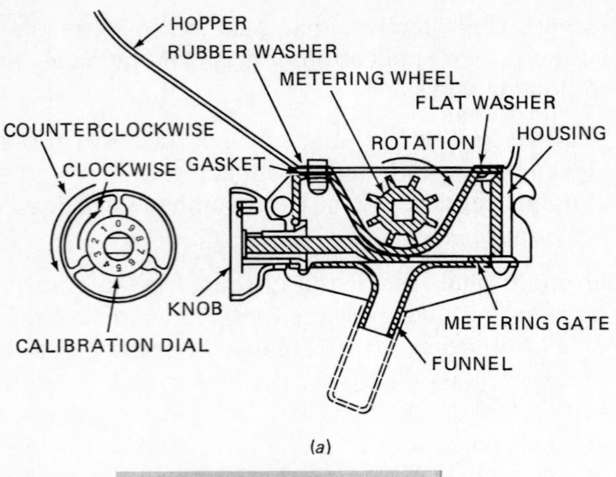


(a)

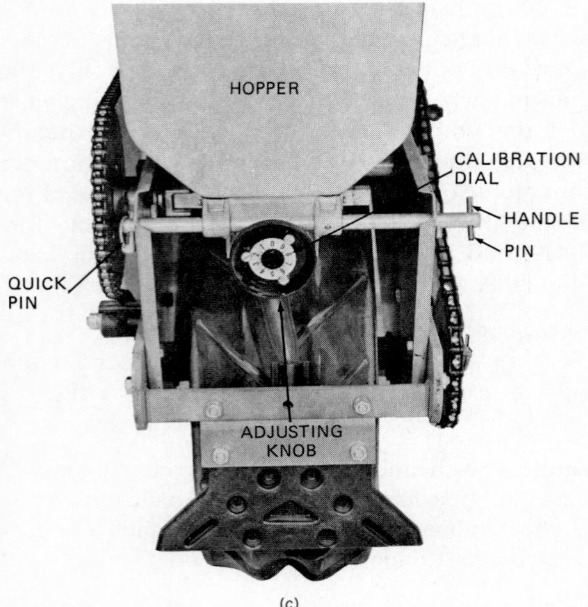


(c)

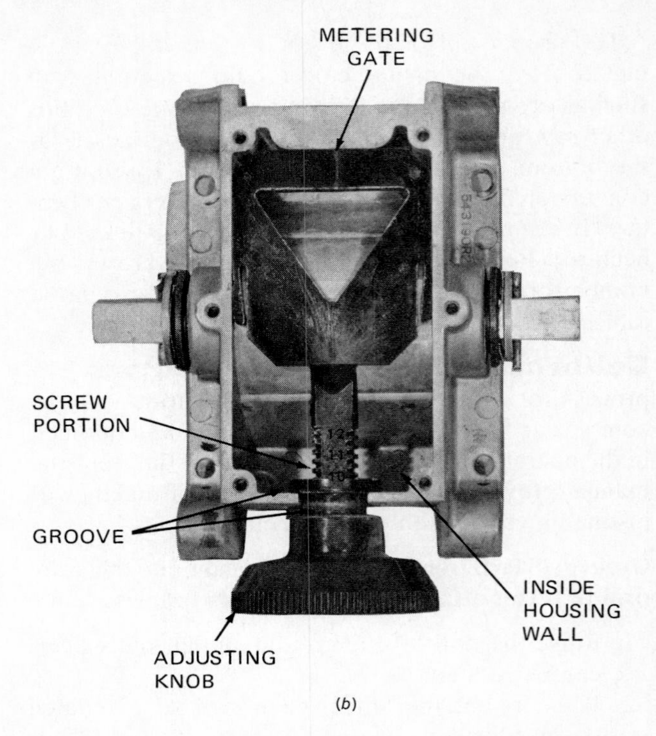


(b)

**Fig. 20-21.** Granular applicator details for adjustment and nomenclature of parts. (*Allis-Chalmers Co.*)

mi/h [8.05 km/h] travel speed for each of eight row spacings. For 7 lb/acre [7.8 kg/ha] in 30-in [76.2 cm] rows, try a setting of gate number 2 and dial number 7.

2. Fill the hoppers with pesticide granules and attach a bag under each applicator to catch the discharge.
3. Drive 1320 ft (80 rods) [402 m] at planting speed (approximately 5 mi/h) [8.05 km/h].
4. Select one of the bags, and record the weight that was distributed.
5. Multiply the ounces [milliliters] delivered by a correction factor for the number of rows per acre found in Appendix C, Table C-7 and divide by 16.

**Example** The planter is set for 30-in [76.2 cm] rows. The weight of one bag of chemicals caught from one row was 8.2 oz [0.23 kg]. From Table C-7 find approximately 13.3 30-in [76.2-cm] rows, 80 rods [402 m] long per acre. Therefore:

$$\text{lb/acre} = \frac{\text{weight of one bag (oz)} \times \text{row factor}}{16 \text{ oz/lb}}$$

$$= \frac{8.2 \times 13.3}{16 \text{ oz}}$$

$$= 6.8 \text{ lb/acre [7.6 kg/ha]}$$

6. Check the remaining bags for uniformity of weight, compared with the bag first selected. If they are uniformly equal, proceed to readjust the metering gate to obtain the desired rate of application.

In this example, the desired rate was 7 lb/acre [7.8 kg/ha]. The calibration check produced 6.8 lb/acre [7.6 kg/acre]. A slightly wider adjustment in the metering gate should be made and the test repeated until the desired amount is delivered.

## MAINTAINING CHEMICAL APPLICATION EQUIPMENT

Equipment will last much longer and give better service if it is properly cleaned and cared for. This applies to use during the application period or season and before storage.

### Maintaining Pesticide Equipment

In addition to preventing damage, pesticide application equipment must be thoroughly cleaned when changing from one chemical to another. The cleaning is necessary to protect the equipment and to avoid damage to crops and livestock from residues left in the sprayer.

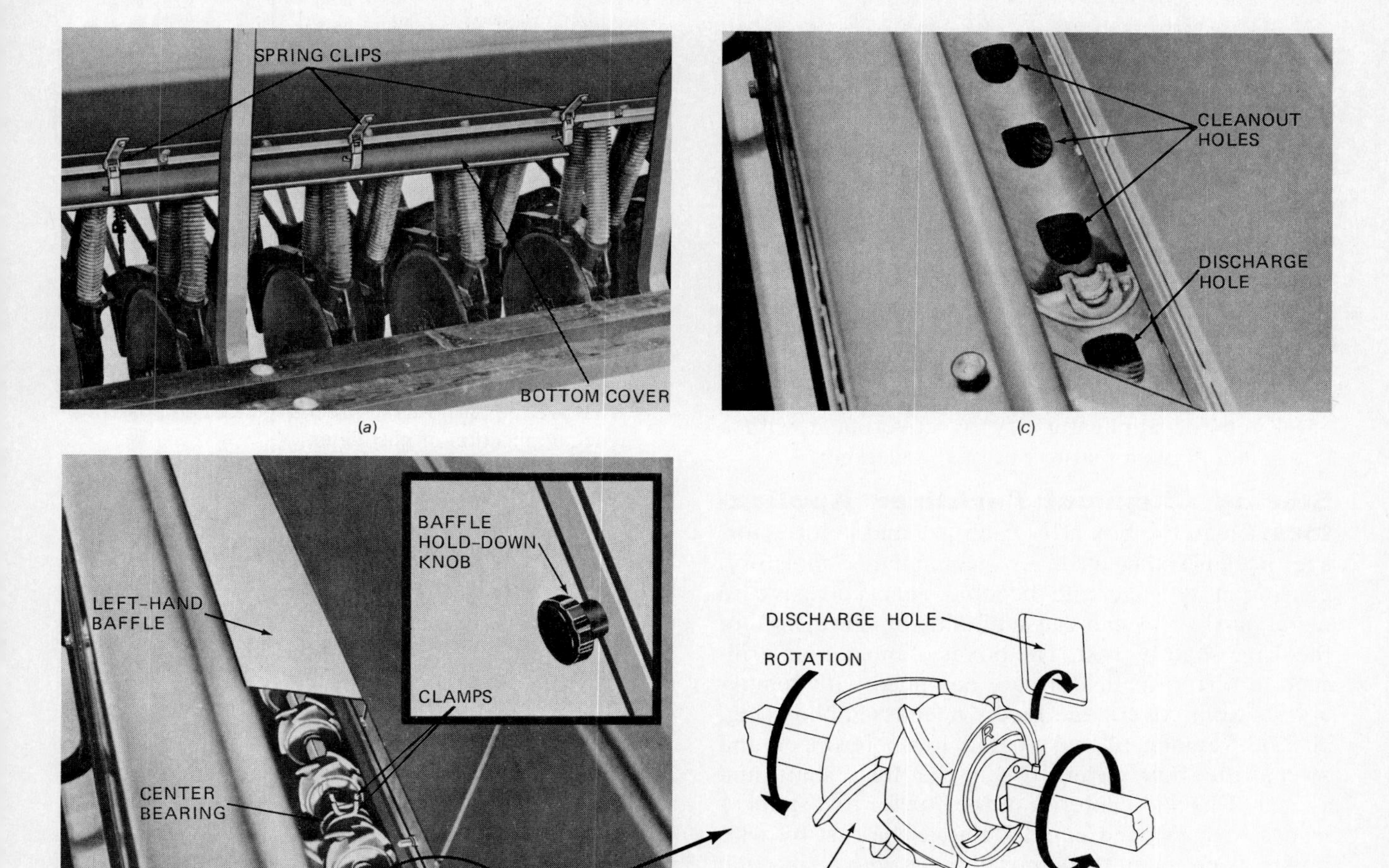


**Fig. 20-22.** Grain drill fertilizer distributor. (*a*) Hinged-bottom cleanout cover, (*b*) feed shaft, (*c*) cleanout with feed shaft removed. (*Deere & Co.*)

When cleaning, servicing, and preparing equipment for storage, always observe all safety precautions. This includes wearing the proper safety clothing and equipment and disposal of residue. Carefully follow the instructions on the label for disposal of remaining chemicals and containers. When cleaning the equipment, collect the residue to prevent it from entering sewers or contaminating water supplies from runoff.

**Cleaning a Sprayer** To clean a sprayer for storage, proceed as follows:

1. Remove the nozzles and clean them separately.
2. Drain the tank and clean the suction and line strainers thoroughly before replacing them.
3. Flush the tank of residue.
4. Add strong soapy water to the tank and wash the inside of the tank thoroughly. Rinse the pump and flush the entire system first with the bypass operating, then with the pressure relief control spring tension completely released.
5. Refill the system with a solution of 1 pint [0.5 L] of household ammonia to 25 gal [95 L] of water. Operate the pump to circulate the solution. Then let stand overnight before draining and flushing the entire system.
6. A roller pump equipped with nylon rollers should be stored in a can of SAE 20 motor oil. (*Note:* Do not store a rubber roller pump in oil.)
7. Clean all nozzles and store in a closed box.
8. Remove the end caps from wet booms and drain before storing.
9. Store polyethylene tanks out of the sun.
10. Clean and repack wheel bearings of trailer-type rigs.
11. Flush exterior parts of equipment with water; use touch-up paint where needed.

**Storing Granular Applicators** Granular chemicals must be kept dry. Any form of moisture can be a problem, causing the granules to harden, which damages the metering system. Protect the entire hopper by covering it with a plastic bag if left in the field overnight. Clean granular applicators as follows:

1. Remove chemical hoppers from the metering unit and clean.
2. Remove the metering wheel from the housing.
3. Clean the metering opening and shaft.
4. Assemble the components in the reverse order.
5. Keep the applicator dry by covering it with a plastic bag and store out of the weather.

**Storing Chemical Fertilizer Applicators** Clean the box after each use and before storage because the fertilizer accumulates moisture, causing it to cake and become very corrosive to metal parts. This can cause binding or "freezing" of the drive shaft by rust. The boxes of most grain drill-type fertilizer applicators are equipped with removable bottom covers and feed wheel assemblies (Fig. 20-22). Remove all caked and loose fertilizer and sweep the box clean. Clean the feed shaft and wheels. Check the interior of the box for rust spots or places with chipped paint. Remove the rust by wire brush. Then, paint the exposed areas with a zinc-rich paint. When dry, coat the inside of the box with diesel fuel.

Remove and clean the fertilizer tubes and store them inside the grain box. Remove the drive chains and clean them with diesel fuel. Soak them in oil before storing.

Lubricate all fittings, bearings, and moving parts. Coat the disks or shovels with oil or grease and release the spring tension by lowering the disks on a board. Place the drill on jacks so that the tires do not touch the ground.

## SUMMARY

Agricultural chemicals are an essential part of modern agriculture. Sustained high yields of crops depend on the use of pesticides and commercial fertilizer.

Chemical application equipment must be assembled and finely tuned to distribute the proper quantity of materials for effective control of weeds and insects and application of commercial fertilizer.

Personal and environmental safety must be observed when servicing, maintaining, or repairing equipment which handles agricultural chemicals. Protection for the eyes, respirators, and suitable clothing must be worn to prevent serious accidents due to freezing, dehydration, inhalation, and skin absorption.

An understanding of the function and operation of components used in sprayers and granular applicators will aid an operator, a salesperson, and the local service center personnel to become more competent. The professional in agricultural power and machinery serves an important role as a person who is able to observe safety regulations, select proper components, calibrate, clean, repair, and properly prepare equipment for storage.

## THINKING IT THROUGH

1. List three reasons why the professional in power and machinery should understand the operation and function of equipment for the application of agricultural chemicals.
2. What are pesticides?
3. What are the purposes of preemergence and postemergence application of pesticides?
4. What are the forms in which pesticides and fertilizers are applied? What equipment is used for each?
5. What are the four "ground equipment" types of sprayers? Identify two machines for each type.
6. What is the characteristic of a positive-displacement pump? Which of the three types of spray pumps fit this classification?
7. Why is a centrifugal pump becoming more commonly used with low-pressure sprayers?
8. List four factors involved in the selection of the tip of a spray nozzle.
9. List one use for each of the six common types of nozzle tips.
10. What are the three types of strainers used in the system of low-pressure sprayers?
11. Identify the purpose of each of the following types of controls for a low-pressure sprayer: (1) main shutoff, (2) pressure relief, (3) unloading, (4) boom selector valve.
12. What are the characteristics of a good material for construction of a spray tank?
13. What personal protection is required when working with agricultural chemicals?
14. How can the amount of wear be evaluated in a nozzle tip? In a roller pump?
15. How does each of the following affect the discharge rate of a spray nozzle? (a) Tip size (b) Density of solution (c) Pressure
16. What are the characteristics of the following spray tips? (a) TeeJet 800462 (b) Delavan LF 4.62 ± 80°
17. Why is it necessary to record the time required to travel 660 ft (40 rods) [201 m] when calibrating a sprayer?
18. Write the formula for calculating the rate of sprayer discharge in gallons per acre [liters per acre].
19. What adjustments must be made in the formula when calculating the field rate in gallons per acre [liters per acre] for band application?
20. What are two methods of controlling the distribution rate of granular materials?
21. Determine the pounds per acre [kilograms per acre] of granular pesticides required to treat 30-in [76.2-cm] rows with a 14-in [35.6-cm] bank at the broadcast rate of 10 lb/acre [112 kg/ha].
22. What material is used to decontaminate a sprayer of chemicals? When is decontamination necessary?
23. Why must a dry chemical applicator for pesticides or commercial fertilizer be protected from moisture and cleaned after each application?

# UNIT VI HARVESTING AND HANDLING AGRICULTURAL PRODUCTS

| Competencies | Agricultural Machinery Service Manager | Agricultural Machinery Parts Person/ Manager | Agricultural Machinery Mechanic | Agricultural Machinery Setup and Delivery Mechanic | Agricultural Tractor and Machinery Mechanic's Helper | Tractor Mechanic | Farmer, Grower, and Rancher |
|---|---|---|---|---|---|---|---|
| Identify various harvesting equipment and component parts | Essential | Essential | Essential | Essential | Desirable | Desirable | Essential |
| Use operator's manuals to determine recommended adjustments/ maintenance | Essential | Desirable | Essential | Essential | Desirable | Desirable | Essential |
| Perform field adjustments on harvesting equipment | Desirable | Not necessary | Essential | Essential | Desirable | Desirable | Essential |
| Determine harvesting losses | Desirable | Desirable | Essential | Not necessary | Desirable | Desirable | Essential |
| Select harvesting equipment for crop and field conditions | Desirable | Not necessary | Desirable | Not necessary | Not necessary | Not necessary | Essential |
| Follow recommended safety precautions during service and operation | Essential | Essential | Essential | Essential | Essential | Essential | Essential |

Essential

Desirable

Not necessary

Mechanized harvesting is essential to produce food and fiber to supply the needs of the world's population. Agricultural producers depend on machines to harvest and handle crops quickly and efficiently with a minimum amount of labor. Through the combined efforts of those engaged in research, development, and production, the productive capacity of farmers and ranchers in the United States has increased dramatically in recent years. Agricultural producers today produce enough food and fiber for themselves *and* 64 other people.

Reliance on machines has created new demands on agricultural producers. They must be able to select equipment that is suited to their needs. They must be able to operate it safely and efficiently. It is important that they understand the machine, its operation, the required maintenance, and recommended adjustments.

There is also a great demand for people trained in sales and service. Producers depend on equipment companies and dealers to provide sales and service. In order to recommend the right machine, a salesperson must understand the crop being harvested, the harvesting conditions, and the operating principles of the machines.

Service workers must be able to identify operating problems and make necessary repairs. In addition, they may be required to make adjustments and recommendations in the field.

Parts workers should be able to discuss machine operation with a customer and recommend special equipment to aid in specific harvesting conditions. They need to understand both the machine and its working parts so that they can supply the correct replacement parts to service people or equipment owners.

In this unit, much emphasis will be placed on using the operator's manual that the manufacturer supplies with the machine. It is important that the operator understand all safety rules and precautions before operating a machine. The operator's manual should be referred to when making all operational adjustments. This will result in better performance and more reliable service. Most operator's manuals include troubleshooting guides that will help locate problems and give suggested adjustments or repairs for the particular make and model.

Upon completion of this unit, you should be able to identify equipment used to harvest many crops produced in the United States. You should be able to describe the operation of the equipment, make common field adjustments, and discuss safety precautions.

It is beyond the scope of this book to discuss all specialized harvesting equipment. It is likely that not all the equipment used in your region is discussed. However, the basic principles presented will make it easier to understand the operation of other machines.

CHAPTER 21

# HAY HARVESTING EQUIPMENT

## CHAPTER GOALS

In this chapter your goals are to:

- Identify the equipment used in hay production
- Use an operator's manual to determine recommended adjustments on hay harvesting equipment
- Make common field adjustments on hay harvesting equipment
- Describe the operation of hay harvesting equipment
- Discuss safety precautions that must be taken when operating hay harvesting equipment

Farmers, ranchers, dairy farmers, and other agricultural producers use hay. Some producers use hay the year round to provide roughage for livestock. Others bale forage crops during the growing season to provide roughage for the winter months. Still others produce hay as a cash crop and sell it to livestock producers.

Hay is produced throughout the United States. The crops used to produce hay vary from region to region, and the harvesting and storage methods also vary. Alfalfa is the leading hay crop in the United States, but many grasses, legumes, and cereal crops such as oats are used to produce hay. Straw and fodder are also baled and called hay.

The quality of the hay depends on the crop used, its stage of maturity, harvesting techniques, and storage. Hay ranges from a low-quality roughage to a nutritious feed that can supply most of the nutrient needs for many classes of livestock.

At one time hay production required a tremendous amount of labor. Hay producers now use modern equipment that has been designed to make hay production and handling a one-person operation. Agricultural machinery manufacturers have produced equipment that has revolutionized the production of hay.

Since hay production has become highly mechanized, manufacturers and dealers need a competent staff to sell and service the equipment. Producers need to be trained as new equipment becomes available so that they can operate it properly and safely. The producer also needs to be able to select the haying equipment that will result in the most economical method of producing and feeding hay.

## CUTTING AND CONDITIONING EQUIPMENT

The cutter bar is used to cut a large percentage of the hay crops grown in the United States. The cutter bar is used on many mowers, mower-conditioners, and windrowers. Multidisk mowers, rotary mowers, and flail mowers are also used.

### Cutter Bar Mowers

The cutter bar, or sickle, is used on many agricultural machines. It is used on combines and forage harvesters as well as hay equipment. The mounted cutter bar mower that attaches to a three-point hitch is widely used in hay production (Fig. 21-1). Other types of cutter bar mowers are the semi-mounted, side-mounted, and trail-type models. Most cutter bar mowers used in hay harvest have a 7-foot (ft) [2.1-meter (m)] or 9-ft [2.7-m] cutting width.

**The Cutter Bar** The complete cutter bar assembly includes the bar, inner and outer shoes, guards, swath or divider board, wear plates, knife holddown clips, and knife assembly (Fig. 21-2). The cutter bar attaches to the mower frame at the yoke, and the knife head attaches the knife assembly to the drive mechanism.

**Knife drive register** When the knife sections are in correct register, they are an equal distance from the centerlines of the guards at each end of the stroke (Fig. 21-3). If the distance is not equal, check the operator's manual for recommended adjustment procedures. Register will not be mentioned in some man-

**Fig. 21-1.** The three-point hitch cutter bar mower. *(International Harvester Co.)*

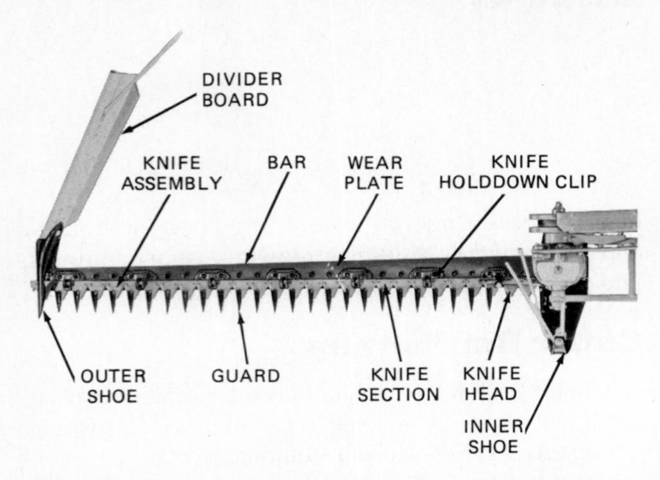


**Fig. 21-2.** The cutter bar assembly. (*International Harvester Co.*)

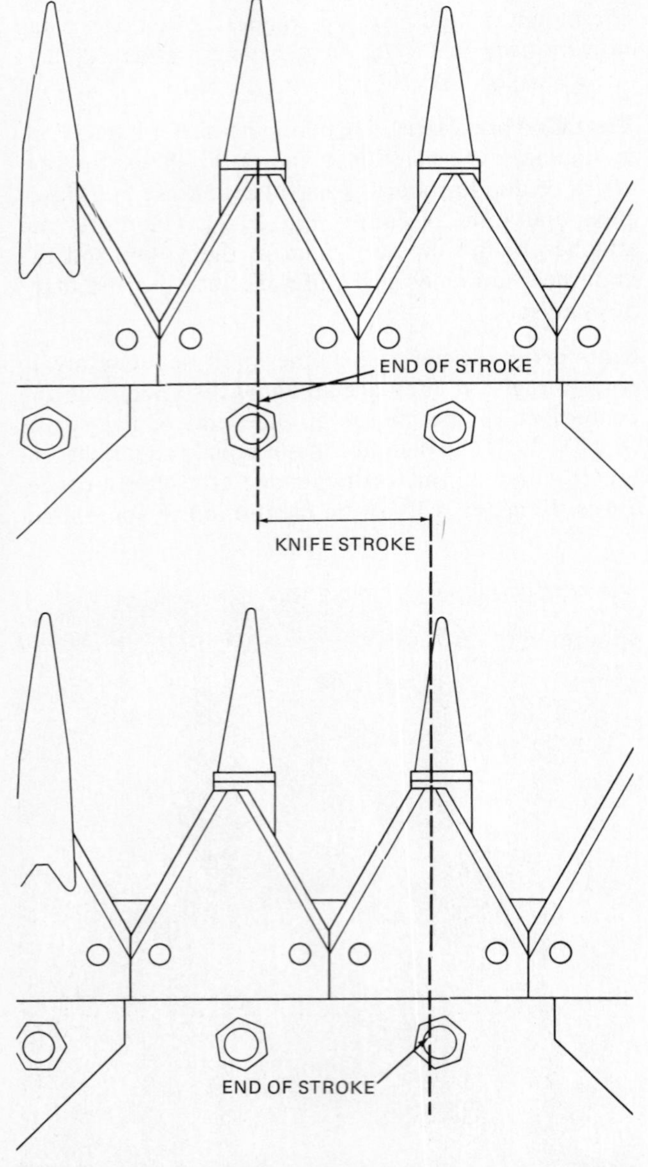


**Fig. 21-3.** Checking knife register on a cutter bar mower. (*International Harvester Co.*)

uals because it is nonadjustable, and the knife sections are in register at any blade operating angle.

**Guard alignment** It is important that the guards be properly aligned so that the knife sections rest on the ledger plates. The crop is cut in a shearing action by the knife sections and the ledger plates (Fig. 21-4). In some guards, ledger plates are held by a rivet and can be replaced when they are worn or broken. Ledger surfaces are an integral part of many newer guards and cannot be replaced.

To check guard alignment, install a new knife assembly or a straight one that is not badly worn. Tighten all the bolts holding the guards on the bar. First set the high guards down by striking at the thickest part in front of the ledger plate surface. Start at the inner shoe and work out. Then set the low guards up to secure the correct knife-to-ledger-plate fit (Fig. 21-5). Retighten all guard bolts.

**Wear plates** Wear plates provide a renewable wear surface at the point where the back of the knife contacts the bar. Wear plates should be adjusted or replaced when there is excessive clearance between the back of the knife and the wear plates (Fig. 21-5). When setting wear plates, be sure that the fronts of the knife sections do not hit the guards. The front edges of all the wear plates should be in line to give the knife a straight bearing surface.

**Knife holddown clips** The knife holddown clips hold the knife sections down against the ledger surfaces. Knife clips can be adjusted by adding or removing shims between them and the bar or by bending them with a hammer (Fig. 21-5). When knife clips are properly adjusted, the knife can be moved forward and backward through the guards without binding.

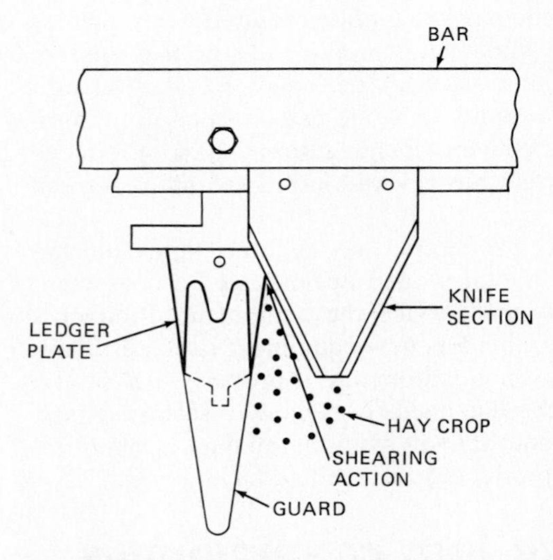


**Fig. 21-4.** The hay crop is cut by a shearing action between the ledger plate and the knife section.

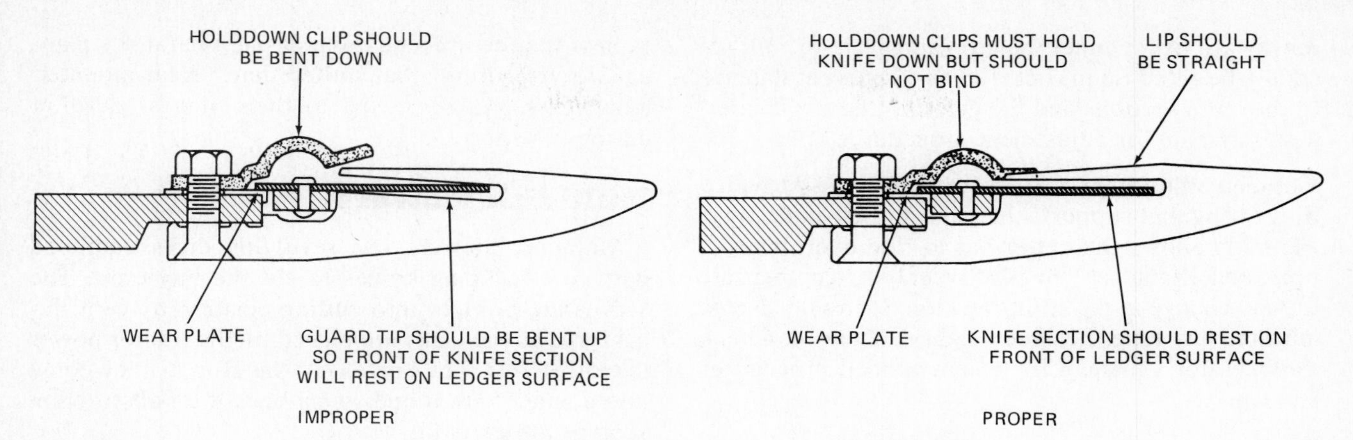


**Fig. 21-5.** Adjust holddown clips and guards for proper cutting. (*Sperry-New Holland*)

**Knife assembly** Before installing a knife assembly, be sure that all knives are tight and properly sharpened (Fig. 21-6). Inspect the knife head for wear and make sure all rivets are tight. The knife should be free from bends or twists. If the knife cannot be straightened, it should be replaced.

It is good procedure to purchase a set of knives and use them in rotation until they all wear out. When this is done, the parts on the cutter bar can be set and they will fit all the knives.

**Cutter bar lead** When the mower is operating, the cutter bar should be at a right angle to the direction of travel. Since there is considerable drag on the bar as it moves through the crop, it tends to move back. To compensate for this movement and to maintain a right angle to the direction of travel, the outer end of the cutter bar should be ahead of the inner end when the mower is stopped. This is called *cutter bar lead.* As a general rule, cutter bar lead is about 1/4 in [6 mm] per 1 ft [0.304 m] of cutter bar length.

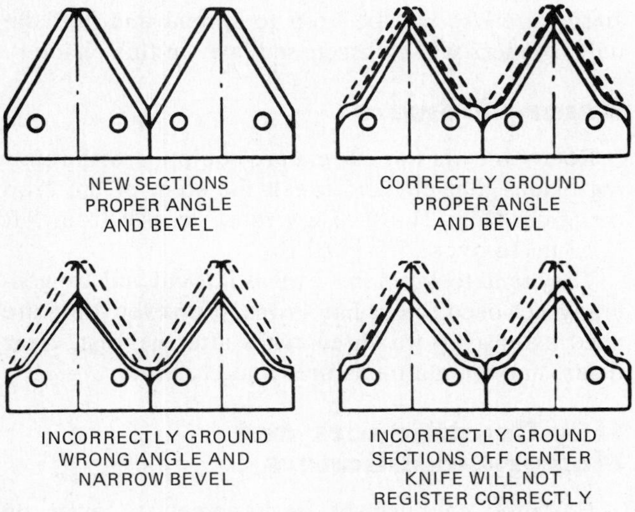

NEW SECTIONS PROPER ANGLE AND BEVEL

CORRECTLY GROUND PROPER ANGLE AND BEVEL

INCORRECTLY GROUND WRONG ANGLE AND NARROW BEVEL

INCORRECTLY GROUND SECTIONS OFF CENTER KNIFE WILL NOT REGISTER CORRECTLY

**Fig. 21-6.** Properly sharpened knife sections. (*Sperry-New Holland*)

To adjust cutter bar lead, first check the operator's manual for the manufacturer's recommended lead. Then place a straight board across the rear of the tractor tires and measure the distance to the board at each end of the bar. If the lead is not correct, consult the operator's manual for recommended adjustment procedures and adjust the lead (Fig. 21-7).

**Shoe adjustment** The inner and outer shoes adjust up and down to control cutting height. When the mower is operated in rough or rocky ground, the shoes should be positioned to raise the bar and protect the guards and knife sections.

**Tilt adjustment** Whenever possible, adjust the tilt so that the cutter bar runs level with the ground. When crops are down, the tips of the guards can be tilted

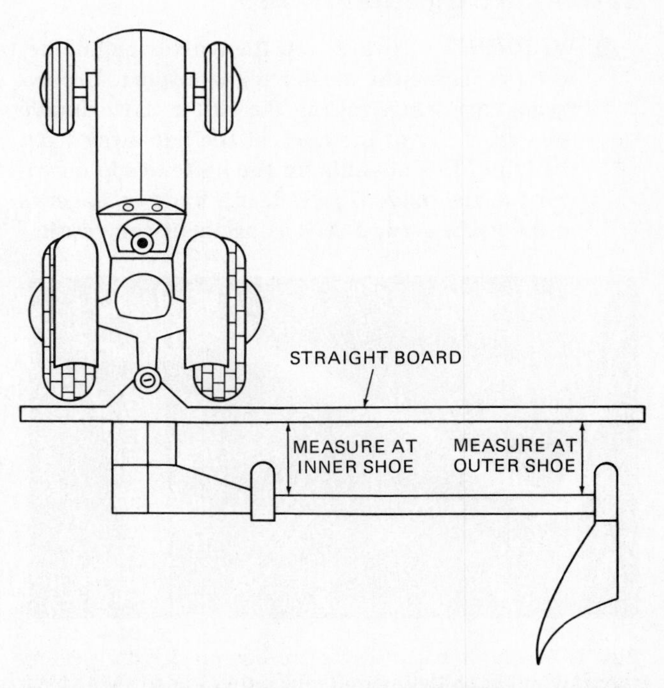


**Fig. 21-7.** Measuring the cutter bar lead during adjustment.

downward to get under the crop and lift it up. Guards should be tilted up in rocky areas to prevent damage to the knife sections and the guards. Check the operator's manual for adjustment procedures.

**Flotation adjustment** Cutter bar flotation is provided by a spring that supports the yoke and the cutter bar. Flotation allows the cutter bar to ride over obstructions and should be checked weekly to compensate for any change in the spring tension. It should also be adjusted any time the shoe height is changed. Check the operator's manual for recommended procedures and setting.

**Breakaway latch** Rear-mounted mowers and trail-type mowers are equipped with a safety latch that allows the cutter bar to swing back if it hits an obstruction such as a tree stump or a large rock. Side-mounted mowers may be equipped with a breakaway or a safety switch which cuts off the tractor power when an obstruction is hit. Breakaways *should not* be adjusted tighter than the adjustment recommended by the manufacturer (Fig. 21-8).

**Field Operation** With the possible exception of the first round, or opening cut on a field, the mower should be operated in a clockwise direction. The field conditions and the crop being harvested will determine ground speed. The operator's manual specifies the manufacturer's recommended normal ground speed range. The top recommended ground speed is usually 7 to 8 miles per hour (mi/h) [11.27 to 12.87 kilometers per hour (km/h)]. Read the operator's manual carefully before operating a mower.

### Transporting the Mower

⚠ WARNING **Always stop the tractor engine before preparing the mower for transport. Use extreme care when raising the cutter bar. Handle the cutter bar at the back of the bar away from the knife. It is possible for the knife to slip downward as the blade is raised. If a hand or finger is in the knife area, a serious accident can result.**

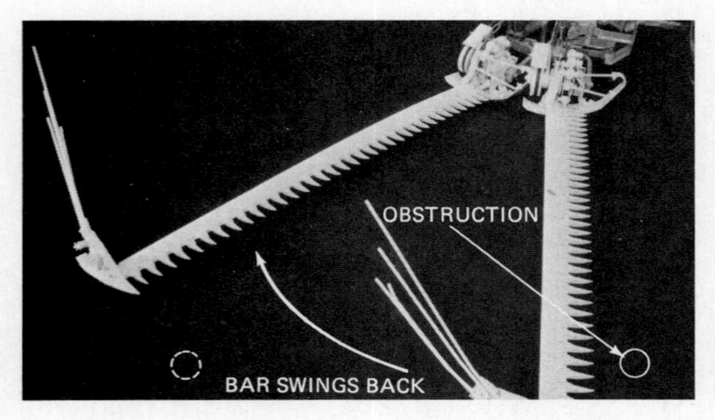


**Fig. 21-8.** Cutter bar mowers are equipped with a safety (breakaway) latch that permits the cutter bar to swing back when an obstruction is hit. (*Sperry-New Holland*).

Follow the recommendations in the operator's manual for securing the cutter bar. Rear-mounted mowers should be raised to their highest position during transport.

## Multidisk Mowers

Multidisk mowers use revolving disks equipped with free-swinging knives to cut the hay crop. The knives are brought into cutting position by centrifugal force as the disks are turned by the tractor power takeoff (PTO). The knives cut the crop as they come into contact with it and swing back if an obstruction such as a rock is hit.

Multidisk mowers are mounted and attached to the three-point hitch on the tractor. They are available in sizes ranging from about 5 ft 3 in [1.6 m] to 7 ft 10 in [2.38 m]. The disk may be suspended under the bar assembly (Fig. 21-9*a*) or mounted on top of the bar assembly (Fig. 21-9*b*).

**Operating Multidisk Mowers** Multidisk mowers are designed to operate between 540 and 620 PTO revolutions per minute (r/min). The disks revolve at high speed (3000 r/min or more) and will cut down and tangle crops with little or no plugging. They can be operated at higher ground speeds than cutter bar mowers. However, ground speed should be no faster than the tractor can be safely operated. Safe speeds will depend on the condition of the field. Ground speed should always be reduced before making sharp turns.

The cutting height of the mower can be adjusted by lengthening or shortening the top link on the three-point hitch. When the link is shortened, the cutting height is lowered. Lengthening the link raises the cutting height.

Multidisk mowers tend to deposit the hay crop in small swaths or windrows. When rapid curing is desired, a *tedder* can be used to spread and fluff the hay. Tedders will be discussed later in this chapter.

## Rotary Mowers

Rotary mowers or cutters are equipped with rotating blades that cut and break the stem of the crop being cut (Fig. 21-10). They range in size from 5 ft [1.52 m] to over 25 ft [7.62 m].

They tend to shred the crop as it is cut and are generally not used to cut hay crops. However, they are used extensively to shred crops after harvest, clear fields, and maintain pastures and orchards.

## Hay Conditioners and Mower-Conditioners

Hay producers use special equipment to lower the labor requirements in hay production and to reduce the time it takes for a crop to cure. Curing is the pro-

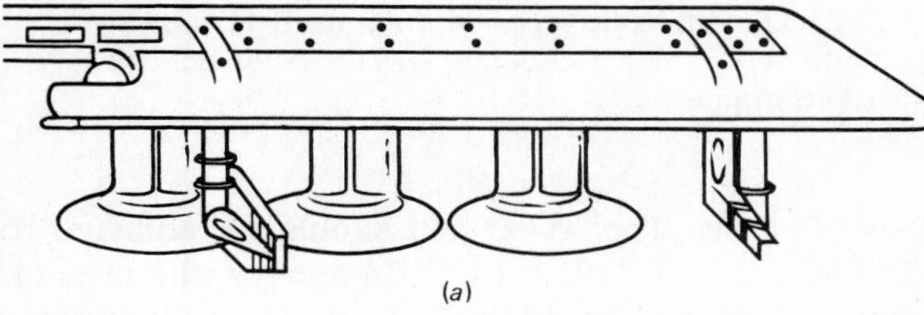

(a)

(b)

**Fig. 21-9.** (*a*) The suspended disk mower. (*Zweegers-PZ.*) (*b*) The bar-type multidisk mower with protective covers raised to show assembly. (*Kuhn Farm Machinery, Inc.*)

cess of allowing the moisture content of the crop to reach a percentage that is safe for storage. Most hay crops in the United States are field-cured. Field-cured hay is baled with a rectangular baler when the moisture content is 15 to 20 percent. When large round balers are used, the moisture content may be as high as 25 percent.

Conditioners are used to crush or crack the plant stems. It is a common practice to condition crops with large stems such as sorghums. Legumes are often conditioned because legume leaves tend to shed easily when normal curing is practiced. When

**Fig. 21-10.** The rotary mower in operation windrowing a hay crop. (*Austin Products-Rhino*)

legumes are conditioned, the stems dry quicker and more leaves can be saved. Conditioning can also reduce the curing time for grass hay crops.

Conditioning can result in a better-quality hay. The leaves are the most nutritious part of the plant. Conditioning reduces leaf loss because the hay can be baled before the leaves become brittle. When the crop cures faster, there is less chance of rain damage and the crop will retain more color, nutrients, and vitamins.

**Types of Conditioners** The hay conditioner which can be operated as an individual unit or hitched behind a tractor equipped with a mower is being replaced with the mower-conditioner (Fig. 21-11). There are actually four types of conditioners. They are the crusher, the crimper, the crusher/crimper, and the flail type.

**Crushers** Crushers have an upper roll that is coated with rubber and a lower steel roll that is ribbed (Fig.

**Fig. 21-11.** Cutting and windrowing hay with a mower-conditioner. (*International Harvester Co.*)

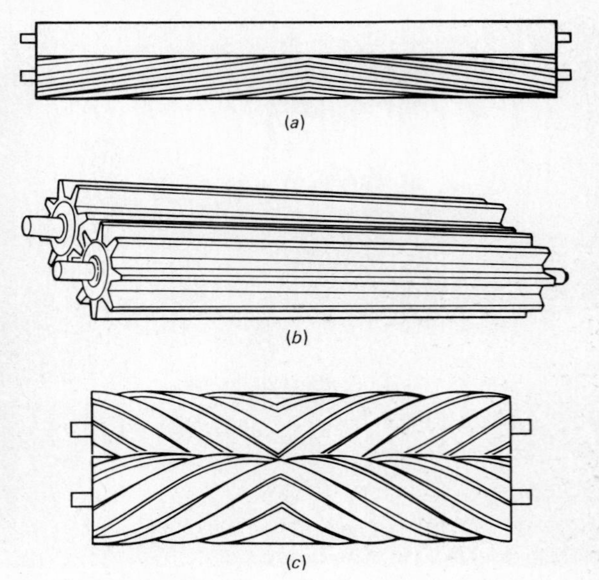

**Fig. 21-12.** Different types of rolls used in mower-conditioners. (*a*) crusher; (*b*) crimper; (*c*) crusher/crimper.

21-12*a*). As the hay crop is fed through the conditioner, it passes between the rollers and the stems are crushed.

The clearance between the conditioning rolls is factory-set, and the clearance must be maintained for maximum roll and drive chain life. The crushing action is determined by the roll tension. The operator's manual will give recommended roll spacing, tension, and adjustment procedures.

**Crimpers** Crimpers are equipped with two corrugated malleable iron rolls (Fig. 21-12*b*). As the crop is fed between the rolls, the stem is cracked at regular intervals. Proper roll clearance, tension, and timing must be maintained.

**Crusher/crimpers** Crusher/crimpers are equipped with two rubber rolls (Fig. 21-12*c*). The stem is both crushed and crimped as it feeds between the rolls. Rollers must have the correct clearance and must be timed to prevent hitting each other. The operator's manual will give procedures for checking and adjusting clearance and timing.

**Flail types** Flail-type mower-conditioners cut and condition the hay crop with rotating flails or knives (Fig. 21-13). The flails are attached to a balanced rotor assembly. Flail-type mowers can be used to shred crops such as corn and milo after harvesting. The rate of conditioning is determined by ground speed and PTO rpm.

Some flail-type conditioners use multidisk mowers to cut the crop. On these units, a flail-type conditioner is positioned behind the cutting unit.

**Field Operation** When a conditioner is operated as a separate harvesting tool, it should travel in the same direction the mower traveled when the hay was cut. The lower roll should be adjusted to within 4 to 7 in [101.6 to 177.8 mm] of the ground. Hay should be conditioned within 15 to 20 minutes after it is cut. Delayed conditioning will result in excessive leaf loss.

Mower-conditioners condition the hay as it is cut. This eliminates one additional field operation and holds leaf loss to a minimum.

**Fluffing board adjustment** The fluffing board or swath deflector helps to control the fluffiness of the windrow or swath. It is usually adjusted to the high position for heavy crops and to the low position for light crops.

**Windrow forming shields** Windrow forming shields can be used to form windrows as the crop is mowed and conditioned. This eliminates the raking of the swath even though a rake may be used to put two or more windrows together before baling the hay. The width of the windrow can be controlled by adjusting the spacing of the forming shields.

**Cutter bar and reel** Most mower-conditioners use a cutter bar and reel to cut the crop. When a cutter bar is used, the components must be properly maintained and adjusted. Refer to the operator's manual and the cutter bar mower section in this chapter for recommended service and adjustment.

**Reel adjustment** The reel should be adjusted according to the condition of the crop. When harvesting tall crops, the reel should be adjusted forward and upward. For short crops, it should be adjusted rearward and downward. The normal reel speed is 1.25 times ground speed.

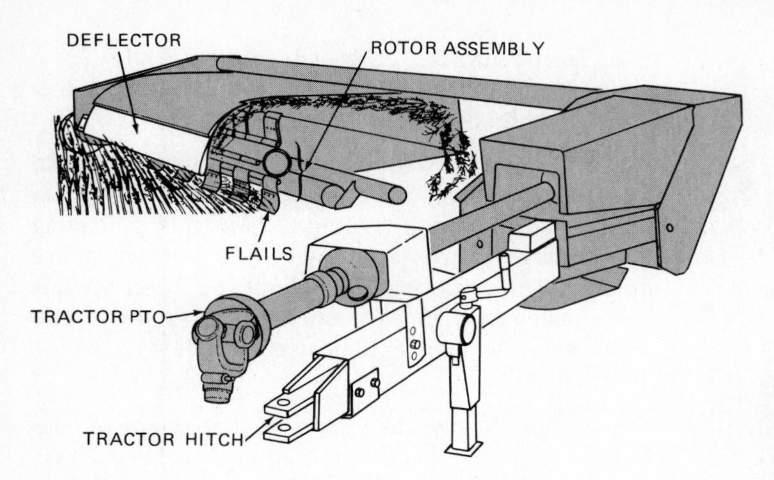


**Fig. 21-13.** The flail-type mower-conditioner cuts and conditions the hay crop with rotating flails or knives.

**Cutting height** The cutting height is adjusted by changing the height of the gauge shoes. The flotation adjustment should be reset when the cutting height is changed. Check the operator's manual for recommendations and procedures.

**WARNING The conditioner rolls operate at very high speed, and they can eject rocks and other objects at high velocity. It is important that all shields be properly maintained and kept in place. Always be certain that no one is behind the conditioner when it is operating.**

### Windrowers

Windrowers (swathers) are used to cut grain crops for threshing with a combine as well as crops to be used for hay (Fig. 21-14).

**Auger-Type Windrowers** Auger-type windrowers use a platform auger to move the crop to the discharge. They are used for hay crops and are especially suited for taller-growing crops. Many auger-type windrowers are equipped with conditioners.

**Fig. 21-14.** A self-propelled auger-type windrower. (*Deere & Co.*)

**Draper-Type Windrowers** Draper-type windrowers use a conveyor belt or belts to move the crop to the discharge. They are designed for windrowing grain crops to be harvested with a combine. They are also used to cut hay crops in areas where both hay and grain are produced. Draper platforms do not handle tall hay crops well.

## WINDROWING AND TEDDING EQUIPMENT

As discussed earlier, the mower-conditioner and the windrower are used extensively to windrow hay crops. However, many crops are swathed or cut to fall evenly over the field and must be windrowed before they are baled or stacked. Windrows may need to be moved after a rain to allow them to dry, and heavy crops may need to be turned to allow the hay near the bottom of the windrow to cure. Also, it is sometimes necessary to put two windrows together. This can be done with a hay rake.

The rakes commonly used today are parallel-bar, finger wheel, and rotary rakes.

### Parallel-Bar Side-Delivery Rakes

Parallel-bar rakes may be PTO-, ground-, or hydraulically driven. They may be rear-mounted to the three-point hitch, they may be the trail type, or they may be front-mounted on the tractor.

A typical parallel-bar rake is shown in Fig. 21-15. It has two parallel plates, or spiders. The spiders are attached at each end of the reel, and the front one faces the rear of the rake. The back spider faces the front of the rake. Parallel bars equipped with teeth are attached to the spiders. Power is delivered to the rear spider, which then turns the bars and the front spider.

**Fig. 21-15.** A parallel-bar side-delivery rake. (*Sperry-New Holland*)

As the bars turn, the teeth stay vertical at all times. When the teeth reach the hay, they move it a short distance. The movement is continued as the next bar comes into contact with the hay. As other bars rotate around, the hay is moved to the side of the rake and a windrow is formed. The stripper rods help pull the hay from the teeth as the reel turns.

**Field Operation** Hay crops should be windrowed when the moisture content is about 50 percent. When hay crops with less than 40 percent moisture are raked, excessive leaf loss can be expected.

Side-delivery rakes do a better job of raking when they are operated in the same direction of travel that the hay was cut. Raking in the reverse direction will usually result in missed hay and a higher leaf loss.

**Setting the teeth** The teeth are the most important part of a side-delivery rake. The teeth should be adjusted to clear the ground by 1 to 2 in [25 to 51 mm]. Normally the teeth should be set to operate just under the top of the stubble. The stubble is that part of the plant left attached to the root system.

The density and shape of the windrow can be changed by adjusting the pitch of the teeth. When teeth have a forward pitch, the windrow will be fluffier. This will allow the hay to cure faster. A rearward pitch of the teeth produces a tighter windrow. Tight windrows may be desirable when the crop is to be used for silage or haylage. (Silage and haylage will be discussed in Chap. 22.) Check the operator's manual for tilt adjustment procedures.

## Finger Wheel Rakes

Finger wheel rakes are made up of a series of individually floating wheels which have tines, or teeth, that extend around each wheel (Fig. 21-16). The wheels turn as the tines come in contact with the hay crop, stubble, and soil.

**Field Operations** Some finger wheel rakes can be adjusted to form one or more windrows. They can also be adjusted for tedding. *Tedding* is scattering and fluffing the crop to allow it to dry or cure more quickly. The position and angle of the finger wheels are adjusted to change the operation of the rake. The operator's manual should be used to adjust the frame and tongue.

**Side raking** To side-rake and form one windrow, the finger wheels are set to overlap each other so that each finger wheel takes over the crop after the preceding one (Fig. 21-17*a*).

**Windrowing** Two windrows can be formed by adjusting the finger wheel angles as shown in Fig. 21-17*b*.

**Tedding and Spreading** Some crops can be tedded or turned with the finger wheel rake. When tedding, the finger wheels are adjusted as shown in Fig. 21-17*c*.

## Radial Rakes

Radial rakes (tedders) are used in certain areas (Fig. 21-18). The rakes have a series of rotary hubs or stars that are equipped with long double-spring tines at each end which pick up the crop and scatter it.

The tedder is used in areas where high humidity and rainfall during the growing season are a problem. The fluffing action allows sun and air to enter the crop so that it will dry or cure in less time.

**Forming Windrows** Radial rakes can be used to form one or more windrows. Some models use windrow cages or shields to form windrows (Fig. 21-19). Others use speed reduction gears to slow the operation of the stars and tines.

# HAY BALERS AND STACKERS

Almost all the hay produced in the United States today is baled or hydraulically stacked. Hay is baled or stacked to make it more convenient to handle, store, and feed.

Many machines have been developed in recent years to reduce the labor required to produce and feed hay. Machines have been developed to handle the rectangular bale. Large round balers and hydraulic stackers are also used on many farms and ranches.

## Rectangular Balers

The rectangular, or square, baler has a pickup attachment that allows it to pick up hay crops that have been windrowed. The balers may be powered by an engine mounted on the baler or by the tractor PTO (Fig. 21-20). Self-propelled balers are also available.

**Fig. 21-16.** A finger wheel rake.

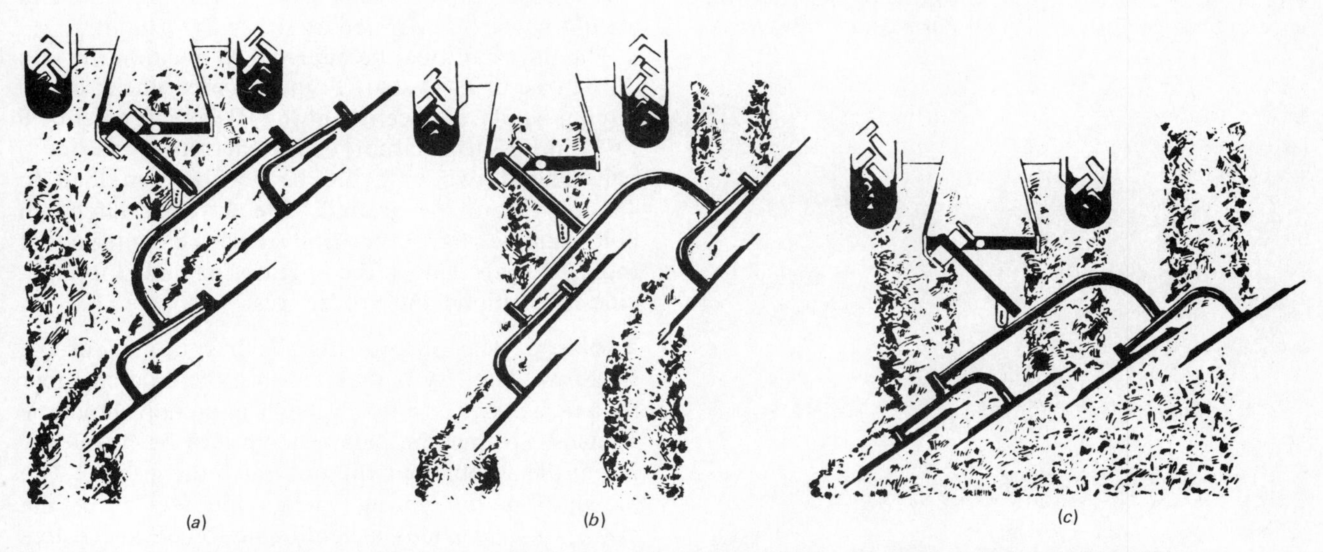


**Fig. 21-17.** Some finger wheel rakes can be adjusted to perform different jobs: (*a*) forming single windrow; (*b*) forming two windrows; (*c*) tedding and fluffing. (*Massey-Ferguson, Inc.*)

Balers are available in different sizes. The size of the baler is usually designated by the tons [metric tons] of hay it can bale in an hour.

Balers are designed to produce different-size bales. The most common bales are 14 × 18 in [356 × 457 mm] and 16 × 18 in [406 × 457 mm]. These bales are normally 36 in [914 mm] long, and the balers can be adjusted to produce longer or shorter bales. Some balers are designed to produce 16 × 24 × 48 in [406 × 610 × 1219 mm] bales. A baler is marketed that produces a 48 × 51 × 96 in [1219 × 1295 × 2438 mm] bale (Fig. 21-21).

The hay in a bale is compressed to a density of 10 to 14 pounds (lb) per cubic foot (ft$^3$) [160 to 225 kilograms (kg) per cubic meter (m$^3$)]. Wire or twine is used to tie the bales so that they can be handled.

**Rectangular Baler Components and Adjustments** A baler has a pickup, a hay compressor (wind guard), an auger or feed rake, feeder

Fig. 21-18. The radial rake can be used to ted or fluff hay. (*Kuhn Farm Machinery, Inc.*)

Fig. 21-19. A radial rake with windrow forming cage being used to form windrows. (*Kuhn Farm Machinery, Inc.*)

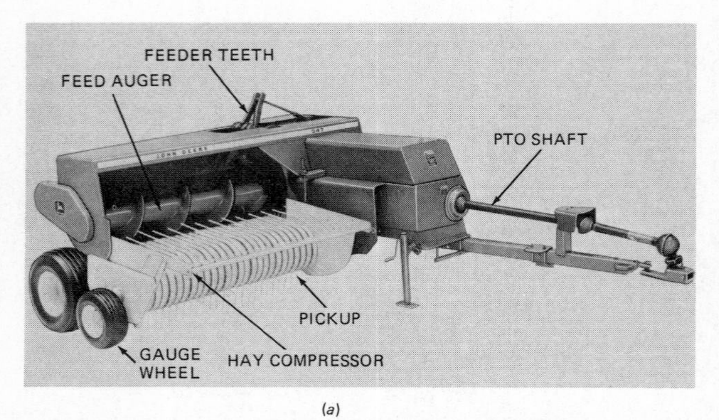


(a)

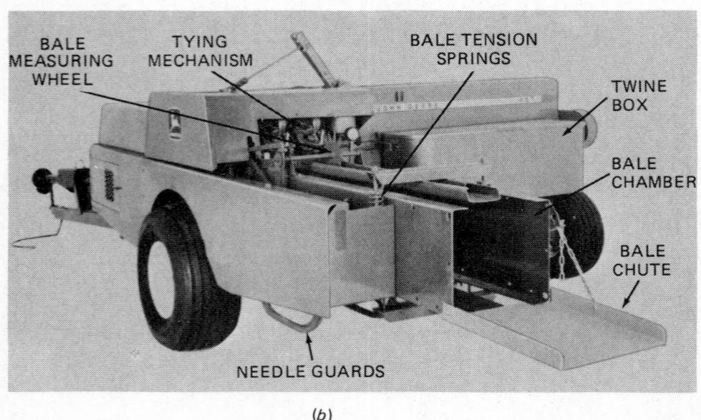


(b)

**Fig. 21-20.** A rectangular baler. (*Deere & Co.*)

**Fig. 21-21.** A rectangular baler that produces 48 × 51 × 96 in [1219 × 1295 × 2438 mm] bales in operation. (*Hesston Corp.*)

teeth, a plunger, knives, a bale chamber, a bale measuring wheel, a tying mechanism, and a bale chute.

## Attaching the Baler to the Tractor

Balers are designed to attach to any tractor that has a drawbar and power takeoff that conforms to American Society of Agricultural Engineers (ASAE) and Society of Automotive Engineers (SAE) standards. Balers operate on 540 r/min power takeoffs and should never be operated at 1000 PTO r/min.

The drawbar must be adjusted to position the end of the tractor PTO shaft 14 in [356 mm] ahead of the hitch pin hole. The center of the PTO shaft must be 6 to 12 in [153 to 305 mm] above the top of the drawbar. The drawbar should be 13 to 17 in [330 to 432 mm] from the ground. The drawbar hitch pin hole must be aligned vertically with the centerline of the PTO shaft. Check the operator's manual for specific instructions and procedures.

**Pickup** The pickup lifts the hay up out of the windrow. The hay is then fed into the bale chamber by a feed auger or a feed rake. The pickup should be adjusted so that the pickup teeth will be below the top of the stubble but will not touch the ground. The pickup flotation spring carries the weight of the pickup and provides a floating action. Some balers are equipped with a gauge wheel on the outside end of the pickup (Fig. 21-20*a*). When a gauge wheel is used, the pickup flotation spring should be adjusted to provide approximately 20 to 25 lb [9 to 11 kg] of weight on the gauge wheel. This will reduce the wear on the gauge wheel. On some balers, the flotation spring must be adjusted each time pickup height is adjusted.

**Hay Compressors** The hay compressors or wind guards hold the hay crop firmly on the pickup teeth (Fig. 21-20*a*). The compressors can be adjusted to feed different hay crops. On some balers, they are adjusted to clear the pickup strippers or straps between the pickup teeth. The manufacturer may specify that the compressors be adjusted to clear the strippers 3 to 4 in [76 to 102 mm] for normal crop conditions. On some balers, the pressure required to lift the compressors off the strippers is adjusted. The operator's manual may specify that approximately 35 lb [241 kPa] of pressure is required to raise the compressors from the strippers.

**Feed Auger and Feeder Tines** Some balers are equipped with feed augers and feeder teeth to move the hay from the pickup assembly into the bale chamber (Fig. 21-20*a*). The feeder teeth can be adjusted to produce uniform bales when baling conditions vary. They are adjusted to shorten or lengthen their stroke. This regulates the distance they move into the bale chamber and controls the placement of the hay in the bale chamber. The feeder teeth must be timed to prevent the plunger from striking them as it moves into the bale chamber.

Other balers use a feeder rake, which consists of tines located on a tine bar to move the hay from the pickup assembly into the bale chamber. The position of the tines on the tine bar can be adjusted to control the placement of the hay in the bale chamber. Check the operator's manual for specific adjustments of the feeder system.

**Plunger** The plunger packs the hay into the bale chamber. As it moves into the bale chamber, the plunger knife passes a stationary knife and cuts the hay that is not completely in the bale chamber (Fig. 21-22). The knife clearance specified in the operator's manual should be maintained, and the knives must be kept sharp.

**Hay Dogs** As the hay is forced into the bale chamber, spring-loaded hay dogs are forced out of the bale chamber. When the plunger starts its return stroke, the hay dogs snap back into position and hold the charge of hay in compression (Fig. 21-22).

**Bale Chamber** The bale is formed in the bale chamber (Fig. 21-22). The density or weight of the bale is controlled by adjusting the upper and lower tension bars in the bale chamber. Hydraulic bale tension attachments are available on some balers.

**Bale measuring wheel** The length of the bales is determined by the adjustment of the stop clamp or bolt on the measuring wheel control arm or trip arm (Fig. 21-20*b*). The bale measuring wheel is turned as the bale is forced out of the bale chamber. The adjustment procedures given in the operator's manual should be followed when adjusting bale length.

**Tying Mechanisms** It is important that a person planning to operate or service rectangular balers understand the operation of wire and twine tying system.

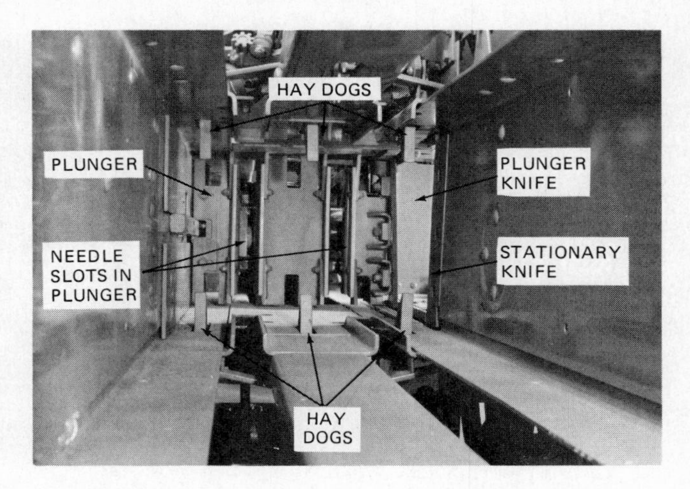


**Fig. 21-22.** The plunger packs the hay in the bale chamber. Hay not completely in the bale chamber is cut by the knives. Spring-loaded hay dogs hold the hay in the bale chamber. (*Sperry-New Holland*)

**Wire tying** In wire tying, the needles are threaded following the specific instructions provided in the operator's manual. After the needles have been threaded, the wire is anchored by the wire grippers (Fig. 21-23*a*). As the bale forms, wire is pulled from the wire box by the force of the hay moving into the bale chamber.

When the bale reaches the desired length, the bale measuring wheel trips the twisting mechanism. The needle moves up, catches the wire at the front end of the bale, carries it up, and positions it at the notch in the shear plate on the opposite side of the anchored wire. The twister shaft begins a clockwise rotation around both wires (Fig. 21-23*b*).

The twister hook completes one revolution and grasps both wires. The wire gripper releases the anchored wire, shears the wire that has been fed up by the needle, and anchors the needle wire for the next tying cycle. The needle moves out of the bale chamber and returns to the home position. The twister hook makes five revolutions twisting the ends of the wire together (Fig. 21-23*c*).

As the bale is forced out of the bale chamber by hay forming the next bale, the twisted knot is pulled off the twister hook (Fig. 21-23*d*). This same process is occurring at the same time at the other wire twisters.

**Twine tying** In twine tying, the needles are threaded following the instructions provided in the operator's manual. After the needles have been threaded, the twine is held in the twine disk by the twine holder (Fig. 21-24*a*). As the bale forms, twine is pulled from the twine box.

When the bale reaches the desired length, the bale measuring wheel trips the tying mechanism. The nee-

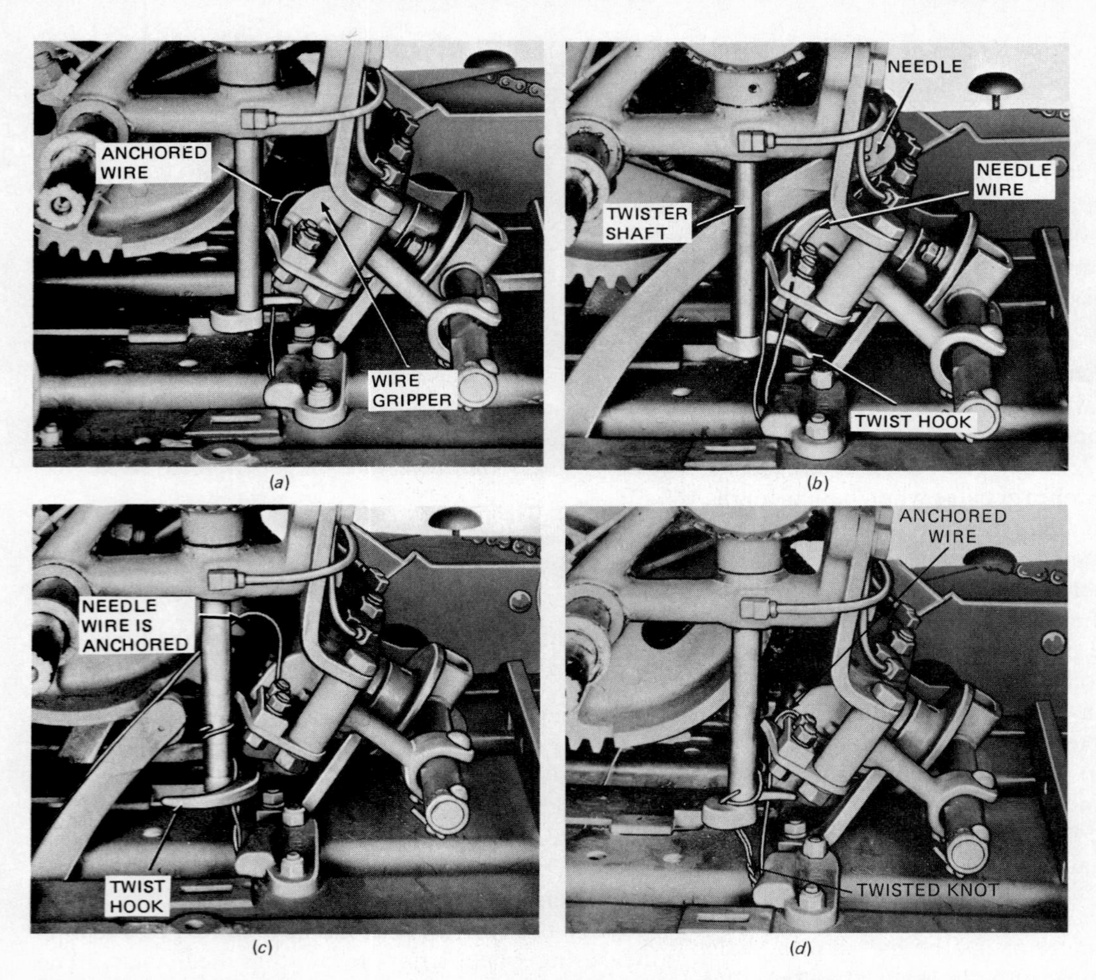


**Fig. 21-23.** (*a*) The wire is anchored by the wire gripper. (*b*) The needle picks up the wire at the front of the bale and positions it in the notch in the shear plate. (*c*) The twister hook completes one revolution and grasps both wires. Wire gripper anchors the needle wire. (*d*) The twisted knot is pulled off the twister hook. (*Deere & Co.*)

dle brings the second strand of twine up through the guide on the knife arm past the billhook and positions it in the twine disk (Fig. 21-24*b*).

The twine holder secures the ends of the twine, and the billhook rotates around both strands (Fig. 21-24*c*). As the billhook turns, it forms a loop in the twine around the hook. As the billhook rotates, the jaws open and the knife advances to cut the twine between the billhook and the twine disk. The needle begins to return to the home position, and the twine disk holds the end of the twine for the next tying cycle (Fig. 21-24*d*).

The billhook jaws close and hold the ends of the twine after it has been cut. The wiper on the knife arm advances to remove the looped twine from the billhook. The jaws continue to hold the cut ends of the twine, and the knot is formed (Fig. 21-24*e*). The knot then drops from the billhook, and the knot is completed (Fig. 21-24*f*). This same process is occurring at the same time at the other knotters.

The twine tying mechanism described here is used by several baler manufacturers. Some knotter systems operate differently. Check the operator's manual for a complete description of the knotter system on the baler you are servicing or operating.

**Checking the tying assembly** The tying assembly can be checked for proper operation by tripping the bale measuring arm and turning the baler through a complete tying cycle by hand. To perform this test, pull the wire or twine out the back of the bale chamber and hold it tight.

If the baler fails to produce a good knot or twist, refer to the troubleshooting section in the operator's manual to determine the recommended adjustments and repairs. Many operator's manuals contain a detailed description of knotter operation and service. A service manual will give more detailed service information.

## Timing a Baler

The feeding mechanism, the plunger, and the tying mechanism must be properly timed to prevent baler damage and assure proper operation. Timing is con-

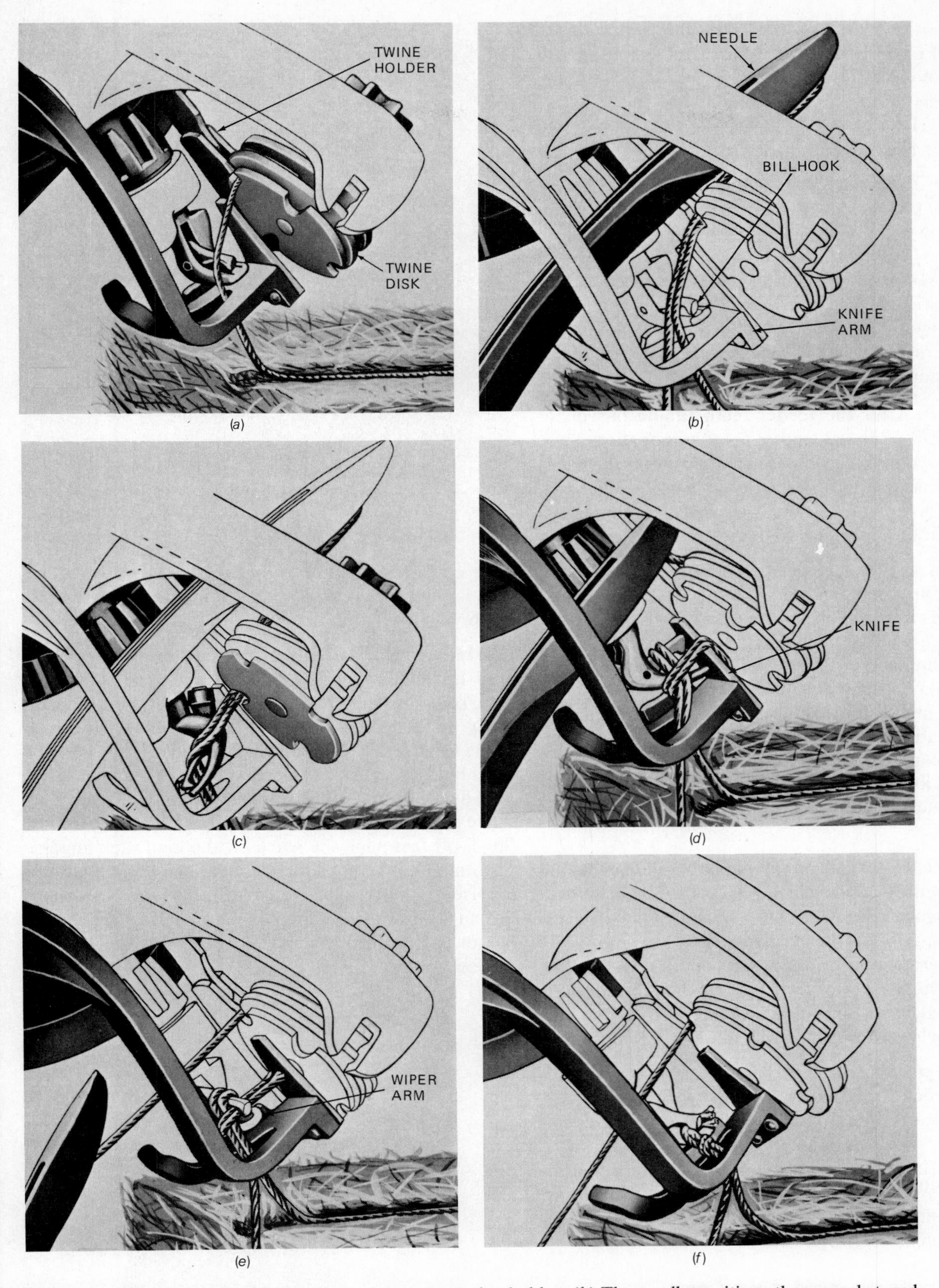


**Fig. 21-24.** (*a*) Twine is held in the twine disk by the twine holder. (*b*) The needle positions the second strand of twine in the twine disk. (*c*) Both ends of the twine are secured in the twine disk, and the billhook rotates around both strands. (*d*) The billhook forms a loop of twine around the hook, and the jaws open to receive the twine. The knife advances to cut the twine between the billhook and the twine disk. (*e*) The billhook jaw closes and holds the ends of the twine. The wiper arm advances to remove the looped twine from the billhook. (*f*) The knot drops from the billhook, and the tie around the bale is completed. (*Deere & Co.*)

trolled by the main drive chain, the feeder drive chain, and the knotter drive gears or chain. If any of these components are removed for service, all timing mechanisms should be checked before the baler is operated. The following checks should be made as the baler is run, by hand, through a complete cycle.

**Feeder Mechanism** Feeder teeth or tine timing is checked by placing the plunger or the main crank in the position indicated in the operator's manual. The position of the mechanism is found by using timing marks or by measuring to determine the location of the feeder teeth. If timing is off, it is corrected by adjusting the position of the feeder drive chain on the feeder drive sprocket. Check the operator's manual for specifications and procedures.

**Plunger** After the feeder teeth timing has been checked and adjusted, move the plunger through a complete cycle. Check to see that the plunger clears the feeder assembly in all the possible adjustment positions. If it does not, adjust the plunger timing as specified in the operator's manual.

**Needles** To check needle timing, trip the bale measuring arm and continue rotating the flywheel, by hand, until the plunger is in the position specified in the operator's manual. Needle timing is then checked by determining the location of the needles as they enter the bale chamber.

Specific measurements are given for determining needle location. If needle location is off, it should be adjusted by repositioning the cluster drive gear or adjusting the knotter/twister drive chain. Check the operator's manual for specific timing and adjustment procedures.

⚠ WARNING **Before making any adjustments or attempting to remove any object from the baler, disengage all power and stop the engine. Balers are equipped with flywheels, and many continue to operate after power is stopped. Always be certain the flywheel has stopped before approaching the baler. All safety precautions listed in the operator's manual should be followed at all times.**

**Baler Operation** The baler should be operated in the same direction the windrow was formed. Ground speed is determined by the type of crop, the size of the windrow, and the condition of the field. The baler should be run at the PTO rpm specified in the operator's manual. When the baler is operated in heavy windrows, a lower gear range should be used on the tractor.

## Bale Wagons and Loaders

Equipment is available to reduce the labor required to handle rectangular or square bales. Bale loaders (Fig. 21-25) can be used to reduce the labor required to load hay on trucks and trailers in the field. Also, bale elevators and conveyors can be used to reduce the labor needed to store and handle bales at the barn.

**Fig. 21-25.** Loading rectangular bales with a bale loader. (*Graves Mfg. Co., Inc.*)

**Automatic Bale Wagons** Both self-propelled and tractor-drawn bale wagons are used to handle square bales (Fig. 21-26). The bale wagon automatically picks up the bales and loads them on the bale table (platform). Side rails and the rear load rack fingers hold the bales in place as they are being stacked and during transport. Hydraulic cylinders are used to raise the table and automatically unload the bales. Push-off feet push the stack off the load rack fingers.

Some bale wagons have a bale conveyor that can be used to load the wagon. With this wagon, the hay is stacked and unloaded manually. The bale conveyor is also used to move the bales from the wagon to the stack.

**Fig. 21-26.** Loading hay on an automatic bale wagon. (*Sperry-New Holland*)

## Round Balers

In recent years, many farmers and ranchers have changed to the large round bale. The major reason for this change is labor. With a large round baler that produces bales that weigh up to 1500 lb [680 kg], the labor required to harvest, handle, and feed hay is greatly reduced (Fig. 21-27).

Large round bales can be stored outside. There is some weather damage, but when the bales are properly made and stored, the loss is about equal to the cost of providing a barn to store hay in. With good management, losses can be held to 10 to 15 percent. However, improperly made bales and improper storage can cause losses as high as 50 percent.

Since the round bales are heavy and bulky, it takes special equipment to handle them. Bales can be moved with trail-type bale movers or with bale movers that attach to the three-point hitch on agricultural tractors (Fig. 21-28). Several bales can be moved at one time with special bale wagons (Fig. 21-29). Special equipment is also being manufactured to assist in feeding round bales.

There are several types of large round balers. Most round balers form the bale in the bale chamber using large flat belts or chains and slats to form the bale. Some form the bale on the ground under the baler. Since most balers form the bale in the bale chamber, they will be discussed in this chapter.

**Fig. 21-28.** Three-point round bale movers are used to move round bales. Two types are shown. *(Sperry-New Holland)*

## Crop Preparation

When a conditioner is used before round baling, the moisture in the stems will escape at about the same rate as the moisture in the leaves and faster curing (drying) will occur. The moisture content of the hay should be at approximately 20 to 25 percent when round baling. Baling at higher moisture content will cause wrapping at the feed roller and spoilage of the hay in the bale.

**Fig. 21-27.** PTO-driven round baler. *(Deere & Co.)*

**Fig. 21-29.** Bale wagon designed to haul and unload several round bales. *(Hay Van Co.)*

Most manufacturers recommend a 3-ft [0.91 m] windrow with a uniform cross section. Uneven windrows are hard to bale and usually result in poorly formed bales.

**Round Baler Components and Adjustments** The primary parts of a large round baler are the pickup and the bale forming belts or chains and slats (Fig. 21-27). Bale density is controlled by springs or by springs and hydraulic cylinders.

**Starting the bale** The windrow is lifted by the pickup assembly. Some balers use floor chains, platform belts, or bottom drum rollers to move the hay into the bale forming chamber. Windguards or compression rollers are used to hold the hay in a flat blanket. The pickup should be adjusted to run as high off the ground as possible and still be able to pick up all the hay in the windrow (Fig. 21-30*a*).

Start the bale with the windrow in the center of the pickup and then drive quickly to the right side and to the left side. This ensures that the hay is fed into the extreme sides of the bale forming chamber.

As the hay enters the bale forming chamber, it is rolled up and over (Fig. 21-30*b* and *c*). The movement of the bale forming belts or chains and slats causes the hay that is moving into the bale chamber to roll onto itself and a bale is started (Fig. 21-30*d*).

**Making the bale** After the bale is started, the baler should be operated at the extreme sides of the windrow. When a crossover is made, it should be made quickly. Crossing over too often will fill the center of the bale chamber, and a "barrel-shaped" bale will be made. Figure 21-31 shows a recommended operational pattern for round balers.

The tension on the forming belts or chains and slats is controlled by large coil expansion springs located on each side of the bale chamber. Some balers also use hydraulic cylinders to assist in bale tension control. As the bale gets larger, the tension increases to form a compact weather-tight outer layer on the bale.

**Unloading the bale** When the bale has reached normal size, a flag or sign will appear on the front of the baler to inform the operator. At this time, forward travel is stopped and a manual, hydraulic, or automatic wrapping system is used to wrap the bale with twine (Fig. 21-32).

The rear gate of the baler is raised to unload the bale (Fig. 21-33). Hydraulic cylinders are used to raise the gate.

> ▲ **WARNING Extreme caution must be taken to prevent injury or damage when unloading a bale. The unloading procedure recommended in the operator's manual should be followed.**

After the bale has been properly unloaded, the rear gate is returned to the operating position and another bale can be formed.

> ▲ **WARNING Before making any adjustments or attempting to remove any object from the baler, disengage all power and stop the engine. All safety precautions in the operator's manual should be followed.**

**Fire safety** There is a high risk of fire when operating round balers. This is due to the flammable nature of some hay crops and the fast-turning rollers, belts, and chains on the baler which generate heat and can cause sparks of static electricity. It is recommended to frequently remove all crop material that accumulates on the ends of rollers, the main drive chain, and the pickup. A periodic check for overheated parts

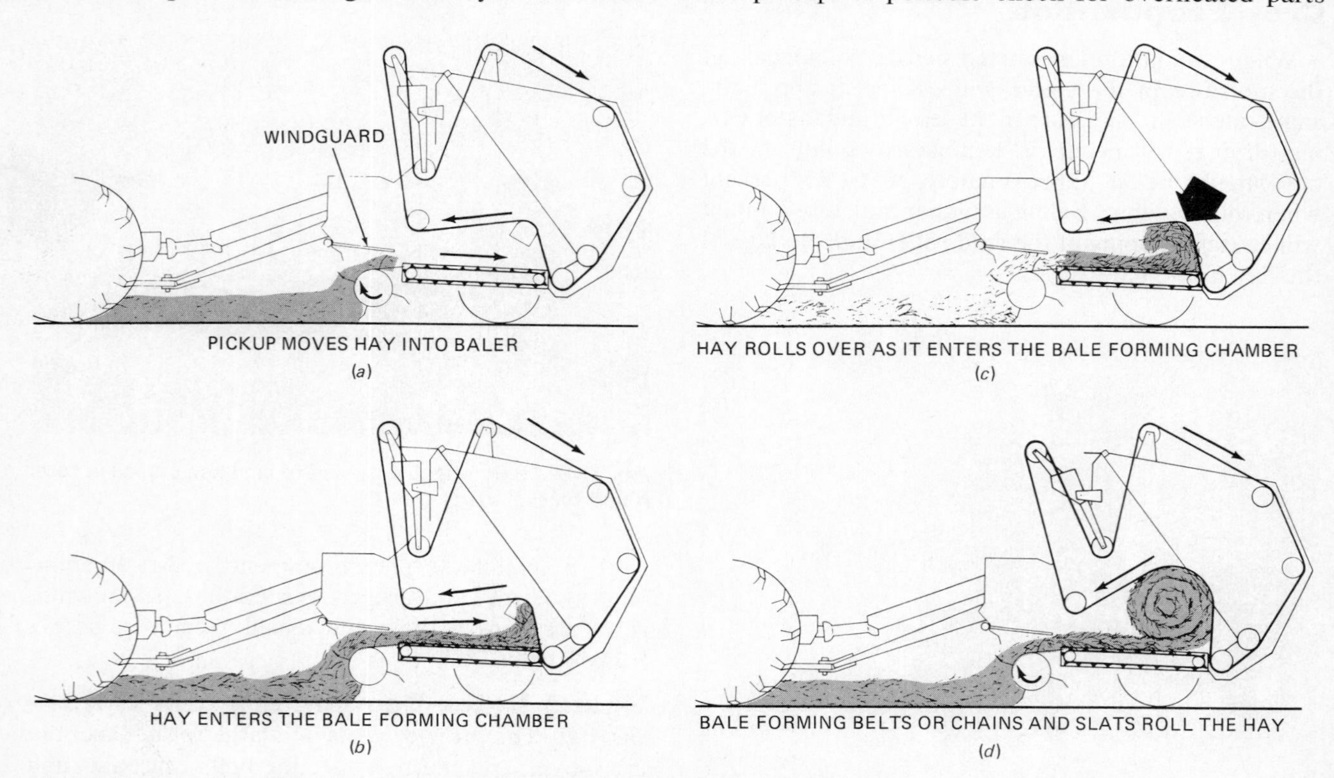


**Fig. 21-30.** (*a*), (*b*), and (*c*) Starting a round bale. (*d*) Forming the core. (*Sperry-New Holland*)

**Fig. 21-31.** Recommended driving pattern for forming round bales. (*Deere & Co.*)

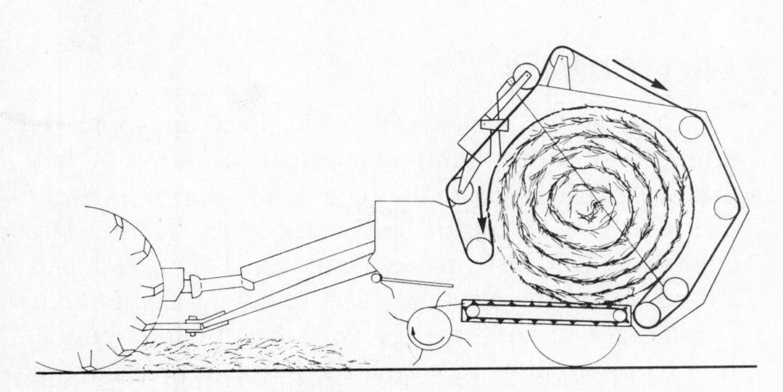

**Fig. 21-32.** When the bale reaches the desired size, it is wrapped with twine. (*Sperry-New Holland*)

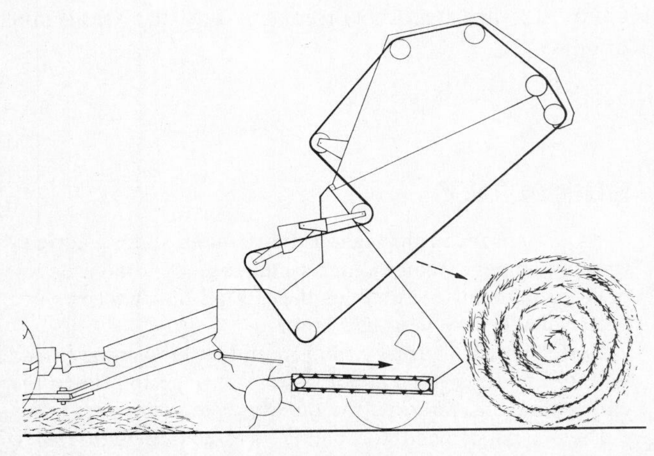

**Fig. 21-33.** The finished bale is unloaded from the baler by raising the rear gate. (*Sperry-New Holland*)

and twine wrapped around parts is also recommended. To limit damage in case of fire, a 2.5-gallon (gal) [9.5-liter (L)] pressurized water fire extinguisher or a 5-lb [2.27-kg] ABC dry chemical fire extinguisher should be mounted on the tractor or baler.

If a fire should occur, the bale should be ejected before attempting to extinguish the fire in the baler.

## Stack Wagons

Another haying system that reduces the labor required to produce and feed hay is the stack method. Hydraulically operated stack wagons are used to produce stacks that weigh from about $1^1/_2$ tons [1.36 metric tons (t)] up to 6 tons [5.44 t].

Some stack wagons are equipped with a pickup similar to the ones used on a baler, an auger to move the hay to the center, and a fan to blow the hay up the discharge duct into the wagon. Others use a rotor equipped with paddles to pick the hay up and blow it into the wagon.

**Stack Wagon Operation** Hay can be stacked when the moisture content is between 25 and 35 percent. The pickup assembly should be adjusted according to the instructions given in the operator's manual. The windrow should be fed into the stack wagon near the center of the pickup.

Stack wagons can be operated at higher ground speeds than balers. However, a loaded stack wagon should not be operated at more than 10 mi/h [16.1 km/h].

**Loading the stack wagon** The operator's manual will give complete loading instructions. When the wagon is filled with loose hay, the canopy is hydraulically depressed to pack the hay (Fig. 21-34). The number of compressions that are necessary to pack the hay completely and form a stack of desired size depends on the crop and its moisture content. High-moisture crops that have a high density will require one or two compressions. Low-density crops that have a low-moisture content will require three to six compressions.

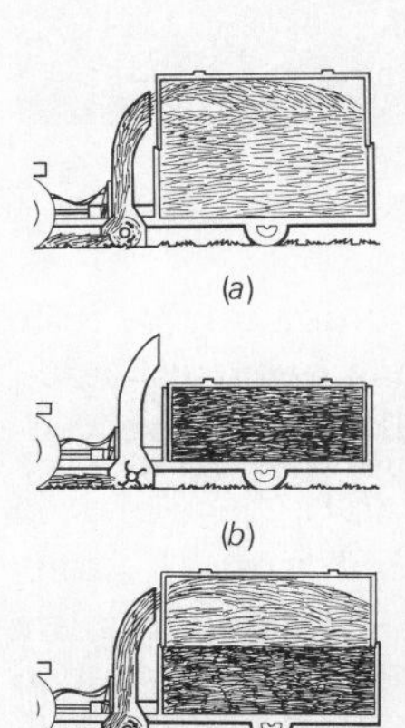


**Fig. 21-34.** Loading the stack wagon to make a stack. (*a*) Loading the wagon. (*b*) Compressing the hay. (*c*) Loading the wagon. (*Hesston Corp.*)

**Fig. 21-35.** A stack mover loading a stack that was made with a hydraulically operated stack wagon. (*Hesston Corp.*)

**Fig. 21-36.** A self-propelled field cuber. (*Deere & Co.*)

**Unloading the stack wagon** Proper unloading will result in an undamaged stack that has been unloaded safely. The operator's manual will give the recommended procedure for unloading stacks.

**Stack Movers** Since stacks weigh from $1\frac{1}{2}$ tons [1.36 t] to 6 tons [5.44 t], special stack movers are needed to move them (Fig. 21-35). Equipment is also available for feeding and grinding stacks.

## Hay Cubers

Self-propelled field cubers are used in some regions to produce tightly compressed cubes of hay that are a little larger than range cubes (approximately $1\frac{1}{2} \times 1\frac{1}{2} \times 4''$) fed to beef cattle (Fig. 21-36). The cubes are easy to process and can be moved and stored with conventional feed handling equipment.

Field-cubed hay must be very dry. Field cubing is limited to climates in which the hay can field-cure to a moisture content of 10 to 12 percent.

Stationary cubers are also used to cube hay. They are popular in alfalfa-producing areas. When hay is cubed in a stationary cuber, driers can be used to dry the hay. By using this method, direct-cut hay can be cubed. Forage harvesters, which will be discussed in Chap. 22, are used to direct-cut hay for stationary cubers.

## SUMMARY

Hay is produced throughout the United States. Agricultural equipment manufacturers and engineers have developed equipment that enables the hay producer to produce better-quality hay with less labor.

Cutter bar, flail, rotary, and multidisk mowers are used to cut hay crops. Hay conditioners are used to reduce the curing time and improve the quality of the hay.

Hay is windrowed for baling or stacking with rakes. Rakes may be parallel bar, wheel type, or rotary type. Tedders are used by some producers to reduce the time it takes for hay crops to cure.

The rectangular or square baler is used to bale hay. Large round balers and stack wagons are replacing the square baler on many farms and ranches. This is mainly due to the labor that they can save.

Equipment is available to reduce the labor required to handle bales. Automatic bale wagons can pick rectangular (square) bales up in the field and stack them outside or in barns. Manufacturers are continually introducing new equipment to handle large round bales.

Hay equipment is complicated, and the machines must be properly maintained and adjusted. The operator's manual should be carefully read before any equipment is operated. It will give service recommendations, adjustments, operating procedures, and safety precautions.

## THINKING IT THROUGH

1. What is the number one hay crop in the United States?
2. What factors affect the quality of hay?
3. List the parts of a cutter bar.
4. How can register be checked on a cutter bar mower?
5. What is cutter bar lead, and how is it adjusted?
6. How is the cutting height adjusted on a cutter bar mower?
7. How does a multidisk mower cut the hay crop?
8. Why are rotary mowers not used extensively to cut hay crops?
9. What is the difference between a crimper and a crusher?
10. When should hay that has been cut with a mower be conditioned?
11. How should the teeth be adjusted on a parallel bar rake?
12. What turns the wheels on a wheel rake?
13. What is a tedder used for?
14. List the major parts of a rectangular baler.
15. How are bale weight and density controlled?
16. How is bale length controlled?
17. Why must the needles on a baler be timed?
18. Why are hay producers using large round balers and stack wagons?

# CHAPTER 22 SILAGE AND HAYLAGE HARVESTING EQUIPMENT

## CHAPTER GOALS

In this chapter your goals are to:

- Describe the different types of forage production
- Discuss the different types of forage harvesting and handling equipment
- Discuss storage facilities for forage crops

Finely chopped forage and grain crops are a popular livestock feed. Chopped forage can be stored in silos for use at a later time, or it can be fed fresh. It can be fed to all ruminant (cud-chewing) animals and is used extensively on dairy farms and in beef feedlots. Many cow-calf beef producers use it to supplement pastures during winter months.

Grain crops such as corn and milo (or maize) can be harvested earlier when they are chopped instead of waiting until the grain has matured enough to combine harvest. This can be a real advantage in sections of the country where the growing season is short. More feed per acre is attained when the whole plant is harvested.

Any forage or grain crop can be chopped. Some farmers chop the crop residue that is left after the grain has been harvested.

Field losses are reduced when forage is chopped, because field curing is eliminated or reduced, depending on the crop being harvested and the crop management system used.

## TYPES OF FORAGE

Chopped forage can be classified by the method of harvesting and feeding the crop.

### Silage

*Silage* is a succulent feed for livestock (primarily beef and dairy cattle) that is preserved by a partial fermentation process. The fermentation takes place while the silage is stored in silos, which will be discussed later in this chapter.

Like hay, silage is produced throughout the United States. The crops used to produce silage vary from region to region, as do the harvesting and storage methods. Corn is the leading silage crop in the United States, but many grasses, legumes, cereal crops, and sorghums are used to produce silage.

The quality of the silage depends on the crop used, its stage of maturity, harvesting techniques, and storage. Well-managed forage and silage production can produce a quality feed that approaches the feed value of grain and is higher in protein than grain.

Silage is made by chopping the crop while it has a high moisture content. Many agricultural producers make silage out of forage and grain crops because it can be handled and fed with automatic equipment.

**Direct-Cut Silage** Direct-cut silage is cut and chopped in one operation while the crop is still standing. Corn is usually harvested by the direct-cut method when the kernels are fully dented. Corn at this stage of maturity will contain 65 to 70 percent moisture. Oats, barley, and other cereal grains may also be direct-cut.

**Wilted Silage** Legumes and grasses may contain 80 to 85 percent moisture. Silage with this much moisture will lose quality and will not be as palatable. The wilt method is often used to harvest these crops. When the wilt method is used, the crop is cut and allowed to field-cure from 1 to 4 hours, depending on the weather conditions. The crop is then chopped when it contains 60 to 65 percent moisture.

**Haylage** Haylage or low-moisture silage is allowed to field-cure until it contains 40 to 60 percent moisture. The production of haylage is increasing in popularity, and it is usually stored in oxygen-limiting structures. It can also be stored in conventional silos.

### Green Chops

Green chops are forage crops that are harvested and fed directly to livestock. This method is used during a crop's growing season. Crops are harvested and fed daily to the animals.

### Stover

The stalks, leaves, husks, and other crop residues left in the field after the corn or milo has been harvested with a combine can be chopped into ensilage, a low-quality silage called stover. Stover silage can provide a second crop which can then, when supplemented with grain, be used to winter beef cows.

## FORAGE HARVESTERS

Forage harvesters are designed to chop the entire crop so that it can be made into silage or fed as green chops. Harvesters may be self-propelled, tractor-drawn (Fig. 22-1), or mounted. Tractor-drawn harvesters may be PTO-driven or powered by an auxiliary engine mounted on the harvester.

## Forage Heads

Harvesters are available with forage heads that are designed to harvest forage crops under different crop management systems. Many harvesters are designed so that forage heads can be interchanged.

**Direct-Cut Head** Forage harvesters can be equipped with cutter bars or flail-type cutters to harvest broadcast or drilled crops such as cereal grains, alfalfa, sorghums, and grasses. The crop is cut and fed into the cutterhead just as the crop is fed into the conditioner on a mower-conditioner.

**Row-Crop Head** Row crops such as corn and milo can be harvested with a forage harvester equipped with a row-crop head. Gathering guides are used to push the crop forward so that it can be fed evenly into the cutterhead (Fig. 22-1).

**Windrow Pickup** Harvesters equipped with a pickup similar to the pickup used on balers are used to pick up forage crops that have been windrowed. The pickup head is used where crops are allowed to wilt or cure before they are chopped (Fig. 22-2).

**Ear Corn Head** Snapper heads (Fig. 22-3) are used to harvest and chop the corn ears and leave the stalk and leaves in the field. The snapper head is used to produce ear corn silage.

**Stover Head** Stover heads can be used to harvest the stalks, leaves, and other forage left in the field after corn or milo has been harvested with a combine (Fig. 22-4).

## Cutterhead

The forage is chopped by the cutterhead. The cutterhead may be designed to cut the crop and throw it up the discharge spout into the forage wagon or truck, or it may be designed just to cut the crop. Fans are provided on forage harvesters with cutterheads designed to cut the crop only. The fan blows the chopped forage into the wagon or truck.

**Knife Cutterhead** The knife cutterhead is a cylinder equipped with knives that chop the forage as it is fed between the knife and the shear bar or stationary knife (Fig. 22-5).

Various methods are used to adjust the length of the cut, which usually ranges from about $^{1}/_{8}$ to $3^{1}/_{2}$ in [3 to 89 mm]. Most haylage is cut from $^{1}/_{4}$ to $^{1}/_{2}$ in [6 to 13 mm]. High-moisture grass or legume silage can be cut from $^{1}/_{4}$ to 1 in [6 to 25 mm].

Many forage harvesters are equipped with automatic or manual sharpeners that can be used to sharpen the knives without removing them from the machine. Electronic metal detectors are also provided on some forage harvesters to prevent metal objects from entering the cutting head.

**Fig. 22-1.** Tractor-drawn forage harvester equipped with a row-crop head being used to chop corn and blow it into a self-unloading forage wagon. (*Sperry-New Holland*)

**Fig. 22-2.** A forage harvester equipped with a pickup attachment harvesting forage after it has been allowed to wilt. (*International Harvester Co.*)

**Fig. 22-3.** Self-propelled forage harvester equipped with ear corn head. (*Deere & Co.*)

**Fig. 22-4.** Self-propelled forage harvester equipped with stover head. (*Deere & Co.*)

**Fig. 22-5.** A knife cutterhead. (*Sperry-New Holland*)

**Flail-type Cutterhead** Some direct-cut forage harvesters use a flail-type cutter (Fig. 22-6). Others use a cutterhead equipped with swinging knives or hammers that chop the crop after it is fed into the machine. Flywheel-type cutterheads are used on some older forage harvesters.

**Blower** Some forage harvesters are equipped with a blower that delivers the chopped forage to the forage wagon or truck. Other cutterheads are designed to throw the crop into the wagon after it is cut.

**Discharge Spout** The discharge spout delivers the chopped forage to the wagon or truck. It is equipped with a swivel deflector that can be operated from the tractor or operator's platform. The deflector allows the operator to direct the forage to any part of the wagon (Fig. 22-6).

## Field Operation

Before the forage harvester is operated, it must be properly adjusted. The cutterhead knives must be sharp and have the proper bevel. The operator's manual will give the manufacturer's recommended adjustments.

Forage harvesters are designed to operate at either 540 or 1000 PTO r/min. The recommended rpm

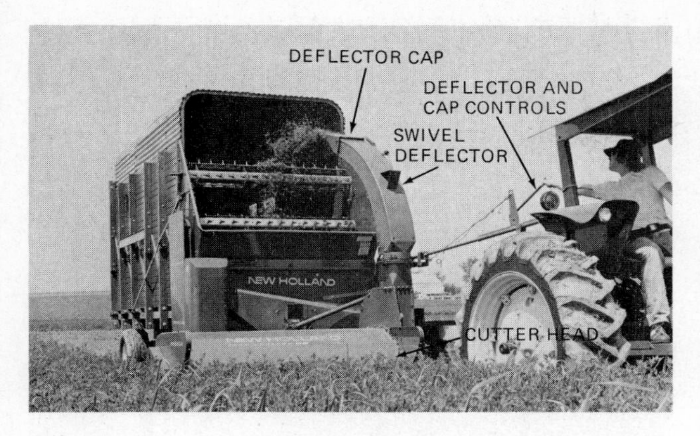


**Fig. 22-6.** A forage harvester equipped with a flail-type direct-cut head. (*Sperry-New Holland*)

should be maintained at all times. When rough field conditions require a reduction in ground speed, a lower gear should be used on the tractor.

> ⚠ **WARNING** **Forage harvesters should not be operated until the operation is completely understood and the operator's manual has been studied. The machine should never be adjusted or repaired before the tractor or engine has been shut off and all moving parts have stopped completely. Failure to do so will cause extensive damage and personal injury.**

## FORAGE WAGONS

Self-unloading forage wagons provide an efficient means of catching the forage as it is chopped and transporting it to the storage or feeding area (Fig. 22-1). Covered wagons are popular because they reduce the wind loss of forage as it is chopped and transported.

When it is necessary to transport forage long distances, dump-type wagons are often used in the field. The dump wagon is filled, and the forage is dumped into a truck for transport.

Forage wagons should be loaded from the rear to the front. The swivel deflector is used to control the loading of the wagon. It is important that forage wagons not be loaded above their capacity. The approximate weight of silage and haylage per cubic foot is given in Table 22-1. Check the operator's manual provided to determine loading recommendations. Also, study the operation and safety information in the manual before operating the wagon.

**TABLE 22-1. Approximate Weight Per Cubic Foot For Silage and Haylage**

| Crop | Weight/ft$^3$ |
|---|---|
| Silage | 26 lb |
| Haylage | 18 lb |

### Unloading Forage Wagons

Forage wagons can unload the forage into a feed bunk, into trench or bunker silos, or into blowers used to fill vertical silos. Unloading is automatic, and the operator's manual will give specific instructions.

### Silos

Silos are structures for storing silage or haylage. The two common types are vertical silos and horizontal silos. New types of structures and storage facilities have been introduced during recent years.

**Vertical Silos** Vertical silos are large vertical cylinders used to store silage. Older-type vertical silos are open-topped and are loaded and unloaded from the top. Their use is very limited.

**Oxygen-limiting silos** Sealed oxygen-limiting silos (Fig. 22-7) are commonly used on farms and ranches today. The silos made of steel and are glass-lined to protect the metal from the acids produced by the silage, and all joints are sealed. Some oxygen-limiting silos are made of concrete and sealed with an interior lining or coating.

An oxygen-limiting silo loads at the top and unloads at the bottom. This makes the structure more useful because silage or haylage can be added at anytime and does not have to be packed.

The bottom unloading feature is well suited to total mechanized feeding systems. Most silo manufacturers also provide feed handling equipment that can be purchased to make feeding more efficient.

**Blowers** Blowers are used to fill vertical silos. Most blowers are PTO-powered, but they can also be powered with electric motors or stationary engines.

Special safety precautions should be taken when operating blowers (Fig. 22-8). The operator is usually operating the unloading wagon and the blower. Both

**Fig. 22-7.** Glass-lined oxygen-limiting silos. (*A. O. Smith Harvestore*)

Fig. 22-8. A blower is used to move the chopped forage into vertical silos. (*Deere & Co.*)

Fig. 22-9. Silage and haylage can be stored in large polyethylene bags. A bagging machine is used to fill the bags which hold 130 to 140 tons [118 to 127 metric tons] of silage. (*AG Bag Corp.*)

may be run by PTO shafts, and both have open shafts, conveyors, and augers that are necessary to move the forage. Be sure to use extreme caution and to follow all the safety precautions given in the operator's manual.

**Horizontal Silos** Horizontal silos provide an economical means of storing silage. Animals can self-feed from this type of silo, or the silage can be loaded and hauled to the feeding site in silage wagons.

**Trench silos** Trench silos are pits dug into the ground. They may have a concrete floor to make feeding easier in wet weather and may be walled with concrete, lumber, or polyethylene to reduce spoilage. Silage is hauled into trench silos and packed. Tractors are commonly used to pack the silage.

**Bunker silos** Bunker silos are aboveground silos that have bunker walls constructed of concrete or lumber. They may also have a concrete floor and may be lined with polyethylene.

**Silage bags** A system for storing and feeding silage that allows producers to store silage near the feeding site is the polyethylene bag system (Fig. 22-9).

A bagging machine is used to place the forage in an airtight plastic bag that is 135 × 8 × 9 feet [41.15 × 2.44 × 2.74 m]. The forage is packed as it is fed into the bag and is protected by a double layer of polyethylene.

Bagged silage can be loaded on a silage wagon and hauled to the feeding site or the animals can self-feed from the bag.

## SUMMARY

A high-quality feed can be attained when forage and grain crops are chopped and properly stored in silos. After the crop is placed in a silo, fermentation takes place and a succulent, highly palatable feed is formed.

Crops to be used for silage are harvested with a high-moisture content. The amount of moisture depends on how the crop is to be managed. Silage is chopped when the crop moisture content is between 60 to 70 percent. When haylage is produced, the crop is allowed to field-cure until the moisture content is 40 to 60 percent.

Forage harvesters can be equipped with different heads to harvest different crops. Row-crop heads, direct-cut heads, pickup heads, ear corn snapper heads, and stover heads are available. Heads can be interchanged on many forage harvesters.

Self-unloading forage wagons and trucks are used to haul the chopped forage from the field to the storage area. Blowers are used to fill vertical silos.

Silos may be vertical or horizontal. Most vertical silos in use today are oxygen-limiting silos and can be unloaded from the bottom. This allows forage to be added to the silo at anytime. Silage is also stored in large plastic bags that can be located near the silage feeding-site.

Forage harvesting and handling equipment must be operated with **extreme caution.** The machines should be fully understood, and the safety precautions given in the operator's manual should always be followed.

## THINKING IT THROUGH

1. Discuss the differences between silage and green chops.
2. Explain the differences between silage and haylage.
3. What crops can be used to produce silage?
4. Discuss the different types of forage heads that are available for forage harvesters.
5. What is the usual length of cut for haylage?
6. Why are electronic metal detectors provided on forage harvesters?
7. What are the advantages of oxygen-limiting silos?
8. Explain the difference between a bunker silo and a trench silo.

# CHAPTER 23 GRAIN HARVESTING EQUIPMENT

## CHAPTER GOALS

In this chapter your goals are to:

- Identify the different types of combines and headers
- Identify the five major systems in a combine
- Explain how a combine works and identify the parts
- Explain the operation of rotary combines
- Use the operator's manual to identify basic combine adjustments and perform these adjustments on a combine
- Discuss safe combine harvesting practices
- Recognize good combining practices and check for harvesting losses

The modern combine is used almost exclusively to harvest the grain and other seed crops grown in the United States today. The combine cuts the crop, feeds the crop to the cylinder, threshes the seed from the seed head, separates the seed from the straw, cleans the seed, and handles the clean seed or grain until it is dumped into a truck or trailer for transport.

Combines are owned and operated on many of the nation's farms and ranches. Agricultural producers use the machine to harvest their own crops and may do some custom-harvesting for neighbors who choose to have their crops harvested. There are combine owners who do custom-harvesting only. They usually operate several machines and provide trucks for hauling the harvested grain to market or to on-farm storage facilities. Custom operators may harvest crops in a specific area or region, or they may have a regular harvest route that they follow each year. Farmers, ranchers, and custom operators provide summer employment opportunities for combine operators and truck drivers.

Agricultural equipment companies and dealers provide employment for people trained in combine operation and service. They need a service staff to sell and service combines. Service people need to be familiar with the operation, maintenance, and adjustment of the machines so that they can work with the farmers, ranchers, and custom operators who run the combines.

## TYPES OF COMBINES

Several types of combines are manufactured to meet the various needs of agricultural producers. The selection of a combine will depend on the crop or crops grown, the terrain, the number of acres to be harvested, and the amount of capital available.

### Self-Propelled Combines

The self-propelled combine is the most popular grain harvesting unit used in modern agricultural production. Many new combines are powered by diesel engines similar to those used on modern agricultural tractors. Others are powered by gasoline engines.

Self-propelled combines provide power for moving the machine through the field and operation (Fig. 23-1). A constant source of power at a predetermined speed (rpm) is delivered to the threshing, cleaning, and separating units. Ground speed can be adjusted to feed the crop evenly into the combine. These systems will be discussed later in this chapter.

The operator on a self-propelled combine sits well above the platform (header) and has a good view of the gathering process. The controls, dials, and gauges that regulate and show the operation of the combine are located within easy reach. Many new combine cabs are equipped with air conditioners, heaters, and electronic devices to monitor the combine operation.

The self-propelled combine is well suited to large farms and ranches, and smaller models can be purchased for the areas where fields are small and fewer acres [hectares] are harvested. Special models are

**Fig. 23-1.** A self-propelled combine equipped with a row-crop head harvesting soybeans. (*Deere & Co.*)

available for harvesting rice, edible beans, corn, soybeans, grass seeds, and other crops.

## Hillside Combines

The land-level self-propelled combine is used where the terrain is level enough to permit good separation and cleaning. In rolling terrain where separation and cleaning become a problem, special hillside combines with an automatic leveling system (Fig. 23-2) or attachments for land-level combines are used.

## Pull-Type Combines

Pull-type combines have the same general features that are found on self-propelled combines, but they are tractor-drawn. The header is located to the right or left of the separating mechanism to allow the unit to be pulled by a tractor. The tractor power requirements for most newer pull-type combines are 80 horsepower (hp) [60 kilowatts (kW)] or more.

## Special Combines

The peanut combine and the castor bean combine are special combines. Another type of combine is the rice special, which has certain features such as a larger engine and oversize flotation tires or tracks that adapt it to the muddy field conditions during rice harvest.

## Types of Cutting Platforms and Heads

Cutting platforms, row-crop heads, corn heads, and windrow pickup platforms or attachments are interchangeable on most combines.

**Fig. 23-2.** A hillside combine in operation. Note that the threshing, separation, and cleaning areas are level and that the cutting platform pivots to follow the land contour. (*International Harvester Co.*)

**Cutting Platforms** Cutting platforms are used to harvest small grain, milo, rice, and other broadcast or row crops where the seed head is cut off the stalk for threshing. They are often referred to as *headers*. Cutting platform operation and adjustment will be discussed later in this chapter.

Flexible cutter bar platforms have been developed. They are especially designed for harvesting soybeans and can be used to harvest small grain. Flexible cutter bar platforms are equipped with long floating divider points that ride independently of the cutter bar on skid shoes. The dividers are designed to lift and separate lodged (fallen) and bushy soybean plants. The cutter bar is designed to flex as it floats on skid shoes that allow it to follow the contour of the ground (Fig. 23-3).

**Row-Crop Heads** Special low-profile row-crop heads are designed to harvest soybeans, sunflowers, and other row crops such as milo. They cannot be used to harvest corn (Fig. 23-1).

The row-crop head is equipped with long low-profile gathers that are designed to get under the crop and pick it up. Rubber gathering belts convey the stalks past individually mounted rotary knives (Fig. 23-4). These units replace the reel and cutter bar on a cutting platform. A trough mounted under the gathering belts catches the soybeans that shatter (fall off) as they pass through the belts into the auger tray.

**Automatic header height control** Automatic header height controls are available on many combines. When a cutting platform or row-crop head is equipped with automatic height control, it can be operated at ground level. The platform cutting height is automatically controlled by sensors (small skid shoes) located under the platform. The sensors are connected to leveling switches which control the operation of an automatic header control valve. The valve controls the flow of hydraulic oil to two single-acting hydraulic cylinders that are located on each side of the platform. The cylinders raise or lower the platform as the ground contour varies.

Cutting height is maintained at the highest ground level. This allows larger platforms and heads to operate at the lowest possible level and keeps them out of

**Fig. 23-3.** A flexible cutter bar platform. (*Deere & Co.*)

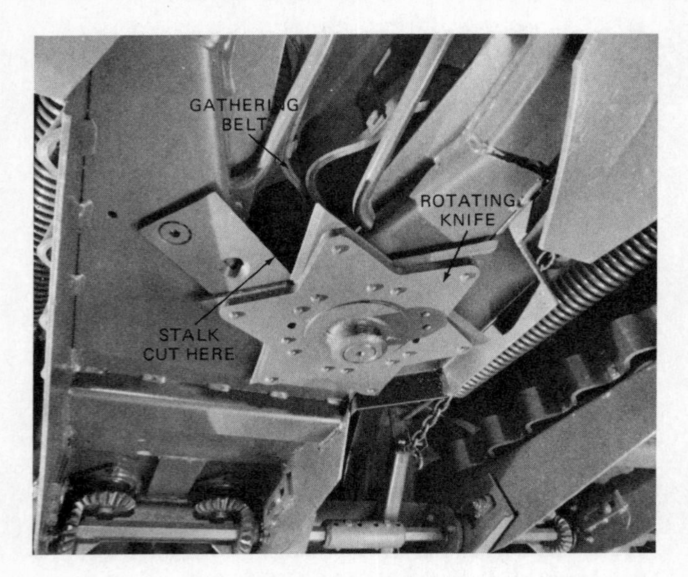


**Fig. 23-4.** The bottom view of a row-crop head showing the individually mounted rotary knife that cuts the stalks in one row. (*Deere & Co.*)

the ground. Automatic header height controls are used extensively when harvesting soybeans and small grains that have lodged. The system has a control switch and can be bypassed for normal header control when automatic height is not needed.

**Corn Heads** The corn head is actually a mechanical corn picker that removes the ear from the stalk (Fig. 23-5). The gathering points guide the corn stalks into the snapping rolls which grab the stalk. The snapping rolls turn in opposite directions and rapidly pull the stalk down. Some corn heads are equipped with stripper plates (deck plates) that are located above the snapping rolls. Stripper plates reduce or eliminate the shelling of corn kernels at the snapping rolls. When an ear of corn reaches the stripper plates or the snapping rolls, it is snapped from the stalk. Gathering chains located above the snapping rolls catch the ears as they are removed and carry them to the cross auger (Fig. 23-6).

**Fig. 23-5.** A self-propelled combine with corn head. (*Allis-Chalmers Manufacturing Co.*)

**Windrow Platforms or Attachments** Windrow platforms or attachments are used to pick up small grain that has been windrowed (Fig. 23-7). Crops may be windrowed to prevent weeds from maturing and to allow crops that ripen unevenly to attain acceptable moisture content at about the same time. Crops are also windrowed when there is considerable moisture at harvest time.

Small grain is windrowed with a windrower (swather). These units were discussed in Chap. 21. Windrowing is usually done about 1 week before the grain is mature enough to combine. After windrowing, the grain is allowed to cure.

The windrowed grain is lifted onto the platform by the windrow platform or a pickup attachment that is

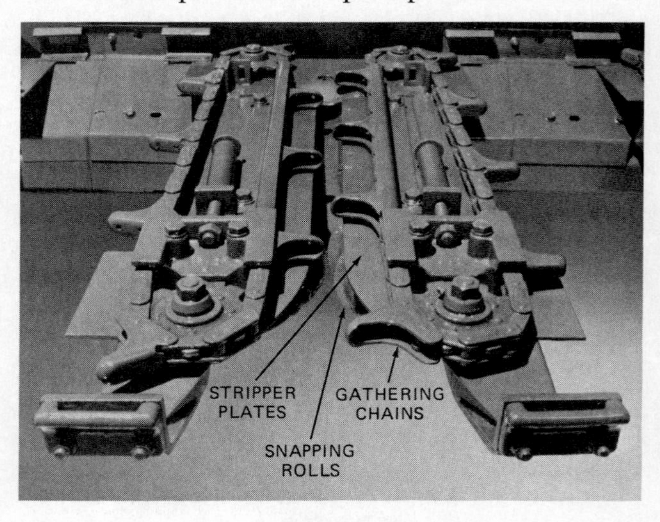


**Fig. 23-6.** A corn head with its gathering points removed showing gathering chains and location of snapping rolls. (*Deere & Co.*)

**Fig. 23-7.** A self-propelled combine with a windrow pickup attachment. (*Sperry-New Holland*)

mounted on a cutting platform. The grain is then handled by the feed auger.

## COMBINE SYSTEMS

Five major functions are performed during the harvesting operation with a combine. These may be classified as (1) cutting the crop and feeding it to the cylinder, (2) threshing the grain from the seed head, cob, or pod, (3) separating the grain from the straw, (4) cleaning the grain, and (5) handling the grain after threshing. These functions are performed as the material is moved through the combine.

### Cutting the Crop and Feeding the Cylinder

The seed head is cut from the stalk by the cutter bar when a cutting platform is used. The cutting takes place between the ledger plate on the guard, which is stationary, and the knife section, which moves. Under normal crop conditions, a slat-type reel that operates at about 1.25 times the ground speed of the combine is used to pull the seed head onto the cutter bar. Special pickup fingers, pickup reels, and other pickup attachments are available for cutting platforms. These may be used for special crops and special conditions where crops have lodged or fallen.

The cutter bar height is adjusted to meet crop conditions. Normally only the seed head is run through the machine when harvesting small grains and milo. Some crops such as soybeans require that the entire stalk be run through the machine.

**CAREFUL** Care should be taken when the cutter bar is operated near ground level to prevent rocks from entering the combine.

**The Reel** The reel has fore and aft adjustment, height adjustment, and speed adjustment. These adjustments are important because they affect grain losses at the cutter bar. In standing grain, the reel should be adjusted so that the slats, in their lowest position, strike just below the lowest grain heads and just slightly ahead of the cutter bar. Pickup reels that are adjusted forward and down may be used to harvest fallen and tangled crops (Fig. 23-8).

**The Platform** After the seed heads have been cut from the stalk, they fall on the platform floor (Fig. 23-9). The seed heads are then moved to the feeder conveyor by platform augers. Draper conveyor belts are used on some rice headers instead of augers. The feeder conveyor moves the seed heads into the cylinder for threshing.

### Threshing the Grain

The seed is removed from the seed head, the kernel is removed from the cob, or the bean is removed from the pod by the cylinder and the concave (Fig. 23-9). The cylinder consists of bars mounted on two hubs that are supported on a shaft that run the width of the threshing unit. Cylinders range from 15 to 22 inches (in) [381 to 558.8 millimeters (mm)] in diameter and from 2 to 5 feet (ft), [610 to 1524 mm] in length. Cylinder speed is adjustable and may range from 200 to 1500 r/min.

**Fig. 23-8.** Sell-propelled combine equipped with a pickup reel. *(Sperry-New Holland)*

The *concave* is located directly under the cylinder. It is a series of bars that are held together by rods or straps and is curved to match the cylinder. It is adjustable up and down to allow the operator to change the cylinder to concave clearance. An angle bar called a *cylinder stripper* is located above the cylinder. It prevents the crop from wrapping around the cylinder.

Three types of cylinder-concaves are available on conventional combines. The type of cylinder used is determined by the crops to be harvested.

**Rasp-Bar Cylinder-Concave** The rasp-bar cylinder is the most common. It can be used to harvest most crops. The cylinder bars are corrugated, and the concave bars are flat and run across the width of the threshing area. The cleats or lugs on the cylinder bars work the seed head against the concave to remove the seed in a rubbing action (Fig. 23-9).

**Spike-Tooth Cylinder and Concave** Spike-tooth cylinder bars have spike teeth attached to them. The concave also has teeth. The teeth on the cylinder bars run between the teeth on the concave (Fig. 23-10). This produces a more aggressive threshing action. The spike-tooth cylinder is used to har-

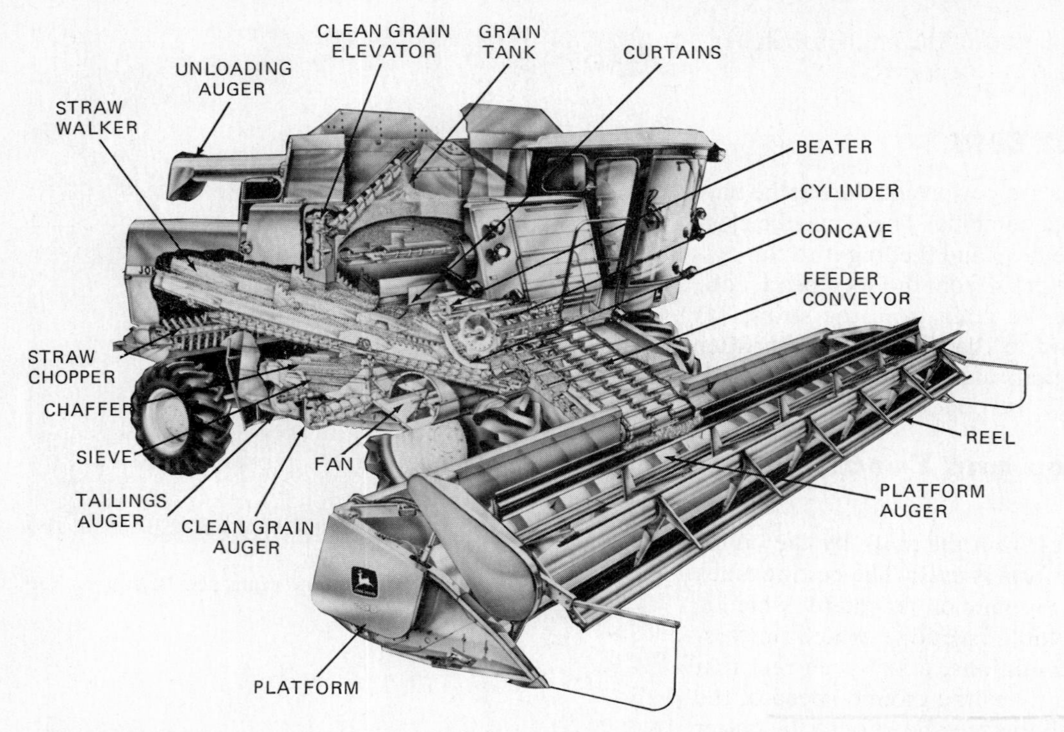


**Fig. 23-9.** A cutaway view of a self-propelled combine. (*Deere & Co.*)

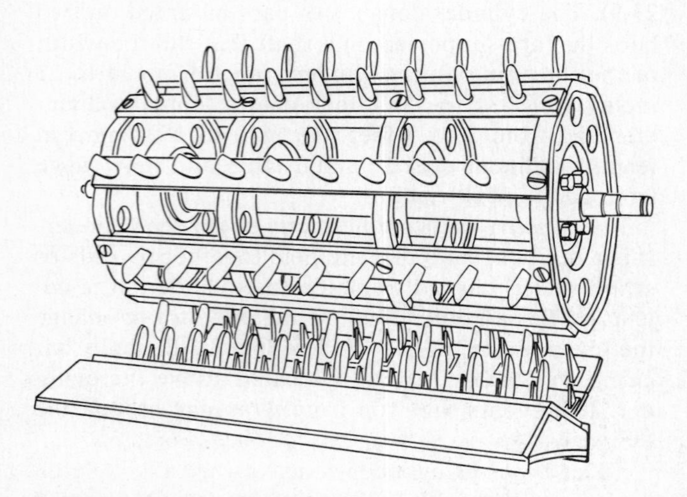

**Fig. 23-10.** A spike-tooth cylinder and concave.

vest rice and other crops difficult to thrash. Rice is also harvested with high-inertia rasp-bar cylinders.

**Angle-Bar Cylinder** The cylinder bars are angle iron covered with rubber. The concave bars are also covered with rubber. This produces a flailing action on the seed being harvested, whereas the rasp-bar cylinder produces a grinding action. The angle bar is used to harvest crops such as alfalfa and clover.

## Separating the Grain

Up to 90 percent of the grain is separated from the straw at the concave. The grain that is suspended in the straw is separated by the beater, finger grate, and straw walker. The beater is a small-diameter drum that is located to the rear and above the cylinder. It is not adjustable, but it may have removable covers or wings that can be removed to expose teeth. The beater slows down the material going off the cylinder and concave and deflects the material down to the front of the straw walker and the finger grates. The beater and the cylinder stripper help prevent straw from wrapping around the cylinder (Fig. 23-9).

**Straw Walkers** The straw walkers or racks may be one piece or multiple type. They shake the straw as they move it out the rear of the combine, and grain suspended in the straw falls through the walker and is carried by the grain pan or grain augers to the cleaning shoe. Curtains are used to help keep the straw in contact with the straw walkers. They help prevent grain from being blown out the back of the combine (Fig. 23-9).

## Cleaning the Grain

The grain that falls through the concave and the grain that is separated from the straw by the beater, finger grates, and straw walker contain chaff, broken straw, and other foreign materials that need to be removed.

**Chaffer** The grain is fed onto the chaffer, which is a flat oscillating rack made up of a series of crosspieces that have overlapping metal lips or louvers. Chaffers available with various shapes of louvers are set at different spacings for different crops. Air from the fan blows up through the chaffer, and the light chaff and straw are suspended while the grain falls through the openings between the louvers (Fig. 23-9). The chaff and straw go out the back of the combine.

**Sieve** The grain and other heavy materials that fall through the chaffer land on the sieve. The sieve is similar to the chaffer except that the openings or louvers are smaller. There are adjustable and nonadjustable sieves available to suit the characteristics of the crop being harvested. Grain and other matter that are small and heavy enough to fall through the sieve are carried by the clean grain auger to the grain tank. Light materials move out the rear of the combine.

The larger materials such as partially threshed seed heads (tailings) that are too heavy to be suspended by the air blast and too large to drop through the sieve, drop off the end of the sieve into the tailings auger and are returned to the cylinder for rethreshing (Fig. 23-9). The chaffer and the sieve are often referred to as the *cleaning shoe*.

### Handling the Grain

The grain handling system includes the clean grain elevator or auger which moves the grain to the grain tank, the grain tank, and the unloading auger. Grain tanks range from 40-bushel (bu) [1.4 cubic meter ($m^3$)] capacity to about 300-bu [10.6-$m^3$] capacity on the larger self-propelled models. Most tanks can be unloaded either on the go or when the combine is stopped. Many combines are equipped with unloading augers which can unload up to 100 bu/min [3.5 $m^3$/min] (Fig. 23-9).

### Rotary-Type Combines

Some manufacturers produce rotary-type combines to thresh grain, corn, soybeans, and other seed crops. The combines are called rotary-type because the crop stays in contact with the cylinder for several full rotations as it moves through the threshing area (Fig. 23-11). These longer cylinders are usually mounted lengthwise "axial" in the combine.

The rotors or cylinders are equipped with rasp bars that rub the grain against the concaves and the rotor cage. Centrifugal action also helps to remove the grain from the seed head, cob, or pod.

Most of the grain is separated at the concave, which is located under the front section of the rotor or rotors. The concave(s) and rotor(s) run parallel with the combine or across the combine. The remainder of the grain is separated through a separating grate, located at the rear section of the rotor, or through a separating cage that is perforated and allows for separation around the full circumference of the threshing area.

Since separation takes place in the threshing unit, no straw walkers are necessary. A beater or paddle wheel is used to feed the straw from the rotor out the rear of the combine.

The principles of the cleaning shoe and grain handling system are not changed. A chaffer is used to remove the chaff and fine straw that falls through with the grain during separation. The sieve finishes the cleaning process, and tailings are returned to the rotor for rethreshing.

## COMBINE OPERATION AND ADJUSTMENT

The operator's manual should be thoroughly read and understood before a combine is operated. Before the combine is started each day, it is important that all servicing including lubrication is performed. Until a worker is thoroughly familiar with a specific combine, the checklist provided in the operator's manual should be carefully followed.

> ⚠ **WARNING When starting a combine engine, be certain that no one is on or around the combine's moving parts. Before the combine is moved, check to see that no one is on or near the combine. Also, check for any obstruction in the path of the combine.**

The separator should be engaged and brought to normal operating speed before the header clutch is engaged if the combine is equipped with separate controls. When the header is adjusted to the desired cutting height, the ground travel can be started by releasing the engine clutch or moving the hydrostatic shifting lever. It is important that the ground speed recommendations given in the operator's manual not be exceeded. A combine with a full grain bin is top heavy and can be overturned easily.

### Determining When to Harvest

When crop drying equipment is not available, a crop should be harvested when the moisture content is at the safe storage level. Most crops are ready to harvest and safe to store when the moisture content of the grain is about 14 percent. Corn can be stored at 17 percent moisture. A local grain terminal operator can be contacted to determine the acceptable moisture content for the crop. If threshed grain feels damp or the kernels can be easily dented with the fingernail, the moisture content is usually too high for safe storage.

It is a common practice to harvest corn with 20 to 30 percent moisture in some areas. When this is the practice, the grain must be dried before it can be stored. Many other crops can also be harvested at up to 20 percent moisture if they are dried before storage.

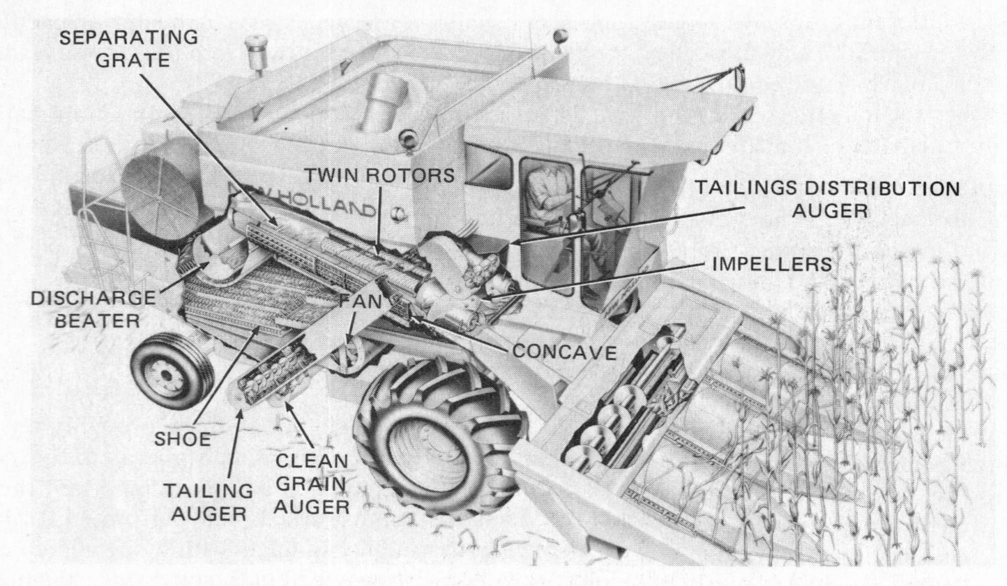


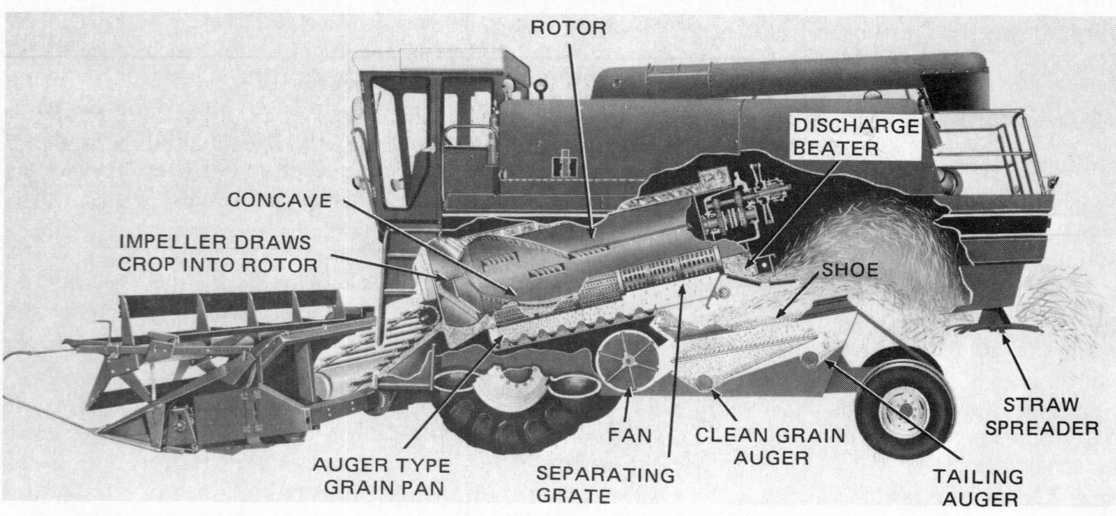


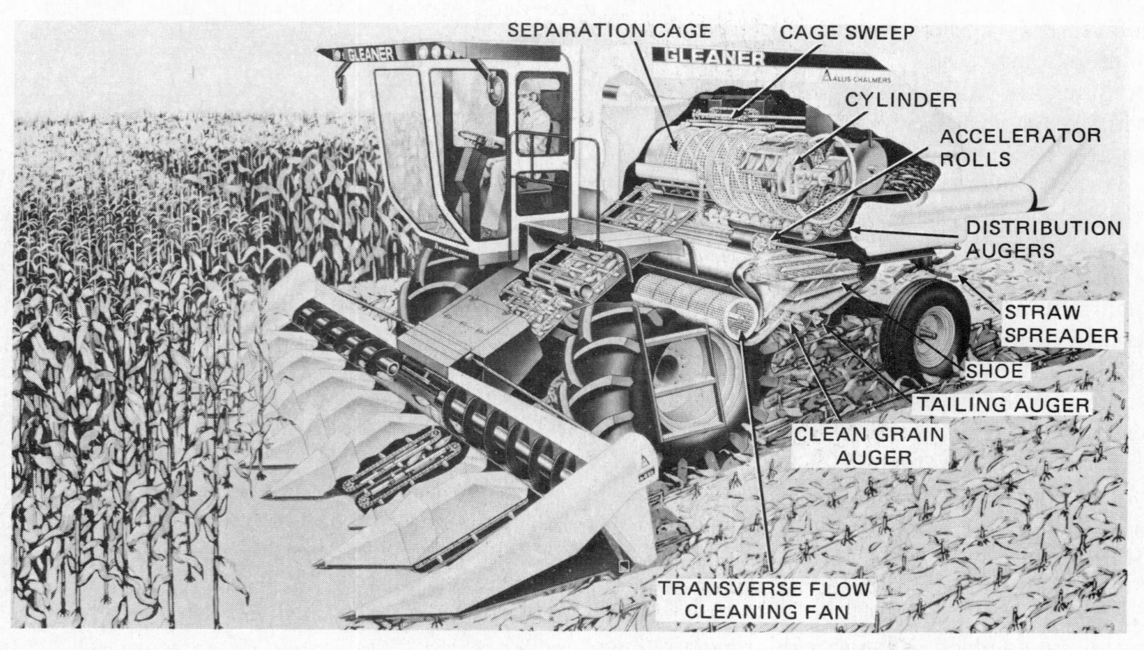


**Fig. 23-11.** Rotary-type self-propelled combines. (*Sperry-New Holland; International Harvester Co.; and Allis-Chalmers Manufacturing Co.*)

## Adjusting the Combine

Combines are designed to operate with the engine in the full-throttle position. Most manufacturers give high and low idle speed specifications and recommend that an accurate speed indicator (tachometer) be used to check and adjust engine speeds.

**Separator Speed** Separators (straw walker, chaffer, and sieve) are designed to operate at a specific speed regardless of the crop being harvested. Separator speed (rpm) should be checked before the harvest season starts. Each combine has a particular shaft that is designated for checking separator speed. The primary countershaft or the cylinder beater cross shaft may be used. An accurate tachometer should be used to check the operating speed with the engine at full governed rpm (Fig. 23-12). Check the operator's manual for specific procedures and specifications.

**Cylinder Speed and Concave Spacing (Clearance)** Basic cylinder and concave adjustments for different crops are given in Appendix D, Table D-1. These are designed to be used as a starting point. The cylinder should be run just fast enough and close enough to the concave to remove the maximum amount of seeds or kernels from the head without breaking up the straw excessively. Cylinder speed and concave spacing should be adjusted alternately until proper threshing is attained. The number of concave bars should be changed as specified when different crops are harvested. The recommended number of concaves is given in Appendix D, Table D-1.

## Chaffer, Sieve, and Fan

Initial chaffer and sieve openings should be set at the maximum openings recommended for the crop being harvested (Appendix D). Figure 23-13 shows a

**Fig. 23-12.** Checking combine separator speed with a tachometer. (*Sperry-New Holland*)

**Fig. 23-13.** Measuring chaffer and sieve opening. (*Allis-Chalmers Manufacturing Co.*)

combine operator checking the amount of opening. If this results in too much trash in the grain tank, increase the air to the shoe. Air volume is controlled by a variable-speed fan or by shutters (choke). If there is still too much trash, the sieve clearance should be reduced. The chaffer clearance may also have to be reduced. Use as much air as possible without blowing clean grain or seed out the back of the combine.

The maximum chaffer and sieve openings that can be used and still have an acceptable amount of trash in the grain tank are recommended. When the clearances are too small, grain will be carried over the shoe and out the back of the combine. Also excessive clean grain may be returned to the cylinder in the tailings.

**Ground Speed** It is important that all threshing and separator adjustments be made for the ground speed at which the combine will be operating. Ground speed should be based on field conditions, yield of the crop, and capacity of the combine. If ground speed is reduced or increased and harvesting conditions remain the same, the threshing, separating, and cleaning performance of the combine will be affected.

A combine is properly adjusted when it puts the most clean whole grain in the grain tank and leaves the least amount of grain in the field. Basic settings should only be used as starting points. The operator should adjust the combine to meet specific crop and harvesting conditions. The ability to do this is developed through operating experience.

## Combine Safety

Operator's manuals include a complete section on safety. Before attempting to service, adjust, or operate a combine, the safety section should be read carefully. One of the primary safety rules is never to lubricate or adjust a combine while it is in motion or the engine is running.

Keep the combine clean. Trash accumulates around the exhaust system and can cause fires. It is a good practice to carry a fire extinguisher on the combine.

Extreme caution should be taken when driving a combine on a public road. Check to see that all lights and safety flashers are properly adjusted and in working condition. They should be turned on, and the slow-moving vehicle emblem should be clean and clearly visible. The combine should be operated at a speed that matches road conditions. High speed is the leading cause of road accidents. Remember that the combine is moving much slower than the other traffic on the road. Watch for other traffic, pedestrians, and obstacles on or above the road.

All the safety precautions listed by the manufacturer should be understood and followed.

**⚠ WARNING Failure to do so may result in extensive combine damage and personal injury.**

# HARVEST LOSSES

Although almost anyone can drive a combine, it takes special skill and knowledge to perform an efficient job of operating a combine. Several types of losses are associated with combining. Some of the major ones are preharvest losses, header losses, threshing losses, separating losses, and cleaning losses.

## Preharvest Losses

Preharvest losses occur before the crop is harvested. They may be due to overmaturity of the crop, weather damage or crop diseases. The combine operator has no control over these losses; the grain is already lost.

## Header Losses

Header losses are caused by header or reel operation. Crops such as soybeans that tend to shatter easily have high header losses. The proper platform and reel adjustment will reduce header losses. In some crops and under some crop conditions special equipment such as pickup reels, pickup fingers, and special-purpose headers are helpful in preventing header losses. Check the operator's manual and visit an implement dealer to determine equipment available for special or unusual conditions.

## Threshing Losses

Threshing losses are due to incomplete removal of the seed from the seed head (underthreshing) or excessive cracking of the seed or grain (overthreshing). Proper cylinder speed, concave spacing, and ground speed all affect threshing losses. Another factor that affects threshing is the stage of maturity of the crop and the moisture content. Good operators know when a crop is mature enough to harvest and when to start in the morning or when to stop at night. These skills are developed through operating experience.

## Separation Losses

Separation losses are losses of threshed grain over the straw walker. They may be caused by too much straw entering the machine at slow cylinder speeds, improper straw walker drive speed, or improper curtain adjustment when threshing lighter seeds. Excessive broken straw may also cause separation losses due to plugging of the holes or slots in the walkers. There are no straw walkers in combines with rotor-type threshing units.

## Cleaning Shoe Losses

Cleaning shoe losses are losses of grain passing over the shoe. They may be caused by too much air from the fan, improper adjustment of the chaffer and sieve, or overthreshing, which results in too much material on the chaffer. Proper adjustment of the fan, chaffer, and sieve and proper threshing will eliminate cleaning losses.

## Checking for Grain Losses

Many operator's manuals give specific instructions for checking combine losses. The few minutes that it takes to check for losses may pay big dividends.

When 19 wheat kernels are found in 1 square foot ($ft^2$) [200 kernels per $m^2$] section of a field, the loss is equal to approximately 1 bu per acre [0.086 $m^3$ per ha]. If 1000 acres [404.7 ha] are harvested at this rate, the loss would amount to 1000 bu [35.24 $m^3$] of wheat. This is equal to a significant amount of money.

Table 23-1 gives the approximate number of kernels equal to 1 bu loss per acre for several crops.

**TABLE 23-1. Approximate Number of Kernels Per Square Foot That Equals One Bushel Loss Per Acre [Kernels per 0.093 $m^3$ Equal to 0.086 $m^3$ Loss per Hectare]**

| Crop | Kernels |
|---|---|
| Barley | 14 |
| Corn | 3 |
| Oats | 11 |
| Rice | 22 |
| Milo | 21 |
| Soybeans | 5 |
| Wheat | 19 |

*Note:* These numbers are approximate and may vary with kernel size and variety.

Consult the operator's manual for additional information regarding other crops.

To check for losses, stop the forward motion of the combine in a section that appears to be average for the field. Allow the combine systems to continue operating until the grain and straw in the machine have been cleaned out. Back the combine up about one machine length. If the combine is equipped with a straw chopper or straw spreader, it should be disconnected when checking grain losses.

Straw spreaders are belt-driven units that are attached under the straw discharge to spread the straw over the full cutting width of the combine. When windrowed straw is wanted, the drive belt is removed and the spreader unit is left on the machine.

Straw choppers cut and shred the straw or other harvest waste materials into small pieces so that they will break down rapidly in the soil. They attach and are driven like a straw spreader.

**Preharvest Losses** Preharvest losses can be determined by checking several 1-ft$^2$ [0.09-m$^2$] areas in the field ahead of the combine (Fig. 23-14). A more accurate count will be made if the number of kernels found in several areas is averaged. If four checks are made and 5, 5, 6, and 4 kernels are found, the average is 5. Referring to Table 23-1, when 19 kernels of wheat are found per square foot, the loss is equal to 1 bu/acre [0.086 m$^3$/ha]. The preharvest loss in this field would be about $^1/_4$ bu/acre [0.009 m$^3$/0.405 ha].

**Header Losses** Header losses can be determined by counting the kernels in several square-foot areas that were under the combine (Fig. 23-14). If four checks are made and 10, 9, 8, and 12 kernels are found, the average is 10. Subtract the preharvest losses, and the header losses average 5 kernels per square foot, or about $^1/_4$ bu/acre [0.009 m$^3$/0.405 ha].

**Separation Losses** If the operator's manual contains a table giving machine losses per square foot behind the separator, total grain loss is easy to calculate. Actual separation and cleaning losses can be determined by subtracting preharvest losses and header losses. For example, the table may state that 76 wheat kernels per square foot is equal to 1 bu [0.035 m$^3$] grain loss per acre [0.405 ha] if the separator is 49 in [1.24 m] wide and the header is 16 ft [4.9 meters (m)] wide.

To determine separation and cleaning losses, count the kernels in several square-foot areas behind the combine (Fig. 23-14). If four checks are made and 31, 29, 26, and 30 kernels are found, the average is 29. When 5 kernels for preharvest losses and 5 kernels for header losses are subtracted, 19 kernels are lost during separation and cleaning. According to the data in the table the 19 kernels amount to about $^1/_4$ bu of wheat per acre [0.009 m$^3$/0.405 ha]. These kernels have gone through the combine and were lost over the straw walker or the cleaning shoe. Kernels in partially threshed heads should not be counted. They are threshing losses.

**Calculating separation losses** When no table is provided in the operator's manual and one is not available, total grain loss can be calculated. Since separator width is much less than the cutting width, the kernels that are lost will be in a rather small area, and an actual count per square foot [0.09 m$^2$] will not predict actual losses. Therefore it is necessary to determine the percentage the separator width is of cutting width. To do this, divide the separator width in inches [meters] by the cutting width in inches [meters]. A 49-in [1.24 m] separator would be approximately 25.6 percent the width of a 16-ft [4.9 m] header.

If an average of 29 kernels is found in the check and 10 are subtracted for preharvest and header losses, 19 kernels are separation losses. To determine separation losses per acre, multiply the number of kernels by the separator percentage of header width. The separation loss would be approximately 5 kernels per square foot or about $^1/_4$ bu/acre [0.009 m$^3$/0.405 ha], which is the same loss that was calculated using the table found in the operator's manual.

**Threshing Losses** Threshing losses can be determined by counting the kernels that are left in partially threshed seed heads (Fig. 23-14). Loose grain or whole seed heads should not be counted. The

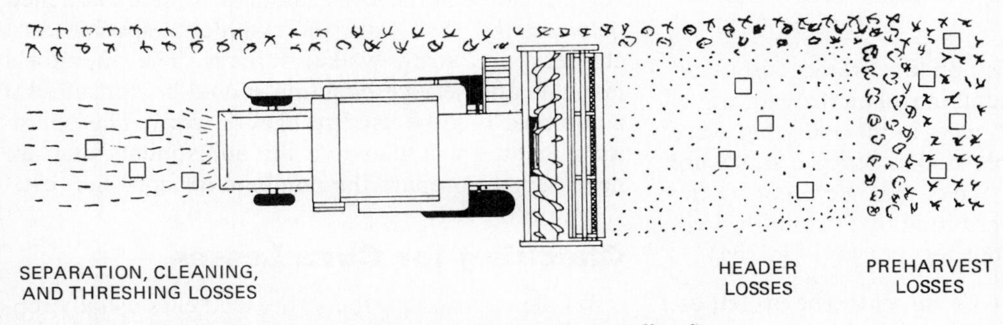


**Fig. 23-14.** Checking for harvest losses. (*Sperry-New Holland*)

threshing losses per acre [0.405 ha] are calculated using the same procedure used to determine total grain loss. Remember that it is easier to use a table to calculate losses if one is available. Total harvest losses should not exceed 3 to 5 percent of yield under normal harvesting conditions.

## Checking for Soybean Losses

Special low-profile row-crop heads have been introduced by combine manufacturers to reduce header and stubble losses in soybeans. These heads are ideal for row crops such as milo and soybeans. They cannot be used for corn (Fig. 23-1).

Soybeans present special harvesting problems. Lodging (falling over) and podding close to the ground are problems that can partially be controlled by variety selection. Poor cultural practices and improper combine adjustment can result in large harvesting losses.

To determine soybean losses, construct a frame that encloses a 10 ft$^2$ [0.93-m$^2$] area with a length that is equal to the combine header width. A 10 ft$^2$ frame for a 12 ft [3.65 m] header would be 10 in [254 mm] wide by 12 ft [3.65 m] long. Use this frame to check for total crop losses and header losses.

Stop the forward motion of the combine in the section of the field where the crop is average. Allow the combine systems to continue to run until the soybeans and stalks have been cleaned out. Back the combine up one machine length and perform the checks for losses.

**Total Crop Losses** Total crop losses include all soybean losses before and during harvest. To check total crop losses, position the frame across the width of the combine in an area behind the combine. Count all soybeans found in the area. Include beans in pods, pods on stalks, etc. Divide the number of beans found by 50 to get the total crop loss in bushels per acre [m$^3$/ha].

If total losses equal 1 bu/acre [0.035 m$^3$/0.405 ha] and the soybeans are producing 33 bu/acre [1.163 m$^3$/0.405 ha], the total crop loss would be 3 percent. A 3 percent loss is acceptable for many combine operators. When the loss is over 3 percent, the cause should be determined and proper adjustments should be made.

**Stubble Losses** Position the frame across an area that was under the combine when forward travel was stopped. Count the beans in all pods that were left attached to the stubble stalks. Stubble stalks are the lower portion of the stalks that were cut by the header. Divide the number found by 50, and this will give you stubble loss in bushels per acre [m$^3$/ha].

**Uncut Stalk Losses** Count all the beans left on lodged or down stalks that were not cut by the header. This number divided by 50 will give you uncut stalk loss in bushels per acre [m$^3$/ha].

**Cut Stalk Losses** Cut stalks are stalks that were cut but lost in the gathering process. The number of beans on cut stalks divided by 50 gives cut stalk loss in bushels per acre [m$^3$/ha].

**Pod Losses** Count the beans in loose pods that were detached from the stalk and lost by the header. This number divided by 50 gives the pod loss in bushels per acre [m$^3$/ha].

**Shatter Losses** Loose beans on the ground are shatter losses. The number of loose beans found divided by 50 equals shatter losses in bushels per acre [m$^3$/ha].

**Preharvest Losses** Preharvest losses can be determined by counting the shattered beans in pods on the ground in an unharvested area just ahead of the combine. Since it is difficult to get the frame into the beans, four stakes and a string may be used to mark off the 10 ft$^2$ [0.93 m$^2$] area. Divide the number of beans found by 50, and this will give you preharvest losses in bushels per acre [m$^3$/ha].

Preharvest shatter and pod losses can be compared with header shatter and pod losses to determine how many beans are being lost by the header. Total machine losses can be determined by subtracting preharvest losses from total crop losses.

This is one method that can be used to check soybean harvesting losses. The method described earlier for grain losses may be used, but it does not identify the types of header losses. Some operator's manuals provide detailed procedures and tables for checking losses.

# HARVESTING CORN

In recent years, much of the nation's corn crop has been harvested with self-propelled combines. This is largely a result of the development of the corn head which can be easily attached to the combine (Fig. 23-5).

Other special equipment has been developed to adapt the standard combine for efficient corn harvesting. Some of these are filler bars that are attached between the rasp bars on cylinders, special corn chaffers, and straw walker screens. The operator's manuals and service bulletins should be consulted if a combine is to be used to harvest corn. The operator's manual will also give the adjustments that are necessary to prepare the combine for corn harvest.

## Checking for Corn Losses

Whole ears and partially threshed ears usually represent the largest loss in threshing corn. Some ears

**TABLE 23-2. Row Length Per 1/100 Acre (0.004-ha)**

| Row Width | | Distance Equal to 1/100 Acre (0.004 ha) in Feet (Meters) | | | | | |
|---|---|---|---|---|---|---|---|
| Inches | (cm) | 4 Rows, ft | (m) | 6 Rows, ft | (m) | 8 Rows, ft | (m) |
| 20 | 50.3 | 65.5 | 14.46 | 43.6 | 13.29 | 32.7 | 9.97 |
| 28 | 71.1 | 46.7 | 14.23 | 31.1 | 9.48 | 23.9 | 7.28 |
| 30 | 76.2 | 43.6 | 13.29 | 29 | 8.84 | 21.8 | 6.64 |
| 38 | 96.5 | 34.5 | 10.52 | 22.9 | 6.99 | | |

are preharvest losses; that is, they are on the ground before the combine enters the field. Others are lost by the corn head, and some kernels are left on the cob after threshing.

**Ear Losses** To determine ear losses, most manufacturers recommend that the ears in 1/100 acre be counted. Many operators's manuals include a table that gives the row length that equals 1/100 acre for different row width or spacing (Table 23-2).

**Preharvest ear losses** Use the table to determine the row length that should be used to mark off 1/100 [0.004 ha] of an acre [0.405 ha] in front of the combine (Fig. 23-15). Carefully pick up all the ears that are on the ground in this area. The ears can be separated by size, or the total ears picked up can be weighed. Each 3/4 pound (lb) [0.34 kilogram (kg)] is equal to 1 bu [0.035 m³] ear loss per acre. Large ears normally weigh about 3/4 lb [0.34 kg]. Therefore each large ear found represents 1 bu loss per acre [0.035 m³/0.045 ha].

**Ear losses behind the combine** To determine ear losses behind the combine, stop the combine in an area that appears to be average for the field. The straw spreader should be disconnected. Carefully pick up all whole, broken, and partially shelled ears in a 1/100-acre [0.004-ha] area behind the combine (Fig. 23-15). Each 3/4 lb [0.34 kg] ear is equal to 1 bu [0.035 m³] loss per acre [0.405 ha].

**Determining corn head ear losses** Subtract the whole ears found in front of the combine (preharvest losses) from the ears found behind the machine that did not go through the threshing unit. This will give corn head losses or the corn that is lost because it was not properly fed into the combine.

**Kernel Losses** Kernel losses can be determined using the methods described for checking separation and header losses for grain. In corn, 2 kernels/ft² [0.09 m²] is equal to a loss of 1 bu/acre [0.086 m³/ha]. Kernels may be lost at the corn head and during separation.

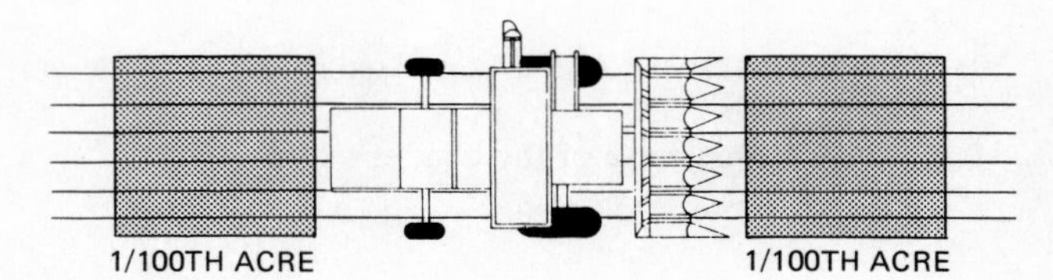


**Fig. 23-15.** Checking for corn ear losses.

**Corn head kernel losses** Corn head kernel losses are determined by counting the kernels found in the area just ahead of the combine after it has been backed up one machine length.

**Separation losses** Separation losses are determined by counting the kernels in an area behind the combine. Most operator's manuals give recommendations and tables for checking kernel losses behind the combine. If a table is not available, the procedure described for calculating separator losses for grain can be used. Remember to subtract corn head kernel losses to determine separator kernel losses.

## SUMMARY

The self-propelled combine is used to harvest the major portion of the nation's grain and seed crops. It is available in standard models and special models for special crops and harvesting conditions.

A combine performs five major functions in the harvesting of a crop. These are (1) cutting the crop and feeding it to the cylinder, (2) threshing the crop, (3) separating the crop from the straw, (4) cleaning the crop, and (5) handling the crop.

A profitable harvest with a combine depends upon proper adjustment of the machine, proper maturity of the crop, and weather conditions. Proper adjustment is attained by studying the operator's manual and through experience.

Four types of losses associated with grain harvesting are preharvest losses, header losses, threshing losses, and separation losses. A good operator will make frequent checks to see that the combine is properly adjusted to keep losses at a minimum.

## THINKING IT THROUGH

1. List three types of combines.
2. List the different cutting platforms and heads available for combines.
3. How do row-crop heads cut the stalk?
4. What crops are automatic header height controls designed for?
5. How are the ears removed from the stalk by corn heads?
6. List reasons for windrowing small-grain crops.
7. What is the purpose of the reel on a cutting platform?
8. What is the recommended reel speed?
9. What two components remove the seed from the seed head?
10. How is the grain or kernel separated from the straw or stalk?
11. What is the purpose of the chaffer? The sieve?
12. Describe the difference in the operating principles of a rotary combine.
13. List the basic adjustments that are made on a combine.
14. Why is it usually necessary to make additional adjustments?
15. What are acceptable harvest losses for most grain crops?
16. What are preharvest losses? How are they checked?
17. Explain header losses and explain how they are checked.
18. What are threshing losses? How are they checked?
19. How do you check for ear losses when harvesting corn?
20. What is the moisture content of most grain when it is ready to combine and store?
21. What is the moisture content of grain that is to be dried following combine harvesting?

# CHAPTER 24 SPECIALIZED HARVESTING EQUIPMENT

## CHAPTER GOALS

In this chapter your goals are to:

- Identify types of cotton harvesting equipment
- Identify types of peanut harvesting equipment
- Identify types of other specialized harvesting equipment

Since many of the machines used to harvest and handle agricultural crops have been developed in recent years, farmers and ranchers can often describe what harvest was like without machines. Crops like corn, cotton, fruits, vegetables, and nuts were harvested almost entirely by hand only a few years ago.

Hand harvesting made the work very hard, required a large labor force, and restricted the number of acres that could be managed by one agricultural producer.

The specialized harvesting machines that are used today are the result of extensive research. Many of the machines started out as the result of efforts made by farmers, ranchers, and other agricultural producers to reduce the amount of labor and time required to harvest their crops.

Agricultural engineers and machinery specialists employed by manufacturers and state agricultural experiment stations work continuously to develop machines that are more efficient and reduce the time and labor required to harvest crops. Crop specialists work with the engineers to develop new varieties of crops that are better suited for mechanical harvesting.

Some fruit and vegetable crops are still hand-harvested. However, research and the development of new machines continue to reduce the number of hand-harvested crops.

## COTTON HARVESTING EQUIPMENT

Only a few years ago much of the nation's cotton crop was harvested by hand. Either the cotton was picked from the boll (picked cotton), or the entire boll was pulled (pulled cotton). Today almost all the cotton grown in the United States is mechanically harvested.

Two basic types of machines are used to harvest cotton. The cotton stripper removes the whole cotton boll from the cotton stalk in a once-over harvesting operation. The cotton picker removes the lint and seed only from the bolls that are matured. Green bolls are left on the stalk and allowed to mature.

### Cotton Strippers

Strippers may be tractor-mounted, pull type, or self-propelled (Fig. 24-1). There are several types of strippers, and a few will be discussed in this chapter. Cotton strippers are usually classified by the type of stripping device used to remove the boll from the stalk.

The stripping units on most cotton strippers can be adjusted for different row spacings. Some four-row units can be adjusted to harvest skip-row cotton (Fig. 24-2).

**Double-Roller Stripper** The double-roller stripper uses two stripping rolls to remove the bolls from the stalk. Some double-roller stripping rolls are equipped with alternating rows of nylon brush bristles (brush combs) and flexible rubberized material (Fig. 24-3). Others use steel stripping rolls. After the bolls are removed, they are moved by augers to the elevator. The undersides of the auger housings are perforated to allow dirt and trash to filter through and drop back on the ground.

The elevator transports the cotton bolls to the basket or cotton trailer. Air is usually used to elevate the cotton. Green bolls are heavier than mature bolls and are separated from the cotton and caught in a green boll box or dropped on the ground.

**Fig. 24-1.** A self-propelled cotton stripper. *(International Harvester Co.)*

Fig. 24-2. A cotton stripper harvesting skip-row cotton. (*Deere & Co.*)

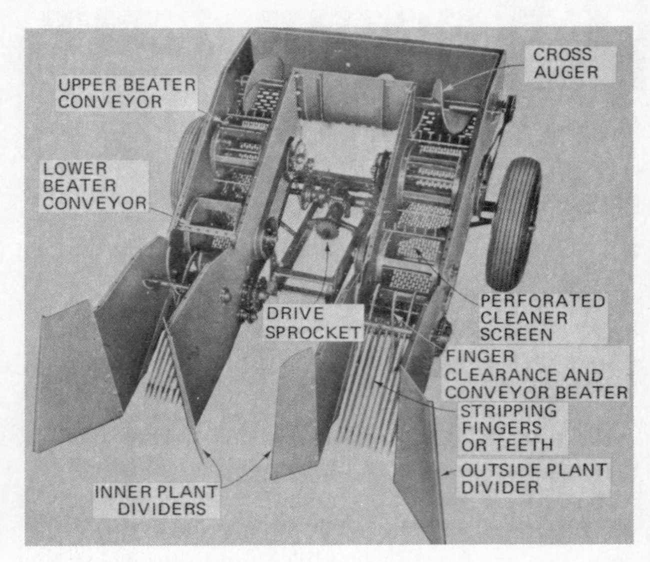


**Fig. 24-4.** The finger-type stripper uses long "fingers" to remove the bolls from the stalks. (*White Farm Equipment Co.*)

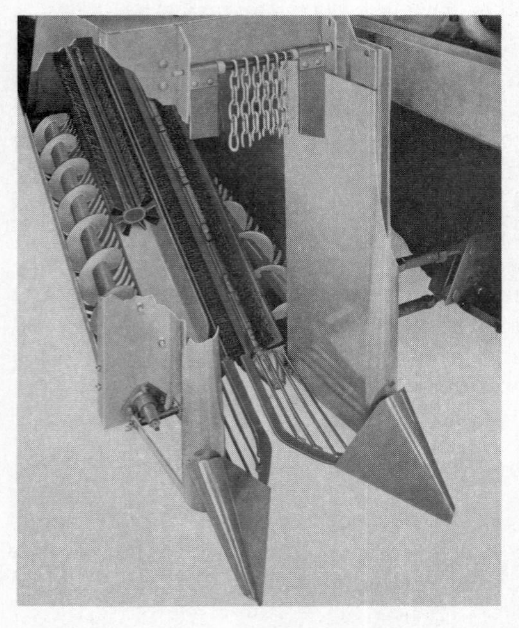

**Fig. 24-3.** A cutaway view of a double-roller type cotton stripper. (*International Harvester Co.*)

**Fig. 24-5.** A self-propelled multirow or broadcast finger-type cotton stripper. (*Allis-Chalmers Manufacturing Co.*)

**Finger-Type Stripper** The finger-type stripper uses long steel teeth or fingers to remove the cotton bolls from the stalk (Fig. 24-4). Finger beaters, which are paddle wheels with fingers on the slats, are often used to move the bolls from the stripper fingers to the elevator system. When finger beaters are used, the bottom of the conveyor is perforated so that dirt and trash will fall through.

**Self-Propelled Multirow Stripper** The multirow-type cotton stripper (Fig. 24-5) uses stripping fingers to remove the cotton bolls from the stalk. After the bolls are removed, they are conveyed by a cross auger over grids that are designed to remove dirt. The cotton is then elevated by a belt to a separating chamber where air provided by two fans is used to separate the lighter mature bolls from the heavier green bolls and other foreign materials such as rocks, clods, and other plant residue.

**Green bolls** The green bolls and other heavy materials are conveyed by an auger into a green boll box. The green boll box is equipped with an indicator light to let the operator know when it is full and needs emptying.

**Mature bolls** Mature bolls are carried by the airstream into the breaker cylinder, which loosens the lint and seed from the bolls. The breaker cylinder also distributes the material which is delivered to the extractor by air (Fig. 24-6).

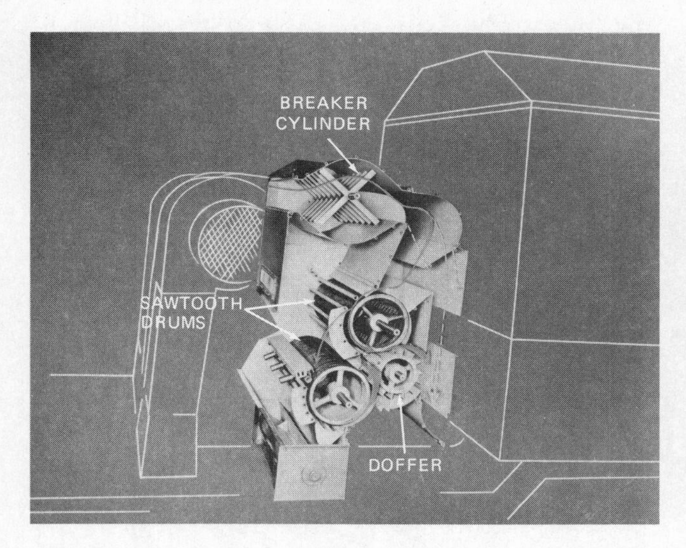


**Fig. 24-6.** The breaker cylinder and extractor unit separate the burrs from the lint in multi-row self-propelled strippers. (*Allis-Chalmers Manufacturing Co.*)

**Lint and seed separation** Lint and seed are separated from the bolls by the extractor. The extractor consists of two steel drums equipped with saw teeth. As the cotton enters the extractor, the lint becomes attached to the saw teeth. The saws carry the lint around to a rotating doffer. The doffer is equipped with brushes that remove the lint from the saws and deliver it back into the airstream which carries the lint to the lint basket (Fig. 24-6).

Burrs, sticks, and other foreign material are separated from the cotton by rod grids. They are deposited back on the ground by burr chutes.

## Cotton Pickers

Cotton pickers are designed to remove the lint from the cotton boll. Unlike strippers, cotton pickers can be used to harvest the crop two or more times and remove only the cotton that is mature and ready to harvest. This allows the cotton farmer to harvest a greater percentage of the cotton produced. Cotton pickers are available in self-propelled and tractor-mounted models.

**Picking Units** One-, two-, and four-row cotton pickers are used (Fig. 24-7). The picking units on some two- and four-row units can be adjusted for various row widths.

**Spindles** Spindles are used to catch lint and pull it from the open boll. Spindles may be tapered or straight. The machine-cut teeth are designed to catch the cotton fiber as they come in contact with it. As the spindles revolve, the fiber wraps around the spindles and is pulled from the boll (Fig. 24-8B, C, D).

Spindles either are mounted on a drum or on a continuous chain belt. The operation of both types is similar, and the drum will be discussed here.

**Fig. 24-7.** Self-propelled cotton pickers remove only the lint cotton. (*International Harvester Co.*)

On the drum-type picker, the spindles are mounted on picker bars which are mounted vertically on the picker drums (Fig. 24-9). Two drums are used, one on each side of the picker unit. The drums turn as the picker moves along the row of cotton. The speed at which the drums turn is synchronized to the travel speed of the picker during first-pick operation so that green bolls are not knocked off the stalk. Pickers can be operated up to about 3.5 miles per hour (mi/h) [5.6 kilometers per hour (km/h)] in first-pick operation. *First-pick* is the first time the cotton is picked. If cotton is picked the second or third time, ground speed can be increased and the picking unit does not have to be synchronized to ground speed.

Both low and high drum pickers are available. High drum pickers are equipped with 20 spindles per bar and are normally used for cotton producing more than 2 bales per acre. Low drum pickers are equipped with 14 spindles per bar and can be used to pick cotton up to 5 feet (ft) [1.5 meters (m)] tall.

**Moisture pads** Moisture pads are used to apply water or a solution of water and special picker wetting agent. The spindles are completely wet down. Moistening helps the cotton to stick to the spindles, aids in doffing or removing cotton from the spindles, and helps keep the spindles clean (Fig. 24-8A).

**Doffers** Rubber-faced doffer plates are used to remove the lint cotton from the spindles. The doffers are located on the outside of the picking unit 180° from where the spindles contact the cotton. As cotton is removed from the spindles, it is carried by a pneumatic (air) pickup to the lint basket (Fig. 24-8E).

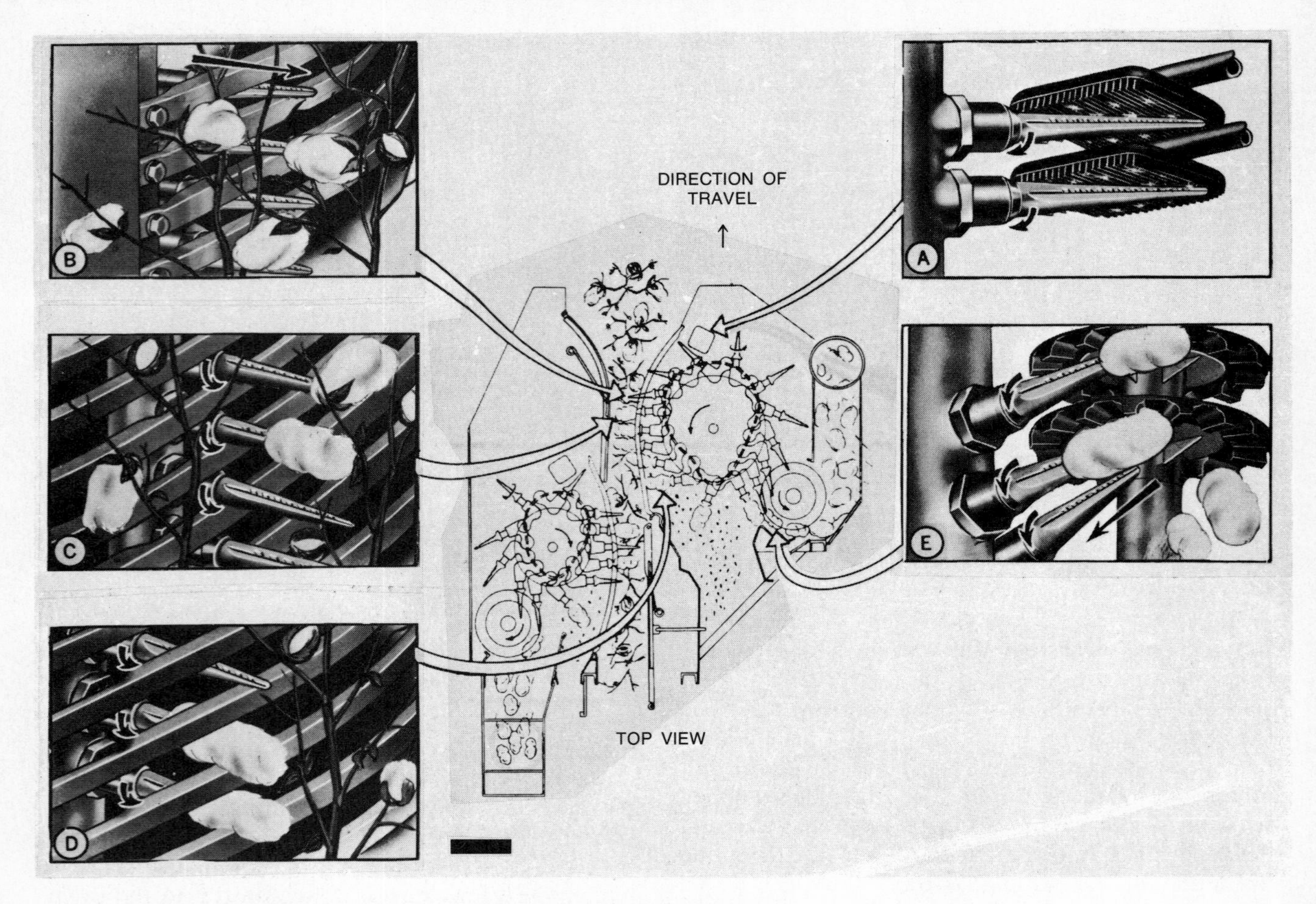


**Fig. 24-8.** The operation of a drum type picker unit. (*Deere & Co.*)

**Conveyor** A fan is used to draw the lint cotton from the picking unit and blow it into the lint basket. Fans and duct systems are designed so that the cotton is handled entirely by air to prevent damage to the lint and seeds. This also prevents the grinding of any trash that is removed with the cotton.

As the cotton is blown into the basket, it is blown against cleaning grates or grids located in the top of the basket or duct. Much of the dirt and trash passes through the grates and is removed from the cotton.

**Lint Basket** Baskets are used on cotton pickers. Newer-model pickers are equipped with baskets that hold up to 3000 pounds (lb) [1360 kilograms (kg)] of lint cotton and can be equipped with compactors that increase the capacity.

Baskets are mounted high and have high lift hinges so that they can be dumped into trailers or compactors. The baskets are hydraulically operated. **WARNING:** Since the baskets are mounted high, special care must be taken when operating cotton pickers to prevent them from turning over.

## Harvesting Cotton

Cotton varieties have been developed for mechanized harvesting. Stripper varieties produce smaller bolls that hold the lint inside the burrs. Picker varieties produce open bolls so that the spindles can catch the cotton fiber and remove it from the boll.

Since cotton stripping is a once-over operation, it is important that the cotton be stripped when the majority of the bolls are mature. Some varieties of cotton will set squares (small bolls) more evenly than others. This means that they also mature more evenly. Recommended varieties vary from region to region, and state agricultural experiment stations can be consulted to determine varieties suited for stripping or picking.

**Cultural Practices** Since pickers and strippers must be able to reach the bolls to harvest the cotton, it is important that the rows be properly formed. A slight elevation at the base of the plant will enable the machine to harvest more of the cotton produced.

It is also important that weeds and grass be controlled in cotton fields. Weeds and grass will reduce harvesting efficiency and lower the grade of the cotton.

**Defoliation** Chemicals are used to cause the leaves to dry up and fall off. This is done before the harvest. The chemicals can be applied with ground

**Fig. 24-9.** Spindles are mounted on picker bars which are mounted vertically on picker drums. (*Deere & Co.*)

**Fig. 24-10.** A module builder which is used to compress cotton in the field. (*Bush Hog Co.*)

**Fig. 24-11.** A module mover bed mounted on a truck. The truck is used to move modules to the cotton gin. (*Bush Hog Co.*)

spray equipment or can be sprayed from an airplane or helicopter equipped for crop spraying.

Always follow the directions supplied with the chemical to determine application time and rate. Also, be sure to follow all recommended safety precautions given for using the chemicals and disposing of the empty containers.

**Machine Operation** The operator's manual should be studied before attempting to operate or adjust a picker or stripper. Also, the safety precautions listed in the manual should be followed.

**Handling Picked Cotton** Cotton trailers are commonly used to transport cotton to the gin after stripping or picking.

**Newer Systems** The increased harvesting capacity of new pickers and strippers has created the need for other handling systems.

**Module system** A module builder is used to compress the cotton on a pallet or on the ground in the field (Fig. 24-10). The cotton is dumped from the picker into the module maker. The compressed cotton *modules* are transported to the gin on large tilting-bed trailers or trucks (Fig. 24-11).

**Rick system** A mechanical rick compactor is used to compress the cotton on the ground in the field. With this method, a long compressed stack of cotton is formed. Front-end loaders are used to load the cotton into conventional cotton trailers or trucks for transport to the gin.

## PEANUT HARVESTING EQUIPMENT

Peanuts develop pods 1 to 3 inches (in) [25 to 76 millimeters (mm)] deep in the soil. Since the peanut grows underground, it must be dug and allowed to dry or cure 3 to 10 days before it is harvested.

### Peanut Diggers

Peanut diggers are designed to cut the taproot and loosen the soil around the peanuts. They also lift the vines and pass them over a shaker that removes the loose soil from the peanuts and roots. The vines and nuts are then deposited on top of the ground in a windrow so that the peanuts will dry.

**Digger-Shaker-Windrowers** The digger-shaker-windrower (Fig. 24-12) digs the peanuts, shakes them, and places them in windrows for curing. The flat plow or knife is set at an angle and is run

**Fig. 24-12.** Peanut digger-shaker-windrowers dig the peanuts, shake them, and place them in a windrow. (*Lilliston Corp.*)

**Fig. 24-13.** Inverters dig the peanuts, shake them, and place them in a windrow with the peanuts exposed on top of the windrow. (*Lilliston Corp.*)

2 to 3 in [51 to 76 mm] deep to cut the taproot and loosen the soil. Steel tines called *radder bars* are attached to the back of the plow. The radder bars move the vines onto the shaker. The shaker elevates the vines and agitates them gently to remove dirt from the pods and roots. As the vines pass over the shaker, forming shields are used to form a windrow.

**Inverters** Inverters dig the peanuts and feed them onto the shaker in the manner described for the digger-shaker-windrower. However, they are equipped with inverter rods that turn the pods up as they pass over the shaker (Fig. 24-13). Peanuts dug with an inverter-type digger are formed into a windrow that exposes the peanuts on the top of the windrow. In other words, the plant is upside down in the windrow.

## Peanut Combines

The peanut combine (Fig. 24-14) picks up the windrowed peanuts and feeds them into a series of cylinders where the pods are picked from the vines. The pods are then separated from the vines and carried to the tank or hopper. Vines, roots, and dirt are discharged out the rear of the combine.

**The Pickup Assembly** The pickup picks the windrowed peanuts up and feeds them into the auger feed assembly (Fig. 24-15). The pickup should be adjusted to run from 1 to 3 in [25 to 76 mm] above the ground. Pickup speed should be adjusted to match ground speed, and the windrow should feed evenly into the combine.

**The Auger-Feeder** After the peanuts are picked up, they are fed into the first cylinder by the auger-feeder. Spring fingers are used to feed the cylinder.

**Fig. 24-14.** Peanut combine harvesting peanuts. Combine is power take-off driven. (*Lilliston Corp.*)

**The Picking Section** A series of cylinders, usually four, pick the peanut pods from the vines (Fig. 24-15). The cylinders are similar to the cylinder used in grain combines, but they are equipped with spring teeth or cylinder springs that comb the vines to remove the pods.

The picking action can be adjusted by controlling the cylinder speed and by positioning the concave stripper spring assemblies.

**Picking cylinder screens and strippers** Picking cylinder screens are located under the picking cylinders.

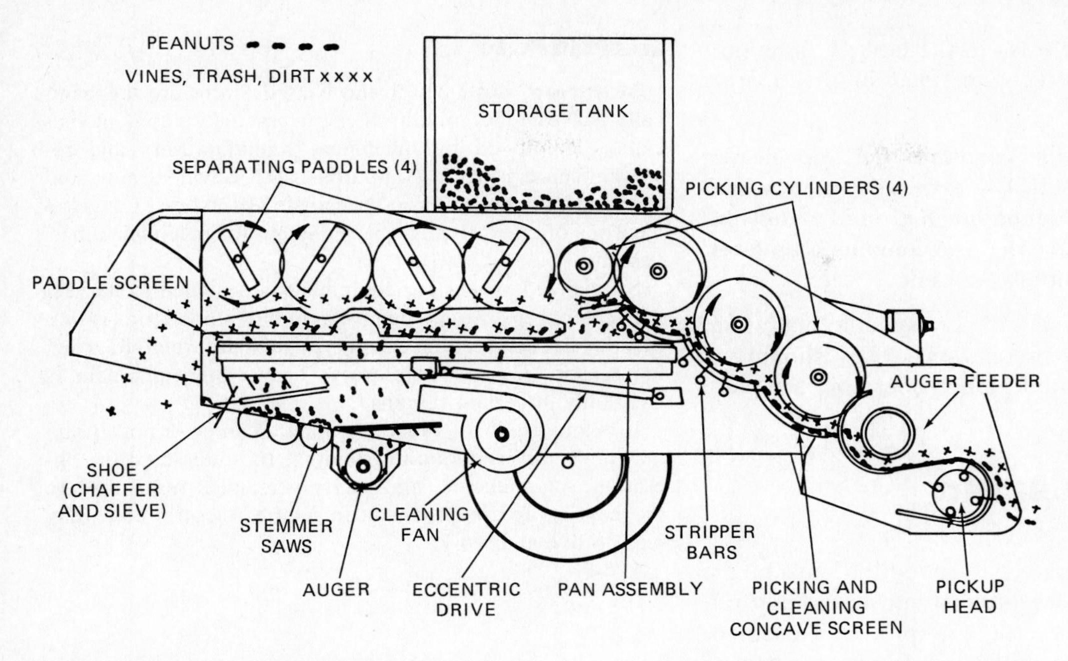


**Fig. 24-15.** Schematic drawing of a peanut combine showing threshing, separation, cleaning, and handling. (*Lilliston Corp.*)

The screens under the front cylinders support the vines and pods as the cylinders move them through the picking section. Separation of the pods from the vines begins at the rear screens, which are open to allow the pods to fall through.

The stripper bars are equipped with spring teeth that engage the cylinder springs to pick the pods from the vines. Stripper springs may be engaged, partially engaged, or not engaged. The springs are engaged when the bars are rotated so that the strippers have maximum contact with the cylinder springs; this results in the highest picking action. The stripper bars are engaged as needed to pick the pods.

**Separation** Peanut combines are equipped with straw walkers or paddle wheels, as shown in Fig. 24-15, to separate the pods from the vines. As the pods are separated, they fall on the cleaning shoe.

Peanut combines are equipped with a chaffer and a sieve. Their operation is very similar to the shoe action in a grain combine. Air is used to suspend the chopped vines and trash that fall on the chaffer, and move them out the back of the combine. The openings in the chaffer are adjustable, and the back openings can be adjusted independently of the front openings.

The sieve is located under the chaffer, and it separates the pods from material that is only slightly larger or lighter than the pods. The openings in the sieve are also adjustable. Both the chaffer and the sieve have a rear lip that can be adjusted to prevent picked nuts from going over the end of the shoe.

**Stemmer Saws** Stemmer saws located under the shoe are designed to remove the stems that remain on the pods. As the stems are removed, they fall through the saw assembly onto the ground.

**Handling System** A cross auger carries the pods to the air lift system (Fig. 24-15). A fan provides the air used to convey picked peanuts into the tank. The air pressure is adjusted by a control valve according to peanut yield.

**Tank** The high-lift storage tank is located above the combine. When the tank is filled, hydraulic rams are used to lift and dump it.

**Field Operation** Pickup and auger-feed speeds are adjusted to meet crop conditions. Cylinder speed and stripper bar position are adjusted to pick the pods from the vines without excessive vine chopping. The highest peanut grade will be attained when the cylinder speed is as slow as possible and the fewest number of stripper bars are engaged.

The cleaning fan should be adjusted so that it floats foreign material out the back of the combine without blowing pods out. The air vanes should be adjusted to begin floating foreign materials at the front of the shoe.

The chaffer and sieve openings should be adjusted wide enough to let the pods fall through. When openings are too small, pods will be carried out the back of the combine. Wide openings will allow excessive foreign material to get into the tank with the peanuts.

**Checking threshing action** Threshing action is checked by examining the vines at the rear of the combine. All pods should be removed from the vines. Also, check for picked pods on the ground behind the combine and excessive trash in the tank. Follow the adjustment procedure recommended in the operator's manual.

**⚠ WARNING Peanut combines get top heavy when the tank is full. Use extreme caution when operating on hillsides or rough ground and avoid sharp turns. Be sure that the combine is on level ground when dumping the tank.**

The operator's manual will give specific operating instructions and safety precautions. They should be read and fully understood before operating the peanut combine.

## OTHER HARVESTING EQUIPMENT

As stated in the chapter introduction, machines have been developed to harvest most agricultural crops. Specialized harvesting equipment is used to harvest potatoes, sweet potatoes, tomatoes, beets, and other vegetables. Sugar beets, sugar cane, fruits, and nuts are also mechanically harvested.

These machines are used extensively in regions where these crops are produced commercially. It is important that people preparing for a career in power and machinery or production agriculture be familiar with the operation and service of the machines used in the region where they plan to live and work. However, since there are a large number of different machines used in the different regions of the United States, it is impossible to discuss them all in this text. Local teachers of vocational agriculture, farm equipment dealers, county extension agents, and agricultural producers can supply operator's manuals and other information concerning the operation and service of the equipment used in your region.

## SUMMARY

Most crops produced in the United States are mechanically harvested. Agricultural engineers and equipment specialists employed by machinery manufacturers and research centers work continuously to develop new and improved machines. Crop specialists also work to develop varieties of crops that are better suited for mechanical harvesting.

Specialized harvesting equipment is designed to harvest particular crops. Since the production of some crops is limited due to such factors as soil type, climate, rainfall, growing season, and markets, some harvesting equipment is used only in certain regions.

It is important for those planning a career in power and machinery or production agriculture to understand the operation, maintenance, and safety principles that relate to the harvesting equipment used in the region where they plan to live and work.

## THINKING IT THROUGH

1. Describe the difference between a cotton stripper and a cotton picker.
2. What is the difference between a double-roller stripper and a finger stripper.
3. Describe the purpose of the spindles, doffers, and moisture pads on a cotton picker.
4. Why is cotton defoliated?
5. Why is it important to plant a certain variety of cotton?
6. Describe the new systems used to transport cotton from the field to the gin.
7. Why is an inverter used to dig peanuts?
8. How does a peanut combine separate the pods from the vines?
9. What is the purpose of the stemmer saws in a peanut combine?

# UNIT VII AGRICULTURAL POWER AND MACHINERY MANAGEMENT

| Competencies | Agricultural Machinery Service Manager | Agricultural Machinery Parts Person/ Manager | Agricultural Tractor and Machinery Mechanic | Agricultural Machinery Setup and Delivery Mechanic | Agricultural Tractor and Machinery Mechanic's Helper | Small-Engine Mechanic | Farmer, Grower, and Rancher |
|---|---|---|---|---|---|---|---|
| Determine the cost of owning and operating agricultural machinery | ● Essential | ● Desirable | ● Desirable | ● Desirable | ● Desirable | ● Desirable | ● Essential |
| Identify methods of reducing machinery costs | ● Essential | ● Desirable | ● Desirable | ● Desirable | ● Desirable | ● Desirable | ● Essential |
| Select machinery on the basis of function, capacity, and size requirements | ● Essential | ● Desirable | ● Desirable | ● Desirable | ● Desirable | ● Desirable | ● Essential |
| Select the size tractor on the basis of horsepower and implement | ● Essential | ● Desirable | ● Desirable | ● Desirable | ● Desirable | ○ Not necessary | ● Essential |
| Use *Nebraska Tractor Test Reports* data in selecting a tractor | ● Essential | ● Desirable | ● Desirable | ● Desirable | ● Desirable | ○ Not necessary | ● Essential |

**Essential**

**Desirable**

**Not necessary**

Managing the use of tractor power and agricultural machinery is of particular concern to the producer of agricultural products. Regardless of the farming enterprise and the products produced, the owner/manager must receive a reasonable financial return on the investment to remain in business. Farming is no different than any other business venture; to remain in business, income must exceed expenses.

Farming includes the management of investments in land, buildings, machinery, raw materials, labor, and operating money. As a reward for overseeing the proper use of these resources, the owner/operator expects to be able to meet all financial obligations and receive a reasonable return (profit). A modern farmer may have from one-fourth to one-half of the total capital investment in machinery and equipment. This value varies with the type of farming enterprise and with yearly changes that result from national economic conditions. Nevertheless, the sum involves a financial investment that represents a major part of crop production cost. Reducing costs of operation is an important way of increasing profit.

The worker in agricultural power and machinery is also involved in the management of a business. As an owner, manager, or representative of a machinery dealership, you will be conducting business affairs and will be financially responsible. By assisting the producer of agricultural products with machinery selection and service, you will need to understand their management practices as well as those of the company you represent.

Your responsibility also requires a basic knowledge of plant, animal, and soil science to be able to communicate intelligently and provide assistance to producers. Your knowledge of the capacities and performance data of machinery can provide the producer with necessary information to make wise machinery purchases. Consequently, tractor and machinery sales depend on your ability to provide accurate information and to promote confidence in the quality of product and services which you represent.

This unit contains the concluding chapters of this book. It is included last, not because agricultural power and machinery management is of lesser importance. On the contrary, it is the capstone of a subject that includes knowledge for decision-making purposes. Proper decisions can only be made when all the information has been assembled and processed. Management therefore requires a thorough understanding of the principles of operation, maintenance, and repair involved in machinery used in production agriculture.

# CHAPTER 25 DETERMINING MACHINERY COSTS

## CHAPTER GOALS

In this chapter your goals are to:

- Determine the cost of owning and operating agricultural machinery
- Identify methods of reducing machinery costs
- Identify how to select and use machinery efficiently

The new models of agricultural tractors have been announced by the Universal Company. As sales manager for the local distributor, you planned and conducted a preview open house for farmers and ranchers in your area. The attendance at the open house was more than you expected, with a lot of interest in the new line. The comments of the people attending varied from discussions about features on the new tractors to general conversations about cropping practices. One person was overheard to say, "We'd have gotten our corn planted in time if we hadn't broken down before the weather turned bad."

Another person was heard saying, "I can't get good help—besides, I can buy horsepower cheaper than I can labor."

"Yes," another commented, "I've got enough horsepower now to handle another 160-acre farm if I could find it."

"But isn't it costing you a lot more to run that big tractor than your old one?" another asks.

As a manager of power and machinery, you, along with other employees of the dealership, were provided the opportunity to hear and heed the comments of owners/managers of local farms. You, in turn, not only made potential sales contacts but identified the emphasis on management, costs, timeliness, and machine capacities. Perhaps one owner summarized the general feeling when he said, "I could use a tractor with more power, but I am not sure I can justify the extra cost."

This is why you have a good feeling—because you can help that person and others with similar needs determine cost to assist in managing their business. The sales potential looks good.

## DETERMINING MACHINERY COSTS

Management of machinery includes an important element of finance. The element is more than how much it costs to purchase a machine (the selling price). Rather, it is how much it costs to own and operate the machine over its useful life, or the period of time that it is considered owned. Ownership costs are locked into the machine when it is purchased. These costs are therefore considered *fixed;* ownership costs are hereafter referred to as fixed cost.

Operating costs are the expenses which occur as the result of actually using the machine. They vary with the amount of annual use and thus are identified as *variable costs* (Fig. 25-1).

### Factors Affecting Fixed Cost

Fixed costs always occur in the ownership of a machine, regardless of how much it is used. Fixed costs are more affected by how long a machine is owned than how much it is used. Fixed cost may vary from 60 to 80 percent of the total machinery cost. In general accounting practices, fixed cost is composed of the following five elements: depreciation, interest, taxes, housing, and insurance.

**Depreciation** The major cost of a machine is its *first cost*. However, a machine will last several years. First cost is therefore amortized (spread out evenly) over a number of years. Purchasing a machine may involve borrowing money and paying it back on in-

**Fig. 25-1.** Management is a decision-making process for the farm owner/manager and local machinery dealer representative. (*Jane Hamilton-Merritt*)

stallment. As a machine gets older, it loses value, or *depreciates*. The *age* of a machine therefore reflects its value. A new machine is worth more than an old one. The more hours a machine is used, the more *wear* accumulates, with a reduction in value. Quality of maintenance affects depreciation. A salesperson, the appraiser, and certainly the manager will evaluate how well a machine has been maintained in order to determine its trade-in value. Another factor influencing depreciation is the degree by which a machine becomes *obsolete* (out-of-date or no longer used). A change of model, an improvement, or a change of process may result in an otherwise good machine having little resale value.

Depreciation is a machinery cost recognized by the U.S. Internal Revenue Service as deductible for income tax purposes. There are three methods to estimate machinery depreciation. They are (1) straight line, (2) declining balance, and (3) sum-of-the-digits.

**Straight-line depreciation** Straight-line depreciation is the most convenient and easiest to figure. It is determined by first assigning an appropriate salvage value to the machine. This is usually 10 to 15 percent of the first cost. The salvage value is then subtracted from the first cost and the remainder divided by the number of years of useful life (Table 25-1). If the useful life is 10 years, the depreciation is 10 percent per year. The following formula is used to figure depreciation by the straight-line method:

$$Ds \text{ (depreciation, straight line)} = \frac{\text{first cost} - \text{salvage value}}{\text{years of useful life}}$$

**Example** A machine was purchased at a cost of $10,000. It has a salvage value of 10 percent and a useful life of 8 years (Table 25-1). What is the annual depreciation, using the straight-line method?

**Solution**

$$\begin{aligned} Ds &= \frac{\text{first cost} - 10\% \times \text{first cost}}{\text{years of useful life}} \\ &= \frac{\$10{,}000 - \$1{,}000}{8 \text{ years}} \\ &= \frac{\$9000}{8 \text{ years}} \\ &= \$1125 \end{aligned}$$

The value of the machine would depreciate at the rate of $1125 each year and at the end of 8 years would still be worth an estimated $1000 in salvage value.

**Declining-balance depreciation** The declining-balance method depreciates a machine at a different amount each year. This method involves a little more figuring, but it depreciates a machine faster when it is new. This can result in a saving of income tax dollars, which could in turn be applied toward the machine's purchase price. Salvage value is not subtracted; rather a percentage (1.5 to 2.0 × the normal straight-line rate) of the remaining value is taken each year until salvage value is reached. The mathematical formula for figuring depreciation on a declining balance is as follows:

$V$ (value, declining balance) $= C \times (1 - r/L)^y$

where $C$ = first cost

$r$ = rate (usually 1.5 or 2.0)

$L$ = years useful life

$y$ = age of machine

**Example** A machine that cost $10,000 new is 3 years old. What is its value?

**Solution**

$$\begin{aligned} V &= C \times (1 - r/L)^y \\ &= \$10{,}000 \times (1 - {}^2/_8)^3 \\ &= \$10{,}000 \times (1 - 0.25)^3 \\ &= \$10{,}000 \times (0.75)^3 \\ &= \$10{,}000 \times 0.42 \\ &= \$4200 \end{aligned}$$

The value of the 3-year-old machine is $4200. If the machine was purchased new, it would be depreciated to the following value in 1 year:

$$\begin{aligned} V &= C \times (1 - r/L)^y \\ &= \$10{,}000 \times (1 - {}^2/_8)^1 \\ &= \$10{,}000 \times 0.75 \\ &= \$7500 \quad \text{at the end of first year} \end{aligned}$$

**Sum-of-the-digits method** The sum-of-the-digits method gives results similar to the declining-balance method. Salvage value may or may not be taken into account at the beginning. To use the system, total the years of useful life. The machine in the previous examples has a useful life of 8 years. Sum the years of life as follows: 8 + 7 + 6 + 5 + 4 + 3 + 2 + 1 = 36.

**TABLE 25-1. Average Useful Life of Selected Machinery**

| | USEFUL LIFE | |
|---|---|---|
| Machine | Years | Hours |
| Combine | 8 | 2,000 |
| Corn picker | 8 | 2,000 |
| Cotton picker | 8 | 2,000 |
| Drill | 8 | 1,000 |
| Forage harvester | 8 | 2,000 |
| Hay machinery | 8 | 2,000 |
| Mower | 8 | 1,000 |
| Planter | 8 | 1,000 |
| Plow | 8 | 2,000 |
| Tractor (wheel) | 10 | 10,000 |
| Tractor (crawler) | 20 | 16,000 |
| Tillage machinery | 8 | 2,000 |

For the first year, take $^8/_{36}$ of the first cost less salvage value ($1000). The second year is $^7/_{36} \times \$9000$, and so on.

**Example (for first year)**

$$Dd \text{ (depreciation, sum-of-the-digits)} = \frac{\left(\text{first cost} - \text{salvage value}\right) \times \text{years}}{\text{sum-of-the-digits}}$$

$$= \frac{(\$10{,}000 - \$1000) \times 8}{36}$$

$$= \frac{\$9000 \times 8}{36}$$

$$= \$2000$$

**Example (for second year)**

$$Dd = \frac{\$9000 \times 7}{36}$$

$$= \$1750$$

A comparative example of the depreciated value of a $10,000 machine using each of the three methods over 8 years is presented in Table 25-2 and Fig. 25-2.

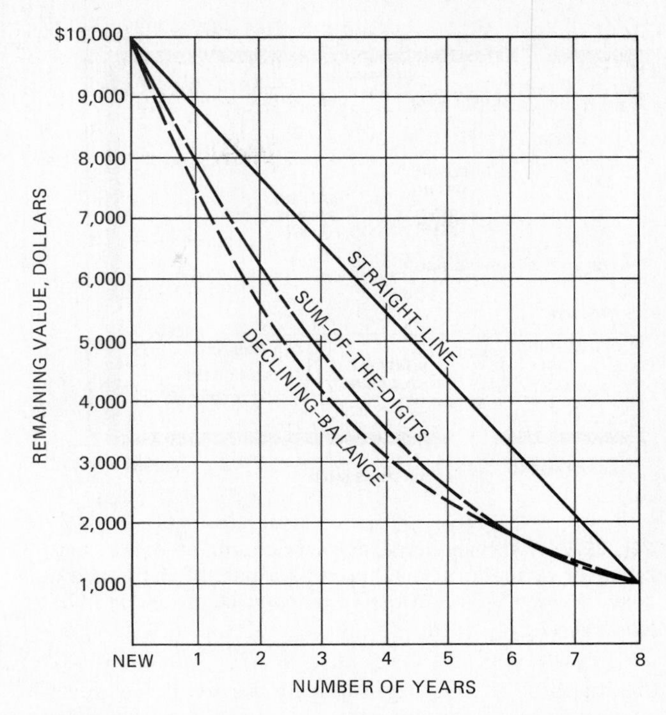


**Fig. 25-2.** Depreciation methods compared for a $10,000 machine.

**Interest** The cost of owning machinery includes a charge for the interest. Interest is considered as an *opportunity cost*. This means that by investing in machinery, you have lost the opportunity for the money to earn interest in a savings account or other interest-earning investment.

The amount of interest charged, including financing charges, is determined by taking one-half the purchase price (first cost) minus 10 percent investment credit plus the salvage value times the interest charge. For example, the purchase price of the machine used in our example was $10,000; one-half this value ($5000) minus 10 percent is $4500 ($5000 − $500 = $4500). Adding the salvage value of $1000 times interest equals $660 (i.e., $4500 + $1000 × 0.12 = $660). A quick method of determining interest cost is as follows: If the salvage value is 10 percent, use 55 percent of purchase value times the interest rate. (If the salvage is figured at 15 percent, substitute 60 percent for 55 percent.) For example, for a 10 percent salvage value, $10,000 × 55 percent × 12 percent interest equals $660.

**Taxes** Machinery is considered personal property and is taxed according to state and local laws. Normally, taxes vary from 1 to 2 percent of the value of

**TABLE 25-2. Comparison of Depreciation Methods ($10,000 machine, with a useful life of 8 years and a salvage value of $1000)**

| Method and Rate | Straight line | | Declining Balance | | Sum-of-the-digits | |
|---|---|---|---|---|---|---|
| Salvage Value | $1000 | | | | $1000 | |
| Beginning Balance | $9000 | | $10,000 | | $9000 | |
| **Year** | **Depreciation** | **Remaining Value** | **Depreciation** | **Remaining Value** | **Depreciation** | **Remaining Value** |
| 1st | $1125 | $8875 | $2500 | $7500 | $2000 | $8000 |
| 2d | 1125 | 7750 | 1875 | 5625 | 1750 | 6250 |
| 3d | 1125 | 6625 | 1406 | 4219 | 1500 | 4750 |
| 4th | 1125 | 5500 | 1055 | 3164 | 1250 | 3500 |
| 5th | 1125 | 4375 | 791 | 2373 | 1000 | 2500 |
| 6th | 1125 | 3250 | 593 | 1780 | 750 | 1750 |
| 7th | 1125 | 2125 | 445 | 1335 | 500 | 1250 |
| 8th | 1125 | 1000 | 335 | 1000 | 250 | 1000 |
| Total Depreciation | $9000 | | $9000 | | $9000 | |
| Salvage Value | | $1000 | | $1000 | | $1000 |
| First 4 years | $4500 | | $6836 | | $6500 | |
| Last 4 years | $4500 | | $3164 | | $3500 | |

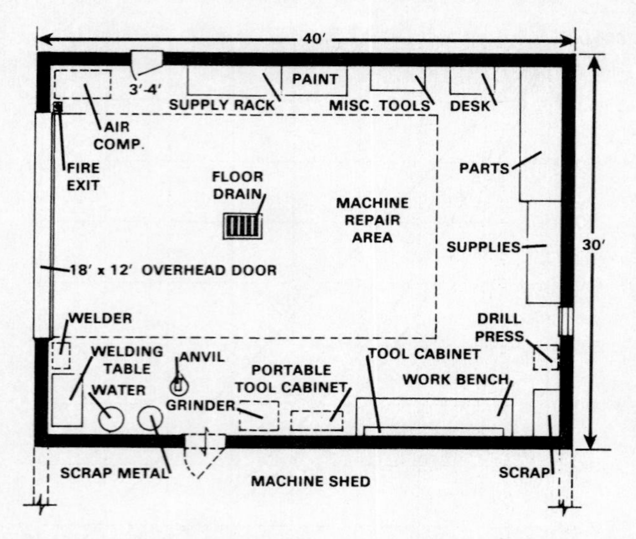


**Fig. 25-3.** A service center for service, maintenance, and repair of agricultural machinery is a part of the housing cost. (*North Dakota State University, Dept. of Agr. Engineering.*)

the machine at the beginning of the year. To determine these costs accurately, you must use the state, county or parish, and local tax rates.

**Housing** The need for machinery housing varies with the climatic conditions of the country. It is generally agreed that machines protected from weather will require fewer repairs. Housing for machinery may include the service center for maintenance and repair (Fig. 25-3). For this reason, an annual charge of 1 to 2 percent of the remaining value of the machine is made for housing.

**Insurance** Even though an insurance policy may not be carried on some of the older and less costly machines, a charge of 0.25 to 0.50 percent is usually added to the remaining value of the machine.

**Fixed costs combined** Except for the depreciation cost, the remaining fixed cost can be figured at a total of 15 percent of the remaining value of the machine. The ownership (fixed cost) for a $10,000 machine is summarized in Table 25-3.

## Factors Affecting Operating Costs

As implied, *operating costs* are those costs which develop as a result of using the machine. The costs include the following: (1) labor to operate the machine, (2) fuel and lubricants, and (3) repairs and maintenance.

**Labor** There is some hesitation to charge the cost of labor to operating costs when you are doing your own work. However, labor is a major operating cost and should always be assigned a value. The cost of labor becomes an important factor when evaluating the size of machine to purchase. The hours saved by using a larger machine often prove to be an important decision-making factor.

When hired labor is used to operate the machine, the actual hourly rate is used to figure labor costs. If the machine is operated by the owner/manager, use a rate equal to the wages he or she would expect to earn.

**Fuel and Lubricants** Tractors and self-propelled machines are powered by internal-combustion engines which require fuel and lubricants to operate. Fuel consumption varies with the power rating of the engine and the type of fuel being used. A properly maintained tractor engine will produce the most power for the fuel used. Alert service department personnel will maintain top tractor operating efficiency for the customers.

The common method of expressing engine fuel consumption is in gallons [liters] consumed per hour. A better term for expressing both engine efficiency and power developed is *horsepower-hours per gallon* [kilowatt-hours per liter], abbreviated as hp·h/gal [kW·h/L]. The average values for hp·h/gal [kW·h/L] which tractor engines develop on an annual basis are identified in Table 25-4.

Since fuel is sold by the gallon [liter], the annual cost for fuel can be estimated using the performance factors in Table 25-4 and the following formula:

**TABLE 25-3. Annual Total Fixed Cost for a $10,000 Machine**

| Year | Remaining Value, Start of Year | Depreciation, Declining Balance | Interest, Taxes, Housing, Insurance (15%) | Total Fixed Costs | Percentage of List Price (Annual Fixed Cost) |
|---|---|---|---|---|---|
| 1 | $10,000 | $2500 | $1500 | $4000 | 40.0% |
| 2 | 7,500 | 1875 | 1125 | 3000 | 30.0 |
| 3 | 5,625 | 1406 | 844 | 2250 | 22.5 |
| 4 | 4,219 | 1055 | 633 | 1688 | 16.9 |
| 5 | 3,164 | 791 | 475 | 1266 | 12.7 |
| 6 | 2,373 | 593 | 356 | 949 | 9.5 |
| 7 | 1,780 | 445 | 267 | 712 | 7.1 |
| 8 | 1,335 | 335 | 200 | 535 | 5.4 |

Annual fuel costs

$$= \frac{\text{price per}}{\text{gallon [L]}} \times \frac{\begin{matrix}\text{maximum} \\ \text{PTO hp [kW]}\end{matrix} \times \begin{matrix}\text{total} \\ \text{hours used}\end{matrix} \times 0.6}{\text{hp·h/gal [kW·h/L]}}$$

where 0.6 = 60 percent of maximum PTO (power takeoff) hp [kW] for average loads on an annual basis.

For example, the machine in our example has a 70-hp [52.2 kW] diesel engine and is used 1000 h/year. Fuel costs average $1/gal [$0.26/L]. What is the annual fuel cost?

**Solution**

$$\text{Annual fuel cost} = \$1 \times \frac{70 \times 1000 \times 0.6}{13}$$
$$= \$3230.77/\text{year}$$

Tractors, trucks, and self-propelled equipment require engine oil, filters, hydraulic and transmission fluid, and greases. An estimate of the lubricating cost of agricultural machinery is approximately 15 percent of the annual fuel cost. In the previous example, total fuel and lubricant cost for the machine's 70-hp [52.2 kW] engine is as follows:

Fuel cost = $3231
Lubricant cost:
$3231 × 0.15 = 485
Total = $3716

*Note:* The above figures are based on diesel fuel at a cost of $1/gal [$0.26/L]. In 1000 h of operation, the fuel and lubrication cost per hour is approximately $3.72/h ($3716/1000 h).

**Repairs and Maintenance** Agricultural machinery must be maintained in the proper operating condition if it is to be reliable. Timeliness of operations is one of the critical factors in achieving maximum crop production. Idleness due to machinery breakdowns can be very costly. It is said that an hour spent in the service center during the slack season getting the machinery ready will avoid a loss of 1 day's time when the machinery is in use.

The best source of information for repair costs is the record book for the machine. Estimates of repair costs are difficult to determine unless actual records have been maintained. Table 25-5 provides the estimated repair cost for various types of machinery.

For example, the $10,000 machine is to be used 1000 h in custom work. What repair costs can be expected? From Table 25-5, repair of hay machinery at 1000 h of use can be expected to be 32.5 percent of the purchase price. Therefore, 32.5 percent of $10,000 (32.5 × $10,000) equals $3250 for repairs in 1000 h. This will average $3.25/h of operation ($3250/1000 h = $3.25).

**TABLE 25-4. Performance Factors for Determining Power Developed per Gallon of Fuel**

| Fuel | Average Performance, Maximum PTO hp [kW], hp·h/gal [kW·h/L] |
|---|---|
| Gasoline | 9.2 [1.8] |
| Diesel | 13.0 [2.6] |
| LP gas | 7.3 [1.44] |

## REDUCING MACHINERY COST

New machinery is expensive to purchase and own. The first year's total fixed cost of owning the new $10,000 machine amounts to $4000. If used an average of 250 h, it would cost $16/h to satisfy the first year's fixed cost. If it is used fewer hours, the cost per hour would be higher; if used more, the cost per hour would be less.

### Ownership Alternatives

There are other ways to obtain machinery rather than to own it. These are to hire custom operators, rent, or lease the necessary equipment. From the owner/manager point of view, the hiring of custom operators has the following advantages:

1. Less capital is tied up in ownership of machinery.
2. Time released by hiring custom work can be devoted to managing another enterprise.
3. The cost of hire may be less than ownership when there is low annual use.
4. Custom operators can increase available machine power to provide more timely operation.

Disadvantages of hiring custom work are:

1. Timeliness of completing the operation depends on the availability of the custom operator.
2. The cost per unit of work may be reduced by farming more acres or doing custom work.

**TABLE 25-5. Estimate of Machinery Repair Costs**

| | HOURS OF USE AT HALF AND FULL LIFE (% OF PURCHASE PRICE) | |
|---|---|---|
| Machine | Half Life | Full Life |
| Combine | 1000 h (9.5%) | 2,000 h (33%) |
| Hay machinery | 1000 h (32.5%) | 2,000 h (80%) |
| Plow, moldboard | 1000 h (32.5%) | 2,000 h (80%) |
| Tractor | 5000 h (29.7%) | 10,000 h (90%) |

Using a machine for custom work spreads the ownership cost over a greater number of hours. For example, the purchaser of a new $10,000 machine expects to use it a total of 1000 h the first year. The cost of operating the machine for 1000 h is determined to be:

| | |
|---|---|
| Total fixed cost | $ 4000 |
| Total fuels and lubricants | 3716 |
| Total repairs and maintenance | 3250 |
| Total labor at $6/h | 6000 |
| | $16,966 |

$$\frac{\$16{,}966}{1000\text{ h}} = \$16.97/\text{h}$$

Capacity of machine = 7.25 acres/h

$$\text{Cost per acre} = \frac{\$16.97\text{ cost/h}}{7.25\text{ acres/h}}$$

$$= \$2.34/\text{acre}$$

Custom work charge = cost plus 50%

$$= \$2.34 + (\$2.34 \times 0.50)$$

$$= \$2.34 + \$1.17$$

$$= \$3.51/\text{acre}$$

Renting and leasing farm equipment are options to owning machinery. Rental arrangements usually involve a short time period of from 1 day to several weeks. Renting machinery is a good option when the need for additional equipment is necessary to achieve timeliness of operation and when one wishes to avoid large capital outlay for short-term use.

Leasing arrangements are usually written for a period of 2 or more years. As in the rental agreement, the owner/manager is actually picking up the fixed costs which the company has invested in the rental or lease equipment. Leasing provides the following advantages: (1) new equipment has high reliability, (2) leasing gives the lessee an opportunity to experiment with size and type of equipment, (3) lease equipment agreements may be written to include the option to purchase at the end of the lease period, (4) limited capital outlay allows the manager to use opportunity cost option to obtain greater return on investment, and (5) leasing provides immediate equipment for completing important operations on time.

Generally, renting and leasing machinery are short-term options—owning provides the lowest machinery cost over the long term.

## Managing Power and Machinery

The owner/manager is concerned with keeping the cost of machinery as low as possible. Ownership costs can be reduced per hour or acre [ha] of machine use by good management practices. For example, normal costs for fuel and repair to operate the machine (used in previous examples) for 1000 h annually amounted to approximately 35 percent of the total cost. Proper tuneup and maintenance of the machine's diesel engine and timely attention to the operating parts can result in reduced operating costs. On the other hand, failure to use competent operators, overloading the machine by excessive travel speeds, and neglecting maintenance of filters and oil changes in the engine and hydraulic system can result in excessive repair costs (Fig. 25-4).

**Oil Analysis** One of the modern management tools that are available to the owner/manager and service personnel of agricultural power and machinery dealerships is laboratory analysis of used oils. Laboratory techniques are used to provide a complete chemical analysis of tractor, truck, and auto engine oil; gear lubricant; and hydraulic oils. The comprehensive evaluations can provide nearly as much information about the proper operation of a power system as a complete teardown and inspection. The heart of the laboratory process is *spectroscopic analysis,* a means of analyzing the chemical components, of the substances contained in used oil.

A sample of used oil is drawn from the crankcase, gear case, or hydraulic reservoir and mailed to the oil analysis laboratory (Fig. 25-5). Upon receipt, the laboratory conducts an extensive analysis of the oil for physical and chemical properties. Upon completion, a report of the analysis is sent by return mail to the sender. In the event the laboratory finds that the oil contains unusual substances, contaminates, or changes in physical characteristics which suggest possible breakdown of the mechanical parts of the engine, the gear system, or hydraulic components, immediate contact is made by telephone.

For example, the oil sample history for a diesel en-

**Fig. 25-4.** Using the operator's manual properly and following the recommended lubrication schedule can reduce machinery costs.

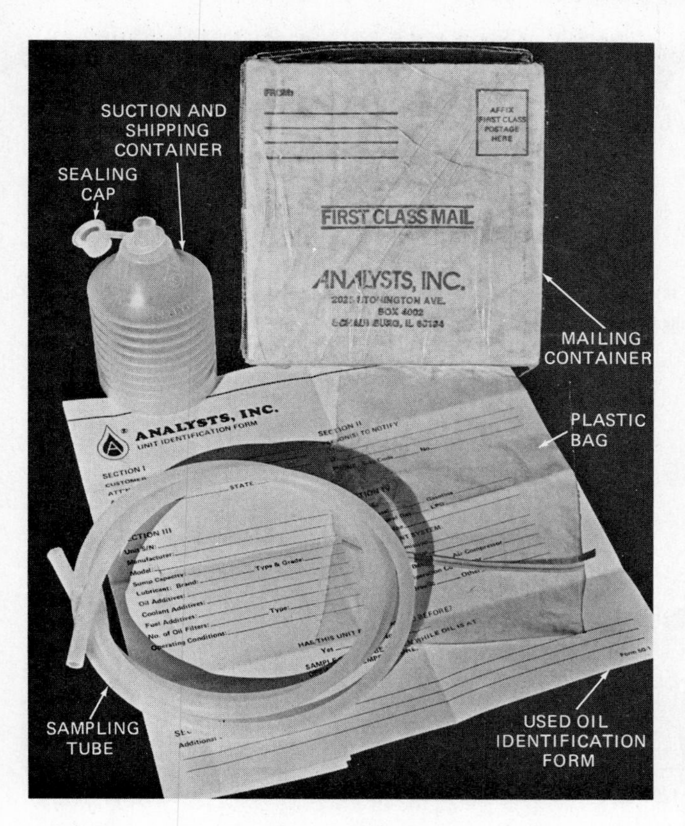


**Fig. 25-5.** A used oil analysis kit is provided by the oil analysis laboratory. The sample is taken from the crankcase of the engine, the gear case, or the hydraulic reservoir by the suction container. When properly sealed, it is packaged and mailed to the laboratory. (*Analysts, Inc.*)

gine is listed in Fig. 25-6. Oil sample analyses 0428 and 1579 indicated no apparent malfunction in the spectrochemical analysis or physical properties tests. However, oil sample 2990 contained excessive amounts of iron, chromium, aluminum, and silicon. High levels of iron and chromium suggest excessive wear of cylinder walls and piston rings. The high silicon level indicates sand and dirt in the oil. In addition, the physical properties tests indicate excessive levels of solids contamination (TS), increased thickening of the oil (V210), and a depletion of the oil additive package (TBN).

A maintenance recommendation for laboratory test 2990 is published suggesting that critical wear conditions and abnormal oil conditions exist and that the following things be done: (1) change oil filters, (2) check air induction system and filters for leaks, (3) check air-fuel ratio, (4) inspect turbocharger or blower for wear, (5) check blowby/compression, and (6) inspect piston rings, liners, and cylinders.

The report further states that the engine was inspected, a defective blower was found, the blower was replaced, and the engine was returned to service.

**Management Decisions** Service personnel are often called upon to assist consumers in obtaining the best performance from their equipment. Special assistance is often given to point out efficiency factors such as operating speeds, hitching, engine tuneup, and control of rolling resistance and wheel slippage.

**Tractor operating speed** Today's tractors have been engineered to operate most efficiently at speeds of 5 to 7 miles per hour (mi/h) [8.05 to 11 kilometers per hour (km/h)]. It is estimated that decreasing the ground speed one-third through excessive loading can result in a 50 percent decrease in the service life of the transmission gear train and final drive.

**Hitching** Improper hitching of tillage equipment can result in increased draft. Also, excessive side draft will often result in abnormal wear of wheel bearings and soil contact tools.

**Engine tuning** An improperly tuned engine can cause a loss of 10 percent or more of its power and consume excessive amounts of fuel.

**Rolling resistance and slippage** Two of the most overlooked causes of power loss in farm tractors are excessive rolling resistance and excessive slippage.

*Rolling resistance:* Resistance of all wheels to rolling in soft soil can cause loss of power or increased draft. Even on level ground, tractor tires sink into soft soil and literally must roll uphill to get out of their own rut. It may take up to 15 percent of the tractor's pulling power to move itself in soft soil conditions. By combining lack of engine tuning (10 percent) and rolling resistance (15 percent), one-fourth of the engine's power is not usable.

*Wheel slippage:* Closely related to power loss due to rolling resistance is wheel slippage. Slippage of the drive wheels should not exceed 8 percent on firm soil and 15 percent on soft soil. This limit is established by engineers because it is not economically desirable to eliminate all slippage of tractor wheels. Excessive wheel slippage usually results in excessive rolling resistance because it causes the wheels to sink deeper into the soil. In terms of horsepower [kW], rolling resistance and wheel slippage could easily result in a 30 percent loss in pulling power at the drawbar.

The adage "Don't spin your wheels" does not necessarily apply to tractor operation. Some slippage is necessary. However, an excessive amount of slippage wastes fuel and power. For example, if a 100-acre [40.5 ha] field can be disk-harrowed in a day with no wheel slippage, 90 acres [36.4 ha] could be tilled with 10 percent slippage. Only 80 acres [32.4 ha] per day could be disked with 20 percent slippage.

An easy way to check how much slippage is occurring in the field is as follows (Fig. 25-7):

1. Place a chalk mark on one of the rear tractor tires at the point where it contacts the ground.
2. Also mark the ground at this point.

VALLEY MACHINERY CO.
SERVICE MANAGER
AG CENTER, USA 00000

ANALYSTS MAINTENANCE LABORATORES, INC.
2450 HASSELL RD. HOFFMAN ESTATES, ILLINOIS 60195 (312) 884-7877
BOX 4002 SCHAUMBURG, ILLINOIS 60194 TELEX: 254088

CUSTOMER NO. : 51000000-0000
UNIT REF. NO. : 5474
SENDER NAME : VALLEY MACHINERY CO.
UNIT IDENT. : TRACTOR #31

MAKE/MODEL : DDV71G
OIL TYPE : CENEX 30W
REPORT DATE : 19Jan'81

MAINTENANCE RECOMMENDATIONS FOR LAB NO. 2990 DIESEL

****IMPORTANT REPORT****IMPORTANT REPORT****IMPORTANT REPORT****

CRITICAL WEAR CONDITIONS
TESTS INDICATE ABNORMAL OIL CONDITION
DRAIN OIL
CHANGE OIL FILTERS
CHECK AIR INDUCTION/ FILTRATION SYSTEM FOR LEAKS
CHECK AIR TO FUEL RATIO
INSPECT TURBOCHARGER OR BLOWER FOR WEAR
CHECK BLOW-BY/ COMPRESSION
INSPECT PISTON,RINGS,LINER/CYLINDERS
RESAMPLE AFTER INSPECTION / CORRECTIVE ACTION
PHONED ****FEED BACK**** INSPECTION REVEALED A DEFECTIVE BLOWER. WEAR
WITHIN THE CYLINDER ASSEMBLY WAS NOT EXCESSIVE THE BLOWER WAS REPLACED
AND THE UNIT WAS RETURNED TO SERVICE

SPECTROCHEMICAL ANALYSIS

| LAB # | IRON | LEAD | COPPER | CHROMIUM | ALUMINUM | NICKEL | SILVER | TIN | SILICON | BORON | SODIUM | PHOS. | ZINC | CALCIUM | BARIUM | MAGNESIUM | TITANIUM | MOLYBDENUM | CADMIUM | ANTIMONY |
|---|---|---|---|---|---|---|---|---|---|---|---|---|---|---|---|---|---|---|---|---|
| 0428 | 62.0 | 10.0 | 3.00 | 2.00 | 0.00 | 0.00 | 0.00 | 0.00 | 7.00 | 0.00 | 36.0 | 840. | 900. | 2800 | 0.00 | 0.00 | 0.00 | 0.00 | 0.00 | 0.00 |
| 1579 | 82.0 | 10.0 | 3.00 | 2.00 | 0.00 | 0.00 | 0.00 | 0.00 | 8.00 | 0.00 | 30.0 | 820. | 960. | 2850 | 0.00 | 0.00 | 0.00 | 0.00 | 0.00 | 0.00 |
| 2990 | 150. | 10.0 | 5.00 | 15.0 | 30.0 | 1.00 | 0.00 | 10.0 | 35.0 | 0.00 | 30.0 | 870. | 980. | 2800 | 0.00 | 0.00 | 0.00 | 0.00 | 0.00 | 0.00 |

OPERATING DATA

| LAB # | MI/HR UNIT | MI/HR OIL | OIL ADD GALS. |
|---|---|---|---|
| 0428 | 2027 | 110 | 1.00 |
| 1579 | 2272 | 245 | 2.00 |
| 2990 | 2700 | 428 | 4.00 |

PHYSICAL PROPERTIES TESTS

| LAB # | SAMPLE DRAWN | FDD %VOL | TS %VOL | W %VOL | V210 SUS | TBN | PH |
|---|---|---|---|---|---|---|---|
| 1428 | 12Oct79 | 0.5 | 1.0 | <0.05 | 67.5 | 5.32 | 6.20 |
| 1579 | 01Nov79 | 0.5 | 2.5 | <0.05 | 68.3 | 3.07 | 7.90 |
| 2990 | 20Dec79 | 1.0 | 5.0 | <0.05 | 88.3 | 1.98 | 7.90 |

UNDERLINED FIGURES INDICATE ABNORMAL VALUES. MAINTENANCE THAT MAY BE REQUIRED IS INDICATED ABOVE UNDER MAINTENANCE RECOMMENDATIONS AND SHOULD BE PERFORMED BY A QUALIFIED MECHANIC. PLEASE ADVISE US OF ANY MAINTENANCE PERFORMED ON THIS UNIT. N/R = TEST NOT PERFORMED
ACCURACY OF RECOMMENDATIONS IS DEPENDENT ON REPRESENTATIVE OIL SAMPLE AND COMPLETE CORRECT DATA ON BOTH UNIT AND OIL. THIS REPORT IS NOT AN ENDORSEMENT OR RECOMMENDATION OF ANY PRODUCT OR SYSTEM. ORIGINAL REPORT MAINTAINED IN ANALYSTS, INC. DATA FILES.



**Fig. 25-6.** Oil analysis lab report. (*Analysts, Inc.*)

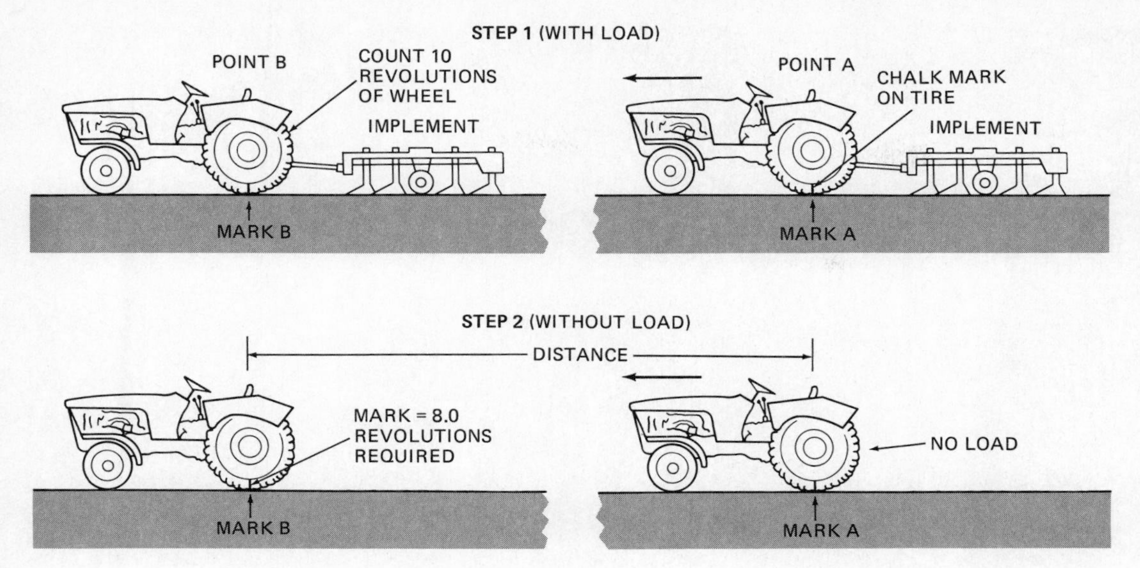


**Fig. 25-7.** Determining the percentage of wheel slippage. *Step 1:* Drive the tractor under load and count 10 revolutions of the rear wheel. *Step 2:* Drive the same distance without load and count the number of wheel revolutions. *Step 3:* Compute the percentage of wheel slippage.

3. Count 10 wheel revolutions as the tractor is driven forward with the implement engaged, traveling at normal operating gear and throttle setting.
4. Mark the ground at this point.
5. Unhitch the implement and count the number of wheel revolutions required to travel the distance at the same gear and throttle setting.
6. Determine the amount of slippage as follows:

$$\text{Percent slippage} = \frac{\text{loaded turns} - \text{no-load turns}}{\text{loaded turns}} \times 100\%$$

**Example**

Loaded turns = 10

No-load turns = 8

$$\text{Percent slippage} = \frac{10 - 8}{10} \times 100\%$$
$$= 0.2 \times 100$$
$$= 20\%$$

In the above example, the slippage of 20 percent is more than the maximum 15 percent recommended. The remedy for correcting the excessive slippage is to add weight to the wheels.

Experienced owners/managers can judge the correct amount of wheel slippage by the imprint of the tread left by the tractor tire when pulling a load. If the imprint is full and unbroken, there is too much weight for load; if the tread marks are badly broken, more weight should be added. The correct amount of weight with 10 to 15 percent slippage is indicated by a slight breaking in the center of the tread (Fig. 25-8).

## Managing Tractor Weight

The management question is, How much weight should be added? Too little weight results in excessive slippage. Too much weight added to the tires causes excessive soil compaction, increased rolling resistance in soft ground, and possible damage to the tractor power train.

Some slippage must occur to avoid excessive compaction and to protect the power train from overload. Although the bearings and gears of a tractor power train are designed to withstand full engine power for a prolonged period of time, a tractor should not be weighted more than will allow a forward travel of 3.5 mi/h [5.6 km/h] at full engine power without a "spinout" (drive wheel slippage without forward travel) to avoid possible damage.

The type of surface the tire travels on affects the pulling ability of a tire. Table 25-6 lists the amount of pull that can be attained for each 100 pounds (lb) [45.4 kg] of weight added to the drive wheels of the tractor when operating on five different types of surface. Use Table 25-7 to make decisions about wheel weights and slippage.

**Methods of Adding Weight** Weight can be added to the drive wheels in several ways:

1. By adding liquid or dry powder to tires and cast-iron wheel weights to wheels
2. By using dual rear wheels with liquid or dry powder in the tires
3. By weight transfer—transferring weight from the implement being pulled to the rear wheels of the tractor

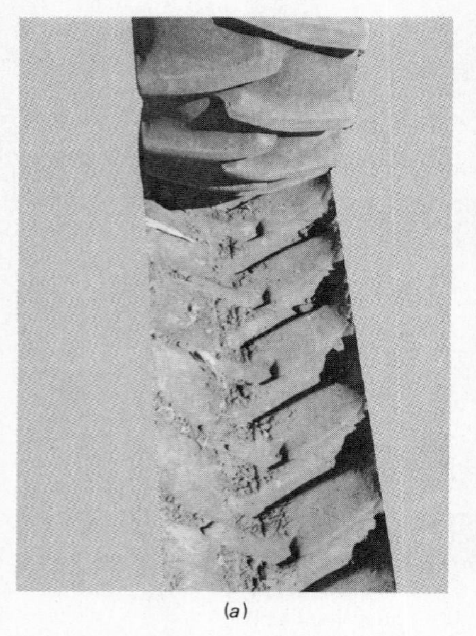
(a)

(b)

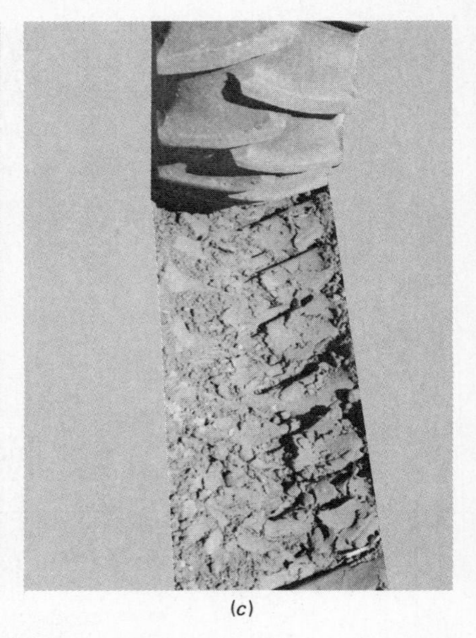
(c)

**Fig. 25-8.** Determining the amount of wheel slippage by studying the tire tread marks in soil. (*a*) Unbroken tread marks mean not enough slippage—too much weight for the load; (*b*) Badly broken tread marks mean too much slippage—add more weight; (*c*) Tread marks slightly broken in the center of the tread mean slippage and weight are correct.

4. By front-end weighting

**Adding wheel weight** The most common way of adding weight to tractor wheels is by filling the tire with liquid or by adding iron weights (Fig. 25-9*a*). A liquid weight consisting of calcium chloride and water is recommended, because the solution has a low freezing point and it weighs 60 percent more per gallon than water. Special equipment is required to fill the tires with liquid. Table 25-8 gives the tire size and the amount of chemical and water to achieve the weight when filled to the 75 percent level. A 25 percent cushion of air is used to help protect the tire from damage. In some instances it will be necessary to reduce the amount of liquid to stay within the weight limits of the tire. Each tractor operator's manual provides a table (see Table 25-9 for an example of one manufacturer's table), which considers the weight of the tractor when computing weight to be added with liquid.

Care must be exercised to prevent tire damage from overweighting. Tire load is the axle weight of the tractor per wheel. Axle weight includes tractor, auxiliary weights, tire ballast, front-end weight transferred under load, and part of the implement weight. The gross weight should not exceed the maximum listed in the operator's manual (Table 25-9).

**Using dual rear wheels** Many owner/managers who have 100 to 130 hp [75 to 100 kW] or more wheel tractors are using dual wheels (Fig. 25-9*b*). Duals provide the following benefits.

1. Less compaction of the soil due to more wheel surface area carrying the weight
2. Increased traction efficiency by decreasing rolling resistance in softer ground (more flotation in soft soil)
3. Added weight from the extra tires' hubs and rims*
4. Better sideway stability for the tractor
5. Lower cost than a set of larger-diameter tires and wheels having the same surface area contact

Tire selection must be based upon the tire manufacturer's data. More than one tire size is available to

**TABLE 25-6. Pulling Power of a Tractor Varies with the Type of Surface on Which the Machine is Operating (For each 100 lb of weight added to the rear wheels, the average drawbar pull will be increased as shown)**

| | Average Pull, lb |
|---|---|
| Concrete road | 66 |
| Dry clay | 55 |
| Sandy loam | 50 |
| Dry sand | 36 |
| Green alfalfa | 36 |

* Do not weight the outer wheels. Inflate outer tires about 4 pounds per square inch (psi) [27.6 kilopascals (kPa)] less than the inner tires. This will reduce the amount of contact on firm ground for better traction and increase contact on softer ground when the inner tires penetrate. This assists by providing greater flotation.

**TABLE 25-7. Percentage of Rear Wheel Slippage and Weight Recommendations**

| % Slippage | Recommendation |
|---|---|
| 0 | Remove weights |
| 5 | Remove weights |
| 10 | Proper weighting |
| 15 | Proper weighting |
| 20 | Add weights |
| 25 | Add weights |

(a)

(b)

**Fig. 25-9.** (*a*) Adding weight to tractor with weights bolted to wheel or tractor. Front weights add stability to steering; rear wheel weights add traction. Front weight can be transferred to the rear by hydraulic weight transfer hitching. (*b*) When properly weighted, dual wheels reduce the rolling resistance and decrease slippage. (*Ford Tractor Operations*)

meet load requirements. Table 25-10 presents the minimum recommended tire size for maximum traction at operating speeds of 4 and 5 mi/h [6.4 and 8.05 km/h]. Selection must also include such factors as ground clearance required for row-crop application, implement clearance in tight turns, and traction versus flotation requirements.

**TABLE 25-8. Tractor Operator's Manual Data for the Recommended Liquid Weight per Tire, 75% Fill**

| Tire Size | 5 lb* Calcium Chloride per Gallon of Water | GALLONS OF WATER ADDED PER TIRE | | Total Pounds Weight of Solution per Tire |
|---|---|---|---|---|
| | | U.S. | Imp. | |
| 5.50-16 | 18 | 3.5 | 2.9 | 47 |
| 6.00-16 | 24 | 4.8 | 4.0 | 64 |
| 6.50-13 | 7.2 | 1.6 | 1.3 | 19.2 |
| 7.50-15 | 32.5 | 6.5 | 5.4 | 87 |
| 7.50-16 | 37 | 7.3 | 6.1 | 98 |
| 7.50-18 | 40 | 8 | 6.7 | 107 |
| 8.3-24 | 50 | 10 | 8.3 | 133 |
| 9.00-10 | 36 | 7.2 | 6.0 | 96 |
| 11.2-28 | 110 | 22 | 18.3 | 293 |
| 12.4-28 | 140 | 28 | 23.3 | 374 |
| 12.4-38 | 185 | 37 | 30.8 | 494 |
| 13.6-24 | 155 | 31 | 25.8 | 414 |
| 13.6-28 | 175 | 35 | 29.2 | 467 |
| 13.6-38 | 230 | 46 | 38.3 | 614 |
| 14.9-24 | 190 | 38 | 31.7 | 507 |
| 14.9-28 | 215 | 43 | 35.8 | 574 |
| 14.9-30 | 230 | 46 | 38.3 | 614 |
| 15.5-38 | 265 | 53 | 44.1 | 707 |
| 16.9-24 | 245 | 49 | 40.8 | 654 |
| 16.9-26 | 260 | 52 | 43.3 | 694 |
| 16.9-30 | 295 | 59 | 49.1 | 787 |
| 18.4-16A | 165 | 33 | 27.5 | 438 |
| 18.4-26 | 320 | 64 | 53.3 | 854 |
| 18.4-30 | 360 | 72 | 60.0 | 960 |

* Freezing point = −53°F slush-free.
*Source:* Ford Tractor Operations

**Transfer of weight** Weight transfer is included as a part of the hydraulic system and hitch of modern tractors. The process utilizes the extra weight of the implement as weight for the tractor. The hydraulic system and weight transfer hitch lifts up on the implement as the pull becomes greater. This action lifts up on the front of the tractor. The result is the transfer of weight to the rear wheels for more traction and less slippage without relying on built-in or added weighting.

**Front-end weighting** Front-end weighting is necessary with tillage implements that require maximum drawbar pull (Fig. 25-9*a*). Every pound [kilogram] of weight added to the front end will add 3 lb [3 kg] to the rear axle weight. This transfer is due to the axle torque reaction, the drawbar/hitch load, the hydraulic reaction of the tractor's draft control system, and the lever action length of the tractor chassis.

Front-end weighting stabilizes the tractor for more responsive steering and safety, especially when hitched to mounted implements.

**TABLE 25-9. Data from Tractor Operator's Manual for Specific Tire Inflation vs. Permissible Load per Tire**

| Front Tire Size | | INFLATION PRESSURES, PSI: 12 | 16 | 20 | 24 | 28 | 32 | 36 |
|---|---|---|---|---|---|---|---|---|
| | | Maximum Permissible Load per Tire, lb* | | | | | | |
| 5.50-16 | 4 ply | | | 655 | 725 | 795 | 860 | |
| 6.60-16 | 4 ply | | | 750 | 835 | 915 | 990 | |
| 6.50-13 | 4 ply | | | 730 | 810 | 885 | | |
| 7.50-15 | 6 ply | | | 1045 | 1160 | 1265 | 1370 | 1470 |
| 7.50-16 | 4 ply | | | 1110 | 1235 | | | |
| 7.50-16 | 6 ply | | | 1110 | 1235 | 1345 | 1455 | 1560 |
| 7.50-18 | 4 ply | | | 1205 | 1340 | | | |
| 9.00-10 | 4 ply | 800 | 1280 | | | | | |

| Rear Tire Size | | INFLATION PRESSURES, PSI: 12 | 14 | 16 | 18 | 20 | 22 |
|---|---|---|---|---|---|---|---|
| | | Maximum Permissible Load per Tire, lb* | | | | | |
| 8.3-24 | 4 ply | | 1055 | 1140 | 1220 | 1300 | 1370 |
| 11.2-28 | 4 ply | 1575 | 1720 | 1860 | | | |
| 12.4-28 | 4 ply | 1890 | 2070 | | | | |
| 12.4-38 | 4 ply | 2185 | 2390 | | | | |
| 13.6-24 | 4 ply | | 2270 | | | | |
| 13.6-28 | 4 ply | | 2430 | | | | |
| 13.6-38 | 4 ply | | 2810 | | | | |
| 14.9-24 | 4 ply | | 2700 | | | | |
| 14.9-28 | 4 ply | | 2880 | | | | |
| 14.9-30 | 6 ply | | 2960 | 3200 | 3430 | | |
| 15.5-38 | 6 ply | | 3160 | 3420 | 3660 | | |
| 16.9-24 | 6 ply | | | 3560 | | | |
| 16.9-26 | 6 ply | | | 3680 | | | |
| 16.9-30 | 6 ply | | | 3920 | | | |
| 18.4-16A† | 6 ply | | 2660 | 2880 | | | |
| 18.4-26 | 6 ply | | | 4370 | | | |
| 18.4-30 | 6 ply | | | 4650 | | | |

* *Do not* exceed the maximum load listed. Do not underinflate or overinflate the tires.
† 18.4-16A 6 ply tires can be used at 10 psi, providing maximum load does not exceed 2180 lb per tire.
*Note:* The above figures are for individual tires only. For maximum permissible loads that can be carried by both front tires or both rear tires, multiply the figures by 2.
*Source:* Ford Tractor Operations

**TABLE 25-10. Minimum Recommended Tire Size for Maximum Traction**

| Drawbar* Horsepower | 4 MI/H Single | | 4 MI/H Dual | | 5 MI/H Single | | 5 MI/H Dual | |
|---|---|---|---|---|---|---|---|---|
| 65 | Tire requirements determined primarily by rear axle weight | | ↑ Duals only necessary for flotation purposes at this speed ↓ | | Tire requirements determined primarily by rear axle weight ↓ | | ↑ Duals only necessary for flotation purposes at this speed ↓ | |
| 75 | | | | | | | | |
| 80 | | | | | | | | |
| 85 | 16.9-34 | 6 PR | | | | | | |
| 90 | 16.9-38 | 6 PR | | | | | | |
| 95 | 18.4-38 | 6 PR | | | | | | |
| 100 | 18.4-38 | 6 PR | | | 16.9-34 | 6 PR | | |
| 110 | 18.4-38 | 8 PR | | | 16.9-38 | 6 PR | | |
| 120 | 20.8-38 | 8 PR | | | 18.4-38 | 6 PR | | |
| 130 | 23.1-34 | 8 PR | | | 18.4-34 | 8 PR | | |
| 140 | 20.8-38 | 10 PR | | | 18.4-38 | 8 PR | | |
| 150 | 24.5-32 | 10 PR | 16.9-34 | 6 PR | 20.8-38 | 8 PR | | |
| 160 | 30.5L-32 | 10 PR | 16.9-38 | 6 PR | 23.1-34 | 8 PR | | |
| 170 | Duals recommended for maximum efficiency at this speed ↓ | | 18.4-38 | 6 PR | 20.8-38 | 10 PR | | |
| 180 | | | 20.8-34 | 6 PR | 24.5-32 | 10 PR | 16.9-34 | 6 PR |
| 190 | | | 18.4-38 | 8 PR | 24.5-32 | 10 PR | 16.9-38 | 6 PR |
| 200 | | | 18.4-38 | 8 PR | 30.5L-32 | 10 PR | 16.9-38 | 6 PR |
| 210 | | | 20.8-34 | 8 PR | Duals recommended for maximum efficiency at this speed | | 18.4-38 | 6 PR |
| 220 | | | 20.8-38 | 8 PR | | | 20.8-34 | 6 PR |
| 230 | | | 23.1-34 | 8 PR | | | 18.4-34 | 8 PR |

* Drawbar hp ranges from 85 to 90 percent of advertised hp.
*Note:* The higher the operating speed, the smaller the tire required at most equivalent drawbar horsepower ratings.
*Source:* Ford Tractor Operations

**Fig. 25-10.** (*a*) Determining rear traction tire inflation by deflection of sidewalls. Tires can be easily damaged by (*b*) running over large objects or (*c*) being punctured or (*d*) cut by sharp fence posts or machinery parts.

**Controlling inflation pressure** Inflation pressure can affect the rolling resistance, wheel slippage, and load-carrying capacity of a tire. Too little inflation not only can decrease rolling resistance but also can cause tire sidewall damage (Fig. 25-10). Table 25-9 presents data on the maximum weight that can be added to each tire when inflated to the pressures listed.

## SUMMARY

The owner/manager of agricultural production operations is a manager of land, labor, and capital. In comparison to land, agricultural machinery costs often exceed $80 to $100/acre. These costs are identified as fixed ownership costs and operating costs.

Fixed costs include depreciation, interest on machinery investment, taxes, housing, and insurance. Depreciation and interest costs represent more than 90 percent of the fixed costs. Fixed costs occur regardless of how much the machine is used per year. Reducing the unit cost (the hourly use cost) of ownership can be accomplished by increasing the annual use of the machine.

Total operating costs increase in nearly direct proportion to machine use. Operating costs include charges made for labor to operate the machine, fuels and lubricants, and repairs and maintenance.

Good management of machinery includes a careful maintenance program, with special attention paid to avoid excessive power losses from wheel rolling resistance and slippage. An untuned engine, excessive rolling resistance, and wheel slippage can easily reduce a tractor's power output by 45 percent and increase costs due to increased fuel consumption. Failure to make timely repairs to avoid unnecessary downtime during critical planting and harvesting time periods can reduce yields and decrease the returns and increase total machinery costs per crop acre.

## THINKING IT THROUGH

1. Briefly explain why it costs money to own a machine. Why are these costs classified as fixed?
2. List the five costs that make up the fixed cost of machine ownership and write a brief explanation of why each must be considered when analyzing the cost of machine ownership.
3. What are the three methods of figuring machinery depreciation? Which is considered the best method for evaluating actual resale value? Why?
4. List the three principal factors that determine operating costs? Why do total operating costs increase with the amount of use?
5. Assume a $20,000 tractor was operated 1000 h/yr. What would be the average yearly repair and maintenance costs for the first 5 years? For 10 years?
6. What is the annual fuel cost for a 100-hp [74.6 kW] diesel tractor operating 1000 h when the fuel costs $1/gal [$0.26/L]? What is the annual cost of oils and lubricants?
7. List the advantages and disadvantages of each of the three alternatives to owning agricultural machinery.
8. Determine the custom rate per acre [ha] you would need to charge if you were to use your new combine 500 h/year for 4 years. Use labor at $6/h and a 50 percent return for profit.
9. How can a regular oil analysis program reduce repair costs?
10. Why does rolling resistance and wheel slippage reduce drawbar power?
11. List three methods of decreasing rolling resistance.
12. What should be done if you found that your tractor wheel turned 10.0 revolutions when loaded but only 7.5 times when unloaded, to cover the same ground?
13. What are two problems that result when too much weight is added to the rear wheels?
14. What are the advantages of using dual drive wheels?

CHAPTER
# 26 SELECTING AND MANAGING MACHINERY

## CHAPTER GOALS

In this chapter your goals are to:

- Identify efficiency factors that are used to select agricultural machinery
- Identify agricultural power requirements
- Be able to use information from the *Nebraska Tractor Test Reports* to select the make and model tractor

Agricultural management practices nearly always involve the selection and management of machinery. The decisions involved include factors which balance the type and size of machine on the one hand with labor and production costs on the other.

Professionals in power and machinery work closely with owners/managers of farms and ranches to select and obtain agricultural production machinery. They provide important information such as performance data and operational features of the equipment so that the proper economic and functional needs can be met.

## SELECTING MACHINERY

Our choices of an automobile are very much influenced by personal preferences. The make, styling, color, and price differences are some of the primary factors which influence our purchasing habits. Some of these same factors influence our choice of an agricultural machine. For example, the color of a machine may be important to some owners/managers because it represents a particular manufacturer. Others may choose a particular make of tractor or combine because they are able to get what they want and need in a machine.

Manufacturers of today's modern agricultural machinery have spent many hours of time and effort and millions of dollars in research, development, and testing to produce a high-quality product. Standardization of such things as control systems, hitches, and the size and speed of the power takeoff (PTO) has made it possible to interchange machines of one make with another without modification.

The capacity of a machine is the primary factor in selecting or purchasing farming equipment. Capacity and size of a machine are often used synonymously. In this chapter, *capacity* of a machine is used to identify its rate of performance. Terms such as bales per hour, acres per hour [ha/h] and other units are used to express capacity. *Size* is identified by such units of measure as width of cut and number of bottoms. The size of a machine may not directly indicate its capacity because a large machine may work more slowly than a smaller one.

### Machine Capacity

Let's assume you wish to purchase a new pickup truck. One of the specifications you look at is the fuel efficiency ratings. These values are published as EPA ratings in miles per gallon [kilometers (km) per liter (L)] of fuel required for highway and city travel, depending on driving habits. This is a measure of capacity of the truck to travel a given number of miles [km] on a gallon [L] of fuel. A pickup is also rated in its weight-carrying ability such as $^1/_2$- or $^3/_4$-ton [0.45- or 0.68-metric ton (t)] size. This rating does not provide a measure of its capacity to carry the load efficiently, because the rate of performance in pounds per gallon [kilograms per liter] of fuel or pounds per mile [kilograms per kilometer] traveled is not stated.

Machine capacity is based on a quantity-time relationship rather than length-number factors. The most common measures of machine capacity are in acres [hectares] per hour and tons [metric tons] per hour. Each of these measures is affected by the rate of travel, the width in feet [meters] of the cutting or harvesting part of the machine, and the effectiveness of the time that it is used.

**Determining Field Travel Speed** One of the first values that must be determined to evaluate machine capacity is the *field travel speed*. Travel speed is measured in miles per hour (mi/h) or kilometers per hour [km/h, where km/h = mi/h × 1.609]. Many machines have speedometers which provide fair estimates. The formula for determining exact field travel speed is as follows:

$$\text{mi/h} = \frac{\text{distance}}{\text{minutes (min)} \times 88}$$

where distance = number of feet traveled
minutes = total minutes to cover distance traveled
88 = 88 feet (ft) [26.8 meters (m)]/min = 1 mi/h [1.61 km/h]

**Example 1** Determine the field travel speed of a 16-ft [4.87 m] field cultivator when pulled across a field $^1/_2$ mile (mi) (80 rods) [0.8 km] long in 6 min (1 mi = 5280 ft [1.6 km]; 1 rod = 16.5 ft [5 m]).

**Solution**

$$\text{mi/h} = \frac{2640 \text{ ft}}{6 \text{ min} \times 88} = 5\ [8.05 \text{ km/h}]$$

The field travel speed of the field cultivator is 5 mi/h [8.05 km/h].

An estimate of field travel speed may be made while walking beside the operating machine. Using a 3-foot [0.9-m] step, count the number of steps taken in 20 seconds (s). Divide this number by 10 to estimate travel speed in miles per hour. This technique is reasonably accurate up to about 5 mi/h [8.05 km/h].

**Example 2** You step 34 three-foot [0.9-m] steps beside a combine in a 20-s period.

**Solution**

$$\text{mi/h} = \frac{34}{10} = 3.4\ [5.47 \text{ km/h}]$$

**Theoretical Field Capacity** The capacity of a machine expresses the rate at which it can do the work it was designed to perform. For example, most tillage tools are rated in the number of acres per hour (acres/h) [hectares per hour (ha/h)] that work can be accomplished. A forage harvester would be rated in its capacity to chop forage crops. Acres per hour [ha/h] would not be a good measure of capacity because forage varies in amount of raw material per acre [ha]. A forage harvester would be rated for field capacity in tons per hour [metric tons (t)/h].

*Theoretical field capacity* (TF cap) expresses the calculated capacity without regard to losses which may occur. TF cap is determined as follows:

$$\text{TF cap, acres/h} = \frac{\text{mi/h} \times \text{width}}{8.25} \qquad \left[\text{ha/h} = \frac{\text{acres/h}}{2.47}\right]$$

where mi/h = travel speed in miles per hour

width = actual width of machine in feet

$$8.25 = \text{constant}\ \frac{43{,}560 \text{ ft}^2/\text{acre}}{5280 \text{ ft/mi}}$$

**Example 3** From the data in Example 1, what is the TF cap of the field cultivator?

**Solution**

$$\text{TF cap, acre/h} = \frac{5 \text{ mi/h} \times 16\text{-ft width}}{8.25} = 9.7\ [3.9 \text{ ha/h}]$$

**Effective Field Capacity** The *effective field capacity* (EF cap) of a machine is an expression of the actual rate at which it can do work, taking into consideration such nonproductive operations as turning at the ends of the field, stopping to add fertilizer or seed, stopping to check performance, and the amount of overlap into previously traveled area. EF cap is determined as follows:

$$\text{EF cap, acres/h} = \frac{\text{acres covered or worked}}{\text{time used in hours}}$$

**Example 4** Assume the field cultivator was able to till a strip 1320 ft [400 m] wide in the 1/2-mi- [0.8-km-] long field in 10 h. What is the EF cap?

**Solution**

$$\text{EF cap, acres/h} = \frac{\text{acres covered}}{10 \text{ h}} \qquad \text{Acres} = \frac{2640 \text{ ft} \times 1320 \text{ ft}}{43{,}560 \text{ ft}^2/\text{acre}} = 80$$

$$= \frac{80 \text{ acres}}{10 \text{ h}} = 8\ [3.2 \text{ ha/h}]$$

**Field Efficiency** The term *field efficiency* (field eff) is used to describe the efficiency of the machine in operation. It is expressed as the difference in percent between the theoretical field capacity and the effective field capacity. The efficiency factor is determined as follows:

$$\text{Field eff, \%} = \frac{\text{EF cap}}{\text{TF cap}} \times 100$$

**Example 5** The values in Examples 3 and 4 for the TF cap and EF cap of the field cultivator were, respectively, 9.7 and 8.0 acres/h [3.9 and 3.2 ha/h]. What is the field efficiency?

**Solution**

$$\text{Field eff, \%} = \frac{8.0 \text{ EF cap}}{9.7 \text{ TF cap}} \times 100 = 82.5\%$$

**Machinery Use Planning** The field efficiency factor is a valuable tool in estimating the productivity capacity of machinery. The data in Table 26-1 provides the field efficiency in percent. These data are used to estimate the effective field capacity in acres per hour [ha/h] when the theoretical capacity is known.

**Example 6** Determine the number of acres [ha] of corn that can be planted with an 8-row planter set for 40-in [101.6-cm] rows, traveling at 5 mi/h [8.05 km/h].

**Solution**

$$\text{Machine capacity, acres/h} = \frac{\text{mi/h} \times \text{width (ft)} \times \text{field efficiency}}{8.25 \times 100\%}$$

where planting = 50 to 85% (use 70%)

**TABLE 26-1. Typical Range of Values for Machinery Performance**

| Operation | Machine | Draft or Power Requirements | Speed or Capacity* | Field Efficiency, % |
|---|---|---|---|---|
| Tillage (primary) | Moldboard or disk plow | See Fig. 26-1 | 3–6 mi/h | 70–90 |
| | Chisel plow (subsoiler) | 100–160 lb/ft in depth | 3–5 mi/h | 70–90 |
| | Lister-bedder | 400–800 lb/bottom | 3–6 mi/h | 70–90 |
| | One-way | 180–400 lb/ft | 4–7 mi/h | 70–90 |
| | Offset-disk | 250–400 lb/ft | 3–6 mi/h | 70–90 |
| | Rotary tiller | 5–10 PTO hp/ft | 1–5 mi/h | 70–90 |
| Tillage (secondary) | Ammonia applicator | 400 lb/knife | 3–5 mi/h | 60–75 |
| | Bed shaper | 15 hp/row | 2–4 mi/h | 70–90 |
| | Disk harrow (tandem) | 100–280 lb/ft | 3–6 mi/h | 70–90 |
| | Field cultivator | 150–500 lb/ft | 3–8 mi/h | 70–90 |
| | Rod weeder | 60–120 lb/ft | 4–6 mi/h | 70–90 |
| | Rotary hoe | 30–100 lb/ft | 5–11 mi/h | 70–90 |
| | Row-crop cultivator | 40–80 lb/ft | 1.5–4 mi/h | 70–90 |
| | Spike-tooth harrow | 30–50 lb/ft | 3–6 mi/h | 70–90 |
| | Spring-tooth harrow | 50–150 lb/ft | 3–6 mi/h | 70–90 |
| Seeding | Planter, with fertilizer, herbicide | 250–450 lb/row | 3–6 mi/h | 50–85 |
| | Drill | 30–150 lb/opener | 2.5–6 mi/h | 65–85 |
| Harvesting | Baler | 2–2.5 hp·h/ton | 2.5–5 mi/h | 70–90 |
| | Combine | 2–2.5 hp·h/ft | 2.0–3.5 mi/h | 65–80 |
| | Cotton picker | 10–15 hp/row | 1.5–3.5 mi/h | 60–75 |
| | Forage harvester, flail | 1.5–2.5 hp·h/ton | 5–10 ton/h | 50–75 |
| | Forage harvester, knife | | | |
| | Green forage | 1.5–2.5 hp·h/ton | 3–4.5 mi/h | 50–70 |
| | Corn silage | 2–5 hp·h/ton | 3–4.5 mi/h | 50–70 |
| | Hay cuber | 15–20 hp·h/ton | 3–4.5 mi/h | 50–70 |
| | Swather-conditioner | 1.5–2.0 hp/ft | 3–6 mi/h | 55–85 |
| Miscellaneous | Fertilizer applicator | | 3–5 mi/h | 60–75 |
| | Sprayer | | | 60–75 |
| | Potato digger | 500–800 lb/row | 1–2 PTO/row plus 500–800 lb/row | 60–80 |

* Metric conversions: lb/ft × 0.0145 = kN/m
hp × 0.746 = kW
hp·h × 0.746 = kW·h
mi/h × 1.609 = km/h

$$= \frac{5 \text{ mi/h} \times 26.7 \text{ ft} \times 70\%}{8.25 \times 100\%}$$

$$= \frac{7000}{825}$$

$$= 11.3 \text{ acres/h } [4.6 \text{ ha/h}]$$

**Selecting Size of Machine** The importance of timeliness in crop production was stressed in the previous chapter. The proper-size machine can make it possible to seed or harvest the crop at the optimum time. For example, the potential yield of corn is reduced 1 to $1^1/_2$ bushels (bu) per acre [0.087 to 0.13 cubic meters/hectare ($m^3$/ha)] for every day of delay after the optimum planting time. In soybeans the loss is approximately $^1/_3$ bu/day [0.0106 $m^3$/day] of delay. In this example, a 10-day delay can result in a loss of 10 to 15 bu [0.35 to 0.52 $m^3$] of corn and approximately 3 bu [0.106 $m^3$] of soybeans per acre.

With the cooperation of the weather, selection of the proper-size planter will allow timeliness of planting. For example, assume the owner/manager wishes to plant 100 acres [40.5 ha] in a 10-h day at a speed of 5 mi/h [8.05 km/h] in 36-in [91.4 centimeters (cm)] rows. What size planter is required?

**Solution**

$$\text{Machine width, ft} = \frac{\text{acres/h} \times 8.25 \times 100\%}{\text{mi/h} \times \text{field eff}}$$

$$\text{Acres/h} = \frac{100 \text{ acres/day}}{10 \text{ h/day}}$$

$$= 10$$

$$= \frac{10 \text{ acres/h} \times 8.25 \times 100\%}{5 \text{ mi/h} \times 70\%}$$

$$= \frac{8250}{350}$$

= 23.6 ft [7.2 m] or 24 ft (24 ft = 8 rows × 36 in)

An 8-row planter will give the desired capacity of 100 acres/day [40.5 ha/day].

## SELECTING SIZE OF TRACTOR

The size of a tractor is identified by its power rating in horsepower [kilowatts]. Manufacturers give each tractor a power rating. This rating is stated as the *maximum rated horsepower [kW] produced at the power takeoff (PTO) and at the drawbar.*

In the previous discussion, the emphasis was placed on fitting the machine to the needs of the production operation to reduce labor costs, obtain timeliness of operations, and complement the system of farming. A machinery system—tractor and machine—must therefore be selected which is compatible with the owner's/manager's needs.

The size of tractor is chosen on the basis of the power requirements of the machine it will power. Machinery power requirements are directly related to the number of acres [hectares] per hour needed to get the job done on time. When selecting a tractor only for drawbar work and light PTO work, power should match the maximum drawbar horsepower [kilowatt] requirements. The machines requiring the greatest drawbar power are the moldboard plow and chisel plow or subsoiler. To achieve the best use of this power, the tractor must be weighted to enable using maximum drawbar power without excessive (more than 15 percent) wheel slippage when running at the desired travel speed. The forage harvester is often used to determine the maximum PTO power requirements.

### Determining Power Needs

Power requirements can be expressed in several ways. The most common ways are (1) number of pounds (lb) [kilonewtons (kN)] of drawbar pull, (2) horsepower-hours (hp·h) per unit [kilowatt-hour (kW·h) per unit], and PTO horsepower [kilowatts]. The power requirements in Table 26-1 are generally in pounds [kg] of pull per foot [m] of width, power per row (or other unit), or horsepower-hours [kW/h] per unit when operating at typical ranges of speed.

Draft values for moldboard plows are presented in graph form in Fig. 26-1. These data take into consideration the effect that soil type has on the draft in pounds [kg] of pull per square inch [cm$^2$] of soil that is moved or turned.

**Determining Pounds of Pull** The amount of drawbar horsepower (dhp) that a machine requires can be calculated using the following formula:

$$\text{dhp} = \frac{\text{draft (lb)} \times \text{speed (mi/h)}}{375}$$

where 375 = constant obtained by dividing 33,000 ft·lb/min by 88 ft/min.

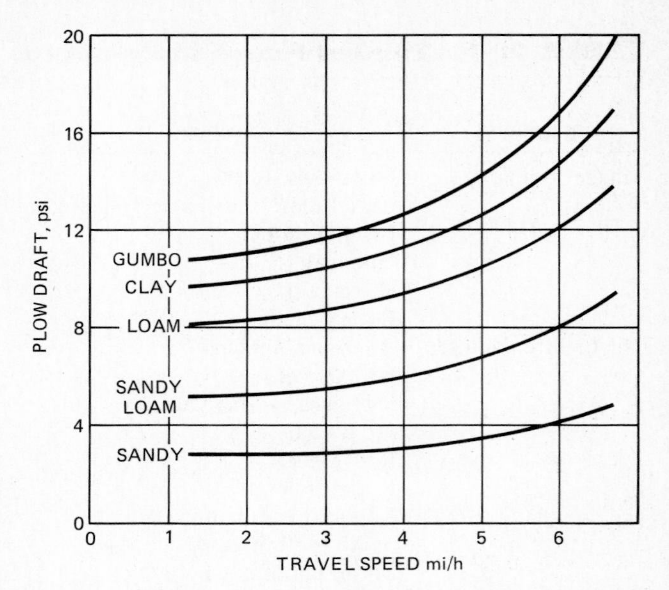


**Fig. 26-1.** Moldboard plow draft requirements according to type of soil and travel speed in mi/h.

The values for the above formula are found in Table 26-1 and Fig. 26-1, depending on the implement.

**Example** Determine the drawbar horsepower [kW] required to operate a 5-bottom, 16-inch (in) [41-cm] moldboard plow plowing loam soil 6 in [15.2 cm] deep at 5 mi/h [8.05 km/h].

**Solution** Pounds of pull required per square inch of turned soil is obtained from Fig. 26-1. This is approximately 11 pounds per square (psi) [75.8 kilopascals (kPa)]. At 5 mi/h [8.05 km/h] the draft per bottom is

Step 1. Pounds draft/bottom = 16-in bottom × 6 in deep × 11 psi [kPa]
= 1056 lb draft/bottom

Step 2. Pounds draft/5 bottoms = 1056 lb × 5 bottoms
= 5280 lb draft for machine

Step 3. $\text{dbh} = \dfrac{\text{draft} \times \text{speed}}{375} = \dfrac{5280 \times 5}{375}$
= 70.4 [52.5 kW]

The required power is 70.4 dhp [52.5 kW]. However, this is *effective* power and does not take into consideration losses in traction (slippage and rolling resistance) and in the gears of the transmission. The data in Table 26-2 provide traction-transmission conversion coefficients for wheel-type tractors. Plowing is a moderately heavy load and is usually performed in untilled (firm) land. The amount of drawbar power required is as follows:

Step 4. $\text{Corrected dhp} = \dfrac{\text{effective dhp}}{\text{coefficient}}$

$= \dfrac{70.4}{0.8} = 88$ [65.6 kW]

In order to achieve the assumed slippage of 8 percent (Table 26-2), the tractor will need to have the following total weight on the rear wheels:

$$\text{Pounds weight rear wheels} = \frac{\text{draft}}{\text{tractor \& transmission coefficient}}$$

Step 5. $$\text{Pounds weight, rear wheels} = \frac{5280\text{ lb}}{0.8} = 6600\text{ lb [2994 kg]}$$

The weight on the rear wheels should be approximately 6600 lb [3000 kg]. This does not take into account weight transfer. Always check the operator's manual for the amount of weight needed when attaching implements to a tractor equipped for weight transfer.

**Determining Horsepower-Hours per Unit** Another way of obtaining power requirements is in horsepower-hours [kW·h] per unit. For some machinery this value (Table 26-1) is more appropriate for determining power requirements of machines which are powered by the tractor's power takeoff (PTO).

For example, the size tractor required to operate a PTO-driven forage harvester to chop corn silage is 2 to 5 hp·h/ton [1.4 to 3.4 (kW·h)/t]. To harvest 20 tons/h [18.1 metric tons/h], assuming 3.5 hp·h/ton [2.4 (kW·h)/t], the PTO horsepower requirement of the tractor would be

$$\begin{aligned}\text{PTO hp} &= \text{tons/h} \times \text{hp·h/ton} \\ &= 20 \times 3.5 \\ &= 75\text{ hp [56 kW]}\end{aligned}$$

The estimate of 75 hp [56 kW] may be conservative. For example, a forage unloading wagon may be pulled behind the harvester, increasing the demand to 5 hp·h/ton, or a 100 PTO hp [74.6 PTO kW] requirement (20 × 5 = 100).

**Determining Tractor Size by PTO Power Data** Power required to operate tillage equipment is also identified by the tractor PTO power. The data in Table 26-3 provide the amount of PTO power required to pull a chisel shank at 4 and 5 mi/h [6.4 and 8.05 km/h] when operating 6 to 12 in [152 to 305 millimeters (mm)] deep for three soil types. For example, the power required for an 11-shank chisel plow operating at 10 in [254 mm] deep at 5 mi/h [8.05 km/h] in firm loam soil is as follows:

11 shanks × 10 PTO hp [7.5 kW]
= 110 PTO hp [82 kW]

Power requirements for the plow disk, disk harrow, and field cultivator are listed in Table 26-4. The data are based on the amount of PTO power required per foot [meter] of width and at a given (average) depth of soil penetration for 4 and 5 mi/h [6.4 and 8.05 km/h] and three soil conditions.

For example, 20-ft [6.1-m] width machines would require PTO horsepower [kilowatts] as follows:

Plow disk: 20 ft × 8.0 hp [6.0 kW]
= 160 PTO hp [119 kW]

Disk harrow: 20 ft × 6.0 hp [4.5 kW]
= 120 PTO hp [89 kW]

Field cultivator: 20 ft × 5.5 hp [4.1 kW]
= 110 PTO hp [82 kW]

Using aids such as these to determine the power requirement is a way to keep machinery working to

**TABLE 26-2. Traction-and-Transmission (T & T) Coefficients for Wheel-type Tractors**

| Surface Condition | LIGHT DRAWBAR LOAD | MEDIUM DRAWBAR LOAD | MODERATELY HEAVY DRAWBAR LOAD | |
|---|---|---|---|---|
| | T & T Coefficient | | T & T Coefficient | Assumed Slip, % |
| Concrete | 0.75 | 0.85 | 0.9 | 4 |
| Firm, untilled field | 0.6 | 0.75 | 0.8 | 8 |
| Tilled, reasonably firm soil | 0.4 | 0.6 | 0.65 | 10 |
| Freshly plowed soil | 0.25 | 0.4 | 0.45 | 10 |

*Note:* Coefficients are based on rolling resistance, wheel slippage, and power train losses.

**TABLE 26-3. Chisel Plow Power Requirements, Tractor PTO Power per Shank,* hp [kW]**

| | 4 mi/h [6.4 km/h] SOIL TYPE | | | 5 mi/h [8.05 km/h] SOIL TYPE | | |
|---|---|---|---|---|---|---|
| Depth, in [mm] | Sandy | Loam | Clay | Sandy | Loam | Clay |
| 6 [152] | 2.5 [1.9] | 5.0 [3.7] | 8.0 [6.0] | 3.0 [2.2] | 6.0 [4.5] | 10.0 [7.5] |
| 7 [178] | 3.0 [2.2] | 5.5 [4.1] | 9.5 [7.1] | 4.0 [3.0] | 6.5 [4.8] | 12.0 [9.0] |
| 8 [203] | 3.5 [2.6] | 6.0 [4.5] | 11.0 [8.2] | 4.5 [3.4] | 8.0 [6.0] | 14.0 [10.4] |
| 9 [229] | 4.0 [3.0] | 7.0 [5.2] | 12.5 [9.3] | 5.0 [3.7] | 9.0 [6.7] | 15.4 [11.2] |
| 10 [254] | 4.5 [3.4] | 8.0 [6.0] | 14.0 [10.4] | 5.5 [4.1] | 10.0 [7.5] | 17.0 [12.7] |
| 11 [279] | 5.0 [3.7] | 9.0 [6.2] | 15.5 [11.6] | 6.0 [4.5] | 11.0 [8.2] | 19.0 [14.2] |
| 12 [305] | 5.5 [4.1] | 9.5 [7.1] | 16.5 [12.3] | 6.5 [4.8] | 12.0 [9.0] | 20.5 [1.9] |

* Equipped with 2 in [51 mm] reversible shovels.

its full capacity to accomplish more work per hour and dollar invested.

## Using *Nebraska* Tractor Test Reports

For over 60 years the tractor testing station at the University of Nebraska has provided performance information on tractors. Tractor testing was established by the state legislature of Nebraska to assure that tractors sold in the state performed in accordance with the manufacturers' claims. As a result, the reports of tests published on each tractor tested serve as the most unbiased data available on power output and economy of operation.

Manufacturers submit new tractors to the testing station for testing. A test number is assigned to the tractor. A final report is prepared which is then made available to the public on a subscription or individual basis upon request. (See Table 26-5 for an example of a test report.) A condensed form of the report is also published in the *Implement and Tractor* Red Book issue. This publication is a reference most agricultural equipment dealerships have available for review.

The testing procedure for all tractors is standardized. This makes it easy and accurate to make comparisons with other tractors of similar capacities. The important data in the report are contained within two major levels of tractor performance, namely "Power Takeoff Performance" and "Drawbar Performance" (Table 26-5). Additional information is presented as lugging ability, sound level, and tire ballast and weight. Particular sections of these test data are important for use in helping to make decisions regarding the purchase of a new tractor to meet expected power needs.

**Power Takeoff Performance** The first line of test data within the power takeoff performance group is headed "Maximum Power and Fuel Consumption." This information shows the tractor's maximum power when operated on the PTO only. No forward motion is involved. Also listed is the amount of fuel used in gallons per hour (gal/h) or liters per hour [L/h] and horsepower-hours per gallon (hp·h/gal) or kilowatt-hours-per liter [kW·h/L]. Horsepower-hours [kilowatt-hours] is an indicator of tractor fuel efficiency. It identifies the power a tractor will develop for each gallon [liter] of fuel used. It is determined as follows:

**TABLE 26-4. Plow Disk, Disk Harrow, and Field Cultivator* Power Requirements Tractor PTO† Power per Foot (0.3 m) Width hp [kW]**

| | | 4 mi/h [6.4 km/h] SOIL TYPE | | | 5 mi/h [8.05 km/h] SOIL TYPE | | |
|---|---|---|---|---|---|---|---|
| Type of Tillage Implement | Average Depth Cut, inches [mm] | Sandy | Loam | Clay | Sandy | Loam | Clay |
| Plow disk | 8 [203] | 3.2 [2.4] | 6.4 [4.8] | 9.6 [7.2] | 4.0 [3.0] | 8.0 [6.0] | 12.0 [9.0] |
| Disk harrow | 5 [127] | 1.9 [1.4] | 4.8 [3.6] | 7.7 [5.7] | 2.4 [1.8] | 6.0 [4.5] | 9.6 [7.2] |
| Field cultivator | ‡ | 1.8 [1.3] | 4.5 [3.4] | 7.4 [5.5] | 2.2 [1.6] | 5.5 [4.1] | 9.2 [6.9] |

* Plow disk values based on firm soil; disk harrow and field cultivator values are for tilled soil.
† For drawbar power, multiply plow disk values by 1.5; for other tools, multiply by 1.8.
‡ Using 1¾-in [44.5-mm] reversible shovels 5 in [127 mm] deep; or 6- to 7-in [152–178 mm] sweep at 3 in [76 mm] deep.

# TABLE 26-5 Nebraska Tractor Test Report with Performance Data and Tractor Specifications

## POWER TAKE-OFF PERFORMANCE

| Power Hp (kW) | Crank shaft speed rpm | Fuel Consumption gal/hr (l/h) | lb/hp.hr (kg/kW.h) | Hp.hr/gal (kW.h/l) | Temperature °F (°C) Cooling medium | Air wet bulb | Air dry bulb | Barometer inch Hg (kPa) |
|---|---|---|---|---|---|---|---|---|
| **MAXIMUM POWER AND FUEL CONSUMPTION** | | | | | | | | |
| **Rated Engine Speed—Two Hours (PTO Speed—1137 rpm)** | | | | | | | | |
| 135.64 (101.14) | 2200 | 9.065 (34.313) | 0.463 (0.281) | 14.96 (2.948) | 178 (81.2) | 59 (15.0) | 75 (23.8) | 29.013 (97.974) |
| **Standard Power Take-off Speed (1000 rpm)—One Hour** | | | | | | | | |
| 129.25 (96.38) | 1935 | 8.102 (30.668) | 0.434 (0.264) | 15.95 (3.143) | 178 (80.9) | 59 (15.0) | 75 (23.8) | 29.010 (97.962) |
| **VARYING POWER AND FUEL CONSUMPTION—Two Hours** | | | | | | | | |
| 120.80 (90.08) | 2304 | 8.482 (32.107) | 0.486 (0.296) | 14.24 (2.806) | 178 (81.1) | 59 (15.0) | 75 (23.9) | ...... |
| 0.00 (0.00) | 2452 | 2.904 (10.992) | ..... | ..... | 171 (77.2) | 58 (14.7) | 74 (23.1) | ...... |
| 62.48 (46.59) | 2384 | 5.634 (21.328) | 0.624 (0.380) | 11.09 (2.185) | 173 (78.3) | 59 (15.0) | 75 (23.9) | ...... |
| 135.22 (100.83) | 2200 | 9.088 (34.403) | 0.465 (0.283) | 14.88 (2.931) | 179 (81.7) | 60 (15.8) | 77 (25.0) | ...... |
| 31.48 (23.48) | 2414 | 4.278 (16.193) | 0.940 (0.572) | 7.36 (1.450) | 171 (77.2) | 60 (15.6) | 76 (24.2) | ...... |
| 92.51 (68.98) | 2356 | 7.203 (27.267) | 0.539 (0.328) | 12.84 (2.530) | 176 (79.7) | 62 (16.4) | 76 (24.7) | ...... |
| **Av 73.75 (55.00)** | **2352** | **6.265** (23.715) | **0.588** (0.358) | **11.77** (2.319) | **175** (79.2) | **60** (15.4) | **75** (24.1) | **29.003** (97.940) |

## DRAWBAR PERFORMANCE

| Power Hp (kW) | Drawbar pull lbs (kN) | Speed mph (km/h) | Crank-shaft speed rpm | Slip % | Fuel Consumption gal/hr (l/h) | lb/hp.hr (kg/kW.h) | Hp.hr/gal (kW.h/l) | Temp. °F (°C) Cool-ing med | Air wet bulb | Air dry bulb | Barom. inch Hg (kPa) |
|---|---|---|---|---|---|---|---|---|---|---|---|
| **Maximum Available Power—Two Hours 8th (5PD) Gear** | | | | | | | | | | | |
| 113.32 (84.50) | 9415 (41.88) | 4.51 (7.26) | 2198 | 7.23 | 8.975 (33.974) | 0.548 (0.334) | 12.63 (2.487) | 180 (82.2) | 71 (21.7) | 81 (26.9) | 28.760 (97.118) |
| **75% of Pull at Maximum Power—Ten Hours 8th (5PD) Gear** | | | | | | | | | | | |
| 94.14 (70.20) | 7236 (32.19) | 4.88 (7.85) | 2326 | 5.19 | 7.866 (29.777) | 0.578 (0.352) | 11.97 (2.358) | 176 (79.9) | 70 (21.0) | 80 (26.6) | 28.837 (97.378) |
| **50% of Pull at Maximum Power—Two Hours 8th (5PD) Gear** | | | | | | | | | | | |
| 64.58 (48.16) | 4805 (21.37) | 5.04 (8.11) | 2364 | 3.61 | 6.248 (23.652) | 0.670 (0.407) | 10.34 (2.036) | 173 (78.1) | 63 (16.9) | 69 (20.3) | 28.855 (97.439) |
| **50% of Pull at Reduced Engine Speed—Two Hours 11th (6PD) Gear** | | | | | | | | | | | |
| 64.89 (48.39) | 4835 (21.51) | 5.03 (8.10) | 1671 | 3.61 | 4.731 (17.910) | 0.505 (0.307) | 13.72 (2.702) | 169 (76.1) | 67 (19.2) | 73 (22.8) | 28.860 (97.456) |
| **MAXIMUM POWER IN SELECTED GEARS** | | | | | | | | | | | |
| 110.18 (82.16) | 15010 (66.77) | 2.75 (4.43) | 2259 | 14.58 | 5th (3PD) Gear | | | 176 (80.0) | 60 (15.6) | 65 (18.3) | 28.850 (97.422) |
| 112.75 (84.08) | 11036 (49.09) | 3.83 (6.17) | 2199 | 8.72 | 7th (4PD) Gear | | | 179 (81.7) | 68 (20.0) | 76 (24.4) | 28.780 (97.186) |
| 116.17 (86.63) | 9646 (42.91) | 4.52 (7.27) | 2201 | 7.23 | 8th (5PD) Gear | | | 179 (81.4) | 67 (19.4) | 73 (22.8) | 28.780 (97.186) |
| 115.20 (85.91) | 7283 (32.40) | 5.93 (9.55) | 2200 | 5.18 | 10th (5DD) Gear | | | 179 (81.7) | 68 (20.0) | 77 (25.0) | 28.780 (97.186) |
| 118.08 (88.05) | 6765 (30.09) | 6.55 (10.53) | 2201 | 4.65 | 11th (6PD) Gear | | | 180 (81.9) | 69 (20.6) | 78 (25.6) | 28.780 (97.186) |
| 114.06 (85.05) | 4267 (18.98) | 10.02 (16.13) | 2200 | 2.95 | 13th (7PD) Gear | | | 179 (81.7) | 69 (20.6) | 78 (25.6) | 28.780 (97.186) |

## LUGGING ABILITY IN RATED GEAR 8th (5PD)

| Crankshaft Speed rpm | 2201 | 1978 | 1763 | 1543 | 1313 | 1101 |
|---|---|---|---|---|---|---|
| Pull—lbs (kN) | 9646 (42.91) | 10422 (46.36) | 10819 (48.12) | 10937 (48.65) | 10471 (46.58) | 9227 (41.04) |
| Increase in Pull % | 0 | 8 | 12 | 13 | 9 | −4 |
| Power—Hp (kW) | 116.17 (86.63) | 111.89 (83.43) | 103.11 (76.89) | 91.00 (67.86) | 74.55 (55.59) | 55.59 (41.46) |
| Speed—Mph (km/h) | 4.52 (7.27) | 4.03 (6.48) | 3.57 (5.75) | 3.12 (5.02) | 2.67 (4.30) | 2.26 (3.64) |
| Slip % | 7.23 | 7.88 | 8.44 | 8.72 | 8.16 | 7.02 |

## TRACTOR SOUND LEVEL WITH CAB

| | dB(A) |
|---|---|
| Maximum Available Power—Two Hours | 81.5 |
| 75% of Pull at Maximum Power—Ten Hours | 82.5 |
| 50% of Pull at Maximum Power—Two Hours | 83.0 |
| 50% of Pull at Reduced Engine Speed—Two Hours | 80.0 |
| Bystander in 16th (8DD) gear | 89.0 |

## TIRES, BALLAST AND WEIGHT

| | | With Ballast | Without Ballast |
|---|---|---|---|
| **Rear Tires** | —No., size, ply & psi (kPa) | Four 18.4-38; 8; inner 16 (110), outer 14 (100) | Four 18.4-38; 8; inner 16 (110), outer 14 (100) |
| Ballast | —Liquid (each inner) | 1200 lb (544 kg) | None |
| | —Cast Iron (each side) | 1020 lb (463 kg) | None |
| **Front Tires** | —No., size, ply & psi (kPa) | Two 11.00-16; 6; 32 (220) | Two 11.00-16; 6; 32 (220) |
| Ballast | —Liquid (each) | None | None |
| | —Cast Iron (each) | 90 lb (41 kg) | None |
| **Height of drawbar** | | 21.5 in (550 mm) | 21.5 in (550 mm) |
| **Static weight with operator** | —rear | 13800 lb (6260 kg) | 9360 lb (4246 kg) |
| | front | 3680 lb (1669 kg) | 3500 lb (1588 kg) |
| | total | 17480 lb (7929 kg) | 12860 lb (5833 kg) |

## 9700 DIESEL

Department of Agricultural Engineering

Dates of Test: May 9 to 20, 1977

Manufacturer: COMPANY.

**FUEL, OIL AND TIME:** Fuel No. 2 Diesel **Cetane No.** 51.8 (rating taken from oil company's typical inspection data) **Specific gravity converted to 60°/60°** *(15°/15°)* 0.8313 **Fuel weight** 6.922 lbs/gal *(0.831 kg/l)* **Oil SAE** 30 **API service classification** SB/SE-CA/CD **To motor** 3.340 gal *(12.643 l)* **Drained from motor** 2.717 gal *(10.281 l)* **Transmission and final drive lubricant** Ford M2C53A **Total time engine was operated** 45.5 hours

**ENGINE:** Diesel **Type** 6 cylinder vertical with turbocharger **Serial No.** *H140627* **Crankshaft** lengthwise **Rated rpm** 2200 **Bore and stroke** 4.4" × 4.4" *(111.8 mm × 111.8 mm)* **Compression ratio** 15.6 to 1 **Displacement** 401 cu in *(6578 ml)* **Cranking system** 12 volt **Lubrication** pressure **Air cleaner** primary and safety paper elements and centrifugal precleaner **Oil filter** full flow paper element **Oil cooler** engine coolant heat exchanger for crankcase oil, radiator for transmission and hydraulic oil **Fuel filter** two parallel paper elements **Muffler** vertical **Cooling medium temperature control** thermostat

**CHASSIS:** **Type** standard with duals **Serial No.** C525432 **Tread width** rear 60" *(1524 mm)* to 88" *(2235 mm)* front 56" *(1422 mm)* to 84" *(2134 mm)* **Wheel base** 109.7" *(2786 mm)* **Center of gravity** (without operator or ballast, with minimum tread, with fuel tank filled and tractor serviced for operation) Horizontal distance forward from center-line of rear wheels 33.3" *(846 mm)* Vertical distance above roadway 44.6" *(1133 mm)* Horizontal distance from center of rear wheel tread 0.4" *(10 mm)* to the right **Hydraulic control system** direct engine drive **Transmission** selective gear fixed ratio with partial (2 range) operator controlled power shift **Advertised speeds mph (km/h)** first 1.5 *(2.4)* second 1.9 *(3.1)* third 2.1 *(3.4)* fourth 2.7 *(4.4)* fifth 3.2 *(5.1)* sixth 4.1 *(6.6)* seventh 4.2 *(6.8)* eighth 4.9 *(7.9)* ninth 5.5 *(8.8)* tenth 6.3 *(10.2)* eleventh 7.0 *(11.2)* twelfth 8.9 *(14.4)* thirteenth 10.4 *(16.8)* fourteenth 13.4 *(21.6)* fifteenth 14.0 *(22.5)* sixteenth 18.0 *(28.9)* reverse 1.9 *(3.0)*, 2.4 *(3.8)*, 5.9 *(9.5)* and 7.6 *(12.3)* **Clutch** single dry plate operated by foot pedal **Brakes** multiple wet disc hydraulically operated by two foot pedals which can be locked together **Steering** hydrostatic **Turning radius** (on concrete surface with brake applied) right 158" *(4.01 m)* left 160" *(4.06 m)* (on concrete surface without brake) right 178" *(4.52 m)* left 180" *(4.57 m)* **Turning space diameter** (on concrete surface with brake applied) right 330" *(8.38 m)* left 334" *(8.48 m)* (on concrete surface without brake) right 370" *(9.40 m)* left 374" *(9.50 m)* **Power take-off** 540 rpm at 1900 engine rpm and 1000 rpm at 1935 engine rpm.

**REPAIRS and ADJUSTMENTS:** Gear shift lever boot was found to be torn and was replaced.

**REMARKS:** All test results were determined from observed data obtained in accordance with SAE and ASAE test code or official Nebraska test procedure. Temperature at injection pump return was 164°F *(73.3°C)*. Six gears were chosen between stability limit and 15 mph *(24.1 km/h)*. #3 cylinder wall was found scored during final inspection. Left brake did not work properly during drawbar tests.

We, the undersigned, certify that this is a true and correct report of official Tractor Test 1243.

LOUIS I. LEVITICUS
Engineer-in Charge

G. W. STEINBRUEGGE, Chairman
W. E. SPLINTER
K. VON BARGEN
Board of Tractor Test Engineers

**Example** From Table 26-5, Test 1243, the tractor developed 135.64 PTO hp [101.15 kW] and used 9.065 gal/h [34.315 L/h] of diesel fuel. What is the horsepower-hours developed per gallon of fuel?

**Solution**

$$\text{hp·h/gal} = \frac{\text{hp developed}}{\text{gal/h}}$$

$$= \frac{\text{hp} \times \text{h}}{\text{gal}}$$

$$= \frac{135.64 \text{ hp} \times 1 \text{ h}}{9.065}$$

$$= 14.96 \ [2.947 \text{ kW·h/L}]$$

For comparison, the average value for 54 different tractors tested in this series was 15.35 hp·h/gal [3.02 kW·h/L] of diesel fuel. This tractor, Model 9700, was approximately 2.5 percent below average.

The second line under "Maximum Power and Fuel Consumption" gives the test data when the engine's speed control (throttle) is adjusted to drive the PTO shaft at a standard speed of 540 or 1000 revolutions per minute (r/min). This tractor is similar to others which produce standard PTO speeds at less than full-rated engine speed. In this tractor the rated engine speed is 2200 r/min. To obtain standard PTO speed of 1000 r/min, the engine was operated at 1935 r/min. Note that the maximum horsepower dropped to 129.25 [96.38 kW] at this speed. This is a horsepower drop of approximately 5 percent—something to consider if the tractor will be doing a lot of heavy PTO work requiring maximum engine output.

The next series of tests are listed under the heading "Varying Power and Fuel Consumption." Six tests with different load settings are run for 2 h each. The last line, No. 7, is an average of all six tests. The values given for gallons per hour [liters per hour] and horsepower-hours per gallon [kilowatt-hours per liter] are a good estimate of the average hourly fuel consumption per year. The Model 9700 tractor, Test 1243, used an average of 6.265 gal [23.715 L] of fuel per hour. Multiplying this figure by the number of hours of annual use will provide a good estimate of the number of gallons [liters] of fuel needed per year.

For this series of tests the average horsepower-hours per gallon for the 54 tractors previously referred to was 12.69 [2.49 kW·h/L]. This tractor produced 11.77 hp·h/gal [2.32 kW·h/L], which is approximately 7 percent under the average.

**Drawbar Performance** An evaluation of the power produced at the drawbar is reported in the series of tests. The first line is listed as "Maximum Available Power" and identifies the horsepower [kW], pounds [kN] of pull, fuel consumption, and fuel efficiency when the tractor drawbar is hitched to a load. The test identifies the maximum amount of pull the tractor will produce when the engine is operating at rated engine speed. For comparison, the average horsepower-hours per gallon of 54 tractors tested was 13.27 [2.6 kW·h/L]. Model 9700 produced 9415 lb [42 kN] of pull at 4.51 mi/h [7.26 km/h] with a performance of 12.63 hp·h/gal, [2.48 kW·h/L] or approximately 5 percent under the average.

The data for the second line are with the drawbar pull set at 75 percent of maximum pull (75% × 9646 lb = 7236 lb [32.19 kN] pull). This test is useful to identify the performance of the tractor under the general run of heavy tillage operations, such as plowing.

The next two lines of tests are "50 Percent of Pull at Maximum Power" and "50 Percent of Pull at Reduced Engine Speed." These two tests are good indicators of how the tractor would perform if used on light loads. For example, these tests provide an evaluation of operating a large tractor on light loads and at reduced engine speeds, while maintaining the same travel speed. In this tractor the transmission was shifted up and the engine rpm reduced to maintain 5-mi/h [8.05 km/h] forward travel. Shifting up dropped the diesel fuel consumed from 6.248 to 4.731 gal/h [23.651 to 17.910 L/h]. This is an increase of approximately 25 percent in fuel efficiency.

Drawbar performance in "Maximum Power in Selected Gears" identifies the horsepower [kW] and pull developed in each of six transmission gear selections. Slippage of drive wheels must not exceed 15 percent. Note that the tractor was equipped with dual rear tires. The total static weight was 17,480 lb [7929 kilograms]. The maximum pull of 15,010 lb [66.27 kN] is 86 percent of its weight.

The drawbar performance data continue with an indication of how well the tractor reacts to overload conditions. "Lugging Ability in Rated Gear" refers to the torque characteristics of the engine and the way the tractor was weighted. Note that the tractor is operated in what is considered *plow gear* (rated gear). The load is increased in steps to reduce the engine rpm. A maximum increase of 13 percent pull was achieved when the engine was "lugged down" to 1543 r/min. The low percentage value of 13 percent is an indicator of rather poor lugging ability.

Sound level for this tractor is reported at various power settings. This tractor was equipped with a cab. Sound levels inside the cab and on the outside 20 ft [6.1 m] from the tractor were measured. The unit of measure is in decibels, A scale [dB(A)]. Sound levels have been established for agricultural equipment operators' exposure in an 8-h period at a maximum of 90 dB(A), without permanent damage to the operator's hearing. The sound level inside the cab of this tractor is below the established maximum level for safety.

**TABLE 26-6 Nebraska Tractor Test Performance Data for Three Agricultural Tractors of Comparable Size**

## 9700 DIESEL
## NEBRASKA TRACTOR TEST 1243

**ENGINE:** **Diesel** Type 6 cylinder vertical with **turbocharger** **Serial No.** *H140627* **Crankshaft** lengthwise **Rated rpm** 2200 **Bore and stroke** 4.4" × 4.4" *(111.8 mm × 111.8 mm)* **Compression ratio** 15.6 to 1 **Displacement** **401 cu in** *(6578 ml)* **Cranking system** 12 volt **Lubrication** pressure **Air cleaner** primary and safety paper elements and centrifugal precleaner **Oil filter** full flow paper element **Oil cooler** engine coolant heat exchanger for crankcase oil, radiator for transmission and hydraulic oil **Fuel filter** two parallel paper elements **Muffler** vertical **Cooling medium temperature control** thermostat

**POWER TAKE-OFF PERFORMANCE**

| | Power Hp (kW) | Crank shaft speed rpm | Fuel Consumption gal/hr (l/h) | lb/hp.hr (kg/kW.h) | Hp.hr/gal (kW.h/l) | Temperature °F (°C) Cooling medium | Air wet bulb | Air dry bulb | Barometer inch Hg (kPa) |
|---|---|---|---|---|---|---|---|---|---|
| | **MAXIMUM POWER AND FUEL CONSUMPTION** | | | | | | | | |
| | **Rated Engine Speed—Two Hours (PTO Speed—1137 rpm)** | | | | | | | | |
| | 135.64 (101.14) | 2200 | 9.065 (34.313) | 0.463 (0.281) | 14.96 (2.948) | 178 (81.2) | 59 (15.0) | 75 (23.8) | 29.013 (97.974) |
| | **Standard Power Take-off Speed (1000 rpm)—One Hour** | | | | | | | | |
| | 129.25 (96.38) | 1935 | 8.102 (30.668) | 0.434 (0.264) | 15.95 (3.143) | 178 (80.9) | 59 (15.0) | 75 (23.8) | 29.010 (97.962) |
| | **VARYING POWER AND FUEL CONSUMPTION—Two Hours** | | | | | | | | |
| | 120.80 (90.08) | 2304 | 8.482 (32.107) | 0.486 (0.296) | 14.24 (2.806) | 178 (81.1) | 59 (15.0) | 75 (23.9) | ...... |
| | 0.00 (0.00) | 2452 | 2.904 (10.992) | ..... | ..... | 171 (77.2) | 58 (14.7) | 74 (23.1) | ...... |
| | 62.48 (46.59) | 2384 | 5.634 (21.328) | 0.624 (0.380) | 11.09 (2.185) | 173 (78.3) | 59 (15.0) | 75 (23.9) | ...... |
| | 135.22 (100.83) | 2200 | 9.088 (34.403) | 0.465 (0.283) | 14.88 (2.931) | 179 (81.7) | 60 (15.8) | 77 (25.0) | ...... |
| | 31.48 (23.48) | 2414 | 4.278 (16.193) | 0.940 (0.572) | 7.36 (1.450) | 171 (77.2) | 60 (15.6) | 76 (24.2) | ...... |
| | 92.51 (68.98) | 2356 | 7.203 (27.267) | 0.539 (0.328) | 12.84 (2.530) | 176 (79.7) | 62 (16.4) | 76 (24.7) | ...... |
| **Av** | **73.75 (55.00)** | **2352** | **6.265 (23.715)** | **0.588 (0.358)** | **11.77 (2.319)** | **175 (79.2)** | **60 (15.4)** | **75 (24.1)** | **29.003 (97.940)** |

## 1086 DIESEL
## NEBRASKA TRACTOR TEST 1247

**ENGINE** **Diesel** Type 6 cylinder vertical with **turbocharger** **Serial No.** 414 TT 2U 104913* **Crankshaft** lengthwise **Rated rpm** 2400 **Bore and stroke** 4.30" × 4.75" *(109.2 mm × 120.6 mm)* **Compression ratio** 16 to 1 **Displacement** **414 cu in** *(6782 ml)* **Cranking system** 12 volt **Lubrication** pressure **Air cleaner** primary and safety paper elements with dust unloader **Oil filter** two full flow spin-on cartridges **Oil cooler** engine coolant heat exchanger for crankcase oil, radiator for transmission and hydraulic oil **Fuel filter** primary and final paper spin-on cartridges **Muffler** underhood **Cooling medium temperature control** thermostat

**POWER TAKE-OFF PERFORMANCE**

| | Power Hp (kW) | Crank shaft speed rpm | Fuel Consumption gal/hr (l/h) | lb/hp.hr (kg/kW.h) | Hp.hr/gal (kW.h/l) | Temperature °F (°C) Cooling medium | Air wet bulb | Air dry bulb | Barometer inch Hg (kPa) |
|---|---|---|---|---|---|---|---|---|---|
| | **MAXIMUM POWER AND FUEL CONSUMPTION** | | | | | | | | |
| | **Rated Engine Speed—Two Hours (PTO Speed—1159 rpm)** | | | | | | | | |
| | 131.41 (97.99) | 2400 | 8.583 (32.489) | 0.455 (0.277) | 15.31 (3.016) | 189 (87.1) | 62 (16.5) | 75 (24.1) | 29.053 (98.109) |
| | **Standard Power Take-off Speed (1000 rpm)—One Hour** | | | | | | | | |
| | 133.67 (99.68) | 2070 | 8.157 (30.877) | 0.425 (0.258) | 16.39 (3.228) | 192 (88.9) | 62 (16.4) | 76 (24.2) | 29.045 (98.081) |
| | **VARYING POWER AND FUEL CONSUMPTION—Two Hours** | | | | | | | | |
| | 115.91 (86.43) | 2490 | 7.973 (30.181) | 0.479 (0.291) | 14.54 (2.864) | 187 (86.1) | 62 (16.4) | 77 (25.0) | ...... |
| | 0.00 (0.00) | 2675 | 2.918 (11.045) | ..... | ..... | 178 (81.4) | 60 (15.6) | 75 (23.9) | ...... |
| | 60.12 (44.83) | 2588 | 5.404 (20.458) | 0.626 (0.381) | 11.12 (2.191) | 182 (83.3) | 60 (15.6) | 76 (24.7) | ...... |
| | 130.52 (97.33) | 2401 | 8.550 (32.367) | 0.456 (0.277) | 15.26 (3.007) | 190 (87.8) | 62 (16.7) | 80 (26.4) | ...... |
| | 30.64 (22.85) | 2638 | 4.258 (16.118) | 0.967 (0.588) | 7.20 (1.417) | 180 (82.2) | 63 (17.2) | 78 (25.8) | ...... |
| | 88.74 (66.17) | 2538 | 6.684 (25.303) | 0.524 (0.319) | 13.28 (2.615) | 186 (85.3) | 64 (18.1) | 81 (27.2) | ...... |
| **Av** | **70.99 (52.94)** | **2555** | **5.965 (22.579)** | **0.585 (0.356)** | **11.90 (2.344)** | **184 (84.4)** | **62 (16.6)** | **78 (25.5)** | **29.043 (98.075)** |

## 4440 DIESEL
## NEBRASKA TRACTOR TEST 1265

**ENGINE** **Diesel** Type 6 cylinder vertical with **turbocharger** **Serial No.** 6466TR-03 024358RG **Crankshaft** lengthwise **Rated rpm** 2200 **Bore and stroke** 4.5625" × 4.75" *(115.9 mm × 120.7 mm)* **Compression ratio** 15.5 to 1 **Displacement** **466 cu in** *(7636 ml)* **Cranking system** 12 volt **Lubrication** pressure **Air cleaner** paper primary and safety elements with dust evacuator **Oil filter** one screw-on cartridge **Oil cooler** engine coolant heat exchanger for crankcase oil, radiator for transmission and hydraulic oil **Fuel filter** two snap-on cartridges **Muffler** vertical **Cooling medium temperature control** 2 thermostats.

**POWER TAKE-OFF PERFORMANCE**

| | Power Hp (kW) | Crank shaft speed rpm | Fuel Consumption gal/hr (l/h) | lb/hp.hr (kg/kW.h) | Hp.hr/gal (kW.h/l) | Temperature °F (°C) Cooling medium | Air wet bulb | Air dry bulb | Barometer inch Hg (kPa) |
|---|---|---|---|---|---|---|---|---|---|
| | **MAXIMUM POWER AND FUEL CONSUMPTION** | | | | | | | | |
| | **Rated Engine Speed—Two Hours (PTO Speed—1002 rpm)** | | | | | | | | |
| | 130.58 (97.38) | 2200 | 8.585 (32.496) | 0.458 (0.279) | 15.21 (2.997) | 188 (86.9) | 57 (13.8) | 75 (23.9) | 28.777 (97.174) |
| | **VARYING POWER AND FUEL CONSUMPTION—Two Hours** | | | | | | | | |
| | 115.03 (85.78) | 2280 | 8.015 (30.339) | 0.485 (0.295) | 14.35 (2.827) | 181 (82.8) | 58 (14.4) | 75 (23.9) | ...... |
| | 0.00 (0.00) | 2356 | 2.963 (11.216) | ..... | ..... | 175 (79.4) | 58 (14.2) | 75 (23.9) | ...... |
| | 58.11 (43.33) | 2299 | 5.375 (20.345) | 0.644 (0.392) | 10.81 (2.130) | 181 (82.8) | 58 (14.4) | 75 (23.9) | ...... |
| | 131.04 (97.72) | 2200 | 8.643 (32.719) | 0.459 (0.279) | 15.16 (2.987) | 189 (87.2) | 58 (14.4) | 75 (23.9) | ...... |
| | 29.15 (21.74) | 2318 | 4.147 (15.699) | 0.991 (0.603) | 7.03 (1.385) | 176 (80.3) | 58 (14.2) | 74 (23.6) | ...... |
| | 86.50 (64.50) | 2285 | 6.658 (25.203) | 0.536 (0.326) | 12.99 (2.559) | 184 (84.4) | 57 (13.9) | 74 (23.6) | ...... |
| **Av** | **69.97 (52.18)** | **2290** | **5.967 (22.587)** | **0.594 (0.361)** | **11.73 (2.310)** | **181 (82.8)** | **58 (14.3)** | **75 (23.8)** | **28.700 (96.916)** |

*(Continued on next page)*

# TABLE 26-6 (Continued)

## 9700 DIESEL
## NEBRASKA TRACTOR TEST 1243
(Continued)

### DRAWBAR PERFORMANCE

| Power Hp (kW) | Drawbar pull lbs (kN) | Speed mph (km/h) | Crank-shaft speed rpm | Slip % | Fuel Consumption gal/hr (l/h) | lb/hp.hr (kg/kW.h) | Hp.hr/gal (kW.h/l) | Temp. °F (°C) Cooling med | Air wet bulb | Air dry bulb | Barom. inch Hg (kPa) |
|---|---|---|---|---|---|---|---|---|---|---|---|
| **Maximum Available Power—Two Hours 8th (5PD) Gear** | | | | | | | | | | | |
| 113.32 (84.50) | 9415 (41.88) | 4.51 (7.26) | 2198 | 7.23 | 8.975 (33.974) | 0.548 (0.334) | 12.63 (2.487) | 180 (82.2) | 71 (21.7) | 81 (26.9) | 28.760 (97.118) |
| **75% of Pull at Maximum Power—Ten Hours 8th (5PD) Gear** | | | | | | | | | | | |
| 94.14 (70.20) | 7236 (32.19) | 4.88 (7.85) | 2326 | 5.19 | 7.866 (29.777) | 0.578 (0.352) | 11.97 (2.358) | 176 (79.9) | 70 (21.0) | 80 (26.6) | 28.837 (97.378) |
| **50% of Pull at Maximum Power—Two Hours 8th (5PD) Gear** | | | | | | | | | | | |
| 64.58 (48.16) | 4805 (21.37) | 5.04 (8.11) | 2364 | 3.61 | 6.248 (23.652) | 0.670 (0.407) | 10.34 (2.036) | 173 (78.1) | 63 (16.9) | 69 (20.3) | 28.855 (97.439) |
| **50% of Pull at Reduced Engine Speed—Two Hours 11th (6PD) Gear** | | | | | | | | | | | |
| 64.89 (48.39) | 4835 (21.51) | 5.03 (8.10) | 1671 | 3.61 | 4.731 (17.910) | 0.505 (0.307) | 13.72 (2.702) | 169 (76.1) | 67 (19.2) | 73 (22.8) | 28.860 (97.456) |

### MAXIMUM POWER IN SELECTED GEARS

| Power Hp (kW) | Drawbar pull lbs (kN) | Speed mph (km/h) | Crank-shaft speed rpm | Slip % | Gear | Cooling med | Air wet bulb | Air dry bulb | Barom. inch Hg (kPa) |
|---|---|---|---|---|---|---|---|---|---|
| 110.18 (82.16) | 15010 (66.77) | 2.75 (4.43) | 2259 | 14.58 | 5th (3PD) Gear | 176 (80.0) | 60 (15.6) | 65 (18.3) | 28.850 (97.422) |
| 112.75 (84.08) | 11036 (49.09) | 3.83 (6.17) | 2199 | 8.72 | 7th (4PD) Gear | 179 (81.7) | 68 (20.0) | 76 (24.4) | 28.780 (97.186) |
| 116.17 (86.63) | 9646 (42.91) | 4.52 (7.27) | 2201 | 7.23 | 8th (5PD) Gear | 179 (81.4) | 67 (19.4) | 73 (22.8) | 28.780 (97.186) |
| 115.20 (85.91) | 7283 (32.40) | 5.93 (9.55) | 2200 | 5.18 | 10th (5DD) Gear | 179 (81.7) | 68 (20.0) | 77 (25.0) | 28.780 (97.186) |
| 118.08 (88.05) | 6765 (30.09) | 6.55 (10.53) | 2201 | 4.65 | 11th (6PD) Gear | 180 (81.9) | 69 (20.6) | 78 (25.6) | 28.780 (97.186) |
| 114.06 (85.05) | 4267 (18.98) | 10.02 (16.13) | 2200 | 2.95 | 13th (7PD) Gear | 179 (81.7) | 69 (20.6) | 78 (25.6) | 28.780 (97.186) |

### LUGGING ABILITY IN RATED GEAR 8th (5PD)

| Crankshaft Speed rpm | 2201 | 1978 | 1763 | 1543 | 1313 | 1101 |
|---|---|---|---|---|---|---|
| Pull—lbs (kN) | 9646 (42.91) | 10422 (46.36) | 10819 (48.12) | 10937 (48.65) | 10471 (46.58) | 9227 (41.04) |
| Increase in Pull % | 0 | 8 | 12 | 13 | 9 | −4 |
| Power—Hp (kW) | 116.17 (86.63) | 111.89 (83.43) | 103.11 (76.89) | 91.00 (67.86) | 74.55 (55.59) | 55.59 (41.46) |
| Speed—Mph (km/h) | 4.52 (7.27) | 4.03 (6.48) | 3.57 (5.75) | 3.12 (5.02) | 2.67 (4.30) | 2.26 (3.64) |
| Slip % | 7.23 | 7.88 | 8.44 | 8.72 | 8.16 | 7.02 |

| TRACTOR SOUND LEVEL WITH CAB | dB(A) |
|---|---|
| Maximum Available Power—Two Hours | 81.5 |
| 75% of Pull at Maximum Power—Ten Hours | 82.5 |
| 50% of Pull at Maximum Power—Two Hours | 83.0 |
| 50% of Pull at Reduced Engine Speed—Two Hours | 80.0 |
| Bystander in 16th (8DD) gear | 89.0 |

| TIRES, BALLAST AND WEIGHT | | With Ballast | Without Ballast |
|---|---|---|---|
| Rear Tires | —No., size, ply & psi (kPa) | Four 18.4-38; 8; inner 16 (110), outer 14 (100) | Four 18.4-38; 8; inner 16 (110), outer 14 (100) |
| Ballast | —Liquid (each inner) | 1200 lb (544 kg) | None |
| | —Cast Iron (each side) | 1020 lb (463 kg) | None |
| Front Tires | —No., size, ply & psi (kPa) | Two 11.00-16; 6; 32 (220) | Two11.00-16; 6; 32 (220) |
| Ballast | —Liquid (each) | None | None |
| | —Cast Iron (each) | 90 lb (41 kg) | None |
| Height of drawbar | | 21.5 in (550 mm) | 21.5 in (550 mm) |
| Static weight with operator | —rear | 13800 lb (6260 kg) | 9360 lb (4246 kg) |
| | front | 3680 lb (1669 kg) | 3500 lb (1588 kg) |
| | total | 17480 lb (7929 kg) | 12860 lb (5833 kg) |

## 1086 DIESEL
## NEBRASKA TRACTOR TEST 1247
(Continued)

### DRAWBAR PERFORMANCE WITH BIAS PLY TIRES

| Power Hp (kW) | Drawbar pull lbs (kN) | Speed mph (km/h) | Crank-shaft speed rpm | Slip % | gal/hr (l/h) | lb/hp.hr (kg/kW.h) | Hp.hr/gal (kW.h/l) | Cooling med | Air wet bulb | Air dry bulb | Barom. inch Hg (kPa) |
|---|---|---|---|---|---|---|---|---|---|---|---|
| **Maximum Available Power—Two Hours 8th (1-HiTA) Gear** | | | | | | | | | | | |
| 108.99 (81.27) | 8049 (35.80) | 5.08 (8.17) | 2400 | 7.30 | 8.494 (32.152) | 0.542 (0.330) | 12.83 (2.528) | 189 (87.2) | 66 (18.9) | 72 (21.9) | 28.960 (97.794) |
| **75% of Pull at Maximum Power—Two Hours 8th (1-HiTA) Gear** | | | | | | | | | | | |
| 89.79 (66.96) | 6192 (27.54) | 5.44 (8.75) | 2516 | 5.23 | 7.434 (28.142) | 0.576 (0.351) | 12.08 (2.379) | 185 (85.0) | 67 (19.4) | 74 (23.3) | 28.935 (97.709) |
| **50% of Pull at Maximum Power—Two Hours 8th (1-HiTA) Gear** | | | | | | | | | | | |
| 62.13 (46.33) | 4113 (18.29) | 5.66 (9.12) | 2576 | 3.68 | 5.908 (22.364) | 0.662 (0.403) | 10.52 (2.072) | 181 (82.8) | 66 (18.9) | 72 (22.2) | 29.035 (98.047) |
| **50% of Pull at Reduced Engine Speed—Two Hours 12th (2-HiDD) Gear** | | | | | | | | | | | |
| 62.13 (46.33) | 4102 (18.24) | 5.68 (9.14) | 1513 | 3.56 | 4.435 (16.790) | 0.497 (0.302) | 14.01 (2.759) | 185 (84.7) | 70 (21.1) | 79 (25.8) | 29.035 (98.047) |

### MAXIMUM POWER IN SELECTED GEARS WITH BIAS PLY TIRES

| Power Hp (kW) | Drawbar pull lbs (kN) | Speed mph (km/h) | Crank-shaft speed rpm | Slip % | Gear | Cooling med | Air wet bulb | Air dry bulb | Barom. inch Hg (kPa) |
|---|---|---|---|---|---|---|---|---|---|
| 78.54 (58.57) | 12137 (53.99) | 2.43 (3.90) | 2536 | 14.10 | 4th (2-LoDD) Gear | 183 (83.9) | 64 (17.8) | 66 (18.9) | 29.020 (97.996) |
| 107.70 (80.31) | 9554 (42.50) | 4.23 (6.80) | 2398 | 8.77 | 6th (3-LoDD) Gear | 187 (86.1) | 65 (18.3) | 70 (21.1) | 28.950 (97.760) |
| 111.56 (83.19) | 8239 (36.65) | 5.08 (8.17) | 2400 | 7.22 | 8th (1-HiTA) Gear | 187 (86.1) | 64 (17.8) | 67 (19.4) | 28.930 (97.692) |
| 110.85 (82.66) | 7149 (31.80) | 5.81 (9.36) | 2399 | 6.08 | 9th (4-LoDD) Gear | 188 (86.4) | 65 (18.3) | 70 (21.1) | 28.950 (97.760) |
| 114.14 (85.12) | 6431 (28.61) | 6.66 (10.71) | 2400 | 5.31 | 10th (1-HiDD) Gear | 190 (87.5) | 68 (20.0) | 74 (23.3) | 28.930 (97.692) |
| 113.91 (84.94) | 6185 (27.51) | 6.91 (11.12) | 2400 | 5.15 | 11th (2-HiTA) Gear | 187 (86.1) | 64 (17.8) | 69 (20.6) | 28.940 (97.726) |

### LUGGING ABILITY IN RATED GEAR 8th (1-HiTA) WITH BIAS PLY TIRES

| Crankshaft Speed rpm | 2400 | 2160 | 1922 | 1684 | 1438 | 1199 |
|---|---|---|---|---|---|---|
| Pull—lbs (kN) | 8239 (36.65) | 9572 (42.58) | 10313 (45.88) | 10805 (48.06) | 10847 (48.25) | 9839 (43.76) |
| Increase in Pull % | 0 | 16 | 25 | 31 | 32 | 19 |
| Power—Hp (kW) | 111.56 (83.19) | 114.65 (85.50) | 108.44 (80.86) | 98.36 (73.35) | 83.96 (62.61) | 64.93 (48.42) |
| Speed—Mph (km/h) | 5.08 (8.17) | 4.49 (7.23) | 3.94 (6.35) | 3.41 (5.49) | 2.90 (4.67) | 2.47 (3.98) |
| Slip % | 7.22 | 8.85 | 9.99 | 11.24 | 11.52 | 9.56 |

| TRACTOR SOUND LEVEL WITH CAB AND RADIAL TIRES | dB(A) |
|---|---|
| Maximum Available Power—Two Hours | 78.5 |
| 75% of Pull at Maximum Power—Ten Hours | 78.5 |
| 50% of Pull at Maximum Power—Two Hours | 80.5 |
| 50% of Pull at Reduced Engine Speed—Two Hours | 77.0 |
| Bystander in 16th (4-HiDD) gear | 86.0 |

| TIRES, BALLAST AND WEIGHT | | With Ballast | Without Ballast |
|---|---|---|---|
| Radial Rear Tires | —No., size, ply & psi (kPa) | Two 20.8R38; 8; 16 (110) | Two 20.8R38; 8; 16 (110) |
| Ballast | —Liquid (each) | 1440 lb (653 kg) | None |
| | —Cast Iron (each) | None | None |
| Bias Ply Rear Tires | —No.; size; ply & psi (kPa) | Two 20.8-38; 8; 18 (120) | Two 20.8-38; 8; 18 (120) |
| Ballast | —Liquid (each) | 1530 lb (694 kg) | None |
| | —Cast Iron (each) | None | None |
| Front Tires | —No., size, ply & psi (kPa) | Two 10.00-16; 6; 32 (220) | Two 10.00-16; 6; 32 (220) |
| Ballast | —Liquid (each) | None | None |
| | —Cast Iron (each) | 20 lb (9 kg) | None |
| Height of drawbar | | 22.5 in (570 mm) | 22.5 in (570 mm) |
| Static weight with operator | —Rear with Radials | 12015 lb (5450 kg) | 9135 lb (4144 kg) |
| | Rear with Bias Ply | 12010 lb (5448 kg) | 8950 lb (4060 kg) |
| | Front | 3625 lb (1644 kg) | 3590 lb (1628 kg) |
| | Total with Radials | 15640 lb (7094) | 12725 lb (5772 kg) |
| | Total with Bias Ply | 15640 lb (7094 kg) | 12540 lb (5688 kg) |

## 4440 DIESEL
## NEBRASKA TRACTOR TEST 1265
(Continued)

### DRAWBAR PERFORMANCE

| Power Hp (kW) | Drawbar pull lbs (kN) | Speed mph (km/h) | Crank-shaft speed rpm | Slip % | Fuel Consumption gal/hr (l/h) | lb/hp.hr (kg/kW.h) | Hp.hr/gal (kW.h/l) | Temp. °F (°C) Cooling med | Air wet bulb | Air dry bulb | Barom. inch Hg (kPa) |
|---|---|---|---|---|---|---|---|---|---|---|---|
| **Maximum Available Power—Two Hours 6th (C-1) Gear** | | | | | | | | | | | |
| 108.59 (80.98) | 8106 (36.06) | 5.02 (8.08) | 2201 | 5.75 | 8.559 (32.401) | 0.549 (0.334) | 12.69 (2.499) | 184 (84.4) | 47 (8.1) | 56 (13.1) | 28.640 (96.713) |
| **75% of Pull at Maximum Power—Ten Hours 6th (C-1) Gear** | | | | | | | | | | | |
| 89.52 (66.76) | 6311 (28.07) | 5.32 (8.56) | 2293 | 4.18 | 7.565 (28.638) | 0.589 (0.358) | 11.83 (2.331) | 181 (82.8) | 36 (2.2) | 45 (7.4) | 28.909 (97.621) |
| **50% of Pull at Maximum Power—Two Hours 6th (C-1) Gear** | | | | | | | | | | | |
| 61.11 (45.57) | 4213 (18.74) | 5.44 (8.75) | 2313 | 2.83 | 6.173 (23.367) | 0.704 (0.428) | 9.90 (1.950) | 179 (81.7) | 43 (6.1) | 53 (11.4) | 28.660 (96.781) |
| **50% of Pull at Reduced Engine Speed—Two Hours 11th (C-3) Gear** | | | | | | | | | | | |
| 60.95 (45.45) | 4203 (18.70) | 5.44 (8.75) | 1400 | 2.87 | 4.450 (16.846) | 0.509 (0.309) | 13.70 (2.698) | 180 (82.2) | 42 (5.3) | 44 (6.7) | 28.680 (96.848) |

### MAXIMUM POWER IN SELECTED GEARS

| Power Hp (kW) | Drawbar pull lbs (kN) | Speed mph (km/h) | Crank-shaft speed rpm | Slip % | Gear | Cooling med | Air wet bulb | Air dry bulb | Barom. inch Hg (kPa) |
|---|---|---|---|---|---|---|---|---|---|
| 101.06 (75.36) | 13392 (59.57) | 2.83 (4.55) | 2263 | 14.78 | 3rd (A-3) Gear | 180 (81.9) | 20 (−6.6) | 22 (−5.5) | 29.380 (99.212) |
| 111.39 (83.06) | 10027 (44.60) | 4.17 (6.70) | 2200 | 7.22 | 5th (B-1) Gear | 185 (85.0) | 46 (7.8) | 54 (12.2) | 28.610 (96.612) |
| 112.56 (83.94) | 8406 (37.39) | 5.02 (8.08) | 2202 | 5.75 | 6th (C-1) Gear | 185 (85.0) | 45 (7.2) | 53 (11.7) | 28.600 (96.578) |
| 111.59 (83.21) | 7736 (34.41) | 5.41 (8.70) | 2200 | 5.15 | 7th (B-2) Gear | 186 (85.6) | 46 (7.8) | 55 (12.8) | 28.620 (96.645) |
| 111.84 (83.40) | 6479 (28.82) | 6.47 (10.42) | 2200 | 4.24 | 8th (C-2) Gear | 186 (85.6) | 47 (8.3) | 56 (13.3) | 28.630 (96.679) |
| 112.41 (83.82) | 5904 (26.26) | 7.14 (11.49) | 2201 | 3.93 | 9th (B-3) Gear | 185 (85.0) | 47 (8.3) | 57 (13.9) | 28.640 (96.713) |

### LUGGING ABILITY IN RATED GEAR 6th (C-1)

| Crankshaft Speed rpm | 2202 | 1980 | 1759 | 1542 | 1315 | 1084 |
|---|---|---|---|---|---|---|
| Pull—lbs (kN) | 8406 (37.39) | 9167 (40.78) | 9662 (42.98) | 10764 (47.88) | 10749 (47.81) | 9799 (43.59) |
| Increase in Pull % | 0 | 9 | 15 | 28 | 28 | 17 |
| Power—Hp (kW) | 112.56 (83.94) | 109.63 (81.75) | 102.09 (76.13) | 98.61 (73.53) | 83.77 (62.47) | 63.79 (47.57) |
| Speed—Mph (km/h) | 5.02 (8.08) | 4.48 (7.22) | 3.96 (6.38) | 3.44 (5.53) | 2.92 (4.70) | 2.44 (3.93) |
| Slip % | 5.75 | 6.56 | 6.86 | 8.01 | 8.15 | 6.86 |

| TRACTOR SOUND LEVEL WITH CAB | dB(A) |
|---|---|
| Maximum Available Power—Two Hours | 78.0 |
| 75% of Pull at Maximum Power—Ten Hours | 78.0 |
| 50% of Pull at Maximum Power—Two Hours | 79.0 |
| 50% of Pull at Reduced Engine Speed—Two Hours | 75.5 |
| Bystander in 16th (D-4) gear | 88.0 |

| TIRES, BALLAST AND WEIGHT | | With Ballast | Without Ballast |
|---|---|---|---|
| Rear Tires | —No., size, ply & psi (kPa) | Four 18.4-38; 6; 14 (100) | Four 18.4-38; 6; 14 (100) |
| Ballast | —Liquid (each inner) | 1115 lb (506 kg) | None |
| | —Cast Iron (each) | None | None |
| Front Tires | —No., size, ply & psi (kPa) | Two 11.00-16; 8; 40 (280) | Two 11.00-16; 8; 40 (280) |
| Ballast | —Liquid (each) | None | None |
| | —Cast Iron (each) | 20 lb (9 kg) | None |
| Height of drawbar | | 22.5 in (570 mm) | 22.5 in (570 mm) |
| Static weight with operator | —rear | 11790 lb (5348 kg) | 9560 lb (4338 kg) |
| | front | 3750 lb (1701 kg) | 3710 lb (1683 kg) |
| | total | 15540 lb (7049 kg) | 13270 lb (6021 kg) |

**TABLE 26-7. Comparison of Annual Fuel Cost of Three Tractors Operated 500 h per Year**

| Model | dhp [kW] | gal/h | Annual Use, h | Total gal | Cost/gal | Annual Fuel Use |
|---|---|---|---|---|---|---|
| 1086 | 70.99 [52.9] | 5.965 | 500 | 2982.5 | $1 | $2982.50 |
| 4440 | 69.97 [52.2] | 5.967 | 500 | 2983.5 | 1 | 2983.50 |
| 9700 | 73.75 [55.0] | 6.265 | 500 | 3132.5 | 1 | 3132.50 |

**Tractor Comparison** *Nebraska Tractor Test Reports* data are especially useful for making comparisons of makes of tractors which meet the calculated power requirement needs. In the previous example of determining drawbar horsepower [kW], a 5-bottom moldboard plow equipped with 16-in [40.6-cm] bottoms and pulled in loam soil at 5 mi/h [8.05 km/h] would require an 88-dhp [65.6-kW] tractor. The power requirement is classified as heavy-duty tillage. Therefore, *Nebraska Tractor Test* data under "Drawbar Performance, 75% of Pull at Maximum Power" would be consulted when seeking appropriate performance information about a tractor to be used for plowing.

Nebraska tractor tests for three tractors have been assembled in Table 26-6. These are Tests 1243, 1247, and 1265. The tractors have turbocharged diesel engines of 401, 414, and 466 cubic inch [6.6, 6.8, and 7.6 liter] displacement, respectively. The horsepower [kW] ratings at 75 percent of maximum pull are 94.14, 89.78, and 89.52 dhp [70.2, 66.9, and 66.76 kW], respectively. All three tractors are within the horsepower [kW] requirements established for 88 dhp [66 kW] and provide a reserve drawbar horsepower of 25 percent for tough spots in plowing.

There is very little difference in performance between the three tractors at the standard PTO speed of 1000 r/min except that Model 1086, Test 1247, produces 16.36 hp·h/gal [3.22 kW·h/L] fuel efficiency, which is the highest of the three and approximately average in rating. None of the three tractors has especially good fuel efficiency in drawbar power performance, although Model 1086 has the highest drawbar horsepower-hours per gallon [kW·h/L].

Perhaps the most significant difference between the three tractors is that the tractor engine for Model 9700, Test 1243, produced a maximum increase in lugging power of 13 percent, compared to 32 percent for Model 1086, Test 1247, and 28 percent for Model 4440, Test 1265.

The weight of 17,480 lb [7929 kg] for tractor Model 9700 was approximately 1850 lb [839 kg] heavier than the other two. The weight difference may have had some effect on the limited torque increase.

Fuel cost for each tractor, based on 500 h of use per year and using the average "PTO varying power and fuel consumption" is summarized in Table 26-7. In this example, the less efficient tractor would cost approximately $150 more to operate for 500 h compared with the other two tractors.

In summary, each of the three tractors would meet the drawbar power needed for tillage. In addition, each tractor would provide an adequate reserve of power for heavy draft situations and up to 30 hp [22.4 kW] for PTO-driven machinery. Model 1086 has superior fuel economy when operating at both 50 and 75 percent of pull at maximum power and at 50 percent of pull at reduced engine speeds. The lugging ability of the engine was outstanding. The increase in lugging power provided, from the increase in torque, is equivalent to another set of transmission gears. Note that the 4440 and 9700 tractors were tested using duals (four), while the 1086 was equipped with single (two) rear tires. Therefore, the tire surface area in soil contact was less for the 1086 than the 4440 even though they were of nearly equal total weight. The result might be more soil compaction with the 1086.

## SUMMARY

As a salesperson in power and machinery, you will have the opportunity to work closely with owners/managers to select machinery for agricultural production operations. Machinery management is the decision-making process which balances the choice of size and type of machine with the cost of ownership and operation.

Selecting the proper type and size of a machine to meet production needs must include an evaluation of timeliness requirements and efficiency of the machine working in the field.

Choosing the proper size and type of tractor is based upon the amount and type of power needed, the operating efficiency, and the characteristics of the field operations. Size of tractor can be determined by calculating the amount of horsepower [kW] required to meet draft or power takeoff needs. *Nebraska Tractor Test Reports* provides a great variety of unbiased data on tractor performance. This information can be invaluable to the owner/manager when making comparisons of tractors for selection purposes.

## THINKING IT THROUGH

1. What are the two terms that are most often used to describe machinery capacity?
2. What is the field travel speed of a 7 × 18 grain drill that is pulled across a field 1/4 mile [0.4 km] long in 3 min?

3. From question 2 what is the theoretical field capacity of the grain drill?
4. The operator of the grain drill in question 2 planted a strip 1320 ft [402.3 m] wide in 8 h. What is the field efficiency of the operation?
5. How large a combine should be purchased if an operator wishes to harvest 90 acres [36 ha] per 10-h day traveling at 6 mi/h [9.7 km/h]?
6. What size tractor is required to pull a 10-ft [3.04 m] offset disk traveling 5 mi/h [8.05 km/h] maximum depth?
7. What size tractor is required to operate a baler that is baling 10 tons/h [9.07 metric tons/h] at an average speed of 4 mi/h [6.44 km/h]?
8. What use can be made of the *Nebraska Tractor Test Reports* by a machinery salesperson? A tractor service manager?
9. Which of the tractor tests found in the *Nebraska Tractor Test Reports* evaluate tractor performance for each of the following?
   (a) Power available to pull heavy tillage tools
   (b) Power at standard PTO speed
   (c) Fuel efficiency when pulling light loads at less than rated engine rpm
   (d) Torque characteristics of the engine
   (e) Amount of fuel required per year
10. What is the efficiency rating of a tractor that develops 60 hp [44.7 kW] on 12 gal [45.4 L] of diesel fuel when operated on the PTO at varying power for 2 h?
11. What data in the *Nebraska Tractor Test Reports* should be evaluated when comparing one tractor with another?

**TABLE A-1. Conversion Factors for U.S. Customary to Metric (SI) Units**

| Quantity | U.S. Customary Unit | Symbol | Multiply by | To Obtain Metric (SI) | Symbol |
|---|---|---|---|---|---|
| **Length** | Inches | in | 25.4 | Millimeters | mm |
| | Inches | in | 2.54 | Centimeters | cm |
| | Inches | in | 0.02540 | Meters | m |
| | Feet | ft | 0.30481 | Meters | m |
| | Yards | yd | 0.9144 | Meters | m |
| | Miles | mi | 1.6093 | Kilometers | km |
| **Area** | Square inches | $in^2$ | 0.000645 | Square meters | $m^2$ |
| | Square feet | $ft^2$ | 0.0929 | Square meters | $m^2$ |
| | Acres | | 4046.8 | Square meters | $m^2$ |
| | Acres | | 0.4047 | Hectares | ha |
| **Force** | Pounds-force | lbf | 4.4482 | Newtons | N |
| | Pounds-force | lbf | 0.004482 | Kilonewtons | kN |
| **Torque** | Pounds-feet | lb•ft | 1.3558 | Newton-meters | N•m |
| | Pounds-inches | lb•in | 0.112985 | Newton-meters | N•m |
| **Mass** (Weight) | Pounds | lb | 0.4536 | Kilograms | kg |
| | Ounces | oz | 28.3495 | Grams | g |
| **Pressure** | Pounds-force per square inches | $lbf/in^2$ | 6.8948 | Kilopascals | kPa |
| | Pounds-force per square feet | $lbf/ft^2$ | 0.04788 | Kilopascals | kPa |
| **Velocity** | Feet per second | ft/s | 0.3048 | Meters/second | m/s |
| | Feet per minute | ft/min | 0.00508 | Meters/second | m/s |
| | Miles per hour | mi/h | 0.4470 | Meters/second | m/s |
| | Miles per hour | mi/h | 1.6093 | Kilometers/hour | km/h |
| **Flow, Volume** | Ounces | oz | 29.574 | Milliliters | mL |
| | Pints | pt | 0.004732 | Cubic meters | $m^3$ |
| | Gallons | gal | 0.003785 | Cubic meters | $m^3$ |
| | Pounds per acre | lb/acre | 1.12 | Kilograms/hectare | kg/ha |
| | Gallons per acre | gal/acre | 9.354 | Liters/hectare | L/ha |
| | Gallons per hour | gal/h | 3.7854 | Liters/hour | L/h |
| | Bushels (USA) | bu | 0.03524 | Cubic meters | $m^3$ |
| | Bushels per acre | bu/acre | 0.087 | Cubic meters/hectare | $m^3ha$ |
| **Rate** | Horsepower-hours per gallon | hp•h/gal | 0.197 | Kilowatt-hours/liter | kW•h/L |
| | Pounds per horsepower-hour | lb/hp•h | 0.6083 | Kilograms/kilowatt-hour | kg/kW•h |
| **Power** | Horsepower | hp | 0.7457 | Kilowatts | kW |
| **Temperature** | Fahrenheit | °F | $\left(\frac{t°F - 32}{1.8}\right)$ | Celsius | °C |

**TABLE B-1. Approximate Plant Population per Acre (Hectare) for Corn Seed Using a 16-cell Seed Plate. (Metric Equivalents are Shown in Parentheses.)**

CHAIN DIAGRAM

DRIVEN 1 DRIVER 2

Left Right

DRIVER 1 (10) (16) DRIVEN 2 (10) (17)

Plant Populations listed are Based on 95 inches (2413 mm) of Travel per Revolution of a 7.60 x 14 or 9.5L x 14 Planter Tires. Values will vary slightly with Air Pressure, Tire Size and Soil Condition.

CHART BASED ON 1 SEED PER CELL

| EFFECTIVE TEETH SPROCKETS | | | | Recommended Speed | SEED SPACING | ROW SPACING | | |
|---|---|---|---|---|---|---|---|---|
| DRIVER 1 | DRIVEN 1 | DRIVER 2 | DRIVEN 2 | mi/h (km/h) | | 30 (762) | 38 (965) | 40 (1016) |
| 16. | 10. | 17. | 10. | 2.1 (3.4) | 4.21 (106.93) | 49666. (122727.) | 39210. (96890.) | 37249. (92046.) |
| 16. | 10. | 16. | 10. | 2.3 (3.6) | 4.47 (113.61) | 46744. (115508.) | 36903. (91191.) | 35058. (86631.) |
| 16. | 10. | 15. | 10. | 2.4 (3.9) | 4.77 (121.19) | 43823. (108289.) | 34597. (85491.) | 32867. (81217.) |
| 16. | 12. | 17. | 10. | 2.6 (4.1) | 5.05 (128.32) | 41388. (102273.) | 32675. (80742.) | 31041. (76705.) |
| 16. | 12. | 16. | 10. | 2.7 (4.4) | 5.37 (136.34) | 38954. (96257.) | 30753. (75992.) | 29215. (72193.) |
| 16. | 12. | 15. | 10. | 2.9 (4.7) | 5.73 (145.43) | 36519. (90241.) | 28831. (71243.) | 27389. (67681.) |
| 16. | 10. | 12. | 10. | 3.0 (4.8) | 5.96 (151.49) | 35058. (86631.) | 27678. (68393.) | 26294. (64973.) |
| 16. | 12. | 14. | 10. | 3.1 (5.0) | 6.13 (155.81) | 34084. (84225.) | 26909. (66493.) | 25563. (63168.) |
| 16. | 14. | 16. | 10. | 3.2 (5.1) | 6.26 (159.06) | 33389. (82506.) | 26360. (65136.) | 25042. (61879.) |
| 16. | 15. | 17. | 10. | 3.2 (5.1) | 6.31 (160.40) | 33111. (81818.) | 26140. (64593.) | 24833. (61364.) |
| 16. | 10. | 11. | 10. | 3.3 (5.3) | 6.51 (165.26) | 32137. (79412.) | 25371. (62694.) | 24103. (59559.) |
| 16. | 11. | 12. | 10. | 3.3 (5.3) | 6.56 (166.63) | 31871. (78756.) | 25161. (62175.) | 23903. (59067.) |
| 16. | 14. | 15. | 10. | 3.4 (5.4) | 6.68 (169.66) | 31302. (77349.) | 24712. (61065.) | 23477. (58012.) |
| 16. | 15. | 16. | 10. | 3.4 (5.5) | 6.71 (170.42) | 31163. (77005.) | 24602. (60794.) | 23372. (57754.) |
| 16. | 16. | 17. | 10. | 3.4 (5.5) | 6.74 (171.09) | 31041. (76705.) | 24506. (60556.) | 23281. (57528.) |
| 16. | 10. | 17. | 17. | 3.6 (5.8) | 7.16 (181.78) | 29215. (72193.) | 23065. (56994.) | 21911. (54144.) |
| 10. | 11. | 17. | 10. | 3.7 (6.0) | 7.41 (188.20) | 28219. (69731.) | 22278. (55051.) | 21164. (52299.) |
| 16. | 10. | 16. | 17. | 3.8 (6.2) | 7.60 (193.14) | 27497. (67946.) | 21708. (53642.) | 20623. (50959.) |
| 16. | 16. | 15. | 10. | 3.9 (6.2) | 7.63 (193.90) | 27383. (67661.) | 21623. (53432.) | 20542. (50760.) |
| 16. | 15. | 14. | 10. | 3.9 (6.2) | 7.67 (194.77) | 27268. (67380.) | 21527. (53195.) | 20451. (50535.) |
| 16. | 12. | 11. | 10. | 3.9 (6.3) | 7.81 (198.31) | 26781. (66177.) | 21143. (52245.) | 20085. (49632.) |
| 16. | 11. | 17. | 17. | 4.0 (6.4) | 7.87 (199.96) | 26559. (65630.) | 20968. (51813.) | 19919. (49222.) |
| 10. | 12. | 17. | 10. | 4.1 (6.6) | 8.08 (205.31) | 25868. (63920.) | 20422. (50464.) | 19401. (47940.) |
| 16. | 10. | 15. | 17. | 4.1 (6.6) | 8.11 (206.02) | 25778. (63699.) | 20351. (50289.) | 19334. (47774.) |
| 16. | 16. | 14. | 10. | 4.1 (6.6) | 8.18 (207.75) | 25563. (63168.) | 20182. (49870.) | 19173. (47376.) |
| 16. | 14. | 12. | 10. | 4.2 (6.8) | 8.35 (212.08) | 25042. (61879.) | 19770. (48852.) | 18781. (46410.) |
| 10. | 11. | 15. | 10. | 4.2 (6.8) | 8.40 (213.29) | 24899. (61528.) | 19657. (48575.) | 18675. (46146.) |
| 16. | 12. | 17. | 17. | 4.3 (7.0) | 8.59 (218.14) | 24346. (60160.) | 19221. (47495.) | 18260. (45120.) |
| 16. | 10. | 14. | 17. | 4.4 (7.1) | 8.69 (220.74) | 24060. (59453.) | 18994. (46936.) | 18045. (44590.) |
| 16. | 11. | 15. | 17. | 4.5 (7.3) | 8.92 (226.62) | 23435. (57908.) | 18501. (45717.) | 17576. (43431.) |
| 10. | 11. | 14. | 10. | 4.5 (7.3) | 9.00 (228.53) | 23239. (57426.) | 18347. (45336.) | 17430. (43069.) |
| 16. | 14. | 11. | 10. | 4.6 (7.4) | 9.11 (231.36) | 22955. (56723.) | 18122. (44781.) | 17216. (42542.) |
| 10. | 12. | 15. | 10. | 4.6 (7.4) | 9.16 (232.68) | 22824. (56400.) | 18019. (44527.) | 17118. (42300.) |
| 10. | 14. | 17. | 10. | 4.8 (7.7) | 9.43 (239.53) | 22172. (54789.) | 17504. (43254.) | 16629. (41092.) |
| 16. | 16. | 12. | 10. | 4.8 (7.8) | 9.54 (242.38) | 21911. (54144.) | 17299. (42746.) | 16434. (40608.) |
| 16. | 12. | 15. | 17. | 4.9 (7.9) | 9.73 (247.22) | 21482. (53083.) | 16959. (41907.) | 16111. (39812.) |
| 10. | 12. | 14. | 10. | 5.0 (8.0) | 9.82 (249.30) | 21303. (52640.) | 16818. (41558.) | 15977. (39480.) |
| 16. | 14. | 17. | 17. | 5.1 (8.1) | 10.02 (254.50) | 20868. (51566.) | 16475. (40710.) | 15651. (38675.) |
| 10. | 15. | 17. | 10. | 5.1 (8.2) | 10.10 (256.63) | 20694. (51136.) | 16337. (40371.) | 15521. (38352.) |
| 16. | 10. | 12. | 17. | 5.1 (8.2) | 10.14 (257.53) | 20623. (50959.) | 16281. (40231.) | 15467. (38220.) |
| 16. | 15. | 11. | 10. | 5.3 (8.5) | 10.41 (264.41) | 20085. (49632.) | 15857. (39183.) | 15064. (37224.) |
| 10. | 11. | 12. | 10. | 5.3 (8.5) | 10.50 (266.61) | 19919. (49222.) | 15726. (38860.) | 14940. (36917.) |
| 16. | 14. | 16. | 17. | 5.4 (8.7) | 10.65 (270.40) | 19641. (48533.) | 15506. (38315.) | 14730. (36400.) |
| 10. | 14. | 15. | 10. | 5.4 (8.7) | 10.69 (271.46) | 19564. (48343.) | 15445. (38166.) | 14673. (36257.) |
| 16. | 15. | 17. | 17. | 5.4 (8.7) | 10.74 (272.67) | 19477. (48128.) | 15376. (37996.) | 14608. (36096.) |
| 10. | 16. | 17. | 10. | 5.4 (8.8) | 10.78 (273.74) | 19401. (47940.) | 15316. (37848.) | 14551. (35955.) |
| 16. | 10. | 11. | 17. | 5.6 (9.0) | 11.06 (280.94) | 18904. (46713.) | 14924. (36879.) | 14178. (35035.) |
| 16. | 11. | 12. | 17. | 5.6 (9.1) | 11.15 (283.28) | 18748. (46327.) | 14801. (36574.) | 14061. (34745.) |
| 16. | 14. | 15. | 17. | 5.7 (9.2) | 11.36 (286.43) | 18413. (45500.) | 14537. (35921.) | 13810. (34125.) |
| 16. | 15. | 16. | 17. | 5.8 (9.3) | 11.41 (289.72) | 18331. (45297.) | 14472. (35761.) | 13748. (33973.) |
| 16. | 16. | 17. | 17. | 5.8 (9.3) | 11.45 (290.85) | 18260. (45120.) | 14415. (35621.) | 13695. (33840.) |
| 10. | 17. | 16. | 10. | 6.1 (9.8) | 12.17 (309.03) | 17185. (42466.) | 13567. (33526.) | 12889. (31850.) |
| 10. | 16. | 15. | 10. | 6.2 (9.9) | 12.21 (310.24) | 17118. (42300.) | 13514. (33395.) | 12839. (31725.) |
| 10. | 15. | 14. | 10. | 6.2 (10.0) | 12.27 (311.63) | 17042. (42112.) | 13454. (33247.) | 12782. (31584.) |
| 10. | 12. | 11. | 10. | 6.3 (10.2) | 12.49 (317.29) | 16738. (41360.) | 13214. (32653.) | 12553. (31020.) |
| 10. | 11. | 17. | 17. | 6.4 (10.3) | 12.60 (319.94) | 16600. (41019.) | 13105. (32383.) | 12450. (30764.) |
| 16. | 17. | 16. | 17. | 6.5 (10.5) | 12.93 (328.35) | 16175. (39968.) | 12769. (31554.) | 12131. (29976.) |
| 16. | 16. | 15. | 17. | 6.6 (10.5) | 12.98 (329.63) | 16111. (39812.) | 12719. (31431.) | 12084. (29859.) |
| 16. | 15. | 14. | 17. | 6.6 (10.6) | 13.04 (331.10) | 16040. (39635.) | 12663. (31291.) | 12030. (29726.) |
| 10. | 16. | 14. | 10. | 6.6 (10.6) | 13.09 (332.40) | 15977. (39480.) | 12613. (31169.) | 11983. (29610.) |
| 16. | 12. | 11. | 17. | 6.7 (10.8) | 13.27 (337.12) | 15753. (38927.) | 12437. (30732.) | 11815. (29196.) |
| 10. | 14. | 12. | 10. | 6.7 (10.9) | 13.36 (339.33) | 15651. (38675.) | 12356. (30533.) | 11738. (29006.) |
| 10. | 12. | 17. | 17. | 6.9 (11.2) | 13.74 (349.02) | 15216. (37600.) | 12013. (29684.) | 11412. (28200.) |
| 16. | 17. | 15. | 17. | 7.0 (11.2) | 13.79 (350.24) | 15164. (37470.) | 11971. (29582.) | 11373. (28103.) |
| 16. | 16. | 14. | 17. | 7.0 (11.3) | 13.90 (353.18) | 15037. (37158.) | 11872. (29335.) | 11278. (27868.) |
| 16. | 14. | 12. | 17. | 7.2 (11.5) | 14.19 (360.54) | 14730. (36400.) | 11629. (28737.) | 11048. (27300.) |
| 10. | 11. | 15. | 17. | 7.2 (11.6) | 14.28 (362.60) | 14647. (36193.) | 11563. (28573.) | 10985. (27145.) |
| 10. | 14. | 11. | 10. | 7.4 (11.9) | 14.57 (370.19) | 14347. (35452.) | 11326. (27988.) | 10760. (26589.) |
| 16. | 17. | 14. | 17. | 7.5 (12.0) | 14.77 (375.25) | 14153. (34972.) | 11173. (27610.) | 10615. (26229.) |
| 16. | 15. | 12. | 17. | 7.7 (12.4) | 15.21 (386.29) | 13748. (33973.) | 10854. (26821.) | 10311. (25480.) |
| 10. | 16. | 12. | 10. | 7.7 (12.4) | 15.27 (387.80) | 13695. (33840.) | 10812. (26716.) | 10271. (25380.) |
| 16. | 14. | 11. | 17. | 7.8 (12.6) | 15.48 (393.31) | 13503. (33366.) | 10660. (26342.) | 10127. (25025.) |
| 10. | 12. | 15. | 17. | 7.9 (12.7) | 15.57 (395.56) | 13426. (33177.) | 10600. (26192.) | 10070. (24883.) |
| 16. | 15. | 11. | 17. | 8.1 (13.0) | 16.59 (421.41) | 12603. (31142.) | 9949. (24586.) | 9452. (23356.) |
| 16. | 16. | 11. | 17. | 8.1 (13.0) | 17.70 (449.50) | 11815. (29196.) | 9328. (23049.) | 8861. (21897.) |
| 16. | 17. | 11. | 17. | 8.1 (13.0) | 18.80 (477.59) | 11120. (27478.) | 8779. (21693.) | 8340. (20609.) |
| 16. | 17. | 10. | 17. | 8.1 (13.0) | 20.68 (525.35) | 10109. (24980.) | 7981. (19721.) | 7582. (18735.) |
| 10. | 17. | 15. | 17. | 8.1 (13.0) | 22.06 (560.38) | 9477. (23419.) | 7482. (18489.) | 7108. (17564.) |
| 10. | 12. | 10. | 17. | 8.1 (13.0) | 23.36 (593.34) | 8951. (22118.) | 7066. (17461.) | 6713. (16588.) |
| 10. | 14. | 11. | 17. | 8.1 (13.0) | 24.78 (629.30) | 8439. (20854.) | 6663. (16464.) | 6329. (15640.) |
| 10. | 15. | 11. | 17. | 8.1 (13.0) | 26.55 (674.25) | 7877. (19464.) | 6218. (15366.) | 5907. (14598.) |
| 10. | 16. | 11. | 17. | 8.1 (13.0) | 28.31 (719.20) | 7384. (18247.) | 5830. (14406.) | 5538. (13685.) |
| 10. | 17. | 11. | 17. | 8.1 (13.0) | 30.08 (764.15) | 6950. (17174.) | 5487. (13558.) | 5212. (12880.) |
| 10. | 17. | 10. | 17. | 8.1 (13.0) | 33.09 (840.56) | 6318. (15613.) | 4988. (12326.) | 4739. (11709.) |

*Courtesy* International Harvester Co.

## TABLE B-2. Seeding Chart for Double-Run Metering

SPROCKET POSITION FOR DRIVE NO.

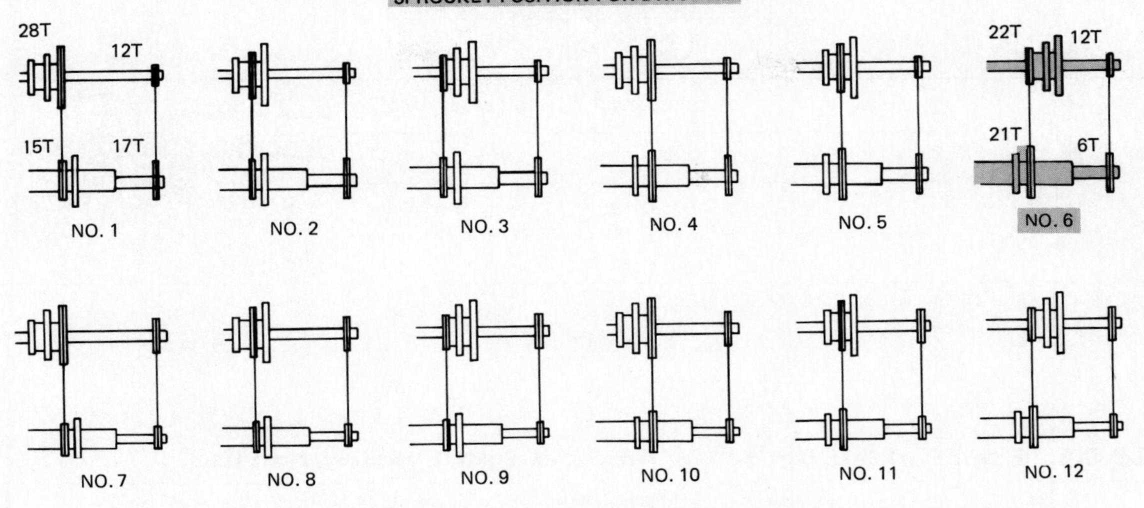


6-in [152 mm] Spacing — Small Side of Feed Wheel (Pounds Per Acre)

| Grain | Wheat | | | Rye | | | Sorghum | | | Flax | | | Buck-wheat | | Sudan Grass | | Millet | | Wheat Grass | | Alfalfa | |
|---|---|---|---|---|---|---|---|---|---|---|---|---|---|---|---|---|---|---|---|---|---|---|
| Lock Lever | 7 | 8 | 10 | 7 | 8 | 10 | 6 | 7 | 8 | 5 | 8 | 10 | 8 | 9 | 8 | 10 | 4 | 6 | 8 | 10 | 2 | 4 |
| Drive No. 1 | 12 | 16 | 24 | 10 | 12 | 19 | 8 | 10 | 13 | 5 | 11 | 17 | 13 | 18 | 13 | 20 | 5.5 | 9 | 1.6 | 4.5 | 5 | 7 |
| 2 | 13 | 18 | 27 | 11 | 14 | 21 | 9 | 11 | 15 | 6 | 13 | 19 | 15 | 20 | 15 | 22 | 6.5 | 10 | 1.9 | 5 | 6 | 8 |
| 3 | 15 | 20 | 31 | 12 | 16 | 24 | 10 | 13 | 17 | 7 | 15 | 22 | 17 | 22 | 17 | 25 | 7.5 | 12 | 2.2 | 5.8 | 7 | 9 |
| 4 | 17 | 22 | 34 | 14 | 18 | 27 | 11 | 15 | 18 | 8 | 17 | 25 | 19 | 25 | 19 | 27 | 8 | 13 | 2.5 | 6.5 | 8 | 10 |
| 5 | 19 | 25 | 38 | 16 | 20 | 30 | 12 | 17 | 20 | 9 | 19 | 28 | 22 | 28 | 21 | 30 | 9.5 | 15 | 2.9 | 7 | 9 | 12 |
| 6 | 21 | 28 | 43 | 18 | 22 | 34 | 14 | 19 | 23 | 10 | 22 | 32 | 25 | 32 | 24 | 34 | 11 | 17 | 3.4 | 8 | 11 | 14 |
| 7 | 24 | 32 | 48 | 20 | 25 | 39 | 16 | 21 | 26 | 12 | 25 | 37 | 28 | 36 | 27 | 38 | 12 | 19 | 3.9 | 9 | 12 | 16 |
| 8 | 27 | 35 | 54 | 23 | 28 | 44 | 18 | 24 | 30 | 14 | 29 | 42 | 31 | 40 | 30 | 43 | 14 | 21 | 4.4 | 10 | 14 | 18 |
| 9 | 31 | 40 | 62 | 26 | 32 | 50 | 20 | 27 | 34 | 16 | 33 | 48 | 35 | 45 | 34 | 48 | 16 | 24 | 5.1 | 12 | 16 | 20 |
| 10 | 34 | 44 | 68 | 29 | 35 | 55 | 22 | 30 | 37 | 18 | 36 | 53 | 39 | 50 | 38 | 53 | 18 | 27 | 5.7 | 13 | 18 | 22 |
| 11 | 38 | 49 | 76 | 33 | 40 | 62 | 25 | 34 | 42 | 20 | 41 | 60 | 44 | 56 | 43 | 59 | 20 | 30 | 6.5 | 14 | 20 | 25 |
| 12 | 43 | 55 | 86 | 37 | 45 | 70 | 28 | 38 | 47 | 23 | 47 | 68 | 50 | 63 | 48 | 67 | 23 | 34 | 7.5 | 16 | 23 | 29 |

6-in [152 mm] Spacing — Large Side of Feed Wheel (Pounds Per Acre)

| Grain | Wheat | | | Oats | | | Barley | | | Soybeans | | | Cow Peas | | | Peas | |
|---|---|---|---|---|---|---|---|---|---|---|---|---|---|---|---|---|---|
| Lock Lever | 11 | 13 | 15 | 14 | 16 | 18 | 12 | 14 | 16 | 14 | 15 | 16 | 12 | 14 | 16 | 18 | 20 |
| Drive No. 1 | 47 | 53 | 62 | 29 | 35 | 47 | 28 | 39 | 49 | 35 | 43 | 54 | 31 | 41 | 56 | 63 | 73 |
| 2 | 53 | 59 | 70 | 33 | 40 | 53 | 32 | 43 | 55 | 39 | 48 | 60 | 35 | 46 | 63 | 70 | 82 |
| 3 | 60 | 68 | 79 | 37 | 46 | 60 | 37 | 48 | 62 | 44 | 54 | 66 | 40 | 52 | 71 | 79 | 93 |
| 4 | 66 | 75 | 88 | 41 | 51 | 66 | 41 | 53 | 68 | 48 | 60 | 72 | 44 | 58 | 78 | 87 | 103 |
| 5 | 74 | 84 | 98 | 46 | 57 | 74 | 47 | 59 | 76 | 53 | 67 | 80 | 50 | 65 | 87 | 98 | 115 |
| 6 | 84 | 95 | 112 | 52 | 66 | 84 | 54 | 67 | 86 | 60 | 76 | 89 | 56 | 74 | 99 | 111 | 131 |
| 7 | 94 | 107 | 126 | 58 | 75 | 95 | 61 | 75 | 97 | 67 | 85 | 100 | 63 | 83 | 112 | 125 | 147 |
| 8 | 105 | 120 | 142 | 65 | 84 | 106 | 69 | 84 | 109 | 76 | 95 | 111 | 71 | 93 | 125 | 140 | 165 |
| 9 | 120 | 137 | 161 | 74 | 96 | 121 | 79 | 95 | 124 | 86 | 108 | 124 | 81 | 106 | 142 | 160 | 187 |
| 10 | 132 | 151 | 178 | 82 | 106 | 133 | 88 | 104 | 136 | 94 | 119 | 136 | 89 | 117 | 156 | 176 | 206 |
| 11 | 147 | 169 | 199 | 92 | 119 | 149 | 99 | 117 | 152 | 106 | 133 | 151 | 100 | 131 | 175 | 197 | 231 |
| 12 | 167 | 192 | 226 | 104 | 136 | 170 | 113 | 132 | 173 | 120 | 151 | 171 | 113 | 149 | 198 | 224 | 262 |

*Courtesy* Deere & Co.

## TABLE C-2. Flat-Even Fan Spray Tip, Angle, Height, Spacing, and Use

| Spray Pattern | Spray Angle Degrees | Nozzle Height, in [mm] 20-in [508 mm] Spacing | 30-in [762 mm] Spacing | Principal Use |
|---|---|---|---|---|
|  | 65 | 21–23 [533–584] | 32–34 [813–864] |  A typical broadcast application |
| | 73 | 20–22 [508–559] | 27–29 [686–737] | |
| | 80 | 17–19 [432–483] | 24–26 [610–660] | |
| | 110 | 10–12 [254–305] | 13–15 [330–381] | |

*Courtesy* Delevan Corp. and Spray Systems Co.

## TABLE C-2. Even Flat-Fan Spray Tip, Angle, Height, Spacing, and Use

| Shape | Spray Pattern | | Principal Use |
|---|---|---|---|
|  | 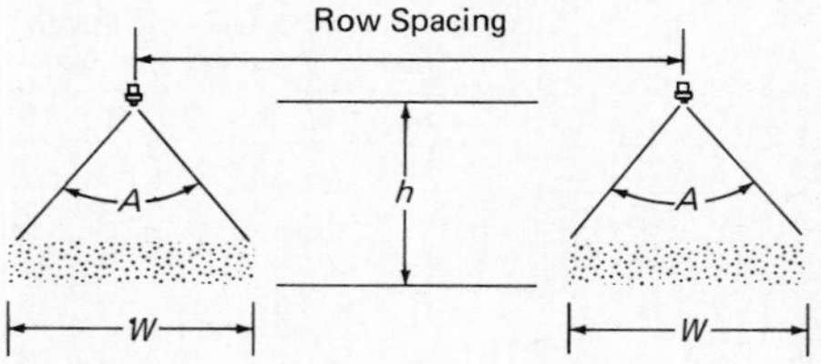 Row Spacing; A; h; W | | 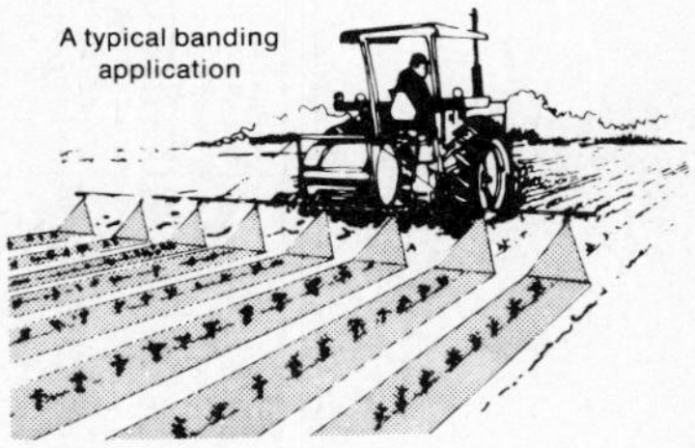 A typical banding application |
| Band Width (*W*), in [mm] | Tip Height (*h*) in [mm] 80° Angle (*A*) | 95° Angle (*A*) | |
| 8 [203] | 5 [127] | 4 [102] | |
| 10 [254] | 6 [152] | 5 [127] | |
| 12 [305] | 7 [178] | 6 [152] | |
| 14 [356] | 8 [203] | 7 [178] | |

*Courtesy* Delevan Corp. and Spray Systems Co.

## TABLE C-3. Flooding Flat-Spray Tips Specification for Speed of Travel, Boom Height, and Tip Spacing

| Nozzle | Speed of Travel, mi/h [km/h] | Tip Spacing, in [cm] 40 [102] | 80 [203] | 120 [305] | Typical Use |
|---|---|---|---|---|---|
| | | Recommended Tip Height, in [cm] | | | |
| 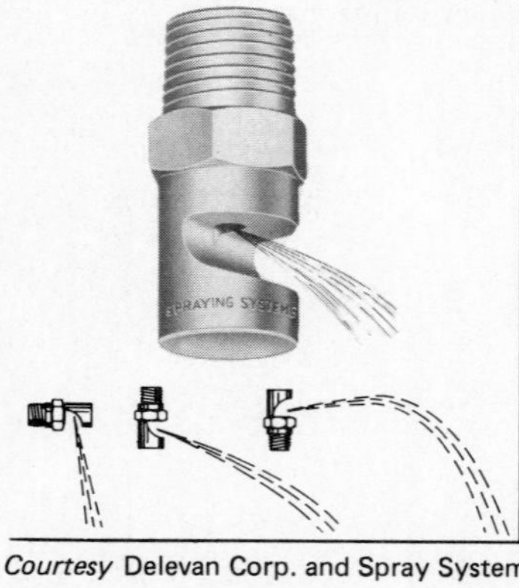 SPRAYING SYSTEMS | 4–20 [6.4–32] | 12–15 [30–38] | 24–30 [61–76] | 36–45 [91–114] | A typical flooding nozzle application.  |

*Courtesy* Delevan Corp. and Spray Systems Co.

**TABLE C-4. Flat-Spray Tip, 73° Series Identification and Performance Data for Delevan and Spray Systems Companies. Size in GPM at 20-60 psi and GPA at 4-10 mi/h with Correction Factor for Density of Solution.**

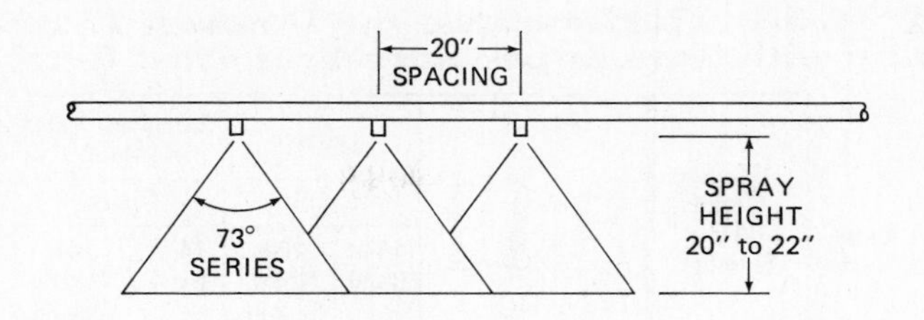


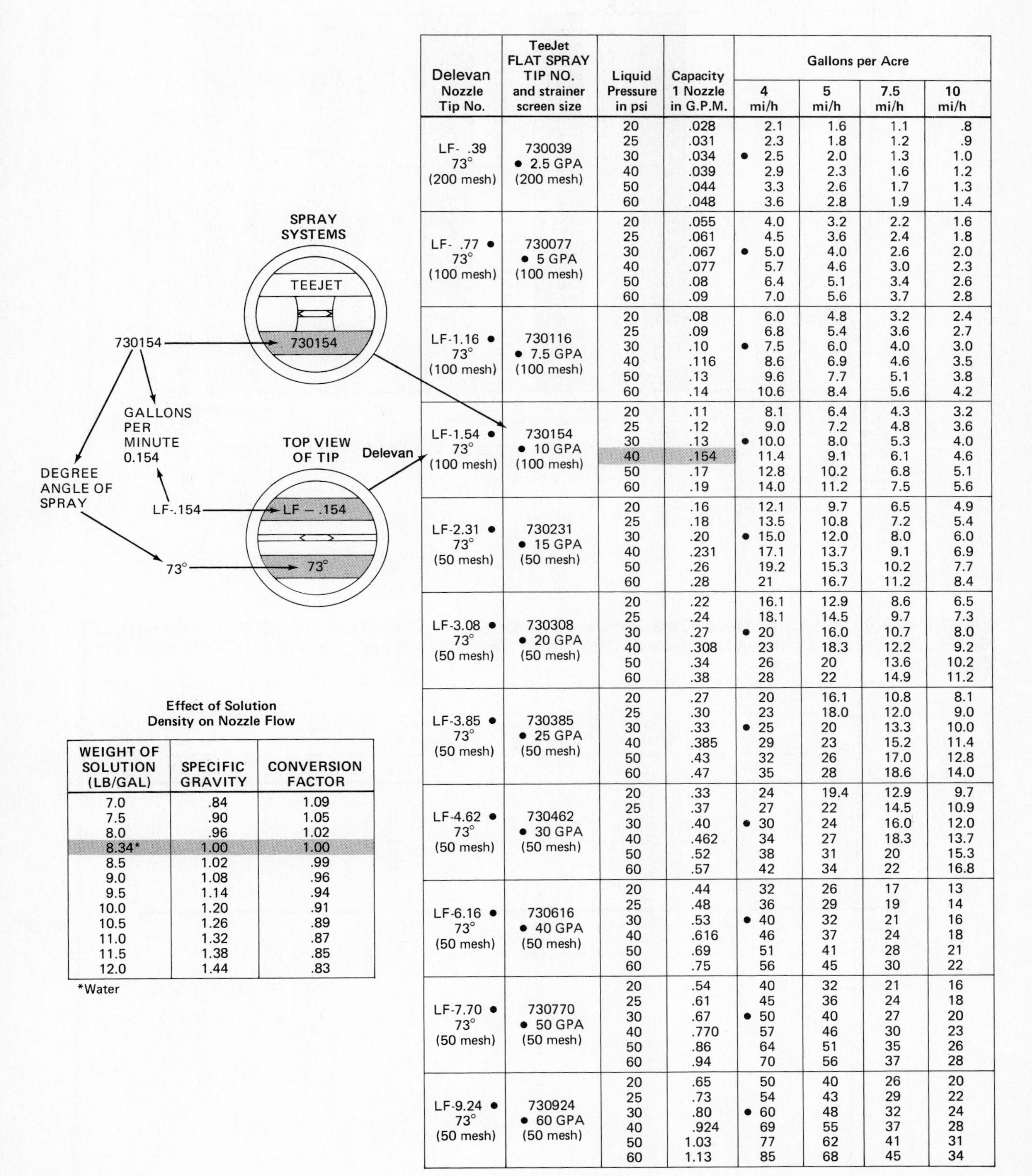


| Delevan Nozzle Tip No. | TeeJet FLAT SPRAY TIP NO. and strainer screen size | Liquid Pressure in psi | Capacity 1 Nozzle in G.P.M. | Gallons per Acre 4 mi/h | 5 mi/h | 7.5 mi/h | 10 mi/h |
|---|---|---|---|---|---|---|---|
| LF- .39 73° (200 mesh) | 730039 • 2.5 GPA (200 mesh) | 20 | .028 | 2.1 | 1.6 | 1.1 | .8 |
| | | 25 | .031 | 2.3 | 1.8 | 1.2 | .9 |
| | | 30 | .034 | • 2.5 | 2.0 | 1.3 | 1.0 |
| | | 40 | .039 | 2.9 | 2.3 | 1.6 | 1.2 |
| | | 50 | .044 | 3.3 | 2.6 | 1.7 | 1.3 |
| | | 60 | .048 | 3.6 | 2.8 | 1.9 | 1.4 |
| LF- .77 • 73° (100 mesh) | 730077 • 5 GPA (100 mesh) | 20 | .055 | 4.0 | 3.2 | 2.2 | 1.6 |
| | | 25 | .061 | 4.5 | 3.6 | 2.4 | 1.8 |
| | | 30 | .067 | • 5.0 | 4.0 | 2.6 | 2.0 |
| | | 40 | .077 | 5.7 | 4.6 | 3.0 | 2.3 |
| | | 50 | .08 | 6.4 | 5.1 | 3.4 | 2.6 |
| | | 60 | .09 | 7.0 | 5.6 | 3.7 | 2.8 |
| LF-1.16 • 73° (100 mesh) | 730116 • 7.5 GPA (100 mesh) | 20 | .08 | 6.0 | 4.8 | 3.2 | 2.4 |
| | | 25 | .09 | 6.8 | 5.4 | 3.6 | 2.7 |
| | | 30 | .10 | • 7.5 | 6.0 | 4.0 | 3.0 |
| | | 40 | .116 | 8.6 | 6.9 | 4.6 | 3.5 |
| | | 50 | .13 | 9.6 | 7.7 | 5.1 | 3.8 |
| | | 60 | .14 | 10.6 | 8.4 | 5.6 | 4.2 |
| LF-1.54 • 73° (100 mesh) | 730154 • 10 GPA (100 mesh) | 20 | .11 | 8.1 | 6.4 | 4.3 | 3.2 |
| | | 25 | .12 | 9.0 | 7.2 | 4.8 | 3.6 |
| | | 30 | .13 | • 10.0 | 8.0 | 5.3 | 4.0 |
| | | 40 | .154 | 11.4 | 9.1 | 6.1 | 4.6 |
| | | 50 | .17 | 12.8 | 10.2 | 6.8 | 5.1 |
| | | 60 | .19 | 14.0 | 11.2 | 7.5 | 5.6 |
| LF-2.31 • 73° (50 mesh) | 730231 • 15 GPA (50 mesh) | 20 | .16 | 12.1 | 9.7 | 6.5 | 4.9 |
| | | 25 | .18 | 13.5 | 10.8 | 7.2 | 5.4 |
| | | 30 | .20 | • 15.0 | 12.0 | 8.0 | 6.0 |
| | | 40 | .231 | 17.1 | 13.7 | 9.1 | 6.9 |
| | | 50 | .26 | 19.2 | 15.3 | 10.2 | 7.7 |
| | | 60 | .28 | 21 | 16.7 | 11.2 | 8.4 |
| LF-3.08 • 73° (50 mesh) | 730308 • 20 GPA (50 mesh) | 20 | .22 | 16.1 | 12.9 | 8.6 | 6.5 |
| | | 25 | .24 | 18.1 | 14.5 | 9.7 | 7.3 |
| | | 30 | .27 | • 20 | 16.0 | 10.7 | 8.0 |
| | | 40 | .308 | 23 | 18.3 | 12.2 | 9.2 |
| | | 50 | .34 | 26 | 20 | 13.6 | 10.2 |
| | | 60 | .38 | 28 | 22 | 14.9 | 11.2 |
| LF-3.85 • 73° (50 mesh) | 730385 • 25 GPA (50 mesh) | 20 | .27 | 20 | 16.1 | 10.8 | 8.1 |
| | | 25 | .30 | 23 | 18.0 | 12.0 | 9.0 |
| | | 30 | .33 | • 25 | 20 | 13.3 | 10.0 |
| | | 40 | .385 | 29 | 23 | 15.2 | 11.4 |
| | | 50 | .43 | 32 | 26 | 17.0 | 12.8 |
| | | 60 | .47 | 35 | 28 | 18.6 | 14.0 |
| LF-4.62 • 73° (50 mesh) | 730462 • 30 GPA (50 mesh) | 20 | .33 | 24 | 19.4 | 12.9 | 9.7 |
| | | 25 | .37 | 27 | 22 | 14.5 | 10.9 |
| | | 30 | .40 | • 30 | 24 | 16.0 | 12.0 |
| | | 40 | .462 | 34 | 27 | 18.3 | 13.7 |
| | | 50 | .52 | 38 | 31 | 20 | 15.3 |
| | | 60 | .57 | 42 | 34 | 22 | 16.8 |
| LF-6.16 • 73° (50 mesh) | 730616 • 40 GPA (50 mesh) | 20 | .44 | 32 | 26 | 17 | 13 |
| | | 25 | .48 | 36 | 29 | 19 | 14 |
| | | 30 | .53 | • 40 | 32 | 21 | 16 |
| | | 40 | .616 | 46 | 37 | 24 | 18 |
| | | 50 | .69 | 51 | 41 | 28 | 21 |
| | | 60 | .75 | 56 | 45 | 30 | 22 |
| LF-7.70 • 73° (50 mesh) | 730770 • 50 GPA (50 mesh) | 20 | .54 | 40 | 32 | 21 | 16 |
| | | 25 | .61 | 45 | 36 | 24 | 18 |
| | | 30 | .67 | • 50 | 40 | 27 | 20 |
| | | 40 | .770 | 57 | 46 | 30 | 23 |
| | | 50 | .86 | 64 | 51 | 35 | 26 |
| | | 60 | .94 | 70 | 56 | 37 | 28 |
| LF-9.24 • 73° (50 mesh) | 730924 • 60 GPA (50 mesh) | 20 | .65 | 50 | 40 | 26 | 20 |
| | | 25 | .73 | 54 | 43 | 29 | 22 |
| | | 30 | .80 | • 60 | 48 | 32 | 24 |
| | | 40 | .924 | 69 | 55 | 37 | 28 |
| | | 50 | 1.03 | 77 | 62 | 41 | 31 |
| | | 60 | 1.13 | 85 | 68 | 45 | 34 |

Effect of Solution Density on Nozzle Flow

| WEIGHT OF SOLUTION (LB/GAL) | SPECIFIC GRAVITY | CONVERSION FACTOR |
|---|---|---|
| 7.0 | .84 | 1.09 |
| 7.5 | .90 | 1.05 |
| 8.0 | .96 | 1.02 |
| 8.34* | 1.00 | 1.00 |
| 8.5 | 1.02 | .99 |
| 9.0 | 1.08 | .96 |
| 9.5 | 1.14 | .94 |
| 10.0 | 1.20 | .91 |
| 10.5 | 1.26 | .89 |
| 11.0 | 1.32 | .87 |
| 11.5 | 1.38 | .85 |
| 12.0 | 1.44 | .83 |

*Water

## TABLE C-5A. Performance for Delevan Flat-Even Fan Tip Band and 40-in Row Spacing with Gallons per Acre Broadcast Rate, and Gallons per Acre Field Rate

| Delevan Nozzle Tip Number | Pressure (psig) | Capacity 1-Nozzle (gal/min) | gal/acre — Broadcast Rate: Application rate on 14-in band (any spacing)* Field Rate: Actual treated field rate (40-in spacing)† | | | | | | | | | |
|---|---|---|---|---|---|---|---|---|---|---|---|---|
| | | | 3 mi/h | | 4 mi/h | | 5 mi/h | | 7.5 mi/h | | 10 mi/h | |
| | | | 14-in Band | 40-in Nozzle Spacing | 14-in Band | 40-in Nozzle Spacing | 14-in Band | 40-in Nozzle Spacing | 14-in Band | 40-in Nozzle Spacing | 14-in Band | 40-in Nozzle Spacing |
| LE - 1 80° (100 mesh) | 20 | .071 | 10.0 | 3.5 | 7.5 | 2.6 | 6.0 | 2.1 | 4.0 | 1.4 | 3.0 | 1.1 |
| | 30 | .087 | 12.3 | 4.3 | 9.2 | 3.2 | 7.4 | 2.6 | 4.9 | 1.7 | 3.7 | 1.3 |
| | 40 | .10 | 14.1 | 5.0 | 10.6 | 3.7 | 8.5 | 3.0 | 5.7 | 2.0 | 4.2 | 1.5 |
| LE - 1.5 80° (100 mesh) | 20 | .11 | 15.6 | 5.4 | 11.7 | 4.1 | 9.3 | 3.3 | 6.2 | 2.2 | 4.7 | 1.6 |
| | 30 | .13 | 18.4 | 6.4 | 13.8 | 4.8 | 11.0 | 3.9 | 7.4 | 2.6 | 5.5 | 1.9 |
| | 40 | .15 | 21 | 7.4 | 15.9 | 5.6 | 12.7 | 4.5 | 8.5 | 3.0 | 6.4 | 2.2 |
| LE - 2 80° (50 mesh) | 20 | .14 | 19.8 | 6.9 | 14.8 | 5.2 | 11.9 | 4.2 | 7.9 | 2.8 | 5.9 | 2.1 |
| | 30 | .17 | 24 | 8.4 | 18.0 | 6.3 | 14.4 | 5.0 | 9.6 | 3.4 | 7.2 | 2.5 |
| | 40 | .20 | 28 | 9.9 | 21 | 7.4 | 17.0 | 5.9 | 11.3 | 4.0 | 8.5 | 3.0 |
| LE - 3 80° (50 mesh) | 20 | .21 | 30 | 10.4 | 22 | 7.8 | 17.8 | 6.2 | 11.9 | 4.2 | 8.9 | 3.1 |
| | 30 | .26 | 37 | 12.9 | 28 | 9.7 | 22 | 7.7 | 14.7 | 5.1 | 11.0 | 3.9 |
| | 40 | .30 | 42 | 14.8 | 32 | 11.1 | 25 | 8.9 | 17.0 | 5.9 | 12.7 | 4.5 |
| LE - 4 80° (50 mesh) | 20 | .28 | 40 | 13.9 | 30 | 10.4 | 24 | 8.3 | 15.8 | 5.5 | 11.9 | 4.2 |
| | 30 | .35 | 49 | 17.3 | 37 | 13.0 | 30 | 10.4 | 19.8 | 6.9 | 14.8 | 5.2 |
| | 40 | .40 | 57 | 19.8 | 42 | 14.8 | 34 | 11.9 | 23 | 7.9 | 17.0 | 5.9 |
| LE - 5 80° (50 mesh) | 20 | .35 | 49 | 17.3 | 37 | 13.0 | 30 | 10.4 | 19.8 | 6.9 | 14.8 | 5.2 |
| | 30 | .43 | 61 | 21 | 46 | 16.0 | 36 | 12.8 | 24 | 8.5 | 18.2 | 6.4 |
| | 40 | .50 | 71 | 25 | 53 | 18.6 | 42 | 14.8 | 28 | 9.9 | 21 | 7.4 |
| LE - 6 80° (50 mesh) | 20 | .42 | 59 | 21 | 45 | 15.6 | 36 | 12.5 | 24 | 8.3 | 17.8 | 6.2 |
| | 30 | .52 | 74 | 26 | 55 | 19.3 | 44 | 15.4 | 29 | 10.3 | 22 | 7.7 |
| | 40 | .60 | 85 | 30 | 64 | 22 | 51 | 17.8 | 34 | 11.9 | 25 | 8.9 |
| LE - 8 80° (50 mesh) | 20 | .57 | 81 | 28 | 60 | 21 | 48 | 16.9 | 32 | 11.3 | 24 | 8.5 |
| | 30 | .69 | 98 | 34 | 73 | 26 | 59 | 20 | 39 | 13.7 | 29 | 10.2 |
| | 40 | .80 | 113 | 40 | 85 | 30 | 68 | 24 | 45 | 15.8 | 34 | 11.9 |
| LE - 10 80° | 20 | .71 | 100 | 35 | 75 | 26 | 60 | 21 | 40 | 14.1 | 30 | 10.5 |
| | 30 | .87 | 123 | 43 | 92 | 32 | 74 | 26 | 49 | 17.2 | 37 | 12.9 |
| | 40 | 1.00 | 141 | 49 | 106 | 37 | 85 | 30 | 57 | 19.8 | 42 | 14.8 |
| LE - 15 80° | 20 | 1.1 | 156 | 54 | 117 | 41 | 93 | 33 | 62 | 22 | 47 | 16.3 |
| | 30 | 1.3 | 184 | 64 | 138 | 48 | 110 | 39 | 74 | 26 | 55 | 19.3 |
| | 40 | 1.5 | 212 | 74 | 159 | 56 | 127 | 45 | 85 | 30 | 64 | 22 |

*Quantity that would be applied to completely cover 1 acre. Approximately one third of this quantity is actually used because of application in 14-in bands on 40-in rows. Use band spray rate capacity columns for exact quantity used per acre of sprayed ground.
†Conversion factors for other than 40-in rows:

| Other Spacing (in) | 18 | 20 | 24 | 28 | 30 | 32 | 34 | 36 | 38 | 42 |
|---|---|---|---|---|---|---|---|---|---|---|
| Conversion Factor | 2.22 | 2.0 | 1.66 | 1.43 | 1.33 | 1.25 | 1.18 | 1.11 | 1.05 | .95 |

*Courtesy* Delevan Corp.

## TABLE C-5B. Plated[1] Bolt (Cap Screw) Torque Specifications by SAE Grade and IFI Metric Property Classes in Pounds-Feet and (Newton-Metre)[2] for Both UNC and UNF Threads

| Bolt Diameter in Inches (mm) | SAE 1 | | | | SAE 5 | | | | SAE 8 | | | |
|---|---|---|---|---|---|---|---|---|---|---|---|---|
| | Torque Minimum | | Torque Maximum | | Torque Minimum | | Torque Maximum | | Torque Minimum | | Torque Maximum | |
| | lbf-ft | (N-m) | lbf-ft | (N-m) | lbf-ft | (N-m) | lbf-ft | (N-m) | lbf-ft | (N-m) | lbf-ft | (N-m) |
| 1/4(6.3) | 4 | (5) | 5 | (7) | 8 | (11) | 9 | (12) | 10 | (14) | 12 | (16) |
| 5/16(8.0) | 10 | (14) | 11 | (15) | 16 | (22) | 18 | (24) | 23 | (31) | 26 | (35) |
| 3/8(10) | 18 | (24) | 20 | (27) | 28 | (38) | 31 | (42) | 38 | (52) | 43 | (58) |
| 7/16(11) | 30 | (41) | 32 | (43) | 45 | (61) | 51 | (69) | 64 | (87) | 72 | (98) |
| 1/2(12) | 44 | (60) | 49 | (66) | 68 | (92) | 77 | (104) | 98 | (133) | 111 | (150) |
| 9/16(14) | 60 | (81) | 68 | (92) | 98 | (133) | 111 | (150) | 140 | (190) | 157 | (213) |
| 5/8(16) | 83 | (113) | 94 | (127) | 136 | (184) | 153 | (207) | 187 | (254) | 213 | (289) |
| 3/4(19) | 148 | (201) | 166 | (225) | 247 | (335) | 272 | (369) | 340 | (461) | 383 | (519) |
| 7/8(20) | 138 | (187) | 154 | (209) | 357 | (484) | 400 | (542) | 553 | (750) | 621 | (842) |
| 1(24) | 212 | (287) | 230 | (312) | 536 | (727) | 604 | (819) | 825 | (1119) | 927 | (1257) |

[1] To obtain torque for *unplated* bolts multiply plated bolt torque by 1.17.
[2] To obtain torque when threads are *lubricated* multiply torque by 0.75.
*Note:* Bolts having lock nuts should be torqued to approximately 50% of amounts listed in above chart.

**TABLE C-6. Fertilizer Drill Calibration Chart Values are in Pounds per Acre with Fertilizer Having a Density of 65**

| ROW SPACING | GEAR BOX SETTING, DRIVE 1 | | | | | | | | | | | | | | | | | | | | |
|---|---|---|---|---|---|---|---|---|---|---|---|---|---|---|---|---|---|---|---|---|---|
| | A1 | B1 | A2 | C1 | B2 | A3 | D1 | C2 | B3 | A4 | C3* | B5 | C4 | D3 | E2 | C5 | D4 | E3 | D5 | E4 | E5 |
| 6 in | 44 | 50 | 55 | 58 | 63 | 66 | 70 | 73 | 75 | 77 | 88 | 100 | 102 | 105 | 109 | 117 | 123 | 131 | 140 | 153 | 175 |
| 7 in | 38 | 43 | 47 | 50 | 54 | 56 | 60 | 63 | 64 | 66 | 75 | 86 | 88 | 90 | 94 | 100 | 105 | 113 | 120 | 131 | 150 |
| 8 in | 33 | 38 | 41 | 44 | 47 | 50 | 53 | 55 | 56 | 58 | 66 | 75 | 77 | 79 | 82 | 88 | 92 | 99 | 106 | 115 | 132 |
| 10 in | 26 | 30 | 33 | 35 | 38 | 40 | 42 | 44 | 45 | 46 | 53 | 60 | 61 | 63 | 66 | 70 | 74 | 79 | 84 | 92 | 105 |

| ROW SPACING | GEAR BOX SETTING, DRIVE 2 | | | | | | | | | | | | | | | | | | | | |
|---|---|---|---|---|---|---|---|---|---|---|---|---|---|---|---|---|---|---|---|---|---|
| | A1 | B1 | A2 | C1 | B2 | A3 | D1 | C2 | B3 | A4 | C3* | B5 | C4 | D3 | E2 | C5 | D4 | E3 | D5 | E4 | E5 |
| 6 in | 142 | 162 | 177 | 189 | 203 | 213 | 227 | 236 | 243 | 248 | 283 | 324 | 331 | 340 | 355 | 378 | 397 | 426 | 454 | 497 | 568 |
| 7 in | 122 | 139 | 152 | 162 | 174 | 182 | 195 | 203 | 208 | 213 | 243 | 278 | 284 | 292 | 304 | 324 | 340 | 365 | 389 | 426 | 486 |
| 8 in | 106 | 122 | 133 | 142 | 152 | 160 | 170 | 177 | 182 | 186 | 213 | 243 | 248 | 255 | 266 | 284 | 298 | 319 | 341 | 372 | 426 |
| 10 in | 85 | 97 | 106 | 114 | 121 | 128 | 136 | 142 | 146 | 149 | 170 | 195 | 199 | 204 | 213 | 227 | 238 | 255 | 272 | 298 | 340 |

*C3 setting is identical to A5, B4, D2, and E1 (not shown).

| Density | Conversion Factor | Density | Conversion Factor |
|---|---|---|---|
| 45 | 1.45 | 65 | 1.00 |
| 50 | 1.30 | 70 | 0.93 |
| 55 | 1.20 | 75 | 0.87 |
| 60 | 1.10 | 80 | 0.81 |

GEAR BOX

DRIVES

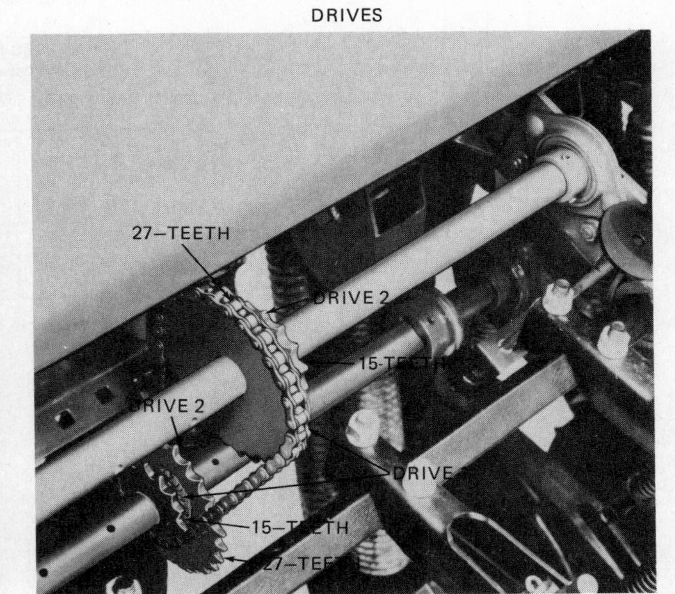


Courtesy Deere & Co.

**TABLE C-7. Calibration Data for Adjusting Metering Gate of Granular Application for Banding in Rows with 5 mi/h Speed of Travel [8.05 km/h] for 80 Rods [402 m]**

| APPROXIMATE RATE OF APPLICATION IN THE ROW, lb/acre, 5 mi/h | | | | | | | | | |
|---|---|---|---|---|---|---|---|---|---|
| Attapulgus Clay Carrier | | | | | | | | Gate No. | Dial No. |
| Row Spacing, in in* | | | | | | | | | |
| 40 | 38 | 36 | 34 | 32 | 30 | 28 | 20 | | |
| 3 | 3.2 | 3.3 | 3.5 | 3.8 | 4.0 | 4.3 | 6.0 | 1 | 8 |
| 4 | 4.2 | 4.5 | 4.7 | 5.0 | 5.3 | 5.7 | 8.0 | 2 | 3 |
| 5 | 5.3 | 5.6 | 5.9 | 6.2 | 6.7 | 7.2 | 10.0 | 2 | 7 |
| 6 | 6.3 | 6.7 | 7.1 | 7.5 | 8.0 | 8.6 | 12.0 | 3 | 1 |
| 7 | 7.4 | 7.8 | 8.3 | 8.8 | 9.3 | 10.0 | 14.0 | 3 | 5 |
| 8 | 8.4 | 8.9 | 9.4 | 10.0 | 10.6 | 11.5 | 16.0 | 4 | 0 |
| 9 | 9.5 | 10.0 | 10.6 | 11.3 | 12.0 | 12.9 | 18.0 | 4 | 4 |
| 10 | 10.5 | 11.1 | 11.8 | 12.5 | 13.3 | 14.3 | 20.0 | 4 | 8 |
| 11 | 11.6 | 12.2 | 13.0 | 13.8 | 14.6 | 15.7 | 22.0 | 5 | 3 |
| 12 | 12.6 | 13.3 | 14.1 | 15.0 | 16.0 | 17.2 | 24.0 | 5 | 7 |
| 13 | 13.6 | 14.4 | 15.3 | 16.3 | 17.3 | 18.6 | 26.0 | 6 | 1 |
| 14 | 14.7 | 15.5 | 16.5 | 17.5 | 18.6 | 20.0 | 28.0 | 6 | 5 |
| 15 | 15.8 | 16.7 | 17.7 | 18.8 | 20.0 | 21.4 | 30.0 | 7 | 0 |
| 16 | 16.8 | 17.8 | 18.8 | 20.0 | 21.3 | 22.9 | 32.0 | 7 | 4 |
| Furadan | | | | | | | | | |
| 7.5 | 7.9 | 8.3 | 8.9 | 9.4 | 10.0 | 10.7 | 15.0 | 1 | 6 |
| 10 | 10.5 | 11.1 | 11.8 | 12.5 | 13.3 | 14.3 | 20.0 | 1 | 8 |
| 20 | 21.0 | 22.2 | 23.6 | 25.0 | 26.6 | 28.6 | 40.0 | 2 | 8 |
| 30 | 31.5 | 33.3 | 35.4 | 37.5 | 39.9 | 42.9 | 60.0 | 3 | 9 |

*Conversion factor for number of 80-rod [402.3 m] rows per acre [ha].

| Number of 80-Rod [402.3-m] Rows per Acre [ha] | |
|---|---|
| Row Width, in [mm] | Number of Rows, acre [ha] |
| 40 [1015] | 10.0 [24.3] |
| 38 [965] | 10.5 [25.9] |
| 36 [915] | 11.1 [27.4] |
| 34 [864] | 11.8 [29.2] |
| 32 [813] | 12.5 [30.9] |
| 30 [765] | 13.3 [32.9] |
| 28 [710] | 14.3 [35.3] |
| 20 [510] | 20.0 [45.4] |

*Courtesy:* Allis-Chalmers Co.

**TABLE D-1. Partial Range Adjustment Chart for Combine Settings (Metric Conversion Table Given Below)**

| Crop | Cylinder Speed RPM | Number of Concaves | Cylinder to Concave Clearance | Chaffer Setting Range: Standard Round End Shallow Tooth 1-1/8" Spacing | Special Round End Shallow Tooth 1-5/8" Spacing | Special Square End Deep Tooth 1-5/8" Spacing | Special Round End Deep Tooth 1-5/8" Spacing | Setting of Adjustable Sieve | Fan Choke Setting |
|---|---|---|---|---|---|---|---|---|---|
| Corn | | | | | | | | | |
| Field Shelled | 400–700 | 3–5 | 1/4"–3/4" | 7/16"–5/8" | 7/16"–5/8" | 7/16"–5/8" | 3/8"–5/8" | 1/2"–5/8" | 5–7 |
| Milo-Maize | 700–1300 | 3–5 | 1/8"–1/2" | 1/2"–3/4" | 5/8"–3/4" | 1/2"–3/4" | 1/4"–1/2" | 1/4"–1/2" | 5–7 |
| Oats | 800–1325 | 2–5 | 1/4"–5/8" | 3/8"–3/4" | 5/8"–1" | 1/2"–3/4" | 1/4"–1/2" | 1/4"–1/2" | 2–7 |
| Rice | 800–1000 | 3–5 | 1/8"–1/2" | 5/8"–3/4" | 5/8"–3/4" | 5/8"–3/4" | 3/8"–1/2" | 1/4"–3/8" | 5–7 |
| Wheat | 800–1325 | 2–5 | 1/8"–1/2" | 1-1/2"–3/4" | 5/8"–1" | 1/2"–3/4" | 1/4"–1/2" | 1/4"–1/2" | 3–7 |

| METRIC CONVERSION – FRACTIONAL INCH TO mm | | | | | | | | | |
|---|---|---|---|---|---|---|---|---|---|
| 1/4" (1.6) | 1/8" (3.2) | 3/16" (4.8) | 1/4" (6.4) | 3/8" (9.5) | 1/2" (12.7) | 5/8" (15.9) | 3/4" (19.1) | 1-1/8" (28.6) | 1-5/8" (41.3) |

# GLOSSARY

**AC** Alternating current.

**Accumulator** A mechanical device for storing hydraulic fluid under pressure.

**After TDC or BDC** The position of the crankshaft in which the connecting rod throw is immediately after top dead center or bottom dead center.

**Air cleaner** A device that removes dirt and other foreign particles from intake air on an internal combustion engine.

**Air-cooled engines** Engines cooled by air circulation. They are usually equipped with a fan or blower, fins, and shrouds to control the air circulation.

**Air-fuel ratio** The ratio of the weight of air to fuel necessary to achieve a combustible mixture.

**Air lock** Air trapped in a diesel or hydraulic system that prevents normal fluid flow.

**Antifreeze** Chemicals added to the cooling system to prevent from freezing. Ethylene glycol base antifreeze also raises the boiling point of the coolant.

**Antifriction bearing** A needle, roller, or ball bearing.

**API oil classification** The American Petroleum Institute's classification of engine oil by type of engine and service conditions.

**Atomize** To break up a liquid (usually fuel) into extremely fine particles.

**Bale mover** A device for moving large round or rectangular bales.

**Beater** A small diameter drum mounted behind the cylinder on a combine that slows down the material coming off the cylinder.

**Before TDC or BDC** The position of the crankshaft in which the connecting rod throw is immediately before top dead center or bottom dead center.

**Bleeding** The process of removing air from a diesel fuel system or a hydraulic system to prevent air locks.

**Bleed screw** A screw or valve located in a filter, pump, or system used to remove air locks (also referred to as a vent plug).

**Blowby** Combustion gases that escape past the rings and piston into the crankcase. Excessive blowby reduces compression.

**Bore** The inside diameter of the cylinder in inches or millimeters.

**Boss** A heavy cast metal section used for extra support.

**Bottom** One of the complete units of a moldboard plow that lifts and inverts the furrow slice.

**Bottom dead center (BDC)** A position of the crankshaft in which the piston is at its lower or innermost position towards the crankshaft. Sometimes referred to as crank dead center.

**Branch center** A regional center maintained by agricultural equipment manufacturers to provide distribution, service, and sales to local dealers.

**Breaker point gap** The distance between the contact surfaces on breaker points when they are in the full open position.

**Breaker points** Contact points located in the distributor that are cam operated. They open and close the primary ignition circuit.

**Calibration** A process of measuring the discharge rate of a chemical applicator or seeder in order to match its performance with a desired output or standards.

**Cam follower bores** The sized holes in the engine block that hold the cam followers in position.

**Cam followers** The part that follows the contour of the cam lobe as the camshaft rotates. They may be solid followers (valve lifters) or hydraulic.

**Camshaft** A shaft with lobes that opens the valves at the proper time in the combustion cycle.

**Carburetor** A device that mixes air and fuel for combustion on spark-ignition engines.

**Center of pull** The hitch point on the tractor located between the center of the wheels and slightly ahead of the rear axle. Sometimes referred to as the center of draft.

**Center of resistance** A point on the implement which is being towed by a tractor that is the balance point of forces, longitudinally, horizontally, and vertically. Sometimes referred to as the center of load.

**Cetane number** The number assigned to diesel fuel to identify its ignition quality.

**Chaffer** A flat oscillating rack with metal overlapping lips or loovers that separates the chaff from the threshed grain or seed in a combine.

**Chamfer** An angle cut or machined on the edge of an otherwise flat surface.

**Chisel plow** A plow consisting of a series of sharp steel points which enter the soil to a depth of 10 to 30 in [254–760mm].

**Choke valve** A disk-shaped valve in a carburetor to restrict the incoming air; it creates a vacuum in the carburetor.

**Cleaning shoe** The chaffer and sieve on a combine.

**Closed-center valving** A hydraulic control valve that does not permit fluid to pass from the pump to the reservoir through its center in the neutral position.

**Clutch** A mechanism for engaging and disengaging the power output of a motor or an engine to the driven machine.

**Clutch pedal free travel** The distance the clutch pedal moves before the clutch release bearing moves the pressure plate release levers.

**Coil** The ignition coil is a voltage transformer that steps battery voltage up to ignition voltage.

**Combine** A machine that cuts, threshes, separates, and cleans seed crops.

**Combustion chamber** The area between a piston at TDC and the cylinder head where combustion occurs.

**Compression gauge** A test instrument used to check compression in a cylinder.

**Compression ignition engines** Diesel engines where fuel ignites as it is mixed with air in the combustion chamber due to heat created by the compression of the air.

**Compression ratio** The total volume of the cylinder divided by its combustion chamber volume.

**Compression rings** Piston rings that seal the combustion chamber.

**Concave** A series of bars (grated area) under the cylinder or rotor on combines.

**Concave spacing** The vertical distance between the concave and the cylinder on a combine.

**Concentric** Located evenly around a common point or center. A valve seat is centered around a valve guide.

**Condenser** An electrical device located in the distributor that serves to prolong breaker-point life and sustain the spark.

**Connecting rod** The connecting link between the piston and the crankshaft.

**Coolant** Usually a mixture of ethylene glycol base antifreeze and water.

**Cooling shroud** A sheet metal component of an air-cooled engine used to direct the flow of air from the flywheel over the cooling fins of the cylinder.

**Corn head** A header designed to remove only the ear from corn for harvest. Used on combines and forage harvesters.

**Cotton pickers** Cotton harvesting machines designed to remove the lint cotton from open bolls.

**Cotton stripper** Cotton harvesting machines designed to remove the whole boll from the cotton stalk.

**Coulter** A device attached to a moldboard plow for cutting a vertical section of the furrow slice to eliminate trash collection.

**Crankshaft** The shaft that runs the length of the engine; it changes the reciprocating motion of the pistons to rotary motion.

**Crocus cloth** An extremely fine abrasive cloth used for polishing.

**Cutter bar mower** A machine used for cutting hay and forage crops that uses a cutter bar equipped with a knife. A cutter bar is also used on other machines such as combines.

**Cutting height** The height at which a crop is cut during harvest.

**Cylinder** A hydraulic actuator which produces linear movement.

**Cylinder (combine)** Bars mounted on a series of hubs on a shaft supported across the threshing area. May be a rasp bar, spike tooth or angle bar.

**Cylinder gauge** A special tool equipped with a dial indicator for checking cylinders for out-of-round and taper.

**Cylinder head** An iron alloy or aluminum casting that bolts on top of the engine block to form the top of the combustion chamber. On liquid-cooled engines, the cylinder head contains cored areas for coolant circulation (water jacket). Valves are in the cylinder head on overhead valve engines.

**Cylinder leakage tester** A test instrument that uses compressed air to check for compression losses in a cylinder.

**Cylinder speed** The rpm of the cylinder in a combine. It is changed to meet specific crop and field conditions.

**DC** Direct current.

**Dead center** The position of the crankshaft in which the connecting rod, journal, main bearing, and piston pin bearing are all in exact alignment with the centerline of the crankshaft.

**Defoliation** Using chemicals to cause the leaves to dry up and fall off cotton stalks to facilitate harvest.

**Deglaze** Removing the glaze from cylinder or liner walls so a new set of rings will seat.

**Depreciation** The loss of value of a machine or piece of equipment due to age, use, or obsolescence.

**Detergent** A chemical added to engine oil to allow it to carry contaminants in suspension.

**Detonation** Explosion of the fuel resulting from spontaneous combustion in the cylinder.

**Dicot** A seed plant having two cotyledons.

**Diesel injection nozzles** Valves that atomize diesel fuel and inject it into the combustion chamber at the proper point in the combustion cycle.

**Diesel injection pump** A mechanically operated pump that supplies diesel fuel under pressure to the injection nozzles at the proper point in the combustion cycle.

**Differential** A drive train mechanism which allows the drive wheels to turn at different speeds when the machine is negotiating a turn.

**Diodes** Electrical devices used in alternators that have the ability to pass electrical current in one direction and resist the flow of current in the other direction.

**Disk harrow** A secondary tillage tool consisting of small diameter disks 16 to 20 in [400–500 mm] in diameter mounted on a spool and operated at an angle.

**Distributor** The ignition distributor opens and closes the primary ignition circuit and delivers the secondary voltage to the cylinders at the proper time in the combustion cycle. It also houses the spark advance system.

**Draper** A continuous conveyor belt used on some harvesting machines to move the crop after it has been cut with a cutter bar.

**Draw bar power** Power which is generated at the draw bar of a farm tractor.

**Dry-type cylinder liner** A sleeve installed in a cylinder to provide a renewable wear surface.

**Dwell** The amount of rotation (in degrees) of the distributor cam that the breaker points remain closed. May be referred to as cam angle.

**Dwell meter** A test instrument that electronically measures the amount of rotation (in degrees) of the distributor cam that the breaker points are closed.

**Dynamic timing** Timing of the ignition, valves, or fuel-injection performed when the engine is operating.

**Dynamometer** A machine used to measure engine power that attaches to the PTO on a tractor.

**Effective machine capacity** The actual or measured performance capacity of a machine.

**Electrolyte** The sulfuric acid and water solution used in storage batteries.

**Endplay** The amount of axial or lengthwise movement between two parts.

**Engine overhaul** A common repair procedure that includes replacement and machining of parts to restore engine performance and reliability.

**Equipment dealers** A company that sells and services agricultural machines and equipment and are often referred to as dealerships.

**Exhaust gas analyzer** An instrument for measuring engine

exhaust gases to determine the amount of carbon monoxide and unburnt hydrocarbons present.

**Exhaust manifold** The casting that bolts to the exhaust port area on the cylinder head or engine block and connects to the exhaust system.

**Exhaust ports** Passages leading from the exhaust valve to the exhaust manifold.

**Exhaust valves** Engine valves controlled by the camshaft that allows for the removal of combustion gases on the exhaust stroke.

**Expansion plug** A hole plug that can be expanded to securely close a hole such as the core holes in cylinder heads and engine blocks.

**Feeler ribbon** A flat, long gauge of known thickness.

**Field cuber** A machine that produces cubes of hay in the field from windrowed hay.

**Field cultivator** A secondary tillage tool used for weed control and seedbed preparation.

**Field curing** Allowing forage crops to reach the desired moisture content in the field.

**Field efficiency** The relationship between the theoretical machine capacity and effective machine capacity expressed in percent.

**Final drive** The last stage in the transmission of power from the engine to the drive wheels or power take-off shaft.

**Finger grates** Small steel rods that extend from the concave to the straw walker in a combine.

**Firing order** The order in which the cylinders fire.

**Fixed costs** Those costs which occur annually regardless of amount of use of a machine, equipment, or buildings.

**Flail** Swinging knives or blades on a rotating horizontal shaft.

**Float** The part of a carburetor which is used to sense the level of fuel in the carburetor bowl.

**Flywheel** A heavy wheel that is part of an internal combustion engine. Attached to the crankshaft, it is used for opposing and moderating by inertia any fluctuations of speed in the rotation of the crankshaft.

**Flywheel ring gear** The gear that is driven by the starter pinion gear to start the engine.

**Forage harvester** A machine that uses a cutter head or flails to chop forage crops into short lengths to be used for silage, haylage, or green chop.

**Force** The cause of the acceleration of the movement of material bodies, such as the force of a hammer against an anvil.

**Four-stroke cycle** An engine which requires two complete revolutions of the crankshaft to complete intake, compression, power, and exhaust.

**Four-way valve** A hydraulic control valve for controlling a double acting cylinder.

**Fuel shutoff knob** The control that shuts off the fuel to the injection nozzles on a diesel engine. Some are electric and shut off fuel when the ignition switch is in the off position.

**Fuel transfer pump** A mechanically or electrically operated in-line pump that is used in many fuel systems to provide a positive flow of fuel.

**Gear backlash** The slack or play between gears as they mesh.

**Gear ratio** The ratio of the number of teeth when one gear is meshed with another gear.

**Governor** A device which maintains a near constant engine speed (rpm) under varying loads.

**Grain drill** A machine which meters grain into rows less than 18 in [457 mm] in width.

**Green chop** Forage crops harvested with a forage harvester and fed directly to livestock.

**Ground speed** The actual operating speed over the ground of a tractor or machine.

**Growler** A test instrument used to test generator and starting motor armatures.

**Guards** The fingerlike projections on a cutter bar that protect the knife. They also separate and guide plants into the knife.

**Harvesting losses** Machine losses during harvest. They vary with the crop and type of harvesting equipment used.

**Hay** Dried grass, legumes, or other forage crops stored for use as livestock feed.

**Hay conditioners** A machine that crushes or cracks plant stems to reduce curing or drying time.

**Haylage** Low moisture silage (40 to 60 percent moisture).

**Head gasket** A gasket that forms a seal between the milled surfaces on the engine block and the cylinder head.

**Horsepower** One horsepower is equal to work performed at a rate of 33,000 foot-pounds per minute.

**Horsepower hours per gallon** Horsepower produced divided by the number of gallons of fuel used per hour.

**Hydraulic motor** A hydraulic actuator which produces rotating movement.

**Hydrodynamic force** Force created by the flow of a liquid; used to transfer power.

**Hydrometer** A float device used to determine the specific gravity of electrolyte in a battery or the level of freeze protection provided by the antifreeze in a cooling system.

**Hydrostatic pressure** A method of transferring power by exerting pressure on a liquid with little or no movement of the liquid.

**Ignition advance** Centrifugal or vacuum systems that automatically advance the point in the combustion cycle where the breaker points open.

**Ignition coil** A transformer for converting low voltage to high voltage to create the ignition spark.

**Ignition coil polarity** The polarity relationship of the primary windings of an ignition coil to that of the battery.

**I head** A type of engine in which the valves are located in the cylinder head.

**Intake manifold** The casting that bolts to the intake port area on the cylinder head or engine block and connects to the carburetor or intake system.

**Intake ports** Passages leading from the intake manifold to the exhaust valve.

**Intake valves** Engine valves controlled by the camshaft that allow air or air and fuel to enter the combustion chamber on the intake stroke.

**Integral plow** A plow which attaches directly to the three-point hitch of a tractor. Usually without transport wheels.

**Intercooler** Coolant or air is used to cool the air forced by the turbocharger into the combustion chamber to reduce the size of the air molecules and allow more air to be forced into the combustion chamber on the intake stroke.

**Interference angle** The angle attained when the valve face and valve seat are ground at slightly different angles.

**Internal combustion engine** A machine in which combustion takes place internally and the expanding gases perform work.

**Journal** The machined surface on which a friction bearing turns.

**Knife assembly** The complete knife used in a cutter bar including the knife sections and knife head.

**Knife drive register** The proper alignment of the reciprocating knife so that knife sections are an equal distance from the centerlines of the guards at each end of the stroke.

**Knife holddown clips** Adjustable clips that attach to the cutter bar to hold the knife sections down against the ledger plates.

**Knocking** A rapping or clattering sound as the result of ignition of the fuel mixture in an engine.

**Knurling** Displacing metal outward to decrease the diameter of a hole such as a valve guide or increase the diameter of a cylindrical component such as a piston.

**Leakoff line** The return line from an injection nozzle to the fuel tank of a diesel engine.

**Ledger plates** The cutting edge on the guard that forms a shear with the knife section.

**L-head** A type of engine in which the valves are located in the cylinder block.

**Line of draft** An imaginary line which extends from the center of pull to the center of resistance.

**Liquid-cooled engines** Engines that are cooled by liquid coolant (water and antifreeze) circulating through the water jacket and radiator.

**LP gas** Liquified petroleum gas; usually propane or a blend of propane and butane.

**Magneto** An electric spark generator for developing a high-voltage ignition spark by means of rotating magnets.

**Main bearings** The friction-type split bearings that provide a renewable wear surface between the crankshaft main journals and the main bearing bore in the engine block.

**Maintenance schedule** A schedule developed by the manufacturer to be used by the owner or operator of an agricultural machine to provide recommended service at recommended intervals.

**Manufacturer's Representatives** People employed by agricultural equipment manufacturers. They usually work out of branch centers in sales, customer relations, or service.

**Maximum diameter** A term used to denote the largest diameter measured on a crankshaft for all the rod journals; main journals. May be referred to as minimum wear.

**Mechanic's helper** A person employed in a dealership that performs the less skilled, entry level jobs, such as equipment set-up, equipment assembly, painting, welding, and may assist a trained mechanic in repair and service of agricultural machines and equipment.

**Microfarad** The unit of measurement used to indicate the capacity of a condenser.

**Micrometer caliper** A measuring instrument which permits measurements to be taken in thousandths of an inch (millimeters).

**Misfire** What happens when the fuel charge in one or more cylinders fails to ignite at the proper point in the combustion cycle.

**Module (cotton)** Cotton compressed in the field on a pallet by a module builder.

**Moldboard** The specially shaped, curved part of a moldboard plow which elevates and inverts a furrow slice.

**Monocot** Any seed plant having a single cotyledon.

**Multidisk mowers** Mowers equipped with revolving discs that have free-swinging knives that are brought into cutting position by centrifugal force.

**Naturally aspirated engines** Engines where air or air and fuel are drawn into the combustion chamber by the low pressure area created by the down stroke of the piston on the intake stroke.

**Nebraska Tractor Tests** Published test results of the Nebraska Agricultural Experiment Station.

**Nozzle tester** A manually operated pump with a pressure gauge used to test diesel injection nozzles.

**Octane rating** The measure of a specific gasoline's ability to resist detonation.

**Ohm** The unit of measurement used to indicate the resistance to the flow of electricity in a circuit or component.

**Ohmmeter** A test instrument used to measure the amount of resistance in ohms in a circuit or component.

**Oil analysis** Spectrochemical analysis for contaminants in engine oil. Used for maintenance management.

**Oil control rings** The lower piston ring or rings that are designed to remove excess oil from the cylinder walls.

**Oil seal** A dynamic seal that prevents oil from leaking around a shaft that rotates.

**Oil sludge** A chemical compound which results from the combination of moisture, acids, and contaminants in engine oil, gear lubricants, and hydraulic fluids.

**Open-center valving** A type of hydraulic valve which allows pump pressure to be diverted directly to the reservoir through an open center when the valve is in the neutral position.

**Operating temperature** The temperature of the coolant in the cooling system when the engine is running.

**Operator's manual** A book explaining operation, safe operation, adjustments, and maintenance; provided by the manufacturer with a machine for customer use.

**Oscilloscope** A test unit that projects a visual reproduction of the ignition system spark action onto a cathode ray tube screen.

**Out-of-round** When a cylinder or journal is worn more on one side than it is on the opposite side.

**Partsperson** People employed by dealerships to order, stock, and provide repair and replacement parts for service departments and agricultural customers.

**Peanut combine** A special combine that uses a series of cylinders equipped with spring teeth to pick the peanut pods from the vines; it also separates and cleans the pods.

**Peanut digger** A machine that utilizes a flat plow or knife to cut the peanut taproot, shake the vines, and place them in a windrow. Some diggers invert the vines and place the peanuts on top of the windrow.

**Pesticide** A chemical used for the control of weeds and/or insects.

**Pickup** A mechanism that uses pickup teeth to pick up a windrowed crop and feed it into a harvesting machine.

**Pilot bearing** The bearing in the flywheel or in the end of the crankshaft that supports the front of the transmission input shaft. Bushings are sometimes used instead of bearings.

**Piston clearance** The clearance between the piston and the cylinder.

**Piston displacement** The volume of the cylinder displaced in cubic inches or cubic centimeters when the piston is moved from the bottom to top dead center.

**Piston head** The top or crown of a piston that forms the bottom part of the combustion chamber.

**Piston lands** The flat surfaces between the ring grooves that support the rings.

**Piston pin** The connecting pin that attaches the piston to the connecting rod.

**Piston pin boss** The heavy part of the skirt that is machined to hold the piston pin.

**Piston pin bushing** The bushing in the eye of the connecting rod that provides a renewable wear surface between the connecting rod and the piston pin.

**Piston pin clearance** The clearance between the piston pin and the piston pin bushing. Also the clearance between the piston pin and the piston pin boss.

**Piston ring groove** The machined areas in the pistons that hold the rings parallel to the cylinder walls.

**Piston skirt** The lower portion of the piston that forms a bearing area with the cylinder wall.

**Plain bearing** A sleeve- or bushing-type bearing, sometimes referred to as a friction bearing.

**Planter** A machine for placing seeds in rows 18 to 42 in [457–1067 mm] in width.

**Plastigage** A gauging plastic string which indicates the distance between parts when crushed between them. A registered trademark of the Dana Corporation.

**Plow disk** A heavy-duty plow consisting of a series of concave- or saucer-shaped disks up to 36 in [914 mm] in diameter spaced 10 to 12 in [254–304 mm] apart on a common spool shaft.

**Point-of-rock** See *Valve overlap.*

**Polarize** Generators are polarized so that the current flow through the field coil is matched to the battery grounding systems used.

**Poppet valve** A valve opened by a cam lobe and closed by a spring. These are used in four-stroke cycle engines and for exhaust valves in some two-stroke cycle engines.

**Power** The rate of doing work (work times time).

**Precleaner** The screen or turbine device that removes large particles of dirt and other foreign matter from intake air before it reaches the air cleaner.

**Precombustion chamber** A small chamber in the cylinder head connected to the combustion chamber by a restricted passage into which the diesel fuel is injected.

**Preharvest losses** Crop losses that occur before harvest.

**Preignition** Ignition of the air-fuel mixture in the combustion chamber before the spark occurs.

**Pressure cap** A special cap on the radiator that controls the pressure in the cooling system and raises the boiling point of the coolant.

**Preventive maintenance** The performance of the routine maintenance services and checks outlined in the operator's manual.

**Primary circuit** The low voltage circuit in an ignition system that functions when the breaker points are closed.

**Prussian blue** A blue die that can be used to coat metal during machining to check the contact area.

**PTO** Power take-off.

**PTO power** Power which is generated at the power take-off shaft of a tractor.

**Pushrods** The rods between the cam followers and the rocker arms that are lifted by the cam lobes to open the valves.

**Rack and pinion** A spur gear mechanism in which the rack is straight and a pinion meshes with the rack to create linear motion.

**Radial rakes** Rakes with a series of rotary hubs or stars equipped with tines that can be used to fluff and windrow hay and forage crops.

**Rain cap** A cap that automatically closes over a vertical exhaust when the engine is not running.

**Reciprocating engine** A piston engine in which the up-and-down motion of the piston is transformed into rotary motion by a crankshaft.

**Recirculating sprayer** A type of weed sprayer which collects overspray and returns it to the supply tank for reuse.

**Rectangular balers** Machines using a reciprocating plunger to press hay crops into a bale chamber where the individual bales are tied with either wire or twine.

**Reed valve** Leaf spring valves used in two-stroke cycle engines to perform the function of opening and closing the crankcase for the intake of the fuel mixture from the carburetor.

**Reel** The cylinder-shaped device used to feed crops into the cutter bar and onto the platform on machines such as combines.

**Resistor** An electrical device for reducing the voltage by offering resistance to flow.

**Restriction indicator** A warning device that may consist of a light on the instrument panel or a colored band to indicate a restriction in the air intake system on engines.

**Rick (cotton)** Cotton compressed in the field by a rick compactor. Ricks are loaded on trucks or trailers with a front-end loader.

**Ring end gap** The distance between the ends of a ring when it is installed in a cylinder.

**Ring ridge** The area in the cylinder directly above the ring travel area.

**Ring side clearance** The clearance between the piston ring and the ring groove in the piston.

**Rocker arm** The lever used to convert the upward movement of the pushrod to the downward movement that opens the valve.

**Rocker arm shaft** The shaft upon which the rocker arms are mounted.

**Rod bearings** Friction-type split bearings that provide a renewable wear surface between the connecting rod and the crankshaft journal.

**Rod cap** The removeable part of the rod head that attaches the connecting rod to the crankshaft.

**Rod eye** The small bore at the top of the connecting rod that attaches the connecting rod to the piston pin.

**Rotor** The distributor rotor attaches to the top of the distributor shaft and carries secondary voltage to the spark plug towers at the proper time in the combustion cycle.

**Rotor-type combine** A combine equipped with rotors or cylinders designed so that the crop stays in contact with the cylinder for several complete revolutions.

**Rotary mower** A mower equipped with rotating blades.

**Rotary tiller** A primary or a secondary tillage tool consisting of a rotating series of knives powered by an internal combustion engine or power take-off.

**Round balers** A machine that rolls the hay crop into large round bales.

**Row crop cultivator** A secondary tillage machine used for cultivating row crops for weed control.

**rpm** Revolutions per minute.

**Running clearance** The clearance between two parts such as a crankshaft journal and a bearing that allows them to turn freely and to lubricate properly.

**Run out** The amount a shaft or other component rotates out of true; wobble.

**Salesperson** A person who sells agricultural equipment and machines to agricultural producers.

**Secondary circuit** The high-voltage circuit in an ignition system that functions when the breaker points are open and ignites the air-fuel mixture in spark-ignition engines.

**Semi-integral plow** A plow which is supported at the front by the tractor lift arms of a three-point hitch and by a tailwheel at the rear.

**Separator speed** The speed (rpm) of the shaft that runs the separator unit on a combine.

**Service manager** The person responsible for assigning work to service persons. Service managers are responsible for scheduling jobs, keeping shop records, supervising work, and dealing with the customers bringing in equipment and machines for service.

**Service manual** A technical manual that gives the manufacturer's recommended specifications and procedures for service.

**Service person** Often referred to as a mechanic. Persons who service and repair agricultural equipment. Many service people specialize in areas such as hydraulics, diesel service, engines, and so on.

**Service records** Records kept by the owner/operator of an agricultural machine to indicate service performed and service intervals.

**Share** The replaceable cutting part of the moldboard.

**Shoes** The hardened steel skids that are used to control cutting height on mowers.

**Side delivery rake** A machine equipped with toothed parallel bars or finger wheels that is used to form windrows.

**Sieve** A flat oscillating rack located under the chaffer on a combine that separates larger particles from the grain or seed.

**Silage** Chopped high moisture forage (60 to 70 percent moisture) that is converted to livestock feed through a fermentation process.

**Silo** A structure for the storage of silage or haylage. May be vertical or horizontal.

**Small engine** An internal combustion engine usually of less than 20 hp and with one or two cylinders, which is self-contained and portable.

**Solid-state ignition** An ignition system which utilizes semiconductors to control the high energy sparks in the spark plugs.

**Spark ignition engines** Engines using highly volatile fuels such as gasoline and LP gas where the air-fuel mixture is ignited by spark created by the ignition system.

**Spark plug gap** The gap between the center and grounded electrode on a spark plug.

**Spark plug heat range** The operating temperature of the firing portion of the spark plug; used for classification purposes.

**Speed ratio** The ratio of speed of rotation of the driven and the driving shafts.

**Stack mover** A device for moving a hydraulically formed hay stack.

**Stack wagon** A machine used to hydraulically form stacks of loose hay or stover.

**Standout** The distance cylinder liners extend past the top of the engine block.

**State of charge** The cranking power of a battery as determined by the specific gravity of the electrolyte.

**Static timing** Valve, ignition, or fuel-injection timing performed when the engine is not operating.

**Storage battery** An electrical device which generates electric energy by chemical reaction.

**Stover** Stalks, leaves, husks, and other materials left in the field after crops such as corn or milo have been harvested with a combine.

**Straw chopper** A belt-driven mechanism that chops the straw as it leaves a combine.

**Straw spreader** A belt-driven mechanism that spreads the straw as it leaves a combine.

**Straw walkers** A rack that moves the straw, cobs, husks, and other material out the back of the combine.

**Stroke** The amount of piston travel from top to bottom dead center expressed in inches or millimeters.

**Subsoiler** A type of chisel plow which penetrates to a depth of up to 32 in [812 mm] for breaking hard layers of subsoil.

**Swashplate** A mechanism which changes rotating motion into linear motion similar to that of a crankshaft. Sometimes called a wobble plate.

**Swath board** The device used to divide the crop cut in one pass from the uncut crop.

**Tailings** Partially threshed seed heads and other larger materials that are too large to fall through the sieve on a combine. They are returned to the cylinder.

**Taper** A condition that occurs when a cylinder or journal is worn more on one end than it is on the opposite end.

**Tedding** The turning and fluffing of a hay or forage crop to reduce curing or drying time.

**Telescoping gauge** A measuring tool for transferring inside measurements to a micrometer caliper.

**Theoretical machine capacity** The calculated capacity of a machine based upon its size and speed of travel.

**Thermostat** The temperature control valve in the cooling system that maintains correct engine operating temperature.

**Thickness gauge** A flat or round gauge of known thickness used to determine the distance between two surfaces or contact areas.

**Third port, loop-scavenged valve** A two-stroke cycle engine in which the intake port of the engine is located in the cylinder in such a position to cause the intake gases to loop the cylinder as the exhaust gases are being scavenged (removed).

**Three-way valve** A hydraulic control valve for the control of a single-acting cylinder.

**Throttle valve** A disk-shaped valve in the carburetor that controls the rate of combustible gases entering the cylinder.

**Timing light (stroboscopic)** A test unit that connects in series in the secondary circuit; it produces flashes of light in unison with the firing of the spark plug. Used to set ignition timing.

**Timing marks** A mark located on the flywheel or crankshaft pulley used to set ignition timing. Also marks on gears that indicate correct alignment.

**Toe-in** The adjustment of the front or steering wheels on a tractor or machine so that they are closer together at the front of the tread than at the back of the tread.

**Tolerance range** The amount of variation allowed by the manufacturer from an exact size or measurement.

**Tool bar** A heavy bar which, when connected to the three-point hitch of a farm tractor, allows planting or cultivating equipment to be attached at variable row spacings.

**Top dead center (TDC)** A position of the crankshaft in which the piston is nearest its cylinder head. Sometimes referred to as head dead center.

**Torque** A lever action which produces or tends to produce rotation.

**Torque wrench** A handtool used to measure the amount of twisting action applied to fasteners and calibrated in foot-pounds, inch-pounds, or newton-meters.

**Total crop losses** Harvesting and preharvest losses combined.

**Transaxle** A drive train component consisting of a transmission and a final drive in one assembly.

**Transmission** A mechanism that controls the rate of release of energy from the driving engine to the source of application.

**Tune-up** The process of making minor adjustments or repairs to maintain or restore engine performance.

**Turbocharged engines** An engine with a device (a turbocharger) that forces air into the cylinders.

**Turbocharger** An exhaust-driven turbine that forces an increased amount of air into the combustion chamber on the intake stroke, thus increasing the engine's power.

**Two-stage clutch** A clutch equipped with a dual pressure plate. There is a separate PTO clutch enclosed in the pressure plate assembly.

**Two-stroke cycle** An engine which requires only one complete revolution of the crankshaft to perform the functions of intake, compression, power, and exhaust.

**Two-way plow** A moldboard plow which can be operated to turn the furrow slice all in one direction by rotating its bottoms.

**U-joint (universal)** A shaft coupling capable of transmitting rotation from one shaft to another even when not in alignment.

**Undersize bearings** Bearings designed for use when a crankshaft is ground to a smaller journal diameter or to compensate for normal wear.

**Universal symbols** Symbols developed by the American Society of Agricultural Engineers to identify gauges, warning lights, adjustments, and controls.

**Vacuum** A condition where the air in a cylinder is at less than atmospheric pressure. It is caused by the downward movement of the piston.

**Vacuum gauge** A test instrument used to check the amount of intake vacuum.

**Valve clearance** The gap between the end of the valve stem and the rocker arm when the valve is closed.

**Valve face** The outer lower edge of the valve head that is machined to contact the valve seat when the valve is closed.

**Valve guides** Sized holes in the cylinder head or block that hold the valves in place and allow them to open and close. Many guides are press fit and can be replaced.

**Valve keepers** The locking devices that fit into the keeper groove on the valve stem to secure the valve spring retainer in place.

**Valve margin** The area between the top of the valve head and the valve face.

**Valve overlap** The amount of crankshaft rotation (in degrees) that both the exhaust and intake valve are open between the intake and exhaust stroke. May be referred to as point-of-rock.

**Valve seat dial indicator** A special tool for checking valve seats for concentricity.

**Valve seat inserts** Hardened seats in the cylinder head that may be removed and replaced.

**Valve seats** The machined surfaces in the cylinder head or engine block that contact the valve face when the valve is closed.

**Valve spring retainer** The disk-shaped retainer that is held in place by the valve keepers; it holds the valve spring in place.

**Valve springs** The coil springs that keep the valves closed.

**Valve timing** The opening and closing of the intake and exhaust valves at the proper point in the combustion cycle. Timing is controlled by the proper alignment of the crankshaft and camshaft gears and proper valve clearance.

**Variable costs** All costs that are associated with machine or equipment operation other than fixed costs.

**Venturi** A mechanical device in a carburetor for increasing the velocity of air through the carburetor without decreasing its volume.

**Viscosity** As used in classifying oils and lubricants, viscosity refers to oil fluidity or its resistance to flow at a given temperature.

**Volatile** A material that evaporates or vaporizes readily.

**Volatility** The property of a fuel such as gasoline to evaporate quickly and at a relatively low temperature.

**Volt** The unit of measurement used to measure electrical pressure.

**Volt-amp tester** A test instrument used to test the voltage, amperage, and resistance in starting and charging circuits.

**Voltmeter** A test instrument used to measure the voltage in a given circuit.

**Water jacket** The cored sections in engine blocks and cylinder heads that allow coolant to circulate in the engine to dissipate heat.

**Wear plate** A renewable wear surface at the point where the back of the knife contacts the cutter bar.

**Welch plug** The plug used to plug the rear camshaft bore; sometimes referred to as the core hole plug.

**Wet-type cylinder liner** A liner installed in engine blocks that are cored so the outside of the liner forms a part of the water jacket.

**Windrow** Hay, forage, or grain crop placed in a fluffy row for pickup with a harvesting machine.

**Windrowers (swathers)** A machine that cuts the crop and places it in a windrow.

**Work** A measurement of the amount of force required to move a body a given distance (force times distance).

# INDEX